Carbon in Electrochemistry

Sheffield, UK

28-30 July 2014

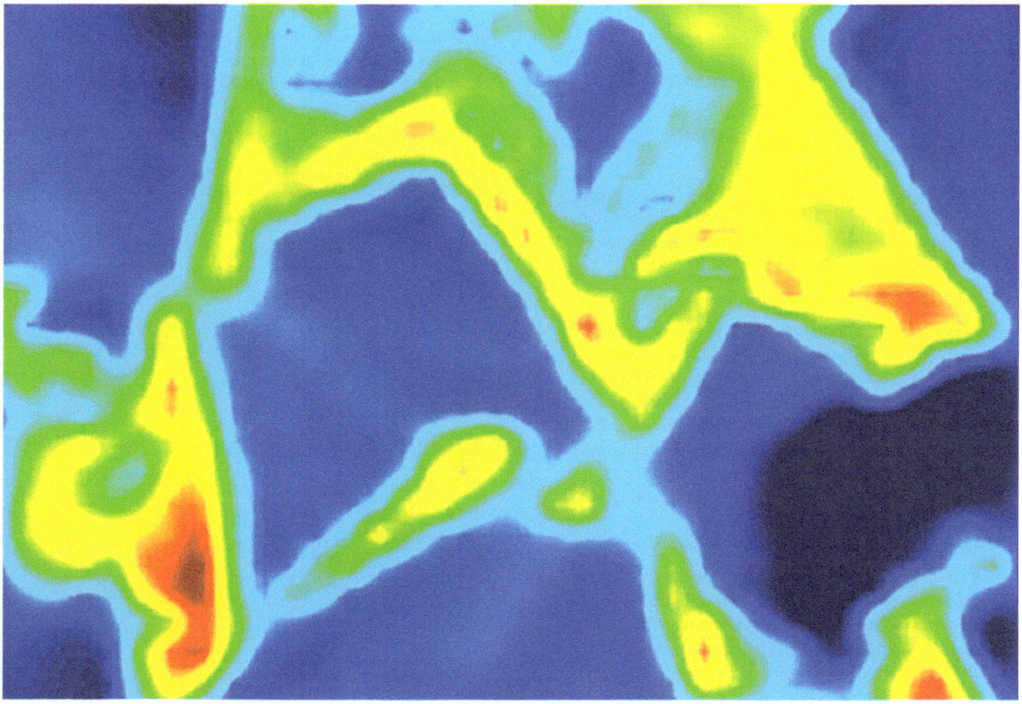

FARADAY DISCUSSIONS

Volume 172, 2014

ROYAL SOCIETY OF CHEMISTRY

The Faraday Division of the Royal Society of Chemistry, previously the Faraday Society, founded in 1903 to promote the study of sciences lying between Chemistry, Physics and Biology.

Faraday Discussions (Print ISSN 1359-6640, Electronic ISSN 1364-5498) is published 8 times a year by the Royal Society of Chemistry, Thomas Graham House, Science Park, Milton Road, Cambridge, UK CB4 0WF.

Volume 172 ISBN:978-1-78262-029-7

2014 annual subscription price: print+electronic £827, US $1542; electronic only £785, US $1465. Customers in Canada will be subject to a surcharge to cover GST. Customers in the EU subscribing to the electronic version only will be charged VAT. All orders, with cheques made payable to the Royal Society of Chemistry, should be sent to RSC Order Department, Royal Society of Chemistry, Thomas Graham House, Science Park, Milton Road, Cambridge, CB4 0WF, UK.
Tel +44 (0) 1223 432398; E-mail www.orders@rsc.org

If you take an institutional subscription to any RSC journal you are entitled to free, site-wide web access to that journal. You can arrange access via Internet Protocol (IP) address at www.rsc.org/ip. Customers should make payments by cheque in sterling payable on a UK clearing bank or in US dollars payable on a US clearing bank.

Printed in the UK

MIX
Paper from responsible sources
FSC® C013604

Faraday Discussions documents a long-established series of Faraday Discussion meetings which provide a unique international forum for the exchange of views and newly acquired results in developing areas of physical chemistry, biophysical chemistry and chemical physics.

Carbon in Electrochemistry

Faraday Discussions

www.rsc.org/faraday_d

A General Discussion on Carbon in Electrochemistry was held in Sheffield, UK on the 28th, 29th and 30th of July 2014.

CONTENTS

ISSN 1359-6640; ISBN 978-1-78262-029-7

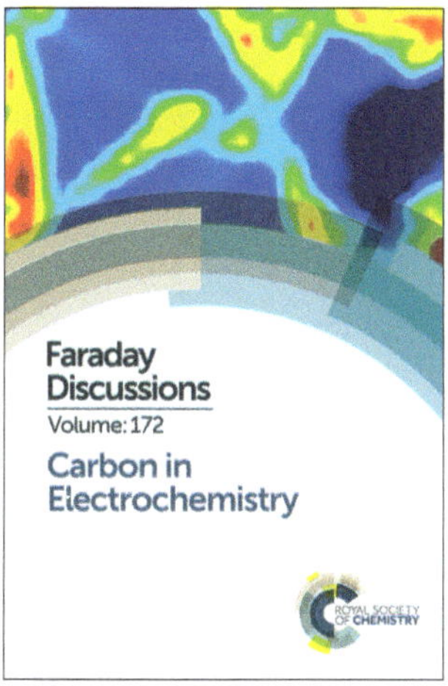

Cover
Image courtesy of Prof. Patrick R. Unwin, University of Warwick, UK.

INTRODUCTORY LECTURE

PAPERS AND DISCUSSIONS

SPONSORS AND EXHIBITORS

ec-lab

Metrohm
Autolab U.K.

bluescientific

SUPPORTING ORGANISATIONS

The
British Carbon Group

CONCLUDING REMARKS

ADDITIONAL INFORMATION

Faraday Discussions

PAPER

Electron transport in all-carbon molecular electronic devices

Richard McCreery,*[ab] Adam Bergren,[b] Amin Morteza-Najarian,[ab] Sayed Youssef Sayed[ab] and Haijun Yan[b]

Received 11th September 2014, Accepted 29th September 2014
DOI: 10.1039/c4fd00172a

Carbon has always been an important electrode material for electrochemical applications, and the relatively recent development of carbon nanotubes and graphene as electrodes has significantly increased interest in the field. Carbon solids, both sp^2 and sp^3 hybridized, are unique in their combination of electronic conductivity and the ability to form strong bonds to a variety of other elements and molecules. The Faraday Discussion included broad concepts and applications of carbon materials in electrochemistry, including analysis, energy storage, materials science, and solid-state electronics. This introductory paper describes some of the special properties of carbon materials useful in electrochemistry, with particular illustrations in the realm of molecular electronics. The strong bond between sp^2 conducting carbon and aromatic organic molecules enables not only strong electronic interactions across the interface between the two materials, but also provides sufficient stability for practical applications. The last section of the paper discusses several factors which affect the electron transfer kinetics at highly ordered pyrolytic graphite, some of which are currently controversial. These issues bear on the general question of how the structure and electronic properties of the carbon electrode material control its utility in electrochemistry and electron transport, which are the core principles of electrochemistry using carbon electrodes.

1. Introduction

Carbon electrodes have been prominent in electrochemistry since the days of Faraday himself, particularly for anodic reactions not possible on mercury due to its relative ease of oxidation. Carbon is essential for operation of the home glucose analyser, which is arguably the most performed chemical analysis in history, certainly in the past several decades. The discovery of the fullerenes, including carbon nanotubes and graphene, significantly raised the profile of carbon materials in areas beyond electrochemistry, as did the development of vapour deposited conducting diamond electrodes. As will be discussed in later

[a]Department of Chemistry, University of Alberta, Canada. E-mail: richard.mccreery@ualberta.ca
[b]National Institute for Nanotechnology, 11421 Saskatchewan Drive, Edmonton, Alberta T6G 2M9, Canada

sections, the electronic properties of carbon, as well as its ability to form a variety of strong bonds with organic molecules, distinguish carbon electrodes from metals and other materials, and underlie many of their useful applications. The Faraday Discussion and associated manuscripts reveal the "many faces" of both sp^2 and sp^3 carbon electrode materials, with broad applications in energy conversion, electroanalysis, electronics, and bioanalytical chemistry. The advent of carbon nanotubes and graphene has stimulated renewed interest in electrochemical applications of novel carbon materials, as reflected in several of the contributed manuscripts. While several of the papers presented at this Faraday Discussion deal directly with fullerenes in electrochemistry, they will not be addressed in this introductory paper. In addition to an overview of some of the special properties of carbon electrodes, this chapter will describe extensions of electrochemical concepts to the area of molecular electronics, where carbon electrodes have both fundamental and practical advantages. The unique property of sp^2 (or sp^3) hybridized carbon as an electronic conductor capable of a wide variety of covalent surface modifications is particularly valuable for providing "contact" between a conductor and an aromatic organic molecule. A practical application of carbon-based molecular electronics in audio processing will be described briefly, followed by a discussion of some recent controversy regarding the electrochemical behaviour of highly ordered pyrolytic graphite electrodes.

2. Special properties of carbon materials as electrodes

As is well known, carbon materials can exhibit high electronic conductivity, either intrinsically in graphite or glassy carbon, or in diamond doped with boron, nitrogen, *etc.*[1,2] Furthermore, the rich covalent chemistry of carbon is the basis of nearly all organic and biochemistry, with a wide variety of bonding to many other elements. The often-cited advantages of both sp^2 and sp^3 carbon solids for electrochemistry include a wide potential window, mainly due to slow kinetics for the oxidation of the carbon itself, as well as low cost and widespread availability in many distinct forms. Metal electrode surfaces can be modified with a variety of chemisorbed and physisorbed molecular mono- and multilayers, but in many cases these modifications are unstable to temperature, time, and potential excursions. The properties of a metal–carbon bond depend strongly on the metal involved, whereas the C–C bond between a conducting sp^2 or sp^3 carbon solid is nearly always strong, and able to survive elevated temperatures and many reactive conditions. An example of a modified carbon electrode is shown in Fig. 1A, for the case of an aromatic molecule bonded to a sp^2 hybridized carbon surface *via* diazonium reduction.[3,4]

As already noted, modified carbon electrodes using glucose oxidase and related cofactors are the basis of the home glucose sensor used by a large fraction of the world's diabetics. The C–C surface bond shown in Fig. 1A is stable to sonication, acid or base treatment, most organic solvents, and temperatures above 400 °C in a vacuum.[3,5–7]

An important carbon fabrication technique for electronics applications is the pyrolysis of photoresist polymers to form "pyrolyzed photoresist films" (PPF).[8–11] The process is shown schematically in Fig. 2, starting with standard silicon wafers

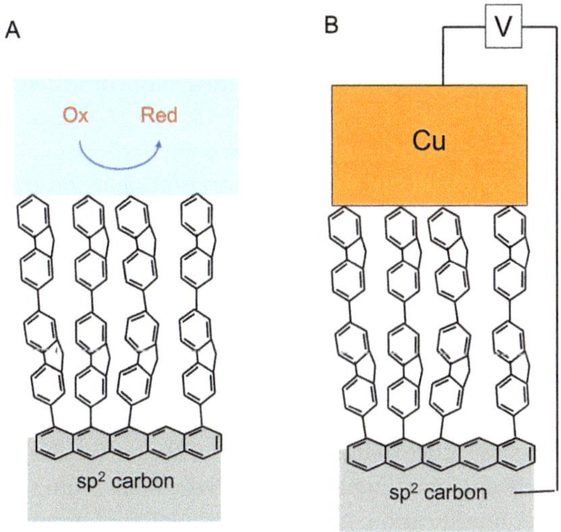

Fig. 1 Schematics of modified sp^2 carbon surfaces consisting of aromatic molecules covalently bonded to edge sites of a conducting carbon solid, used as a modified electrode in solution (A) or in a molecular junction (B).

and commercial photoresist materials. The photoresist is often a phenolic resin (*e.g.* "novolac") which is designed to spin coat evenly on silicon, often coated with native or thermal oxide. A photosensitizer permits patterning of the resist by conventional photolithography if desired, then the wafer is heat treated in a 5%

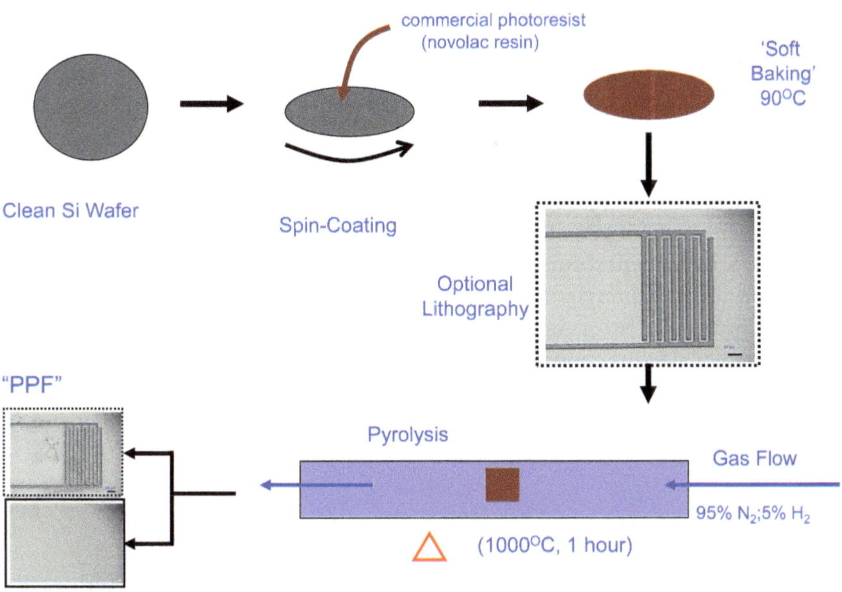

Fig. 2 Formation of pyrolyzed photoresist film (PPF) from a positive photoresist to make either "blanket" or patterned sp^2 hybridized carbon films on Si/SiO_x surfaces. The surface exhibits an rms roughness of <0.4 nm, measured by AFM.

H_2–95% N_2 atmosphere. The result is a very flat (<0.5 nm rms by AFM), mainly sp^2 carbon surface, with conductivity similar to that of glassy carbon. PPF has been used to make a variety of carbon nanostructures, many of which have been used for electrochemistry.

The electrochemical properties of PPF are similar to those of glassy carbon, but its low oxide coverage results in weaker adsorption, and the observed electron transfer rates are generally lower than those of glassy carbon.[10] The reduction of diazonium reagents on PPF results in modified electrodes with high coverage, but often forms multilayers by subsequent radical attack of the initial modification layer.[12,13] While the layer thickness can be controlled by the electrochemical conditions, it should be verified with AFM "scratching".[14] Since the PPF surface is disordered and contains a mixture of edge and basal sites, the bonding geometry is difficult to determine, although the more rapid reaction of diazonium reagents on step edge defects[13] implies that the majority of the surface bonding is between sp^2 edges on the PPF and an aromatic ring from the diazonium reagent, as shown in Fig. 1A. While modification of the basal plane of graphitic surfaces is possible,[15,16] the resulting disruption of three aromatic rings likely makes the reaction slower than on edge plane.

Electron-beam deposited carbon is another relative newcomer to the field, and can be useful in electronics applications.[17–20] E-beam deposition is widely used for metals and refractory materials, and is based on an energetic electron beam striking a target in high vacuum ($<10^{-6}$ Torr). Atoms and clusters are generated at the target, which travel to a sample surface where they deposit, possibly with subsequent diffusion.[21,22] When graphite rods are used as a target, the resulting beam of carbon atoms and clusters forms a flat, conformal film on the sample (designated e-carbon, or e-C). Since carbon atoms and clusters are very reactive, diffusion is unlikely after deposition. Surprisingly, the Raman spectrum of a diazonium-derived molecular layer shows no change in relative intensities after e-C deposition, implying minor modifications to the molecular layer, if any.[20] Although the conductivity of e-C is much lower than that of PPF, the e-C layer may be sufficiently thin (<20 nm) that its ohmic resistance is often negligible along the axis perpendicular to the layer surface. E-beam deposition of either Cu or e-C "top contacts" completes the molecular junction (MJ), as shown schematically in Fig. 1B. A thin layer of e-C (2–5 nm) may also be deposited on the molecular layer to prevent subsequent penetration of metals such as Au into the molecular layer.[20] For e-C, Cu, or Au deposition, no changes were observed in the Raman spectrum after deposition, although Ti and Pt caused significant changes in relative and absolute peak intensities, indicating damage of the molecular layer.[23] The "all carbon" molecular junction, consisting of an organic molecular layer between PPF and e-C (or two films of e-C), has some special advantages due to the stability of carbon and its resistance to oxidation and electromigration, which cause failure of PPF/molecule/Cu devices at high bias.[20,24]

3. Carbon-based molecular electronics

The MJ structure in Fig. 1B indeed turns the molecular layer into a circuit component,[5,6,25,26] but in contrast to the modified electrode, there is no solvent, mobile ions, ionic double layer, or mass transport. Furthermore, there is no reference electrode as in conventional electrochemistry, hence the relevant potential scale is usually referred to the vacuum level. An electron at the normal hydrogen electrode (NHE) potential has a Fermi energy of −4.5 V relative to vacuum,[27] while a typical molecular HOMO level has an energy of −6 V relative to vacuum. The relationship between redox potentials and the vacuum level is rather complex, but the relative energies of molecular orbitals compared to the contact Fermi level are critical to determining junction behaviour. For the current discussion, suffice it to say that comparisons of energetics in a MJ with the electrochemical potential scale should be made carefully, since there is no reference electrode in the MJ. The parallel fabrication of PPF/nitroazobenzene/Cu/Au devices is shown in Fig. 3, taking full advantage of the ability to pattern PPF.[28] In the case shown, 36 "chips" can be fabricated on one 4″ silicon wafer, then each chip may be modified electrochemically and have its top contact deposited to yield 32 identical MJs/chips.

Fig. 4A shows the current density/voltage (*JV*) curves for all 32 junctions on one chip, obtained at 1000 V s^{-1}, with the applied bias stated as PPF relative to Cu (*i.e.* PPF is +). The steep curve occurs with the molecular layer absent, due to the direct contact between the two electronic conductors (PPF and Cu). If molecules behaved like resistors and were governed by Ohm's law, a linear dashed line would be observed with its slope proportional to the inverse of the resistance. If a 3.5 nm thick layer of nitroazobenzene (NAB) is present between the PPF and Cu, the 32 overlaid sigmoidal curves are observed, with a reproducibility of current density of 10–20% across the 32 devices. A plot of the same data on a log scale in

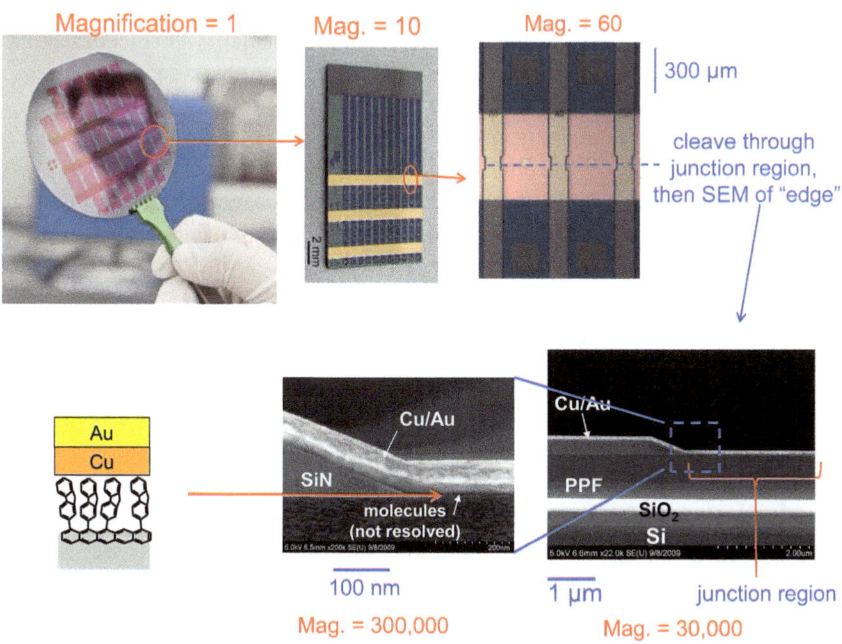

Fig. 3 Images of fabrication steps for PPF/molecule/Cu molecular junctions from a photoresist pattern on SiO_x. Increasing magnification clockwise from upper left photograph.

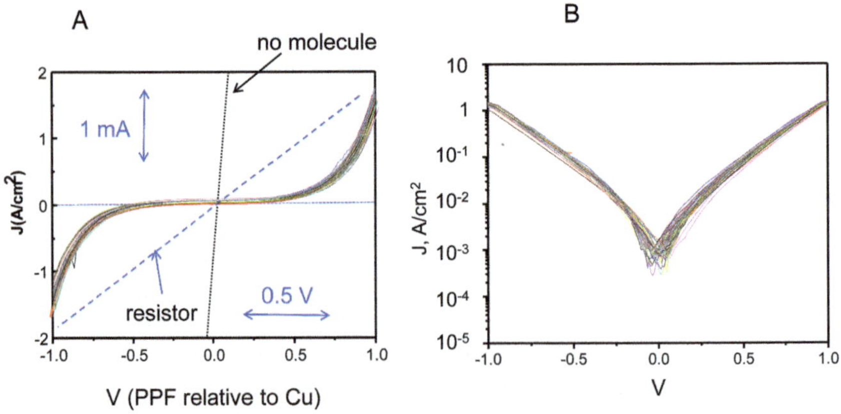

Fig. 4 Current density (*J*) *vs.* voltage curves for PPF/NAB/Cu molecular junctions containing an NAB multilayer 3.5 nm thick. (A) Shows an overlay of 32 curves from one "chip", compared to a device without an NAB layer; (B) is the same data on a logarithmic *J* scale.

Fig. 4B exhibits linear regions above ~0.2 V, indicating exponential dependence of current on applied bias.

The *JV* curves are weakly dependent on temperature, with no change of shape or current density between 5 K and 200 K, and an Arrhenius slope of 37 meV between 250 and 350 K.[29] They were also scan rate independent, at least between 0.1 and 10 000 V s^{-1}, except for a capacitive current observed at high scan rates. The *JV* curves were unchanged after >10^9 cycles to $|J| > 1$ A cm^{-2}, even at 100 °C. We have reported extensively on the characteristics of MJs corresponding to Fig. 1B but with variations in molecular structure, molecular layer thickness, top contact composition, and temperature.[6,25,29–33] The MJ current at a particular bias is exponentially dependent on molecular layer thickness (*d*), as shown in Fig. 5A, and alkanes differ significantly from NAB in the slope of the ln *J* *vs.* *d* plot, generally designated as β, the attenuation coefficient, with units of nm^{-1} ($Å^{-1}$ is also common). A β of 2.5 nm^{-1} indicates that the observed current density decreases by a factor of 1/*e* for each 0.4 nm of molecular layer thickness, while an alkane junction would show a similar attenuation with a thickness increase of 0.12 nm.

Based on a long history of electron transfer in donor–acceptor molecules and through molecular monolayers on electrodes in solution, the distance dependence observed in Fig. 5A and in a variety of molecular junctions has been attributed to quantum mechanical tunnelling, depicted for the current MJs in Fig. 5B. The electron density in the sp^2 carbon extends beyond the carbon surface due to the wave-like behaviour of electrons. There is a finite probability of finding an electron at some distance (*d*) from the carbon surface, and this probability generally decreases exponentially with *d*. In a vacuum, the decrease is very rapid, with the probability decreasing by a factor of *e* (ref. 25) over a distance of 1 nm. However, if the vacuum is replaced by a molecule or metal oxide, the attenuation is more gradual, and β depends on the properties of the material involved. If *d* is small enough, there is a finite overlap between the electron density from the carbon surface and the Cu contact, thus permitting electron transport across the

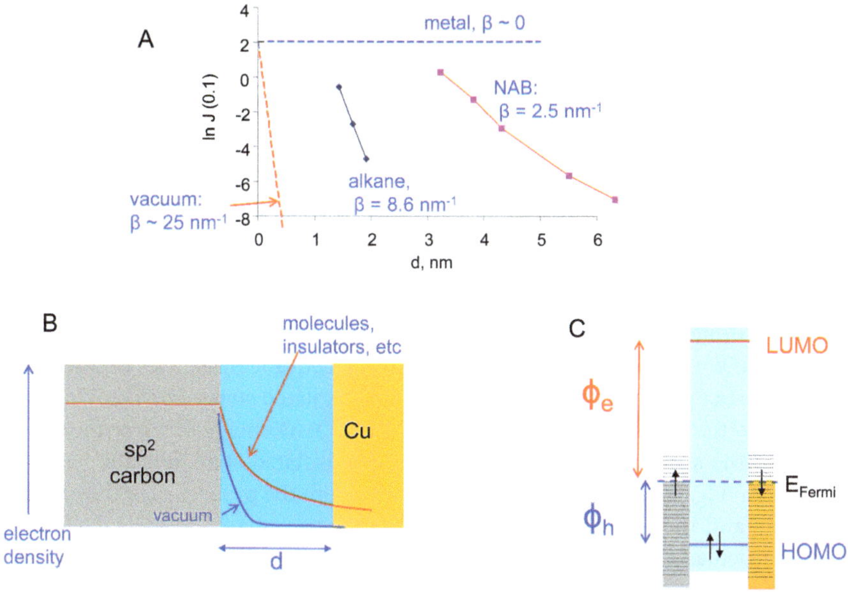

Fig. 5 (A) Plots of ln(*J*) observed at a 0.1 V bias *vs.* the thickness of molecules between the two conducting contacts. Dashed lines are predicted for a vacuum or metal, while experimental results for an alkane series and NAB[29,34] are shown with their slopes. (B) Schematic of electron density distribution outside of an sp^2 carbon surface, with a Cu electrode positioned a distance *d* from the carbon. (C) Simplified schematic of tunnelling barriers for electron (Φ_e) or hole (Φ_h) tunnelling across a molecular junction.

molecular layer in response to an applied bias. Electron tunnelling with its exponential distance dependence has been reported for a wide range of systems,[5,22,35–45] including single molecule as well as "ensemble" MJs, and it is reassuring that a β of 8–9 nm^{-1} is often reported for aliphatic molecules in homogeneous, electrochemical, and solid state electron transport.

Off-resonant tunnelling mechanisms always involve an energy barrier, with higher or thicker barriers decreasing the probability of a tunnelling event. The often assumed barriers for a MJ are shown in Fig. 5C, with Φ_e indicating the electron tunnelling barrier, assumed to equal the difference in energy between the Fermi level of the contacts and the lowest unoccupied molecular orbital (LUMO). An analogous but less intuitive hole tunnelling barrier, Φ_e is mediated by the highest occupied molecular orbital (HOMO). In both cases, transport would be favourable if the orbital involved were at the same energy as the Fermi level, and the displacement of the orbital energy from E_{Fermi} represents the corresponding tunnelling barrier. By classical tunnelling theory, β should be proportional to the square root of the barrier height;[46–48] hence the slope of plots such as Fig. 5A should depend strongly on the HOMO or LUMO levels of the molecules involved. We tested this hypothesis by making an extensive series of 400 MJs of the type shown in Fig. 1B, with seven aromatic structures in addition to the alkanes.[22,34] The aromatic molecules had structures similar to NAB, but were chosen to have a range of HOMO energies spanning 2.3 eV, and a range of LUMO energies spanning 2.7 eV. These variations in orbital energies should have caused large variations in current density, since the barrier height occurs in the exponential portion of the tunnelling current expression. Surprisingly, the seven aromatic molecules had β coefficients which were statistically equal (2.7 ± 0.6 nm^{-1}), and the absolute current densities were also quite similar, indicating a barrier height for the aromatic molecules of 1.3 ± 0.2 eV.[29,34] Furthermore, the insensitivity of the observed tunnelling barrier to the orbital energies of the free molecules was confirmed by ultraviolet photoelectron spectroscopy, which indicated a barrier of 1.4 ± 0.2 eV for the aromatic series.

This unexpected (in fact, disappointing) result is a consequence of the insufficiency of free molecule properties for predicting the behaviour of molecules bonded to one or more conductors. The Mott–Schottky limit is often used in semiconductor electronics for a case where the electronic properties of two solid phases are not perturbed significantly from their isolated states when they are put together to form an interface. In this limit, the HOMO energy of a free molecule would be the same in a MJ as in the molecule's isolated state, and the barriers depicted in Fig. 5C could be readily determined from the properties of the carbon electrode and the free molecule. However, if the two phases interact electronically, particularly by charge transfer across the interface, the classical literature used the terms "strong electronic coupling" and also the "Bardeen limit". Such electronic interactions have been described extensively for interfaces between two semiconductors, a metal and semiconductor, or a metal and organic layer, and can result in apparent changes in work function and alterations in orbital energies of the system.[43,49–53] Bonding an aromatic molecule to the surface of PPF creates a conjugated, covalent bond between two aromatic systems, which are likely to interact electronically. If an electron withdrawing group is placed in the molecular layer, partial electron transfer occurs from the graphene to the molecular layer, which in turn makes the local electrostatic potential more negative. This electrostatic shift increases the energies of the HOMO or LUMO of the molecule relative to the contact Fermi level, thus decreasing Φ_h and increasing Φ_e. Similarly, partial charge transfer from the molecule to the carbon causes an electronic shift in the opposite direction, increasing Φ_h and decreasing Φ_e. As a result, the changes in free molecule HOMO and LUMO energies from donating or withdrawing groups are counteracted by a "levelling effect" which compresses the expected tunnelling barriers to similar values, *i.e.* 1.3 ± 0.2 eV. These results indicate strongly that a MJ must be treated as a complete electronic system consisting of both contacts and the molecular layer, rather than basing expectations on the properties of isolated materials.

Although the levelling effect frustrated attempts to vary tunnelling current by changes in molecular structure, it was observed for PPF/molecule/Cu devices with thicknesses less than 6 nm, in which tunnelling is the dominant transport mechanism. We next considered transport "beyond tunnelling" where MJ thickness was extended beyond 6 nm, and other transport mechanisms might be possible. Three examples are shown in Fig. 6, which is an attenuation plot for PPF/molecule/e-C/Au devices similar to that of Fig. 1B, but with e-C as the top contact and a wider thickness range.

The alkane and aromatic devices with $d < 6$ nm have the same β values as the copper devices, with a similar "levelling" effect. However, extension of anthraquinone (AQ) and NAB to $d > 6$ nm showed some significant differences. β for AQ remains at 3.0 nm^{-1} for d up to 9 nm, where the current decreased below the

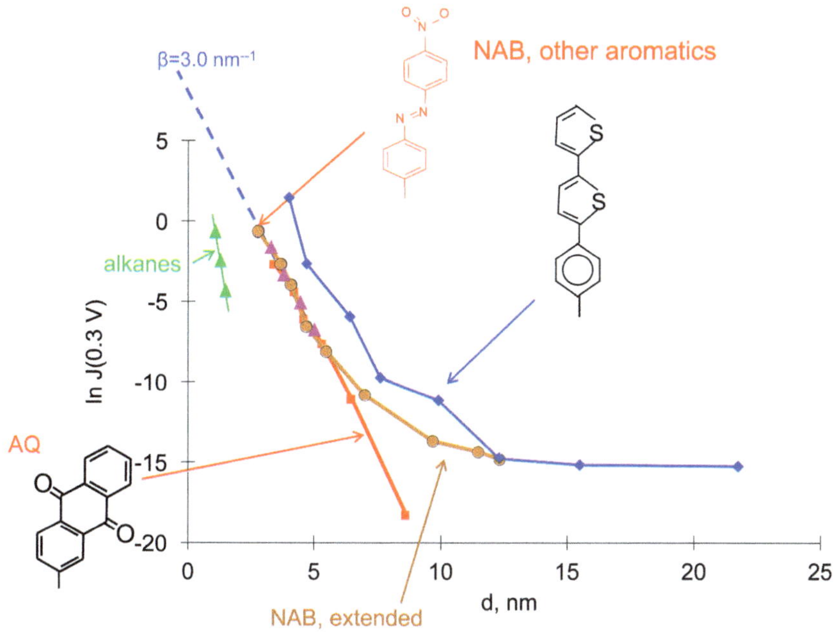

Fig. 6 Attenuation plot of ln(*J*, 0.3 V) *vs.* molecular layer thickness for alkanes,[22] NAB, anthraquinone, and bis-thienylbenzene[30] between PPF and Cu (alkane) and e-Carbon/Au electrodes (NAB, AQ, BTB).

detection limit for the 0.3 V bias. NAB exhibited a curvature of about 6 nm, with ~400 times higher current density than AQ for $d = 9$ nm. Bisthienylbenzene (BTB) shows three regions in the attenuation plot, $\beta = 3$ nm^{-1} below $d = 8$ nm, $\beta = 1.0$ nm^{-1} for $d = 8$–13 nm, and $\beta \approx 0$ nm^{-1} for $d > 15$ nm. A detailed analysis of the BTB case, including temperature dependence, concluded that three mechanisms were operative, including tunnelling for short distance, activated hopping for $d > 15$ nm and $T > 200$ K, and possibly field ionization for $d = 12$–16 nm. Therefore, there are likely several transport mechanisms "beyond tunnelling", and they may have significant roles to play in practical applications of molecular electronics.

4. Audio processing with carbon-based molecular junctions

The tunnelling behaviour evident in Fig. 4A may have commercially useful properties in the near future, in a rather unexpected electronic application. Electric guitar output is often modified before amplification to improve sound quality and introduce unusual audible effects. "Analog distortion pedals" intentionally distort the guitar output to generate harmonics which are perceived as a "warmer" sound. The harmonic distribution of any musical instrument has a strong effect on its sound quality, so electronic modifications are quite common in popular music. The tunnelling behaviour of Fig. 4A is rarely observed in conventional silicon electronic devices, and in fact is usually avoided to reduce

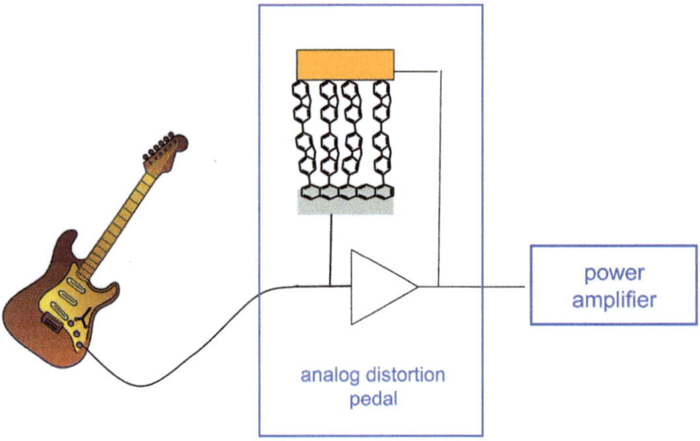

Fig. 7 Schematic of an audio distortion circuit incorporating a molecular junction. The triangle represents the amplifier itself, which could be a transistor or operational amplifier. The distortion introduced by the molecular junction changes the harmonic distribution of the signal, thus modifying the "sound".

"leakage" current. However, the nonlinear *JV* behaviour of a MJ may be well suited for generating harmonics, using an arrangement similar to that shown in Fig. 7.

A MJ combined with a conventional transistor or operational amplifier introduces harmonics into the signal which are clearly audible. Since the *JV* curve in Fig. 4A for the MJ is not possible with silicon (to our knowledge), the distribution or harmonics should be distinct from what is currently possible. Furthermore, the MJ behaviour is easily varied by changing its thickness, so a range of distortion should be possible using different MJs, or with the associated conventional circuit components combined with an MJ. A commercial prototype of an analog distortion pedal which incorporates an "all carbon" MJ is currently under evaluation, and commercial availability is planned for early 2015.

5. Highly ordered pyrolytic graphite (HOPG) revisited

HOPG has long been used as an X-ray monochromator, due to its well-defined interplanar spacing of 3.35 Å. Following an early investigation by Randin and Yeager regarding the low double layer capacitance of HOPG,[54] the electrochemical behaviour of HOPG as an electrode in solution was explored extensively, starting in the 1980s.[55–60] One motivation was to use HOPG, as a structurally well-defined form of graphitic carbon, as a model for widely used glassy carbon and polycrystalline graphite electrodes. As described in several reports and reviews, a general picture of what factors control the electrochemical reactivity of graphitic electrodes emerged, based on the properties of HOPG and variations in electrochemical kinetics observed at the basal and edge planes of HOPG and with a variety of surface modifications. To briefly review these conclusions, it became generally accepted that the basal plane of HOPG exhibited 3–5 orders of magnitude lower k^o values for outer-sphere redox reactions such as $Ru(NH_3)_6^{+3/+2}$ and $IrCl_6^{-3/-4}$, compared to the edge plane. Certain redox systems such as aquated

$Fe^{+3/+2}$ and $V^{+2/+3}$ were very sensitive to C=O groups on the carbon surface,[61,62] which rarely exist on the basal plane, if at all. Catechol oxidation was very sensitive to hydrogen bonding sites on the electrode surface, including those present during adsorption of the catechol itself. Removal of such sites nearly totally prevents catechol oxidation, and the effect was attributed to proton coupled electron transfer catalysed by surface hydrogen bonding sites.[63,64] A related observation was that anthraquinone-2,6-disulfonate (AQDS) reduction was small or undetectable on HOPG basal plane, but strong on edge plane, leading to the proposal that AQDS adsorption occurs mainly at step edges, possibly due to the electronic disturbance introduced into the surface by defects.[65–67] The initial proposal that AQDS adsorbed solely to edge plane defects was later amended when it was observed that AQDS adsorption observed voltammetrically corresponded to ~30 times the step edge density observable with STM.[65] This observation led to the proposal that electronic disturbances caused by step edges caused adsorption over a much larger area than the defect itself. The broad conclusion from these experiments is that basal plane HOPG is less active than edge plane regarding electrode kinetics and adsorption, but also that the nature of the kinetic heterogeneity varies significantly depending on the particular redox mechanism(s) involved.[2,68]

Subsequent reports from various laboratories using different techniques supported the generally accepted kinetic and electrochemical heterogeneity of HOPG, including electrochemical deposition of metals at step edges,[69,70] Raman spectroscopy of enhanced adsorption at edge sites,[71] faster electrode kinetics at edge sites compared to basal plane,[72–79] enhanced electrochemiluminescence at edge plane,[59] and nucleation of diazonium derived phenyl radicals at step edges.[13] However, the advent of scanning electrochemical microscopy (SECM) and scanning electrochemical cell microscopy (SECCM) permitted monitoring electrode activity on a microscopic scale, and some of the results challenged the accepted model. In particular, "close to reversible" kinetics for $Fe(CN)_6^{-3/-4}$ and other outer sphere systems were reported on HOPG basal plane[80] using SECCM, in contrast to the very slow kinetics (ΔE_p > 700 mV for $Fe(CN)_6^{-3/-4}$) reported previously on "validated" HOPG basal plane surfaces.[66,67] The widely studied catechol dopamine was reported to have very slow kinetics on basal plane HOPG using macroscopic voltammetry,[58] but later reports using SECM[81] and SECCM[82–84] reported fast kinetics on "pristine" basal plane HOPG. Furthermore, the observed electrode kinetics and AQDS adsorption on basal plane HOPG were claimed to be independent of the grade of HOPG or the density of step edges,[83,85] leading to the conclusion that defects were not controlling the observed electrochemical behaviour. A more detailed review of the many reported electrochemical results reported on HOPG basal plane is beyond the scope of the present article, but suffice it to say that there are two incompatible viewpoints: one that basal plane HOPG has low reactivity for outer sphere electron transfer reactions, anthraquinone adsorption, and dopamine oxidation, provided there is a low density of defects; and a second that such processes occur rapidly on basal plane HOPG and are not affected by step edges or other defects.

The discrepancy between macroscopic and microscopic observations on basal plane HOPG stimulated a brief reconsideration of $Fe^{+2/+3}$ kinetics in the authors' laboratory, almost two decades after the original reports.[61,62] Using the "low defect" HOPG which is not available commercially, a series of cleaved surfaces were examined with voltammetry of 1 mM $Fe^{+2/+3}$ in 1 M $HClO_4$, using the "inverted drop" cell to avoid mechanical strain on the sample. The experiments were done using ultrapure water (18 MΩ, <10 ppb TOC), with less than 1 minute exposure of the cleaved surfaces to air, in a different lab (and different country) from the 1990s experiments, with the results shown in Fig. 8. The inverted drop area was not controlled, producing the variation in peak current, but the electrode kinetics clearly vary significantly for surfaces cleaved by different methods. When the cleaving method prescribed in 1992 (ref. 67) was used repeatedly, four successive cleaves produced ΔE_p > 1000 mV for $Fe^{+2/+3}$.

There are at least two possible reasons for a scientific disagreement; a problem with the experimental results or a difference in interpretation. An example of the latter is the proposal that step edges or defects and their effects on the density of electronic states are responsible for certain electrochemical behaviour on HOPG, such as adsorption or faster electron transfer kinetics. The complexity of the electronic and chemical properties of carbon materials in their variety of allotropes and crystallinity will undoubtedly result in variations in behaviour, and it will be a long time before the underlying materials properties which determine electrochemical reactivity are fully understood. However, the discrepancy regarding HOPG basal plane reactivity is more fundamental, in that different laboratories have reported many results at opposite extremes, *e.g.*, from "very slow" to "reversible" in the case of the electrode kinetics of $Fe(CN)_6^{-3/-4}$. Several established laboratories report discordant results using similar experimental conditions, in some cases even the same source of HOPG. Obviously, since the

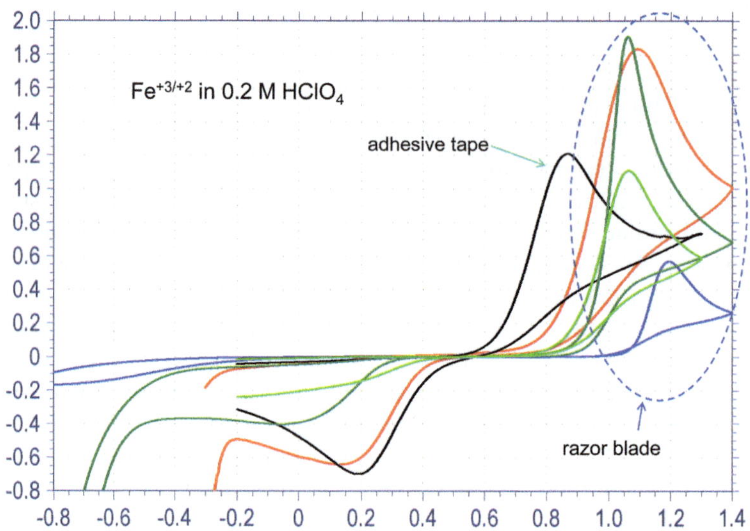

Fig. 8 Cyclic voltammograms (0.2 V s^{-1}) on five basal plane HOPG surfaces obtained from the same HOPG sample with various cleaving techniques, including removal of several layers using adhesive tape. The solution contained $Fe^{+3/+2}$ in 0.2 M $HClO_4$ in ultrapure water (18 MΩ resistivity, TOC < 10 ppb), and was placed on the HOPG surface as an inverted drop with uncontrolled diameter within one minute of cleaving the sample in air. The blue curve showing the largest ΔE_p was obtained with the technique used previously,[66,67] and the green and red curves by a razor blade inserted between graphite layers to expose a cleaved surface.

interpretation and any scientific conclusions are based on experimental results, it is difficult to proceed if those results differ as much as claimed for HOPG basal plane electrochemistry.

Our early conclusion that HOPG surfaces varied significantly in defect density was based on the experimental observation of adsorption, capacitance, and ΔE_p ($Fe(CN)_6^{-3/-4}$, 1 V s^{-1}) for a series of HOPG surfaces. For the ungraded or highest commercial grades available at the time, the AQDS capacitance varied from <1 to 69 pmol cm^{-2}, the capacitance from 0.6 to 6.5 μF cm^{-2}, and ΔE_p from 92 to >1000 mV.[67] Therefore, we selected surfaces for closer study which had the lowest adsorption and largest ΔE_p as being the most representative of low-defect HOPG basal plane behaviour. Recall that HOPG is far from a single crystal, with both a finite in-plane microcrystallite size (~ a few μm) and "turbostratic" disorder, with crystallites rotated relative to each other around the axis perpendicular to the basal plane surface. The effect of such disorder on electronic properties is an active and complex research area, and will likely remain so for a long time. Step edges or other defects may have a large effect on electronic properties, through changes in density of electronic states, scattering centres, grain boundaries, *etc.* Suppose that these electronic disturbances underlie the variation in electron transfer rate and adsorption, and that the electronic changes depend strongly on defects. Since catechols can be "self-catalyzed" by their own adsorption and subsequent proton coupled electron transfer,[63,64] a defective surface would appear reactive once catechol (or its quinone) adsorption occurs. Furthermore, Ta and McDermott[86] reported that using voltammetry to measure AQDS adsorption may be complicated by the interaction of both adsorption itself and the varying electrochemical reactivity near defects. As an indication of the strong electronic consequences of defects, consider the example of basal plane graphite modified by chemisorption of H atoms, to form single sp^3 point defects in an otherwise sp^2 hybridized surface.[87] STM in an UHV environment revealed a "charge redistribution" which extended ~5 nm away from the point defect, corresponding to an area of ~78 nm^2. Taking the carbon number density as 3.8×10^{15} cm^{-2},[88] there are ~3000 carbon atoms in the surface graphene layer within 5 nm of the sp^3 centre. Stated differently, ~2×10^{-12} mol cm^{-2} of defects would result in charge redistribution of the entire basal surface. Such electronic perturbations do not necessarily result in changes in adsorption or electrode kinetics, but they do indicate the sensitivity of basal plane HOPG to a low density of defects. As pointed out in early reports, mechanical strain (*e.g.* from adhesive tape) during cleaving significantly reduced the probability of obtaining a low defect surface.[67]

The discussion about the electrochemical reactivity of basal plane HOPG will likely continue for some time, given the disparity in experimental results. The only way this author can explain such disparity is by invoking electronic disturbances which vary with the source of HOPG and the history of each individual basal surface. Without some type of "selection" procedure, the results can vary greatly, using the same sample and identical procedure. Either the "selection" procedure is fundamentally flawed, or there are possibly unknown characteristics of HOPG which are not captured in a simple, crystalline model, and are not controlled experimentally.

Future perspectives

It is clearly evident from the many interesting papers and associated discussion in this volume that the electrochemistry and electronic properties of carbon materials remain active research fields even after many decades of modern research. The advent of fullerenes, particularly carbon nanotubes and graphene, have rekindled the already active research on electrochemistry at carbon electrodes. Nearly all of the allotropes of elements other than carbon have been known for more than a century, yet a new allotrope of carbon was discovered in 1985, and a vast new research area emerged as a result. The combination of a metallic or semiconducting solid material with the rich covalent chemistry of carbon underlies both the attractive electrochemical properties of carbon materials, and their application in solid state molecular electronic devices. The widespread applications of carbon electrodes described in this Faraday Discussion range from nanoscale phenomena at interfaces in both solution and solid state, to large scale industrial applications in supercapacitors and electrocatalysis. Like many vibrant areas of scientific research, our current knowledge of the electronic and electrochemical behaviour of carbon materials is really the "tip of the iceberg", and the possibilities for future development are both extensive and intriguing.

Acknowledgements

This work was supported by the Natural Sciences and Engineering Research Council of Canada, the National Institute for Nanotechnology, the University of Alberta, and the Province of Alberta. The National Institute for Nanotechnology is operated as a partnership between the National Research Council and the University of Alberta and is jointly funded by the Government of Canada, the Government of Alberta, and the University of Alberta. Research from the authors' laboratory preceding 2006 was funded by the National Science Foundation (USA), the Air Force Office of Scientific Research (USA) and Ohio State University.

References

1 R. L. McCreery, in *Electroanalytical Chemistry*, ed. A. J. Bard, Dekker, N.Y., 1991, vol. 17, pp. 221–374.
2 R. L. McCreery, *Chem. Rev.*, 2008, **108**, 2646–2687.
3 D. Belanger and J. Pinson, *Chem. Soc. Rev.*, 2011, **40**, 3995–4048.
4 J. Pinson and F. Podvorica, *Chem. Soc. Rev.*, 2005, **34**, 429–439.
5 R. L. McCreery and A. J. Bergren, *Adv. Mater.*, 2009, **21**, 4303–4322.
6 R. McCreery, J. Wu and R. J. Kalakodimi, *Phys. Chem. Chem. Phys.*, 2006, **8**, 2572–2590.
7 T.-C. Kuo, Ph.D. thesis, The Ohio State University, 1999.
8 R. Kostecki, B. Schnyder, D. Alliata, X. Song, K. Kinoshita and R. Kotz, *Thin Solid Films*, 2001, **396**, 36–43.
9 R. Kostecki, X. Song and K. Kinoshita, *Electrochem. Solid-State Lett.*, 1999, **2**, 465–467.
10 S. Ranganathan and R. L. McCreery, *Anal. Chem.*, 2001, **73**, 893–900.
11 S. Ranganathan, R. L. McCreery, S. M. Majji and M. Madou, *J. Electrochem. Soc.*, 2000, **147**, 277–282.

12 J. K. Kariuki and M. T. McDermott, *Langmuir*, 2001, **17**, 5947–5951.
13 J. K. Kariuki and M. T. McDermott, *Langmuir*, 1999, **15**, 6534–6540.
14 F. Anariba, S. H. DuVall and R. L. McCreery, *Anal. Chem.*, 2003, **75**, 3837–3844.
15 P. Allongue, M. Delamar, B. Desbat, O. Fagebaume, R. Hitmi, J. Pinson and J. M. Saveant, *J. Am. Chem. Soc.*, 1997, **119**, 201–207.
16 P. M. Kirkman, A. G. Güell, A. S. Cuharuc and P. R. Unwin, *J. Am. Chem. Soc.*, 2013, **136**, 36–39.
17 T. P. DeAngelis, R. W. Hurst, A. M. Yacynych, H. B. Mark Jr, W. R. Heineman and J. S. Mattson, *Anal. Chem.*, 1977, **49**, 1395–1398.
18 J. S. Mattson and C. A. Smith, *Anal. Chem.*, 1975, **47**, 1122–1125.
19 J. J. Blackstock, A. A. Rostami, A. M. Nowak, R. L. McCreery, M. Freeman and M. T. McDermott, *Anal. Chem.*, 2004, **76**, 2544–2552.
20 H. Yan, A. J. Bergren and R. L. McCreery, *J. Am. Chem. Soc.*, 2011, **133**, 19168–19177.
21 A. P. Bonifas and R. L. McCreery, *Nano Lett.*, 2011, **11**, 4725–4729.
22 A. P. Bonifas and R. L. McCreery, *Nat. Nanotechnol.*, 2010, **5**, 612–617.
23 A. M. Mahmoud, A. J. Bergren and R. L. McCreery, *Anal. Chem.*, 2009, **81**, 6972–6980.
24 S. Ssenyange, H. Yan and R. L. McCreery, *Langmuir*, 2006, **22**, 10689–10696.
25 R. McCreery, H. Yan and A. J. Bergren, *Phys. Chem. Chem. Phys.*, 2013, **15**, 1065–1081.
26 F. Anariba, J. Steach and R. McCreery, *J. Phys. Chem. B*, 2005, **109**, 11163–11172.
27 A. J. Bard and L. R. Faulkner, *Electrochemical Methods*, Wiley, New York, 2nd edn, 2001.
28 J. Ru, B. Szeto, A. Bonifas and R. L. McCreery, *ACS Appl. Mater. Interfaces*, 2010, **2**, 3693–3701.
29 A. J. Bergren, R. L. McCreery, S. R. Stoyanov, S. Gusarov and A. Kovalenko, *J. Phys. Chem. C*, 2010, **114**, 15806–15815.
30 H. Yan, A. J. Bergren, R. McCreery, M. L. Della Rocca, P. Martin, P. Lafarge and J. C. Lacroix, *Proc. Natl. Acad. Sci. U. S. A.*, 2013, **110**, 5326–5330.
31 S. Y. Sayed, A. Bayat, M. Kondratenko, Y. Leroux, P. Hapiot and R. L. McCreery, *J. Am. Chem. Soc.*, 2013, **135**, 12972–12975.
32 J. A. Fereiro, R. L. McCreery and A. J. Bergren, *J. Am. Chem. Soc.*, 2013, **135**, 9584–9587.
33 A. M. Mahmoud, A. J. Bergren, N. Pekas and R. L. McCreery, *Adv. Funct. Mater.*, 2011, **21**, 2273–2281.
34 S. Y. Sayed, J. A. Fereiro, H. Yan, R. L. McCreery and A. J. Bergren, *Proc. Natl. Acad. Sci. U. S. A.*, 2012, **109**, 11498–11503.
35 S. M. Lindsay and M. A. Ratner, *Adv. Mater.*, 2007, **19**, 23–31.
36 H. T. Nicolai, M. Kuik, G. A. H. Wetzelaer, B. de Boer, C. Campbell, C. Risko, J. L. Bredas and P. W. M. Blom, *Nat. Mater.*, 2012, **11**, 882–887.
37 P. A. Van Hal, E. C. P. Smits, T. C. T. Geuns, H. B. Akkerman, B. C. De Brito, S. Perissinotto, G. Lanzani, A. J. Kronemeijer, V. Geskin, J. Cornil, P. W. M. Blom, B. De Boer and D. M. De Leeuw, *Nat. Nanotechnol.*, 2008, **3**, 749–754.
38 H. B. Akkerman and B. De Boer, *J. Phys.: Condens. Matter*, 2008, **20**, 013001.
39 H. B. Akkerman, R. C. G. Naber, B. Jongbloed, P. A. van Hal, P. W. M. Blom, D. M. de Leeuw and B. de Boer, *Proc. Natl. Acad. Sci. U. S. A.*, 2007, **104**, 11161–11166.

40 L. Luo, S. H. Choi and C. D. Frisbie, *Chem. Mater.*, 2011, **23**, 631–645.
41 S. H. Choi, C. Risko, M. C. R. Delgado, B. Kim, J.-L. Bredas and C. D. Frisbie, *J. Am. Chem. Soc.*, 2010, **132**, 4358–4368.
42 S. H. Choi, B. Kim and C. D. Frisbie, *Science*, 2008, **320**, 1482–1486.
43 O. Yaffe, Y. B. Qi, L. Scheres, S. R. Puniredd, L. Segev, T. Ely, H. Haick, H. Zuilhof, A. Vilan, L. Kronik, A. Kahn and D. Cahen, *Phys. Rev. B: Condens. Matter Mater. Phys.*, 2012, **85**, 045433.
44 A. Vilan, O. Yaffe, A. Biller, A. Salomon, A. Kahn and D. Cahen, *Adv. Mater.*, 2010, **22**, 140–159.
45 I. Ron, I. Pecht, M. Sheves and D. Cahen, *Acc. Chem. Res.*, 2010, **43**, 945–953.
46 J. G. Simmons, *J. Appl. Phys.*, 1964, **35**, 2655–2658.
47 J. G. Simmons, *J. Appl.Phys.*, 1963, **34**, 2581–2590.
48 J. G. Simmons, *J. Appl. Phys.*, 1963, **34**, 1793–1803.
49 J. Hwang, A. Wan and A. Kahn, *Mater. Sci. Eng., R*, 2009, **64**, 1–31.
50 A. Salomon, T. Boecking, O. Seitz, T. Markus, F. Amy, C. Chan, W. Zhao, D. Cahen and A. Kahn, *Adv. Mater.*, 2007, **19**, 445–450.
51 D. Cahen, A. Kahn and E. Umbach, *Mater. Today*, 2005, **8**, 32–41.
52 F. Li, Y. Zhou, F. Zhang, X. Liu, Y. Zhan and M. Fahlman, *Chem. Mater.*, 2009, **21**, 2798–2802.
53 C. Tengstedt, W. Osikowicz, W. R. Salaneck, I. D. Parker, C.-H. Hsu and M. Fahlman, *Appl. Phys. Lett.*, 2006, **88**, 053502.
54 J.-P. Randin and E. Yeager, *J. Electroanal. Chem.*, 1972, **36**, 257–276.
55 H. Gerischer, *J. Phys. Chem.*, 1985, **89**, 4249–4251.
56 H. Gerischer, R. McIntyre, D. Scherson and W. Storck, *J. Phys. Chem.*, 1987, **91**, 1930.
57 R. McIntyre, D. Scherson, W. Storck and H. Gerischer, *Electrochim. Acta*, 1987, **32**, 51–53.
58 R. J. Bowling, R. T. Packard and R. L. McCreery, *J. Am. Chem. Soc.*, 1989, **111**, 1217–1223.
59 R. J. Bowling, R. L. McCreery, C. M. Pharr and R. C. Engstrom, *Anal. Chem.*, 1989, **361**, 2763–2766.
60 R. Bowling, R. T. Packard and R. L. McCreery, *Langmuir*, 1989, **5**, 683.
61 P. Chen, M. Fryling and R. L. McCreery, *Anal. Chem.*, 1995, **67**, 3115.
62 C. A. McDermott, K. Kneten and R. L. McCreery, *J. Electrochem. Soc.*, 1993, **140**, 2593.
63 S. DuVall and R. L. McCreery, *J. Am. Chem. Soc.*, 2000, **122**, 6759–6764.
64 S. DuVall and R. L. McCreery, *Anal. Chem.*, 1999, **71**, 4594–4602.
65 M. T. McDermott and R. L. McCreery, *Langmuir*, 1994, **10**, 4307.
66 K. K. Cline, C. A. McDermott and R. L. McCreery, *J. Phys. Chem.*, 1994, **98**, 5314.
67 M. T. McDermott, K. Kneten and R. L. McCreery, *J. Phys. Chem.*, 1992, **96**, 3124.
68 R. L. McCreery and M. T. McDermott, *Anal. Chem.*, 2012, **84**, 2602–2605.
69 E. C. Walter, B. J. Murray, F. Favier, G. Kaltenpoth, M. Grunze and R. M. Penner, *J. Phys. Chem. B*, 2002, **106**, 11407–11411.
70 R. M. Penner, M. J. Heben, T. L. Longin and N. S. Lewis, *Science*, 1990, **250**, 1118–1121.
71 K. G. Ray and R. L. McCreery, *Anal. Chem.*, 1997, **69**, 4680–4687.
72 T. J. Davies, M. E. Hyde and R. G. Compton, *Angew. Chem., Int. Ed.*, 2005, **44**, 5121–5126.
73 R. R. Moore, C. E. Banks and R. G. Compton, *Anal. Chem.*, 2004, **76**, 2677–2682.

74 T. J. Davies, R. R. Moore, C. E. Banks and R. G. Compton, *J. Electroanal. Chem.*, 2004, **574**, 123–152.
75 C.-Y. Lee and A. M. Bond, *Anal. Chem.*, 2009, **81**, 584–594.
76 F. Armstrong, A. Bond, H. Hill, I. Psalti and C. Zoski, *J. Phys. Chem.*, 1989, **93**, 6485–6493.
77 F. Armstrong, A. Bond, H. Hill, N. Oliver and I. Psalti, *J. Am. Chem. Soc.*, 1989, **111**, 9185–9189.
78 N. L. Ritzert, J. Rodríguez-López, C. Tan and H. D. Abruña, *Langmuir*, 2013, **29**, 1683–1694.
79 W. Li, C. Tan, M. A. Lowe, H. D. Abruna and D. C. Ralph, *ACS Nano*, 2011, **5**, 2264–2270.
80 S. C. S. Lai, A. N. Patel, K. McKelvey and P. R. Unwin, *Angew. Chem., Int. Ed.*, 2012, **51**, 5405–5408.
81 S. Lhenry, Y. R. Leroux and P. Hapiot, *Anal. Chem.*, 2012, **84**, 7518–7524.
82 A. N. Patel, K. McKelvey and P. R. Unwin, *J. Am. Chem. Soc.*, 2012, **134**, 20246–20249.
83 A. N. Patel, S.-y. Tan, T. S. Miller, J. V. Macpherson and P. R. Unwin, *Anal. Chem.*, 2013, **85**, 11755–11764.
84 A. N. Patel, S.-y. Tan and P. R. Unwin, *Chem. Commun.*, 2013, **49**, 8776–8778.
85 G. Zhang, P. M. Kirkman, A. N. Patel, A. S. Cuharuc, K. McKelvey and P. R. Unwin, *J. Am. Chem. Soc.*, 2014, **136**, 11444–11451.
86 T. C. Ta, V. Kanda and M. T. McDermott, *J. Phys. Chem. B*, 1999, **103**, 1295–1302.
87 P. Ruffieux, M. Melle-Franco, O. Gröning, M. Bielmann, F. Zerbetto and P. Gröning, *Phys. Rev. B: Condens. Matter Mater. Phys.*, 2005, **71**, 153403.
88 N. Jung, A. C. Crowther, N. Kim, P. Kim and L. Brus, *ACS Nano*, 2010, **4**, 7005–7013.

Faraday Discussions

PAPER

Single layer graphene as an electrochemical platform†

Nicole L. Ritzert,[a] Wan Li,[b] Cen Tan,[a] Gabriel G. Rodríguez-Calero,[a] Joaquín Rodríguez-López,[a] Kenneth Hernández-Burgos,[a] Sean Conte,[a] Joshua J. Parks,[ab] Daniel C. Ralph*[b] and Héctor D. Abruña*[a]

Received 4th April 2014, Accepted 9th May 2014

DOI: 10.1039/c4fd00060a

Over the past decade, there has been a great deal of interest in graphene with regards to its electrochemical behavior. Previous studies have focused on understanding fundamental processes such as charge transfer and molecular transport at the graphene–electrolyte interface as well as on applications of graphene in electronic, optical, and mechanical systems. We present illustrative examples of large area, single layer graphene platforms for applications such as optical and sensing devices as well as microfluidic systems. Three examples of graphene modified with thin polymer films are discussed. We have explored the use of graphene as an electrochemical platform for surface-generated electrogenerated chemiluminescence (ECL) using poly-$[Ru(v\text{-}bpy)_3]^{2+}$, where v-bpy is 4-vinyl, 4′-methyl 2,2′-bipyridine, as a model system. We found that while graphene can sustain ECL conditions, there was film degradation during ECL, as demonstrated by a decrease in ECL intensity upon potential cycling even in the presence of a graphene coating ("graphene blanket"). Using poly 3,4-ethylenedioxythiophene (EDOT), we demonstrate a facile method of fabricating electrochromic electrodes from large area graphene. The oxidation of NADH at graphene was catalyzed using an electrodeposited layer of 3,4-dihydroxybenzaldehyde as an effective redox mediator. In addition, we describe the fabrication and characterization of a microfluidic device based on a solution-gated field effect transistor which was able to detect changes of 60 mV per pH unit change in an inverted cell design. On the other hand, a 29 mV shift in the Dirac point per unit pH change was measured with our microfluidic devices, and a *ca.* 10% FET conductance change was measured when we continuously changed the pH in

[a]Department of Chemistry and Chemical Biology, Cornell University, Ithaca, NY, 14853, USA. E-mail: hda1@cornell.edu; Fax: +1-607-255-9864; Tel: +1-607-255-4720

[b]Laboratory of Atomic and Solid State Physics, Department of Physics, Cornell University, Ithaca, NY, 14853, USA. E-mail: ralph@ccmr.cornell.edu; Fax: +1-607-255-6428; Tel: +1-607-255-9644

† Electronic supplementary information (ESI) available: Cyclic voltammograms for deposition of 3,4-DHB on graphene; cyclic voltammograms of poly-$[Ru(v\text{-}bpy)_3]^{2+}$ on ITO in acetonitrile and in phosphate buffer; cyclic voltammograms of poly-$[Ru(v\text{-}bpy)_3]^{2+}$ on ITO before and after ECL with tripropylamine; Raman spectra of ITO/poly-$[Ru(v\text{-}bpy)_3]^{2+}$/graphene before and after ECL; cyclic voltammograms of poly-$[Ru(v\text{-}bpy)_3]^{2+}$ on ITO before and after ECL with $C_2O_4^{2-}$; cyclic voltammograms of immobilized 3,4-DHB at different sweep rates. See DOI: 10.1039/c4fd00060a

solution from 6.91 to 7.64 in the microfluidic channel, demonstrating local microfluidic pH sensing (albeit non-Nerstian) in real time.

Introduction

During the past decade since it was isolated through simple exfoliation,[1,2] graphene has demonstrated interesting optical, mechanical, and electronic properties.[3–6] Graphene has also found use in electrochemical systems, from sensing devices to fundamental studies of its charge transfer properties.[7–11] Large area, single layer graphene[12–15] is interesting because unlike graphene flakes, it can be used as the sole electrode material, removing the need for a conductive substrate and thus allowing for measurements on graphene without complications from the underlying substrate. Single layer graphene also offers the opportunity for simultaneously probing two interfaces in proximity. Recent debate has focused on any differences in charge transfer processes at the edges and basal planes of graphene[16,17] and the effects of impurities introduced during growth and processing on the purported catalytic properties of graphene.[18]

Our group has fabricated and electrochemically characterized large area, single layer graphene electrodes.[12,19–21] We have used these electrodes to probe phenomena such as controlling mechanical and chemical defect sites[19] and measuring the 2D surface diffusion coefficient of electroactive adsorbates.[20] To demonstrate the utility of large area, single layer graphene in different electrochemical platforms, we have expanded on our previous studies by demonstrating optical and sensing devices based on single layer graphene electrodes. In this paper, we discuss graphene as a protective layer for thin polymer films during electrogenerated chemiluminescence (ECL), as an electrode material for electrochromic devices based on electropolymerized, conducting polymer films, and as a substrate for electrochemical sensing of biomolecules. We also describe the fabrication and characterization of a working microfluidic device based on electrochemically-gated single layer graphene (FET) transistors.

Electropolymerizable materials, such as those based on conducting, organic polymers, are of great interest in electrochromic displays, electrochemical energy storage, organic light emitting diodes, flexible electronics, and photo-voltaic devices because of their long-term stability, low oxidation potential, high contrast ratio, fast charge/discharge rates, tunable electronic structure, and fast switching times.[22–34] Graphene has been proposed as an alternative to commonly used support materials such as indium tin oxide (ITO) due to graphene's extraordinary electrical, physical, optical, and mechanical properties.[1,3,4,35–37]

We have developed methodologies for electropolymerizing electrochromic compounds, *viz.*, 3,4-ethylenedioxythiophene (EDOT), onto large area graphene electrodes through the use of electrochemical techniques such as cyclic voltammetry, chronoamperometry, and galvanostatic methods. Electropolymerization provides advantages in terms of sample preparation, cost, and processing, especially when compared to other chemical and physical deposition methods. The use of electrochemical techniques provides control over material properties such as thickness, electronic structure, chemical structure, and morphology by simply changing the deposition parameters and conditions. In addition, by using electropolymerization to modify the graphene electrodes, we ensure the conformal coating of the electrode material.

ECL based on thin organic films is useful in devices such as optical displays and small-molecule sensors.[38–40] Briefly, ECL is produced when anions and cations of an electroactive, luminescent species that are electrogenerated in close proximity react, provided that sufficient energy is provided to form an excited electronic state, as shown below for annihilation ECL:

$$R^{\bullet -} + R^{\bullet +} \rightarrow R^* + R \tag{1}$$

$$R^* \rightarrow R + h\nu \tag{2}$$

Because these reactions produce visible light, the required energy is typically *ca.* 2 to 3 eV.[38] Coreactant ECL is induced by a single potential step or a linear potential sweep. A coreactant is used to generate a very reactive species that provides the energy needed to excite the emitting species. Proposed mechanisms are available for a variety of coreactants.[38–40] One example is the tris(2,2′-bipyridyl) ruthenium(II)/oxalate system, in which a reducing reagent is formed upon sweeping the potential to oxidizing conditions:[41]

$$[Ru(bpy)_3]^{2+} \rightarrow [Ru(bpy)_3]^{3+} + e^- \tag{3}$$

$$[Ru(bpy)_3]^{3+} + C_2O_4^{2-} \rightarrow [Ru(bpy)_3]^{2+} + C_2O_4^{\bullet -} \tag{4}$$

$$C_2O_4^{\bullet -} \rightarrow CO_2^{\bullet -} + CO_2 \tag{5}$$

$$[Ru(bpy)_3]^{2+} + CO_2^{\bullet -} \rightarrow [Ru(bpy)_3]^{+} + CO_2 \tag{6}$$

$$[Ru(bpy)_3]^{3+} + [Ru(bpy)_3]^{+} \rightarrow [Ru(bpy)_3]^{2+*} + [Ru(bpy)_3]^{2+} \tag{7}$$

Coreactant ECL is useful in systems where electrolyte media, such as aqueous solutions, are unstable at potentials necessary to generate the reactive species or if one of the electrogenerated species is unstable.

Polymer films can degrade during ECL due to effects such as dissolution caused by enhanced solubility of the charged film and mechanical breakdown of the film due to incorporation of solvent or counter ions.[38] Particularly, highly reactive radicals, which may react with the film, are formed during coreactant ECL. In the following studies, we considered using graphene as a protective layer at the film–electrolyte interface during ECL of polymer films to prevent/mitigate mechanical loss, *e.g.*, dissolution, of the film. Large area graphene has already been studied as a protective layer for metals and semiconductors under thermally and electrochemically oxidizing conditions[42–44] although passivation of metals such as copper was limited to several hours. We investigated ECL at ITO electrodes modified using poly-$[Ru(v\text{-}bpy)_3]^{2+}$, where v-bpy is 4-vinyl, 4′-methyl 2,2′-bipyridine, as a model system since the ECL of surface-immobilized $[Ru(bpy)_3]^{2+}$ is well-documented[45–49] and films of electropolymerized $[Ru(v\text{-}bpy)_3]^{2+}$ can be readily grown.[50] We first considered coreactant ECL in order to probe electron transfer processes across graphene, requiring interaction of a solution species, *viz.*, the coreactant, through the graphene layer with immobilized $[Ru(v\text{-}bpy)_3]^{2+}$ films.

Another useful area of polymer-modified electrodes is in biosensing. Polymer films have been used as antifouling agents as well as providing a soft platform for antibody detection.[51–53] In our studies, we considered sensors based on nicotinamide adenine dinucleotide (NADH) because NADH is involved in over 200 known enzymatic reactions,[54] making it a popular analyte, and glassy carbon modified with polymerized 3,4-dihydroxybenzaldehyde (DHB) catalyzes its oxidation.[55] NADH oxidation has already been documented at graphene-composite electrodes.[7,8,56,57] Graphene offers the ability to easily pattern electrodes in different shapes, including coating 3D structures and flexible substrates, that may be advantageous in designing new biosensor configurations. In addition, and in the specific case of DHB as a surface modifier, it needs to be pointed out that such films only grow on carbon surfaces.

In addition to polymer-modified graphene platforms, we report our effort to construct a graphene sensing platform on which we integrate the solution-gated field effect transistor into a microfluidic system to achieve local pH sensing. Graphene is a two dimensional material that is fully comprised of surfaces. The electrical conductance of graphene is thus sensitive to the changes of charge environment induced by analyte reaction or adsorption on the graphene surface, and is often used as a readout for various graphene-based sensing devices. The high charge carrier mobility of graphene imparts lower noise and thus better sensing performance in graphene-based sensing device.[58] In solution-gated graphene field effect transistors,[59,60] the graphene surface is directly exposed to the solution and the gate voltage is applied electrochemically to graphene through the Debye layer at the graphene–electrolyte interface. Due to the relatively large capacitance of the Debye layer, changes of the gate voltage induced by a chemical change in solution are expected to lead to significant changes of the graphene charge carrier density, and consequently the graphene conductance. Solution-gated field effect transistors have been extensively studied for pH sensing.[59–62] A moderate response of 6–20 mV shift in the Dirac point per unit pH change was typically observed. The corresponding change of graphene conductance was found to be 3–13% per unit pH change, when compared to the conductance at pH 7 at gate voltages close to the Dirac point.

Experimental methods

Materials

All chemicals were used as received unless otherwise noted. Tripropylamine (TPrA) (≥98%), 3,4-ethylenedioxythiophene (EDOT), tetrabutylammonium hexafluorophosphate ($TBAPF_6$) (for electrochemical analysis, ≥99.0%), tetrabutylammonium perchlorate (TBAP) (for electrochemical analysis, ≥99.0%), anisole (anhydrous, 99.7%), Tris buffer, and acetonitrile (MeCN) (anhydrous, 99.8%), were from Sigma-Aldrich (St. Louis, MO). Strem Chemicals (Newbury Point, MA) was the supplier of tris(2,2′-bipyridyl)ruthenium(II) chloride hexahydrate, $[Ru(bpy)_3]^{2+}$ (min. 98%). Purified $[Ru(v\text{-}bpy)_3](PF_6)_2$ was from an available batch. Reduced nicotinamide adenine dinucleotide (NADH) (grade II, 98%) disodium salt was from Roche Diagnostics (Indianapolis, IN). Sodium oxalate ($Na_2C_2O_4$), sodium acetate, sodium phosphate monobasic, sodium phosphate dibasic, sodium nitrate, acetone, isopropanol, and dichloromethane were all analytical reagent grade and from Mallinkrodt-Baker (Phillipsburg, NJ). 3,4-DHB (97%) was from Pfaltz and Bauer (Waterbury, CT). Silicon wafers with a 300 nm layer of thermally grown oxide (Si/SiO_2) were from Silicon Quest International (San Jose, CA), and 1.1 mm glass slides coated with indium tin oxide (ITO) (10 Ω cm^{-2}) were from Nanocs Inc. (New York, NY). Water (18 MΩ cm) purified using a Millipore (Billerica, MA) Milli-Q system was used to clean all glassware and to prepare all aqueous solutions.

In our studies, "large area" refers to the size of the graphene sheet so that the graphene electrode was between 2 and 4 cm^2 at the end of processing. Graphene was grown on 0.025 mm thick copper foil (99.8%, Alfa-Aesar, Ward Hill, MA) using the chemical vapor deposition (CVD) method as described previously.[12,20,63] Graphene was grown using Cu foil initially *ca.* 2 $\times$ 8 cm^2. Then, the foil was cut into smaller, rectangular pieces with dimensions between 1.5 and 2 cm along each side. A layer of poly(methyl methacrylate) (PMMA) was added to the graphene before etching the copper in order to provide support during transfer. After etching the copper and transferring to water at least six times, graphene–PMMA was transferred to Si/SiO_2, ITO, modified ITO (*vide infra*), or a glass microscope slide and dried with nitrogen. To remove the PMMA, the chips were immersed first in anisole for two to three hours, then in 50/50 (v/v) dichloromethane–acetone for at least 12 hours, and finally in isopropanol for at least two hours. The quality of the single layer graphene was verified by the 2D and G peaks at 2700 and 1600 cm^{-1}, respectively, in the Raman spectra,[64,65] obtained using a Renishaw (Gloucestershire, UK) InVia Confocal Raman microscope with excitation wavelength of 488 nm. Raman spectra also indicated that the graphene samples contained some disorder, shown by a small D peak at 1350 cm^{-1}.

Electrochemical and ECL measurements

All electrochemical measurements were performed in a three-electrode cell configuration with graphene as the working electrode. Electrical contact to graphene was made using gold evaporated on the graphene or using copper tape and indium wire directly on the graphene. Auxiliary and reference electrodes were a gold or platinum wire and an Ag/AgCl/1 M KCl electrode inside a salt bridge (3% agar with 0.2 M potassium nitrate), respectively. All potentials are reported *vs.* Ag/AgCl unless noted otherwise. A silver wire quasi reference electrode (Ag QRE) was used in nonaqueous solutions. No calibration of the QRE was made because the potential was usually stable over the time required for each measurement. Electrodes were placed in a homemade Teflon cell with a Plexiglas base, and an o-ring was used to expose an area of 0.5 cm^2 graphene to solution. For ECL measurements, the base was an aluminum piece with a hole to allow any emitted light to pass through the bottom of the cell. In this configuration, any ECL emission was detected through the ITO electrode on the bottom of the cell opposite to the solution (Fig. 1).

ECL measurements were performed using an EG&G Princeton Applied Research (Princeton, NJ) Model 173 Potentiostat-Galvanostat with a Model 175 Universal Programmer and Model 176 Current Follower. Data were recorded using a custom LabVIEW (National Instruments, Austin, TX) program and a National Instruments USB-6210 DAQ device. Emission was detected using a Hamamatsu Model R928 photomultiplier tube (PMT) operated at −750 V using a Model 556 high-voltage power supply from EG&G Ortec (Oak Ridge, TN). The PMT

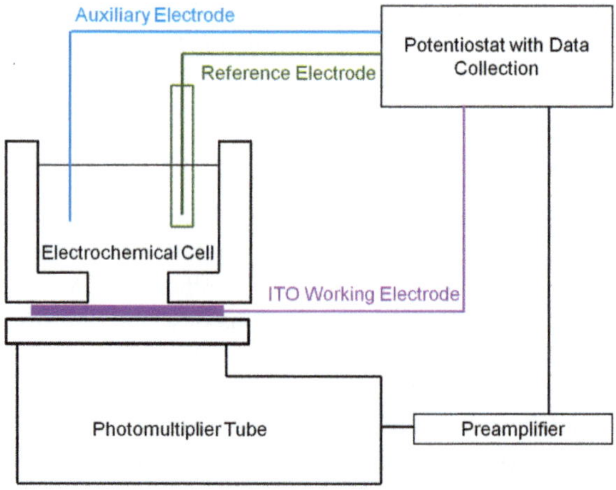

Fig. 1 Diagram of equipment for detecting ECL emission. ECL generated in the electrochemical cell is detected by the photomultiplier tube placed directly below the cell, opposite to the electrolyte solution.

was connected to the DAQ device using an Ithaco (Ithaca, NY) Model 1211 current preamplifier.

Electrochromic measurements were performed at room temperature using a Model HSV-100 or HABF1510m potentiostat (Hokuto Denko). All other electrochemical measurements were performed using a Model 900 potentiostat (CH Instruments, Austin, TX).

Polymer film growth for electrochemical measurements

Films of poly-$[Ru(v\text{-}bpy)_3]^{2+}$ were made through electroreductive polymerization[50] by immersing an ITO electrode into a 0.5 mM solution of $[Ru(v\text{-}bpy)_3](PF_6)_2$, in 0.1 M $TBAPF_6$ in acetonitrile and cycling the potential between −0.2 V and −1.8 V *vs.* the Ag QRE, just past the second reduction wave of poly-$[Ru(v\text{-}bpy)_3]^{2+}$, which is the second ligand reduction. The potential was typically cycled at a scan rate of 100 mV s^{-1}. After electropolymerization, the electrode was removed from the $[Ru(v\text{-}bpy)_3]^{2+}$ solution and rinsed with acetone. ITO–poly-$[Ru(v\text{-}bpy)_3]^{2+}$, *i.e.*, the ITO modified with poly-$[Ru(v\text{-}bpy)_3]^{2+}$, was then immersed in fresh electrolyte solution or graphene was transferred over the polymer layer as described previously.

3,4-DHB films were grown through electrooxidative polymerization[55] by cycling the potential of graphene on Si/SiO_2 between −0.2 and +0.8 V *vs.* Ag/AgCl for 5 cycles in a solution of 1 mM 3,4-DHB in 0.1 M Tris buffer, pH 8.21, with 0.1 M $NaNO_3$ (Fig. S1†). After rinsing with water, the electrode was immersed in fresh buffer or buffer containing NADH.

Microfluidic device fabrication and characterization

Fabrication of graphene field effect transistor. For the fabrication of graphene field effect transistors, metal electrodes made of 5 nm Ti and 200 nm Au were first patterned on a fused silica substrate using photolithography and electron beam evaporation (Fig. 2a; step 1). CVD graphene was then transferred onto the fused silica substrate to make electrical contact with the metal electrodes using the same processes as those for ECL sample preparation, except that the PMMA supporting layer was kept on top of the graphene (Fig. 2a; step 2). In the next step, (Fig. 2a; step 3) the graphene was patterned into a rectangular stripe using photolithography and oxygen plasma. Since the PMMA was not removed in the earlier step, the oxygen plasma patterned graphene by first removing the PMMA and then the underneath graphene. This fabrication scheme avoided the direct contact of photoresist with graphene and was capable of obtaining cleaner graphene surfaces. In the last step, the photoresist and PMMA on graphene strips were removed by heated 1165 solution, which consists mostly of *N*-methyl-2-pyrrolidone, for 1 hour at 70–80 degrees. The devices were then rinsed with isopropanol and blown dry with N_2 (Fig. 2a; step 4). Fig. 2b shows the AFM image of a graphene sample we patterned with the above method. The CVD graphene surface remained smooth and clean after the patterning step.

Microfluidic solution-gated graphene FET construction and measurement schemes. For the microfluidic solution-gated graphene FET devices, PDMS microfluidic channels 200 μm in width and 40 μm in height with multiple inlet ports and one outlet reservoir were fabricated using standard methods. A closed, sealed reservoir was placed along the channel for reference electrode insertion (Fig. 3a). To construct the microfluidic device, the polydimethylsiloxane (PDMS) block was treated with oxygen plasma and then bonded to a freshly fabricated graphene FET. The two compartments were carefully aligned to ensure that the microfluidic channel was right above the graphene stripe (Fig. 3c, inset). A platinum wire used as a reference electrode was punched into the sealed reservoir and further sealed with epoxy (Fig. 3b and c).

During measurement, buffer solutions of known pH were injected into the microfluidic channel *via* the inlet port at a controlled flow rate of $<\sim 15\ \mu L\ min^{-1}$. A gate voltage (0–0.5 V) was applied to the graphene FET through the Pt reference

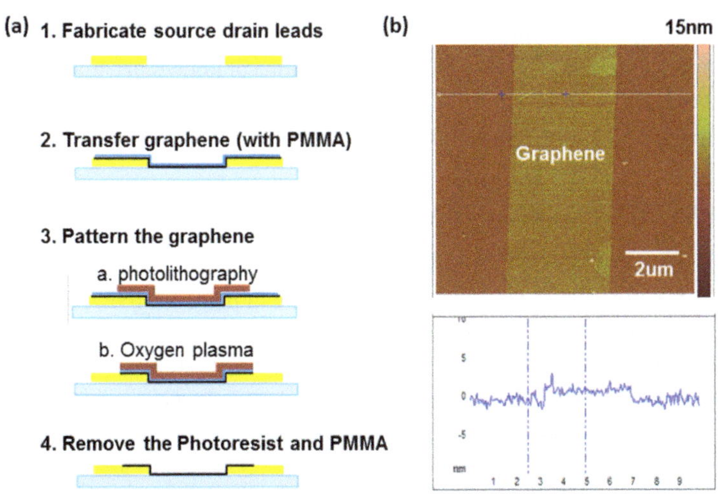

Fig. 2 Graphene FET fabrication. (a) Fabrication steps for graphene FET. (b) AFM characterization of graphene surface after patterning.

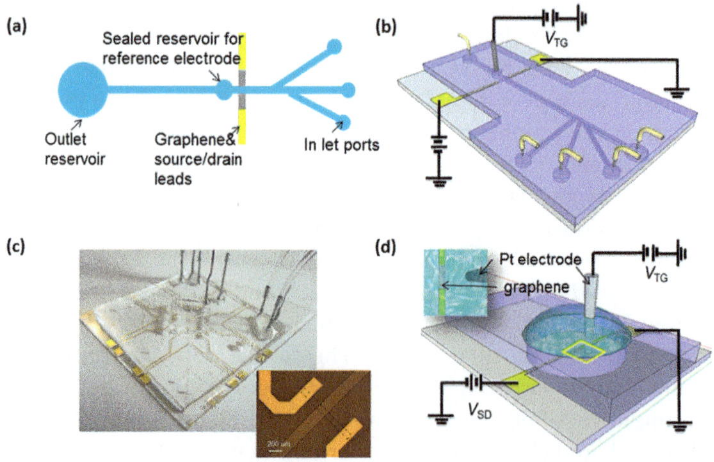

Fig. 3 Solution-gated graphene FET fabrication. (a) Path of the microfluidic channel. (b) Structure of the microfluidic graphene FET and measurement scheme. (c) Optical image of an assembled microfluidic graphene FET. (d) Structure and measurement scheme of the solution-gated FET in an inverted cell structure.

electrode, while the conductance of graphene was monitored through the metal electrodes with constant source–drain voltage of 5 or 10 mV.

Fig. 3b illustrates the structure of the microfluidic device together with our electrical measurement scheme. Fig. 3c shows the optical image of the microfluidic device after construction. To make good use of a single piece of CVD graphene and the fabrication resources, 8 devices were fabricated in parallel processes on a quarter of a 4 inch fused silica substrate.

Solution-gated graphene FET in inverted cell structure for system characterization. We also constructed solution-gated graphene FETs in an inverted cell structure to more easily characterize the pH sensitivity of our solution-gated graphene FETs. This kind of device was constructed by placing a PDMS open reservoir (6 mm in diameter) directly on top of the graphene FET. During measurement, the reservoir was filled with solutions of known pH, and the Pt wire was directly inserted to the reservoir. Fig. 3d illustrates the structure of the microfluidic device together with the electrical measurement scheme.

Results and discussion

Graphene as a protective blanket for ECL platforms

For ECL, ITO was used as the supporting substrate in place of Si/SiO_2 because ITO is conductive, allowing facile polymer growth, and transparent, allowing for light transmission. Measurements were performed on ITO modified with poly-$[Ru(v\text{-}bpy)_3]^{2+}$ and/or graphene in solutions of two different coreactants, tripropylamine (TPrA) and $C_2O_4^{\ 2-}$. The surface coverage, Γ, of poly-$[Ru(v\text{-}bpy)_3]^{2+}$ was determined using the charge passed by the $Ru^{2+/3+}$ couple in a cyclic voltammogram of the film immersed in fresh electrolyte (Fig. S2a†). The area is related to the charge, Q, of an electroactive species immobilized on an electrode. From Q, the surface coverage is calculated from the following equation:

$$Q = nFA\Gamma \tag{8}$$

where n is the number of electrons passed per redox active group ($n = 1$ for poly-$[Ru(v\text{-}bpy)_3]^{2+}$), A is the area of the electrode ($0.5\ cm^2$) and F is Faraday's constant (96 485 C per mole e^-). Values of Γ were typically between 1.5×10^{-10} and 1.5×10^{-9} mol cm^{-2}, as controlled by the number of potential cycles, corresponding to 2 to 18 monolayers with a thickness between 2.8 and 25.6 nm, assuming that each monolayer is 8.3×10^{-11} mol cm^{-2} and is 1.42 nm thick.[50]

ECL with TPrA. We started our studies of coreactant ECL at graphene using TPrA[66] in 0.2 M phosphate buffer, pH 7.0. Visible, orange light was observed at single layer graphene on Si/SiO_2 immersed in a solution of $[Ru(bpy)_3]^{2+}$ and TPrA (Fig. 4a) with an applied potential of +1.4 V *vs.* Ag/AgCl for at least 10 minutes, indicating that graphene is stable under these conditions. The ECL intensity was similar at bare ITO and ITO/graphene (Fig. 4b and c), indicating that graphene does not significantly quench ECL.

Fig. 5a shows ECL as a function of potential for ITO/poly-$[Ru(v\text{-}bpy)_3]^{2+}$ exposed to TPrA. Upon potential cycling, the ECL intensity decreased, with a 75% loss in signal observed in the second cycle. Chronoamperometry indicates that emission was no longer detected after 60 s (Fig. 5b), indicating that the film is unstable under our experimental conditions. Indeed, *ca.* 80% of the charge for the Ru^{2+}/Ru^{3+} couple in the immobilized film was lost after the first cycle (Fig. S3†), which could be due to film dissolution or that the film simply became inactive, as has been reported in the presence of Ru^{1+} species.[45] These results are in contrast to annihilation ECL in acetonitrile, in which emission was observed for at least 20 minutes of a poly-$[Ru(v\text{-}bpy)_3]^{2+}$.[45]

With a single layer of graphene placed between the poly-$[Ru(v\text{-}bpy)_3]^{2+}$ and electrolyte solution, *i.e.*, ITO/poly-$[Ru(v\text{-}bpy)_3]^{2+}$/graphene, ECL emission decreased upon potential cycling, similar to the results for the bare ITO/poly-$[Ru(v\text{-}bpy)_3]^{2+}$, with a 50% decrease in signal upon the second cycle (Fig. 5c). Therefore, single layer graphene offers little protection in the presence of TPrA. Raman spectra (Fig. S4†) show the characteristic G and 2D peaks near 1600 and 2700 cm^{-1}, indicating that graphene was present before and after ECL. These

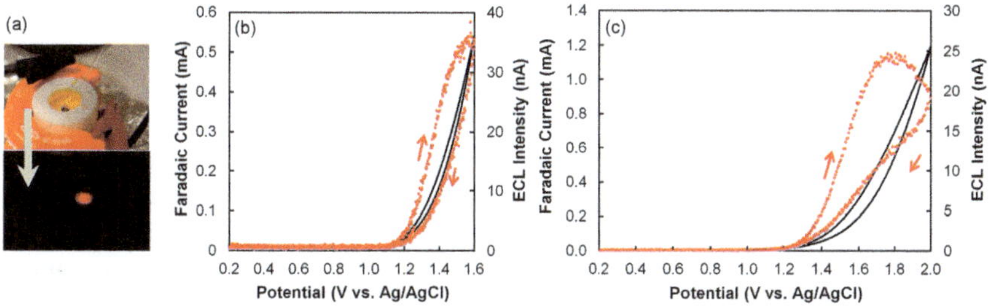

Fig. 4 Solution-based ECL with TPrA. (a) Image of ECL of graphene on Si/SiO_2 in electrochemical cell with 0.2 mM $[Ru(bpy)_3]^{2+}$ and 0.1 M TPrA. Applied potential, +1.4 V *vs.* Ag/AgCl. (b) Faradaic current (black) and ECL emission (red) of bare ITO exposed to a solution of 0.2 mM $[Ru(bpy)_3]^{2+}$ and 0.1 M TPrA as a function of applied potential. (c) Faradaic current (black) and ECL emission (red) of ITO covered with single layer graphene exposed to a solution of 0.2 mM $[Ru(bpy)_3]^{2+}$ and 0.1 M TPrA as a function of applied potential. Scan rate, 50 mV s^{-1}; supporting electrolyte, 0.2 M phosphate buffer, pH 7.0; electrode area, 0.5 cm^2.

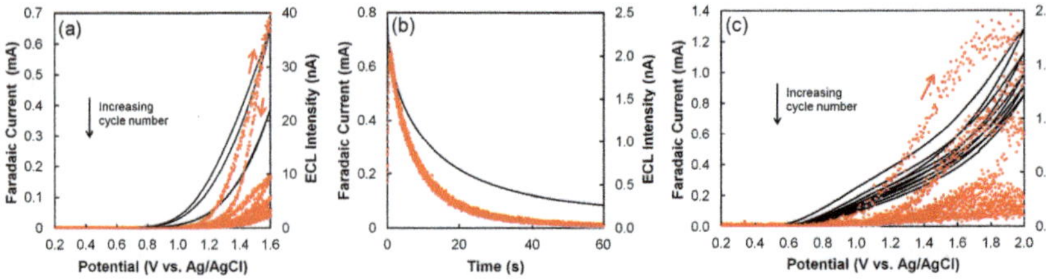

Fig. 5 (a) Faradaic current (black line, first and seventh cycles shown) and ECL emission (red) as a function of potential of ITO modified with poly-$[Ru(v\text{-}bpy)_3]^{2+}$ exposed to a solution of 0.13 M TPrA. (b) Faradaic current (black) and ECL emission (red) as a function of time for ITO/poly-$[Ru(v\text{-}bpy)_3]^{2+}$ exposed to a solution of 0.13 M TPrA. Applied potential, +1.2 V *vs.* Ag/AgCl. (c) Faradaic current (black) and ECL emission (red) as a function of potential of ITO/poly-$[Ru(v\text{-}bpy)_3]^{2+}$/graphene in 0.13 M TPrA. Scan rate, 50 mV s^{-1}; supporting electrolyte, 0.2 M phosphate buffer, pH 7.0; electrode area, 0.5 cm^2.

results suggest that loss in ECL intensity is due to the film itself and that graphene did not decompose during ECL.

ECL with $C_2O_4^{2-}$. Because poly-$[Ru(v\text{-}bpy)_3]^{2+}$ became inactive during ECL with TPrA, we pursued $C_2O_4^{2-}$ as the coreactant.[46–48] These measurements were performed in 0.1 M acetate buffer, pH 4.6, which is at a lower pH than the TPrA measurements at pH 7.0. These conditions should be less oxidizing, and therefore less aggressive, at the same applied potentials. In addition, $[Ru(bpy)_3]^{2+}$ immobilized in Nafion films exhibited ECL for over 1500 cycles using $C_2O_4^{2-}$ as a coreactant in aqueous solution.[47]

Cyclic voltammetry and chronoamperometry (Fig. 6a and b) of ITO/poly-$[Ru(v\text{-}bpy)_3]^{2+}$ exposed to $C_2O_4^{2-}$ showed that a 25% loss in ECL intensity occurred upon the second cycle and that ECL emission was detected without a diminishing signal for over two minutes under applied potential at +1.4 V *vs.* Ag/AgCl. In addition, 80% of the original charge in the Ru^{2+}/Ru^{3+} couple remained after ECL (Fig. S5†). These results demonstrate that the films are significantly more stable with $C_2O_4^{2-}$ as the coreactant. However, as with TPrA, we observed a decrease in ECL intensity even when single layer graphene was transferred on top of the film (Fig. 6b and d).

It is interesting to note that the cyclic voltammetric profile is different for the bare polymer film and for that covered with graphene over the same potential range (Fig. 6a and c). The peak shape in the cyclic voltammogram of the graphene-covered film is reminiscent of a mass transport-limited reaction such as a solution species diffusing to an array of ultramicroelectrodes. It is possible that the coreactant diffuses to the underlying polymer film exposed to solution through physical holes in the graphene. However, the fact that the ECL intensities of the bare and covered films are of the same order of magnitude suggests that ECL probably originates from more than just sites with holes. Perhaps the coreactant diffuses through the holes and becomes sandwiched between the graphene and polymer film. Some questions that we are exploring are: does ECL occur only at sites where bare polymer is exposed through holes in the graphene or does ECL occur over the entire surface? Does electron transfer occur through single layer graphene or does the graphene mediate electron transfer?

Although graphene seemed to offer little improvement as a protective layer in ECL, it may be interesting to consider graphene as a contact material in

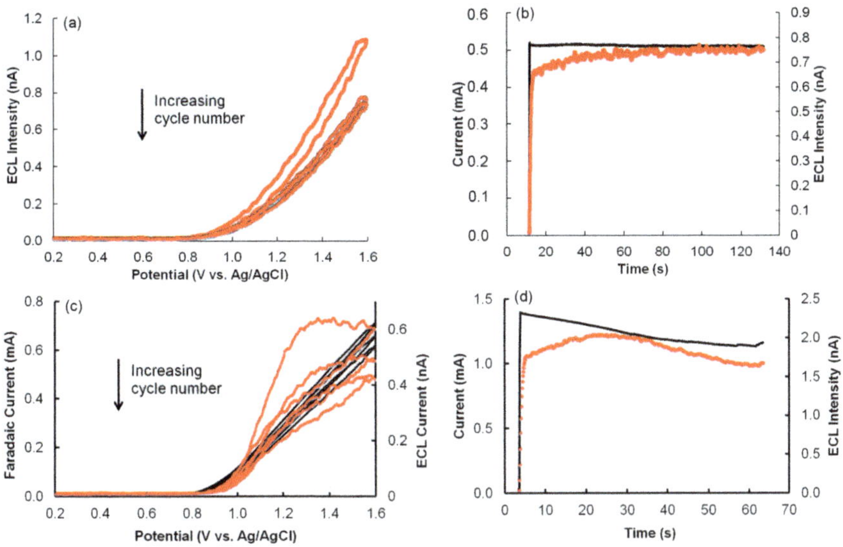

Fig. 6 (a) ECL emission as a function of potential of ITO modified with poly-$[Ru(v\text{-}bpy)_3]^{2+}$ exposed to a solution of 50 mM $C_2O_4^{2-}$. (b) Faradaic current (black) and ECL emission (red) as a function of time for ITO/poly-$[Ru(v\text{-}bpy)_3]^{2+}$ exposed to a solution of 50 mM $C_2O_4^{2-}$. Applied potential, +1.4 V *vs.* Ag/AgCl. (c) Faradaic current (black) and ECL emission (red) as a function of potential of ITO/poly-$[Ru(v\text{-}bpy)_3]^{2+}$/graphene in 50 mM $C_2O_4^{2-}$. (d) Faradaic current (black) and ECL emission (red) as a function of time for ITO/poly-$[Ru(v\text{-}bpy)_3]^{2+}$/graphene exposed to a solution of 50 mM $C_2O_4^{2-}$. Applied potential, +1.4 V *vs.* Ag/AgCl. Scan rate, 25 mV s^{-1}; supporting electrolyte, 0.1 M acetate buffer, pH 4.6; electrode area, 0.5 cm^2.

luminescence devices. It has been demonstrated that processes such as metal deposition affect the behavior of electroluminescent devices, possibly due to film degradation during the deposition process.[67] Large area graphene may offer a more gentle way of introducing the top contact electrode in these systems.

Electropolymerization of EDOT on graphene for electrochromic applications

Electropolymerization of materials onto graphene electrodes has similarities to electropolymerization onto conducting substrates in general (Fig. 7A). A 20 mM solution of EDOT was cycled between 0.0 V *vs.* Ag/Ag^+ and +0.9 V *vs.* Ag/Ag^+. The first cycle shows that the monomer onset potential is +0.7 V *vs.* Ag/Ag^+. Subsequent cycles show a decrease in the onset potential and an increase in the pseudocapacitive current. Both features are characteristic of the electropolymerization of these types of monomers. Our results indicate that the monomers used for these experiments can be anodically electropolymerized onto graphene electrodes (Fig. 7B). A cyclic voltammogram of the PEDOT/graphene composite in 0.1 M TBAP/MeCN solution shows the characteristic peaks of a PEDOT film-modified electrode, where the major doping peak can be observed around 0 V *vs.* Ag/Ag^+ and the major de-doping peak can be observed at −1.1 V *vs.* Ag/Ag^+. Similar anodic and cathodic current responses were observed for other monomer solutions studied, *e.g.*, polythiophene.

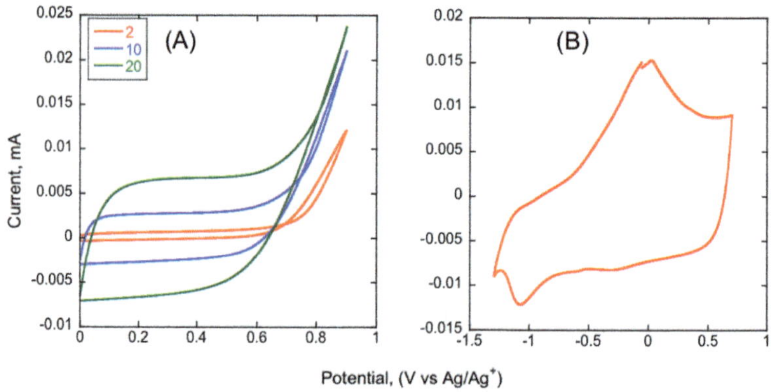

Fig. 7 (a) Cyclic voltammogram for electropolymerization in 10 mM EDOT in at different cycle numbers. (b) Cyclic voltammogram of the film modified graphene electrode in fresh electrolyte solution. Sweep rate, 20 mV s^{-1}; supporting electrolyte, 0.1 M TBAP/MeCN.

During the electrochemical cycling, color changes, *i.e.*, electrochromism, were observed at the electrode surface (Fig. 8). Upon application of positive potentials, the PEDOT/graphene model system faded from blue to a colorless film, and this color change could be reversed by scanning in the negative direction. In Fig. 9, a film-modified graphene electrode is presented. In this figure, the electrode setup used is illustrated and includes the electrical contacts, graphene monolayer and polymer (blue).

NADH oxidation at 3,4-DHB-modified graphene

To test large area graphene as an electrochemical biosensor, oxidation of NADH was characterized at bare single layer graphene and graphene modified with an

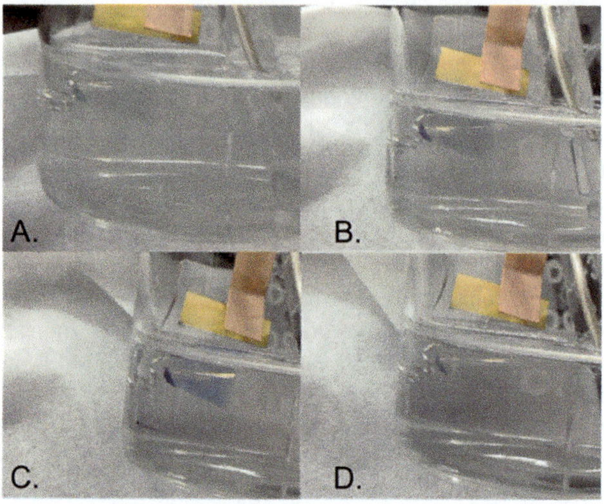

Fig. 8 Electrochemical cycling of a PEDOT/graphene composite in 0.1 M TBAP/MeCN. (a) 0 min, (b) 1 min, (c) 2 min and (d) 3 min of electrochemical cycling.

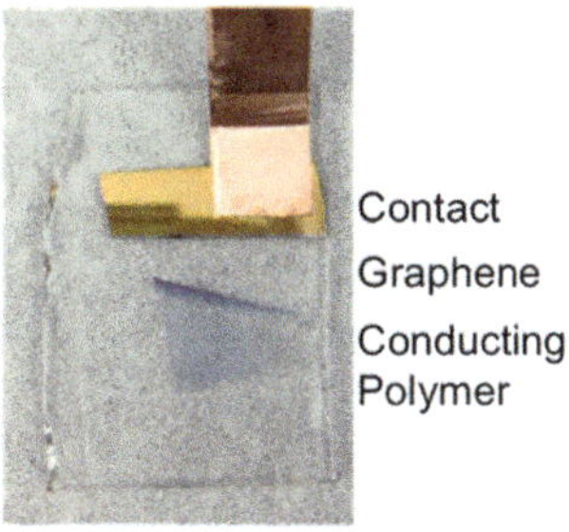

Fig. 9 Image of the film modified graphene electrode supported on a glass slide.

electropolymerized layer of 3,4-DHB. 3,4-DHB-modified graphene exhibited well-behaved surface anodic and cathodic waves, corresponding to the two electron oxidation and reduction, respectively, of the quinone functionalities, in 0.1 M acetate or Tris buffer with 0.1 M $NaNO_3$ at different pH values ranging from 1.05 to 10.53 (Fig. 10). A plot of the peak current *vs.* the sweep rate was linear (Fig. S6†), indicating that the 3,4-DHB films were immobilized to graphene. Typical surfaces coverages were *ca.* 5×10^{-12} mol cm^{-2}, calculated using eqn (8) with $n = 2$. The formal potential, $E^{0\prime}$, was estimated as the average of the cathodic and anodic peak potentials. Similar to results reported by Pariente *et al.*,[55] the slope of $E^{0\prime}$ *vs.* pH changed above pH 6.82, which was attributed to the first deprotonation of the quinones. For pH values below 7, $E^{0\prime}$ shifts negatively by 53 mV per pH unit (Fig. 10b), corresponding to a 59 mV per pH unit expected for two electrons transferred per two protons, in agreement with the literature.[55] ΔE_{peak} increased with increasing pH, suggesting that the electron transfer kinetics become slower as the concentration of protons decreases. It should be noted that the DHB films desorb upon potential cycling, indicated by a decrease of *ca.* 15% in the surface coverage after 30 minutes in phosphate buffer. This desorption was attributed to the increased solubility of the deprotonated, charged polymer.[55]

One anodic wave at +0.47 V *vs.* Ag/AgCl was observed at bare graphene immersed in 0.5 mM NADH in pH 7.5 phosphate buffer (Fig. 11), corresponding

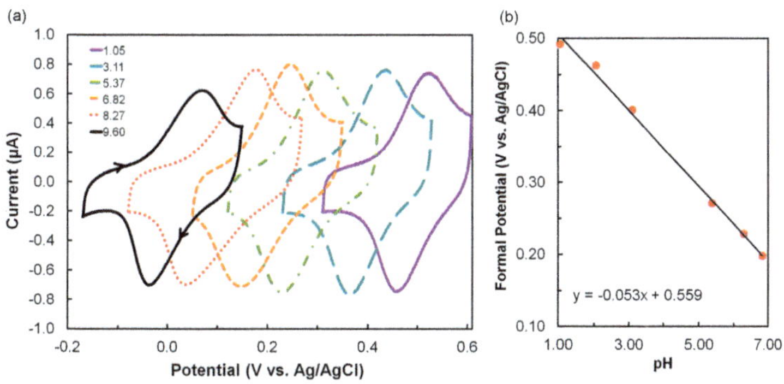

Fig. 10 (a) Cyclic voltammograms of 3,4-DHB-modified graphene electrode on Si/SiO_2 in buffer at various pH values. Sweep rate, 100 mV s^{-1}; supporting electrolyte, 0.1 M acetate or Tris buffer with 0.1 M $NaNO_3$; electrode area, 0.5 cm^2. (b) Corresponding plot of formal potential *vs.* pH value.

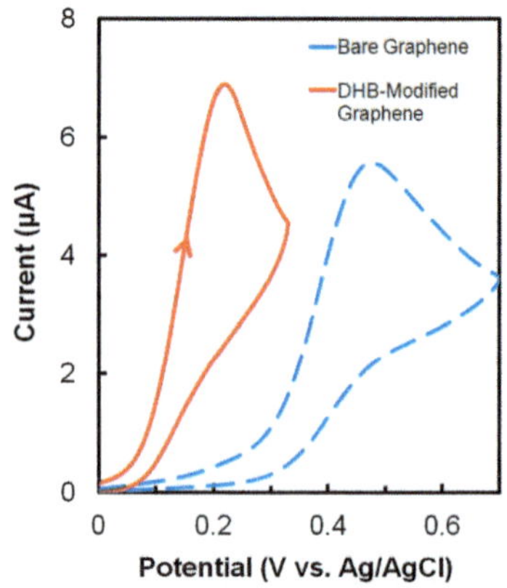

Fig. 11 Typical cyclic voltammograms of 0.5 mM NADH at bare graphene and 3,4-DHB-modified graphene. Sweep rate, 5 mV s^{-1}; supporting electrolyte, pH 7.5 phosphate buffer; electrode area, 0.5 cm^2.

to the oxidation of NADH to NAD^+, similar in potential to that previously observed.[56,57] At 3,4-DHB-modified graphene, this wave shifted negative by 250 mV and the peak current increased about 20%, indicating that NADH oxidation at graphene is catalyzed in the presence of 3,4-DHB. A similar enhancement was observed in NADH oxidation activity was observed for chemically-reduced graphene modified with methylene green.[56]

pH sensing with graphene microfluidic devices

pH sensitivity of solution-gated FETs in the inverted cell structure. We constructed solution-gated graphene FETs in an inverted cell structure to study the intrinsic pH sensitivity of our microfluidic devices. Fig. 12a shows the conductance of an inverted-cell device as a function of the gate voltage in bulk volumes of phosphate buffer solutions of different pH values. For each fixed pH condition, a "V" shaped, p- to n-type transition was observed as the Pt top-gate voltage V_{TG} was scanned, and the conductance curves shifted to more negative potentials at higher pH values. A good linearity is observed when the measured Dirac point gate voltage, as defined by the conductance minimum, is plotted as a function of the pH value, for data obtained across different buffer solutions over the pH range of 5.9 to 8 (Fig. 12b). Notably, a pH sensitivity of *ca.* 60 mV shift of Dirac point voltage per unit pH change was observed for our solution gated graphene FET device. While this is very close to the anticipated (ideal) Nernstian value, the value is 2–3 times larger than the typical value obtained in previous studies.[59–62]

We emphasize that Fig. 12b includes experimental data measured in two groups of buffer solutions with different buffer types and buffer capacity. All data points fall on the same line, indicating good reliability of the solution-gated graphene FETs as a pH sensor. We further plot the measured conductance of graphene as a function of pH for gate voltage at 0.3 V and 0.5 V. The graphene conductance changed by −40% per unit pH and 45% per unit pH, respectively, for the two fixed gate voltages, compared to the measured conductance at pH 6.88. This value is again approximately 3–4 times larger the previously reported values.

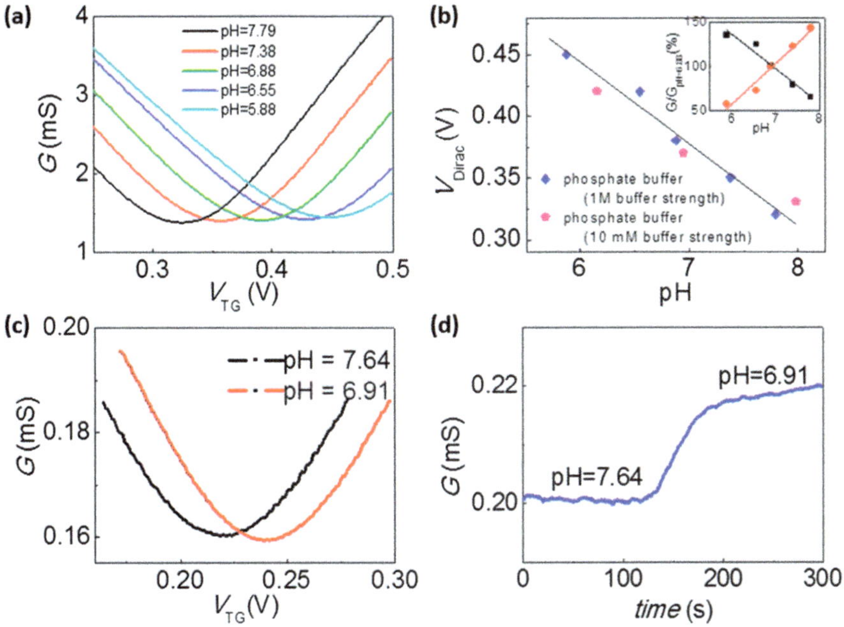

Fig. 12 pH sensitivity of the solution-gated graphene FETs. (a) The gate-dependent FET conductance measured with an inverted-cell device in phosphate buffers at different pH. (b) The Dirac point gate voltage as a function of pH for two different groups of buffer solutions in the inverted cell. Inset: conductance as a function of pH value at gate voltage of 0.3 V (black) and 0.5 V (red). (c) The gate-dependent FET conductance measured with a microfluidic device at two different pH values. (d) Continuous monitoring of the conductance of a microfluidic device when solution pH changes from 7.64 to 6.91.

pH sensitivity of solution-gated FETs in microfluidic devices. Integrating the solution-gated FETs to in a microfluidic system enables us to detect the local pH value in the system. Fig. 12c shows the gate-dependence data obtained with our microfluidic device at two different pH values. The Dirac point voltage changed by 21 mV when the solution pH value changes from 7.64 to 6.91, corresponding to *ca.* 29 mV per unit pH change. This value is smaller than what we obtained with solution-gated FETs in the inverted cell structure, but is still larger than the typical values from previous study. We are not certain as to why the pH sensitivity of the microfluidic device (*ca.* 21 mV per pH unit) is significantly smaller than that of the "inverted cell" (*ca.* 60 mV per pH unit). We believe that the pH sensitivity of graphene likely arises as the result of electrostatic interactions and having oxygenated species, such as phenols and quinones, on the surface of the graphene. It is possible that the relative importance of these two contributions could give rise to mixed potentials. However, since the graphene used in both the inverted cell and the microfluidic device was grown using similar methods, there could also be something else giving rise to the difference in pH sensitivity between the two devices.

Fig. 12d shows our continuous measurement of graphene FET conductance when we changed the solution in microfluidic channel from pH 7.64 buffer to pH 6.91 buffer during a time period of 2 min. We clearly observed that the FET conductance evolves from a plateau at 0.20 mS to another plateau at 0.22 mS during the time frame. This experiment further demonstrated that the graphene FET is capable of monitoring the pH change in microfluidic channels in real time.

Conclusions

We presented several examples of single layer, large area graphene electrodes as a platform for optical and biosensing devices. Graphene can be readily modified using various polymer films deposited through electrochemical methods. These electropolymerization processes significantly improve the prospects for using graphene in applications such as those that require a transparent electrode with electropolymerized materials. Simple modification of graphene with polymers can readily be used to improve its pH sensing properties and catalysis of biologically-relevant reactions. These results suggest that modifying graphene itself may offer an alternative to trying to modify other electrodes with graphene, such as composite electrodes made with graphene flakes. We demonstrated a working microfluidic device based on electrochemical gating of field effect transistors that is more responsive to pH changes than those reported. Future studies will involve combining microfluidic devices with polymer-modified graphene electrodes. In addition, we are currently replacing ITO with graphene to make all-graphene devices for ECL applications.

Acknowledgements

The authors acknowledge the National Science Foundation Centers for Chemical Innovation-Phase I: Center for Molecular Interfacing (0847926) for financial support and the Cornell Center for Materials Research (DMR 1120296) and Cornell NanoScale Science and Technology Facility (ECS 0335765) for use of instrumentation. N.L.R. and C.T. gratefully acknowledge the Integrative Graduate Education and Research Traineeship (IGERT) Program in the Nanoscale Control of Surfaces and Interfaces, supported under the National Science Foundation Award DGE-0654193. G.G.R., K.H.B., S.C., and C.T. gratefully acknowledge the Jung-Hyun Oh-SSCP for financial support for the electrochromic experiments. The authors thank Prof. Paul McEuen's group and Kevek Innovations for help and providing facilities for graphene growth.

References

1 K. S. Novoselov, A. K. Geim, S. V. Morozov, D. Jiang, Y. Zhang, S. V. Dubonos, I. V. Grigorieva and A. A. Firsov, *Science*, 2004, **306**, 666–669.
2 K. S. Novoselov, D. Jiang, F. Schedin, T. J. Booth, V. V. Khotkevich, S. V. Morozov and A. K. Geim, *Proc. Natl. Acad. Sci. U. S. A.*, 2005, **102**, 10451–10453.
3 A. K. Geim and K. S. Novoselov, *Nat. Mater.*, 2007, **3**, 183–191.
4 A. K. Geim, *Science*, 2009, **324**, 1530–1534.
5 S. Pang, Y. Hernandez, X. Feng and K. Müllen, *Adv. Mater.*, 2011, **23**, 2779–2795.
6 J. S. Bunch, S. S. Verbridge, J. S. Alden, A. M. van der Zande, J. M. Parpia, H. G. Craighead and P. L. McEuen, *Nano Lett.*, 2008, **8**, 2458–2462.

7 K. R. Ratinac, W. Yang, J. J. Gooding, P. Thordarson and F. Braet, *Electroanalysis*, 2011, **23**, 803–826.
8 Y. Shao, J. Wang, H. Wu, J. Liu, I. A. Aksay and Y. Lin, *Electroanalysis*, 2010, **22**, 1027–1036.
9 D. Chen, L. Tang and J. Li, *Chem. Soc. Rev.*, 2010, **39**, 3157–3180.
10 D. A. C. Brownson, D. K. Kampouris and C. E. Banks, *Chem. Soc. Rev.*, 2012, **41**, 6944–6976.
11 M. Pumera, *Electrochem. Commun.*, 2013, **36**, 14–18.
12 W. Li, C. Tan, M. A. Lowe, H. D. Abruña and D. C. Ralph, *ACS Nano*, 2011, **5**, 2264–2270.
13 A. T. Valota, I. A. Kinloch, K. S. Novoselov, C. Casiraghi, A. Eckmann, E. W. Hill and A. W. Dryfe, *ACS Nano*, 2011, **5**, 8809–8815.
14 A. G. Güell, N. Ebejer, M. E. Snowden, J. V. Macpherson and P. R. Unwin, *J. Am. Chem. Soc.*, 2012, **134**, 7258–7261.
15 A. Ambrosi and M. Pumera, *J. Phys. Chem. C*, 2013, **117**, 2053–2058.
16 S. C. S. Lai, A. N. Patel, K. McKelvey and P. R. Unwin, *Angew. Chem., Int. Ed.*, 2012, **51**, 5405–5408.
17 A. N. Patel, M. G. Collignon, M. A. O'Connell, W. O. Y. Hung, K. McKelvey, J. V. Macpherson and P. R. Unwin, *J. Am. Chem. Soc.*, 2012, **134**, 20117–20130.
18 A. Ambrosi and M. Pumera, *Nanoscale*, 2014, **6**, 472–476.
19 C. Tan, J. Rodríguez-López, J. J. Parks, N. L. Ritzert, D. C. Ralph and H. D. Abruña, *ACS Nano*, 2012, **6**, 3070–3079.
20 J. Rodríguez-López, N. L. Ritzert, J. A. Mann, C. Tan, W. R. Dichtel and H. D. Abruña, *J. Am. Chem. Soc.*, 2012, **134**, 6224–6236.
21 N. L. Ritzert, J. Rodríguez-López, C. Tan and H. D. Abruña, *Langmuir*, 2013, **29**, 1683–1694.
22 L. Groenenedaal, F. Jonas, D. Freitag, H. Pierlarzik and J. R. Reynolds, *Adv. Mater.*, 2000, **12**, 481–494.
23 B. C. Thompson and J. M. J. Fréchet, *Angew. Chem., Int. Ed.*, 2008, **47**, 58–77.
24 A. J. Heeger, S. Kivelson, J. R. Schrieffer and W.-P. Su, *Rev. Mod. Phys.*, 1988, **60**, 781–850.
25 H. Sirringhaus, T. Kawase, R. H. Friend, T. Shimoda, M. Inbasekaran, W. Wu and E. P. Woo, *Science*, 2000, **290**, 2123–2126.
26 G. Gustafsson, Y. Cao, G. M. Treacy, K. Klavetter, N. Colaneri and A. J. Heeger, *Nature*, 1992, **357**, 477–479.
27 P. Novák, K. Müller, K. S. V. Santhanam and O. Haas, *Chem. Rev.*, 1997, **97**, 207–282.
28 G. Heywang and F. Jonas, *Adv. Mater.*, 1992, **4**, 116–118.
29 A. O. Patil, A. J. Heeger and F. Wudl, *Chem. Rev.*, 1988, **88**, 183–200.
30 L. Groenendaal, G. Zotti, A. Pierre-Henri, S. M. Waybright and J. R. Reynolds, *Adv. Mater.*, 2003, **15**, 855–879.
31 S. E. Burkhardt, G. G. Rodríguez-Calero, M. A. Lowe, Y. Kiya, R. H. Hennig and H. D. Abruña, *J. Phys. Chem. C*, 2010, **114**, 16776–16784.
32 S. Conte, G. G. Rodríguez-Calero, S. E. Burkhardt, M. A. Lowe and H. D. Abruña, *RSC Adv.*, 2013, **3**, 1957–1964.
33 I. F. Perepichka, D. F. Perepichka, H. Meng and F. Wudl, *Adv. Mater.*, 2005, **17**, 2281–2305.
34 R. J. Mortimer, *Chem. Soc. Rev.*, 1997, **26**, 147–156.

35 A. H. Castro Neto, F. Guinea, N. M. R. Peres, K. S. Novoselov and A. K. Geim, *Rev. Mod. Phys.*, 2009, **81**, 109–162.

36 S. Park and R. S. Ruoff, *Nat. Nanotechnol.*, 2009, **4**, 217–224.

37 F. Schwierz, *Nat. Nanotechnol.*, 2010, **5**, 487–496.

38 *Electrogenerated Chemiluminescence*, ed. A. J. Bard, Marcel Dekker, New York, 2004.

39 W. Miao, in *Handbook of Electrochemistry*, ed. C. G. Zoski, Elsevier, Oxford, UK, 2007, pp. 541–590.

40 M. M. Richter, *Chem. Rev.*, 2004, **104**, 3003–3036.

41 I. Rubenstein and A. J. Bard, *J. Am. Chem. Soc.*, 1981, **103**, 512–516.

42 S. Chen, L. Brown, M. Levendorf, W. Cai, S.-Y. Ju, J. Edgeworth, X. Li, C. W. Magnuson, A. Velamakanni, R. D. Piner, J. Kang, J. Park and R. S. Ruoff, *ACS Nano*, 2011, **5**, 1321–1327.

43 M. Schriver, W. Regan, W. J. Gannett, A. M. Zaniewski, M. F. Crommie and A. Zettl, *ACS Nano*, 2013, **7**, 5763–5768.

44 A. C. Nielander, M. J. Bierman, N. Petrone, N. C. Strandwitz, S. Ardo, F. Yang, J. Hone and N. S. Lewis, *J. Am. Chem. Soc.*, 2013, **135**, 17246–17249.

45 H. D. Abruña and A. J. Bard, *J. Am. Chem. Soc.*, 1982, **104**, 2641–2642.

46 Y. S. Obeng and A. J. Bard, *Langmuir*, 1991, **7**, 195–201.

47 I. Rubenstein and A. J. Bard, *J. Am. Chem. Soc.*, 1980, **102**, 6641–6642.

48 I. Rubenstein and A. J. Bard, *J. Am. Chem. Soc.*, 1981, **103**, 512–516.

49 X. Zhang and A. J. Bard, *J. Phys. Chem.*, 1988, **92**, 5566–5569.

50 H. D. Abruña, P. Denisevich, M. Umaña, T. J. Meyer and R. W. Murray, *J. Am. Chem. Soc.*, 1981, **103**, 1–5.

51 P. Harder, M. Grunze, R. Dahint, G. M. Whitesides and P. E. Laibinis, *J. Phys. Chem. B*, 1998, **102**, 426–436.

52 R. L. C. Wang and H. J. Kreuzer, *J. Phys. Chem. B*, 1997, **101**, 9767–9773.

53 A. Rastogi, S. Nad, M. Tanaka, N. Da Mota, M. Tague, B. A. Baird, H. D. Abruña and C. K. Ober, *Biomacromolecules*, 2009, **10**, 2750–2758.

54 D. L. Nelson and M. M. Cox, *Lehninger Principles of Biochemistry*, W. H. Freeman, New York, 4th edn, 2005, ch. 13, pp. 489–520.

55 F. Pariente, E. Lorenzo and H. D. Abruña, *Anal. Chem.*, 1994, **66**, 4337–4394.

56 H. Liu, J. Gao, M. Xue, N. Zhu, M. Zhang and T. Cao, *Langmuir*, 2009, **25**, 12006–12010.

57 A. Ambrosi and M. Pumera, *J. Phys. Chem. C*, 2013, **117**, 2053–2058.

58 F. Schedin, A. K. Geim, S. V. Morozov, E. W. Hill, P. Blake, M. I. Katsnelson and K. S. Novoselov, *Nat. Mater.*, 2007, **6**, 652–655.

59 P. K. Ang, W. Chen, A. T. S. Wee and K. P. Loh, *J. Am. Chem. Soc.*, 2008, **130**, 14392–14393.

60 Y. Ohno, K. Maehashi, Y. Yamashiro and K. Matsumoto, *Nano Lett.*, 2009, **9**, 3318–3322.

61 J. Ristein, W. Zhang, F. Speck, M. Ostler, L. Ley and T. Seyller, *J. Phys. D: Appl. Phys.*, 2010, **43**, 345303.

62 Z. Cheng, Q. Li, Z. Li, Q. Zhou and Y. Fang, *Nano Lett.*, 2010, **10**, 1864–1868.

63 X. Li, W. Cai, W. J. An, S. Kim, J. Nah, D. Yang, R. Piner, A. Velamakanni, I. Jung, E. Tutuc, S. Banerjee, L. Colombo and R. S. Ruoff, *Science*, 2009, **324**, 1312–1314.

64 A. C. Ferrari, J. C. Meyer, V. Scardaci, C. Casiraghi, M. Lazzeri, F. Mauri, S. Piscanec, D. Jiang, K. S. Novoselov, S. Roth and A. K. Geim, *Phys. Rev. Lett.*, 2006, **97**, 187401.
65 C. Casiraghi, S. Pisana, K. S. Novoselov, A. K. Geim and A. C. Ferrari, *Appl. Phys. Lett.*, 2007, **91**, 233108.
66 J. K. Leland and M. J. Powell, *J. Electrochem. Soc.*, 1990, **137**, 3127–3131.
67 A. A. Gorodetsky, S. Parker, J. D. Slinker, D. A. Bernards, M. H. Wong, G. G. Malliaras, S. Flores-Torres and H. D. Abruña, *Appl. Phys. Lett.*, 2004, **84**, 807–809.

Faraday Discussions

PAPER

Boron doped diamond biotechnology: from sensors to neurointerfaces

C. Hébert,[a] E. Scorsone,[a] A. Bendali,[c] R. Kiran,[ad] M. Cottance,[b] H. A. Girard,[a] J. Degardin,[c] E. Dubus,[c] G. Lissorgues,[b] L. Rousseau,[b] P. Mailley,[e] S. Picaud[c] and P. Bergonzo*[a]

Received 14th March 2014, Accepted 14th April 2014
DOI: 10.1039/c4fd00040d

Boron doped nanocrystalline diamond is known as a remarkable material for the fabrication of sensors, taking advantage of its biocompatibility, electrochemical properties, and stability. Sensors can be fabricated to directly probe physiological species from biofluids (*e.g.* blood or urine), as will be presented. In collaboration with electrophysiologists and biologists, the technology was adapted to enable structured diamond devices such as microelectrode arrays (MEAs), *i.e.* common electrophysiology tools, to probe neuronal activity distributed over large populations of neurons or embryonic organs. Specific MEAs can also be used to build neural prostheses or implants to compensate function losses due to lesions or degeneration of parts of the central nervous system, such as retinal implants, which exhibit real promise as biocompatible neuroprostheses for *in vivo* neuronal stimulations. New electrode geometries enable high performance electrodes to surpass more conventional materials for such applications.

A Introduction

Boron doped diamond (BDD) electrodes are extremely promising in the field of biomedical applications as they exhibit a unique combination of properties. Doped diamond synthesis has become a major focus of research and development: undoped diamond exhibits a large band gap and is an electrical insulator at room temperature. When doped with boron, diamond results in an electrode with remarkable electro-analytical properties, which can be used to detect species in solution before oxygen and hydrogen evolution interfere with the analysis. Thus, conductive diamond electrodes take electrochemical detection to new levels, and extend their usefulness to analytes which are not measurable with conventional

[a]*CEA, LIST, Diamond Sensors Laboratory, F-91191, Gif-sur-Yvette, France. E-mail: philippe.bergonzo@cea.fr*
[b]*ESIEE, Université Paris Est, Noisy le Grand, France*
[c]*INSERM, U968, Institut de la Vision, Paris, France*
[d]*MED-EL Research Center, Feodor-Lynen-Srasse 35, 30625, Hannover, Germany*
[e]*CEA-LETI-DTBS Minatec, 17 rue des Martyrs, 38054, Grenoble*

electrode materials. This opens the field of direct analytical detection with diamond, and in addition to applications where analytes can be detected in biofluids, reverse current spikes can be used to stimulate neural tissues.

There is a real interest in the development of novel neuroprostheses and brain-machine interfaces: these applications witnessed an exponential growth since the success of cochlear implants and of deep brain stimulation for Parkinsonian patients.[1] Neuroprostheses can either be surface electrodes, meaning the electrodes are simply in contact with the glial surface of the neuronal tissues such as the cortex, or penetrating electrodes, reaching deeper neuronal structures in the brain. Although cortical implants have resulted in a function gain after a few months, this was unfortunately frequently followed by a complete loss of activity.[2] This loss of functionality could be attributed to a major gliosis occurring around the prostheses or its electrodes in the months following implantation.[3] Another challenge for neuroprostheses is to increase the resolution of individual electrodes to stimulate more precisely discrete neuronal areas. These objectives motivate the development for new biocompatible materials that limit glial reactions and improve direct interactions with neurones.

Recently, the concept of retinal prostheses was validated in clinical trials, showing that such prostheses can enable blind patients to read short words, identify contrasting objects or follow lanes on the ground.[4,5] Those retinal prostheses aiming at restoring vision in patients having lost their photoreceptors are either placed in the subretinal space or on the epiretinal side, in direct contact with either the outer or the inner limiting membrane, both produced by glial Muller cells. In the first configuration, the subretinal implant will stimulate retinal bipolar cells (neurones normally postsynaptic to photoreceptors),[5] whereas epiretinal implants target retinal ganglion cells (spiking neurones that send visual information to the brain *via* their axon through the optic nerve).[4] Although these implants have already restored some visual functions in patients,[4,5] an increase in electrode resolution is required to further improve the restored visual performances, to achieve face recognition, text reading or independent locomotion. Different 3D designs have already been proposed to reach this objective using either pillars[6] or wells[7] on the implants.

Among potential new biomaterials, diamond has attracted great attention for its use in neuroprostheses because boron doped diamond exhibits semiconductive properties. Diamond biocompatibility has already been demonstrated with osteoblast cultures[8] and even embryonic cortical neurons.[9,10] Embryonic neurons were found to grow selectively on protein patterns stamped on diamond surfaces,[9] but they could not grow directly on a polished polycrystalline diamond layer or on a nanocrystalline diamond layer unless these layers were coated with peptides or with dispersed nanodiamond particles.[10] The use of diamond as electrodes has further been demonstrated by recording neuronal activities using cell lines.[11]

B Boron doped diamond electrodes and microelectrode arrays for biosensing

BDD electrodes possess unique electrochemical properties, namely, a low capacitive background current, a wide potential window in aqueous media, poor adsorption of polar molecules, and corrosion resistance in harsh environments, making them highly promising as electrochemical sensors compared to other conventional electrode materials.[12–14] BDD can be grown using microwave plasma enhanced chemical vapour deposition (CVD) techniques on substrates from 2 to 4 inches using 2.45 GHz plasma excitation. Such layers have demonstrated a bio-inertness and long term stability that makes them ideal candidates for biomedical applications.[15] Due to its carbon nature, BDD also opens the way to the immobilization of functional groups such as DNA, proteins, enzymes *etc.* on the electrode surface,[16] thereby bringing high selectivity to the sensing process.

We developed a novel process for the fabrication of diamond microelectrodes that involves the selective growth of diamond over silicon substrates, followed by the deposition of metal contacts and passivation layers. This process includes the initial fixing of detonation diamond nanoparticles over a pre-oxidized 4 inch silicon wafer using a protocol described elsewhere.[17] Then, an aluminum hard mask consisting of a disc 100 μm in diameter was selectively deposited over the areas where the BDD microelectrodes have to be grown. Photolithography enabled us to partially pattern the Al mask, which was used to protect areas of diamond nanoparticles from being etched by a reactive ion etching (RIE) process in oxygen. The Al hard mask was finally removed and the diamond electrodes were grown. The dimensions of the resulting diamond disc electrodes were typically 300 nm in thickness and 100 μm in diameter. Then, an assembly of Ti (50 nm)/Pt (150 nm) metal tracks was deposited over the substrate, with a metal ring going around the edge of the diamond disc in order to take an electrical contact from the diamond electrode. Then, a silicon nitride (Si_3N_4) passivation layer was deposited by CVD over the substrate. Finally, an opening was made over the diamond electrodes by local etching of the Si_3N_4 layer by RIE with SF_6 gas. Typical dimensions range between 10 and 100 μm (Fig. 1).

Further to the use of BDD electrodes for the electrochemical detection and quantification of *e.g.*, explosive traces in real media like sea water,[18] the novel ability to fabricate diamond microelectrode arrays has further enabled the possibility to detect redox active compounds directly in unprocessed media such as urine, wine, milk, blood *etc.*,[19] *e.g.*, uric acid (UA) in real urine (Fig. 2).[20] In fact, conventional BDD electrodes are prone to fouling when used in biological (urine, blood plasma) and synthetic fluids. Kiran *et al.*[19] have recently proposed an electrochemical treatment where a train of short cathodic and/or anodic pulses are applied to clean fouled electrodes. This technique can be used to retrieve the lost reactivity, characterized by the electron transfer rate k_0 of the boron doped diamond electrodes, thereby enhancing their reusability over long periods of

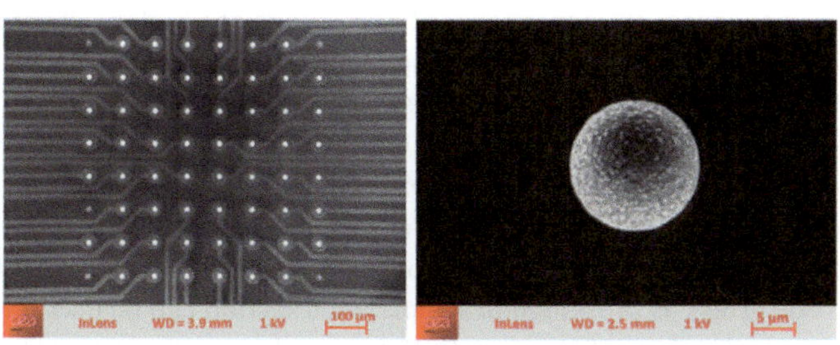

Fig. 1 Image of typical BDD diamond microelectrode arrays.

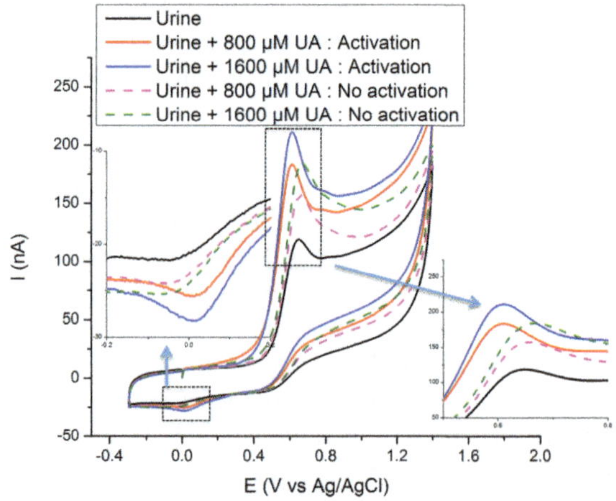

Fig. 2 Cyclic voltammograms of urine diluted 2 fold, and that of diluted urine containing added uric acid (800–1600 μM), scanned at 20 V s^{-1}. The insets provide details of the oxidation and reduction peaks, displaying the interests of the activation process.

measurements without degradation of the signal, thus significantly extending the field of monitoring and surveying applications. With this approach, uric acid detection in real urine samples was made feasible, with good discrimination of ascorbic acid. Human urine samples were collected from volunteers as real samples for direct BDD uric acid analysis, which was compared to the spectrophotometric method. Measured values using the proposed model were observed to be very close to the spectrophotometric results with a maximum difference of 13%. The interest in continuous monitoring of UA in urine is particularly for patients admitted to intensive care units, where the early diagnostic of acute renal failure can have a major impact on the survival rate of those patients.[21] In this context, electrochemical detection techniques are seen as a promising alternative to conventional optical methods due to their good sensitivity, fast measuring time, portability, low power consumption and cost effectiveness, thus enabling, for example, direct bedside monitoring.

C Boron doped diamond microelectrode arrays for recording of neuronal signals

Electrophysiology is the study of the electrical properties of cells and tissues, which involves the measurement of the voltage change of a biological entity. It is a powerful approach, not only to study the electrical activity of animal cells to understand the workings of the nervous system, brain, hypothalamus, *etc.*, but also to diagnose and treat nervous system disorders. The electrical activity of neurons can be measured directly using extracellular microelectrode arrays. The potential changes in the vicinity of the electrode, caused by currents flowing across the neuronal membranes of multiple neurons, can be detected. The same microelectrode arrays as presented in the previous section can be used in order to record the invoked activity of cell signals. For example, arrays of 64 pixel diamond microelectrodes have successfully been used by Maybeck *et al.*[22] for recording of the cardiomyocyte-like cell line HL-1. The study demonstrated that boron doped nanocrystalline diamond (BNCD) electrodes were able to detect cardiac cell action potentials as well as gold electrodes, while electrodes performed up to a factor of four times better than planar metal electrodes of the same diameter. BNCD electrodes survived the mechanical stresses of contractile cells without the defects commonly observed in other nanostructured materials after cell culture (loss of platinum black, removal of CNTs,[23] and tipping of pillars[24]). This enabled us to conclude that the diamond layer remained on the surface during cell contraction by the lack of thinning-induced defects after multiple cultures and the contractile tension generated in mature HL-1 cells (the cells did not pull themselves off of the surface), thus demonstrating that the diamond MEA's physical robustness makes it a durable and easy to handle device for repeated or long-term use, with promise for recording electrical signals from biological samples in addition to its established role in electrochemistry.

D Boron doped diamond microelectrode arrays for neurostimulation and implants

The main constraint where *in vivo* tissue stimulation is concerned is that the microelectrode array must conform to the size and shape of the tissues to be stimulated. This requires specific developments to enable the fabrication of devices that, further to providing the advantages of previous ones, must also be flexible and exhibit a long connecting foil, enabling the electrical signals to be taken from outside the body to the *in vivo* neuronal tissues. This is particularly demanding in the case of retina implants since the curvature is particularly high and the surgery constraints impose very strict specifications. In spite of this, the past decade has seen an explosion of research efforts into retinal prostheses aimed at restoring sight to patients blinded by retinitis pigmentosa or age-related macular degeneration. In addition to classical approaches, novel technologies are being explored, often in the context of newly founded commercial enterprises. Also, for the devices to offer a real breakthrough to blind people, theoretical studies have demonstrated that the ability to reach stimulation structures of 600 pixels would be of great benefit as it should be sufficient for patients to read.[25] This extreme number is not reachable today with current technologies. Several approaches have been proposed, with systems offering typically 60[26–28] to 49[29–31] pixel electrode arrays, with often partial clinical success in short term tests.

In a similar approach as for the fabrication of diamond MEAs used for recording applications, here, nanocrystalline diamond was processed using standard nanotechnology approaches for the fabrication of pixelized microelectrode arrays, but the structure was grown on a sacrificial layer (*e.g.* SiO_2). This process is illustrated through a 3D approach, in Fig. 3. In brief, after diamond electrodes are deposited on such a sacrificial layer, the standard processing of a soft material such as polyimide is made possible on the substrate. Feed lines to contact each electrode can then be embedded in several layers of soft polyimide, to lead, after removal of the sacrificial layer, to a flexible microelectrode array with high flexibility for retinal stimulation as shown in Fig. 4A. The preparation of such implants has been made feasible with polyimide as well as with parylene. We typically fabricated matrices of 64 independent pixels for the stimulation of

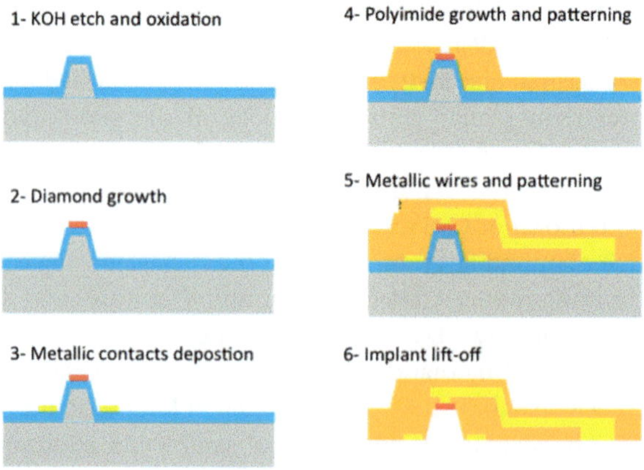

Fig. 3 Process for the fabrication of 3D nanocrystalline diamond implants.

tissues, which were evaluated on laboratory animals in the form of preliminary prototypes of diamond implantable electrode devices (Fig. 4).

E Non-planar diamond electrode matrices towards the improvement of spatial resolution

Furthermore, a study driven by the Vision Institute in Paris rendered the specifications more challenging, as Djilas *et al.*[7] demonstrated the benefit of fabricating 3D wells where a small group of retinal cells could be stimulated, in order to improve the stimulation precision. In fact, the major challenge in visual

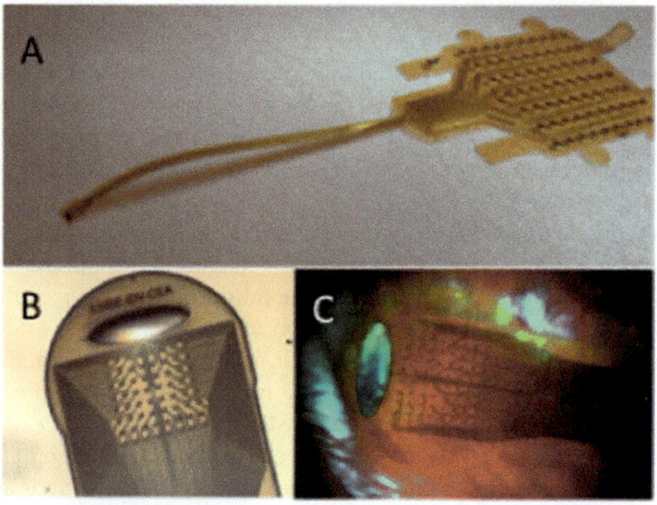

Fig. 4 (A) A view of a non-3D diamond MEA implant for retinal stimulation. The left part consists of 64 diamond electrodes as shown in (B). (C) An *in vivo* MEA array as implanted in P23H rats at the Paris Vision Institute.

rehabilitation using neuroprostheses is to improve the spatial resolution of the stimulated area of each individual electrode, which is a direct consequence of the density of independent pixels. It has been demonstrated that the electrode resolution can be improved using bipolar stimulations between two neighbouring electrodes rather than with a configuration based on a stimulating electrode and a distant common ground.[32] Using a retinal implant where each pixel exhibits a 3D geometry, one could further improve such bipolar stimulations by locally maintaining neurons between both stimulating electrodes[33]. The success of such 3D designs implies that the residual blind retina remains sufficiently plastic to shape itself around the electrodes. This plasticity of the blind residual retina was demonstrated using pillars penetrating the tissue or cavities filled with cells.[34] However, the neuronal nature of the cells surrounding pillars or located in the cavities was not fully assessed, since cell immunolabelling could not be performed within this approach. In addition, bipolar stimulation would require two wires per pixel, thus significantly increasing the complexity of the implant design when the number of pixels is increased. An alternative approach proposed the use of a returning ground grid surrounding each of the individual stimulating electrodes, thus enabling improved localisation of the stimulating current providing the grid material exhibits a low impedance.[34] Electrical stimulations have to be constrained to small tissue areas, as proposed in three-dimensional arrays aimed at confining the current within a small volume around the stimulating electrode.[6] A previous study by our group validated *in vivo* a model of 3D well shaped electrodes, demonstrating the favourable integration of the retina cells within such cavities when designed at optimized sizes.[7] This modelling was based on the hypothesis that the neuronal tissue would shape itself into the 3D wells located within the implant.

To address these challenges, Rousseau *et al.*[35] have proposed novel approaches enabling the synthesis of 3D shaped nanodiamond microelectrode arrays. The goal is to achieve an implant with each electrode placed at the bottom of a typically 30 μm deep recess and a ground located at the surface of the implant. To fabricate this 3D structure, we have chosen to use silicon moulds fabricated by KOH etching of silicon <100> substrates. A specific design with compensation structures was added to obtain these moulds. Initially a simple structure without electrodes was made for histological testing as well as to check the plasticity of the retina. A silicon wafer was oxidized in a furnace at 1050 °C, in a wet environment with a gaseous mixture of hydrogen and oxygen, for 50 min, in order to get 500 nm of silicon dioxide. A photolithography step was done to define the patterns for KOH etching. Silicon dioxide not covered by photoresist was etched away in a solution of buffered HF (10%). Then, the wafer was placed in KOH solution and kept at 80 °C for approximately 30 min so as to obtain a structure 30 μm high. After cleaning, the wafer was oxidized once more under the same conditions to obtain a thick silicon dioxide layer (1 μm). Selective growth of diamond could then be initiated, as described above, by spreading a colloidal solution of diamond nanoparticles in deionised water onto the substrate, followed by spin coating.[36] Nevertheless, due to the topology of the 3D substrate used here, spin coating cannot be used, and an alternative approach based on electrostatic grafting of the diamond nanoparticles was used instead.[37] This latter method is based on electrostatic interactions between the diamond nanoparticles and a substrate coated with a polyelectrolyte. Thus, the substrates were dipped into poly diallyldimethylammonium chloride (PDDAC) polymer aqueous solution. The substrates were then washed with water and dried prior to immersion in the colloidal solution of nanoparticles. Finally, the substrates were washed again with water and dried under N_2 flux, resulting in 3D structures completely covered by diamond nanoparticles. A short diamond growth was then performed to fix the diamond nanoparticles on the surface. The patterning of the diamond electrodes was ensured from the direct diamond nanoparticle patterning before growth.[38] A 500 nm thick aluminium film was sputtered over the substrate and patterned by photolithography, followed by chlorine plasma etching so as to define the electrode areas. The diamond particles not covered by the aluminium film were etched away by classical RIE in a gas mixture of argon and oxygen with a power of 200 W for 20 min. Lastly, the aluminium protecting the nanoparticles was removed by wet etching and the wafer was placed in a SEKI 6500 CVD diamond growth reactor to selectively grow the diamond electrodes on a 4-inch substrate. Next, polyimide 2611 was poured on the wafer and spin coated to obtain a 10 μm thick layer. The polyimide was cured at 450 °C under nitrogen for 6 h. Then, a metal mask (aluminium, 500 nm) was deposited onto the substrate by sputtering. Photolithography with thick photoresist (AZ 4562) was carried out to define the shape of the implant. The aluminium was then etched away by RIE in chlorine and the polyimide layer was etched by RIE with oxygen plasma. The aluminium was removed and the sacrificial layer (a thick silicon dioxide layer) was etched by HF. The implants were finally rinsed and dried.

A complete surgical procedure was developed to implant, in the sub-retinal position, a complete 3D diamond implant. 3D structures were implanted in P23H rats in the sub-retinal position for 11 weeks. The retina was then explanted along with the implant, and cell labelling was performed. Cell nuclei were labeled by the DAPI stain (blue), while glial cells and retinal bipolar neurons were immunolabelled by the GFAP antibody (grey) and the Goα and PKCα antibodies (green and red), respectively. In Fig. 5, the ×20 magnification view of the retina/implant wholemount allows visualization of the presence of the nuclei stain in all 25 cavities. Confocal analysis was carried out to check the plasticity of the retina. In Fig. 5, the blue cell nuclei are well localised within the implant cavities,

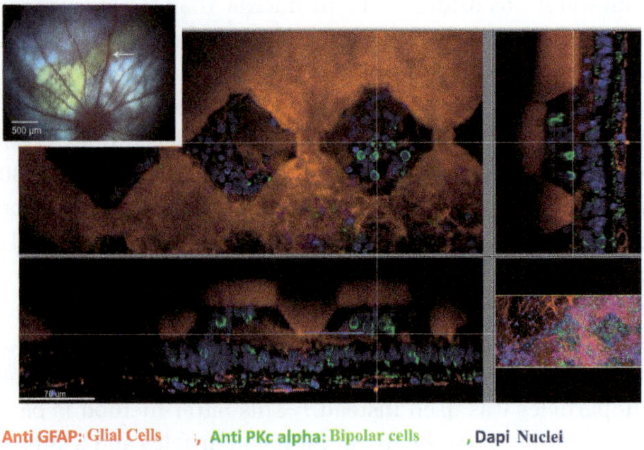

Fig. 5 Result of the histology study on the plasticity of the retina.

demonstrating that the retina has followed the shape of the implant. Cell labelling revealed that a small group of bipolar cells is present inside the recess.

Further tests were carried out in order to check the biocompatibility of the BDD diamond electrode with the retina. A 3D implant with diamond on the whole surface was fabricated and implanted in a rat subretinal region. As before, cell labeling was performed after several months. The confocal image (Fig. 6) shows that no specific reaction occurs on the tissue and no glial proliferation is visible on the surface. No significant difference is observed here when comparing the sample with our reference (a polyimide implant). Moreover, some bipolar cells can be observed inside each cavity.

F Boosting diamond's electrochemical properties for implants using nanostructured electrodes

Although diamond exhibits very attractive properties when used as an electrode material, particularly in terms of its electrochemical window, biocompatibility, and resilience to fouling, the material does not match its competitors in terms of impedance and double layer capacitance, which limits the amount of injectable charges.[39] In fact, values for flat diamond electrodes are generally expected to be around 5 $\mu F\ cm^{-2}$, which are extremely low values, comparable to that of flat Pt, although black Pt, which exhibits a 3D texture, can display capacitance values up to 100 $\mu F\ cm^{-2}$.

One way to increase the double layer capacitance of the material is to increase its specific surface area. We propose the use of vertically aligned carbon nanotubes with high surface areas as a template onto which boron doped diamond is grown. The resulting composite was found to exhibit a double layer capacitance as high as 0.58 $mF\ cm^{-2}$ and very low impedance when compared to planar diamond electrodes in a phosphate buffered saline solution (Fig. 7 and 8). The influence of the CNTs' length as well as the thickness of the diamond coatings were shown to play a very significant role on the electrode performance as discussed by Hebert *et al.*[40] The entire approach relies on the ability to control the seeding and nucleation stage to avoid the etching of the CNT during the first initial stages of growth.

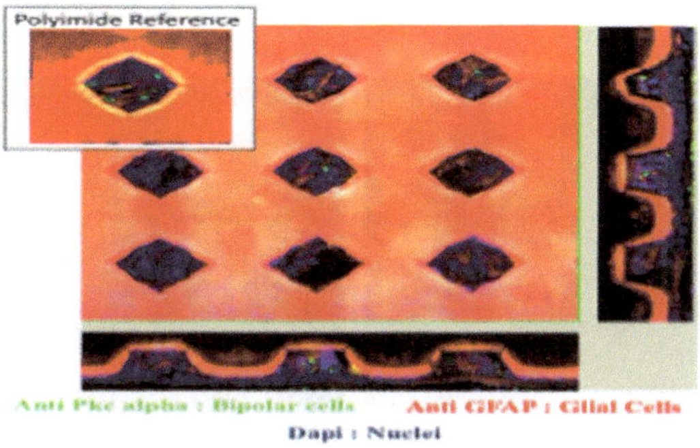

Fig. 6 Biocompatibility test of the BDD material.

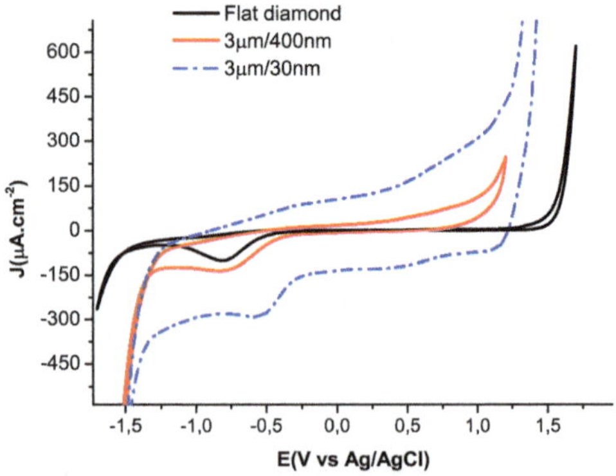

Fig. 7 Cyclic voltammetric (100 mV s^{-1}) electrochemical characterization of BDD coated vertically aligned CNTs in PBS (electrode surface 1 cm^2).

It is to be noted that such structures exhibit mechanical resilience that is close to that of diamond and far better than CNTs, which are known to be very fragile. As an example, scratching experiments performed with stainless steel tweezers showed that it was possible to not damage the diamond tipped structures. We are currently in the process of implementing such very highly structured electrodes in the shape of microelectrode arrays for neurophysiology evaluation as well as histology experiments.

Conclusions

From diamond planar electrodes to highly structured surfaces, diamond can take a wide range of forms which are applicable for neurointerfacing and particularly for stimulation purposes. Recent work has been dedicated to the development of retinal interfacing structures for the stimulation of the retina in blind patients. The great progress over the last few years will hopefully enable the validation of interest in diamond and its convincing development. In neuroprostheses, focal stimulation is a real challenge, especially for retinal implants, and we believe this challenge can be successfully overcome by taking advantage of diamond's properties. As presented, we have recently investigated a new approach to obtain focal stimulation for sub-retinal implants, based on the fabrication of a 3D implant

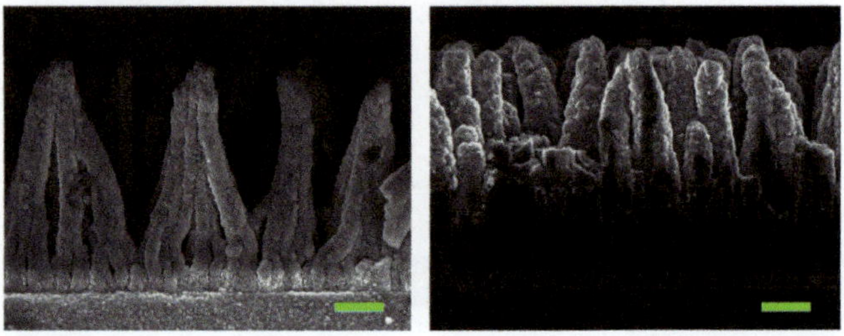

Fig. 8 Cross section of vertically aligned CNT scaffolds covered with diamond (scale bar 1 μm).

with electrodes localized at the bottom of a cavity, and a common return ground electrode placed at the surface of the implant. This geometry has previously been modelled to evaluate the efficiency of this 3D implant. With this structure, the stimulation current is confined within the cavity. Furthermore, we have worked on a new material for long term *in vivo* stability, namely BDD material. A specific process was optimised to realize a specific implant for histological tests (3D and 3D with diamond). A complete process was also developed to include BDD electrodes on a soft implant. During *in vivo* experiments, we have seen that the retina is soft enough to follow the shape of the implant. Using specific cell labelling of the tissues, confocal pictures showed that bipolar cells were present in each cavity of the implant. The tests have so far revealed that BDD material is well accepted by the tissue. A complete surgical procedure for a 3D soft implant was tested to place this implant in the sub-retinal region. The combination of the 3D shape and the BDD electrode's electrochemical performance is expected to lead to significant progress for the next generation of implants. The ultimate challenge for diamond compared to competing materials is the fact that nanocrystalline diamond still exhibits a relatively low capacitance, much lower than that of *e.g.* black Pt. With the advance of creating a nano-textured material at the hundreds of nanometres scale, recent progress opens up the route for diamond to match the performances of other materials, while it surpasses them with its extreme stability and biocompatibility.

Acknowledgements

We gratefully acknowledge the funding authorities that have supported this work, namely the Agence Nationale de la Recherche Française *via* the MEDINAS ANR07TECSAN014 project, as well as the AVIESAN agency for ITS-ITMO IMPLANTS, and more recently the FP7 European commission *via* the NEURO-CARE project no. 280433.

References

1 M. A. Lebedev and M. A. L. Nicolelis, Brain-machine interfaces: past, present and future, *Trends Neurosci.*, 2006, **29**, 536.

2 W. H. Dobelle, M. G. Mladejovsky and J. P. Girvin, Artificial vision for the blind: electrical stimulation of visual cortex offers hope for a functional prosthesis, *Science*, 1974, **183**, 440–444.

3 E. M. Maynard, E. Fernandez and R. A. Normann, A technique to prevent dural adhesions to chronically implanted microelectrode arrays, *J. Neurosci. Methods*, 2000, **97**, 93.

4 M. S. Humayun, J. D. Dorn, A. K. Ahuja, A. Caspi, E. Filley, G. Dagnelie, J. Salzmann, A. Santos, J. Duncan, L. Dacruz, S. Mohand-Said, D. Eliott, M. J. McMahon and R. J. Greenberg, Preliminary 6 month results from the argus II epiretinal prosthesis feasibility study, *Conf. Proc. IEEE Eng. Med. Biol. Soc.*, 2009, **1**, 4566–4568.

5 E. Zrenner, K. U. Bartz-Schmidt, H. Benav, D. Besch, A. Bruckmann, V. P. Gabel, F. Gekeler, U. Greppmaier, A. Harscher, S. Kibbel, J. Koch, A. Kusnyerik, T. Peters, K. Stingl, H. Sachs, A. Stett, P. Szurman, B. Wilhelm

and R. Wilke, Subretinal electronic chips allow blind patients to read letters and combine them to words, *Proc. R. Soc. London, Ser. B*, 2011, **278**, 1489–1497.
6 A. Butterwick, P. Huie, B. W. Jones, R. E. Marc, M. Marmor and D. Palanker, Effect of shape and coating of a subretinal prosthesis on its integration with the retina, *Exp. Eye Res.*, 2009, **88**, 22–29.
7 M. Djilas, C. Oles, H. Lorach, A. Bendali, J. Degardin, E. Dubus, G. Lissorgues-Bazin, L. Rousseau, R. Benosman, S. H. Ieng, S. Joucla, B. Yvert, P. Bergonzo, J. Sahel and S. Picaud, Three-dimensional electrode arrays for retinal prostheses: modeling, geometry optimization and experimental validation, *J. Neural Eng.*, 2011, **8**, 046020.
8 L. B. L. Grausova, A. Kromka, S. Potocky, M. Vanecek, M. Nesladek and V. Lisa, Nanodiamond as promising material for bone tissue engineering, *J. Nanosci. Nanotechnol.*, 2009, **9**, 3524–3534.
9 C. G. Specht, O. A. Williams, R. B. Jackman and R. Schoepfer, Ordered growth of neurons on diamond, *Biomaterials*, 2004, **25**, 4073.
10 A. Thalhammer, R. J. Edgington, L. A. Cingolani, R. Schoepfer and R. B. Jackman, The use of nanodiamond monolayer coatings to promote the formation of functional neuronal networks, *Biomaterials*, 2010, **31**, 2097–2104.
11 P. Ariano, A. Lo Giudice, A. Marcantoni, E. Vittone, E. Carbone and D. Lovisolo, A diamond-based biosensor for the recording of neuronal activity, *Biosens. Bioelectron.*, 2009, **24**, 2046.
12 T. Yano, D. A. Tryk, K. Hashimoto and A. Fujishima, *J. Electrochem. Soc.*, 1998, **145**, 1870–1876.
13 R. Tian and J. Zhi, *Electrochem. Commun.*, 2007, **9**, 1120–1126.
14 G. M. Swain, *Anal. Chem.*, 1993, **65**, 345–351.
15 M. Panizza and G. Cerisola, *Electrochim. Acta*, 2005, **51**, 191–199.
16 C. Agnès, S. Ruffinatto, E. Delbarre, A. Roget, J.-C. Arnault, F. Omnès and P. Mailley, New one step functionalization of polycrystalline diamond films using amine derivatives , *IOP Conf. Ser.: Mater. Sci. Eng.*, 2010, **16**, 012001.
17 E. Scorsone, S. Saada, J. C. Arnault and P. Bergonzo, *J. Appl. Phys.*, 2009, **106**, 014908.
18 J. De Sanoit, E. Van Hove, P. Bergonzo and P. Mailley, Electrochemical diamond sensors for TNT detection in water, *Electrochim. Acta*, 2009, **54**, 5688–5693.
19 R. Kiran, E. Scorsone, J. de Sanoit, J.-C. Arnault, P. Mailley and P. Bergonzo, Boron Doped Diamond Electrodes for Direct Measurement in Biological Fluids: An *In Situ* Regeneration Approach, *J. Electrochem. Soc.*, 2013, **160**, H67–H73.
20 R. Kiran, E. Scorsone, P. Mailley and P. Bergonzo, Quasi-Real Time Quantification of Uric Acid in Urine Using Boron Doped Diamond Microelectrode with *in Situ* Cleaning, *Anal. Chem.*, 2012, **84**, 10207–10213.
21 N. Lameire, W. Van Biesen and R. Vanholder, Nephrology, dialysis, transplantation: official publication of the European Dialysis and Transplant Association, *European Renal Association*, 1999, **14**, 2570–2573.
22 V. Maybeck, R. Edgington, A. Bongrain, J. O. Welch, E. Scorsone, P. Bergonzo, R. B. Jackman and A. Offenhäusser, Boron-doped nanocrystalline diamond microelectrode arrays monitor cardiac action potentials, *Adv. Healthcare Mater.*, 2014, **3**(2), 283–9.

23 Y. Hanein, presented at IBN-2, Forschungszentrum Juelich GmbH, Juelich, Germany (6 Apr 2011), unpublished results.
24 D. Brüggemann, unpublished results.
25 J. Sommerhalder, B. Rappaz, R. de Haller, A. P. Fornos, A. B. Safran and M. Pelizzone, *Vision Res.*, 2004, **44**, 1693–1706.
26 G. J. Chader, J. Weiland and M. S. Humayun, Artificial vision: needs, functioning, and testing of a retinal electronic prosthesis, *Prog. Brain Res.*, 2009, **175**, 317–332.
27 A. Horsager, G. M. Boynton, R. J. Greenberg and I. Fine, Temporal interactions during paired-electrode stimulation in two retinal prosthesis subjects, *Invest. Ophthalmol. Visual Sci.*, 2011, **52**, 549–557.
28 R. A. Fernandes, B. Diniz, R. Ribeiro and M. Humayun, Artificial vision through neuronal stimulation, *Neurosci. Lett.*, 2012, **519**(12), 122–128.
29 O. Zeitz, M. Keseru, R. Hornig and G. Richard, Artificial sight: recent developments, *Klin. Monatsbl. Augenheilkd.*, 2009, **226**, 149–153.
30 Velikay-Parel, D. Ivastinovic, T. Georgi and R. Hornig, Perceptual Threshold And Neuronal Excitability As Long-term Safety Evaluation In Retinal Implants, *Acta Ophthalmologica*, 2010, **88**(s246), 0.
31 M. Keseru, M. Feucht, N. Bornfeld, T. Laube, P. Walter, G. Rossler, M. Velikay-Parel, R. Hornig and G. Richard, Acute electrical stimulation of the human retina with an epiretinal electrode array, *Acta. Ophthalmol.*, 2012, **90**(1), 1–8.
32 S. Joucla and B. Yvert, *PLoS One*, 2009, **4**(3), e4828.
33 D. Palanker, P. Huie, A. Vankov, R. Aramant, M. Seiler, H. Fishman, M. Marmor and M. Blumenkranz, Migration of Retinal Cells through a Perforated Membrane: Implications for a High-Resolution Prosthesis , *IOVS*, 2004, **45**(9), 3266–3270.
34 D. Palanker, A. Vankov, P. Huie and S. Baccus, Design of a high-resolution optoelectronic retinal prosthesis, *J. Neural Eng.*, 2005, **2**, S105–20.
35 P. Bergonzo, M. Bonnauron, G. Lissorgues, L. Rousseau, E. Scorsone, Patent IPC: A61N1/05; C30B1/00; C30B29/04, Application number: FR20100054550 20100609, 2011.
36 E. Scorsone, S. Saada, J. C. Arnault and P. Bergonzo, *J. Appl. Phys.*, 2009, **106**, 014908.
37 H. A. Girard, S. Perruchas, C. Gesset, C. Chaigneau, M. Vieille, J.-C. Arnault, P. Bergonzo, J. Boilot and T. Gacoin, Electrostatic grafting of diamond nanoparticles: a versatile route to nanocrystalline diamond thin films, *ACS Appl. Mater. Interfaces*, 2009, **1**, 2738–2746.
38 A. Bongrain, E. Scorsone, L. Rousseau, G. Lissorgues and P. Bergonzo, Realisation and characterisation of mass-based diamond micro-transducers working in dynamic mode, *Sens. Actuators, B*, 2011, **154**(2), 142–149.
39 D. J. Garrett, K. Ganesan, A. Stacey, K. Fox, H. Meffin and S. Prawer, Ultra-nanocrystalline electrodes: optimization towards neural stimulation applications, *J. Neural Eng.*, 2012, **9**(1), 016002.
40 C. Hébert, J.-P. Mazellier, E. Scorsone, M. Mermoux and P. Bergonzo, *Carbon*, 2014, **71**, 27–33.

Faraday Discussions

PAPER

Creating and testing carbon interfaces – integrating oligomeric phthalocyanines onto single walled carbon nanotubes

Volker Strauß,[a] Almudena Gallego,[b] Gema de la Torre,[b] Thomas W. Chamberlain,[d] Andrei N. Khlobystov,[de] Tomás Torres[bc] and Dirk M. Guldi*[a]

Received 10th April 2014, Accepted 11th April 2014
DOI: 10.1039/c4fd00063c

Two oligomers, that is, *para*-phenylenevinylene with either pendant Pd(II)Pc or Zn(II)Pc units, have been synthesized and their interactions with SWCNTs and peapods – C_{60}@SWCNT – have been probed by complementary spectroscopy and microscopy. Spectroscopy assisted mainly in shedding light onto electronic interactions in the ground and excited state, while microscopy enabled insights into the degree of individualizing SWCNTs/C_{60}@SWCNTs and their stabilization.

Introduction

Nanosized carbon based materials offer exciting opportunities in a wide palette of areas of nanotechnology, since they possess novel electrical, magnetic, thermal, optical, mechanical, chemical, and morphological properties.[1,2] Carbon nanotubes (CNTs) are cylinders constituted of one – single walled carbon nanotubes (SWCNTs) – or more – multiwalled carbon nanotubes (MWCNTs) – graphene sheets with diameters from 0.4 to 5 nm and from 1.4 to 200 nm, respectively, and lengths up to half-meter.[3] In the past two decades, CNTs have triggered considerable attention owing to their unprecedented and interesting features. The band gap and optical properties of semiconducting CNTs are widely tunable with their diameters enabling specific light absorption.[4] The latter span from the ultraviolet to the infrared region. More notable are the electronic properties: CNTs reveal

[a]*Department of Chemistry and Pharmacy Interdisciplinary Center for Molecular Materials (ICMM) Friedrich-Alexander-Universitaet Erlangen-Nuernberg, Egerlandstr. 3, 91058, Erlangen, Germany. E-mail: dirk.guldi@fau.de*

[b]*Departamento de Química Orgánica, Universidad Autónoma de Madrid, Cantoblanco, 28049, Madrid, Spain*

[c]*Instituto Madrileño de Estudios Avanzados (IMDEA)–Nanociencia, c/ Faraday 9, Cantoblanco, 28049, Madrid, Spain*

[d]*School of Chemistry, University of Nottingham, Nottingham, NG7 2RD, UK*

[e]*Nottingham Nanoscience & Nanotechnology Centre, University of Nottingham, University Park, Nottingham, NG7 2RD, UK*

ballistic charge conduction, field effect mobilities on the order of 10^5 cm^2 V^{-1} s^{-1} at room temperature, electrical conductivities up to 1.7×10^6 S cm^{-1} and current carrying capability or ampacity up to 10^9 A cm^{-2}.[5]

In SWCNT bundles, fluorescence is hardly ever observed. Metallic SWCNTs, which are statistically present in bundles, are the inception for photoexcited carriers in semiconducting SWCNTs to relax along nonradiative pathways.[6] On the contrary, fluorescence measurements with individual SWCNTs have revealed distinct fluorescing transitions for more than 30 different SWCNTs.

The combination of SWCNTs with electron donors or acceptors generates active materials, which can produce electrical energy when irradiated.[7] In contrast to empty fullerenes and endohedral metallofullerenes, SWCNTs are, however, very elusive species when it comes to the characterization of their metastable states. This problem mainly arises from the polydispersive nature of SWCNT samples and the inevitable presence of SWCNTs in bundles of different sizes. An interesting aspect for electronic applications is the manipulation of the electronic properties of semiconducting SWCNTs, which is at the forefront of current investigations.[8] Charge transfer/doping is a leading example of a potential strategy toward this aim, which is expected to substantially increase the density of free charge carriers and thereby enhance the electrical and thermal conductivities. Notable are contributions in the area of field effect transistors.[9] Along the same lines, the effects that are generated when replacing carbon by boron, nitrogen or silicon are carefully studied in terms of transport properties of SWCNTs. A final example of charge transfer interactions includes a tunable photosensor. An alternative approach rests on the meta-stable doping of SWCNTs, triggered by light. In particular, the use of photoexcited state electron donors or electron acceptors generate transient charge carriers in the conduction and valence bands of SWCNTs, respectively. One way to achieve this goal relies on the use of versatile protocols to covalently attach electron donors or electron acceptors to SWCNTs.[10] However, the impact that the covalent functionalization exerts on the electronic structure is noticeable and affects transport properties, *etc.* Thus, efforts are directed to a non-covalent strategy, that is, the implementation of multifunctional groups without compromising the electronic and structural features of SWCNTs.

What renders working with SWCNTs so difficult is that all methods used to synthesize them in reasonable quantities produce SWCNTs with a very broad range of electronic, thermal, and structural properties that change depending on the different kinds.[11] A viable strategy implements charge transfer/doping. Our approach, on one hand, facilitates the integration of a molecular handle, while, on the other hand, preserving the SWCNT original electronic structure. Such interactions implicitly require the physical adsorption of suitable molecules onto the SWCNT sidewalls. We have performed systematic and complementary investigations by means of microscopy and spectroscopy, which have shed light onto mutual interactions between semiconducting SWCNTs and our dye-containing surfactants, towards the realization of two major milestones. The first milestone is the establishment of a versatile methodology to achieve water-soluble SWCNTs for processing them under environmentally friendly conditions.[12,13] In fact, microscopy demonstrated the benefits of tightly interacting π-systems towards the successful debundling and suspension. The second milestone is the establishment of a protocol to achieve *p*-doped SWCNTs for the

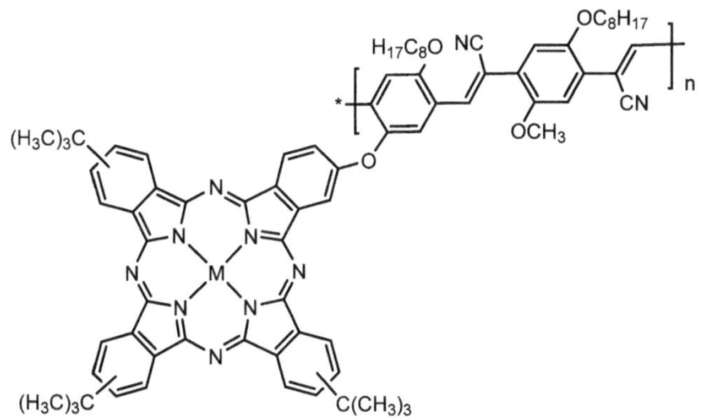

Scheme 1 PPV-Zn(II)Pc **1a** and PPV-Pd(II)Pc **1b**.

integration into novel electronic devices. Enhancing the *p*-type character is also achieved by trapping C_{60} inside of SWCNT – C_{60}@SWCNT.[14] In fact, preliminary experiments point to the successful shift of electron density from SWCNT to C_{60}.[15]

Oligomers as templates for electroactive components are similarly susceptible towards chemical changes. Simply incorporating electron withdrawing functionalities such as cyano groups into the oligomer backbone will help to transform the oligomer to feature *n*-type instead of *p*-type character.[16] These experiments should be contrasted to work on electron donating polyethylene imines, which transform *p*-type SWCNT into *n*-type SWCNT.[17]

In this contribution, we opted for two phthalocyanine (Pc)-containing oligomers, that is, PPV-Zn(II)Pc **1a** and PPV-Pd(II)Pc **1b** (Scheme 1), as excited state electron donors and immobilize them onto SWCNTs and C_{60}@SWCNTs. C_{60}@SWCNTs were chosen owing to the intrinsic charge transfer from SWCNTs to encapsulated C_{60} as a means to facilitate immobilizing PPV-Zn(II)Pc **1a** and PPV-Pd(II)Pc **1b**, on one hand, and individualizing not only C_{60}@SWCNTs but also SWCNTs, on the other hand.

Experimental section

General

Chemicals were purchased from Sigma Aldrich and used as received without further purification. IR spectra were recorded with a Bruker Vector 22 spectrophotometer. MALDI-TOF MS and HRMS spectra were recorded with a Bruker Reflex III. NMR spectra were recorded with a Bruker AC-300 instrument. Column chromatography was carried out on silica gel Merck-60 (230–400 mesh, 60 Å) and TLC on aluminum sheets pre-coated with silica gel 60 F254 (E. Merck).

9,16,23-Tri-*tert*-butyl-2-[4-(2-ethyl)hexyloxy-2,5-bis(hydroxymethyl)phenoxy] phthalocyaninato palladium(II) (mixture of regioisomers) (3)

A mixture of 4[(4′-(2-ethyl)hexyloxy-2′,5′-bis(methylacetoxy)]phenoxyphthalonitrile (**2**)[18] (200 mg, 0.41 mmol), 4-*tert*-butyphthalonitrile (227 mg, 1.23 mmol) and

DBU (0.08 mL) in *N,N*-dimethylethanolamine (4 mL) was heated at 100 °C under argon atmosphere until complete dissolution of the materials. Then, $PdCl_2$ (95 mg, 0.54 mmol) was added and the mixture was refluxed for 20 h. After cooling, methanol (10 mL) was added and the mixture was refluxed again for further 2 h and poured into water (30 mL). The resulting dark blue solid was filtrated and washed with water, mixtures of methanol–water (1 : 2) and (1 : 1), and methanol. Afterwards, the solid was dried and redissolved in tetrahydrofuran and purified by column chromatography (SiO_2) using $CHCl_3$–hexane–dioxane (1 : 2 : 0.1) as eluent. Phthalocyanine **3** was obtained as a blue solid (79 mg, 18%). δ_H (500 MHz; $CDCl_3$; Me_4Si) 0.5–2.0 (42H, m, CH, CH_2,CH_3), 4.2 (2H, d, $ArOCH_2$), 5.1–5.6 (4 H, m, $ArCH_2OH$), 6.8–7.0 (2 H, m, ArH), 7.7–8.2 (4 H, m, Pc–H) and 8,6–9.2 (8 H, m, Pc–H); *m/z* (MALDI, DCTB) 1074–1066. ($C_{60}H_{64}N_8O_4Pd$ requires 1066). λ_{max} (tetrahydrofuran)/nm (log ε) 332 (3.8), 596 (3.7) and 660 (4.5). FTIR spectrum shows the corresponding O–H wide band at 3410 cm^{-1}.

9,16,23-Tri-*tert*-butyl-2-[4-(2-ethyl)hexyloxy-2,5-diformyl)phenoxy] phthalocyaninato palladium(II) (mixture of regioisomers) (4)

To a well stirred solution of 1-hydroxy-1,2-benziodoxole-3(1H)-one-1-oxide (IBX) (50 mg, 0.17 mmol) in DMSO (60 mL), compound **3** (60 mg, 0.056 mmol) was added at once. The solution was stirred at room temperature for 24 h and then, poured over brine (80 mL) and extracted with diethyl ether (3 × 30 mL). The organic layer was separated, washed with a saturated solution of sodium bicarbonate (2 × 30 mL), brine (2 × 30 mL) and dried over sodium sulfate. After filtration of the drying agent, the solvent was evaporated and the blue solid was purified on a BiobeadsR© column using THF as eluent, affording aldehyde **4** as dark blue solid (39 mg, 65%). δ_H (500 MHz; $CDCl_3$;Me_4Si) 0.9–2.1 (42 H, m, CH, CH_2, CH_3), 4.3 (2 H, m, $ArOCH_2$), 7.3–8.8 (14 H, m, Pc–H, ArH), 10.68 (1 H, bs, CHO) and 11.00 (1 H, bs, CHO); *m/z* (MALDI, DCTB) 1070–1062 ($C_{60}H_{60}N_8O_4Pd$ requires 1062). λ_{max} (tetrahydrofuran)/nm (log ε) 331 (3.9), 595 (3.8) and 659 (4.5). FTIR spectrum reveals a strong band at 1610 cm^{-1} (νC-O).

Synthesis of PPV-PdPc oligomer 1b *via* Knoevenagel reaction

Diformylderivative **4** (30 mg, 0.028 mmol), 2,5-bis(cyanomethyl)-4-(2-ethylhexyloxy)anisol,[18] (9 mg, 0.029 mmol) and anhydrous potassium *tert*-butoxide (5 mg, 0.04 mmol) were dissolved in a mixture of dry tetrahydrofuran (6 mL) and dry *tert*-butanol (3 mL). The solution was heated at 60 °C and then a solution of tetrabutylammonium hydroxide (9 mL, 0.1 M in methanol) was added at once, turning the solution a dark green colour. After heating for 24 h, further anhydrous potassium *tert*-butoxide (5 mg, 0.04 mmol) and tetrabutylammonium hydroxide solution (9 mL, 0.1 M in methanol) were added. The reaction was maintained until consumption of the starting diformyl compound (*ca.* 24 h). Then, the solvents were evaporated and the residue was triturated in methanol (20 mL) and acidified with a drop of acetic acid in an ultrasound bath. The solid obtained was purified on a BiobeadsR© column, with dichloromethane as solvent. The main fraction (first to elute) was separated and the solvent eliminated. After washing with a mixture of hexane–acetone (1 : 1) the green compound was dried in vacuum affording 16 mg of 3 as a green solid. δ_H (500 MHz; THF-d_8; Me_4Si) 0.6–2.3 (m, CH, CH_2, CH_3, $C(CH_3)_3$), 5.0–5.3 (2 × m, $ArOCH_2$, $ArOCH_3$), 8.0–8.2 (m, ArH, vinylH), 9.0–9.5 (m, Pc–H). λ_{max} (tetrahydrofuran)/nm (log ε) 327 (4.2), 600 (3.7) and 661 (4.2). FTIR spectrum shows a band at 2231 cm^{-1} (νC-N).

Synthesis of C_{60}@SWCNTs

C_{60}@SWCNTs were produced from arc discharge SWCNT (Helix Material Solutions Inc., USA) and C_{60} (SES Research, USA). As received SWCNTs (20 mg) were annealed in air at 370 °C for 10 min to remove the end caps and residual amorphous carbon from the nanotube interior, and then stored at 300 °C to avoid the presence of water in the sample. A 3-fold excess of C_{60} (60 mg) was added to the nanotubes and the resultant material was sealed in a quartz tube under reduced pressure (6×10^{-6} mbar) and heated at 550 °C for 3 days. The material was then washed with CS_2 (20 mL) to remove any fullerenes adsorbed on the surface of the nanotubes and filtered through a PTFE membrane (pore diameter = 0.2 μm).

Spectroscopy and Microscopy

Steady state absorption measurements were carried out with either a Lambda 2 UV/Vis/NIR-spectrometer (Perkin Elmer) or a Cary 5000 UV/Vis/NIR-spectrometer (Varian). Steady state fluorescence measurements were performed with a FluoroMax®-3 spectrofluorometer (Horiba). All spectra were corrected for the instrument response. Transmission electron microscopy was performed with a TEM 912 Omega (Zeiss) at an acceleration voltage of 80 kV. Femtosecond transient absorption studies were carried out with a CPA 2101 (Clark-MXR Inc.) coupled to a Helios transient absorption pump/probe system (Ultrafast systems) with 387 nm laser pulses (1 kHz, 150 fs pulse width). Raman spectra were recorded using a FT-Raman RFS 100 system (Bruker) with a Ge detector using a 1064 nm Nd-YAG laser for excitation. High resolution transmission electron microscopy analysis was performed using a JEOL-2100F field-emission gun microscope. The imaging conditions were carefully tuned by lowering the accelerating voltage of the microscope to 100 kV and reducing the beam current density to a minimum. Nanotube samples (2–3 mg) were dispersed in 2 mL methanol using an ultrasonic bath and deposited onto Lacey carbon-film-coated TEM copper grids. All experiments were performed at room temperature in either solid state or in dimethylformamide (DMF) spectrophotometric grade (99.8%, Sigma-Aldrich). CoMoCAT® SWCNTs SG 65, enriched with (6,5)-SWCNTs were obtained from SIGMA ALDRICH (Purity ≥ 77% carbon as SWCNT, diameter 0.7–0.9 nm, > 90% semi-conducting).

Results and Discussions

Synthesis

The synthesis of the *o*-PPV oligomer with pendant Pd(II)Pc units (**1b**) by Knoevenagel condensation procedures requires the preparation of the corresponding diformyl- and dicianomethylene-functionalized precursors (Scheme 2). On one hand, the synthesis of the Pc-containing co-monomer was carried out by initial mixed condensation reaction between phthalonitrile **2**[18] and 4-*tert*-butylphthalonitrile, followed by hydrolysis of the acetate moieties to hydroxymethyl groups to afford **3** and further oxidation with IBX to give the diformyl compound **4**. On the other hand, the synthesis of the dicyanomethylene co-monomer **5** was

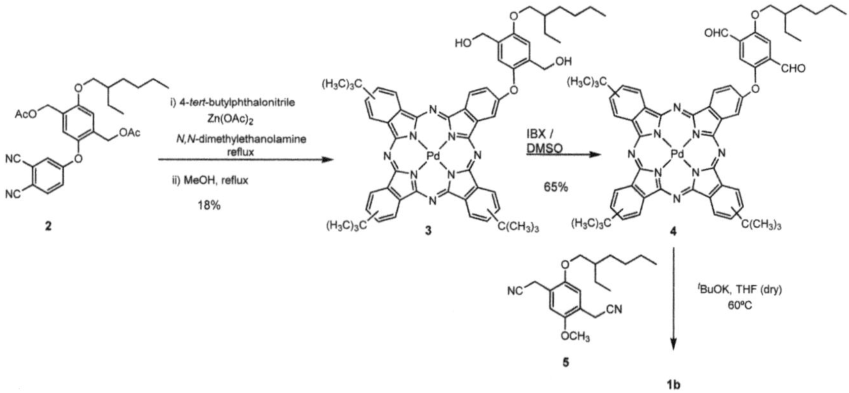

Scheme 2 Synthesis of Pd(II)Pc-containing oligomer – PPV-Pd(II)Pc (**1b**)

carried out as previously described.[19,20] Polymerization of co-monomers **3** and **4** was accomplished following similar conditions to that previously reported for related Zn(II)Pc-containing *o*PPV oligomers **1a**, that is, condensation in the presence of potassium *tert*-butoxide and tetrabutylammonium hydroxide during long reaction times (48h).

1b is very soluble in polar solvents such as THF and DMF. FT-IR showed a characteristic peak at 2231 cm^{-1}, which is assigned to the cyano moieties. ^1H-NMR in THF-d_8 showed broad signals for the aromatic and also for the CH_3O and -CH_2O protons.

Photophysical and microscopic characterization

When compared to the known PPV-Zn(II)Pc **1a**, both the absorption and the fluorescence features of PPV-Pd(II)Pc **1b** are considerably blue shifted by 14 nm (Fig. 1). In particular, the Q-bands shift from 676 to 662 nm and from 686 to 672 nm. At first glance, the fluorescence patterns of the two oligomers are equivalent. A closer look reveals, however, that the fluorescence quantum yield is – due to an increased spin orbit coupling – significantly reduced to 0.5 % in PPV-Pd(II)Pc **1b** when compared to PPV-Zn(II)Pc **1a**. For example, photoexcitation into its Soret-band at 350 nm fails to produce any appreciable fluorescence. Only upon direct excitation into the Q-band fluorescence is observed.

Recently, the interactions of PPV-Zn(II)Pc **1a** with SWCNTs were tested with HiPco SWCNTs.[16] In the current work we used two other samples of SWCNTs, that is, CoMoCAT SWCNTs and C_{60}@SWCNTs. The first are enriched with semiconducting SWCNTs to a degree of up to 90%. Typically, CoMoCAT SWCNTs exhibit very low defect rates and are – due to their small diameter of ~0.8 nm – comparatively flexible. On the other hand, C_{60}@SWCNTs have larger diameters of ~1.4 nm and the incorporation of C_{60} adds some degree of stiffness.

For the production of **1b**/SWCNT and **1a**/SWCNT hybrids we followed a sequential enrichment procedure. To this end, we started with a 1×10^{-5} M solution of the respective oligomers to disperse maximum amounts of SWCNTs/C_{60}@SWCNTs. A decolorization, which sets in as soon as the mixture was stirred, points to rather strong interactions between the oligomers and SWCNTs or C_{60}@SWCNTs. To avoid the latter, SWCNTs/C_{60}@SWCNTs were pre-dispersed in

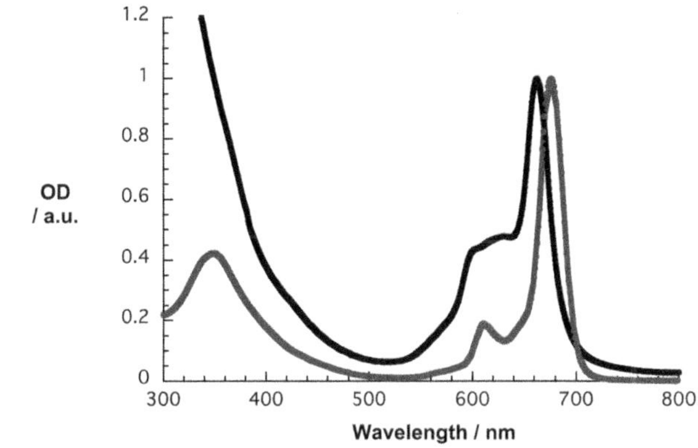

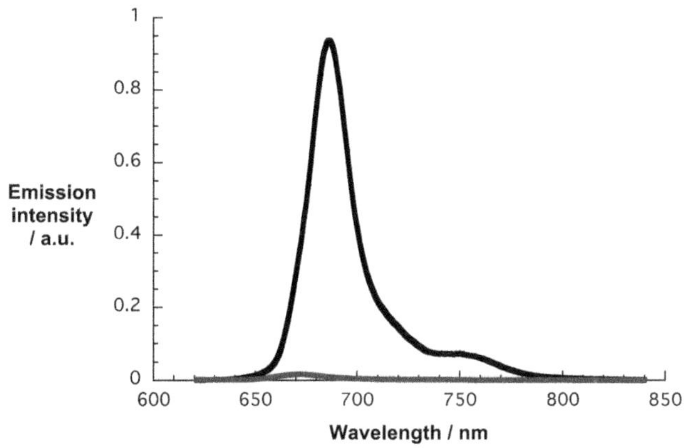

Fig. 1 Upper part – normalized absorption spectra of 1×10^{-5} M solutions of PPV-Pd(II)Pc **1b** oligomer (black spectrum) and PPV-Zn(II)Pc **1a** oligomer (dark grey spectrum) in DMF at room temperature. Lower part – fluorescence spectra of 1×10^{-5} M solutions of PPV-Pd(II)Pc **1b** oligomer (dark grey spectrum) and PPV-Zn(II)Pc **1a** oligomer (black spectrum) in DMF at room temperature upon photoexcitation at 600 nm.

a bare minimum of DMF. The pre-dispersed SWCNTs/C_{60}@SWCNTs were added stepwise to the oligomer solutions followed by 10 min of ultrasonication. Steady state absorption and fluorescence spectroscopy were employed to monitor the progress of the enrichment. The steps were repeated until no further changes were observed in the optical spectra (Fig. 2).

Firstly, the typical SWCNT transitions emerge in the near infrared parts of the absorption spectra. Upon applying prolonged ultrasonications, these bands sharpen and blue-shift. As a matter of fact, we consider this as an indicator for debundling of SWCNTs. Secondly, upon adding SWCNTs/C_{60}@SWCNTs to either of the oligomer solutions, a sequential decrease of the Q-band absorptions is discernable with the concomitant formation of new absorption bands at 698 and 684 nm for the Zn(II)Pc and Pd(II)Pc, respectively. From previous studies, we infer that this band is due a charge-transfer state.[16,21,22]

A look at the fluorescence spectra reveals that the successive addition of SWCNTs to solutions of either PPV-Zn(II)Pc **1a** or PPV-Pd(II)Pc **1b** induces a

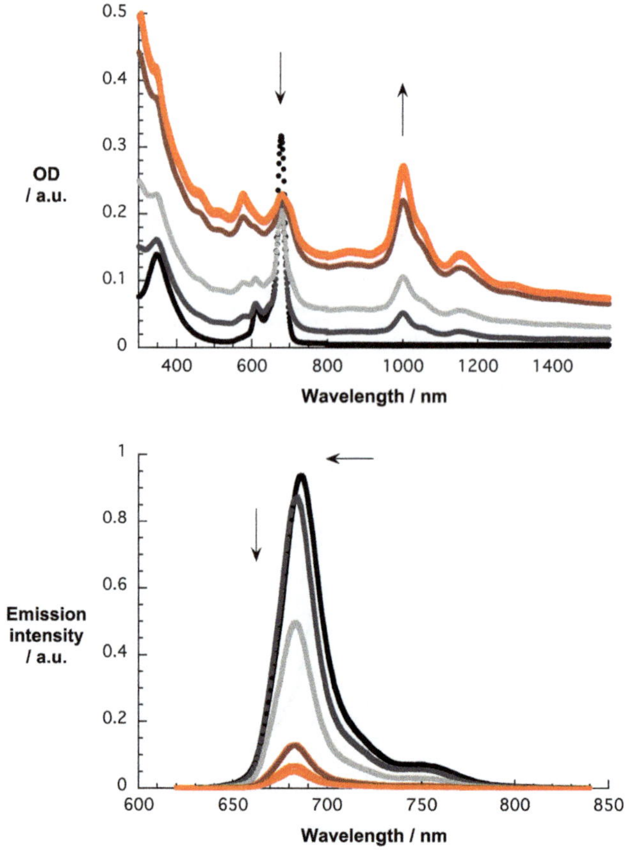

Fig. 2 Upper part – absorption spectra recorded during the sequential enrichment of a 1 × 10^{-5} M PPV-Zn(II)Pc **1a** solution with CoMoCAT SWCNTs in DMF at room temperature. Lower part – fluorescence spectra recorded during the sequential enrichment of a 1 × 10^{-5} M **1a** solution with CoMoCAT SWCNTs in DMF at room temperature upon photoexcitation at 600 nm.

significant quenching of the phthalocyanine related fluorescence. From the latter we deduce the formation of stable charge transfer hybrids. Nevertheless, we cannot rule out the presence of aggregates of **1a** or **1b**, in which the phthalocyanines stack. To this end, interactions with SWCNTs and C_{60}@SWCNTs induce the break up of the aggregates.

Likewise, the enrichment procedure with C_{60}@SWCNTs resulted in the same trends as noted for CoMoCAT SWCNTs. Nevertheless, the absorption features are drastically different (Fig. 3). C_{60}@SWCNT, for instance, shows two broad sets of absorptions, namely one at ~1000 nm and one at ~1800 nm.[23] The latter relates to S_{11} transitions, while the earlier is assigned to the corresponding S_{22} transitions.[20] Striking is the intensity of the charge transfer band at ~700 nm, which develops and intensifies upon sequential addition of C_{60}@SWCNTs. Overall, the rather weak and broad nature of the C_{60}@SWCNT centered absorptions point to poorly individualized C_{60}@SWCNTs. We hypothesize that the stiffness and the diameter of C_{60}@SWCNTs compared to CoMoCAT SWCNTs might prevent a better

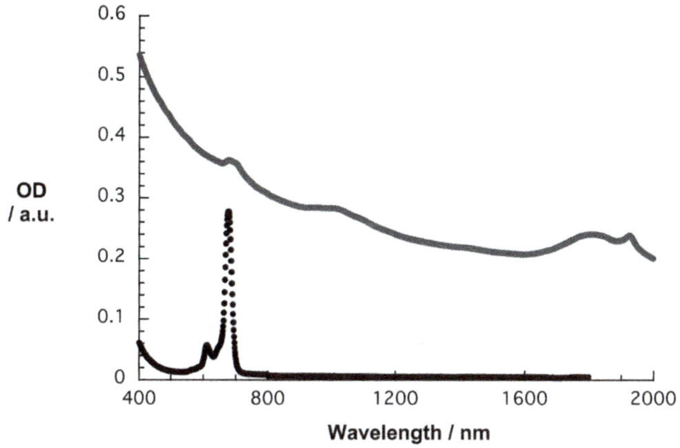

Fig. 3 Absorption spectra recorded before (black spectrum) and after (dark grey spectrum) the sequential enrichment of a 1×10^{-5} M PPV-Zn(II)Pc **1a** solution with C_{60}@SWCNTs in DMF at room temperature.

individualization. Comparative absorption and fluorescence titration assays with predispersed C_{60}@SWCNTs in DMF and small amounts of, for example, solid **1b** corroborate the aforementioned. Representative absorption spectra for C_{60}@SWCNTs and **1b** are shown in Fig. 4. Following the addition of first amounts of **1b** the charge transfer feature at 684 nm is discernable. In the presence of additional amounts of Pd(II)Pc, the 662 nm absorption of free Pd(II)Pc is noted and the overall ratio between the two absorptions, that is 684 nm absorption *versus* 662 nm absorption, reflects the equilibration in the complex formation. In line with the strong fluorescence quenching, which was observed during the above described enrichment process, the increase in Pd(II)Pc fluorescence is barely detectable throughout the course of the titration.

To gather additional information on the electronic interactions we performed complementary Raman spectroscopic analyses. The Raman specimen were

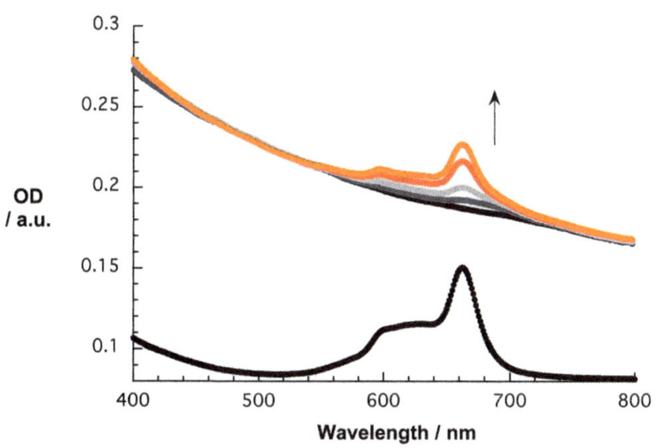

Fig. 4 Titration assays of C_{60}@SWCNTs with PPV-Pd(II)Pc **1b** in DMF at room temperature.

prepared by drop-casting the hybrid suspensions onto gold coated object slides and the solvent was evaporated under reduced pressure. All spectra were recorded upon excitation at 1064 nm, baseline corrected, and normalized with respect to the G^+-band. Comparison of the spectra before and after functionalization revealed a slight down-shift of the respective G^+-bands. It is known from doping experiments with SWCNTs that shifting of RBM and G^+-modes indicate charge transfer effects.[24] In the present case, the shifts towards smaller frequencies denote a shift of charge density from **1a** to SWCNTs.

When turning to the radial breathing modes (RBM), further insights into possible electronic interactions are gathered. In Fig. 5, the normalized RBM region for **1a**/C_{60}@SWCNT and **1b**/C_{60}@SWCNT and pristine C_{60}@SWCNT are shown. Importantly, the RBM mode at 167 cm^{-1} correlates with that of C_{60}@SWCNTs featuring diameters of ~1.4 nm as, for instance, in (16,3) or (14,6) SWCNTs.[25] Importantly, the RBM undergoes a slight shift towards higher

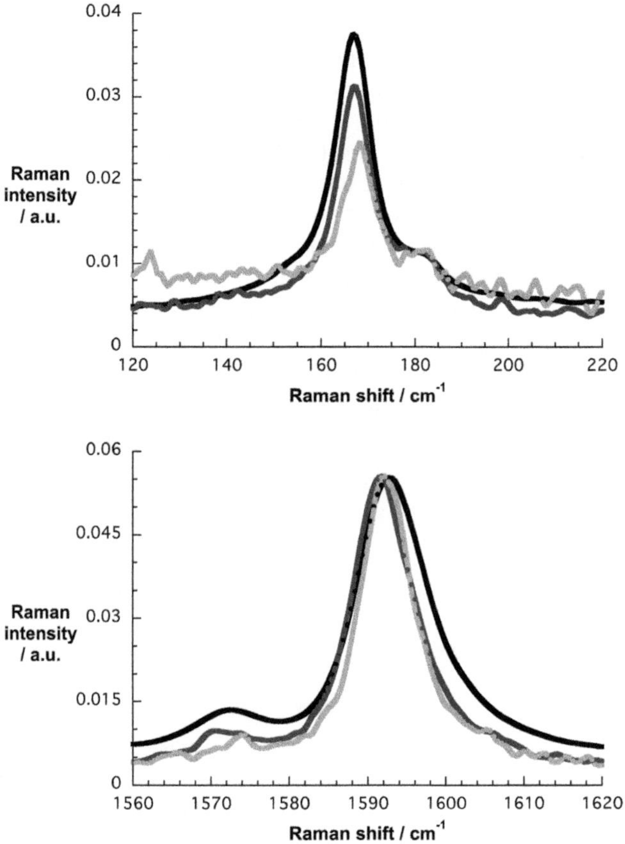

Fig. 5 Upper part – normalized solid state Raman spectra (λ_{ex} = 1064nm) of pure C_{60}@SWCNT (black spectrum), **1b**/C_{60}@SWCNT (dark grey spectrum), and **1a**/C_{60}@SWCNT (light grey spectrum) with particular emphasis on the RBM region. Lower part – normalized solid state Raman spectra (λ_{ex} = 1064nm) of pure C_{60}@SWCNT (dark spectrum), **1b**/C_{60}@SWCNT (dark grey spectrum), and **1a**/C_{60}@SWCNT (light grey spectrum) with particular emphasis on the G-band region.

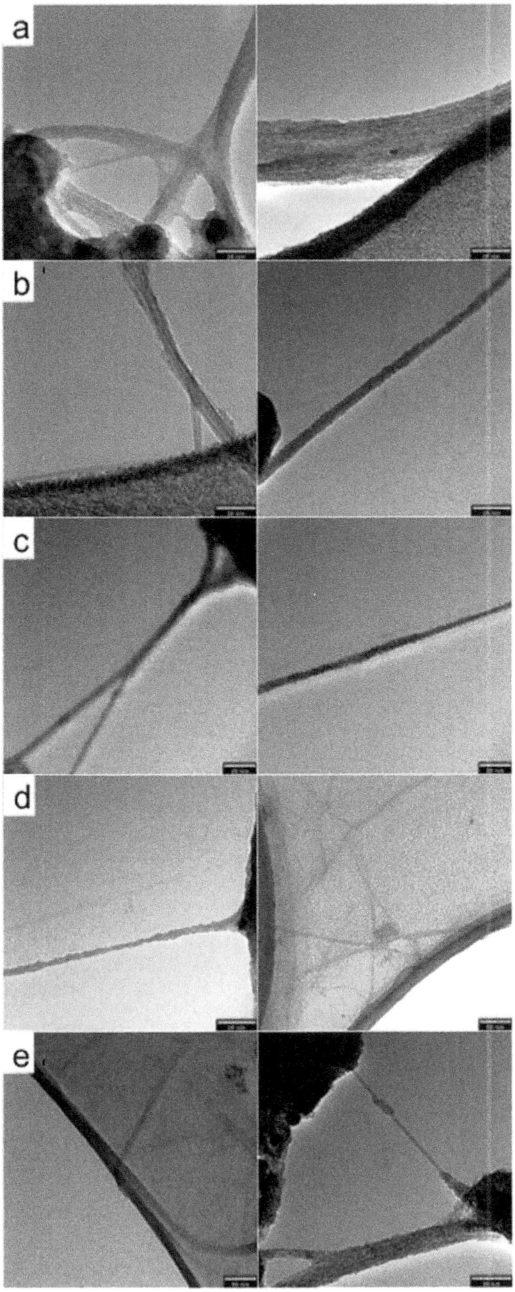

Fig. 6 a) TEM images of C_{60}@SWCNT bundles. Scale bars are 25 nm. b) TEM images of a small bundle of partially wrapped **1b**/C_{60}@SWCNT on the left and a fully PPV-Pd(II)Pc **1b** functionalized C_{60}@SWCNT on the right. Scale bars are 20 and 25 nm. c) TEM images of a small bundle and an individualized **1b**/SWCNT on the left and right. respectively. Scale bars are 25 nm. d) TEM images of an individualized **1a**/SWCNT on the left and a corresponding overview image on the right. Scale bars are 25 and 50 nm. e) TEM images of an individualized partially wrapped **1a**/C_{60}@SWCNT on the left and a corresponding overview image on the right. Scale bars are 50 and 20 nm.

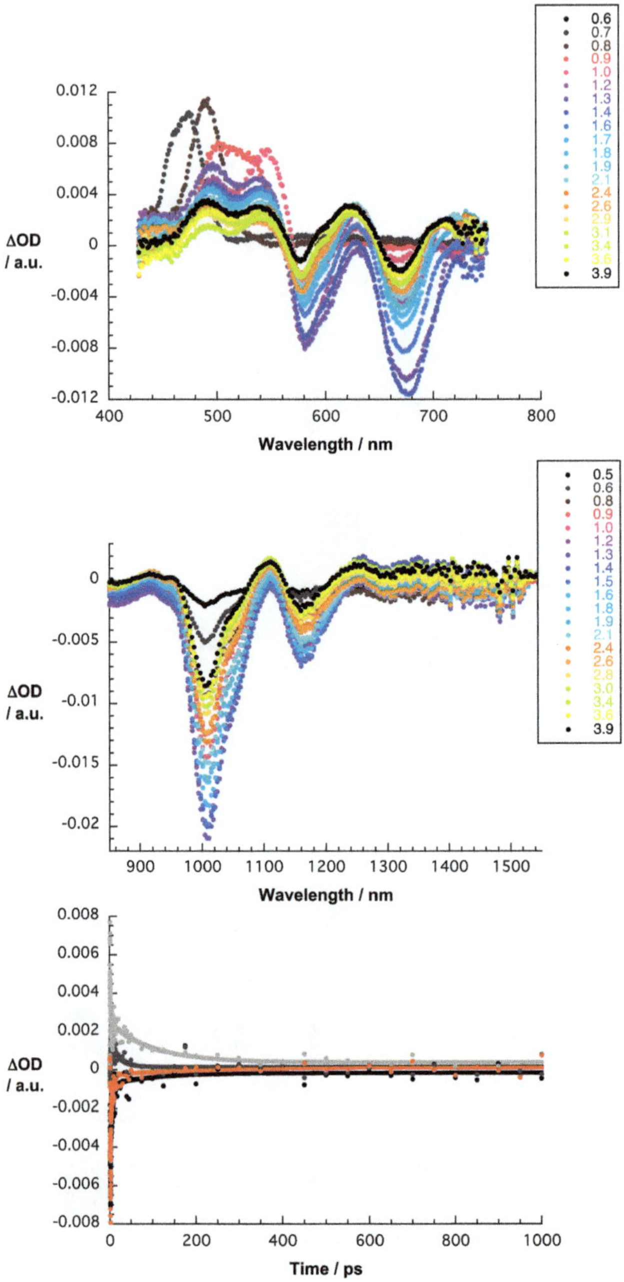

Fig. 7 Upper part – differential absorption spectra (visible) obtained upon femtosecond pump probe experiments (387 nm) of **1b**/SWCNTs in DMF with several time delays between 0.5 and 3.9 ps at room temperature. Central part – differential absorption spectra (near infrared) obtained upon femtosecond pump probe experiments (387 nm) of **1b**/SWCNTs in DMF with several time delays between 0.5 and 3.9 ps at room temperature. Lower part – time absorption profiles of the spectra shown in the upper and central parts at 1165 (black spectrum), 1110 (dark grey spectrum), 580 (red spectrum), and 540 nm (light grey spectrum) monitoring the excited state decay.

frequencies upon immobilization of the oligomers and decrease in intensity.[26] Both of these effects are typically observed in charge transfer hybrids containing SWCNTs.

Microscopic characterizations by means of transmission electron microscopy (TEM) were performed with all possible combinations of PPV-Zn(II)Pc **1a**/PPV-Pd(II)Pc **1b** and SWCNTs/C_{60}@SWCNTs – Fig. 6. To this end, dilute solutions of the hybrids were dropcasted onto Lacey carbon/Cu grids. The solutions were soaked by capillary forces to obtain individualized SWCNTs/C_{60}@SWCNTs crossing over the holes of the Lacey film. By this method, we are able to corroborate individual and freestanding SWCNTs/C_{60}@SWCNTs, which is crucial in order to visualize the amorphous oligomer coating.[27]

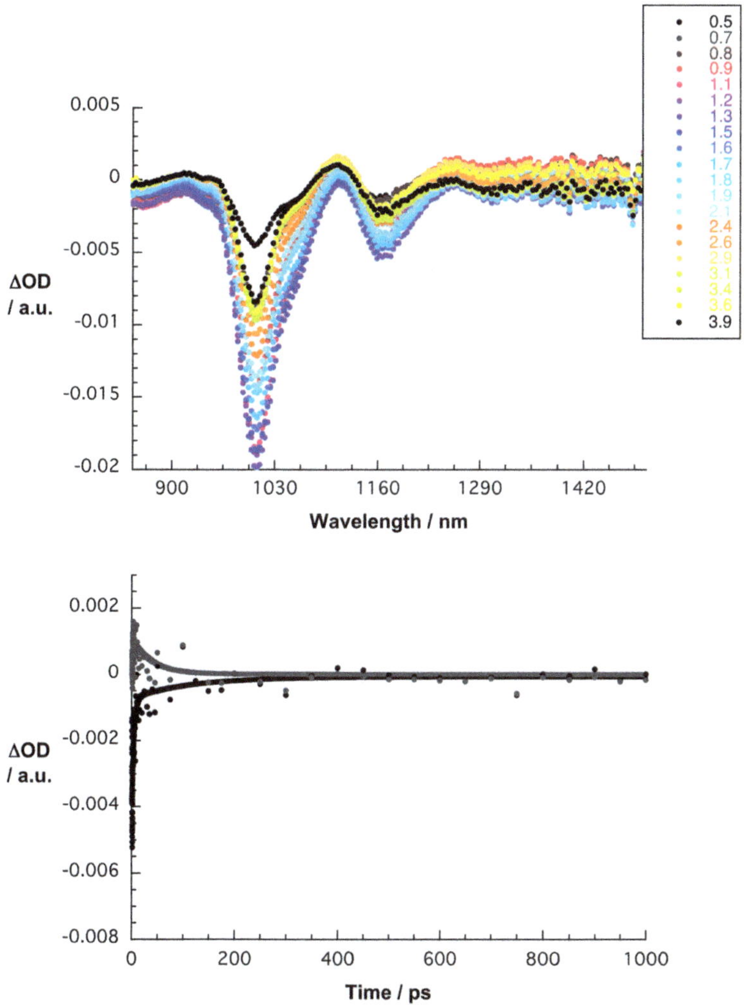

Fig. 8 Upper part – differential absorption spectra (near infrared) obtained upon femtosecond pump probe experiments (387 nm) of **1a**/SWCNTs in DMF with several time delays between 0.5 and 3.9 ps at room temperature. Lower part – time absorption profiles of the spectra shown in the upper part at 1165 (black spectrum) and 1110 nm (dark grey spectrum) monitoring the excited state decay.

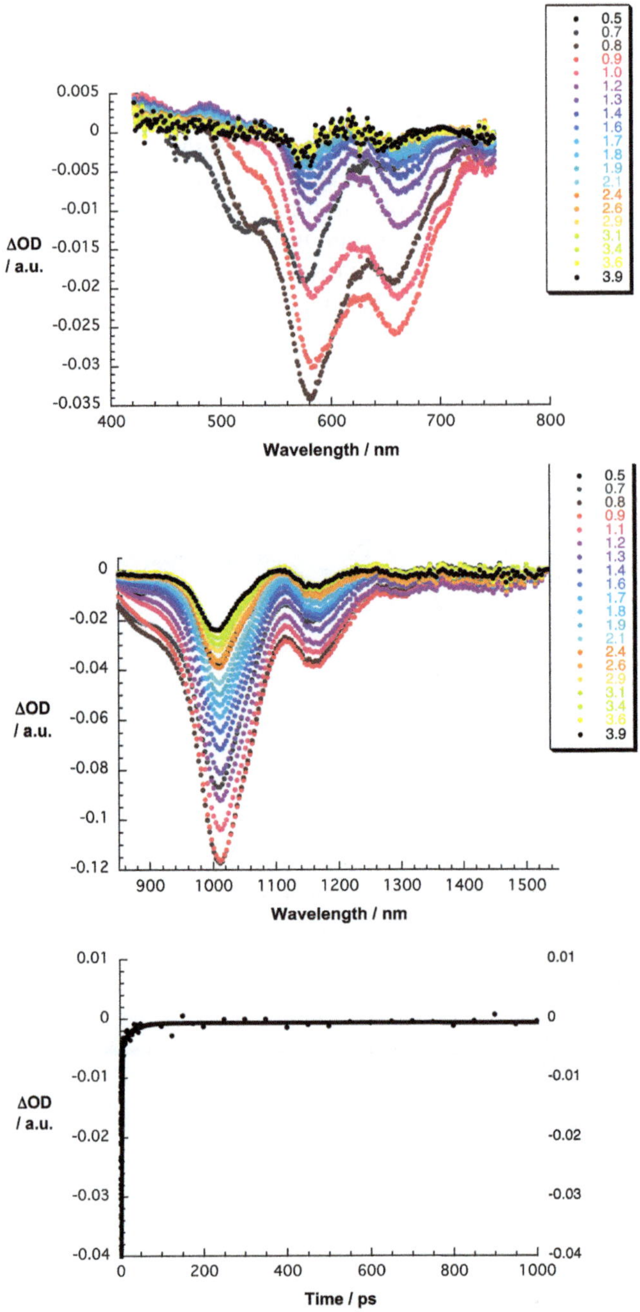

Fig. 9 Upper part – differential absorption spectra (visible) obtained upon femtosecond pump probe experiments (387 nm) of **1b**/SWCNT thin films with several time delays between 0.5 and 3.9 ps at room temperature. Central part – differential absorption spectra (near infrared) obtained upon femtosecond pump probe experiments (387 nm) of **1b**/SWCNT thin films with several time delays between 0.5 and 3.9 ps at room temperature. Lower part – time absorption profile of the spectra shown in the central parts at 1159 nm (black spectrum) monitoring the excited state decay.

As for the oligomer wrapped SWCNTs/C_{60}@SWCNTs, throughout the examined areas predominantly small bundles of and/or loosely entangled SWCNTs/C_{60}@SWCNT were found. We commonly observe the tangling of individualized SWCNTs and smaller bundles during the preparation of TEM samples. In stark contrast only large and straighten bundles (~20–50 nm) were seen for TEM grids prepared with non-functionalized SWCNTs/C_{60}@SWCNTs. Therefore, we deduce an efficient debundling of SWCNTs/C_{60}@SWCNT upon wrapping with **1a**/**1b**. Independent confirmation for the debundling came from overview images. SWCNTs and C_{60}@SWCNTs on Lacey films reveal reasonable individualization for the probed hybrids. Representative images of individualized SWCNTs/C_{60}@SWCNT are shown in Fig. 6. In each of the images, a relatively dense

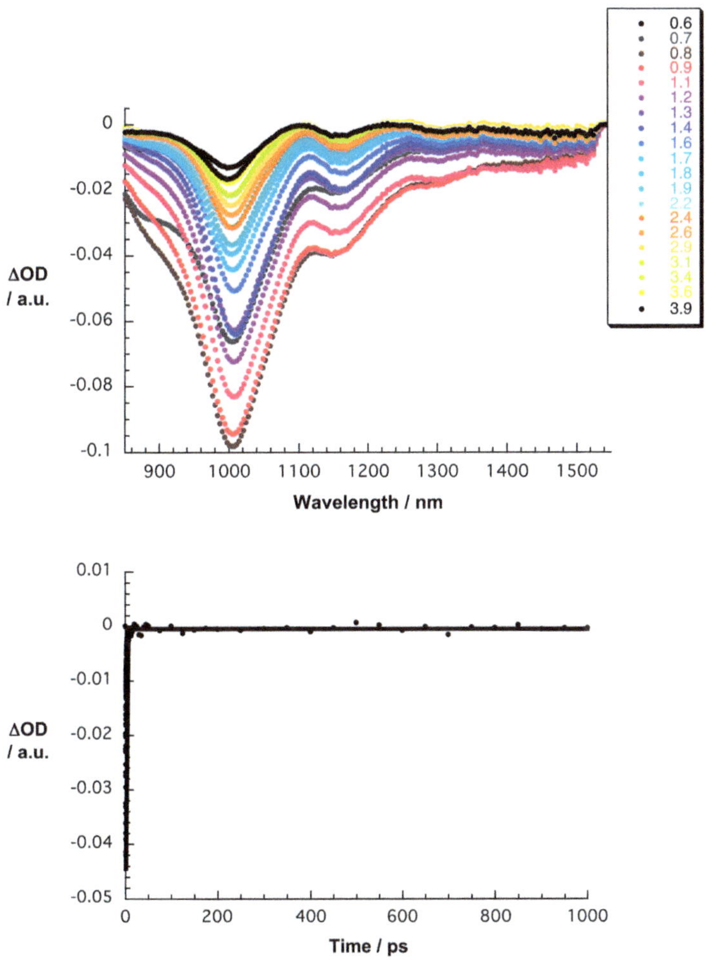

Fig. 10 Upper part – differential absorption spectra (near infrared) obtained upon femtosecond pump probe experiments (387 nm) of **1a**/SWCNT thin films with several time delays between 0.5 and 3.9 ps at room temperature. Lower part – time absorption profile of the spectra shown in the upper part at 1159 nm (black spectrum) monitoring the excited state decay.

amorphous coating around SWCNTs is noted. In the case of C_{60}@SWCNTs, the amorphous **1a**/**1b** coating hampers, however, the visualization of the encapsulated fullerenes. It is only in the references, that is, in the absence of **1a**/**1b**, that the fullerene arrays are recognizable. In the latter, well-defined and large sized bundles of C_{60}@SWCNTs predominate, while **1a**/**1b** coated C_{60}@SWCNTs are rather randomly ordered.

Finally and as a complement to the aforementioned fluorescence titration experiments, we turned to pump probe experiments. For the first time, we compared solution based experiments with those performed in solid state films. As to solution based experiments, 387 nm photoexcitation of **1b**/SWCNTs – Fig. 7

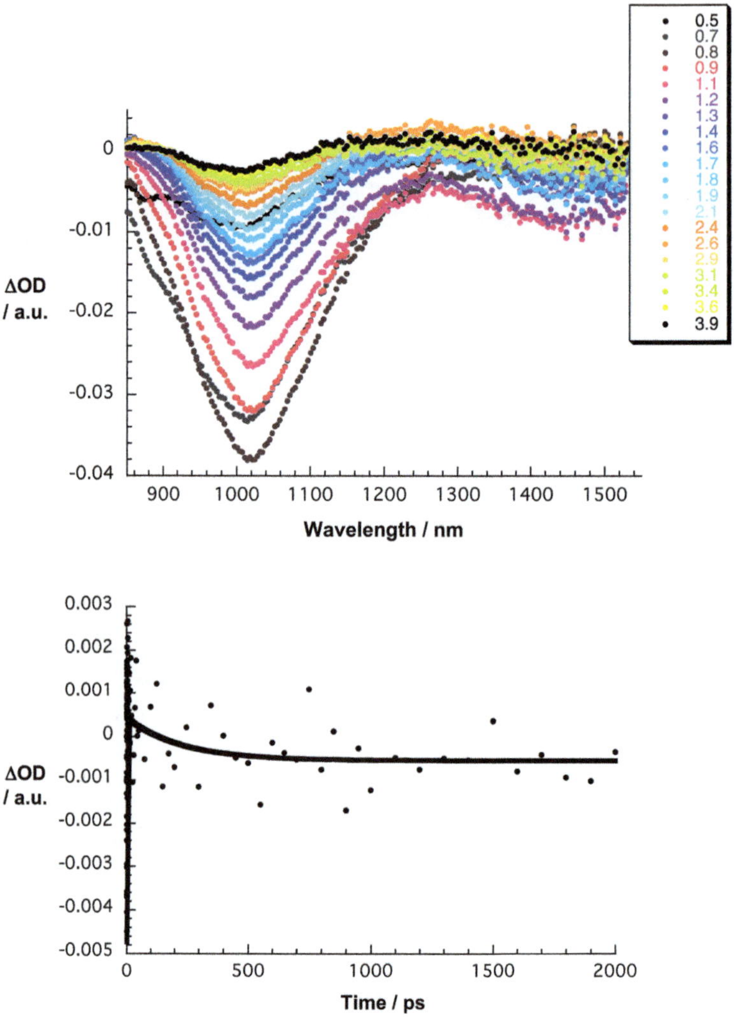

Fig. 11 Upper part – differential absorption spectra (near infrared) obtained upon femtosecond pump probe experiments (387 nm) of **1a**/C_{60}@SWCNT thin films with several time delays between 0.5 and 3.9 ps at room temperature. Lower part – time absorption profile of the spectra shown in the upper part at 1159 nm (black spectrum) monitoring the excited state decay.

– results in differential absorption changes that include maxima at 495, 542, 630, 730, 1110, 1256, and 1345 nm as well as minima at 580, 677, 1007, and 1164 nm. Notably, the minima reflect the ground state absorption maxima and, in turn, attest the photoexcitation of **1b**/SWCNTs. In contrast to reference experiments with SWCNTs, whose transient features are subject to a biexponential decay but to no appreciable red or blue shifts, photoexcited **1b**/SWCNTs reveal a 5 ps lasting blue shift of all the excited state characteristics. The maxima shift to 490, 622, 713, 1106, 1250, and 1340 nm, while the minima are now seen at 575, 667, 1004, and 1159 nm. Implicit is in line with previous reports, electron injection into the conduction bands of SWCNTs and oxidation of **1b**. On the longer time scale, that is, with 135 ps, the SWCNT conduction band electrons and holes on **1b** recombine to populate the ground state. The same findings were gathered for **1a**/SWCNTs – Fig. 8. For example, the 1007 and 1164 nm minima transform *via* 1004 and 1159 nm intermediates with lifetimes of 115 and 65 ps to the ground state.[28]

In thin films – Fig. 9 and 10 – upon photoexciting **1b**/SWCNTs 486, 624, 730, 1116, and 1273 nm maxima together with 582, 661, 1009, and 1161 nm minima were noted. Following the trend in suspensions the maxima and minima blue shift to 480, 612, 701, 1102, and 1244 nm as well as 576, 652, 1005, and 1155 nm, respectively, before reinstating the baseline. Considering lifetimes of 46 and 25 ps, we conclude that the electron injection into the conduction bands of SWCNTs, on one hand, and the population of the ground state, on the other hand, renders to be faster in the films than in solution. Likewise, the kinetics, by which the 1003 and 1161 nm minima in **1a**/SWCNTs convert upon photoexcitation to 999 and 1059 nm minima and to the baseline, are faster than in suspension. Here, the corresponding lifetimes are 40 and 4 ps.

Finally, we noted for **1a**/C_{60}@SWCNTs the immediate formation of a 1260 nm maximum and minima at 1019 and 1465 nm in the near infrared – Fig. 11.[29] Again, these features transform with a few ps into blue shifted maxima and minima. Most dominant is the 1250 nm maximum and the 1004 nm minimum, whose formation dynamics give rise to 5 ps lifetime. The latter correlate with the electron injection into the conduction bands of SWCNTs. Once formed, its decay sets in with a 160 ps lifetime to yield the ground state.[30]

Conclusions

Two Pc containing oligomers, that is, PPV-Zn(II)Pc **1a** and PPV-Pd(II)Pc **1b** as excited state electron donors have been synthesized and probed in terms of immobilization onto SWCNTs and C_{60}@SWCNTs. Importantly, the latter was chosen owing to the intrinsic charge transfer from SWCNTs to encapsulated C_{60}. Such a charge transfer facilitates the immobilization of **1a** and **1b** and, in turn, the individualization of not only C_{60}@SWCNTs but also SWCNTs. Notably is that the larger radii of C_{60}@SWCNTs, when compared to SWCNTs, renders the stability of the resulting C_{60}@SWCNT suspensions rather difficult. Still, sizeable electronic communications were found in absorption and fluorescence assays, which were complemented by Raman studies. Pump probe experiments on the femto- and picosecond time scales corroborated finally that the ground state charge transfer leads upon photoexcitation to the electron injection into conduction bands of SWCNTs and C_{60}@SWCNTs.

Acknowledgements

Financial support is acknowledged to the Spanish MEC and MICINN (CTQ2011-24187/BQU and PRI-PIBUS-2011-1128). Generous funding by the DFG is greatly appreciated.

References and Notes

1 D. M. Guldi and N. Martín, *Carbon Nanotubes and Related Structures*, Wiley-VCH, 2010.
2 G. Bottari, G. de la Torre, D. M. Guldi and T. Torres, *Chem. Rev.*, 2010, **110**, 6768–6816.
3 R. Zhang, Y. Zhang, Q. Zhang, H. Xie, W. Qian and F. Wei, *ACS Nano*, 2013, **7**, 6156–6161.
4 V. Sgobba and D. M. Guldi, *Chem. Soc. Rev.*, 2009, **38**, 165–184.
5 X. Wang, N. Behabtu, C. C. Young, D. E. Tsentalovich, M. Pasquali and J. Kono, *Adv. Funct. Mater.*, 2014, DOI: 10.1002/adfm.201303865.
6 S. M. Bachilo, M. S. Strano, C. Kittrell, R. H. Hauge, R. E. Smalley and R. B. Weisman, *Science*, 2002, **298**, 2361–2366.
7 G. M. A. Rahman, A. Troeger, V. Sgobba, D. M. Guldi, N. Jux, M. N. Tchoul, W. T. Ford, A. Mateo-Alonso and M. Prato, *Chem. – Eur. J.*, 2008, **14**, 8837–8846.
8 K. Dirian, M. Á. Herranz, G. Katsukis, J. Malig, L. Rodríguez-Pérez, C. Romero-Nieto, V. Strauss, N. Martín and D. M. Guldi, *Chem. Sci.*, 2013, **4**, 4335–4353.
9 Z. Wang, S. Mohammadzadeh, T. Schmaltz, J. Kirschner, A. Khassanov, S. Eigler, U. Mundloch, C. Backes, H.-G. Steinrück, A. Magerl, F. Hauke, A. Hirsch and M. Halik, *ACS Nano*, 2013, **7**, 11427–11434.
10 C. Oelsner, M. A. Herrero, C. Ehli, M. Prato and D. M. Guldi, *J. Am. Chem. Soc.*, 2011, **133**, 18696–18706.
11 J. Prasek, J. Drbohlavova, J. Chomoucka, J. Hubalek, O. Jasek, V. Adam and R. Kizek, *J. Mater. Chem.*, 2011, **21**, 15872–15884.
12 C. Romero-Nieto, R. García, M. Á. Herranz, L. Rodríguez-Pérez, M. Sánchez-Navarro, J. Rojo, N. Martín and D. M. Guldi, *Angew. Chem., Int. Ed.*, 2013, **52**, 10216–10220.
13 C. Romero-Nieto, R. García, M. Á. Herranz, C. Ehli, M. Ruppert, A. Hirsch, D. M. Guldi and N. Martín, *J. Am. Chem. Soc.*, 2012, **134**, 9183–9192.
14 D. A. Britz and A. N. Khlobystov, *Chem. Soc. Rev.*, 2006, **35**, 637–659.
15 L. Kavan, L. Dunsch and H. Kataura, *Carbon*, 2004, **42**, 1011–1019.
16 J. Bartelmess, C. Ehli, J.-J. Cid, M. García-Iglesias, P. Vázquez, T. T. Torres, D. M. Guldi, M. Garcia-Iglesias and P. Vazquez, *Chem. Sci.*, 2011, **2**, 652–660.
17 Z. Li, V. Saini, E. Dervishi, V. P. Kunets, J. Zhang, Y. Xu, A. R. Biris, G. J. Salamo and A. S. Biris, *Appl. Phys. Lett.*, 2010, **96**, 033110.
18 J.-J. Cid, C. Ehli, C. Atienza-Castellanos, A. Gouloumis, E.-M. Maya, P. Vázquez, T. Torres and D. M. Guldi, *Dalton Trans.*, 2009, 3955–3963.
19 T. Ahn, M. S. Jang, H.-K. Shim, D.-H. Hwang and T. Zyung, *Macromolecules*, 1999, **32**, 3279–3285.
20 N. S. Cho, D.-H. Hwang, J.-I. Lee, B.-J. Jung and H.-K. Shim, *Macromolecules*, 2002, **35**, 1224–1228.
21 L. Brinkhaus, G. Katsukis, J. Malig, R. D. Costa, M. Garcia-Iglesias, P. Vázquez, T. Torres and D. M. Guldi, *Small*, 2013, **9**, 2348–2357.

22 J. Malig, N. Jux, D. Kiessling, J.-J. Cid, P. Vázquez, T. Torres and D. M. Guldi, *Angew. Chem., Int. Ed.*, 2011, **50**, 3561–3565.
23 The small peak at 1923 nm is due to an artefact stemming from minor amounts of decomposed DMF during ultrasonication. This peak increases with prolonged ultrasonication.
24 A. G. S. Filho, A. Jorio, G. G. Samsonidze, G. Dresselhaus, R. Saito and M. S. Dresselhaus, *Nanotechnology*, 2003, **14**, 1130–1139.
25 Y. F. Wang, X. W. Cao, S. F. Hu, Y. Y. Liu and G. X. Lan, *Chem. Phys. Lett.*, 2001, **336**, 47–52.
26 M. Kalbac, L. Kavan, M. Zukalova and L. Dunsch, *J. Phys. Chem. B*, 2004, **108**, 6275–6280.
27 The lack of long term stability for **1b**/C_{60}@SWCNTs and **1a**/C_{60}@SWCNTs hampered any meaningful analysis, besides that the fact the intrinsic bleaching in the near infrared region mirrors the ground state absorption.
28 The visible lacks notable transient absorption features.
29 The lack of long term stability for **1b**/C_{60}@SWCNTs hampered any meaningful analysis, besides that the fact the intrinsic bleaching in the near infrared region mirrors the ground state absorption.
30 Individualized SWCNTs/C_{60}@SWCNTs are more soluble than small bundles and, in turn, are more likely to remain in solution.

Faraday Discussions

PAPER

Multifunctional structural energy storage composite supercapacitors

Natasha Shirshova,[a] Hui Qian,[a] Matthieu Houllé,[b] Joachim H. G. Steinke,[a] Anthony R. J. Kucernak,[a] Quentin P. V. Fontana,[c] Emile S. Greenhalgh,[a] Alexander Bismarck[a] and Milo S. P. Shaffer*[a]

Received 27th March 2014, Accepted 14th May 2014

DOI: 10.1039/c4fd00055b

This paper addresses the challenge of producing multifunctional composites that can simultaneously carry mechanical loads whilst storing (and delivering) electrical energy. The embodiment is a structural supercapacitor built around laminated structural carbon fibre (CF) fabrics. Each cell consists of two modified structural CF fabric electrodes, separated by a structural glass fibre fabric or polymer membrane, infused with a multifunctional polymeric electrolyte. Rather than using conventional activated carbon fibres, structural carbon fibres were treated to produce a mechanically robust, high surface area material, using a variety of methods, including direct etching, carbon nanotube sizing, and carbon nanotube *in situ* growth. One of the most promising approaches is to integrate a porous bicontinuous monolithic carbon aerogel (CAG) throughout the matrix. This nanostructured matrix both provides a dramatic increase in active surface area of the electrodes, and has the potential to address mechanical issues associated with matrix-dominated failures. The effect of the initial reaction mixture composition is assessed for both the CAG modified carbon fibre electrodes and resulting devices. A low temperature CAG modification of carbon fibres was evaluated using poly(3,4-ethylenedioxythiophene) (PEDOT) to enhance the electrochemical performance. For the multifunctional structural electrolyte, simple crosslinked gels have been replaced with bicontinuous structural epoxy–ionic liquid hybrids that offer a much better balance between the conflicting demands of rigidity and molecular motion. The formation of both aerogel precursors and the multifunctional electrolyte are described, including the influence of key components, and the defining characteristics of the products. Working structural supercapacitor composite prototypes have been produced and characterised electrochemically. The effect of introducing the necessary multifunctional resin on the mechanical properties has also been assessed. Larger scale demonstrators have been produced including a full size car boot/trunk lid.

[a]*The Composites Centre, Imperial College London, South Kensington Campus, London, SW7 2AZ, UK. E-mail: m.shaffer@imperial.ac.uk; Tel: +44 (0)20 7594 5825*

[b]*Nanocyl S.A., Rue de l'Essor 4, 5060 Sambreville, Belgium*

[c]*Cytec Industrial Materials, Heanor, Derbyshire, UK*

1 Introduction

Despite the progress in the development of electrical energy storage devices, the quest for more efficient, safer systems, with improved specific properties, continues, driven by increasing energy requirements. There is a particular need for volume and/or weight savings in mobile products, such as portable electronics or hybrid/electric vehicles, and aircraft. A particularly promising strategy relies on the development of multifunctional energy storage devices, which offer net savings by combining two conventionally independent sub-systems. One approach simply involves the assembly of two separate components, for example, shaping a battery to fit within a car door. A more challenging but potentially beneficial approach is to develop truly multifunctional devices, built out of multifunctional materials. This paper discusses the possibility of creating a structural material that is also an electrochemical energy storage device. The structural performance, here, is not just about robustness, but involves carrying significant mechanical loads in service. This concept was motivated by the observation that carbon-based materials are both the leading structural reinforcement and a common electrode component, and that both structural composites and electrochemical devices are typically laminates. The question is the extent to which these functions can be unified.

Many different electrical energy storage mechanisms could be considered. However, here the focus is on the development of multifunctional supercapacitors, since they offer useful performance, for example in load-levelling of high power applications, without involving significant changes of electrode shape or volume. Electrochemical double layer capacitors consist of two high surface area electrodes separated by a thin electrically-insulating but ionically-conducting electrolyte. To create a structural supercapacitor, at least two multifunctional components are required; a structural reinforcement/electrode, and a structural separator/electrolyte. Both aspects are discussed in more detail below.

Supercapacitor electrodes must have a high, ionically-accessible surface area and good conductivity, as well as thermal and (electro)chemical stability. There are a number of publications[1–4] which discuss possible materials and methods of their synthesis and modification, however, carbon in a variety of activated forms remains the most common choice. Structural carbon fibres are widely used in composite systems that exploit their high strength and stiffness,[5] and good electrical conductivity.[2] However, structural carbon fibres have very low surface areas, whilst industrial activated carbon fibres are usually poorly graphitised and lack the required properties. Instead, a means of increasing the effective surface area of structural carbon fibres is required. A number of possible modification methods can be considered, including physical and chemical activation, introduction of carbon nanotubes and impregnation with carbon aerogel, as discussed and compared further below.

The multifunctional electrolyte may be even more challenging, since it must combine high ionic conductivity with excellent rigidity and strength. Unfortunately, it is generally accepted that ionic conductivity and mechanical performance have an inverse relationship, since ionic conductivity requires molecular motion and rigidity does not.[6–10] Nevertheless, it is possible to obtain electrolytes with an unexpectedly good balance of performance by designing a bicontinuous structure[6] where one phase is responsible for the mechanical properties and a second phase provides ionic conduction. A more conventional, simple molecular interpenetrating network (molecular mixture) does not adequately isolate the two property requirements, and the phase separation should occur on a longer length scale, most likely defined by the other components in the device. The most effective existing examples use epoxy as the structural component and an ionic liquid based electrolyte as a second phase.[6,7,11,12] Both of these choices excel at their attributed role. Epoxies have high mechanical performance in combination with chemical and thermal stability[13,14] whereas ionic liquids have good ionic conductivity, a wide electrochemical window, negligible vapour pressure and high thermal stability.[15–17] Moreover, there is great variety in the types of both materials; for example, epoxy resins may contain two or more oxirane groups and a wide range of hardeners are available, and ionic liquids can combine a wide variety of cations and anions. Systems can therefore be tuned to produce the desired phase behaviour required to manifest the complementary properties, although formulation can be difficult, as illustrated below.

This paper illustrates the development of a structural supercapacitor system based on laminates of activated structural carbon fibre fabric electrodes sandwiching a structural glass fibre separator filled with (structural) electrolyte. The results show a comparison of different activation methods, and various approaches to developing an optimised, processable system. The data highlight the importance of bicontinuous microstructures for both electrodes and electrolytes. A range of proof of concept demonstrators have been fabricated, and the prospects for further progress are discussed.

2 Experimental

2.1 Materials

Plain weave carbon fibre fabric (HTA, Toho Tenax) was used for modification and as the electrodes of the structural supercapacitors (Tissa Glasweberei AG 862.0200.01) with an areal density of 200 g m^{-2} and a thickness of 0.21 mm, and a glass fibre fabric (plain weave, 200 g m^{-2}, 842.0200.01, TISSA Glasweberei AG) was used as the separator. The resorcinol–formaldehyde resin AX2000 was kindly provided by the INDSPEC Chemical Corporation.

Three fully formulated, commercially available, high performance epoxy resins with trademarks MVR444 and toughened MTM57 and VTM266 were prepared at Cytec Industrial Materials. All chemicals, including solvents for extraction were used as received: poly(ethylene glycol)diglycidyl ether (PEGDGE, M_n = 526, Aldrich), triethylenetetramine (TETA, technical grade, Aldrich), 1-ethyl-3-methylimidazolium bis(trifluoromethylsulfonyl)imide (EMITFSI, purity ≥98%), bis(trifluoromethane)sulfonimide lithium salt, LiTFSI, (puriss., ≥99.0% (^{19}F-NMR) Aldrich), KCl (Sigma-Aldrich), EDOT (Sigma-Aldrich), NaPSS (M_w 70 kDa, Sigma), $Fe(NO_3)_3 \cdot 9H_2O$ (Sigma), $FeCl_3$ (Sigma), cobalt acetate ($Co(CH_3COO)_2 \cdot 4H_2O$, Sigma-Aldrich), ethanol (puriss. >96%, Sigma-Aldrich).

2.2 Modification of the carbon fibres

The modification of carbon fibres has been described in detail elsewhere, for example the chemical/physical activation of carbon fibres,[18,19] the grafting of

CNTs onto carbon fibres,[20] as well as the preparation of structural electrolytes based on commercial epoxy resins.[6,10] CAG modification of the carbon fibres was performed as follows: resorcinol–formaldehyde (RF) polymer was prepared using a commercially available RF resin (AX2000, INDSPEC Chemical Corporation). The AX-2000 resin contains about 73.1 wt% resorcinol at a R : F molar ratio of 2 : 1. Potassium hydroxide (KOH) was used as the catalyst (C) and the R : C molar ratio was 50 : 1. Additional formaldehyde (37 wt% solution) was added to the mixture to keep the R : F ratio at 1 : 2. The weight percentage of RF in the mixture was controlled to be 25%, 40% or 50% by adjusting the quantity of distilled water diluent. The mixture was tightly sealed and stirred for 2 h. RF solution was infused into the carbon fibre using the resin infusion under flexible tooling (RIFT) method at room temperature. The specimens were then cured for 24 h, dried, placed into a chamber furnace (Lenton ECF 12/30) with a gas tight Inconel metal retort (Lenton 'A105') and carbonised under an inert atmosphere (constant N_2 flow at 0.5 l min^{-1}) at 800 °C for 30 min.

As carbon fabric is capable of withstanding high temperatures, especially under non-oxidizing atmospheres, it is possible to perform a direct growth of carbon nanotubes by catalytic chemical vapor deposition (CCVD). This growth is achieved by using a previously deposited catalytic phase on the surface of the carbon fabric. Previous work[20,21] has already shown that such a process is able to provide homogeneous coverage of the carbon fabric by a layer of CNTs, giving rise to a hybrid composite with a high specific surface area. The grafting of CNTs onto the CF was performed as follows: the fabric was heated at 80 °C in a ventilated oven (Binder) for 2 h before any treatment, to remove any adsorbed moisture that would prevent proper grafting of the CNTs over the carbon surface. A precursor solution of metallic salts was prepared using iron nitrate ($Fe(NO_3)_3 \cdot 9H_2O$) and cobalt acetate ($Co(CH_3COO)_2 \cdot 4H_2O$) dissolved in a mixture of distilled water and ethanol. The carbon fabric was impregnated with the catalyst precursor solution and then left to dry in open air at room temperature overnight. After complete drying, pieces of the fabric were placed in a tubular oven (Carbolite) and heated under an inert atmosphere (N_2) to the reaction temperature used for the growth process (750 °C). Once heated to a suitable temperature, a mixture of gaseous reactants was introduced into the reaction chamber to initiate the CNT-growth process. Nitrogen, ethylene and hydrogen were used in volumetric proportions of 50 : 40 : 10 with a total gas flow of 4 l min^{-1}. Accurate feeding of the gas flows was realised with mass flow controllers (Bronkhorst). Ethylene acts as a carbon source while hydrogen prevents excessive self-cracking of ethylene and promotes the catalyst lifetime. Nitrogen is added to the gas stream to dilute the reactive gases and ensure a homogeneous growth of CNTs on all the carbon fibres. Finally, the reaction was continued for a defined time (5 to 20 min) until the desired yield of growth was achieved (from 5 wt% up to more than 20 wt%). Once CCVD synthesis was completed, the fabrics were left to cool to room temperature under an inert atmosphere.

Sizing of the carbon fabric with CNTs was also performed as follows: the fabric was impregnated with a specifically designed formulation (Aquacyl™ custom, Nanocyl) containing 3 wt% of NC7000™ carbon nanotubes (Nanocyl) as well as a polymeric dispersing agent. The thickness of impregnation was controlled using a bar coater of 200 μm (BYK) to ensure homogeneous deposition of the sizing solution over the fabric surface, and the excess solution was removed. The so-coated fabric was dried at room temperature for 2 h, and then placed in a ventilated oven (Binder) at 120 °C for 2 additional hours to guarantee complete evaporation of water. After drying, the fabric was removed and cooled down to RT before a second impregnation step. The second stage of impregnation was performed using the same protocol, again using a wet-coating thickness of 200 μm. The wet sample was then dried at RT for 2 h and at 120 °C for 2 extra hours. The fabric was transferred to another oven (Bouvier-Technofour) that was heated up to 450 °C at a rate of 2 °C min^{-1} under a steady flow of air (1 l min^{-1}). Heating to 450 °C triggers the complete decomposition of the dispersing agent while the carbon nanotubes remain unharmed by the heat treatment. This thermal decomposition engenders the formation of a carbonaceous deposit that helps to anchor the CNTs onto the CF surface. Sized fabric was recovered after complete cooling.

2.3 Further modification of the CAG-modified carbon fibres

The CAG modified carbon fibres were further modified by PEDOT using two methods.[22,23] For the first (PEDOT 1), 0.3 g NaPSS (M_w 70 kDa, Sigma) was dissolved in 15 ml H_2O followed by the addition of 0.21 g EDOT (Sigma). The components were homogenised using an ultrasonic probe for 15 s to form a milky emulsion and 14.8 g $Fe(NO_3)_3 \cdot 9H_2O$ in 10 ml H_2O was added to the solution. The mixture was vigorously stirred using a magnetic stirrer for 5 s. The carbon fibres were submerged in the formed mixture for 5–10 min and then placed onto a PTFE coated film for 24 h. Afterwards, the carbon fibres were placed into 0.1 M HCl for 1 week, followed by washing with water and ethanol, and drying until constant weight was reached.

For the second route (PEDOT 2), equal amounts of EDOT and $FeCl_3$ (0.25 g) were dissolved in 5 ml EtOH and mixed. Carbon fibres were submerged in this solution for 5 min and then left at room temperature (21 °C) for the reaction to proceed. After 1 h, the carbon fibres were washed in EtOH to remove unreacted reagents and dried until constant weight was reached.

2.4 Characterisation of the modified carbon fibre fabric

Morphology and surface characterization of CAG modified carbon fibre fabric. The microstructure of the carbon fibre fabric was characterized using a field emission gun scanning electron microscope (SEM) (Gemini LEO 1525 FEG-SEM, Carl Zeiss NTS GmbH), operating at 5 kV, without metal coating. The specific surface area and pore size of the modified carbon fibre was studied using a Micromeritics TriStar 3000 analyzer (Micromeritics UK Ltd.) by pure N_2 adsorption. The data were analysed using the Brunauer–Emmett–Teller (BET) and the Barrett–Joyner–Halenda (BJH) methods.

Electrochemical characterization of the modified carbon fibre fabric. Electrochemical tests were performed on the carbon fibre fabric tows (~3k extracted fibre bundles, around 3.5 cm long) before and after modification at ambient temperature, using a three-electrode cell (platinum wire counter electrode, silver–silver chloride (Ag–AgCl) reference electrode) filled with 3 M KCl aqueous electrolyte. Cyclic voltammetry (CV) experiments were conducted between −0.2 and +0.2 V, using a potentiostat SI 1287 (Solartron Instruments). This scan range was chosen for this work as it provided representative capacitance results and allowed for quick comparison between different modified carbon fibres. The specific capacitance was calculated from the current $(I_a - I_c)/2$ at 0 V, *i.e.* the average of the current from the anodic and cathodic part of the curves.

Fabrication of the structural supercapacitors. To manufacture the structural supercapacitors, different matrix–electrolyte mixtures of PEGDGE : TETA (90 wt%, molar ratio 3 : 1) and the ionic liquid EMITFSI (10 wt%) were used. Mixtures were degassed before infusion using resin infusion under flexible tooling (RIFT) at 1 bar. The lay-up for producing the structural supercapacitors consisted of two carbon fabric electrodes, which sandwiched two pieces of glass fabric as the separator (mirrored at the midplane to ensure the laminates were symmetrical, for mechanical testing).[18] Copper tape (with conductive adhesive, 542-5511, RS components) was applied to the carbon fibre fabric to form the current collector for electrochemical characterization. The laminates were cured at 80 °C for 24 h.

Characterisation of the structural supercapacitors

Mechanical properties. The in-plane shear properties of the structural supercapacitors were investigated by tensile testing[24] of a ±45° laminate according to ASTM standard D3518. The composite specimens were dry cut to dimensions of 150 × 25 mm. Glass-fibre composite end-tabs were then applied to both ends of each specimen, resulting in a gauge length of 90 mm. A biaxial strain gauge (FCA-5-11, Techni Measure) was adhered to the sample, using cyanoacrylate glue, at the midspan of the specimens, to measure the strains in both the longitudinal and transverse directions. The tests were conducted at room temperature using an INSTRON 4505 universal testing machine, equipped with a 1 kN load cell, at a crosshead speed of 2 mm min^{-1}. A minimum of five measurements were conducted for each type of specimen. The fibre volume fraction of the reinforcements in the composites was measured[25] according to Method II of ASTM standard D3171.

Electrochemical properties. Charge–discharge experiments were performed at room temperature using a 0.1 V step voltage applied for 10 s and calculations were performed as described elsewhere.[18]

2.5 Characterisation of the structural electrolyte

To study the morphology of the structural electrolyte, EMITFSI and LiTFSI were extracted from the cured formulations as follows. Pre-weighed structural electrolyte, about 0.5–1.0 g, was placed in a glass vial filled with ethanol which was changed twice a day for at least one week, and subsequently dried in an oven at 70 °C under reduced pressure, to a constant weight.

SEM images. SEM images of extracted samples were recorded using a Jeol JSM 5610 LV scanning electron microscope with an accelerating voltage of 15 kV, after gold sputter coating for 120 s in an argon atmosphere.

Ionic conductivity. Ionic conductivity was determined using impedance spectroscopy (Solartron 1260) in the frequency range 10^5–0.1 Hz.

3 Results and discussion

3.1 Comparison of methods for carbon fibre modification

Woven carbon fibre mats were modified using a number of different treatments and compared to identify the one with the most promising performance (Fig. 1);

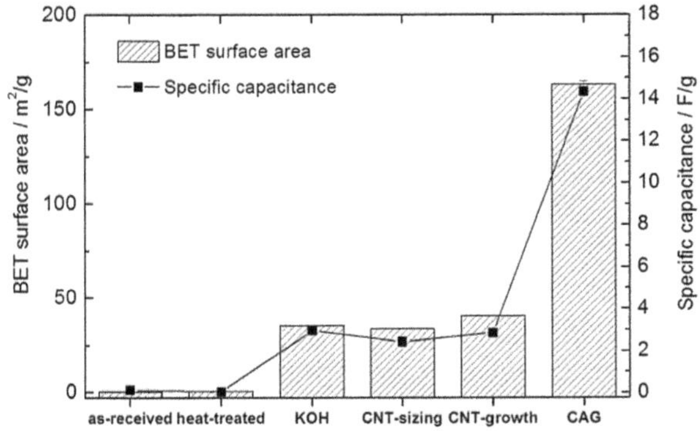

Fig. 1 The effect of different types of carbon fibre modification on the BET surface area and specific capacitance of the CF–GF–CF composite. (Electrolyte used – 3 M KCl; the specific capacitance was calculated from cyclic voltammetry, range −0.2 V to +0.2 V, scan rate 5 mV s^{-1}.)

the goal is to provide a maximum increase in electrochemical surface area whilst retaining the mechanical properties of the pristine carbon fibres. The examples shown have relatively little deterioration of mechanical performance,[19,20,26] unlike many types of activation[19,27–30] or nanotube grafting.[21] Carbon fibres annealed under an inert atmosphere are included as a control since many methods include heat treatment, either for growing nanotubes or removing sizing or surfactants. The most straightforward approach is simply to activate the surface of the structural fibres using an etching process. After a study of various activation chemistries, chemical activation of the carbon fibres using KOH was found to produce the best results[19] (Fig. 1). In this case, the surface of the carbon fibres appear etched (Fig. 2), creating mesopores[19] responsible for the increase of the surface area and specific capacitance (Fig. 1). However, the increase is limited to a relatively small volume, since the fibre core must remain broadly intact for reinforcement purposes. The addition of nanostructures to the carbon fibre surface, therefore, offers better prospects for increasing the capacitance.

The introduction of carbon nanotubes (CNTs) onto carbon fibres is known to increase their surface area;[20] intrinsically the use of CNTs is appealing as, in principle, they offer high surface areas and excellent mechanical performance. There are examples of pure CNT fibres with promising performance, but currently only very small amounts of material are available.[31–33] The combination of CNTs with carbon fibres to form hierarchical composites is very promising for a variety of structural and functional purposes,[21] and is already commercialised in some cases. As well as providing the necessary electrode performance, the addition of transverse CNTs onto unidirectional fibres may inhibit matrix-dominated failure processes prevalent in carbon fibre composites; tackling matrix dominated failures is likely to be particularly important in a structural electrical energy storage context, since multifunctional electrolytes/matrices are likely to be mechanically less robust than conventional resins. CNTs can be dispersed at low concentrations in the matrix, or attached to the primary carbon fibres by growth or post-growth coating processes. Since connection to the primary carbon fibre electrodes is key, these latter options are most relevant; two possible examples have been

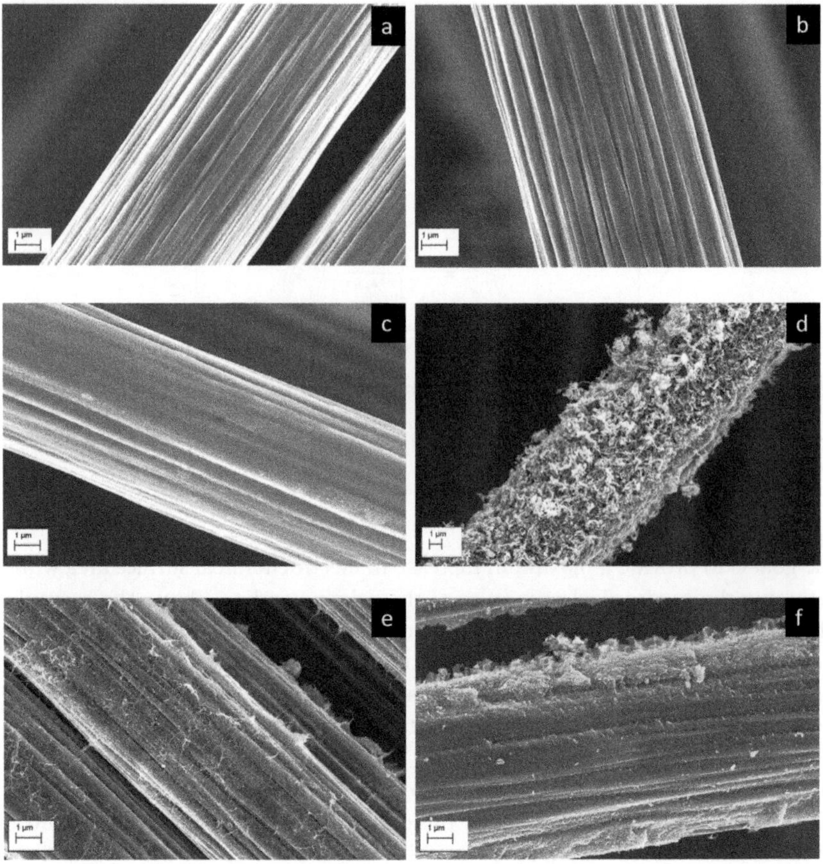

Fig. 2 Scanning electron micrographs of (a) as-received, (b) heat-treated, (c) KOH activated carbon fibres, (d) CNT grafted carbon fibres, (e) CNT sized carbon fibres, (f) CAG modified carbon fibres.

tested here. In the first case, the carbon fibres were impregnated with a CNT sizing, followed by pyrolysis of the surfactant/binder. In the second, the CNTs were grown directly from the surface of the carbon fibres, using CCVD. The impregnation process is simpler to implement, and relates to a large number of possible variants, such as electrophoretic deposition or chemical grafting. The disadvantage from a mechanical point of view is that the CNTs typically lie predominantly parallel to the carbon fibre surface and likely offer less transverse contribution to the matrix properties. The initial improvement using the CNT sizing was very minor (Fig. 1); however, to remove the binder/surfactant, to allow the electrolyte access to the hybrid electrode and to ensure a good electrical connection, the CNT impregnated carbon fibres were subjected to pyrolysis. A SEM image (Fig. 2d) shows a homogeneous coverage of the carbon fibres with a layer of CNTs, consistent with the observed increase in surface area and capacitance (Fig. 1). CVD growth of CNTs from the surface of carbon fibres, initiated by the use of a pre-deposited metal catalyst, avoids the presence of a (pyrolyzed) binder and may improve CNT anchoring. Again, a uniform covering of carbon fibres forms but with a more three dimensional microstructure than for the pyrolysed system, leading to a further improvement in the accessible surface area. Etching of the carbon fibre by the metal catalyst during growth tends to degrade the primary mechanical properties but the effects can be limited by careful choice of parameters.[21]

Whilst CNTs are intrinsically promising, the embodiments discussed have either limited volume, or relatively low CNT packing density, or both. The net contribution of the CNT layer is, therefore, low. A simple geometric model indicates that a significant increase in accessible area is possible, in principle, using small diameter nanotubes at high packing densities (Fig. 3). The nanotubes cannot be too small, or they will tend to bundle and leave insufficient space to accommodate the electrochemical double layer; they cannot be too long as they will limit the primary carbon fibre packing efficiency. An optimum will exist for nanotubes with diameters of ~3–5 nm, lengths <10 μm, and a relatively high packing density. Fig. 3 is plotted for an areal packing density of around 15%; higher packing density might be possible, but space is needed for the electrolyte and electrochemical double layer. Unfortunately, despite progress in preparing dense small diameter CNT arrays for vias in microelectronics,[34] the conditions for preparing such a structure grafted on carbon fibres remain elusive.

As an alternative, a three-dimensional carbon aerogel (CAG) can be formed around the carbon fibres,[26] filling all the available free volume with a bicontinuous, electrically-connected structure with high surface area and low density. CAG is formed *via* a sol–gel process between organic precursors, for example resorcinol (RF) and formaldehyde (FA), in the presence of a catalyst, resulting in a highly crosslinked organic gel, which after pyrolysis forms a highly porous carbon network.[35,36] CAG is often used as a pure supercapacitor electrode material. In addition, the combination with non-structural, low density, short carbon fibre felts has been shown to minimise shrinkage, simplify handling, and maximise electrochemical performance.[37,38] The unexpected combination with high loadings of continuous structural fibres retains these advantages, and introduces

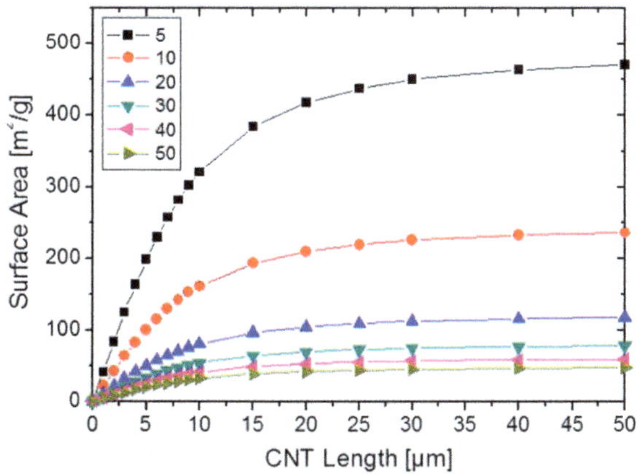

Fig. 3 Simple geometric estimate of the potential surface area of the CNT modified carbon fibres as a function of CNT length, for a range of different diameters (given in nm). These data are for a 15% surface coverage.

additional mechanical benefits through the formation of a monolithic supportive matrix network.

Of the methods studied, the CAG-modified carbon fibres showed the greatest potential for improving the surface area and electrochemical double layer capacitance, whilst retaining the structural performance of the primary fibres. The method increased the surface area and specific capacitance by three orders of magnitude over the initial structural fibres, as well as fivefold and fourfold, respectively, over the next best approach, using KOH activated carbon fibres (Fig. 1). In principle, the physical and mechanical characteristics of the CAG are determined by the geometry and length scales of the structures, which can be modified[39–41] by varying synthesis parameters, including the concentration of catalyst, concentration of resorcinol, ratio of resorcinol to formaldehyde and temperature.

3.2 Effect of resorcinol–formaldehyde concentration on the properties of the modified carbon fibres

In the context of structural supercapacitors, the volume fraction of primary reinforcing fibres should be kept high (>40 vol% of the woven carbon laminates), and thus the volume available for loading with CAG is limited. Since the volumetric capacitance of CAG is related to the CAG density and amount of active material,[1] it was decided to vary the concentration of RF in the reaction mixture[42] keeping all other parameters constant, including the R : F precursor to catalyst ratio, as well as temperature and time. A set of CAF–CF laminates were prepared, infused using varying RF concentration, and the performance assessed. A steady increase of the BET surface area was observed as the concentration of RF was increased from 20 wt% to 50 wt%, as a result of the increase in the volume of both mesopores (from 20 to 40 wt%) which can be seen as an increase in the hysteresis between the adsorption and desorption isotherms (Fig. 4) and micropores (from 40 to 50 wt%). The literature data describing the dependence of the BET surface area upon RF concentration is not consistent, ranging from not affected,[43] to

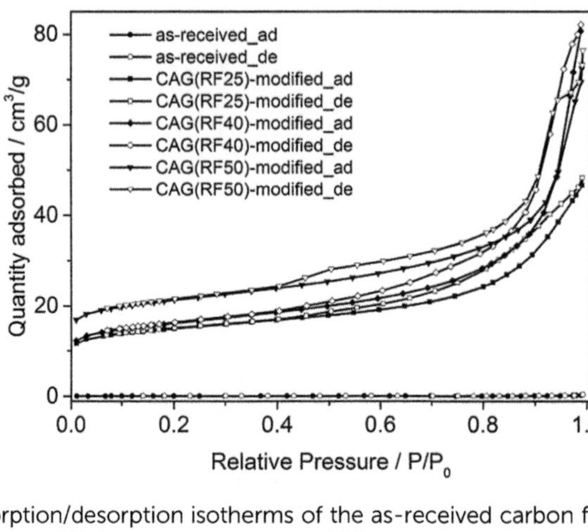

Fig. 4 N_2 adsorption/desorption isotherms of the as-received carbon fibres and carbon fibres infused using different concentrations of RF.

decreased[35,42] and increased.[44,45] These variations can be attributed to other experimental parameters such as the type and amount of catalyst, or the carbonisation conditions (temperature and time). However, a detailed comparison of the literature data is outside of the scope of this paper.

As expected, the CAG loading in the carbon fibre fabrics increased with increasing infused RF concentration. The decrease of the RF concentration from 40 wt% to 25 wt% led to a decrease of the CAG loading in the resulting modified carbon fibres and an associated reduction of the specific capacitance (Table 1).[41] The slightly greater than proportionate loss of CAG loading may be due to dilution of the reaction mixture, leading to a reduction of the polycondensation rate[43] and the yield of the reaction. The decrease in the volume of mesopores, associated with a decrease of the RF concentration, may also contribute to the loss of capacitance.[43,44] The increase of the RF concentration to 50 wt% led to an increase of the CAG loading and an increase of the surface area, due to a reduction of pore size (Fig. 5) but it did not increase the specific capacitance. Clearly, many of the smaller pores formed at higher RF concentrations are too small to allow effective electrolyte access, and the more open 40 wt% structure is more favourable. The more open structure likely offers further advantages in improved ion transport.

The change in pore size observed is consistent with previous investigations into the formation mechanism of these CAG microstructures.[40,46] There are two possibilities for phase separation, *i.e.* the formation of a porous structure during resorcinol–formaldehyde polycondensation, depending on the RF : W ratio and concentration of catalyst in the sol. One of them is nucleation–growth, which upon an increase of resorcinol concentration may become a spinodal decomposition.[40,46] It is reasonable to assume that for the system studied, the transition occurs between RF concentrations of 40 wt% and 50 wt%, leading to the observed changes in the morphology. Similar findings were reported by Schwan and Ratke[40] on an example of resorcinol–formaldehyde polycondensation catalysed using sodium carbonate and nitric acid. They observed that even small variations in the water content led to significant differences in the mechanical performance of the CAG formed, which was attributed to the change in the phase separation mechanism. However, no mechanism transition was reported by Szczurek *et al.*[44] and a steady decrease of the surface area, as well as porosity, was reported for CAG produced from cured resorcinol–formaldehyde resins synthesised with higher RF concentrations. It was also reported[35] that increases in RF : W led to increased pore polydispersity which has a negative effect on the double layer capacitance.[47]

Table 1 Effect of the RF content in the sol on the properties of the resulting CAG modified carbon fibres

RF concentration, wt%	CAG loading, wt%	BET surface area, $m^2\ g^{-1}$	Pore width, nm	Specific capacitance,[a] $F\ g^{-1}$
25	6.0	51.7 ± 0.5	5.2	3.8 ± 0.1
40 (initial conc.)	9.1	56.1 ± 0.5	7.3	7.4 ± 0.3
50	12.9	73.2 ± 0.8	5.5	4.0 ± 0.2

[a] Electrolyte – 3 M KCl; specific capacitance was calculated from CV measurements using the three electrode setup in the voltage range −0.2 to 0.2 V at a scan rate of 5 mV s^{-1}.

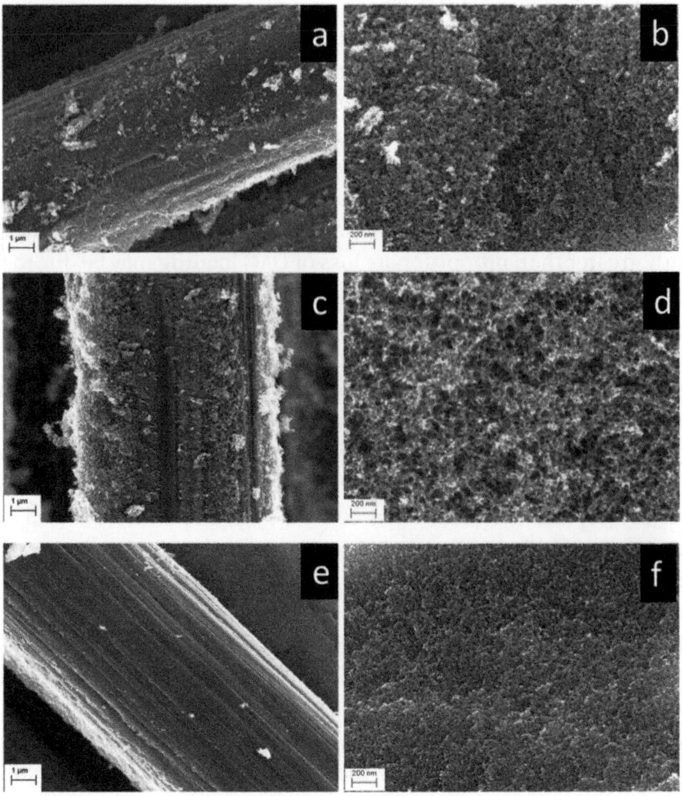

Fig. 5 Scanning electron micrographs of CAG modified carbon fibres prepared using different amounts of RF: (a and b) 25 wt%, (c and d) 40 wt%, and (d and e) 50 wt%. For each sample, two different magnifications are presented.

Fig. 6 shows typical cyclic voltammograms (CVs) of CAG modified carbon fibres in 3 M KCl solution over a potential range of −0.2 V to +0.2 V. All curves exhibit a nearly rectangular shape consistent with the expected capacitive behaviour. The specific capacitance from cyclic voltammetry was calculated from the current $(I_a - I_c)/2$ at 0 V, *i.e.* the average of the current from the anodic and cathodic part of the curves. CAG modified carbon fibres prepared from the mixture containing 40 wt% RF displayed the highest current density as a result of the presence of larger pores.

3.3 Multifunctional performance of the composites based on CAG-modified carbon fibres

Composite electrodes manufactured using CAG-modified carbon fibres were considered to have the properties required for structural supercapacitors, *i.e.* high strength and electrochemical surface area. Structural supercapacitor laminates containing 40 wt% CAG-modified carbon fibres have been demonstrated, using woven glass fibre separators and a multifunctional electrolyte matrix.[26] Using the same multifunctional polymer electrolyte, which was comprised of PEGDGE and

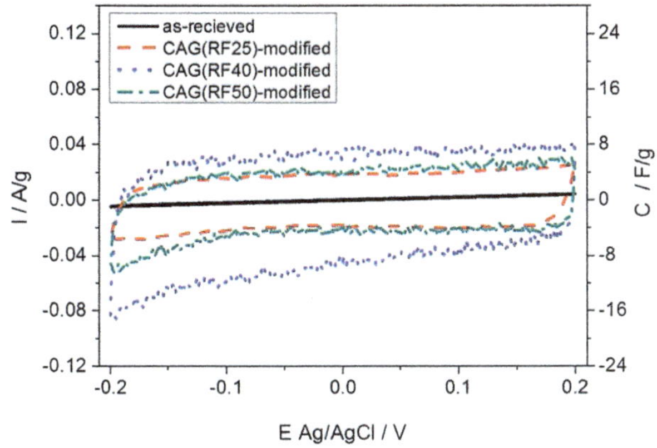

Fig. 6 Cyclic voltammetry response for CAG–GF–CAG composites in 3 M KCl depending on the amount of the resin in the sol formulation. Scan rate 5 mV s^{-1}.

EMITFSI, structural supercapacitor composites were manufactured using CAG-modified carbon fibres with different RF content in the initially infused sol.

The specific capacitance of the composites was calculated from charge–discharge experiments with a step voltage of 0.1 V and charging time of 60 s, and overall the values (Table 2) follow the trend observed for the liquid electrolyte measurements (Table 1). However, there is a very significant decrease in the apparent specific capacitance which may be attributed to changes in wettability, segregation of the multifunctional electrolyte components at the electrode interface, or the availability of ions. Furthermore, the ionic conductivity of the polymer electrolyte is much lower than that of the liquid aqueous electrolyte; the ionic conductivity of PEGDGE epoxy containing 10 wt% ionic liquid electrolyte is 2.8×10^{-2} mS cm^{-1}, compared to above 90 mS cm^{-1} for 3 M KCl. As discussed further below, improvements in the multifunctional electrolyte are required.

Mechanically, the primary structural fibres, which are thought to be relatively unaffected by the CAG modification, determine the in-plane tensile performance of these structural devices; the tensile performance is, therefore, expected to be excellent, since the composite performance is "fibre-dominated". A more interesting, less obvious question is whether the matrix-dominated properties will be

Table 2 Effect of the resin content in the sol formulation on the electrochemical properties and in-plane shear response from tensile testing of the structural supercapacitor devices based on CAG-modified carbon fibres[a]

RF concentration, wt%	CAG loading, %	C_g, mF g^{-1}	ESR, kΩ cm^2	Shear strength τ_{12}^{m}, MPa	Shear modulus G_{12}^{chord}, MPa	Vol. reinforcement, %
0	—	12.5	120.8	5.36 ± 0.07	276 ± 14	47.2
25	6.0	73.1	82.1	6.02 ± 0.25	686 ± 68	42.0
40	9.1	90.5	79.2	9.78 ± 0.08	1172 ± 112	42.0
50	12.9	88.0	80.2	7.94 ± 0.60	1379 ± 116	42.0

[a] Electrolyte – PEGDGE with 10 wt% IL; separator – GF.

improved or degraded. Given the limited amount of material available, it was decided to focus on matrix-dominated composite tests, as a more demanding and informative evaluation of the different systems; specifically, an in-plane shear test was chosen to evaluate the effect of the properties of the CAG-modified carbon fibres on the fibre/matrix interface. In this study, CAG can be potentially considered to be a matrix reinforcement.[26] Indeed, the introduction of CAG significantly improves the in-plane shear modulus, which rises monotonically with CAG (reinforcement) content (Table 2). However, the relative increase is smaller at higher CAG loadings, suggesting that the coarse 40 wt% RF-derived CAG structure, with thicker struts, is relatively more effective at stiffening. The in-plane shear strength of the manufactured structural supercapacitors was less affected by the presence and amount of CAG, although it begins to decrease at the highest CAG loading.

3.4 Further improvement of the CAG modified carbon fibres

The data presented above confirm that the CAG-modified carbon fibres prepared from the sol containing 40 wt% RF showed the most promising results, so far. Whilst further optimisation is clearly possible, there does appear to be a balance between CAG-content, mechanical performance and electrochemical accessibility. Within the available volume, there will be an optimum microstructure for pure electrochemical double layer capacitance somewhere near the current system. To improve the performance even further, additional modification of the CAG-modified carbon fibres could be performed using one of the available methods, including physical (oxidation) or chemical (KOH) activation, or the introduction of pseudo-capacitive components, such as metal oxides.[4] However, all these methods require high temperature treatment, making the introduction of a pseudo-capacitive component based on conducting polymer an attractive alternative.[3,4,23,48] Among the most commonly studied conducting polymers, in this context, are polypyrrole (PPy),[48,49] polyaniline (PANI),[50] poly(3-methylthiopene) (PMET)[51] and poly(3,4-ethylenedioxythiopene) (PEDOT).[52,53] In this research, PEDOT was chosen[23] because of its high conductivity and good electrochemical and mechanical properties, as well as environmental and thermal stability.

There are a number of methods reported in the literature for the polymerisation of EDOT,[22,52,54] including electrochemical polymerisation,[52] room temperature emulsion (co)polymerisation of EDOT with NaPSS in the presence of $Fe(NO_3)_3$ (Scheme 1a)[22] and room temperature polymerisation in $FeCl_3$–methanol solution (Scheme 1b).[23] For preliminary studies, the two last methods have been explored, as shown in Scheme 1.

Both methods still require optimisation, but preliminary studies showed that a two fold improvement in the specific capacitance can be achieved when CAG-modified carbon fibres are subjected to EDOT polymerisation (Fig. 7). However, further optimisation is required in order to maximise the contribution of the PEDOT.

3.5 Multifunctional electrolyte

The discussion so far has concentrated on the development of structural electrodes *via* the modification of carbon fibres. The next crucial component for

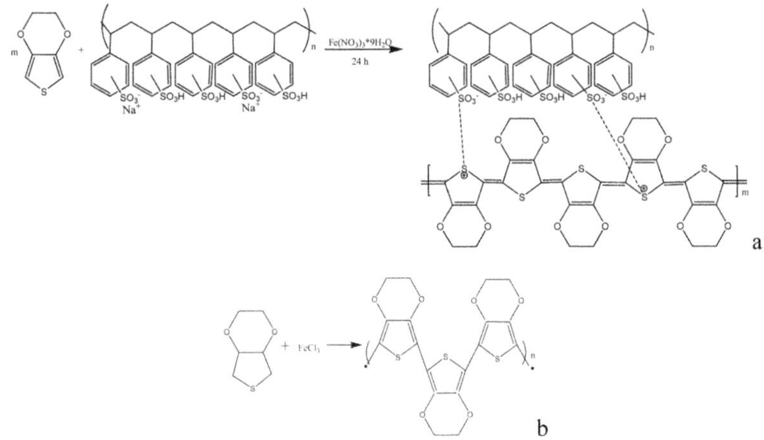

Scheme 1 Reactions used to synthesise PEDOT; (a) (co)polymerisation of EDOT with NaPSS in the presence of $Fe(NO_3)_3$, (b) polymerisation of EDOT in $FeCl_3$–methanol solution.

structural supercapacitors, is a structural electrolyte which should have high ionic conductivity (>1 mS cm^{-1}) and good mechanical properties (Young's modulus > 1 GPa).[6] As mentioned above, the ionic conductivity and mechanical properties are inversely related, making this multifunctional requirement particularly challenging.[6–10] This demanding relationship is especially obvious when high performance structural polymers are used, such as epoxies, where the architecture is already optimised for stiffness. Nevertheless, it has been possible[6] to prepare electrolytes based on commercial epoxy resins (with the trade names MVR444, VTM266 and MTM57) with an ionic conductivity as high as 0.8 mS cm^{-1} and a Young's modulus of 0.18 GPa, by developing a bicontinuous structure

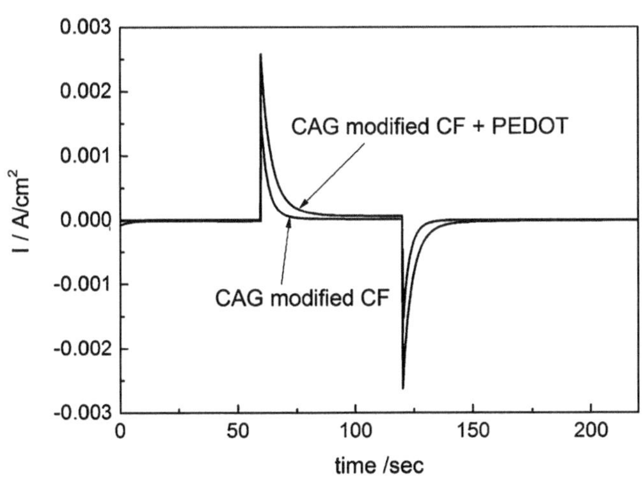

Fig. 7 Charge–discharge curves of the CAG modified carbon fibres before and after modification with PEDOT 2. Step voltage 0.1 V. Electrolyte 3 M KCl.

where one phase is responsible for the mechanical strength and another for ionic conductivity. By varying the composition of the epoxy and electrolyte, it is possible to tune the morphology of the resulting structural electrolyte from traditional bicontinuous networks, in the case of MVR444 and VTM266 (Fig. 8a and b), to more complicated structures, in the case of MTM57 (Fig. 8c and d). The changes in morphology affect the properties of the structural electrolytes. For example, as a finer structure is formed, the ionic conductivity of the structural electrolyte decreases. The microstructure presented in Fig. 8b is an example of a desirable morphology, as it possesses a large scale continuous epoxy network (based on MVR444) and pores big enough for high ionic mobility. One of the benefits of using a commercial epoxy is that it allows for straightforward scale up, first to large scale composite demonstrators, and ultimately to production, as long as the addition of the electrolyte does not lead to changes in processing behaviour, for example in the curing kinetics and/or film forming ability of the epoxy resin. However, the formulation which produced the promising microstructure in Fig. 8b had a very short shelf life, most likely due to reactions between the imidazolium ions and the epoxy resin,[55,56] making it impossible to use it for structural supercapacitor manufacturing. As an alternative, formulations based on MTM57, which resulted in more complex microstructures, were studied in the composite devices.

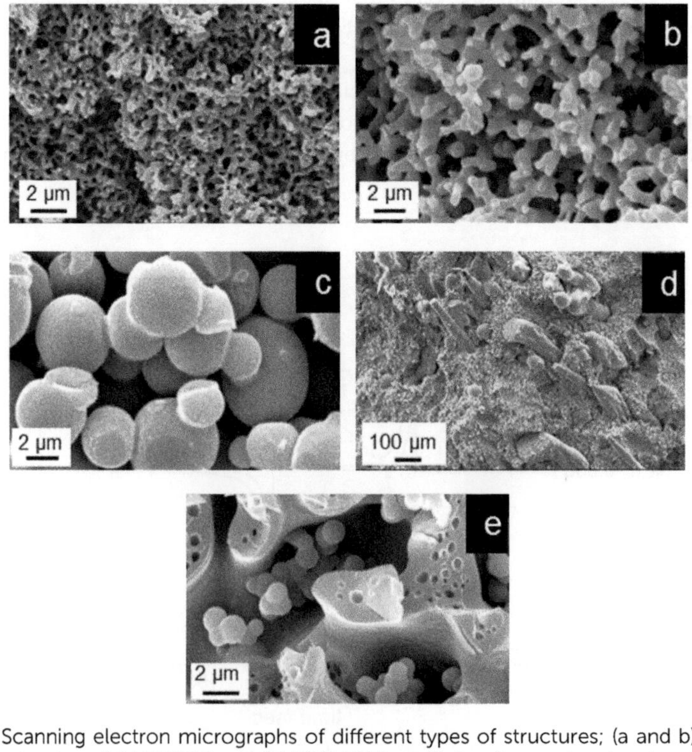

Fig. 8 Scanning electron micrographs of different types of structures; (a and b) bicontinuous structure ((a) VTM266 and (b) MVR444 based), (c–e) complex structure (MTM57 based). Note that (c and d) show the same sample at different magnifications.

Detailed studies of the effect of electrolyte composition on the ionic conductivity and mechanical performance of the structural electrolyte based on MTM57 will be published in a separate manuscript;[10] however, an example of processability issues is described here. A key issue for practical use is the need to spread the resin formulation to form a stable film that can subsequently be infused or 'pre-pregged' with the carbon fibre weave. Although formulations based on MTM57 blended with ionic liquid based electrolyte had a wider processability window (time/temperature stability) compared to MVR444 formulations, films cast from MTM57 based formulations had a tendency to reticulate,† (Fig. 9a). To overcome this shortcoming, the effect of adding rheology modifiers or thixotropes was studied, using for example hydrophilic fumed silica (Cab-o-sil M5), hydrophobic fumed silica modified with polydimethylsiloxane (Aerosil R202) and "neutral" mixed mineral (Garamite 1958) additives, introduced at 1 wt% to the MTM57. The film forming properties of the uncured formulations that were cast and left for 1 h were imaged (Fig. 9). The addition of a thixotrope improved film forming ability and only minimal reticulation was observed for formulations containing hydrophobic and hydrophilic thixotropes.

However, the introduction of 1 wt% thixotrope not only affected the film forming ability but also the morphology and properties of the resulting structural electrolytes. Fig. 10 shows SEM images of the cured 50/MTM57 formulations containing different thixotropes. The addition of the Aerosil R202 led to the formation of a microstructure consisting mainly of closed pores (Fig. 10a and b), which resulted in a decrease in ionic conductivity values from 0.42 mS cm^{-1} for the 50/MTM57 structural electrolyte to 0.004 mS cm^{-1}. The presence of Cab-o-sil and Garamite also resulted in changes in the morphology of the 50/MTM57 (Fig. 10c and d) but the ionic conductivity was less affected and was still 0.13 mS cm^{-1} and 0.3 mS cm^{-1} for 50/MTM57 containing Cab-o-sil and Garamite, respectively. Nanoparticulate fillers are known to modify the ionic conductivity of polymer electrolytes by providing alternative conduction paths or varying the free volume. In addition, small particles are known to act as Pickering stabilisers for

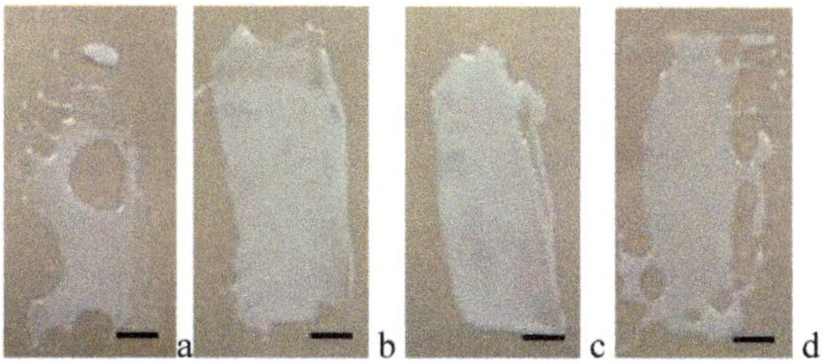

Fig. 9 Images of the films of uncured formulations based on 50/MTM57 after 1 h upon addition of thixotropes, drawn down by casting knife; (a) neat, (b) Aerosil, (c) Cab-o-sil, and (d) Garamite. Scale bar is 1.3 cm.

† The term "reticulation", here, is used to describe a process of rupture or dewetting of the resin film from the substrate and can proceed to the extent of forming discrete droplets, rather than the intended continuous film.

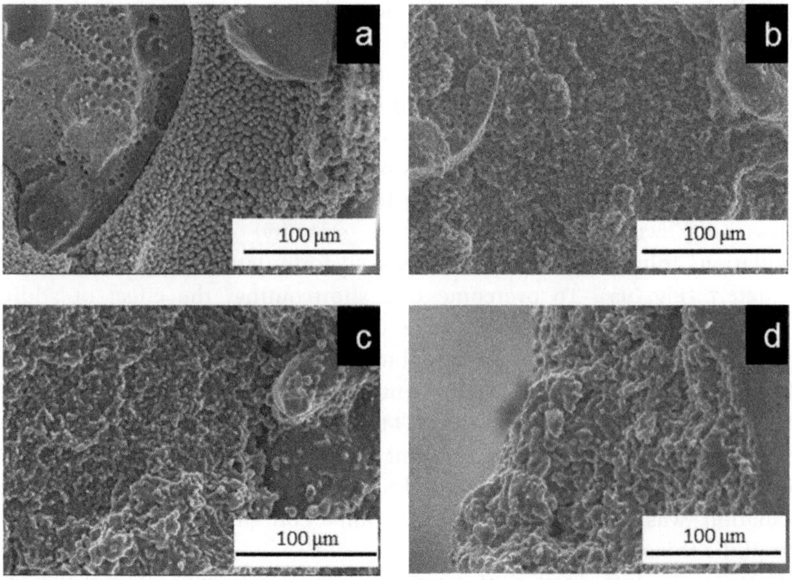

Fig. 10 Scanning electron micrographs of 50/MTM57 based structural electrolytes with and without different thixotropes; (a) neat, (b) Aerosil, (c) Cab-o-sil, and (d) Garamite.

emulsions, with an ability to stabilise different curvature interfaces, depending on the wettability.[57,58] The hydrophobic Aerosil is likely to promote epoxy as the continuous phase, as is indeed observed. Even though the addition of the hydrophilic thixotrope had a positive effect on the film forming ability, and only resulted in a small decrease in the ionic conductivity of the resulting structural electrolyte, the overall microstructure became less ordered. This change may be attributed to the influence of the particulate filler distribution on the phase separation during the curing process. Previous developments showed that substitution of a small amount (1 wt%) of ionic liquid by an organic electrolyte with high dielectric constant also results in a significant change of the microstructure of the resulting structural electrolyte.[10] Thus, further formulation of these structural electrolyte systems is likely to be complicated, in general, by the potential to disturb the specific phase separating behaviour required. The phase behaviour in the presence of the (modified) reinforcing fibres must also be considered.

3.6 System issues

Once the individual multifunctional components of the structural supercapacitor are developed, there are still many other issues to resolve relating to assembly, integration, manufacturing, machining/finishing, current collection, and ownership. As an example of a potential component implementation for structural supercapacitors, a variety of demonstrators have been manufactured using a pre-preg route. Due to the high viscosity of the uncured formulations of epoxy-based electrolyte, resin infusion was not applicable. Carbon fibres were separated using two glass fibre fabrics (GFs) to balance the composite in the midplane. Using the pre-preg route, it was possible to manufacture not only flat laminates but also complex shapes, including components for a full size trunk lid for a Volvo S80, where LEDs were powered by the structural supercapacitors, as presented in Fig. 11. Copper tape with conductive adhesive was chosen as a current collector and was introduced onto the carbon fibre surface prior to composite manufacturing. The copper tape does not have to be used as a long strip and it is efficient even when it is attached to the carbon fibres in a small rectangular shape with a soldered wire connection (Fig. 11b). The spacing and geometry of the copper contacts is fundamentally determined by the in-plane conductivity of the carbon fibre electrodes and the performance of the electrochemical system (*i.e.* the current densities generated). Despite the effectiveness of the copper contacts, there is a concern regarding the long term stability of the copper in the presence of the ionic liquid based electrolytes; different current collector materials or modification of the copper system will be needed.

Given that it will not be always possible to manufacture structural supercapacitors as a single unit, it is important to consider the connectivity and machining of the fabricated structural supercapacitors. Preliminary data showed that it is possible to drill multiple holes and trim edges without causing a short-circuit and/or compromising the electrochemical performance of the structural supercapacitor. It was also possible to build stacks containing a number of single structural supercapacitor cells by separating them using layers of glass fibre/structural matrix (MTM57) pre-preg. Among many others, one question is how best to encase the manufactured structural supercapacitor without compromising its electrochemical performance, to protect the electrolyte component from the environment, specifically moisture. The most straightforward approach is to use a conventional structural glass fibre pre-preg material, as long as there is no interdiffusion between the structural electrolyte and the conventional matrix of the glass fibre pre-preg.

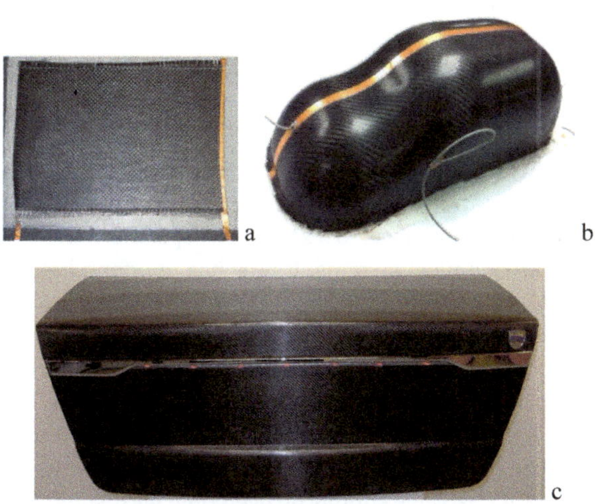

Fig. 11 Structural supercapacitor flat laminate (a), model car body (b), and trunk lid with added chrome trimming and lit LED lights (c).

4 Conclusions

Composite materials intrinsically combine the characteristics of their components; they are an ideal platform for the creation of multifunctional materials. There is an exciting opportunity to save (or redistribute) weight and volume, by combining electrical energy storage within the basic construction of a physical system. For example, in the context of hybrid/electric transport, electrical energy can be stored in the bodywork, as illustrated by the demonstrator above (Fig. 10). Similar possibilities exist in other electric vehicles (trains, trams, bikes) and in aerospace, as well as in portable electronics. In order to make these multifunctional structural power systems work effectively, multifunctional reinforcements and matrices must be developed and combined. One encouraging opportunity is the use of carbonaceous materials as both high strength/stiffness reinforcements and as electrochemical electrodes. The laminated nature of many fibre composites also facilitates many energy storing embodiments, although possibilities also exist at other length scales, such as at the single fibre level.[59] This discussion paper presents a strategy for preparing structural supercapacitors from a laminated architecture. For this system, the surface area of the carbonaceous layer must be maximised whilst retaining the mechanical performance of the primary carbon fibres. After comparing several methods, the use of a carbon aerogel matrix was found to be most effective, supplying the desired surface area, and maximising the utilisation of the available space. This same aerogel also offers a mechanical performance advantage by reinforcing the electrolyte matrix phase, and addressing matrix-dominated composite failures that are otherwise amplified by the compromises of the multifunctional matrix. The bicontinuous nanostructured nature of the aerogel appears to be a characteristic motif in multifunctional materials. For the electrode, the scale of the porosity is defined by the electrochemical double layer that must be accommodated. Other requirements associated with the electron/ion transport and mechanics do not appear limiting at present, although further improvements can be anticipated, particularly if hierarchical features are introduced. The multifunctional electrolyte matrix is even more demanding as it combines more inherently conflicting properties. However, the bicontinuous motif again provides encouragement. In this case, the characteristic length scale is larger, most likely defined by the diameters of the primary fibres that must be supported.

Now that the concept is demonstrated, the challenge is to increase absolute performance to practical levels; in the case of structural supercapacitors, power density remains the central target. It's worth noting, however, that the multifunctional system does not have to exceed the specific performance of both structural and electrochemical systems. It only has to offer a better performance than the combined weight of the two conventional solutions. Whilst laminated structural supercapacitors seem promising, there are many other architectures and electrochemical systems that might be considered. If suitable performance can be obtained, practical manufacturing routes will be needed for real component testing and eventual deployment. Many of the multifunctional components will bring their own challenges to formulation and processing, as highlighted above; however, these obstacles should be overcome with suitable development. Further investigation of machining, connectivity, power systems design, and through life service issues will need to be considered. Bringing such multifunctional systems to reality requires a multi-disciplinary collaboration between electrochemists, materials chemists, composite engineers and electrical power systems engineers. It is important to build mutual understanding of the requirements and challenges through dialogue and discussion.

Acknowledgements

Funding from the EU as part of the FP7 project "Storage" (Grant Agreement no. 234236) and the MAST2.2 program funded by the UK Ministry of Defence is gratefully acknowledged. The authors would like to thank Mr Richard Shelton from Cytec Industrial Materials for help with sample preparation, and the wider "Storage" team for the fabrication of the Volvo boot.

References

1 H. Pröbstle, M. Wiener and J. Fricke, *J. Porous Mater.*, 2003, **10**, 213–222.
2 A. G. Pandolfo and A. F. Hollenkamp, *J. Power Sources*, 2006, **157**, 11–27.
3 Y. Zhang, H. Feng, X. Wu, L. Wang, A. Zhang, T. Xia, H. Dong and X. L. L. Zhang, *Int. J. Hydrogen Energy*, 2009, **34**, 4889–4899.
4 G. Wang, L. Zhang and J. Zhang, *Chem. Soc. Rev.*, 2012, **41**, 797–828.
5 D. D. Edie, *Carbon*, 1998, **36**, 345–362.
6 N. Shirshova, A. Bismarck, S. Carreyette, Q. P. V. Fontana, E. S. Greenhalgh, P. Jacobsson, P. Johansson, M. J. Marczewski, J. Scheers, G. Kalinka, A. Kucernak, M. S. P. Shaffer, J. H. G. Steinke and M. Wienrich, *J. Mater. Chem. A*, 2013, **1**, 15300–15309.
7 K. Matsumoto and T. Endo, *Macromolecules*, 2008, **41**, 6981–6986.
8 J. F. Snyder, E. D. Wetzel and C. M. Watson, *Polymer*, 2009, **50**, 4906–4916.
9 S. Wang and K. Min, *Polymer*, 2010, **51**, 2621–2628.
10 N. Shirshova, A. Bismarck, E. S. Greenhalgh, P. Jacobsson, P. Johansson, M. J. Marczewski, G. Kalinka, M. S. P. Shaffer, J. H. G. Steinke and M. Wienrich, 2014, submitted.
11 K. Matsumoto and T. Endo, *J. Polym. Sci., Part A: Polym. Chem.*, 2011, **49**, 1874–1880.
12 K. Matsumoto and T. Endo, *Macromolecules*, 2009, **42**, 4580–4584.
13 J.-P. Pascaut and R. J. J. Williams, *Epoxy Polymers*, Wiley-VCH, 2010.
14 P.-L. Kuo, W.-J. Liang and T.-Y. Chen, *Polymer*, 2003, **44**, 2957–2964.
15 M. Galiński, A. Lewandowski and I. Stępniak, *Electrochim. Acta*, 2006, **51**, 5567–5580.
16 A. Lewandowski and A. Świderska-Mocek, *J. Power Sources*, 2009, 601–609.
17 N. D. Khupse and A. Kumar, *Indian J. Chem., Sect. A: Inorg., Bio-inorg., Phys., Theor. Anal. Chem.*, 2010, **49**, 635–648.
18 N. Shirshova, H. Qian, M. S. P. Shaffer, J. H. G. Steinke, E. S. Greenhalgh, P. T. Curtis, A. Kucernak and A. Bismarck, *Composites, Part A*, 2013, **46**, 96–107.
19 H. Qian, H. Diao, N. Shirshova, E. S. Greenhalgh, J. G. H. Steinke, M. S. P. Shaffer and A. Bismarck, *J. Colloid Interface Sci.*, 2013, **395**, 241–248.
20 H. Qian, A. Bismarck, E. S. Greenhalgh and M. S. P. Shaffer, *Composites, Part A*, 2010, **41**, 1107–1114.

21 H. Qian, E. S. Greenhalgh, M. S. P. Shaffer and A. Bismarck, *J. Mater. Chem.*, 2010, **20**, 4751–4762.
22 X. Zhang, D. Chang, J. Liu and Y. Luo, *J. Mater. Chem.*, 2010, **20**, 5080–5085.
23 L. Jin, T. Wang, Z. Q. Feng, M. K. Leach, J. Wu, S. Mog and Q. Jiang, *J. Mater. Chem. B*, 2013, **1**, 1818–1825.
24 ASTM-D3518, ASTM: West Conshohocken, PA, 1995.
25 ASTM-D3171, ASTM: West Conshohocken, PA, 2000.
26 H. Qian, A. R. Kucernak, E. S. Greenhalgh, A. Bismarck and M. S. P. Shaffer, *ACS Appl. Mater. Interfaces*, 2013, **5**, 6113–6122.
27 J. C. Lee, B. H. Lee, B. G. Kim, M. J. Park, D. Y. Lee, I. H. Kuk, H. Chijng, H. S. Kang, H. S. Lees and D. H. Ahn, *Carbon*, 1997, **35**, 1479–1484.
28 F. Severini, L. Formaro, M. Pegoraro and L. Posca, *Carbon*, 2002, **40**, 735–741.
29 W. Li, S. Y. Yao, K. M. Ma and P. Chen, *Polym. Compos.*, 2013, **34**, 368–375.
30 L. Liu, H. Chen and D. Pan, *Fibers Polym.*, 2012, **13**, 587–592.
31 C. Choi, J. A. Lee, A. Y. Choi, Y. T. Kim, X. Lepró, M. D. Lima, R. H. Baughman and S. J. Kim, *Adv. Mater.*, 2014, **26**, 2059–2065.
32 J. A. Lee, M. K. Shin, S. H. Kim, H. U. Cho, G. M. Spinks, G. G. Wallace, M. D. Lima, X. Lepró, M. E. Kozlov, R. H. Baughman and S. J. Kim, *Nat. Commun.*, 2013, **4**, 1970.
33 C. L. Pint, N. W. Nicholas, S. Xu, Z. Sun, J. M. Tour, H. K. Schmidt, R. G. Gordon and R. H. Hauge, *Carbon*, 2011, **49**, 4890–4897.
34 S. Esconjauregui, M. Fouquet, B. C. Bayer, C. Ducati, R. Smajda, S. Hofmann and J. Robertson, *ACS Nano*, 2010, **4**, 7431–7436.
35 S. A. Al-Muhtaseb and J. A. Ritter, *Adv. Mater.*, 2003, **15**, 101–114.
36 R. W. Pekala, *J. Mater. Sci.*, 1989, **24**, 3221–3227.
37 J. C. Lytle, J. M. Wallace, M. B. Sassin, A. J. Barrow, J. W. Long, J. L. Dysart, C. H. Renninger, M. P. Saunders, N. L. Brandell and D. R. Rolison, *Energy Environ. Sci.*, 2011, **4**, 1913–1925.
38 R. Petričević, M. Glora and J. Fricke, *Carbon*, 2001, **39**, 857–867.
39 W. Li, G. Reichenauer and J. Fricke, *Carbon*, 2002, **40**, 2955–2959.
40 M. Schwan and L. Ratke, *J. Mater. Chem. A*, 2013, **1**, 13462–13468.
41 R. Saliger, U. Fischer, C. Herta and J. Fricke, *J. Non-Cryst. Solids*, 1998, **225**, 81–85.
42 A. Szczurek, G. Amaral-Labat, V. Fierro, A. Pizzi and A. Celzard, *Sci. Technol. Adv. Mater.*, 2011, **12**, 035001.
43 H. Tamon, H. Ishizaka, M. Mikami and M. Okazaki, *Carbon*, 1997, **35**, 791–796.
44 E. A. Oyedoh, A. B. Albadarin, G. M. Walker, M. Mirzaeian and M. N. M. Ahmad, *Chem. Eng. Trans.*, 2013, **32**, 1651–1656.
45 A. M. ElKhatat and S. A. Al-Muhtaseb, *Adv. Mater.*, 2011, **23**, 2887–2903.
46 D. W. Schaefer, R. Pekala and G. Beaucage, *J. Non-Cryst. Solids*, 1995, **186**, 159–167.
47 S. Kondrat, C. R. Pérez, V. Presser, Y. Gogotsi and A. A. Kornyshev, *Energy Environ. Sci.*, 2012, **5**, 6474–6479.
48 L. Yuan, B. Yao, B. Hu, K. Huo, W. Chen and J. Zhou, *Energy Environ. Sci.*, 2013, **6**, 470–476.
49 H. An, Y. Wang, X. Wang, L. Zheng, X. Wang, L. Yi, L. Bai and X. Zhang, *J. Power Sources*, 2010, **195**, 6964–6969.
50 K. S. Ryu, K. M. Kim, N. G. Park, Y. J. Park and S. H. Chang, *J. Power Sources*, 2002, **103**, 305–309.

51 C. Arbizzani, M. Mastragostino and F. Soavi, *J. Power Sources*, 2001, **100**, 164–170.
52 A. S. Saraç, G. Sönmez and F. Ç. Cebeci, *J. Appl. Electrochem.*, 2003, **33**, 295–301.
53 G. P. Pandey and A. C. Rastogi, *Electrochim. Acta*, 2013, **87**, 158–168.
54 R. Starbird, C. A. García-González, I. Smirnova, W. H. Krautschneider and W. Bauhofer, *Mater. Sci. Eng., C*, 2014, **37**, 177–183.
55 B. G. Soares, S. Livi, J. Duchet-Rumeau and J.-F. Gerard, *Macromol. Mater. Eng.*, 2011, **296**, 826–834.
56 M. S. Heise and G. C. Martin, *Macromolecules*, 1989, **22**, 99–104.
57 V. O. Ikem, A. Menner and A. Bismarck, *Langmuir*, 2010, **26**, 8836–8841.
58 B. P. Binks and S. O. Lumsdon, *Phys. Chem. Chem. Phys.*, 1999, **1**, 3007–3016.
59 S. Leijonmarck, T. Carlson, G. Lindbergh, L. E. Asp, H. Maples and A. Bismarck, *Compos. Sci. Technol.*, 2013, **89**, 149–157.

Faraday Discussions

PAPER

Electro-deposition and re-oxidation of carbon in carbonate-containing molten salts

Happiness V. Ijije, Richard C. Lawrence, Nancy J. Siambun,† Sang Mun Jeong,‡ Daniel A. Jewell,§ Di Hu and George Z. Chen*

Received 21st March 2014, Accepted 6th May 2014
DOI: 10.1039/c4fd00046c

The electrochemical deposition and re-oxidation of solid carbon were studied in CO_3^{2-} ion-containing molten salts (*e.g.* $CaCl_2$–$CaCO_3$–LiCl–KCl and Li_2CO_3–K_2CO_3) at temperatures between 500 and 800 °C under Ar, CO_2 or N_2–CO_2 atmospheres. The electrode reactions were investigated by thermodynamic analysis, cyclic voltammetry and chronopotentiometry in a three-electrode cell under various conditions. The findings suggest that the electro-reduction of CO_3^{2-} is dominated by carbon deposition on all three tested working electrodes (Ni, Pt and mild steel), but partial reduction to CO can also occur. Electro-re-oxidation of the deposited carbon in the same molten salts was investigated for potential applications in, for example, direct carbon fuel cells. A brief energy and cost analysis is given based on results from constant voltage electrolysis in a two-electrode cell.

Introduction

Electrochemical reduction of the carbonate ion (CO_3^{2-}) to solid carbon in molten salts has been known to be possible since the mid-1960s.[1,2] It then gained only a limited interest,[3,4] possibly because of the ready availability of various forms of carbon from fossil resources and biomass. However, in recent years renewed interest has grown fairly fast,[5–11] due largely to the need to reduce CO_2 emissions to avoid detrimental climate change, and, more importantly, to utilise this waste gas as a carbon source for the synthesis and processing of materials and chemicals, such as steel carburisation and the production of so called solar fuels.[8,12] In

Department of Chemical and Environmental Engineering, and Energy and Sustainability Research Division, Faculty of Engineering, University of Nottingham, University Park, Nottingham NG7 2RD, UK. E-mail: george.chen@nottingham.ac.uk

† Current address: School of Engineering and Information Technology, Universiti Malaysia Sabah, 88999 Kota Kinabalu, Sabah, Malaysia.

‡ Current address: Department of Chemical Engineering, Chungbuk National University, 410 Seongbongno, Heungduk-gu, Cheongju-si, Chungcheongbuk-do, 361-763, Republic of Korea.

§ Current address: Coogee Titanium, Unit 4/25, Agosta Drive, Laverton North, VIC Australia 3026.

fact, solar fuel production can be regarded as an energy storage process that is very suitable for remote and seasonal purposes. Yet, considering its high energy content, carbon production from CO_2 is far better for energy storage ($\Delta H^o = -25.97$ MJ L^{-1} at 20 °C for a carbon powder packed to 0.792 g mL^{-1} in apparent density) in comparison with, for example, methanol production from CO_2 and H_2O ($\Delta H^o = -16.71$ MJ L^{-1} at 20 °C for liquid CH_3OH of 0.792 g mL^{-1} in density).[13] This simple thermodynamic comparison has formed the basis of our research programme since 2006.[14] Herein we summarise our findings on the electro-deposition of solid carbon from molten salts under CO_2 containing atmospheres,[8,11] and report for the first time our investigation on the electrochemical oxidation of electro-deposited carbon in molten salts.

The molten salt suitable for the electrolytic production of carbon from CO_2 should be able to dissolve the O^{2-} ion which is a product of carbon deposition, and also helps to absorb CO_2 and convert it to the CO_3^{2-} ion. Because of their low costs and environmental impacts, both chloride and carbonate salts, and in particular their mixtures, were investigated in this laboratory.[8,11] It has been observed that electro-deposition of carbon can proceed in all tested molten salt combinations, as long as Li^+ ions are present.[1–4] In other cases, deposition is not impossible, but occurs at much lower rates.[15] This phenomenon may be accounted for by the relative deposition potential of Li, Na, or K metal compared with that of carbon, as expressed by the following reactions:

$$2M_2CO_3 = 4M + 2CO_2 + O_2 \text{ (M = Li, Na or K)} \tag{1}$$

$$3M_2CO_3 = C + 3M_2O + 2CO_2 + O_2 \tag{2}$$

Note that reaction (2) is actually the sum of the following reactions (3) and (4) × 2:

$$M_2CO_3 = C + M_2O + O_2 \tag{3}$$

$$M_2CO_3 = M_2O + CO_2 \tag{4}$$

This combination aids electrochemical analysis, since both reactions (1) and (2) have the same anodic reaction (5) below:

$$2CO_3^{2-} = 2CO_2 + O_2 + 4e\ (E^o = 0\text{ V}) \tag{5}$$

The potentials of reactions (1) and (2) against reaction (5), *i.e.* a hypothetical reference electrode of $CO_3^{2-}/CO_2–O_2$, are listed in Table 1[13] which shows that in

Table 1 Deposition potentials (*vs.* $CO_3^{2-}/CO_2–O_2$) of alkali and alkaline earth metals *via* reaction (1), and carbon *via* reaction (2) in their own molten carbonate salts at 600 °C

Molten salt	Alkali metal	Carbon
Li_2CO_3	−2.964 V	−1.719 V
Na_2CO_3	−2.546 V	−2.551 V
K_2CO_3	−2.612 V	−3.083 V
$CaCO_3$	−3.033 V	−1.349 V
$BaCO_3$	−3.069 V	−1.992 V

molten alkali carbonates, carbon deposition is thermodynamically preferred to Li deposition, but more difficult than Na or K deposition, which is in line with experimental observations of the authors and others.[1–4,15] (With reference to CO_3^{2-}/CO_2–O_2, in the following discussion, the standard potential, E^o, provided next to the electrode reaction, is for the reaction occurring in pure molten Li_2CO_3 at 600 °C.)

Previous studies of the electro-deposition of carbon were mostly carried out in molten alkali metal salts. However, thermodynamic calculations show that it is also possible to use molten alkaline earth metal salts. It can be seen in Table 1 that carbon deposition is also preferred to alkaline earth metal deposition in both molten $CaCO_3$ and $BaCO_3$. However, $CaCO_3$ decomposes at temperatures only slightly above its melting point (825 °C). $MgCO_3$ is even worse and decomposes at temperatures below 350 °C before melting, which is the reason why Table 1 does not contain data for $MgCO_3$. $BaCO_3$ is thermally more stable but it is expensive to use, whilst barium salt toxicity is also a concern. Nevertheless, in this work, it was found that electro-deposition of carbon could be achieved in the molten mixture of $CaCl_2$ and $CaCO_3$ (84 : 16 in a molar ratio) at temperatures higher than 730 °C. To lower the working temperature, mixtures of chlorides were used in some cases to dissolve $CaCO_3$ and enable electro-deposition of carbon.

Experimental

Details of chemicals, materials, experimental equipment, set up and procedures can either be found in previous publications from this laboratory,[8,11,17] or are specified in this paper.

Results and discussion

Fig. 1a presents the cyclic voltammogram (CV) of a Ni wire working electrode in the quaternary mixture of $0.30CaCl_2$–$0.17CaCO_3$–$0.43LiCl$–$0.10KCl$ in a molar ratio. The CV was recorded under argon, instead of CO_2, so that if carbon deposition does occur, it can be attributed without doubt to the reduction of the CO_3^{2-} ion. The most prominent feature of the CV in Fig. 1a is an almost symmetrical and large reduction current peak C1. To confirm carbon deposition, the Ni wire working electrode was removed from the molten salt bath after recording two consecutive CVs similar to that in Fig. 1a. A carbon coating was clearly seen on the Ni wire as evidenced by the photograph in Fig. 1b. According to Table 1, peak C1 can be attributed to carbon deposition because its potential was less negative than those of the other two reduction peaks (C2 and C3). The potentials of C2 and C3 relative to that of C1 suggest K and Li deposition, respectively. Deposition of Ca requires a more negative potential and hence did not appear on the CV.

To enhance the carbon deposition, electrolysis was carried out in a two electrode cell under different CO_2 partial pressures. The cathode was a mild steel rod of 5.0 mm dia., and the anode was a SnO_2 rod of 10.0 mm dia. which functioned as an inert anode.[16] A high cell voltage (4.0 V) was applied to gain a high deposition rate. The molten salt bath temperature was also increased to 579 °C for the same reason. The electrolysis current was observed to increase with the CO_2 partial pressure as exemplified in Fig. 2a, in agreement with the reduction (6) and regeneration (7) of CO_3^{2-} ions as follows:

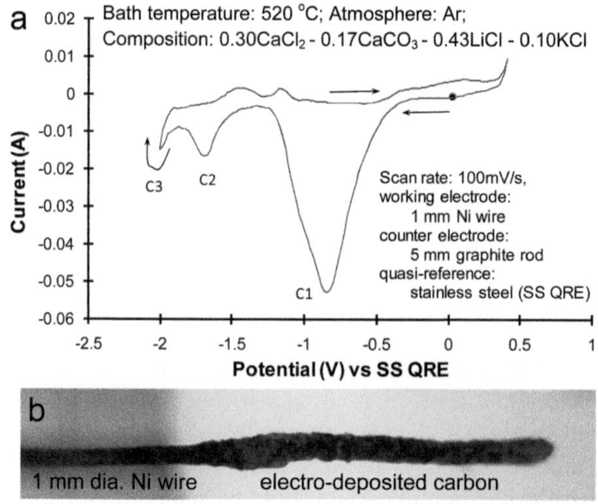

Fig. 1 (a) The cyclic voltammogram (CV) of a nickel wire in a quaternary molten salt bath. Experimental conditions are indicated with the bath composition in a molar ratio. The potential scan started cathodically from 0 V as indicated by the round solid spot on the CV. (b) A photograph showing the Ni wire working electrode (1 mm dia.) with deposited carbon after recording two consecutive CVs.

$$CO_3^{2-} + 4e = C + 3O^{2-} \ (C1, E^o = -1.719 \text{ V}) \quad (6)$$

$$CO_2 + O^{2-} = CO_3^{2-} \quad (7)$$

Note that the bath compositions differ slightly in Fig. 1a and 2a, which was not intentional, but was due to the inconvenience of preparing the multicomponent molten salt bath. To compare the results, electrolysis was terminated at 1 h, and the cathode was removed from the bath. Fig. 2b displays a photograph of the cathodes, showing an increasing thickness of the deposited carbon coating with increasing CO_2 partial pressure.

It is interesting to note in Fig. 2b that the surface of the deposited carbon was covered by small craters that were strong evidence of gas bubbles being formed on the cathode during electrolysis. This was thought to be CO formation *via* reaction (8). If not escaping from the molten salt, CO may undergo further changes on the electrode *via* reactions (9) and/or (10):

$$CO_3^{2-} + 2e = CO + 2O^{2-} \ (E^o = -0.947 \text{ V, A2}) \quad (8)$$

$$CO + 2e = C + O^{2-} \ (E^o = -1.442 \text{ V, A1}') \quad (9)$$

$$2CO + O^{2-} = CO_3^{2-} + C \quad (10)$$

Note that (10) = (9) − (8), and it is a chemical reaction. However, the observed small craters in Fig. 2b seem to indicate slower kinetics of reactions (9) and (10), likely because CO was in the gas phase. Also, reactions (8) and (10) suggest that the O^{2-} ions could prevent or reduce the formation of CO bubbles, which agrees

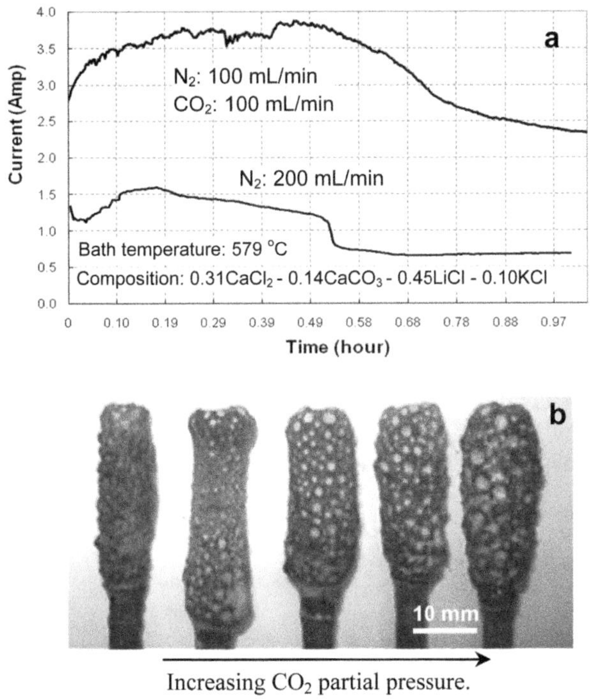

Fig. 2 (a) Current–time plots recorded during constant voltage (4.0 V) electrolysis in a quaternary molten salt bath with a mild steel cathode (5 mm dia. rod) and a SnO_2 anode (10 mm dia. rod) under different atmospheres as indicated. (b) The mild steel rod cathodes with carbon deposits collected after 1 h of electrolysis. The electro-deposition on each cathode was carried out consecutively in the same bath, but under different ratios of N_2 and CO_2 flow rates (mL min^{-1}, from left to right): 200 : 0 (pure N_2); 150 : 50, 100 : 100, 50 : 150 and 0 : 200 (pure CO_2). The total pressure in the electrolyser was 1 atm.

with the absence of craters on the carbon coatings deposited from carbonate dominated baths, as discussed below.

The fairly symmetrical shape of peak C1 in Fig. 1a implies that Ni may have some catalytic effect on CO_3^{2-} reduction. It was also observed that the current of the symmetrical peak C1 decreased significantly on the 2nd potential cycle, which may result from the Ni surface being covered by the deposited carbon and hence losing its catalytic effect. On the other hand, the absence of a re-oxidation counterpart of peak C1 suggests the irreversibility of the carbon deposition in the chloride dominated molten salt bath. This could be related to the CV being recorded under Ar, and that there were too few O^{2-} and/or CO_3^{2-} ions in the chloride dominated molten salt bath to assist carbon oxidation *via* reaction (11) or (12) below:[6]

$$C + 2O^{2-} = CO_2 + 4e \text{ (A1, } E^{o} = -1.488 \text{ V)} \tag{11}$$

$$C + 2CO_3^{2-} = 3CO_2 + 4e \text{ (A3, } E^{o} = -1.025 \text{ V)} \tag{12}$$

It was then thought that if the CVs were recorded in a carbonate dominated molten salt bath under CO_2, carbon deposition and re-oxidation of the deposited carbon could both be facilitated. To confirm this, the molten eutectic mixture of Li_2CO_3–K_2CO_3 (molar ratio: 62 : 38) was used to record CVs.[11] Note that the Ni wire was replaced by a Pt wire which can offer a greater stability at more positive potentials to allow studies of all possible anodic reactions. Also, an alumina membrane Ag/AgCl reference electrode was used to offer a more stable reference potential.[16]

Typical CVs obtained in Li_2CO_3–K_2CO_3 are presented in Fig. 3. The potential window is about 1.90 V between the current onset potentials of peaks C1 (−1.65 V *vs.* Ag/AgCl) and A4 (0.25 V *vs.* Ag/AgCl) measured in Fig. 3a. This is slightly wider than that of pure Li_2CO_3 (1.72 V, Table 1), but it is understandable considering the influence of K_2CO_3. The CVs present a significant reduction peak C1, and several re-oxidation peaks as highlighted in Fig. 3b. To confirm that these anodic peaks are indeed due to re-oxidation of the deposited carbon, the potential was held at the cathodic limit (−2.0 V) for 10 s before the scan was reversed. In response, all anodic peaks increased significantly in current, whilst peak A1′ was engulfed in

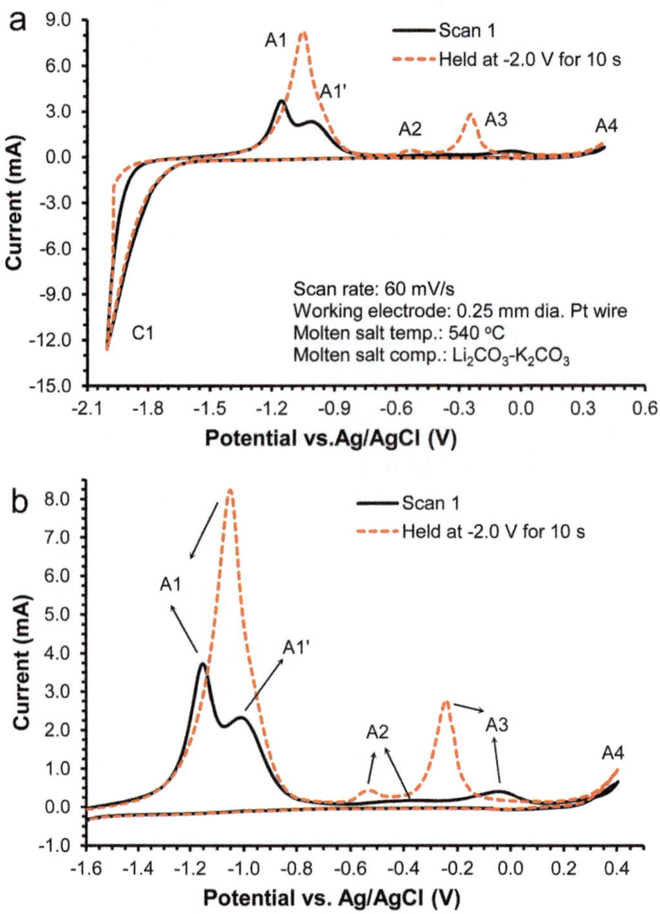

Fig. 3 CVs of a Pt wire (0.25 mm dia.) in molten Li_2CO_3–K_2CO_3 (62 : 38 molar ratio) at 540° C; scan rate: 100 mV s^{-1}. (a) Full potential range. (b) Enlargement of the anodic branches of the CVs in (a).

A1. This phenomenon can be explained as follows: more carbon was deposited when the potential was held at −2.0 V and hence contributed to greater re-oxidation currents. According to their relative potentials, peaks A1 and A3 can be attributed to reactions (11) and (12), respectively. Peaks A1′ and A2 were likely the reversals of reactions (9) and (8), respectively, considering their potentials relative to that of A1 for reaction (11). Interestingly, both peaks A2 and A3 shifted negatively after holding the potential scan at −2.0 V for 10 s. It is very likely that when CO is produced after peak A1′ *via* the reverse of reaction (9), it adsorbs on the carbon surface, impeding reaction (12) and hence there is a more positive potential for peak A3. Thus, the absence of A1′ on the CV after holding the potential at −2.0 V means a lowered influence of CO, and hence the negative shift of A3. If peak A2 corresponds to the reverse of reaction (8), its negative shift may result from a smaller number and size of the CO nuclei on the deposited carbon.

The oxidation current, A4, at the anodic potential limit (0.5 V) may be attributed to the discharge of either the O^{2-} ion to form O_2 (13), or of the ${CO_3}^{2-}$ ion to form O_2 and CO_2 (5).

$$2O^{2-} = O_2 + 4e\ (E^o = -0.463\ V) \tag{13}$$

If reaction (13) had occurred, its current should have appeared at potentials between A3 and A4 on the CV. However, for reaction (13) to proceed, a significant amount of O^{2-} ions should be present near the electrode. This was unlikely to be the case, considering that the O^{2-} ions, including those already present in the molten carbonate salts and those produced during carbon deposition *via* reaction (6), should have mostly been consumed in the course of the potential scan from A1 to A3. Thus, A4 should correspond to reaction (5).

Having obtained a basic understanding of the CV, further analysis was carried out to study the feasibility of employing the carbon deposition and dissolution processes for electrochemical energy storage and reuse. This effort started in this work by comparing the charges passed during the cathodic and anodic potential scans when recording the CVs. Some of the results obtained are listed in Table 2.

As it can be seen in Table 2, the Q^+/Q^- ratio increased with a decreasing potential scan rate. In line with this trend, it was found that holding the potential for 10 s at the cathodic end could further increase the Q^+/Q^- ratio to close to 1. These observations imply that carbon deposition may not proceed well on the

Table 2 The ratios of anodic (Q^+) and cathodic (Q^-) charges measured from CVs (*cf.* Fig. 3) as a function of potential scan rate[a]

Scan rate (mV s^{-1})	Q^- (C)	Q^+ (C)	Q^+/Q^-
100	3.575	1.776	0.50
80	6.220	3.151	0.51
60	6.524	3.362	0.52
40	10.75	5.737	0.53
20	25.56	14.41	0.56
10	73.56	47.70	0.65

[a] Note: all measurements were made using a fresh Pt working electrode in the same molten salt bath.

fresh Pt electrode, but becomes more efficient (although not necessarily quicker) after the Pt electrode is covered by the deposited carbon. In other words, carbon deposition on the Pt surface may proceed in competition with other reactions whose products do not remain on the electrode (*e.g.* CO formation), but these side reactions encounter higher kinetic barriers on the carbon surface. This understanding is practically important because it basically means that for the electrodeposition of carbon, the cathode material can be any metal or alloy as long as it is cathodically stable in the molten salt.

To further study the electro-oxidation of the deposited carbon, potentiostatic control was applied to produce a sufficient amount of carbon deposit for further analyses. In particular, chronopotentiometry was carried out to confirm reactions (11) and (12) as the main anodic reactions on the deposited carbon. In these experiments, the working electrode was a 5 mm dia. mild steel rod, whilst graphite, SnO_2 and stainless steel were all tested as the counter electrode material, and similar results were obtained. However, stainless steel was proven to be the most convenient choice for various carbonate dominated baths, although minor anode dissolution and consequent contamination of the deposited carbon were observed.[11] Thus, the following discussion will focus on data obtained from using the stainless steel counter electrode.

Fig. 4 compares the chronopotentiometric plots recorded during anodic oxidation (dissolution or galvanostatic discharge) of the electro-deposited carbon obtained under potentiostatic control. Both pure Li_2CO_3 and Li_2CO_3–K_2CO_3 were tested for comparison. On all plots in Fig. 4, there are two clearly distinguishable plateaux at potentials corresponding approximately to those of reactions (11) and (12). In Fig. 4b and c, a third potential plateau is shown at a more positive potential. The potential difference between the lowest and highest plateaux is about 1.75 V which can be explained by polarisations in addition to the theoretical value of 1.488 V between reactions (5) and (11). Polarisation is evident from a comparison between Fig. 4a and b. The potential difference between the first and second plateaux is about 0.40 V in Fig. 4a, but it increases to about 0.78 V in Fig. 4b. The larger current and lower temperature for recording Fig. 4b would have contributed to increasing both the ohmic and charge transfer polarisations. Thus, the third plateau can be attributed to reaction (5), *i.e.* electro-oxidation of the CO_3^{2-} ion.

Similar to the CVs in Fig. 3 which indicate no clear presence of direct anodic discharge of the O^{2-} ion, *i.e.* reaction (13), the plots in Fig. 4 also show only a barely visible inflexion of the rapidly increasing current between the 2nd and 3rd plateaux. This is understandable because the O^{2-} activity must be significantly lower than that of the CO_3^{2-} ion in the molten carbonate salts used. Also, these experiments were carried out under CO_2 which could also help convert O^{2-} to CO_3^{2-}.

For energy storage, a high Q^+/Q^- ratio approaching 1 is desirable. Cyclic voltammetry has revealed promising results as shown in Table 2. More analyses were carried out on carbon coatings deposited under potentiostatic control and discharged (re-oxidised) *via* chronopotentiometry. Some of the preliminary findings are presented in Table 3.

The Q^+/Q^- ratio was found to be a complex function of both deposition and re-oxidation variables. The deposition potential (and voltage if in a two-electrode cell) plays the most significant role. With an increasingly negative deposition

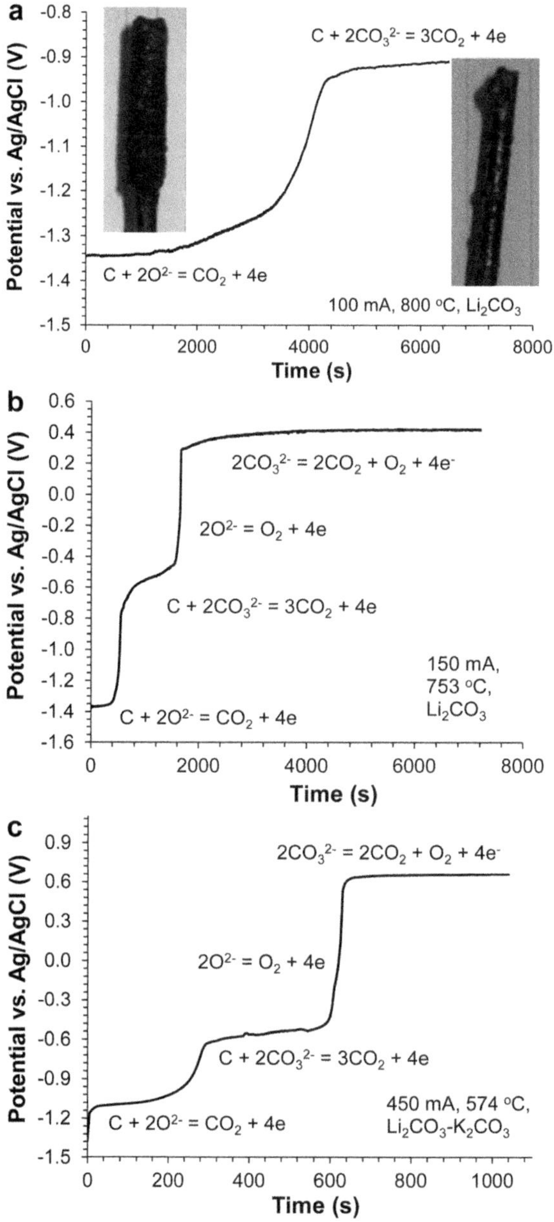

Fig. 4 Chronopotentiometric plots for anodic oxidation of the electro-deposited carbon in the same molten salt under CO_2. (a) Li_2CO_3 at 800 °C, oxidation: 100 mA, deposition: −1.8 V *vs.* Ag/AgCl, 3600 s. (b) Li_2CO_3 at 753 °C, oxidation: 150 mA, deposition: −2.1 V for 600 s. (c) Li_2CO_3–K_2CO_3 (molar ratio: 62 : 38) at 574 °C, oxidation: 450 mA, deposition: −2.6 V for 600 s. The photographs in (a) show the working electrode (5 mm dia. mild steel rod) with the carbon deposit (left), and after anodic oxidation (right). Counter electrode: 10 mm diameter graphite rod.

potential, the Q^+/Q^- ratio decreases quickly. This trend can be explained by parasitic reactions (*e.g.* CO formation) in competition with carbon deposition. Temperature change has a relatively small effect on the Q^+/Q^- ratio.

Increasing the deposition charge has a negative effect on the Q^+/Q^- ratio, whilst decreasing the re-oxidation current is beneficial. This can be related to the fact that the electro-deposited carbon is composed of various nanoparticulates, instead of a dense continuum.[11] Thus, carbon loss from the electrode with a thick carbon coating is highly likely during electro-oxidation, particularly considering the disturbance from CO_2 bubbles leaving the electrode at high re-oxidation currents.

It was observed in some cases that the overall Q^+/Q^- ratio would be higher if it were higher at the end of the first re-oxidation plateau (*cf.* Fig. 4 and Table 3). The first plateau corresponds to reaction (11) which produces one CO_2 molecule at the expense of four electrons. At the second plateau, reaction (12) also involves four electrons, but creates three CO_2 molecules. Thus, CO_2 disturbance would be more significant at the second plateau than the first one, likely contributing to a greater non-electrochemical loss of carbon and hence a lower overall Q^+/Q^- ratio. Unfortunately, reaction (11) consumes O^{2-} ions supplied to the electrode surface *via* diffusion which limits the maximum reaction rate. On the contrary, the $CO_3^{\ 2-}$ ion in reaction (12) is a component of the molten carbonate salt and hence cannot be depleted at the electrode surface. Thus, to promote reaction (11) and mitigate the impact from CO_2 disturbance during re-oxidation (discharge), a well-balanced engineering design of the bath composition and electrode and cell structures is crucial.

Prospects

According to Fig. 4, the oxidation of carbon should ideally proceed along the lowest potential plateau for reaction (11), which requires sufficient activity of the O^{2-} ion. This is, however, difficult, if not impossible, to achieve in molten Li_2CO_3 or mixed Li_2CO_3–K_2CO_3 under a CO_2 containing atmosphere. Thus, future work should include studies of carbon deposition and dissolution in molten salts with added alkali or alkaline earth metal oxides (Li_2O and CaO). Apparently, at sufficiently high activities of the O^{2-} ion, reaction (11) ($E^o = -1.488$ V) may change to the reverse of reaction (6) ($E^o = -1.719$ V) at a more negative potential. This is desirable from the viewpoint of carbon oxidation in a battery or fuel cell. On the other hand, however, because a high O^{2-} activity in the molten salt does not

Table 3 The Q^+/Q^- ratio as a function of the carbon deposition potential (*E vs.* Ag/AgCl), deposition charge (Q^-), and re-oxidation current (*I*), measured in molten Li_2CO_3–K_2CO_3 (62 : 38, molar ratio) at 700 °C under CO_2

E (V)	Q^- (C)	*I* (mA)	Q^+/Q^-, 1st plateau	Q^+/Q^-, 2nd plateau
−1.68	1017	150	0.26	0.83
−1.68	973.6	300	0.17	0.48
−1.68	1301	450	0.18	0.34
−1.98	1764	300	0.16	0.32
−2.28	2318	300	0.15	0.28

favour electro-deposition of carbon *via* reaction (6) (nor the reduction of O_2 gas in a direct carbon fuel cell), a careful balance is needed.

In addition to energy storage through electrochemical deposition and dissolution, the process of, and products from, indirect electro-reduction of CO_2 to carbon can have other applications. For example, the deposited carbon can be used directly for making the electrodes of supercapacitors,[10] or, after a proper treatment, for pollutant absorption. The process itself may be used to convert CO_2 gas back to O_2 gas, and hence support life in space, undersea, or in mines. It is thus worth making an estimation of the cost of the electro-deposition itself. In this work, a very high current efficiency (*e.g.* 95%) has been achieved in carbonate dominated baths. However, a voltage of or higher than 4.0 V was found to be necessary for electrolysis in the two-electrode cell,[11] which is much higher than the thermodynamically predicted voltage as shown in Table 1. A possible reason is the relatively low activity of O^{2-} ions in the carbonate dominated bath under CO_2 that invokes reaction (5) instead of (13) as the anode reaction. Thus, it is reasonable to predict that the electrolysis voltage can be reduced to below 3.0 V if either Li_2O or CaO is added to the molten salts. Nevertheless, even assuming a cell voltage of 3.0–4.0 V and a 95% current efficiency, the energy consumption of the electrolysis can be derived as 28.2–37.6 W h per kg-C, which can be translated to a cost of less than $5 per kg-C considering a 20% margin for heating and pre- and post-processing. For comparison, the current market price for supercapacitor grade activated carbon falls in the range of $20–40 per kg-C.

Summary

The electrochemical deposition and dissolution of carbon can be achieved in carbonate containing molten salts. The deposition becomes faster when CO_2 is present in the atmosphere above the molten salts. The cathodic process is dominated by carbon deposition at high current efficiency (up to 95%), whilst CO formation and other reactions may also occur. It was observed that carbon deposition occurs on the deposited carbon more efficiently than on a freshly prepared metal electrode. Electro-re-oxidation of the deposited carbon can be achieved with the aid of either the preferred O^{2-} ion, or the CO_3^{2-} ion at a more positive potential. Both chronopotentiometry and cyclic voltammetry revealed a fairly high ratio of deposition and dissolution charges, but also a large difference between the carbon deposition and dissolution potentials. A brief analysis of the process suggests addition of the O^{2-} ion in the form of either Li_2O or CaO to the molten salts may not only be beneficial for decreasing the electrolysis voltage by avoiding direct anodic oxidation of the CO_3^{2-} ion, but also enable anodic oxidation of the deposited carbon at more negative potentials.

Acknowledgements

This research received financial support from the Royal Society (2007–2009, Brian Mercer Feasibility Award), the EPSRC (2010–2015, Doctor Training Grant, EP/F026412/1, and EP/J000582/1), and the University of Nottingham (2011–2014, Dean of Engineering Scholarships).

Notes and references

1 Y. K. Delimarskii, O. V. Gorodiskii and V. F. Grishchenko, *Dokl. Akad. Nauk SSSR*, 1964, **156**, 650–651.

2 M. D. Ingram, B. Baron and G. J. Janz, *Electrochim. Acta*, 1966, **11**, 1629–1639.

3 H. Kawamura and Y. Ito, *J. Appl. Electrochem.*, 2000, **30**, 571–574.

4 L. Massot, P. Chamelot, F. Bouyer and P. Taxil, *Electrochim. Acta*, 2003, **48**, 465–471.

5 H. Groult, B. Kaplan, F. Lantelme, S. Komaba, N. Kumagai, H. Yashiro, T. Nakajima, B. Simon and A. Barhoun, *Solid State Ionics*, 2006, **177**, 869–875.

6 K. Le Van, H. Groult, F. Lantelme, M. Dubois, D. Avignant, A. Tressaud, S. Komaba, N. Kumagai and S. Sigrist, *Electrochim. Acta*, 2009, **54**, 4566–4573.

7 S. Licht, B. Wang, S. Ghosh, H. Ayub, D. Jiang and J. Ganley, *J. Phys. Chem. Lett.*, 2010, **1**, 2363–2368.

8 N. J. Siambun, H. Mohamed, D. Hu, D. Jewell, Y. K. Beng and G. Z. Chen, *J. Electrochem. Soc.*, 2011, **158**, H1117–H1124.

9 Q. S. Song, Q. Xu, Y. Wang, X. J. Shang and Z. Y. Li, *Thin Solid Films*, 2012, **520**, 6856–6863.

10 H. Yin, X. Mao, D. Tang, W. Xiao, L. Xing, H. Zhu, D. Wang and D. R. Sadoway, *Energy Environ. Sci.*, 2013, **6**, 1538–1545.

11 H. V. Ijije, C. Sun and G. Z. Chen, *Carbon*, 2014, **73**, 163–174.

12 M. Tahir and N. S. Amin, *Energy Convers. Manage.*, 2013, **76**, 194–214.

13 *HSC Chemistry 6, v. 6.12*, Outotec Research Oy, 2007.

14 G. Z. Chen, *A feasibility study of solar-electrochemical cells for capturing atmospheric carbon*, Application for The Royal Society 2007 Brian Mercer Feasibility Award, October 2006.

15 I. A. Novoselova, N. F. Oliinyk, S. V. Volkov, A. A. Konchits, I. B. Yanchuk, V. S. Yefanov, S. P. Kolesnik and M. V. Karpets, *Phys. E*, 2008, **40**, 2231–2237.

16 R. Barnett, K. T. Kilby and D. J. Fray, *Metall. Mater. Trans. B*, 2009, **40**, 150–157.

17 H. Wang, N. J. Siambun, L. P. Yu and G. Z. Chen, *J. Electrochem. Soc.*, 2012, **159**, H740–H746.

Faraday Discussions

The many faces of carbon in electrochemistry: general discussion

Robert Dryfe, George Zheng Chen, Domnik Bayer, Martin Jonsson-Niedziolka, Richard McCreery, Julie Macpherson, Héctor Abruña, Yury Gogotsi, Chi-Chang Hu, François Béguin, Laurence Hardwick, John Foord, Matteo Duca, Elizabeth Howe, Katherine Holt, Manuel Alvarez-Guerra, Patrick R. Unwin, Thomas Rabbow, Milo Shaffer, Jet-Sing Lee, Elzbieta Frackowiak, Fulian Qiu, Boris Dyatkin, Philip A. Ash, Keith Stevenson, Dirk Guldi and Philippe Bergonzo

DOI: 10.1039/C4FD90025A

George Zheng Chen opened the discussion of the Introductory Lecture by Richard McCreery: In your list of electrode capacitance for different carbons, the values differ significantly. Would you please comment on the causes of such differences, and also explain how you have measured the capacitance of the respective carbon materials? Particularly, what electrolyte have you used in the measurements?

Richard McCreery replied: We used either 0.1 M $HClO_4$ or 1 M KCl in water for capacitance measurements, and used cyclic voltammetry, where the charging current is proportional to the capacitance and the scan rate. HOPG has an unusually low capacitance.

George Zheng Chen opened the discussion of the paper by Héctor Abruña: It is well know that conducting polymers can be used for various electrochromic purposes, such as displays and smart windows. What are the improvements of your graphene–PEDOT hybrid in comparison with previous work? Also, if your graphene–PEDOT hybrid is to be used for smart windows, what is the light transmittance?

Héctor Abruña responded: You are certainly correct in that there is well-established literature on the use of conducting polymers for electrochromic applications. Part of the intent with the graphene–PEDOT hybrid that we have prepared was to move away from ITO-type electrodes, which have been traditionally employed in these applications. Graphene offers the advantages of being conductive and highly transmissive. (Single-layer graphene is about 97–98% transmissive.) In addition, it can make a conformal coating. In terms of light transmission, since the graphene itself has very low absorbance, the transmittance of the composite will be dominated by the thickness of the PEDOT film. One can readily make a calibration curve of transmittance *vs.* polymer thickness so as to adjust and optimize the PEDOT film thickness to the particular application/need.

Richard McCreery asked: Have the oxygen containing functional groups on graphene been characterized, for example by XPS? You invoke them to explain the dependence of electronic properties on pH, but has anyone established the identity of the oxygen containing functionalities?

Héctor Abruña answered: We have not done XPS characterization of the oxygen functionalities on the graphene surface. My sense is that such measurements are likely to be difficult because the signal levels are likely to be low and broad. We invoke their presence as a (reasonable) way to rationalize the pH dependence.

Julie Macpherson queried: There are different types of oxygenated functional groups that will form on the graphene surface. For pH FET sensing is there a preferred oxygen functionality for designing an optimal pH sensor?

Héctor Abruña responded: Although we do not have any direct information concerning the various oxygen functionalities that may be present on the graphene surface, my sense is that carboxylic acid groups are likely to be the best for pH sensing. I base this on the fact that such groups (as opposed to hydroxyl or aldehyde groups) are more likely to respond rapidly and reversibly to changes in pH.

Katherine Holt asked: What is the required density/coverage of surface functional groups to allow for pH sensing in FET devices? Has anyone calculated or determined such a value?

Héctor Abruña replied: While to my knowledge nobody has calculated or determined this, one can estimate the value. The saturation coverage of quinones on HOPG is of the order of 1.5×10^{-10} mol cm^{-2}.[1] Given the voltammetric profiles and S/N considerations, we would estimate (clearly speculative on our part) that a surface coverage of 1–5% should be sufficient to provide pH dependent behavior.

1 M. T. McDermott and R. L. McCreery, *Langmuir*, 1994, **10**, 4307.

Katherine Holt remarked: With regards to the pH sensitivity of functional groups on graphene, our paper[1] discusses the pH dependent behaviour of graphene nanoflakes with carboxylic acid edge groups. The oxygen functionalities of these nanoflakes are well defined, which allows us to ensure the pH dependent behaviour can be attributed to the carboxylic acid functionalities. We haven't used the graphene nanoflakes for pH sensing but it is possible they would be a good candidate material for this application.

1 M. M. Lounasvuori *et al.*, Electrochemical characterisation of graphene nanoflakes with functionalised edges, *Faraday Discuss.*, 2014, DOI: 10.1039/c4fd00034j.

Héctor Abruña added: Graphene-type materials, such as graphene nanoflakes with a high density of oxygen functionalities, would appear to be good candidates for such an application. My concern would be how they would be supported on a conducting substrate to which one could make contact.

Milo Shaffer communicated to Richard McCreery and Héctor Abruña: The issue of oxide structures on carbon has been studied in various ways. Recently, I heard an excellent presentation by Neil Wilson (Warwick) describing a detailed study of atomic oxygen adsorption on graphene, associated with atomic resolution microscopy and *in situ* ARPES measurements. He showed the complexity of the resultant structures and the fact that a couple of O atom% were sufficient to destroy the band structure of the graphene seen in ARPES. It is interesting to compare this result with the estimate given by Richard McCreery for the maximum 1% functionalisation before the whole layer is affected. In then considering Hector Abruña's devices, and his estimate that >1% functionalisation is required for a successful pH sensor, are the key properties of graphene lost too soon? Or is the associated change in properties relevant to the device operation?

Héctor Abruña communicated in response: I was not aware of the study by Neil Wilson, thanks for bringing it up. The issue of 'oxygenated groups' on carbon surfaces has been (and continues to be) the subject of numerous studies. The 1% estimate that I gave was based more on S/N considerations for measurement. In the statement, "the fact that a couple of O atoms were sufficient to destroy the band structure of the graphene", what would this correspond to in terms of surface coverage? As carbon surfaces go, graphene is likely to be the most affected in terms of the consequences of having oxygenated species. A detailed study in which the surface coverage, and defect density, are controlled would be of great utility. However, this is likely to be very difficult.

Richard McCreery communicated in reply: I agree on most points. Everyone should be careful to distinguish between electronic disturbances and actual chemically distinct sites. Outer sphere systems should be sensitive to electronic effects, which may extend well beyond the site of the defect itself. An electrocatalytic process, however, may depend on specific functional groups present only at the defect. An example is aquated $Fe^{+3/+2}$, which is catalyzed by surface C=O groups.

Patrick Unwin asked: How does the response in Fig. 11 of the paper for bare graphene compare to that for HOPG? Additionally, what happens to the response on the graphene electrode in subsequent cycles? I ask this as the electrochemical oxidation of NADH can produce products which foul the electrode. Is this seen on graphene?

Héctor Abruña answered: While we have not done an exhaustive study of changes in the response for NADH oxidation during consecutive scans, there are some observations that we have made. For the oxidation of NADH at bare graphene, we did observe a decay in the magnitude of the response with consecutive scans. For graphene surfaces modified with 3,4-DHB, we did not observe dramatic changes. While somewhat speculative on our part, we believe that this could be due to depletion of NADH within the diffusion layer, as well as potential blocking/inhibiting effects for the formation of non-reactive products.

Philip A. Ash commented: We observe the onset of NADH oxidation at a bare pyrolytic graphite edge electrode at approximately +300 mV *vs.* the standard hydrogen electrode (SHE), *i.e. ca* +60 mV *vs.* Ag/AgCl. A range of other carbon electrodes have been demonstrated as NADH-based biosensors, and tend to report similar operating potentials as those stated for Prof. Abruña's graphene.[1]

NADH oxidation to NAD^+ can be summarised as involving the transfer of two e^-s and one H^+. The overall process has been proposed to follow an ECE mechanism involving at least one radical intermediate after removal of the first electron. In the case of biological NADH-dependent enzymes, side reactions involving any such radical intermediate or intermediates are avoided by rapid proton transfer and removal of the second electron. In the biological case, NADH oxidation is often coupled to highly regio- and stereospecific reduction of a substrate, with NADH effectively acting as a hydride donor that is required stoichiometrically during the course of the reaction and must be either regenerated or replenished.[2] Therefore, any process that leads to formation of biologically 'inactive' forms of NAD^+ (for example due to dimerisation or reactions leading to immobilisation on an electrode) must be avoided in any practical NADH biosensor or NAD^+ regeneration system. With this in mind it is surprising that you see no decrease in activity towards NADH oxidation with repeated potential cycles at your graphene electrode. It is possible that this is due to NADH depletion in the diffusion layer during your measurements limiting the overall extent of the reaction at the highest applied potentials, rather than the graphene electrodes being intrinsically inert to either reaction with or adsorption of NADH. The difference in activity between the mediator-coated and bare graphene surfaces could be that the adsorbed mediator limits the extent of NADH adsorption and reactive immobilisation, or that the semi-labile mediator (as indicated by the stated decrease in surface coverage with successive potential cycles) transfers two electrons simultaneously (or at least in quick succession) thus avoiding any long-lived radical intermediates.

You quote an overpotential of roughly 1 V for NADH oxidation – $E(NAD^+/NADH)$ at pH 7 is approximately −400 mV *vs.* SHE at a 1000-fold excess of NADH, *i.e. ca* −640 mV *vs.* Ag/AgCl. These operating potentials are quite high in biological terms. Do you know what the activities of the bare or modified graphene electrodes are with respect to other possible substrates that might be present in a realistic biosensor? NADH-based biosensors sometimes incorporate a NAD^+-dependent enzyme that produces NADH during substrate oxidation, and it is oxidation of this NADH by-product that is detected by the biosensor. It is therefore very important that the electrode oxidises exclusively NADH at the operating potential of the biosensor. An alternative strategy might be to use a more selective electrocatalyst that works at a lower overpotential to improve the selectivity of a NADH-based biosensor by minimising any unwanted oxidations.

1 See, for example: *Chapter 4: NAD(P)-Based Biosensors* in *Bioelectrochemistry: Fundamentals, Experimental Techniques and Applications*, ed. P. N. Bartlett, John Wiley and Sons, 2008.
2 Reeve *et al.*, *Chem. Commun.*, 2012, **48**(10), 1589–1591.

Elizabeth Howe asked: The graphene used in your study was used after being transferred with PMMA. How do you verify whether all the PMMA has been removed by your washing procedure?

Héctor Abruña responded: As mentioned in the manuscript, in order to remove the PMMA, the chips were immersed in anisole for two to three hours, in 50/50 (v/v) dichloromethane–acetone for at least 12 hours, and finally in iso-propanol for at least two hours. The quality of the single layer graphene was verifed by the 2D and G peaks at 2700 and 1600 cm^{-1}, respectively, in the Raman spectra. In some cases we also obtained AFM and TEM images that showed very low levels of PMMA residues.

Richard McCreery enquired: You described some Raman results in your manuscript. Does the Raman spectrum of your graphene change at all when you modify the graphene with polymers or other reagents?

Héctor Abruña responded: We see changes that we can relate to the nature of the films being deposited. In addition, the intensities of the bands associated with graphene itself (*i.e.* G and 2D) tend to be attenuated.

Richard McCreery continued: The ratio of the 2D peak intensity to that of the G peak in Raman is often used to indicate graphene structural changes. Does this ratio remain reasonably constant when your graphene is modified with polymers, etc?

Héctor Abruña answered: While we have not done an exhaustive/extensive study of this, we generally find that while there is some attenuation of intensities, the ratio remains relatively similar.

Katherine Holt said: A paper last year[1] reported that the hydrophobicity/hydrophilicity of graphene is strongly affected by the adsorption of atmospheric contaminants like hydrocarbons, which adsorb almost instantaneously on exposure to the laboratory atmosphere. Is this something that has been noted by those working on graphene electrochemistry and is it likely to effect the performance of graphene in electrochemical devices?

1 Z. Li *et al.*, Effect of airborne contaminants on the wettability of supported graphene and graphite, *Nature Materials*, 2013, **12**, 925.

Héctor Abruña replied: We have not carried out studies addressing this point. However, we would estimate that if there was strong adsorption of such molecules over a significant fraction of the surface, the redox response of redox molecules would be affected. We have not observed such behavior.[1]

1 N. L. Ritzert, J. Rodríguez-López, C. Tan and H. D. Abruña, *Langmuir*, 2013, **29**, 1683–1694.

Julie Macpherson opened the discussion of the paper by Philippe Bergonzo: Fig. 7 of your paper shows that for your heavily doped diamond (1% doping) there is an oxygen wave present, occurring about −0.8 V in the planar surface CV. This is commonly associated with the presence of non diamond carbon (NDC) impurities. For your proposed applications does NDC presence matter?

Philippe Bergonzo replied: No, I don't think NDC carbon impurities matter.

Julie Macpherson continued: When you grow the heavily doped diamond films on structured surfaces, such as the CNT study described in the paper, does the NDC content increase compared to growth on planar surfaces – *i.e.* is it harder to control BDD quality on structured surfaces? It was difficult to tell from the data presented in Fig. 7 as the background currents had also increased due to an increased surface area of the electrode.

Philippe Bergonzo responded: According to the thickness of diamond deposited on the CNT structure, we observe a varying window, probably due to the (inverse) variation in NDC content.[1]

1 This is further discussed in C. Hébert *et al.*, *Carbon*, 2014, **71**, 27–33.

Richard McCreery remarked: Do electrochemical properties, particularly electron transfer kinetics, track the doping density? Has there been a systematic study of k^{o} as a function of boron content in BDD?

Philippe Bergonzo replied: I don't know if such a systematic study has been conducted on all forms of diamond. For the material we grow, we indeed have published such a study.[1]

1 E. Vanhove *et al.*, High reactivity and stability of diamond electrodes: The influence of the B-doping concentration, *Physica Status Solidi a*, 2009, **206**, 2063–2069.

Richard McCreery said: You showed three reversible redox systems on boron doped diamond electrodes. One might expect a variable density of states in the BDD, possibly resulting in the dependence of kinetics on the standard potential of the redox system. Has anyone studied the variations in electrode kinetics with the E^{o} of the redox system?

Philippe Bergonzo responded: Not to my knowledge. It could be interesting. But again one has to do it on well controlled materials before they can generalize this to 'diamond'.

Julie Macpherson commented: This is a comment related to the previous question Rick McCreery asked about the relationship between boron content, sp^2 content and the density of states in diamond. When boron doped diamond (BDD) is doped above the metallic threshold $>1\times10^{20}$ B atoms cm^{-3}, we find that the intrinsic rate constant of electron transfer, k^{o}, for simple outer sphere redox species, scales with the boron content of the material, which in turn, we find, controls the local density of states (LDOS).[1] This was elucidated on polycrystalline BDD, which contains differently doped crystallographic grains and negligible sp^2 carbon. All grains were doped above the metallic threshold. A factor of ~4 difference in the measured k^{o} values across the electrode surface was determined using scanning electrochemical microscopy (SECM), where more highly doped grains showed faster k^{o} values (for example, for $FcTMA^+$ we measure k^{o} values of 0.097 ± 0.04 cm s^{-1} on high doped grains and 0.022 ± 0.008 cm s^{-1} on lower doped ones); these k^{o} values are high enough for CVs at macroelectrodes in stationary solution to look reversible, at typical scan rates *e.g.* 50 mV s^{-1}. Scanning electrochemical cell microscopy (SECCM) was used to measure the local capacitance in the differently doped regions, from which a LDOS value could be extracted. The more highly doped regions showed a factor of ~4 increase in LDOS compared to the less doped areas; values of 1.7 ± 0.7×10^{20} cm^{-3} eV^{-1} and 6.3 ± 2.0×10^{20} cm^{-3} eV^{-1} were obtained for the polycrystalline BDD (at 0.0 V *vs.* Ag/AgCl) employed in the study. Notably, our 'metal-like' BDD had a LDOS 2–3 orders of magnitude smaller than for a metal, which is typically *ca.* 10^{23} cm^{-3} eV^{-1}. Note, when sp^2 carbon is present, its affect on the LDOS, the electrical properties and the k^{o} values should also be considered. The presence of sp^2 carbon should also be taken into consideration when interpreting the k^{o} values for inner sphere couples, as sp^2 regions will have different surface chemistry with respect to the BDD surface.

1 This work was published in Patten *et al.*, *Angew Chemie Intl.*, 2012, **51**, 7002–7006.

John Foord remarked: You showed this rather nice *in situ* pulsed electrochemical cleaning process for diamond electrodes. Have you tried it out on other materials where electrode fouling can be more problematic?

Philippe Bergonzo responded: We tried it on Pt, and for deposited thin Pt layers it seems to work, but the layers are progressively altered, something that is not observed on diamond. I suspect such a cleaning process may also work on bulk Pt materials or on other highly 'bulky' materials.

Keith Stevenson asked: You discussed the concept of self-cleaning of the electrodes and I was curious as to the potential dependence of self cleaning mechanism. Did you employ symmetrical pulses during the redox process?

George Zheng Chen also asked: When the electrode was cleaned by application of short cathodic and/or anodic pulses, were the pulses symmetrical? What potentials were selected, and would these potentials affect the cleaning result?

Philippe Bergonzo answered: Indeed they were symmetrical cathodic and anodic. We don't drive a voltage, but a current, typically from 10 to 100 mA maximum. More information is available in the following patent.[1]

1 J. de Sanoit, R. Kiran and E. Scorsone, *US Pat.* 2013313120, 2013.

Keith Stevenson continued: As a follow up question, do you see a change in the behavior with a change in chloride ion concentration? Are bubbles observed due chlorine gas generation?

Philippe Bergonzo responded: We never saw any modification of the solution during the cleaning process. There are no bubbles. We believe this is the case because we go too fast; bubbles don't have time to form.

Patrick Unwin enquired: What does the voltammetry look like at the individual electrodes in these microelectrode arrays?

Philippe Bergonzo replied: We have published this.[1] We did a statistical study of how many of the microelectrodes were not responding as perfect diamond, and indeed always found close to 15% faulty microelectrodes. We incriminated metallic residues from the final lithographic processing step, as this seemed to be improved after better cleaning. Thus it's not due to the diamond itself but to processing contamination. A minor issue as better cleaning can solve it.

1 R. Kiran *et al.*, *Sensors*, 2012, **12**, 7669–7681.

Julie Macpherson remarked: Given that sp^2 carbon is present in your highly doped boron doped diamond films, do you see any evidence for preferential etching of the sp^2 regions (*e.g.* at grain boundaries) when you apply extreme potentials?

Philippe Bergonzo answered: We don't use potentials but currents. And indeed their could be high surface potentials but I don't know how to evaluate them. What we have observed is that we have never seen any etching. Some colleagues (*e.g.* C.E. Nebel at IAF Freiburg) have observed diamond etching applying several 10s of V (tip −20 or -30V). But we have never observed such an etching at the resulting 'potentials' of our activation treatments. For example, depositing IrO_x mediating spheres on the nanocrystalline diamond surface, we have observed closely how they behave, following their populations and sizes grain by grain, but we have never observed any etching on the diamond.

Katherine Holt said: It seems to be the case that sometimes the advantages of certain electrode materials (like diamond) need to be compromised to get the best out of them in a device. For example in the applications of diamond electrodes requiring interfacing with cells, is it not the case that the cells may not interact or grow well at the interface of the best quality planar diamond? So the diamond needs to be nanostructured, which introduces sp^2 content and defects, and therefore its electrochemical properties (wide potential window etc.) might be compromised?

Philippe Bergonzo responded: Nanocrystalline diamond is heavily structured at the nanoscale, and we have studied a lot of cell lines from neurons to organ cells or osteoblastes or cardiomyocytes, and they all grow extremely well on nanocrystalline diamond. In other words, I don't think we grow the "best quality planar diamond", and indeed we believe it works best without.

Katherine Holt continued: Does the surface termination of the diamond have any effect on the ability of cells to attach to it or the electrochemical response?

Philippe Bergonzo answered: Indeed. But not quite on the attachment. However, we have observed that O– or H– termination has a effect on the promotion or the reduction of glial to neuron cell growth. As such, we in fact deposited a patent application on this as it is a nice way to pattern the neuron growth on carbon materials with respect to localizing glial cells that are also required to build a neural network. This patent has been published recently,[1] where more details are provided.

1 S. Picaud *et al.*, *Eur. Pat.* 2747158 (A1), 2014.

Richard McCreery asked: Diamond electrodes are generally recognized as having longer lifetimes *in vivo*, probably due to lower adsorption of 'impurities', *etc.* But biological redox systems such as dopamine may also require chemical interactions with the electrode surface during electrocatalysis. Does the weak tendency to adsorb work against the need for catalysis? Does the stability of diamond become a disadvantage for the electrochemical detection of biological molecules?

Philippe Bergonzo responded: I don't think so, as electron transfer rates are still very high, reaching values close to 0.1 cm s^{-1}. However, this is only obtained on very highly doped diamond, typically in the 1% at. ratio range, thus it is likely that we are talking about diamond with a bit of sp^2 carbon.

Patrick Unwin commented: We have recently compared the voltammetric response for dopamine electro-oxidation on a variety of carbon electrodes.[1] There is a clear difference in the response between freshly cleaved HOPG (different grades) and boron doped diamond (BDD), with the process apparently much faster on HOPG and other graphite electrodes than on BDD. The comparison is interesting as HOPG and BDD have a broadly similar density of electronic states (DOS), yet the faster process on the graphite basal surface suggests stronger adsorption and electronic coupling. Such comparisons between graphite basal surfaces and diamond would be beneficial for other systems, given the similar DOS, but different structure and surface terminations (sp^2 *vs.* sp^3). It should be noted that although the electro-oxidation process is slower on BDD, the BDD surface is much less prone to blocking by dopamine oxidation products at relatively high dopamine concentration (100 μM), although at lower concentration (1 μM) HOPG shows minimal blocking and a fast response that is clearly seen due to the low background currents characteristic of pristine HOPG.

1 A. N. Patel, S. Tan, T. S. Miller, J. V. Macpherson and P. R. Unwin, *Anal. Chem.*, 2013, **85**(24), 11755–11764

Richard McCreery asked: In your experience with diamond electrodes for analysis *in vivo*, are biological redox systems detectable, given the often weak activity of diamond surfaces for the electrocatalysis involved in biological redox systems?

Julie Macpherson replied: Yes, biological redox systems are detectable, but given the electrocatalytic inert nature of the sp^3 diamond electrode surface, the detection potentials are typically pushed out compared to other carbon electrodes such as glassy carbon, carbon nanotubes, HOPG (EPPG, BPG) and classical metal electrodes such as Pt and Au, as we show, for example, in ref. 1 and 2. However, if this is not perceived to be a problem for the measurement of interest, diamond electrodes are well-suited towards *in vivo* measurements given their reduced fouling, bio-compatibility and mechanical robustness.

1 Patel *et al.*, *Anal. Chem.*, 2013, **85**, 11755–11764.
2 A. Güell *et al.*, *PCCP*, 2010, **12**, 10108–10114.

Matteo Duca said: Considering the interest in hybrid sp^2–sp^3 carbon systems, do you believe that it is possible to control the BDD coating to prepare a partially-covered BDD–CNT composite using the approach described in your paper (Fig. 8)?

Philippe Bergonzo answered: Probably. This was partially discussed in ref. 1. However we always aimed at fully covering (no pinhole) diamond layers. When the layers are too thin, we see that the window decreases, and thus the advantages of the diamond layer for neurointerfacing as a passivating layer are lost. But for other applications it could be interesting indeed.

1 C. Hébert *et al.*, *Carbon*, 2014, **71**, 27–33.

Jet-Sing Lee asked: How much more effective are these boron doped diamonds compared to conventional materials for biosensing? I respect that conventional materials are more prone to fouling; are your materials using diamond stable enough to make them cost efficient for its enhanced properties?

Philippe Bergonzo responded: Boron doped diamond is costless. It is as cheap as any carbon based material to fabricate. We grow diamond from methane and water. The cost is definitely not an issue. Indeed we have shown that diamond is extremely stable *in vivo*, more than any other conventional carbon materials, and it is also much cheaper than any precious metal.

Richard McCreery opened the discussion of the paper by Dirk Guldi: One would expect a significant difference in the photoactivity of 'peapod' materials made with semiconducting compared to metallic nanotubes. How do you select the tubes used for making 'peapod' structures, *i.e.* what sort of pre-screening is carried out?

Dirk Guldi replied: There are indeed significant differences in terms of behaviour between semiconducting and metallic carbon nanotubes and between empty and filled carbon nanotubes. No attempts were, however, made to pre-screen the samples or to separate the different types of carbon nanotubes. Still, to guarantee unambiguous results we took great care to individualise carbon nanotubes and to ensure the stabilities of the corresponding dispersions. In addition, enriched semiconducting carbon nanotubes were used.

Yury Gogotsi asked: What is the percentage/degree of filling of the nanotubes by fullerenes and how does it affect the adsorption?

Dirk Guldi responded: The degree of filling is nearly 100%. In addition, filled carbon nanotubes, that is, peapods, require a slightly different work-up procedure than empty carbon nanotubes. The samples were quantitatively checked for filling *vs.* non filling by TEM.

Katherine Holt remarked: In your difference absorption spectra (for example Fig. 7 in the paper) you see very clear loss peaks (minima) corresponding to loss of the ground state absorptions. However only in a few cases do you see any corresponding gain peaks (maxima) in the difference spectra. Why is this?

Dirk Guldi answered: In general, the negative regions in the differential absorption spectra correlate to a loss of oscillator strength relative to the reference spectrum, that is, the ground state absorption. Overall, the spectra are correlated with those obtained in spectroelectrochemical reduction and oxidation experiments performed with filled and non filled carbon nanotubes. One may expect, for example, upon one electron reduction, a bleaching of the ground state absorption accompanied by a transient absorption red-shifted relative to the latter. The fact that this is not always observable relates to the fact that we are dealing with a rather broad distribution of carbon nanotubes and, in turn, a superimposition of many spectra. As such, the bleaching seems in most cases more intense than the new absorption.

Katherine Holt asked: For the proposed applications for your functionalised carbon nanotubes would you require them to be dispersed in solution or deposited as a film?

Dirk Guldi responded: Ideal for an application would be carbon nanotubes deposited as films – 2D or 3D. One way to use the deposition would be to employ them in field effect transistors, in which the individual carbon nanotubes are electronically and spatially well isolated from each other.

Katherine Holt continued: To clarify the role of charge transfer in the dispersion of the functionalised nanotubes – if you tried to separate the nanotubes with a surfactant for which charge transfer does not take place between surfactant molecule and the nanotube, you will be unable to disperse the CNT?

Dirk Guldi replied: Just the bare separation is even possible for surfactants, which lack any redox features. However, the stability of the resulting dispersions is governed by increasing the activation barrier for the desorption of the surfactants from the surface of the carbon nanotubes. In other words, the redox features of the surfactants/charge transfer between the surfactants and the carbon nanotubes assist in raising the activation barrier.

François Béguin opened the discussion of the paper by Milo Shaffer: Your system is characterized by low power and energy density. The low energy density is related to the electrode construction with a very small amount of active material. Would it be possible to include activated carbon cloth in the structural composites? Activated carbon cloths from phenolic resin or viscose are currently available; these materials display a very high specific surface area and have good mechanical properties. Therefore, it is possible to believe that they could be substituted with a part of the structural fibers without greatly altering the mechanical properties. Could that be a future direction to overcome the low performance of your system? The low power density is essentially due to the low electrical conductivity of ionic liquid electrolyte. Which kind of alternative electrolyte could be considered to improve this parameter?

Milo Shaffer responded: Traditional activated carbon fibres are not suitable for structural supercapacitors as the intrinsic fibre properties are poor. They are not based on structural graphitised carbon, and have low strength, stiffness, as well as very poor electrical conductivity. They have a tendency to fragment during handling, increasing the risk of device shorts. Whilst they could be hybridised with conventional structural fibres, the reduction in structural fibre content would damage mechanical performance. The aerogel approach is the most attractive, since it uses all of the matrix space to provide an electrochemically active electrode volume. The structural fibre content can therefore be fully maintained, whilst redeploying the remaining space that is always present, so a second function can be integrated. In due course, other fibres may become available that offer good strength and surface area simultaneously. Aligned carbon nanotube fibres are already promising in this context, and although not yet available in commercial quantity, efforts are underway to scale-up production.[1] The relatively low performance of the structural capacitors in the current study is due to the very poor conductivity of the basic cross-linked electrolyte used. Already we have produced a very significantly improved structural electrolyte based on a bicontinuous epoxy/ionic liquid system, as outlined in the paper, and in more detail elsewhere.[2] Once integrated with the aerogel/carbon fibre electrodes, significant performance improvements can be anticipated.

1 N. Behabtu *et al.*, *Science*, 2013, **339**(6116), 182–186.
2 N. Shirshova *et al.*, *J. Mat. Chem. A*, 2013, **48**, 15300.

Yury Gogotsi said: It is important to think of applications when deciding whether to use a multifunctional energy-storing composite or a supercapacitor film on a strong structural composite plate.

Milo Shaffer replied: As indicated in the paper, the combination of two fully independent mechanical and electrical systems by simple adhesion, to form a multifunctional structure, may offer some advantages particularly in usable system volume. However, the scope for weight saving is very limited, on the whole relating only to packaging. Conceptually, it is also a much less interesting or revealing challenge. Multifunctional materials, in which at least some of the atoms provide at least two traditionally orthogonal functions, are more demanding but potentially much more rewarding. The motivation to balance conflicting materials requirements may inspire new understanding and new materials solutions of broad relevance.

Yury Gogotsi continued: What is better in terms of strength and capacitance per weight – using your multifunctional composite or a strong structural composite combined with a thin film supercapacitor?

Milo Shaffer responded: As noted in the previous answer, the multifunctional materials approach has much greater potential to save both weight and volume. The materials challenges are, of course, also greater.

Richard McCreery asked: You see a very large increase in capacitance (approx. 100×) when the carbon aerogel is added to other carbon materials. Is this all due to the increase in surface area, or is there also a component due to formation of oxygen functional groups? The latter are important to activated carbon, for example.

Milo Shaffer answered: In the examples within the paper, the capacitance of the electrodes closely follows the BET surface area measurements; the as grown carbon nanotubes form part of this series, even though their synthesis does not involve intentional oxygen-containing precursors. Some oxygen groups will be present but only at a low level. It is also worth noting that these examples have mostly mesoporosity accessible to the electrolyte. The consistency between surface area and these various electrodes suggests that surface groups contribute little, in this case. Deliberate attempts to increase the effective capacitance by introducing pseudocapacitive species is of interest, as illustrated in the paper by the polymerisation of PEDOT onto the aerogel framework, although there are a wide range of possibilities.

George Zheng Chen commented: Regarding the influence of the oxygen content in nano-structured carbon materials on their specific capacitance, I shall mention that when presented in an appropriate concentration in surface functional groups, oxygen does contribute to the so-called pseudo-capacitance that involves electron transfer reactions. The cyclic voltammograms of such materials do not necessarily present current peaks because these oxygen containing surface groups are connected by highly conjugated graphene layers, which means that the electrons transferred to and from these groups are also likely partly or fully delocalised and hence do not give rise to current peaks. Our experience is that the oxygen content should not be higher than 10% to benefit the specific capacitance. Too high an oxygen content may disrupt the surface conjugation of the carbon material. You may find more explanations on the relation between pseudo-capacitance and delocalised valance electrons in my recent paper (which is free to access on the journal's webpage).[1]

1 G. Z. Chen, Understanding supercapacitors based on nano-hybrid materials with interfacial conjugation, *Prog. Nat. Sci.-Mater. Int.*, 2013, **23**(3), 245–255.

Richard McCreery enquired: Oxygen functional groups are often responsible for the degradation of high surface area carbon materials. Since the aerogels are made from oxygen-rich precursors, are they prone to corrosion more than other carbon materials?

Milo Shaffer answered: It is true that RF aerogels are formed by the condensation of oxygen-containing precursors; however, many other activated carbons are similarly derived from carbohydrate sources, usually with other heteroatoms present. The oxygen content can be adjusted by the carbonisation temperature, which for our aerogels is around 800 °C. Other cross-linking chemistries could be introduced to the same concept/architecture if necessary, though the RF system is considered favourable for supercapacitor electrodes.[1]

1 A.G. Pandolfo and A.F. Hollenkamp, *J. Power Sour.*, 2006, **157**, 11–27.

Richard McCreery continued: Has anyone tried H-termination to stabilize carbon aerogels? It is not a simple process, but should remove oxygen surface groups and make stable C–H bonds.

Milo Shaffer replied: Controlled termination of the surface chemistry is an important process for controlling electrolyte interactions, and the associated effects on electrolyte wetting and stability.[1] In the context of aerogels, thermal activation or passivation with oxides have been considered.[2] Introducing hydrogen terminations would be worth exploring for these monolithic aerogels. We have explored high temperature activated radical reactions[3] that might be applied, without the complications of plasma systems.

1 A. G. Pandolfo and A. F. Hollenkamp, *J. Power Sour.*, 2006, **157**, 11–27.
2 J. Biener *et al.*, *Energy Environ. Sci.*, 2011, **4**, 656–667.
3 R. Menzel *et al.*, *Chem. Sci.*, 2010, **1**, 603–608.

Thomas Rabbow asked: Did you try to combine first a chemical activation to achieve a high active surface area and second the aerogel coating to achieve a high total surface area? Thereby I would not expect a higher BET surface area, but if the chemical activation would persist for a long enough duration, I could imagine that the capacitor performance increases (faster charging and discharging, lower internal resistance).

Milo Shaffer answered: We did try various combinations of the strategies for improving the structural electrode surface area/capacitance, which are outlined in the paper. As yet, we have found no significant advantage. Introducing the RF after an initial fibre activation is likely to block the surface pores created in the first step. Activation after the aerogel formation ought to offer some advantage, but relatively less than might be anticipated from the increased BET surface area, due to the formation of inaccessible pores.[1]

1 A. G. Pandolfo and A. F. Hollenkamp, *J. Power Sour.*, 2006, **157**, 11–27.

Fulian Qiu commented: When a double layer supercapacitor is fully charged, its open circuit potential will gradually decay due to self discharge processes. These processes could include a localized short circuit through the separator membrane, thermo-desorption, back reaction at the substrates, reactions involving impurities on carbon particles and surface morphology changes due to strain–stress under the charged conditions.

Fulian Qiu asked: What are the advantages of composite supercapacitors over the activated carbon coated metal mesh supercapacitor?

Milo Shaffer replied: The point is that the primary carbon fibres are truly structural, of the type used for a wide range of structures under high mechanical load – for example, airframes, chassis, sports equipment, *etc.* Thus, in principle, the composite can be truly multifunctional – both a mechanical structural component and an electrical energy storage device.

George Zheng Chen remarked: Would you please comment on the differences between your laminated structural supercapacitors and the cable-type structural supercapacitors that can be woven or embedded into different structures?

Milo Shaffer responded: There are a wide range of possible architectures for structural supercapacitors, with the cells implemented at the single fibre, tow, or laminate level. In general, using a finer scale design offers improved rate performance at the expense of complicating the connections and increasing the risk of shorts. Twisted yarn supercapacitors and batteries have been produced previously, with a focus on flexible systems, for example, to integrate in textiles. In our structural supercapacitors, the structural requirements are more demanding; good stiffness, in particular, demands good fibre alignment, and is less consistent with existing cable approaches. On the other hand, structural supercapacitor architectures designed around predominantly tensile loading of unidirectional fibres would be attractive. The preparation of multifunctional matrices which balance both ionic conductivity and stiffness are challenging, implying that matrix dominated failures are likely to be limiting. Purely tensile loading in the fibre direction should offer a simpler route to good multifunctional performance.

Héctor Abruña said: The processes that you described are likely to significantly add to the cost of the materials and thus the devices. In the long run, will the improved performance compensate the additional costs?

Milo Shaffer replied: Of course, the balance of cost to benefit will ultimately determine the take up in commercial application. However, it is an early stage to predict either the costs of any finalised system or the extent of the benefits available. Neither the materials systems nor processes are settled. We are at the stage of a proof of concept, exploring the fundamental constraints and compromises involved. Different applications will have very different requirements. The value of weight saving will vary dramatically between, for example, the aerospace and automotive sectors. However, as an indication, even in the cost-sensitive mass produced car market, there is intense interest in purely structural carbon fibre composites to replace metals, particularly for hybrid and electric vehicles. Once carbon fibres are implemented, the marginal cost of structural supercapacitors may be modest. There may even be savings, since a secondary electrical energy system will be redundant. The work has also generated a new type of hierarchical composite which has wider potential significance than structural capacitors alone, since there are indications of improved absolute mechanical performance and a range of other opportunities.

Elzbieta Frackowiak asked: In your experiments concerning multifunctional composites for a rigid energy storage device, you were using a very narrow range of potential, ± 0.2 V, that is quite far from the voltage applied for a capacitor operating in practice. What is the maximum voltage which you reached in the two solutions used, *i.e.*, KCL and ILs?

Additionally, could you please answer what is the benefit of using a step voltage (0.1 V) method for capacitance estimation? Are you sure that capacitance, ESR and BET surface area values (Table 1 and 2) are measured with so high accuracy?

Milo Shaffer responded: Broadly, our test regime was driven by the challenges of making a true structural supercapacitor. Many of the samples, indeed all the structural ones, in the paper still have a high ESR and low power density, for reasons explained in the text. In order to test all samples under consistent conditions in a reasonable timeframe, we adapted our approach, as you mention. Once we have plausibly practical values, which we are approaching with our next generation architecture, then of course it will be important to test under a range of more realistic operating conditions. For non-aqueous electrolytes, where we hope to broaden the voltage window significantly, a new challenge will arise: how to process and seal the composite such that it remains sufficiently dry. However, we have plans and commercial collaborators in place to address this requirement.

Boris Dyatkin remarked: Your publication describes two approaches of grafting nanotubes onto carbon fibers. The first involves catalytic growth, with the carbon nanotubes stemming directly from the carbon structure. The second involves the decomposition and carbonization of a solvent containing suspended nanotubes. Does the latter approach increase charge transfer resistance in the system, given that the electrical contact between the two carbon structures is not an ideal, defect-free graphitic interconnect? What is the structure of the carbonized residue that forms the interconnect?

Milo Shaffer replied: We have not explored this question in detail, due to the modest capacitances identified, but found no evidence of such a problem. The high temperature of the carbonisation, the very small distances involved, and the network forming ability of the nanotubes are likely to limit the influence of the amorphous carbon resistance. Note that in the sizing (second) approach, the nanotubes lie predominantly parallel to the carbon fibre surface, due to capillarity effects, implying a large number of contact points per nanotube.

Jet-Sing Lee asked: I understand that for your material its structure is important for applications where weight is important. You show a capacitance of ~15 F g^{-1} for your CAG material. Is this not still lower than reported activated carbons that can reach up to ~200 F g^{-1}, therefore less can be used? You say an aim of your material is to address mechanical issues but from my knowledge, carbon and other materials that operate by electric-double layer capacitance are already robust with stabilities of up to 500,000 cycles.

You also mentioned that you are now modifying your CAG with conducting polymers to enhance its performance. How stable are these new composites, as polymeric supercapacitors that work *via* pseudocapacitance are known to have low stabilities?

Milo Shaffer answered: The capacitance mentioned only appears lower because it is normalised to the combined mass of the carbon fibre and CAG (hybrid) material. The capacitance of the CAG normalised to its own mass is within the usual range for these materials. The carbon fibres contribute to current collection, but little to the capacitance. The concept uses the necessary matrix volume within the composite to accommodate CAG and hence electrical energy storage, at the same time as maintaining the structural performance of the primary fibres.

The pseudocapacitive components have the same advantages and limitations as in conventional supercapacitors, but can contribute to the overall performance of this novel multifunctional system. The motivation, here, is not at all the simple mechanical robustness (stability) of a pure electrochemical electrode. The point is that these multifunctional composite devices may potentially have sufficient mechanical performance to be truly structural, in the sense of carrying the load needed, for example, to act as a wing spar, car door, or aerofoil.

Robert Dryfe opened the discussion of the paper by George Chen: Please could you clarify the motivation behind the study. If the objective is electrochemical energy storage, the utility of the approach seems limited because the efficiency (based on the anodic/cathodic charge ratios given in Table 2) seems low. Repeated cycling would yield <1% charge recovery within 10 cycles.

George Zheng Chen responded: The data in Table 2 are derived from cyclic voltammograms recorded at different potential scan rates, and show clearly that the Q^+/Q^- ratio increases with decreasing scan rate. This suggests some kinetic difficulties. In the text following Table 2, we describe that if the potential is held at the cathodic end for a short while (*e.g.* 10 s), the Q^+/Q^- ratio can increases to close to 1. In practice, the charging and discharging are more likely carried out under a steady state, and we can expect improvement in charge efficiency upon optimisation of process variables. In fact, Table 3 contains a Q^+/Q^- ratio of 0.83 when the deposition and dissolution were carried out under a constant current of 150 mA. In addition to fundamental understanding, there are however engineering issues in terms of what type of electrode/cell designs will be optimal to avoid the deposited carbon falling off the electrode when the deposit becomes thicker. Other factors, such as formation of CO gas, may also reduce the charge efficiency. I shall point out that for seasonal (and long term) energy storage, carbon presents many merits over other options, particularly its high energy density and convenience for transport and storage. Cost is another important factor affecting commercial development which favours the molten salt approach, not to mention that there is the tax benefit for capture and utilisation of CO_2.

Robert Dryfe continued: Following on from the previous question, do you have any idea of where the 'missing' charge is going, *i.e.* what are the side reactions that lead to the loss of charge recovery?

George Zheng Chen answered: The side reactions include the formation of CO gas, and possibly alkali metals, both of which may leave the cathode and are not recoverable. The deposited carbon, if sufficiently thick, may also fall off the cathode, accounting partly for the missing charge.

Matteo Duca said: Since you observe that "carbon deposition occurs on the deposited carbon more efficiently than on a freshly prepared metal electrode" (line 35, page 11 of your paper), have you tried using graphite as a cathode instead of metals?

George Zheng Chen replied: We have tested using graphite as the cathode material, but it suffered from lithium intercalation which agrees with what was reported previously.[1,2]

1. G. Z. Chen *et al.*, *High Temp. Mater. Processes*, 1998, **2**, 459–469.
2. G. Z. Chen *et al.*, *J. Electroanal. Chem.*, 1998, **446**, 1–6.

Yury Gogotsi asked: Is there any fundamental reason why you can't use a lower temperature for electrochemical CO_2 conversion?

George Zheng Chen responded: There are indeed fundamental reasons. From thermodynamic analysis, electrochemical reduction of CO_2 should proceed at room temperature at a fairly low voltage ($\Delta G = 394.363$ kJ mol^{-1} $\Rightarrow$ $\Delta E = 1.022$ V at 273 K).

However, the mechanistic and kinetic factors play more important roles here. It is difficult to find a molecular or ionic acceptor that can work in low temperature electrolytes to capture and stabilise the O^{2-} ion that is discharged from direct reduction of CO_2, and then transport the O^{2-} ion for discharging at the anode.

Nevertheless, I shall mention that working at elevated temperatures for electrolysis is not a big technical or energy concern because the industry is familiar with using molten salts at 1173–1273 K for the extraction of aluminium which has always been a profitable business.

Martin Jonsson-Niedziolka enquired: When calculating efficiency, what part of the energy consumption goes into maintaining the high temperature needed for the process?

George Zheng Chen explained: Energy consumption for supplying the electrolysis current is far greater than that for heating the molten salt to maintain the working temperature. Heat loss is obviously inevitable, but can be minimised *via* the use of appropriate thermal insulation materials and also reactor design. Particularly, the Joule heat resulting from the electrolysis current passing through the molten salt should be considered in balancing the heat loss. In our laboratory, the thermal energy loss from our furnace, which is far from optimal for heat insulation, has been estimated to be about 5% or less of the total energy consumption. On the other hand, I shall mention that an important prospect of our work is to utilise electricity and heat that can be generated from sunlight or other renewable sources, or that from the excess generation of existing power plants.

Manuel Alvarez-Guerra commented: I think these types of approaches for CO_2 valorisation can be particularly interesting as a future option for using excess energy from intermittent renewable sources (*e.g.* solar or wind energy), allowing the balancing of fluctuations in production and demand cycles, which is usually considered to be essential for intermittent renewable energies to be used on a large scale.

Manuel Alvarez-Guerra asked: In your paper you have some results of the electrolysis under different CO_2 partial pressures using mild steel as cathode. Did you find important changes in the performance and deposition rates depending on the nature of the working electrode used?

George Zheng Chen answered: In the early stage of carbon deposition, it depends strongly on the nature of the working electrode, except for graphite which suffers from lithium intercalation. However, after the electrode surface is fully covered by carbon, the deposition become very much independent of the substrate materials. Also, we have not found notable differences between samples obtained from using different cathode materials, but we have not analysed the carbon layer immediately next to the surface of the cathode.

Manuel Alvarez-Guerra continued: You carried out the experiments in a two electrode undivided cell. Did you try using a divided cell with a membrane?

George Zheng Chen replied: Thank you for your question. The use of a membrane or separator in our cell has been considered, but this option increases the cell resistance and hence energy consumption. It also remains challenging in terms of membrane materials. Ideally, the membrane should be made from an ionic conductor that is stable in molten salts at elevated temperatures. However, all known high temperature ion conducting materials (*e.g.* zirconium stabilised yttrium oxide or sodium beta alumina) are ceramic types and vulnerable to thermal shock and molten salt attack. Porous ceramic materials are an alternative, but these would be less effective in terms of separation of the anolyte and anodic products from the catholyte and cathodic product. However, there are still engineering innovations that can help improve the process efficiency and product quality, and these are under investigation in our laboratory and possibly others.

Héctor Abruña remarked: Data comes from the automotive industry like Ford and GM.

George Zheng Chen answered: Thank you. I have checked a few recent review articles[1] on PEMFCs, and it seems that in some laboratories, the power density has reached beyond 1 W per sq. cm, which implies a current density of over 2 A per sq. cm at a cell voltage of 0.5 V.

1 *e.g.* H. Zhang and P.K. Shen, *Chem. Rev.*, 2012, **112**, 2780–2832.

Domnik Bayer said: Your paper presenting a direct carbon energy storage is appealing. This implies the reversible reduction of CO_2 and re-oxidation of deposited carbon at a reasonable efficiency. My question relates to the system cost in terms of power specific (€ per kW installed) and energy specific costs (€ per kWh stored). These numbers would make comparison with existing storage solutions easier (*e.g.* battery technologies). Carbon production from CO_2 might also be interesting, however, if there are no additional special features of the electrochemically generated carbon, it will pricewise compete against standard carbon black which is available at a sales price of ~1000 US$ per metric ton.

George Zheng Chen responded: We have not yet constructed a full device (*e.g.* direct carbon fuel cell) to test and evaluate the energy efficiency of the carbon in great confidence. For discussion, we have estimated the cost of our electrolytic carbon to be $3–5 per kg-C, which includes the electricity, heat and post-electrolysis processing costs based on our laboratory findings, but not yet the benefit from the carbon tax. Also, for energy storage, the cost for post-electrolysis processing may not be needed. With further research and technical improvement, the process efficiency can increase further, and the cost may be driven below $3 per kg-C. The energy content in pure carbon is about 9.2 kWh per kg-C (enthalpy change of carbon combustion at 600 °C), whilst the direct carbon fuel cell can achieve over 70% in energy efficiency, leading to a delivery of 6.4 kWh per kg-C. Thus, $0.5 per kWh in energy stored is not impossible. For material uses, there are many options such as electrode materials for supercapacitors and batteries. Specifically, the molten salt approach reduces the carbon footprint, whilst all existing processes for production of various carbon materials strain the environment with some levels of carbon emission.

François Béguin stated: Simply based on the energy consumption of electrolysis, you estimate the cost of your carbon to be less than $5 per kg, and you compare it to the current market price of activated carbon for supercapacitors ($20 to 40 per kg). The relatively high cost of supercapacitor carbons is due to the post-treatments required to reach a high purity. Therefore, the estimated cost of your material should include all the washing treatments which are necessary to reach the same purity as for the supercapacitor carbons.

George Zheng Chen responded: I fully appreciate your comment. Our estimation has been based on the preparation procedures and materials/chemicals used in our laboratory, including washing with acid, which is of course subject to correction when the process is scaled up for commercial purposes. Our intention to report our preliminary estimate is to provide a starting point for more comprehensive cost analyses so that potential stakeholders, researchers and business managers alike, can have a scale for comparison.

Chi-Chang Hu said: Is it possible to control the microstructures (*e.g.* porosity, crystallinity, doping, *etc.*) of carbon by means of electrochemical deposition?

George Zheng Chen replied: Your question is also mine. According to our results obtained so far, there is some degree of controllability *via* manipulating the temperature and voltage (or the cathode potential) as we have described in ref. 1. I hope our ongoing research can provide more accurate answers, at least in terms of the controlled formation of a particular structure.

1 H. V. Ijije, C. Sun and G. Z. Chen, *Carbon*, 2014, **73**, 163–174.

Laurence Hardwick asked: Have you investigated what type or form of carbon you produce electrochemically from CO_2 – say for example with Raman?

George Zheng Chen answered: We have carried out some basic spectroscopic and microscopic characterisations of the electro-deposited carbon, and reported the findings in ref. 11 of our paper. However, we have not yet analysed our samples by Raman, although you may find ref. 3 and 10 of the paper to be relevant.

Milo Shaffer remarked: What is the structure of the carbon deposit? Does its integrity, structure, or electrical conductivity/contact resistance limit the thickness of coating that might be deposited effectively? From the images that you showed, it appears that the deposit is porous, containing either voids due to gas evolution, or trapped electrolyte. If graphitic carbons form, is (undesirable) intercalation favoured?

George Zheng Chen responded: The deposited carbon layer is fairly porous and formed by packed nano-particulates of carbon in various morphologies. The mechanism for these nano-particulates to be fairly stably attached to the electrode surface is not clear, and deserves further investigation. My initial explanation is that it is related to the insolubility of carbon in the molten salt. Thus, once formed, the carbon nano-particulates would precipitate to the nearest solid surface such as that of the cathode. It is not to say that carbon deposition would not produce a crystalline structure *via* the nucleation and growth mechanism, but the fast deposition rate, by product (CO gas and alkali metal) formation and gas bubbling on the nearby anode, could all disturb the growth of a crystalline deposit. The contact resistance between these nano-particulates, if any, is inevitable, but is also very well compensated by the increase of the overall surface area when more carbon is deposited, and hence does not affect the growth of the deposit. As for intercalation, it will surely happen to graphitic carbon, as we know it is possible to produce carbon nanotubes by intercalation of lithium or sodium into graphite in molten salts.[1] Although intercalation may not occur to amorphous carbon or the carbon nano-particulates, alkali carbides may still form, although this was found to be insignificant.

1 *cf.* (a) Chen *et al.*, *High Temp. Mater. Processes*, 1998, **2**, 459–469; (b) Chen *at al.*, *J. Electroanal. Chem.*, 1998, **446**, 1–6.

Faraday Discussions

PAPER

Effects of structural disorder and surface chemistry on electric conductivity and capacitance of porous carbon electrodes†

Boris Dyatkin and Yury Gogotsi*

Received 21st March 2014, Accepted 6th May 2014
DOI: 10.1039/c4fd00048j

This article reports on changes in electric double layer charge storage capacity as a function of surface chemistry and graphitic structure of porous carbon electrodes. By subjecting 20 nm to 2.0 μm sized carbide-derived carbons (CDCs) synthesized at 800 °C to high-temperature vacuum annealing at 700–1800 °C, we produce three-dimensional internal surface architectures with similar pore sizes and volumes but divergent surface chemistry and wall graphitization. Annealing increases carbon ordering and selectively removes functional groups, and both transformations affect conductivity and wettability. Contrary to an expected increase in gravimetric capacitance, we demonstrate no increases in charge storage despite increased conductivity and pore accessibility. At the same time, annealing improves the charge/discharge rates in EMIm-TFSI ionic liquid electrolyte. The annealing process eliminates faradaic reactions that limit the voltage window, but potentially accelerates catalytic breakdown of the ions themselves. We therefore corroborate the theory that surface groups and defects in the graphitic structure act as dopants that allow facile movement of ions into pores, improve screening in the superionic state, and affect the quantum capacitance contribution from the carbon structure.

1 Introduction

Electric double layer capacitors (EDLCs), also known as supercapacitors and electrochemical capacitors (ECs), are attracting substantial interest as electrical energy storage solutions for hybrid electric vehicles and other systems.[1] The primary charge storage harvesting mechanism, which relies on the electrosorption of ions from an electrolyte onto a porous electrode,[2] offers essential advantages such as extended cyclability, stable operation in a wide temperature range,[3] versatile geometric configuration, and scalability.[4] Porous carbon

A.J. Drexel Nanomaterials Institute & Department of Materials Science and Engineering, 3141 Chestnut Street, Philadelphia, PA 19104, USA. E-mail: gogotsi@drexel.edu; Fax: +1 2158951934; Tel: +1 2158956446

† Electronic supplementary information (ESI) available: Additional materials characterization, electrochemical performance, and quantum capacitance discussions. See DOI: 10.1039/c4fd00048j

electrodes can provide a tuneable, electrically conductive structure with over 3000 $m^2\ g^{-1}$ in accessible surface area,[5] extending their gravimetric power densities above 10 kW kg^{-1}.[6] However, the small operating potential of traditional supercapacitors (<2.7 V) has so far capped their energy densities under 10 W h kg^{-1},[1] hindering their implementation in automotive and grid storage applications. Most fundamental materials research in the field has therefore focused on tuning the carbon structure and electrolyte composition to maximize capacitance and minimize electrolyte breakdown at high potentials.

Recent research efforts have detailed significant progress optimizing porous carbon structures to accommodate ions. Previous work has repeatedly shown that a tailored pore structure that matches its diameter with the ion that it confines maximizes capacitance due to ion desolvation.[7] Diameter deviations of ± 0.05 nm have been shown to drastically (40%+) affect capacitance.[8] Moreover, molecular simulations predict a superionic state[9] of conductive pore walls that, under an applied critical potential, effectively screen similar charges and eject counterions; this process further maximizes capacitance.[10] While single-layer graphene exhibits a theoretical 2630 $m^2\ g^{-1}$ specific surface area (SSA),[11] it cannot offer a sufficient aerial energy density that would make it practical as a standalone material in most power applications.[12] Conversely, activated and carbide-derived carbons (CDCs)[13] provide 1000–3000 $m^2\ g^{-1}$ SSA and can be easily scaled up and integrated into various device-specific applications using configurations such as polymer-bonded flexible films.[14]

CDC materials, which are synthesized by halogen etching of various carbides (SiC, TiC, B_4C, Ti_3AlC_2, *etc.*),[15] produce highly tuneable porous structures that can be tailored by varying Cl_2 treatment temperatures (300–1200 °C) and precursor materials.[16] The resulting structures exhibit robust surface chemistry, including C–Cl, C–H, and C–NH_2 groups present as a result of halogen treatment and subsequent H_2/NH_3 annealing.[17] Various hydroxyl and carbonyl groups originate from exposure of the material to air.[18] Additionally, structural defects and lattice strain points are present[19] in the carbon architecture. The structure is composed of a mostly amorphous carbon, with predominant short-range sp^2 bonding.[20] The material's graphitization increases with synthesis temperature, since thermally-induced mobility facilitates growth of crystallites in the [100] direction and enhances stacking along the [002] direction.[21] The CDC's electric conductivity improves as a result: as delocalized π-bonding becomes more prevalent, the graphitic segments' size and concentration increase and push the material's electron transport mode from semiconducting to metallic behaviour.[22]

While CDC materials have previously demonstrated gravimetric capacitances exceeding 100 F g^{-1},[5] and novel organic electrolytes and room-temperature ionic liquids (RTILs) have expanded operating voltage limits,[23] the fundamental relations between carbon structure and corresponding ion behaviour are not fully understood. Standard CDC synthesis convolutes pore diameter, graphitization, and surface chemistry, making it difficult to decouple the standalone effects of each. The thermodynamically limited 4.0–5.6 V windows[24] of organic electrolytes and RTILs are reduced to 2.3–2.7 V due to potential-induced breakdown reactions of molecules with functional groups on pore surfaces.[25] The reactions also convolute pure double layer charge accumulation with pseudocapacitive contributions from surface redox processes.[26] While defunctionalization increases hydrophobicity,[24] it is unclear how this affects the ion dynamics, the carbon–

electrolyte interface,[27] and subsequent charge transfer resistance. This critical knowledge can optimize the dynamics of ions with divergent size, hydrophobicity, and charge distributions ($EMIm^+$, $TFSI^-$, NEt_4^+, BF_4^-, *etc.*) entering and exiting narrow pores.[28]

Recent mathematical modelling efforts predict that the pores' fundamental structural properties significantly impact interactions with electrolytes and subsequently affect double layer charge storage.[29] The traditional approach has simulated their behaviour according to the Stern model;[30] however, coupled with an energetically-favourable confinement in pores,[31] ion configuration in internal surface carbons is impacted by surface chemistry and roughness of the latter. The presence of defects (functional groups and lattice deformations) disrupts sp^2-inherent π bonding symmetry, magnifying the p_z-controlled density of states and subsequently enhancing the carbon material's quantum capacitance (C_Q) by over 300%.[32] Impurity-driven enhanced C_Q reduces the screening length of similarly charged ions from 30 nm to ion-sized dimensions, while pristine graphene requires 0.66 nm to isolate ions in the superionic state.[32] At the same time, excessive n- and p-type dopant impurities[33] interfere with ballistic conductivity along the graphitic plane and may maximize resistance losses in the system.[34] While maximization of C_Q is important for electrodes composed of few-layer graphene, its impact on total capacitance is diminished when compared to double-layer induced capacitance (C_{DL}), with an increase in the number of stacked graphite layers.[35] Furthermore, it is unclear how these simulations, which model carbon as an infinite graphene surface, translate to a polycrystalline material that best describes the CDC structure. In particular, narrow graphene ribbons, which can be assumed to make up the majority of pore surfaces, have lower intrinsic carrier concentrations and impeded electron transport behaviour.[36] The correlation of these simulations with 3-dimensional porous architectures is complicated by divergent pore wall thicknesses, pore tortuosity, and the heterogeneous nature of surface morphology. The latter has an additional (but difficult to empirically quantify) factor of enhancing the electric field at points of surface edges and extreme curvature, potentially boosting capacitance by over 250%.[37] In addition to changing the quantum capacitance contribution,[32] functional groups with different surface charges also affect the ions' geometrical orientation and interface with the pore surface,[38] impacting charge storage density.[39]

Similar fundamental problems impede the integration of novel carbons as high-performance electrodes in electroanalytical chemistry. Glassy carbon (GC) electrodes have been traditionally used to determine redox potentials, reaction kinetics, and concentrations of various aqueous systems (As/As^{3+}, NO_2^-, tryptophan, ferrocene, *etc.*).[40] Peak currents have been shown to correlate with the active species' concentrations, with theoretical limits typically precluding measurements below 10^{-7} M.[40] Increasing accessible surface areas significantly magnifies these currents (10^3 fold increase from μA to mA range).[40] Various electrodes composed of carbon functionalized graphene,[41] highly ordered pyrolytic graphite (HOPG),[42] and single walled carbon nanotubes (SWCNTs)[43] have been developed to transition from a low-SSA semi-infinite planar diffusion behaviour to a thin layer cell with convergent diffusion.[40] These materials are commonly functionalized with Au nanoparticles to maximize electrical conductivity, limit overpotential reactions, and reduce mass transport limitations.[42] GC electrodes typically face problems with fouling and irreversibility, as 2nd cycle performance decreases current intensity by 35%.[40] Graphitic electrodes have shown enhanced electron transport rates, improved cyclabilities, and higher sensitivities to reactions.[42] This further underscores the need for optimized porous materials that are electrochemically stable and contribute no additional redox peaks to highly sensitive systems. Depending on the electrochemical process in question (Nernstian, quasi-reversible, or irreversible),[41] different surface areas may be required from the electrode material to ensure that redox peaks can be properly identified.

The objective of our work is to partially deconvolute the key properties of double layer charge storage and provide more fundamental information regarding ion dynamics, charge storage density, and electrochemical stability in confined porous environments. We rely on high-temperature (700–1800 °C) vacuum annealing (10^{-3} Pa) to transform the surface chemistry and carbon ordering of pore surfaces. This technique has been previously effectively used to modify structural ordering of the material and its surface area without compromising narrow pore size distributions.[44] The procedure allows us to develop passivated, stable 3-dimensional porous architectures with different degrees of functionalization and graphitization to understand the effects of those parameters on key charge storage characteristics. By correlating key structural parameters and corresponding electrochemical behaviour, we can compare materials with different defect concentrations and derive the effects of hydrophobicity, unstable oxygen species, and graphitization-induced conductivity on the dynamics and the double layer configuration of the electrolyte. We rely on different CDC particle sizes to demonstrate effects of vacuum annealing at different scales and pore lengths. Our goal is to provide key information regarding the "ground state" configuration that maximally resembles the structures implemented in simulated models in order to correlate experimental evidence with predicted behaviour.

2 Experimental approach

Synthesis and annealing

Carbide precursors. TiC (Alfa Aesar, 99.5% purity, 2 μm diameter flakes) was selected for the microparticle system.[45] SiC (MTI crystal, 99% purity, <30 nm diameter) was selected for the nanoparticle system.[46]

Halogen etching. Carbide powders were loaded into quartz boats and placed in a tube furnace. Using a previously described procedure,[47] the samples were purged with Ar, heated under Ar to 800 °C, and subjected to flowing Cl_2 gas for 6 h. They were subsequently cooled to 600 °C under Ar and treated with H_2 for 3 h.[17] After removal from the furnace, samples were dried at 80 °C under a low vacuum (0.5 Pa) for 24 h to remove any trapped adsorbed species.

Vacuum annealing. CDC materials were loaded into graphite crucibles and placed in a graphite vacuum furnace (Solar Atmospheres).[48] The furnace was outgassed for 24 h at room temperature, during which time the pressure decreased to approximately 10^{-3} Pa. The samples were subsequently heated at a rate of 10 °C min^{-1} to the following temperatures: 700 °C, 1200 °C, 1400 °C, and 1800 °C.[44] They were held at those conditions for 8 h each and subsequently cooled to room temperatures under the same low pressure.

Materials characterization

Gas sorption and porosimetry. Adsorption/desorption isotherms were collected using N_2 gas at −196 °C using a Quadrasorb (Quantachrome Instruments). The Brunauer–Emmett–Teller (BET) method was used to calculate the specific surface area in the P/P_0 0.05–0.10 adsorption range.[49] Quenched Solid Density Functional Theory (QSDFT) model for slit-shaped carbon pores was used to calculate SSA, pore volume, and the pore size distribution.[50] DFT analyses were performed using only the adsorption branches. QuadraWin software was used for all calculations.

Structural characterization. Raman spectra of the materials were collected using a Renishaw inVia Spectrometer and a 514 nm Ar laser excitation (5% power, ∼5 μm beam diameter). Spectra were deconvoluted using an 1800 grooves cm^{-1} grating. WIRE 3.2 (Renishaw Instruments) software was used for peak fitting. Transmission Electron Microscopy (TEM) was done using a JEM 2100 (JEOL Instruments) with a 200 kV accelerating voltage. Thermogravimetric analysis (TGA) was carried out using a Q50 TGA Analyzer (TA Instruments); samples were dried at 150 °C for 1 h (in an Al_2O_3 crucible on a Pt pan) and heated at a 2 °C min^{-1} rate in a flowing air atmosphere to 750 °C. WIRE 3.2 software was used for dM/dT peak deconvolution.

Surface chemistry. X-Ray Photoelectron Spectroscopy (XPS) analysis was performed using a Kratos AXIS surface analysis system (Ultra DLD). CasaXPS Version 2.3.16 RP 1.6 software was used to calculate elemental analysis and deconvolute functional group bands. Dynamic Water Vapour Sorption (DVS) was performed using a Q5000 SA Analyzer (TA Instruments); samples were placed in Pt-coated pans and the relative humidity of the chamber was increased while the temperature was held at a constant 25 °C. Inert TGA analysis was carried out using similar parameters as the air TGA tests, except that an Ar atmosphere was used to heat the samples to 800 °C and a 2 h preliminary purge time was used.

Intrinsic material conductivity. CDC particles were rolled into films typical of those used for supercapacitor electrodes. Powders were mixed with PTFE (60 wt% solution in water, Sigma Aldrich) polymer in a 95 : 5 ratio, and the slurry was blended using ethanol at 25 °C. The subsequent films were rolled to maintain a 6.7 mg cm^{-2} mass loading for each material, producing 95 μm thick microparticle films and 150 μm thick nanoparticle films. For 4-point probe measurements, 4 cm × 4 cm pieces of each film were cut and a 4-point probe (2.09 mm separation) was used in conjunction with a DC power source (Keithley Instruments). Intrinsic conductivity derived from applied voltage (V), measured current (I), and film thickness (t) was calculated using the following equation:

$$\sigma = \frac{I\ln(2)}{V\pi\, t} \tag{1}$$

Electrochemical testing

Electrode setup. 1.5 cm × 1.5 cm squares (2.25 cm^2) were cut from main electrode films, resulting in identical masses for working (WE) and counter (CE) electrodes. After 48 h drying at 80 °C under 0.5 Pa, the samples were placed in an Ar glovebox (Vacuum Atmospheres, <1 ppm content of H_2O and O_2). The cells were assembled into a standard "sandwich" configuration,[51] using 2 layers of separator (Celgard) and a carbon-coated aluminium current collector (Exopack). Ag/AgCl wire was implemented as a quasi-reference electrode.[52] C-Clamps were used to maximize compression between electrodes and current collectors.

Electrochemical testing. All experiments were conducted at room temperature (~25 °C). 1-Ethyl-3-methylimidazolium bis(trifluoromethylsulfonyl)imide (EMIm-TFSI, IoLiTec Ionic Liquids Technologies Inc.) room temperature ionic liquid was used in a solvent-free state. A VMP3 Potentiostat (Bio-Logic Instruments) was used for cyclic voltammetry (CV), galvanostatic cycling (GCPL), electrochemical impedance spectroscopy (EIS), and square wave chronoamperometry (CA). CV sweep rates varied between 2 mV s^{-1} and 1000 mV s^{-1}, while galvanostatic cycling was conducted to test various current densities between 0.1 A g^{-1} and 6.0 A g^{-1}. Values were normalized by mass and accessible SSA (derived from gas sorption). Samples were held at open circuit potential (OCP) for 1 h prior to the beginning of each experiment series, and an OCP 10 minute segment was used between different experiments. The same sequence of tests was used for each material. From cyclic voltammetry, capacitance was derived from the voltage window (V_w), measured current, (I), sweep rate (dV), working electrode mass (m_{WE}) and vertex potentials (V_0 and V_f), and calculated using the following equation:

$$C_{sp} = \int_{V_0}^{V_f} \frac{2I}{m_{WE}\, V_w} dV \tag{2}$$

Capacitance derived from galvanostatic cycling was derived from the slope of the voltage line (dV/dt) using the following equation:

$$C_{sp} = \frac{2I}{m_{WE}\, dV/dt} \tag{3}$$

Discharge segments of each cycle were used for calculations for both techniques. Impedance spectroscopy was collected using a 10 mV dampening oscillation in the 200 kHz to 0.01 Hz range. Series resistance (R_s) was derived for the value of $Z'(\omega)$ at which $Z''(\omega) = 0$. Time constant was derived from the −45° impedance angle frequency. Ionic resistance was derived from the $Z'(\omega)$ range between the 1 Hz range and the −45° Z angle (mid-frequency regime). The Randles–Sevcik coefficient (x) was calculated from both CV and GCPL techniques using the following equation relating peak current (I), charge transfer coefficient (n), Faraday constant (F), SSA (A), molar IL concentration (c), diffusion constant (D), temperature (T), gas constant (R), and the scan rate (ν):

$$I = 0.4463(nF)^{1.5} Ac\left(\frac{D}{RT}\right)^{0.5} \nu^{x} \tag{4}$$

Chronoamperometry was conducted by holding a cell at a given potential for 30 min, measuring current, and subtracting the current measured at 0.0 V using the same conditions.

3 Results and discussion

Structural transformations

The main processes that occurred during vacuum annealing were, in order of prevalence, removal of functional groups from pore surfaces, graphitic reorganization and increase in size of ordered sp^2 domains, and limited pore restructuring. The procedure removed 10–15% of the mass of annealed CDC, with the amount of lost material proportional to the annealing temperature. It is expected that this loss accounts for chemical groups, attached carbons, and any species chemisorbed or trapped in tight pores.

Pore structure changes. The TiC–CDC microparticles and SiC–CDC nanoparticles underwent very similar pore structure transformations under the influence of vacuum annealing. The initial materials had SSA values of 1682 m^2 g^{-1} and 2001 m^2 g^{-1}, respectively, and pore distributions showcase a predominant 0.61 nm pore diameter for TiC–CDC and 0.85 nm for SiC–CDC. When annealed at 700 °C, the SSA of the microparticles increased by 15%, while the pore diameter increased marginally to 0.67 nm. Higher annealing temperatures decreased SSA by 19% (1371 m^2 g^{-1} after a 1400 °C anneal), with the micropore distribution remaining relatively unchanged and an increase in mesoporosity. Fig. 1 shows the pore size distributions and changes in accessible SSA and pore volume. The SiC–CDC nanoparticles reported similar responses to annealing temperatures: the SSA increased by 55% to 3100 m^2 g^{-1} (1.01 nm pores) after 700 °C treatment and decreased by 30% to 1401 m^2 g^{-1} (0.79 nm pores) after a 1400 °C anneal. While the systems (synthesized at 800 °C) in their initial, non-annealed states exhibited similar structural and chemical properties, the effects of high temperature vacuum treatment differed for materials with divergent initial structures.[44] Low-temperature annealing removed functional groups and attached carbon atoms from outer pore walls, opening closed channels and increasing SSA, while the high-temperature restructuring merges some pores, collapsing small branches and decreasing the cumulative surface area. The ESI† includes a comprehensive summary of annealing effects on porosity (Table S1†) and demonstrates effects of 1400 °C annealing on TiC–CDC synthesized at 1000 °C, for comparison (Fig. S1(a) in ESI†). For all annealed materials, the micropore structure remained generally intact. The increase in mesoporosity for materials annealed at high temperatures was expected to enhance the mobility of ions during charging and discharging,[53] improving rate handling abilities.

Changes in graphitic structure. The high-temperature annealing resulted in increased graphitization, as sp^2 domains became larger and more ordered. However, this was only observed for materials annealed at 1100 °C and above. Treatment at 700 °C removed surface functional groups, yet did not provide sufficient energy-dependent mobility for the carbon atoms to induce significant graphitization. Raman spectroscopy (Fig. 2) shows a temperature-dependent narrowing of both D and G bands, as well as the emergence of the resonant 2D and D + G bands. These observations confirm the expected increase in graphitic ordering.[54] The material retains a generally amorphous carbon structure before and after annealing, and its three-dimensional architecture (featuring small interlocked sp^2 domains) impedes the formation of long-range, defect-free graphene sheets on pore wall surfaces. As a result, the full widths at half maxima (FWHM) of the D band are more appropriate than I_D/I_G ratios (compared in Fig. S2 in ESI†) to describe graphitization.[21] The appearance of the D′ band at the ~1610 cm^{-1} for both materials after an 1800 °C anneal suggests the eventual collapse of the CDC particles to non-porous, polycrystalline structures; this is described well by the porosimetry data from Fig. 1.

Air TGA analysis (Fig. S3 in ESI†) shows an increase in oxidation temperature for samples annealed at higher temperatures, corroborating the increased

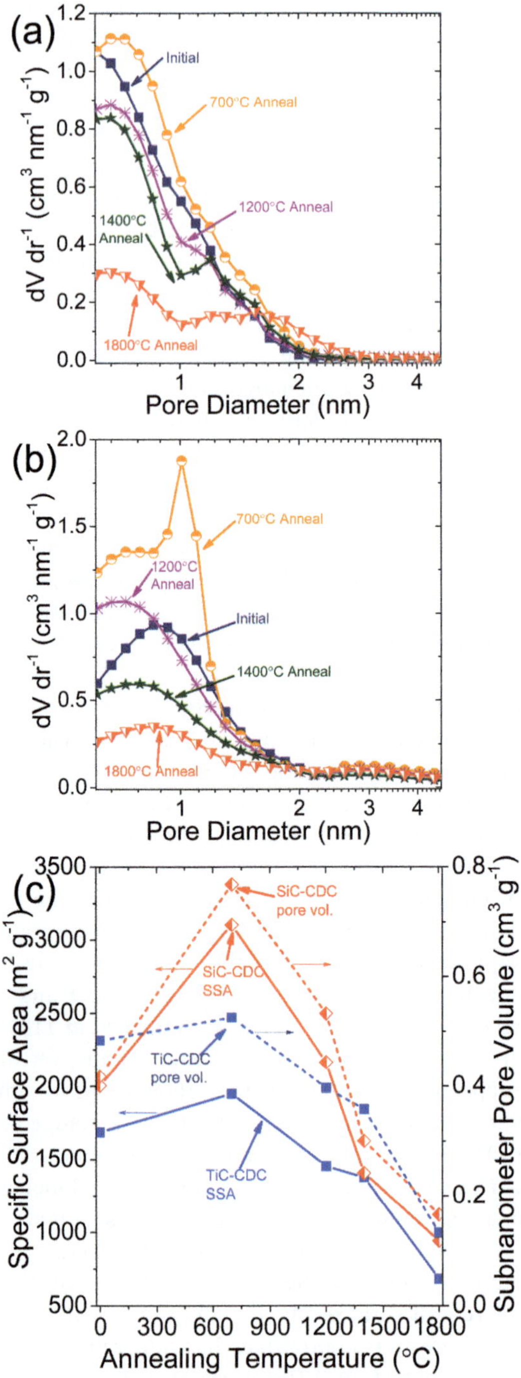

Fig. 1 (a) Pore size distributions for TiC–CDC microparticles in their initial state and annealed at 700 °C, 1200 °C, 1400 °C, and 1800 °C. (b) Pore size distributions for SiC–CDC nanoparticles annealed using the same conditions. Similar trends are observed. (c) Changes in surface area and cumulative pore volume from all pores <1 nm in diameter for the two systems.

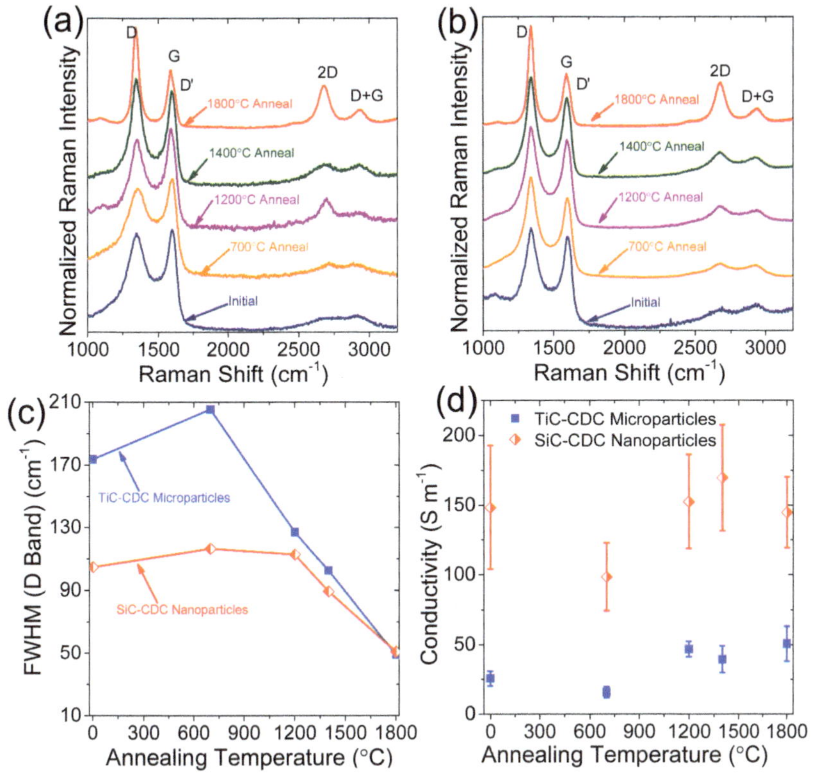

Fig. 2 Raman spectroscopy offset spectra for the (a) microparticle and (b) nanoparticle systems. As annealing temperature increases, the resonant modes become more prominent and (c) the full width half maxima of the main bands become narrower, indicating increases in ordering. Low-temperature annealing (700 °C) appears to slightly decrease ordering, as chemical species removal occurs without giving carbon thermal energy to re-arrange itself into sp^2 domains. (d) Effects of annealing on electric conductivity of PTFE-bound electrode films. Resistance values measured using the four-point probe were very small for nanoparticles, so deviations translated to more significant distributions than the calculated conductivity values for microparticles.

prevalence of ordered carbon. Analysis of the FWHM of the oxidation profiles showed the broadest oxidation profile for the 700 °C samples, further confirming a large dispersity and a generally disordered structure of both microparticles and nanoparticles annealed at that temperature. Previous experiments correlated narrower d*M*/d*T* peaks with a larger structural defect concentration in MWCNTs (inter-wall faults, broken walls, *etc.*).[55] However, the narrowing of the peaks shown for CDCs, which have a more heterogeneous structure, more likely corresponds with an increase in uniformity in the annealed materials. The G-band shifted to the left by 7 cm^{-1} (for microparticles) and 5 cm^{-1} (for nanoparticles) after an 1800 °C anneal (Fig. S2 in ESI†). This effect has been correlated with an increase in p doping (or minimization of n doping) of carbon's electron density of states in graphene before,[56] but it is unclear whether these minor shifts can be assigned to such transformations.

The effects of annealing and increased graphitization on intrinsic electric conductivity are shown in Fig. 2(d). Conductivity generally increases with annealing temperature for both CDC materials, with nanoparticles generally exhibiting greater electron transport due to improved interparticle connectivity.[57] Both materials exhibited a 33–38% conductivity drop after a 700 °C anneal, caused in part by a lack of increase in ordering and a significant increase in pore volume, which impedes electron transfer.[58] Annealing at 1400 °C increased conductivity of TiC–CDC by 14 S cm^{-1} and SiC–CDC by 21 S cm^{-1}. The 16–50 S m^{-1} microparticle conductivities and 98–168 S m^{-1} nanoparticle conductivities were smaller than the values reported for monolithic, chemically reduced graphene oxide with high sp^2 ordering.[59]

Surface chemistry changes. Increasing the annealing temperature defunctionalized the surfaces of the CDC materials. The elemental analysis of the materials, which is shown in Table 1, shows that elements such as oxygen, nitrogen, and chlorine steadily decreased in atomic % after the 700 °C anneal and were almost completely eliminated after the 1400 °C treatment. The SiC–CDC material, which had a lower relative amount of starting surface chemistry, did not experience such drastic defunctionalization, which could be related to both its higher SSA (increasing the prevalence of contaminants adsorbed from air) and its particle aspect ratio, which could facilitate re-functionalization. TGA analysis in an Ar environment (Fig. S3 in ESI†) shows that the functional group content, which constituted 7–9% of the total CDC mass in its initial state, decreased to <3% after the high-temperature treatment.

It should be noted that neither analysis can accurately quantify the hydrogen content, which is assumed to be present on the surfaces of as-produced samples. Functional group analysis revealed predominantly C–OH, C=O, and C–Cl bonds, with trace amounts of acid, ether, and C–O–Cl groups. Chlorine was found only on the insides of particles (after Ar sputtering), so its presence could be limited to the XPS-induced opening of previously closed channels. For both systems, the materials did not become noticeably functionalized during exposure to air after annealing, indicating that the treatment produced passivated and stable structures.

The defunctionalization process influenced the hydrophobicity of the materials, as shown in Fig. S4 in ESI.† As the number of surface species decreased,

Table 1 Atomic percentages of key chemical species identified for the two systems using X-ray Photoelectron Spectroscopy (XPS). Entries indicating a range of values signify amounts found outside of particles (before Ar sputtering) and inside (after sputtering away several nm)

	TiC–CDC microparticles			SiC–CDC nanoparticles		
	Initial CDC	700 °C annealed	1400 °C annealed	Initial CDC	700 °C annealed	1400 °C annealed
Carbon	92.3	96.1	99.7	97.9	98.1	97.4
Oxygen	4.1	1.4	0.0	1.7	1.4	1.7
Nitrogen	2.3	2.1	0.0	0.0	0.0	0.0
Chlorine	1.3	0.0–0.3	0.0–0.3	0.0–0.4	0.0–0.5	0.0–0.9
Titanium/silicon	0.0	0.0–0.7	0.0	0.0	0.0	0.0

water vapour uptake commenced at higher relative humidity states; this suggests adsorbance dominated by capillary action and not attraction from chemical groups.[60] This effect was more pronounced for microparticles than for nanoparticles. While the trend was still observed for the latter system, the low pore lengths diminished the energy limitation of water adsorption on CDC surfaces. Given that pore wetting significantly affects the mobility of ion dynamics,[61] and ions are likely to be more mobile in hydrophobic pores, it is expected that these transformations will facilitate electrolyte mobility.[62]

Summary. Vacuum annealing modified several material properties of microparticle and nanoparticle CDCs. The following transformations have the most significant impact on the systems' electrochemical behaviour:

1. Pores retain their overall diameter distribution. At 700 °C annealing temperatures, SSA and pore volume increase, but the accessible surface area starts to decrease (and pores merge or collapse) for materials annealed at or above 1100 °C.

2. Materials become more ordered and contain fewer surface functional groups after annealing at higher temperatures. Conductivity and hydrophobicity increase as a result, although these properties are partially convoluted with porosity.

Electrochemical behaviour: effects of annealing on capacitance

Gravimetric capacitance. Effects of vacuum annealing on gravimetric capacitances of both materials are shown in Fig. 3. The microparticle sample (Fig. 3(a)) shows a 19% increase in capacitance after the 700 °C anneal. This is explained in part by the 16% increase in SSA and a slightly larger pore diameter, which minimizes the ion-sieving effect that is somewhat present in the initial CDC material (based on the resistive nature of the discharge curve) and hinders its capacitance. The capacitance of the 1400 °C annealed microparticle material decreased by over 30%, despite retaining 81% of its SSA and sufficiently large pores. The decreasing capacitance trend was even clearer for the SiC–CDC nanoparticle system (Fig. 3(b)): with no ion sieving limitations, capacitance decreased slightly for the 700 °C annealed material (despite gaining 1000 $m^2 g^{-1}$) and by over 20 F g^{-1} for the 1400 °C annealed material. While certain inflection points in the cyclic voltammograms suggest possible side reactions (and, consequently, faradaic contributions to capacitance) present in both initial CDC systems, they noticeably diminish after a 700 °C anneal and completely disappear in materials annealed at 1400 °C. The "butterfly"-shape of cyclic voltammograms observed for TiC–CDC after the 1400 °C anneal was also found for the nanoparticle sample. This corresponds to improved electric conductivity of both materials and suggests voltage-dependent capacitance.[63]

Capacitance retention (Fig. 3(c)) indicates that the microparticle CDC system best handled high sweep rates after undergoing 700 °C annealing; however, above a 250 mV s^{-1} scan rate, the 1400 °C material performs the best. The nanoparticle system clearly improved rate handling abilities at low and high sweep rates after 1400 °C annealing. Analysis of capacitance derived from galvanostatic cycling (Fig. 3(d)) shows the best performance for materials annealed at 700 °C, underscoring the importance of high SSA and accessible pore diameter to handle high current loads.[64]

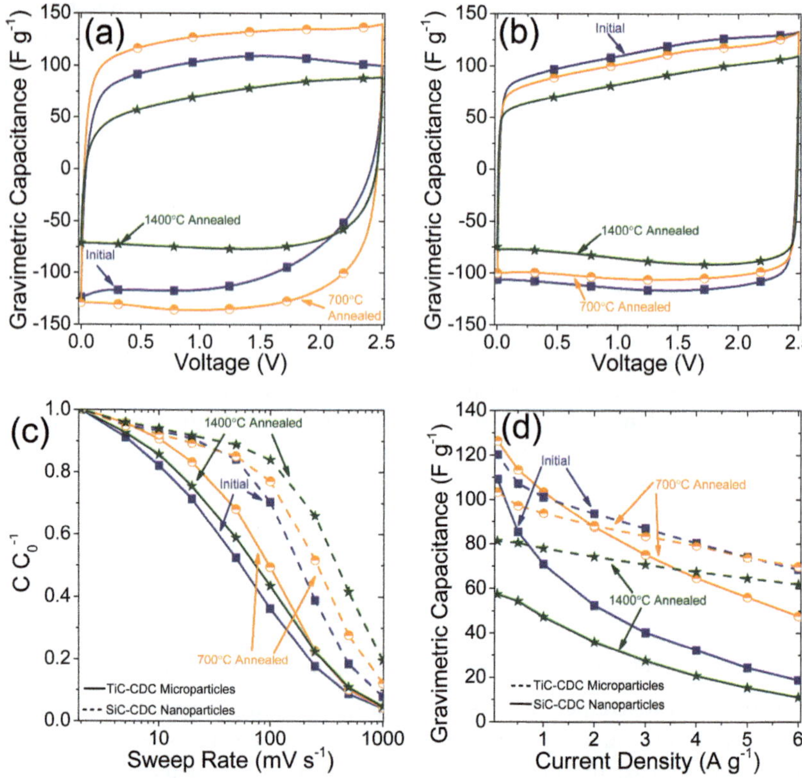

Fig. 3 Cyclic voltammograms of (a) TiC–CDC microparticles and (b) SiC–CDC nanoparticles in their initial, 700 °C annealed, and 1400 °C annealed states, collected at 5 mV s^{-1} and normalized by WE mass. (c) Capacitance retention of the 2 systems and their defunctionalized and graphitized versions in the 2 mV s^{-1} to 1000 mV s^{-1} range, normalized to the maximum capacitance recorded for each sample at 2 mV s^{-1}. (d) Capacitance values derived from galvanostatic cycling at current densities (normalized by WE mass) between 0.1 A g^{-1} and 6.0 A g^{-1}.

Capacitance normalized by surface area of carbon. Considering that the amount of accessible surface area is different for each sample, effects of annealing on aerial capacitance are shown in Fig. 4. For the TiC–CDC microparticles, both sweep rate and current density comparisons show that the 700 °C annealed material performs best, although cyclic voltammograms indicate a convergence in stored charge at 500 mV s^{-1} sweep rates. The SiC–CDC nanoparticle system indicates that the 1400 °C material stores the largest amount of charge per unit of area (Fig. 4(b)). However, this is partially convoluted with rate handling, as current density comparisons (Fig. 4(d)) show that the initial material exceeds the annealed counterpart's capacitance per surface area at very low scan rates.

Given the convergent appearance of the rate handling response to increased sweep rates, it is likely that the initial material stores more charge per m^2 than the 1400 °C annealed CDC below the minimum 2 mV s^{-1} tested rate. Although the BET-derived SSA values were used for normalization of aerial capacitances, DFT-

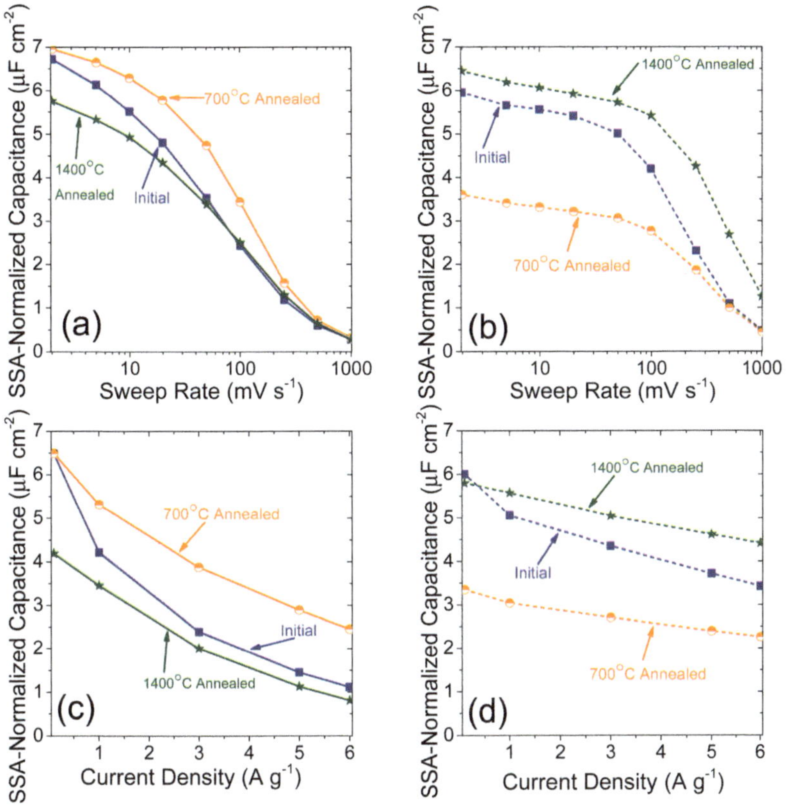

Fig. 4 Capacitance changes with increasing sweep rates derived from cyclic voltammetry normalized by SSA (BET value) for (a) microparticles and (b) nanoparticles. Galvanostatic cycling shows a similar trend for (c) microparticles, with the large pore diameters of the 700 °C annealed system providing greatest ion mobility. The higher cyclic voltammetry aerial capacitances of the 1400 °C annealed systems are shown to depend on its improved rate handling, as the initial and annealed materials show similar performance when galvanostatic cycling (d) is executed at 0.1 A g^{-1}.

derived SSA values would produce very similar trends: the SSA differences between the two are insignificant. The BET equation consistently calculated slightly higher m^2 g^{-1} values than the DFT simulation (16–75 m^2 g^{-1} greater SSA for microparticles and 44–240 m^2 g^{-1} for nanoparticles). The corresponding 3–12% changes would slightly affect the absolute F m^{-2} numbers, but the trends will remain consistent.

Discussion. Since the capacitance per area is different for each material, and the pore diameters exhibit insignificant changes, the removal of functional groups and the increase in graphitization change the relative contributions from quantum capacitance, the charge confinement in the superionic state, and the ion mobility. It is also likely that the annealing process produces different types of pore morphologies for microparticles with longer pores and nanoparticles with shallow pores. For instance, 700 °C annealing may create more mono- or divacancies in the microparticle system (disorder indicated by Raman spectroscopy and TGA), while graphitization-induced strain in the sp^2 carbon network may produce more hillocks and Stone–Wales type transformations for the 1400 °C annealed nanoparticles. Given that the types and amounts of those defects and curvature changes may influence the quantum capacitance of the materials,[32] additional theoretical analysis and structural characterization of pore morphology is needed to identify and quantify these contributions. The defects affect charge storage density by changing both the screen lengths (controlling ion spacing) and ion-pore distance (*via* steric forces).[29] The impact of quantum capacitance on overall capacitance is dependent on interlayer carbon thickness.[65] While nanoparticles are expected to have thinner pore walls, the number of graphitic layers separating each channel has not, to date, been accurately determined. The concentration of these defects has proven to be significant in these CDC systems, however, as the expected benefit of increased charge screening facilitated by conductive pore walls did not translate into enhanced gravimetric capacitance.

It is unlikely that the decrease in gravimetric capacitance for the annealed materials (or, conversely, a higher C_{sp} for the initial CDCs) is due to significant changes in contributions from faradaic reactions of the ionic electrolyte with functional groups.[66] A more direct observation is the positive correlation between increased surface defunctionalization and improved rate handling capabilities. Defect-free pores facilitate rapid $EMIm^+$ and $TFSI^-$ movement, as previously reported for similar systems.[61] This approach should therefore be considered for development of high-power systems to minimize resistance and facilitate charge and discharge. It should be noted that improved graphitic ordering and electrode conductivity are somewhat convoluted with this relationship: nanoparticles, although exhibiting relatively less defunctionalization, showed the greatest increases in capacitance retention.

The effects of pore defunctionalization depend on properties of the electrolyte. Previous experiments evaluated TiC–CDC microparticles synthesized at 1000 °C and annealed at 1400 °C. The treatment increased SSA and hydrophobicity, which translated to capacitance-impeding resistance in Na_2SO_4 (in H_2O) that was less evident for NEt_4BF_4 (in AN). This data is shown in Fig. S1 in ESI.†

Summary. The following key effects of annealing-induced defunctionalization and graphitization were observed for CDC materials:

1. Microparticles exhibit a slightly higher gravimetric capacitance after 700 °C annealing (with a higher SSA) but lower after 1400 °C annealing. Annealed nanoparticles show decreases in gravimetric capacitance regardless of surface area changes.

2. Although aerial capacitance of nanoparticles increases after 1400 °C annealing, low charge/discharge rates show similar or slightly higher values for the initial material.

3. Defect removal and increased conductivity achieved after high-temperature annealing improve rate handling and capacitance retention.

4. Surface defects, which are removed or transformed during annealing, significantly contribute to total charge storage. Annealing modifies these defects differently for microscale and nanoscale systems, uniquely affecting quantum capacitance.

Electrochemical behaviour: ionic resistance

Impedance spectroscopy results. Nyquist impedance plots for the TiC–CDC microparticles and SiC–CDC nanoparticles are shown in Fig. 5. Microparticles (Fig. 5(a)) show different impedance behaviour when comparing the initial material and its 700 °C annealed counterpart. This divergence stemmed primarily from sieving effects: the larger pore diameter of the annealed material decreases ionic resistance and produces a more capacitive behaviour at low frequencies.[67] The 1400 °C annealed material features a lager R_s and a higher ionic resistance. While electrode assembly could affect the interfacial resistance between the electrode and current collector,[51] the absence of a significant semi-circular region at high frequencies suggests that the intrinsic material structure is mostly responsible for impedance of ions. A combination of the smaller ion diameter (0.05 nm larger than the initial material) and the presence of annealing-induced defects most likely increases ionic resistance.

As the annealing temperature increases, the time constant decreases and impedance becomes more capacitance-dominated. The Bode impedance and phase angle response to frequency are shown in Fig. S4 in ESI.† Nanoparticles (Fig. 5(b)) indicate a decrease in ionic resistance, which corroborates the previously discussed improved rate handling abilities of these materials. The time constant monotonically decreases with higher annealing temperatures, as it does with the microscale system. The real capacitance (C') normalized by SSA showcases the same trends that CV and GCPL showed for aerial charge storage densities. Table 2 includes the key parameters regarding the system's ionic resistance and capacitance derived from EIS.

Charge/discharge dynamics. The diffusion limitations of the systems that calculated the Randles–Sevcik equation showed a correlation between annealing temperature and prevalence of capacitance-dominated behaviour (Table 3). The maximum coefficient value for the microparticles was obtained on the 700 °C annealed CDC, mostly because of the larger pore diameter that facilitates ion

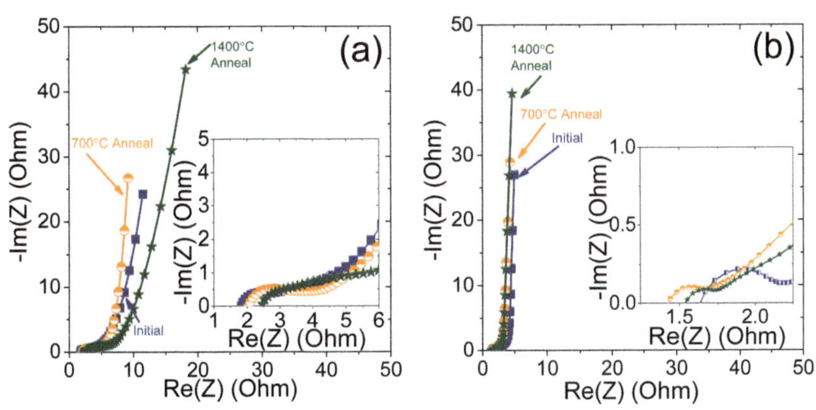

Fig. 5 Nyquist impedance plots derived from EIS for the (a) TiC–CDC microparticles and (b) SiC–CDC nanoparticles. Insets in each plot zoom in on the high-frequency region to properly describe electrical resistance and effects of the carbon-current collector interface.

Table 2 Electric and ionic resistance parameters derived from impedance spectroscopy. The time constant is derived from the frequency value at which impedance angle = −45.0°. Series resistance (R_s) is derived from the real impedance value at which $-\mathrm{Im}(Z) = 0\ \Omega$. C' is calculated using the following equation: $C'(\omega) = -Z''(\omega)\ [wZ(\omega)^2]^{-1}$. Values are shown normalized by the materials' BET SSA

	TiC–CDC microparticles			SiC–CDC nanoparticles		
	Initial CDC	700 °C annealed	1400 °C annealed	Initial CDC	700 °C annealed	1400 °C annealed
R_s, Ω	1.84	2.04	2.44	1.63	1.42	1.52
τ, s	26.5	23.2	19.3	13.3	10.1	7.4
Ionic resistance, Ω	3.39	2.17	4.27	1.02	0.69	0.48
Z (0.01 Hz), °	−64.6	−70.9	−67.3	−79.6	−81.5	−83.2
C'_{sp} (0.01 Hz), F m^{-2}	0.24	0.27	0.18	0.25	0.15	0.28

Table 3 Randles–Sevcik coefficient derived from cyclic voltammetry and galvanostatic cycling for the microparticle and nanoparticle systems. The value represents the slope of the line of log(I) *vs.* log(dV/dt). Values approaching 1 indicate classic double layer behaviour, while values approaching 0.5 indicate significant diffusion limitations. R-squared correlations were 0.979–0.998 for all data sets

	TiC–CDC microparticles			SiC–CDC nanoparticles		
	Initial CDC	700 °C annealed	1400 °C annealed	Initial CDC	700 °C annealed	1400 °C annealed
CV-derived	0.698	0.808	0.707	0.884	0.916	0.938
GCPL-derived	0.711	0.774	0.765	0.855	0.893	0.919

mobility. This was not the case for the nanoparticles: the 1400 °C material exhibited minimal diffusion-limited behaviour despite having a smaller pore diameter. This further suggests the possible differences in the structure and defects that are present on the surfaces of the two materials.

While the ionic resistance decreases for the nanoscale particles from 1.02 Ω (initial CDC) to 0.48 Ω (1400 °C annealed), the same process increases this resistance from 3.39 Ω to 4.27 Ω for microparticles. It is possible that the annealing-induced morphology developed on the surface of large TiC–CDC particles increases Warburg impedance and hinders ion diffusion at semi-infinite length scales.[68,69] These impediments could include out-of-plane graphite sheets that increase the mean free path, increased interparticle separation in films due to repulsion between similarly-functionalized particles, and formation of bottlenecks at pore openings. Extensive finite element analysis modelling and small-angle X-ray scattering (SAXS)[45] may shed more light on the prevalence of these features.

Summary. Annealing-induced graphitization and defunctionalization produce the following key changes in the behaviour of EMIm-TFSI ions:

1. Annealed materials exhibit lower time constants and a general decrease in diffusion limitations and ionic resistance. Pore diameter and any ion sieving effects have an equally significant impact and are coupled with these observations.

2. Types and quantities of defects on pore surfaces, which depend on both particle size and annealing temperature, affect ion dynamics. Mathematical modelling and further investigation of the structure are required for system optimization.

Electrochemical behaviour: charge distribution and symmetry

The quasi-reference 3-electrode configuration allowed CV measurements of both working and counter electrodes. Performances for the two systems (collected using a 5 mV s^{-1} sweep rate) showing the contribution of each electrode to total capacitance are shown in Fig. 6. Identical trends are observed for nanoparticles and microparticles: in its initial state, CDC materials store charge predominantly in the working electrode, while the counter electrode extends to a significantly lower potential (*vs.* Ag/AgCl). It should be re-iterated that the masses and thicknesses of the electrodes for each sample are almost identical. This suggests that the materials' structure (both pore diameter and surface chemistry) facilitates $TFSI^-$ anion's entry over the $EMIm^+$ cation.

After the 700 °C vacuum annealing treatment, both the microscale and nanoscale materials experience a negative shift in potential, with the samples' charge distribution becoming even more asymmetric. Nanoscale and microscale materials annealed at 1400 °C show near-perfect charge distribution symmetry, with the equilibrium potential shifting closer towards 0.00 V *vs.* Ag/AgCl. The "butterfly" shape is evident for the working electrodes for both microscale and nanoscale powders annealed at this condition. Evaluating the capacitance contributions for each system, however, shows the almost-symmetrical behaviour that was expected. Table S3 in ESI† provides a quantitative summary of effects of annealing on symmetry.

Discussion. The changes in symmetry cannot be attributed to porosity changes alone: while materials annealed at 700 °C have the largest pore diameters

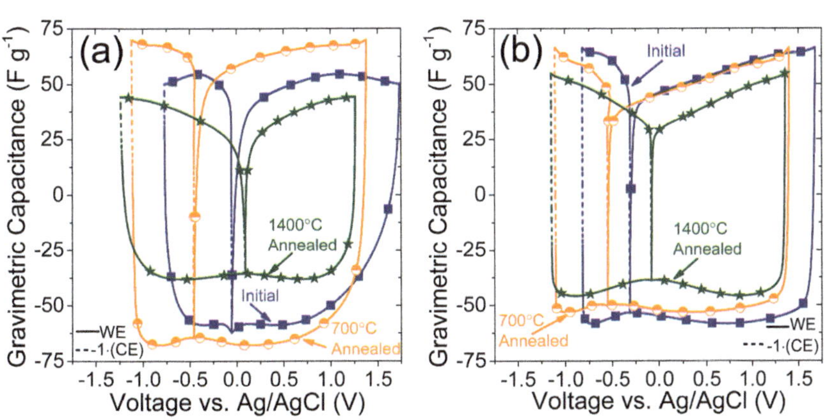

Fig. 6 Comparison of operating potential distribution and capacitance contributions (normalized by mass) of the working and counter electrodes of (a) microparticles and (b) nanoparticles collected using cyclic voltammetry at 5 mV s^{-1}. For counter electrodes, the negative of the current is shown to facilitate comparison of the symmetry. Quasi-reference electrode potential was identical for each case. Open circuit potentials fluctuated between 6 mV and 60 mV for each system.

(expected optimal access to both ions), they respond most asymmetrically to EMIm-TFSI electrosorption. Therefore, surface chemistry and morphologically-induced structural defect transformations govern charge distribution. These changes also control ion sieving: the TiC–CDC initial and 700 °C annealed microparticles showcase noticeable resistance-induced deviation from the ideal rectangular shape of the discharge voltammograms of the working electrodes. Annealing transformations also influence contributions from faradaic reactions, which are completely absent for both WE and CE after a 1400 °C anneal. The 700 °C annealed micro- and nanoparticles exhibit inflection points during the charge cycle of the CE at −1.00 V; it is unclear whether this is evidence of voltage-dependent capacitance or electrochemical breakdown. This behaviour is not observed for the 1400 °C systems.

Summary. Surface chemistry has a significant effect on electrolyte ion and, subsequently, charge distribution in symmetrical cells. A defunctionalized surface makes it equally favourable for $EMIm^+$ and $TFSI^-$ to enter and exit pores.

Electrochemical behaviour: voltage window analysis

While the majority of CV and GCPL experiments were conducted in the 0.00–2.50 V potential range, final experiments assessed the systems' operations with vertex potentials of 2.75 V, 3.00 V, and 3.25 V. Voltammograms of these measurements performed at 2 mV s^{-1} are shown in Fig. 7. The behaviour at vertex potentials suggests that electrochemical breakdown occurs above 2.50 V for all microscale and nanoscale samples, regardless of defunctionalization or ordering. The materials annealed at 700 °C showcase the highest breakdown and the lowest overall coulombic efficiencies. However, this is partially convoluted with the large accessible surface areas of these materials, providing more ion reaction and decomposition sites. Analysis of parasitic currents and voltammograms shows that the 1400 °C annealed systems (with fewest functional groups) are the most stable ones under high potentials, but still show signs of electrolyte breakdown above 2.50 V.

Separate analysis of the working and counter electrodes (Fig. S5 in ESI†) suggests that breakdown occurs for both $EMIm^+$ and $TFSI^-$ ions. Excessive vertex potentials show different charge distribution asymmetry than those shown in Fig. 6 and likely indicate changes in pore structure and chemistry (such as gas evolution and grafting of new chemical species). Analysis of respective coulombic efficiencies of the microparticle electrodes (Table S3 in ESI†) consistently shows greater resistance and breakdown associated with the positive electrode and, subsequently, the predominance of anion decomposition.

Significant contributions from impurities in the IL (trapped water, synthesis byproducts, *etc.*) are possible but unlikely: no decomposition is observed below the 2.50 V initial threshold. Furthermore, the differences between materials annealed at different temperatures account for over 30% of breakdown-induced contributions to capacitance, suggesting the overwhelming significance of surface chemistry effects compared to any present impurities. These differences are less significant for the annealed SiC–CDC systems, further corroborating the difference in the amounts and types of defects produced in nanoscale carbons as a result of annealing.

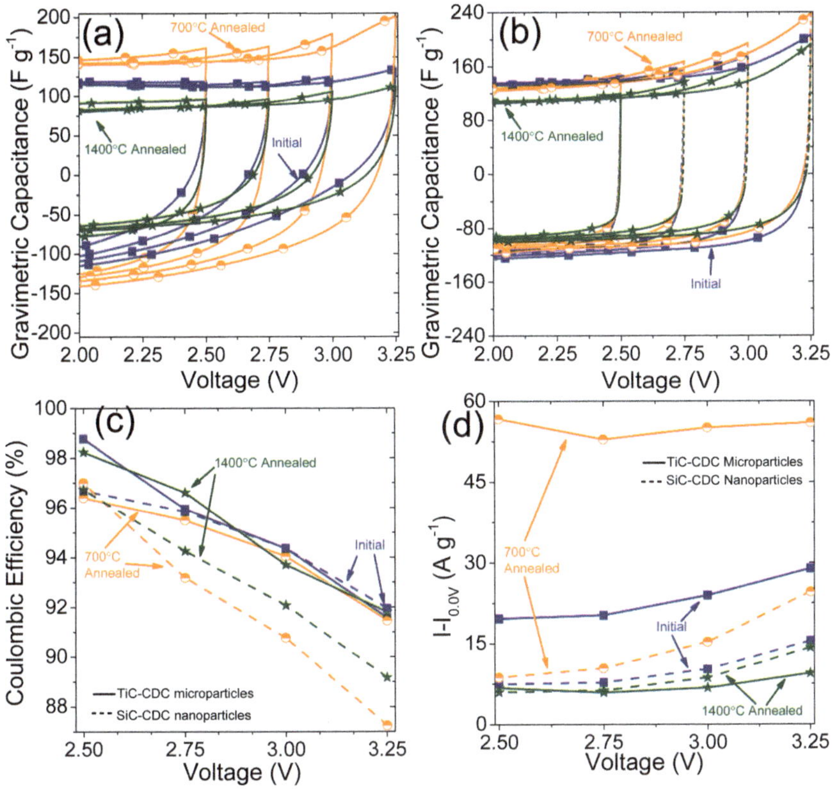

Fig. 7 Vertex potentials shown for cyclic voltammograms collected at 2 mV s^{-1} for (a) microparticles and (b) nanoparticles. The first cycle is shown for each system. (c) Coulombic efficiencies calculated from above CV results, showcasing increases in breakdown-related reactions increasing as the operating potential is increased from 2.50 V to 3.25 V. (d) Net parasitic currents (currents measured at 0.0 V subtracted) normalized by mass for each system for square wave amperommetry conducted at 2.50 V, 2.75 V, 3.00 V, and 3.25 V.

Discussion. Although high-temperature annealing slightly decreases electrolyte decomposition, pure defunctionalization is still insufficient for ILs to operate in their theoretical 4.0 V window.[70] The differences in magnitude of the breakdowns between materials annealed at different temperatures can be attributed to different mechanisms affecting the reactions. These processes probably depend on ion confinement, pore wall conductivity (derived from graphitization), and chemical species present at the ion–carbon interface. Reactions with –OH and –COOH groups may yield to dimerization-induced breakdown of $EMIm^+$ (releasing H_2 that blocks pores and degrades charge storage).[71] Alternative simulations predict a potential-induced breakdown of the S–C bonds and subsequently the C–H bonds (both of which are present in the $TFSI^-$ anion) under applied potentials.[72]

While both processes require overcoming substantial activation energy barriers, a conductive, accessible surface (*i.e.*, the structure of annealed pores) may catalyse these processes and increase the dominance of their contribution to electrolyte decomposition. Furthermore, the three-dimensional confinement of ions in pores changes the interaction and distortion of the ions' electron cloud, creating a very different environment than the conditions that have been used for simulation and tests with GC electrodes. Defunctionalizing the surface will not completely solve the problem of low voltage windows, and a combination of materials (introducing specific stable groups) and device design (asymmetrical electrodes) is required.

Summary. Extending the voltage window of materials with different surface chemistries and graphitic ordering showed the following key properties:

1. Defunctionalization slightly reduces breakdown. This is also convoluted with pore size and carbon surface defects.

2. Cations and anions likely decompose *via* different processes and at different rates, as observed by increased asymmetry between the positive and negative electrodes.

3. As surface-chemistry related breakdown behaviour diminishes, ion dimerization catalysed by conductive pore walls may dominate decomposition in identical voltage windows.

Conclusions

We presented experimental analysis to assess the existing predictions that anticipate an increase in double layer charge storage densities and facilitated ion screening for materials with defunctionalized, conductive pore walls. We found that vacuum annealing, which created these properties for carbon electrodes with internal surfaces, did not yield expected higher gravimetric capacitances. Our primary explanation of this phenomenon is that surface defunctionalization and passivation removes chemical defects and subsequently decreases the carbon walls' quantum capacitance contributions and ion screening, reducing double layer capacitance as a result. The magnitude of these changes depends on the carbon structure, including degree of sp^2 ordering, accessible surface area, and pore length and tortuosity. As expected, we observe fewer faradic reactions in defunctionalized materials under applied potentials. However, they remain difficult to classify and quantify, so their pseudocapacitive impact on total charge storage remains an open question.

Charge and discharge behaviour at high loads suggests that surface defunctionalization, increased mesoporosity, and graphitization facilitate ion mobility in and out of pores, improving rate handling ability. Diffusion limitations decreased with increased annealing temperatures independently of pore diameter or the presence of mesopores. This effect is likely a combination of the presence of an EMIm-TFSI favourable defect-free surface and conductive pore walls. Although surface defects have been shown to enhance quantum capacitance in static modes, there is little knowledge of dynamic effects of quantum capacitance and charge screening on ion mobility.

We found that charge distribution between the positive and negative electrodes is strongly affected by surface chemistry, with the robust surface species that are present in typical non-annealed CDCs favouring $TFSI^-$ ion electrosorption. A defunctionalized, defect-free surface with increased sp^2 ordering creates a more mutually favourable ion–carbon interface and improves the system's symmetry. Effects of hydrophobicity are different for every electrolyte (see

Fig. S1(c) and S1(d) in ESI† to compare performance of aqueous and organic systems).

When we extended the operating voltage window beyond the conventional 2.50 V limit, we found that defunctionalization of pore walls slightly decreased electrochemical breakdown. However, we also noticed that the severity of decomposition is correlated with accessible surface area, which should be considered for future electrode design. Some of the potential-induced breakdown reactions were likely accelerated by the conductive pore walls that acted as catalysts. It remains unclear whether the theoretical 4.00+ V operating limit of ILs can ever be reached in confined pore environments, but pure defunctionalization will not accomplish this goal.

Although a wide array of characterization techniques was used for this study, it remains difficult to identify every aspect of the pore structure that subsequently affects electrochemical performance. The treatment produces heterogeneous morphological effects that cannot be calculated using conventional means. This may partially explain the divergent effects of annealing on aerial capacitance for microscale and nanoscale systems. TEM imaging (Fig. S6 in ESI†) showed the formation of various graphitic barrels,[73] ribbons, and protruding features. These defects have as significant of an effect on quantum capacitance as surface functionalities.[32] The trade-off between quantum and double layer contributions to total charge storage depends on the pore wall thickness,[35] which needs to be properly calculated. ESI† contains a discussion of relative quantum capacitance contributions from the six tested materials. For subsequent optimization of porous carbons for electrochemical energy storage, extensive mathematical modelling is required to properly assess effects of defects. This knowledge can be subsequently used to select a combination of thermal treatments and chemical grafting with appropriate functional groups to increase capacitance end electrochemical stability.

Acknowledgements

Research was funded by the NSF grant CNS-0960061. TEM imaging was conducted at Drexel's Centralized Research Facility (CRF). TGA was conducted in the Polymers and Composites Laboratory (Department of Chemical and Biological Engineering, Drexel University). The authors gratefully acknowledge the assistance of Dr K. M. Cook with XPS analyses, F. W. Richey with DVS measurements, and O. Mashtalir (Drexel University) with TEM imaging. The authors thank M. R. Lukatskaya, K. B. Hatzell, and M. Sipics (Drexel Unviversity) for valuable comments on the manuscript. The authors wish to additionally acknowledge Prof. P. Simon (Université Paul Sabatier) and Prof. V. Presser (Leibniz-INM, Saarbrücken) for helpful discussions, as well as Dr. J. Chmiola and Prof. S. Osswald (NPS) for producing initial experimental results that inspired this study.

References

1 P. Simon and Y. Gogotsi, *Acc. Chem. Res.*, 2013, **46**, 1094–1103.
2 B. E. Conway, *Electrochemical Supercapacitors: Scientific Fundamentals and Technological Applications*, Springer, New York, 1999.

3 R. Lin, P.-L. Taberna, S. Fantini, V. Presser, C. R. Perez, F. Malbosc, N. L. Rupesinghe, K. B. K. Teo, Y. Gogotsi and P. Simon, *J. Phys. Chem. Lett.*, 2011, **2**, 2396–2401.

4 P. Simon and Y. Gogotsi, *Nat. Mater.*, 2008, **7**, 845–854.

5 F. Béguin, V. Presser, A. Balducci and E. Frackowiak, *Adv. Mater.*, 2014, **26**, 2219–2251.

6 L. L. Zhang, Y. Gu and X. S. Zhao, *J. Mater. Chem. A*, 2013, **1**, 9395–9408.

7 P. Simon and Y. Gogotsi, *Philos. Trans. R. Soc. London, Ser. A*, 2010, **368**, 3457–3467.

8 R. Lin, P. L. Taberna, J. Chmiola, D. Guay, Y. Gogotsi and P. Simon, *J. Electrochem. Soc.*, 2009, **156**, A7–A12.

9 S. Kondrat, N. Georgi, M. V. Fedorov and A. A. Kornyshev, *Phys. Chem. Chem. Phys.*, 2011, **13**, 11359–11366.

10 S. Kondrat, C. R. Perez, V. Presser, Y. Gogotsi and A. A. Kornyshev, *Energy Environ. Sci.*, 2012, **5**, 6474–6479.

11 Y. Zhu, S. Murali, M. D. Stoller, K. J. Ganesh, W. Cai, P. J. Ferreira, A. Pirkle, R. M. Wallace, K. A. Cychosz, M. Thommes, D. Su, E. A. Stach and R. S. Ruoff, *Science*, 2011, **332**, 1537–1541.

12 Y. Gogotsi and P. Simon, *Science*, 2011, **334**, 917–918.

13 F. Beguin and E. Frackowiak, *Supercapacitors: Materials, Systems, and Applications*, Wiley-VCH, Weinheim, Germany, 2013.

14 A. G. Pandolfo and A. F. Hollenkamp, *J. Power Sources*, 2006, **157**, 11–27.

15 V. Presser, M. Heon and Y. Gogotsi, *Adv. Funct. Mater.*, 2011, **21**, 810–833.

16 *Carbon nanomaterials*, ed. Y. Gogotsi and V. Presser, CRC Press, 2nd edn, 2013.

17 C. Portet, D. Kazachkin, S. Osswald, Y. Gogotsi and E. Borguet, *Thermochim. Acta*, 2010, **497**, 137–142.

18 C. R. Perez, S.-H. Yeon, J. Segalini, V. Presser, P.-L. Taberna, P. Simon and Y. Gogotsi, *Adv. Funct. Mater.*, 2013, **23**, 1081–1089.

19 Z. G. Cambaz, G. Yushin, S. Osswald, V. Mochalin and Y. Gogotsi, *Carbon*, 2008, **46**, 841–849.

20 J. C. Palmer, A. Llobet, S. H. Yeon, J. E. Fischer, Y. Shi, Y. Gogotsi and K. E. Gubbins, *Carbon*, 2010, **48**, 1116–1123.

21 S. Urbonaite, L. Halldahl and G. Svensson, *Carbon*, 2008, **46**, 1942–1947.

22 P. M. Vora, P. Gopu, M. Rosario-Canales, C. R. Perez, Y. Gogotsi, J. J. Santiago-Aviles and J. M. Kikkawa, *Phys. Rev. B: Condens. Matter Mater. Phys.*, 2011, **84**, 155114.

23 E. Frackowiak, *J. Braz. Chem. Soc.*, 2006, **17**, 1074–1082.

24 A. M. O'Mahony, D. S. Silvester, L. Aldous, C. Hardacre and R. G. Compton, *J. Chem. Eng. Data*, 2008, **53**, 2884–2891.

25 P. Kurzweil and M. Chwistek, *J. Power Sources*, 2008, **176**, 555–567.

26 D. Hulicova-Jurcakova, M. Seredych, G. Q. Lu and T. J. Bandosz, *Adv. Funct. Mater.*, 2009, **19**, 438–447.

27 H. Zhou, P. Ganesh, V. Presser, M. C. F. Wander, P. Fenter, P. R. C. Kent, D.-e. Jiang, A. A. Chialvo, J. McDonough, K. L. Shuford and Y. Gogotsi, *Phys. Rev. B: Condens. Matter Mater. Phys.*, 2012, **85**, 035406.

28 T. Batchelor, J. Cunder and A. Y. Fadeev, *J. Colloid Interface Sci.*, 2009, **330**, 415–420.

29 C. Merlet, B. Rotenberg, P. A. Madden, P.-L. Taberna, P. Simon, Y. Gogotsi and M. Salanne, *Nat. Mater.*, 2012, **11**, 306–312.

30 D. Qu, *J. Power Sources*, 2002, **109**, 403–411.
31 C. Merlet, C. Péan, B. Rotenberg, P. A. Madden, B. Daffos, P. L. Taberna, P. Simon and M. Salanne, *Nat. Commun.*, 2013, **4**, 2701.
32 B. C. Wood, T. Ogitsu, M. Otani and J. Biener, *J. Phys. Chem. C*, 2014, **118**, 4–15.
33 K. Tahara, T. Iwasaki, S. Furuyama, A. Matsutani and M. Hatano, *Appl. Phys. Lett.*, 2013, **103**, 143106.
34 C. Punckt, F. Muckel, S. Wolff, I. A. Aksay, C. A. Chavarin, G. Bacher and W. Mertin, *Appl. Phys. Lett.*, 2013, **102**, 023114.
35 E. Uesugi, H. Goto, R. Eguchi, A. Fujiwara and Y. Kubozono, *Sci. Rep.*, 2013, **3**, 1595.
36 T. Fang, A. Konar, H. Xing and D. Jena, *Appl. Phys. Lett.*, 2007, **91**, 092109.
37 J. Vatamanu, Z. Hu, D. Bedrov, C. Perez and Y. Gogotsi, *J. Phys. Chem. Lett.*, 2013, **4**, 2829–2837.
38 O. Borodin, W. Gorecki, G. D. Smith and M. Armand, *J. Phys. Chem. B*, 2010, **114**, 6786–6798.
39 A. I. Frolov, K. Kirchner, T. Kirchner and M. V. Fedorov, *Faraday Discuss.*, 2012, **154**, 235–247.
40 B. R. Kozub, N. V. Rees and R. G. Compton, *Sens. Actuators, B*, 2010, **143**, 539–546.
41 C. Punckt, M. A. Pope and I. A. Aksay, *J. Phys. Chem. C*, 2013, **117**, 16076–16086.
42 Y. Guo, S. Guo, Y. Fang and S. Dong, *Electrochim. Acta*, 2010, **55**, 3927–3931.
43 R. C. Rodriguez, A. B. Moncada, D. F. Acevedo, G. A. Planes, M. C. Miras and C. A. Barbero, *Faraday Discuss.*, 2013, **164**, 147–173.
44 J. Chmiola, S. Osswald and Y. Gogotsi, *Carbon*, 2012, **50**, 4880–4886.
45 G. Laudisio, R. K. Dash, J. P. Singer, G. Yushin, Y. Gogotsi and J. E. Fischer, *Langmuir*, 2006, **22**, 8945–8950.
46 C. Portet, G. Yushin and Y. Gogotsi, *J. Electrochem. Soc.*, 2008, **155**, A531–A536.
47 Y. Gogotsi, A. Nikitin, H. Ye, W. Zhou, J. E. Fischer, B. Yi, H. C. Foley and M. W. Barsoum, *Nat. Mater.*, 2003, **2**, 591–594.
48 J. McDonough, A. Frolov, V. Presser, J. Niu, C. Miller, T. Ubieto, M. Fedorov and Y. Gogotsi, *Carbon*, 2012, **50**, 3298–3309.
49 S. Brunauer, P. H. Emmett and E. Teller, *J. Am. Chem. Soc.*, 1938, **60**, 309–319.
50 G. Y. Gor, M. Thommes, K. A. Cychosz and A. V. Neimark, *Carbon*, 2012, **50**, 1583–1590.
51 M. D. Stoller and R. S. Ruoff, *Energy Environ. Sci.*, 2010, **3**, 1294–1301.
52 G. Snook, A. Best, A. Pandolfo and A. Hollenkamp, *Electrochem. Commun.*, 2006, **8**, 1405–1411.
53 Y. Wang and Y. Xia, *Adv. Mater.*, 2013, **25**, 5336–5342.
54 A. C. Ferrari and J. Robertson, *Phys. Rev. B: Condens. Matter*, 2000, **61**, 1495–14107.
55 D. K. Singh, P. K. Iyer and P. K. Giri, *J. Appl. Phys.*, 2010, **108**, 084313.
56 J.-H. Zhong, J.-Y. Liu, Q. Li, M.-G. Li, Z.-C. Zeng, S. Hu, D.-Y. Wu, W. Cai and B. Ren, *Electrochim. Acta*, 2013, **110**, 754–761.
57 P. Ganesh, P. Kent and V. Mochalin, *J. Appl. Phys.*, 2011, **110**, 073506.
58 J. Sánchez-González, A. Macías-García, M. F. Alexandre-Franco and V. Gómez-Serrano, *Carbon*, 2005, **43**, 741–747.
59 L. L. Zhang, X. Zhao, M. D. Stoller, Y. Zhu, H. Ji, S. Murali, Y. Wu, S. Perales, B. Clevenger and R. S. Ruoff, *Nano Lett.*, 2012, **12**, 1806–1812.
60 P. Kim, Y. Zheng and S. Agnihotri, *Ind. Eng. Chem. Res.*, 2008, **47**, 3170–3178.

61 K. Pinkert, M. Oschatz, L. Borchardt, M. Klose, M. Zier, W. Nickel, L. Giebeler, S. Oswald, S. Kaskel and J. Eckert, *ACS Appl. Mater. Interfaces*, 2014, **6**, 2922–2928.
62 J. Lemus, J. Palomar, F. Heras, M. A. Gilarranz and J. J. Rodriguez, *Sep. Purif. Technol.*, 2012, **97**, 11–19.
63 O. Kimizuka, O. Tanaike, J. Yamashita, T. Hiraoka, D. N. Futaba, K. Hata, K. Machida, S. Suematsu, K. Tamamitsu, S. Saeki, Y. Yamada and H. Hatori, *Carbon*, 2008, **46**, 1999–2001.
64 A. B. Fuertes, G. Lota, T. A. Centeno and E. Frackowiak, *Electrochim. Acta*, 2005, **50**, 2799–2805.
65 H. Ji, X. Zhao, Z. Qiao, J. Jung, Y. Zhu, Y. Lu, L. L. Zhang, A. H. MacDonald and R. S. Ruoff, *Nat. Commun.*, 2014, **5**, 3317.
66 A. Balducci, R. Dugas, P. L. Taberna, P. Simon, D. Plée, M. Mastragostino and S. Passerini, *J. Power Sources*, 2007, **165**, 922–927.
67 J. Segalini, B. Daffos, P. L. Taberna, Y. Gogotsi and P. Simon, *Electrochim. Acta*, 2010, **55**, 7489–7494.
68 X.-m. Liu, R. Zhang, L. Zhan, D.-h. Long, W.-m. Qiao, J.-h. Yang and L.-c. Ling, *New Carbon Mater.*, 2007, **22**, 153–158.
69 S. Prabaharan, R. Vimala and Z. Zainal, *J. Power Sources*, 2006, **161**, 730–736.
70 A. B. McEwen, H. L. Ngo, K. LeCompte and J. L. Goldman, *J. Electrochem. Soc.*, 1999, **146**, 1687–1695.
71 T. Romann, O. Oll, P. Pikma, H. Tamme and E. Lust, *Electrochim. Acta*, 2014, **125**, 183–190.
72 O. Borodin, W. Behl and T. R. Jow, *J. Phys. Chem. C*, 2013, **117**, 8661–8682.
73 A. Perkson, J. Leis, M. Arulepp, M. Kaarik, S. Urbonaite and G. Svensson, *Carbon*, 2003, **41**, 1729–1735.

Faraday Discussions

PAPER

Ionic liquid based EDLCs: influence of carbon porosity on electrochemical performance

Asa Noofeli, Peter J. Hall and Anthony J. R. Rennie*

Received 31st March 2014, Accepted 7th May 2014
DOI: 10.1039/c4fd00057a

Electrochemical double layer capacitors (EDLCs) are a category of supercapacitors; devices that store charge at the interface between electrodes and an electrolyte. Currently available commercial devices have a limited operating potential that restricts their energy and power densities. Ionic liquids (ILs) are a promising alternative electrolyte as they generally exhibit greater electrochemical stabilities and lower volatility. This work investigates the electrochemical performance of EDLCs using ILs that combine the bis(trifluoromethanesulfonyl)imide anion with sulfonium and ammonium based cations. Different activated carbon materials were employed to also investigate the influence of varying pore size on electrochemical performance. Electrochemical impedance spectroscopy (EIS) and constant current cycling at different rates were used to assess resistance and specific capacitance. In general, greater specific capacitances and lower resistances were found with the sulfonium based ILs studied, and this was attributed to their smaller cation volume. Comparing electrochemical stabilities indicated that significantly higher operating potentials are possible with the ammonium based ILs. The marginally smaller sulfonium cation performed better with the carbon exhibiting the largest pore width, whereas peak performance of the larger sulfonium cation was associated with a narrower pore size. Considerable differences between the performance of the ammonium based ILs were observed and attributed to differences not only in cation size but also due to the inclusion of a methoxyethyl group. The improved performance of the ether bond containing IL was ascribed to electron donation from the oxygen atom influencing the charge density of the cation and facilitating cation–cation interactions.

1 Introduction

Electrochemical double layer capacitors (EDLCs) are a category of supercapacitors; electrical energy storage devices that have the ability to operate at substantially higher rates than similar sized electrochemical cells.[1] They typically

Dept. of Chem. & Bio. Eng., University of Sheffield, Sheffield, S1 3JD, UK. E-mail: a.rennie@sheffield.ac.uk; Fax: +44 (0)114 222 7501; Tel: +44 (0)114 222 8257

exhibit long cycle lives (>500 000 cycles), display relatively high energy efficiencies and a high degree of reliability.[1–7] For these reasons, among others, they have become the subject of much research and are the theme of a growing number of reviews.[2–12] To date, EDLCs have mainly been employed in applications where their useful traits allow for the design of systems that require less servicing/replacement than those based on batteries.[4–7,9]

EDLCs store energy through the physical separation of charges at the interface between electrodes and an electrolyte, and are particularly useful when dealing with systems where variations in power demand leads to a large mismatch between the average and maximum power required. They are typically charged/discharged within seconds and can display values of specific power up to 10 kW kg^{-1}.[6,7,10]

An application where this is useful is in electric vehicles where there are rapidly changing demand profiles, resulting in batteries being discharged at rates that cause reduced operating lifetimes.[4,6] Employing EDLCs to cater for peak demand, alongside batteries or fuel cells, not only allows for the design of smaller, lighter systems but can increase the lifetime of the system and enable the efficient storage of high charging currents generated by regenerative braking systems.[13–15]

However EDLCs would be a more viable option for many applications if greater specific energy/energy densities could be achieved; this in turn is restricted by the operating voltage of the device. Operating voltage is defined by the electrochemical stability of the electrolyte, which limits not only the amount of energy that can be stored, but also the maximum rate at which energy can be delivered.

Eqn (1) shows that the energy stored by an EDLC (E) varies with the capacitance (C) of the device, and with the square of the operating voltage (V):[1]

$$E = \frac{1}{2}CV^2 \tag{1}$$

The maximum power (P), that can be delivered by an EDLC also varies quadratically with the operating voltage and inversely with the equivalent series resistance (ESR) of the device, as shown in eqn (2):[1]

$$P = \frac{V^2}{4(\text{ESR})} \tag{2}$$

Eqn (1) also shows that the majority of the useful energy (as well as higher quality energy[16]) is delivered at elevated potentials, and by considering the relationships in eqn (1) and (2), it is evident that increases in operating potential enhance both the energy, and power, delivered by EDLCs.

Commercially available devices are limited by decomposition of the electrolyte, which is typically acetonitrile or propylene carbonate based.[3,5,10,12,17] Ionic liquids (ILs) can possess relatively wide electrochemical stability windows (ESWs) allowing for significantly increased operating potentials. ILs are salts usually consisting of poorly interacting anions and cations; due to these weak interactions they display relatively low melting temperatures (<100 °C).[17–21] In addition, ILs typically possess several other physical properties that are desirable in EDLC electrolytes, such as non-flammability, high thermal and chemical stability, and due to their inherent low volatility can be considered to be less hazardous than organic solvent based electrolytes.[17,22] Physical properties other than the ESW of ILs play a crucial role in their performance as electrolytes and vary considerably with the structure of their constituent ions. In particular, viscosity (which in turn influences conductivity) affects not only the capacitance of a device, but also the maximum power output.[23] High conductivities lead to a lower cell resistance, which remains a challenge in the case of ILs, as their conductivity tends to be at least an order of magnitude lower than that of conventional electrolytes.

In this study we investigate the effect that cation structure has on the performance of IL based EDLCs using various cations paired with the bis(trifluoromethanesulfonyl)imide ([Tf_2N]) anion. The partially delocalized charge in the [Tf_2N] anion imparts some flexibility into the structure and results in weak interactions with cations.[24] This generally results in ILs with relatively low viscosities, wide ESWs and it has been shown that ILs containing this anion can form a stable protective layer passivating the surface of aluminium current collectors.[25]

The most widely studied cations are pyrrolidinum or imidazolium based.[17] When coupled with the [Tf_2N] anion, pyrrolidinium based electrolytes are typically stable in the region of 3.5 V; however, their relatively high viscosity results in poor performance at high rates.[17,23,26–28] Imidazolium [Tf_2N] electrolytes, on the other hand, tend to be less viscous at room temperature but are less electrochemically stable and are limited to operating potentials around 3.2 V.[17,20,23,29,30] For example, the conductivity of 1-butyl-3-methylpyrrolidinum [Tf_2N] is in the region of 2.2 mS cm^{-1} (ref. 24) whereas 1-ethyl-3-methylimidazoium [Tf_2N] is 3.4 mS cm^{-1}.[29] There is therefore a compromise between the quantity of energy stored in an IL based EDLC and the rate at which this energy can be delivered.

In this work we report on the performance of ILs consisting of sulfonium and ammonium based cations coupled with the [Tf_2N] anion as EDLC electrolytes. Triethyl sulfonium [Tf_2N] exhibits a remarkably high conductivity of 7.1 mS cm^{-1} (ref. 31) and has been shown to exhibit relatively high values of specific capacitance when compared with a carbonate based electrolyte.[32] In the sulfonium based ILs used in this work the cation volume is smaller than that of the anion. To investigate the influence of carbon pore size on performance, bulkier ammonium cation based ILs were employed for comparative purposes.

The physical properties of ILs, in particular the viscosity/conductivity, can be altered through the inclusion of ether linkages in the alkyl side chains of cations.[33] It has been proposed that the inclusion of ether bonds results in increased specific capacitances in EDLCs using mesoporous carbon electrodes by adding a degree of flexibility into the anion and alteration of the charge density.[34] Therefore to explore this behaviour we compare the performance of butyl triethyl ammonium [N_{2224}] [Tf_2N] with that of *N*,*N*-diethyl-*N*-methyl-*N*(2-methoxyethyl)-ammonium [$N_{221(1O2)}$] [Tf_2N].

As a substantial number of reports have shown that the pore characteristics of electrodes strongly influence the performance of EDLCs,[22,28,35–38] we also compare the behaviour of these ILs when combined with electrodes of differing average pore size (d) in the small mesopore range (*i.e.* $2 < d < 10$ nm).

2 Experimental

2.1 Carbon materials

Porous carbon materials were produced by carbonisation and physical activation of resorcinol–formaldehyde xerogels in a manner similar to that used in previous investigations.[23,38,39] The polycondensation of an aqueous solution of resorcinol, R (Aldrich, 99%) with formaldehyde, F (Aldrich, 37 wt%, stabilised with 10–15 wt% methanol) was catalysed by the addition of sodium carbonate, C (Aldrich, Na_2CO_3, >99.5%). R/F molar ratio was kept at 0.5 and R/W was 0.10 g cm^{-3}. R/C ratios of 200, 250, 300 and 350 were used to control the pore characteristics of the resultant gels. Gelation occurred over the following 3 days at a temperature of 85 °C. Organic hydrogels then underwent solvent exchange with acetone before being dried at 80 °C under vacuum.

Xerogels were carbonised in a tube furnace under flowing argon with a final temperature of 850 °C being maintained for 180 min. Physical activation under flowing CO_2 was then carried out in a tube furnace at a temperature of 850 °C for 60 min. The activated carbon materials were ball milled under argon for 60 min and stored under vacuum at 80 °C prior to use in electrodes/characterisation.

Nitrogen adsorption/desorption isotherms at −196 °C were obtained using a Micromeritics TriStar II 3020 instrument; samples were degassed at 200 °C for 10 h before analysis. Porosity distribution was evaluated using the original DFT model for nitrogen on carbon with slit shaped pores.

Electrodes were manufactured by mixing the carbon material with polymer binder (KynarFlex® 2801) in an 85–15 ratio by mass. An homogeneous slurry of this mixture with acetone was spread to a wet film thickness of 250 μm on 15 μm thick aluminium foil using a micrometer adjustable gap paint applicator. These sheets were dried under vacuum, cut into individual electrodes, and paired with those of identical mass to produce symmetrical cells.

2.2 Ionic liquids

The ionic liquids used in this study were purchased from Io-Li-Tec GmbH (Germany) and had a stated purity >99%. Fig. 1 illustrates the structure of the ions

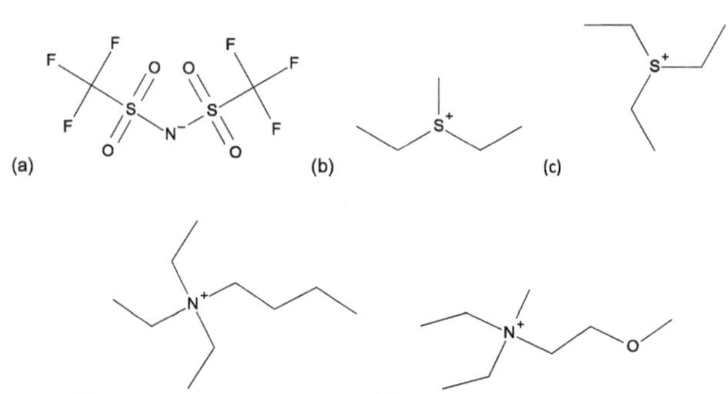

Fig. 1 Schematic structure of the constituent ions in the ionic liquids under study, (a) bis(trifluoromethanesulfonyl)imide [Tf_2N], (b) diethyl methyl sulfonium [S_{221}], (c) triethyl sulfonium [S_{222}], (d) butyl triethyl ammonium [N_{2224}], (e) *N*,*N*-diethyl-*N*-methyl-*N*-(2-methoxyethyl)ammonium [$N_{221(1O2)}$].

contained in these ILs along with the abbreviated name used throughout this report.

Prior to characterisation and cell assembly, the ILs were heated under rapid stirring for several hours in an argon filled glovebox (H_2O < 0.1 ppm, O_2 < 0.1 ppm). The moisture content of ILs dried using this technique was determined to be less than 7 ppm using Karl Fischer titration (KF899 Coulometer, Metrohm). Viscosity was measured using a Bohlin Gemini (Malvern Instruments Ltd.) rheometer with 50 mm diameter stainless steel conical plates (°1) using a gap size of 20 μm for shear rates from 10 to 1000 s^{-1}.

Electrochemical tests were carried out using a Solartron Analytical 1470E Multichannel Potentiostat/Galvanostat with 1455A Frequency Response Analysers. The electrochemical stability window of the ILs was determined using linear sweep voltammetry at 50 mV s^{-1} in a three electrode setup using a glassy carbon working electrode (3 mm dia.), Ag wire pseudo reference electrode and Pt wire counter electrode. Swagelok®-type symmetrical cells using non-porous electrodes were used to identify suitable operating potentials in a manner similar to that used in previous investigations. Linear sweep voltammetry was performed between 0 and 4.5 V at 50 mV s^{-1}, with operating potentials identified using a cut-off current of 30 μA cm^{-2}.

2.3 EDLC assembly and characterisation

Two-electrode button cells (2016) were assembled using stainless steel spacers, paired electrodes, and glass fibre filter paper separator (Whatman, GF/F). The separator was soaked with the electrolyte under study and the cell components were then placed under vacuum in the glovebox antechamber for *ca.* 5 min to encourage the impregnation of the electrolyte. Cells were then crimped closed inside the glovebox.

Electrochemical impedance spectroscopy (EIS) of symmetrical EDLCs was performed at a potential bias of 0.1 V (*vs* OCP), using a 10 mV perturbation over the frequency range 100 kHz to 10 mHz. Cells were also cycled galvanostatically between 0 V and the operating voltage at various rates between 0.1 and 10 A g^{-1} (where *IR* drop permitted) using an Arbin SCTS instrument. Specific capacitance values are based on the mass of active material and are expressed on a three-electrode basis. All measurements were performed in triplicate at 25 °C.

3 Results & discussion

3.1 Viscosity of ionic liquids

Low viscosity is a desirable property in EDLC electrolytes as lower viscosities are associated with higher specific capacitances and lower cell resistances.[23] Viscosities determined for each of the ILs in this work are given in Table 1 along with literature values of conductivity and cation volumes calculated using the Molinspiration Property Calculation Service.[40] (For comparison, the volume of the [Tf_2N] anion is 148 $Å^3$.)

The viscosity of [S_{221}][Tf_2N] was determined to be 38.4 mPa s, which is in broad agreement with that found in literature (lit.,[41] 36 mPa s). However, reports indicate that a significantly lower viscosity would be associated with [S_{222}][Tf_2N] (lit.,[31,42] 30–33 mPa s) which was not observed in this study; in this case,

Table 1 Physical properties[a] of the ionic liquids used in this work

IL	η [mPa s]	σ [mS cm^{-1}]	V_c [Å^3]	ESW [V]
$[S_{221}][Tf_2N]$	38.4	5.8 (ref. 41)	116	3.8
$[S_{222}][Tf_2N]$	38.6	7.1 (ref. 31)	133	2.8
$[N_{2224}][Tf_2N]$	140.3	1.3 (ref. 43)	196	3.9
$[N_{221(1O2)}][Tf_2N]$	66.5	2.6 (ref. 45)	171	4.7

[a] η represents viscosity (at 25 °C), σ conductivity (at 25 °C), V_c cation volume,[40] ESW electrochemical stability window determined in a three electrode cell (cut-off current density 40 μA cm^{-2}).

$[S_{222}][Tf_2N]$ exhibited a viscosity almost equal to that of $[S_{221}][Tf_2N]$. This may arise from differences in moisture content.

In contrast, the ammonium based ILs display relatively high viscosities resulting from their appreciably larger cation volumes. The viscosity of $[N_{2224}][Tf_2N]$ determined experimentally was 140.3 mPa s which also compares favourably with reported values (lit.,[43] 144.8 mPa s).

The ether group present in the $[N_{221(1O2)}]$ cation results in significantly lower viscosity (as reported for other ILs that incorporate ether-bonds in alkyl side chains[33,34]). Electron donation from the oxygen atom diminishes the charge of the cation resulting in weaker anion–cation interactions.[30] The viscosity observed for $[N_{221(1O2)}][Tf_2N]$ of 66.5 mPa s agrees with the majority of published values (lit.,[44,45] 64 mPa s at 24 °C, 69 mPa s) however it is noted that this is substantially different from the value reported by Sato *et al.* (120 mPa s (ref. 30)). Again, this disparity may arise from differences in moisture content.

From the properties given in Table 1 it can be anticipated that the sulfonium based ILs will be associated with increased values of specific capacitance, reduced resistances and better rate performance when compared with the ammonium ILs.

3.2 Electrochemical stability of ionic liquids

Table 1 also reports the ESW determined using linear sweep voltammetry in a three-electrode cell, with a cut-off current density of 40 μA cm^{-2}. These conditions result in narrower ESWs when compared with the majority of those reported in the literature. For comparison, the ESW observed under these conditions for a 1 mol l^{-1} solution of tetraethylammonium tetrafluoroborate ($TEABF_4$) in propylene carbonate was 4 V.

ESWs are extremely sensitive to the conditions used to define them. The point at which the current density is deemed large enough to indicate significant electrolyte decomposition is fairly arbitrary. Several flaws in using this method to determine the operating voltage of devices have been raised, and alternative approaches explored.[46–49] As the ESW tends to be asymmetric with respect to the open circuit potential of symmetric EDLCs, the operating potential that can be used without appreciable electrolyte decomposition is often substantially lower than the full ESW.

Operating potentials for the symmetric cells used in this work were identified using symmetrical two electrode cells in a manner similar to that employed in previous investigations.[34,38] (The operating potential determined under these conditions for a 1 mol l^{-1} solution of $TEABF_4$ in propylene carbonate was 2.1 V.) It is noted that higher operating potentials are achievable for these ILs, perhaps through the use of asymmetric electrode loadings to make fuller use of the ESW. The conservative values used herein minimise the possibility of surface area being rendered inactive by species generated by electrolyte decomposition.

Fig. 2 shows that there is little difference between the behaviour of the sulfonium based ILs. However it is worth noting that $[S_{221}][Tf_2N]$ is slightly more stable than $[S_{222}][Tf_2N]$ in both the three electrode and two electrode methods. This could suggest that the smaller cation volume of $[S_{221}][Tf_2N]$ influences the charge density of the ion resulting in slightly stronger ion–ion interactions.

A typical organic solvent-based electrolyte (1 mol l^{-1} $TEABF_4$ in propylene carbonate) is demonstrably less stable with increasing potential than the ILs as illustrated in Fig. 2. It is also evident that the ammonium based ILs have greater electrochemical stability than the sulfonium ILs which can be mainly attributed to the difference in cation volumes.

The methoxyethyl group in the side chain of $[N_{221(1O2)}]$ reduces the positive charge density at the ammonium centre when compared with $[N_{2224}]$. For smaller cations, it has been reported that this results in a diminished ESW[33,45] and this may be expected to occur in the case of $[N_{221(1O2)}][Tf_2N]$. However $[N_{221(1O2)}][Tf_2N]$ exhibits the widest ESW of the ILs studied using the three electrode method. It appears that the inclusion of the methoxyethyl group influences the degree of ionic coordination and enhances the stability of the IL.[45] In contrast with the ESWs in Table 1, Fig. 2 indicates that in a two-electrode configuration the ether bond reduces electrochemical stability, and the broader ESW seen in three electrode experiments does not result in an increased operating potential.

3.3 Characteristics of carbon materials

Pore characteristics of the activated carbon materials were investigated using nitrogen adsorption/desorption isotherms at −196 °C. Characteristics determined from the isotherms are given in Table 2. As the R/C ratio increases it can be seen that the specific surface area determined by both the BET and DFT methods increases. There is no substantial change in the micropore volume, however it increases slightly with R/C ratio.

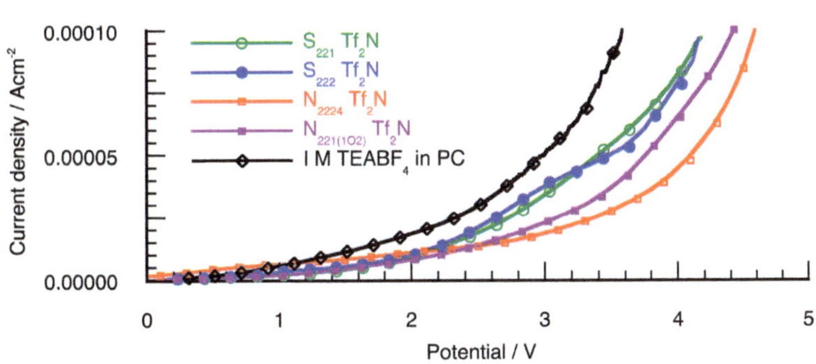

Fig. 2 Electrochemical stability of ionic liquids and a typical EDLC electrolyte determined in a symmetrical two-electrode cell at a sweep rate of 50 mV s^{-1}.

Table 2 Specific surface areas and pore characteristics calculated from nitrogen sorption isotherms at 77 K

RC	S_{BET}[a] [m^2 g^{-1}]	S_{DFT}[b] [m^2 g^{-1}]	V_{tot}[c] [cm^3 g^{-1}]	V_{mic}[d] [cm^3 g^{-1}]	V_{meso}[e] [cm^3 g^{-1}]	d_{BJH}[f] [nm]	d_{DFT}[g] [nm]
200	670	530	0.48	0.21	0.18	4.3	5.4
250	720	580	0.53	0.22	0.21	5.1	6.3
300	750	620	0.62	0.23	0.27	6.3	7.2
350	790	630	0.67	0.24	0.32	7.5	8.1

[a] Specific surface area calculated using the BET method. [b] Specific surface area determined using DFT model. [c] Total pore volume calculated at P/P_0 > 0.99. [d] Micropore volume determined using the *t*-plot method. [e] Mesopore volume from DFT model. [f] Modal pore width estimated from BJH pore size distribution. [g] Modal pore width estimated from DFT pore size distribution.

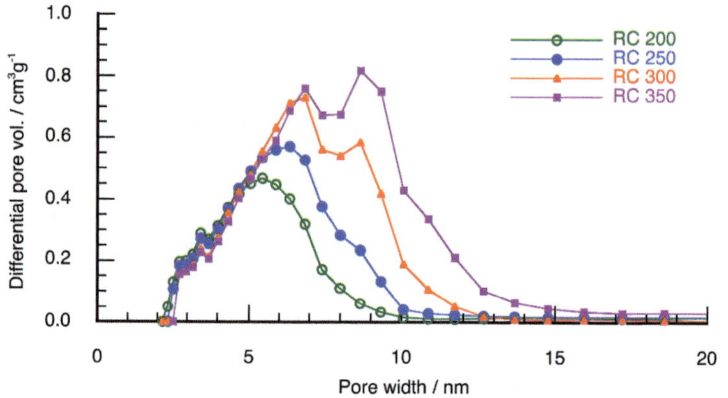

Fig. 3 Mesopore size distribution of resorcinol–formaldehyde derived activated carbon materials determined by the DFT model.

Mesopore volume increases markedly with increasing R/C ratio. This is illustrated in Fig. 3, which shows the pore size distribution determined using the DFT model for each of the carbon materials in the small mesopore region. As the R/C ratio increases, the mode of the distribution shifts to larger pore sizes. Estimates of the modal pore width attributed to each material from the DFT model indicate that increasing the RC ratio by 50 in this region results in an increase in the average pore width of roughly 0.9 nm. This trend was also observed when applying the BJH model to the desorption branch of the isotherms, as indicated by the values of d_{BJH} given in Table 2.

3.4 EDLC Performance – sulfonium based ionic liquids

Firstly, considering the behaviour of cells assembled using sulfonium based ILs, where the cation volume is smaller than that of the anion, Fig. 4(a) shows Nyquist plots produced using each of the carbon materials paired with $[S_{221}][Tf_2N]$. These spectra show that the cells do not behave as ideal capacitors, (*i.e.* vertical response at low frequencies) however at lower frequencies the response can be described as

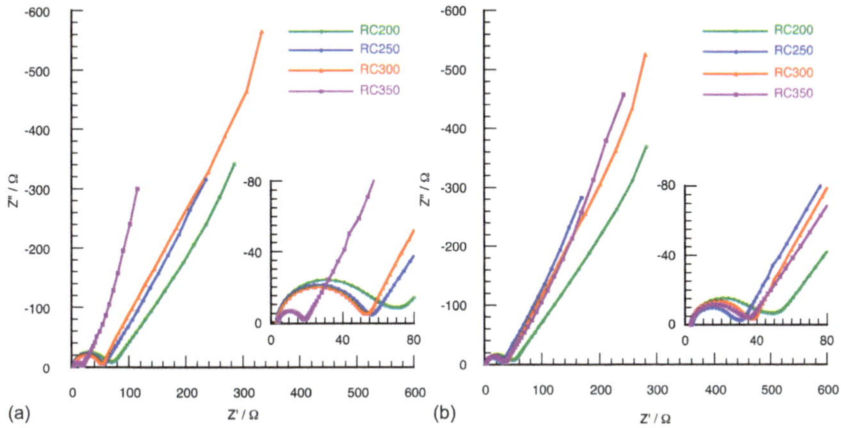

Fig. 4 Nyquist plots obtained using electrochemical impedance spectroscopy for sulfonium based ionic liquids with controlled porosity carbons including magnified high-frequency regions (inset). (a) $[S_{221}][Tf_2N]$ and (b) $[S_{222}][Tf_2N]$.

approaching ideal as the R/C ratio, and therefore pore width, increases. Increased pore widths are also associated with reduced specific capacitances as determined from the spectra at 10 mHz. Cell characteristics obtained from the spectra are summarised in Table 3.

The inset of Fig. 4(a) shows a magnified section of the spectra at higher frequencies and their intersection with the real axis (R_s) varies little with the carbon used. It can also be seen that the increased pore width from RC200 to RC350 has an influence on the diameter of the semicircle that appears in the high frequency region. This diameter is influenced by ionic mobility and represents a measure of the ionic resistance (R_i). As the pore size increases, R_i and equivalent series of resistance (ESR) are seen to decrease. This trend is also seen in values of resistance (R_{dc}) determined from the *IR* drop using 10 ms pulses at a current density of 2 mA cm^{-2}.

Table 3 Cell characteristics for sulfonium based ionic liquids combined with controlled porosity carbon materials[a]

IL	RC	c_s [F g^{-1}]	R_s [Ω]	R_i [Ω]	ESR [Ω]	$c_{0.2}$ [F g^{-1}]	$c_{2.0}$ [Fg^{-1}]	R_{dc} [Ω]
$[S_{221}][Tf_2N]$	200	27.4	3.4	55.3	41.1	39.1	17.1	39.6
	250	26.8	3.2	48.3	37.9	43.0	23.0	37.7
	300	18.1	3.3	46.3	36.9	32.2	17.4	32.2
	350	24.3	4.2	13.6	17.0	38.7	27.5	10.0
$[S_{222}][Tf_2N]$	200	22.8	3.9	35.9	31.9	35.4	16.4	26.1
	250	26.1	4.2	22.7	24.2	40.5	24.5	19.8
	300	18.0	3.0	30.9	29.0	28.8	16.5	25.5
	350	19.2	3.4	28.2	26.1	29.5	17.5	21.1

[a] c_s represents the specific capacitance (determined using EIS at 10 mHz), R_s the series resistance, R_i ionic resistance, ESR equivalent series resistance (determined at 1 kHz), $c_{0.2}$ specific capacitance determined at a current density of 0.2 A g^{-1} (and $c_{2.0}$ at 2.0 A g^{-1}) and R_{dc} represents the resistance determined using pulses of current at 2 mA cm^{-2}.

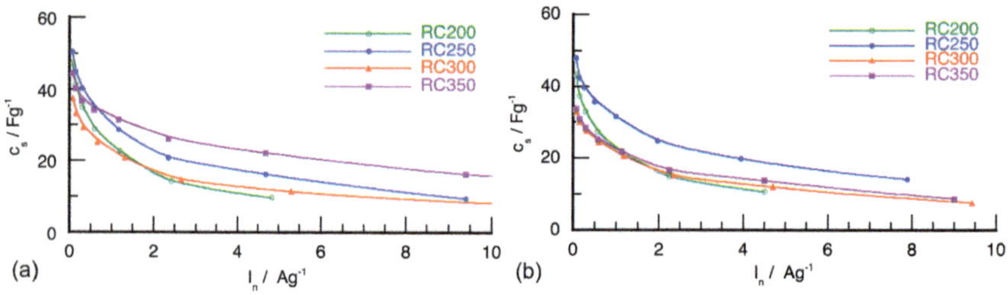

Fig. 5 Specific capacitance determined at different rates of constant current discharge between 0 and 2.5 V for cells using sulfonium based ILs. (a) $[S_{221}][Tf_2N]$ and (b) $[S_{222}][Tf_2N]$.

Increased resistances resulting from smaller pore sizes are also evident in Fig. 5(a), where the variation of specific capacitance with rate is illustrated. At rates greater than *ca.* 5 A g^{-1} for RC200, the *IR* drop was greater than the operating voltage of the cell. The curves associated with each carbon material do not display any obvious trend indicating that there are complicated relationships between the migration of ions, the packing of these ions at the electrode surface, and the pore characteristics of the electrode. Specific capacitances determined during galvanostatic discharge at rates of 0.2 A g^{-1} ($c_{0.2}$) and 2.0 A g^{-1} ($c_{2.0}$) are given in Table 3. Unlike the specific capacitances determined using EIS, the values of $c_{0.2}$ and $c_{2.0}$ in Table 3 are noticeably higher when RC250 is coupled with $[S_{221}][Tf_2N]$ when compared with RC200 and RC300.

Considering the spectra in Fig. 4(b), which show the behaviour of cells produced with $[S_{222}][Tf_2N]$, it can again be seen that the cells do not display ideal capacitor behaviour and that there is no clear correlation between low frequency behaviour and variations in pore width.

From the detail in the inset of Fig. 4(b), and the characteristics in Table 3, again it is seen that there is little variation in the values of R_s. However, unlike the cells using $[S_{221}][Tf_2N]$, there is no clear trend between the values of R_i, ESR or R_{dc} with the carbon material used. Specific capacitance is greatest when $[S_{222}][Tf_2N]$ is paired with RC250 and, moreover, R_i, ESR and R_{dc} are lowest for this combination. This supports the idea that the performance of ionic liquid based EDLCs can be improved by utilising carbon materials with carefully controlled pore characteristics.[37,38] The highest values of R_i, ESR and R_{dc} were observed with the carbon displaying the narrowest pore size. This is due to the increased fraction of vacant pore volume in these electrodes, as a substantial fraction of pores are too narrow to accommodate electrolyte ions.

Fig. 5(b) shows that the combination of RC250 with $[S_{221}][Tf_2N]$ unambiguously exhibits superior performance when compared with the other carbon materials.

Comparing the characteristics of cells using $[S_{221}][Tf_2N]$ with those using $[S_{222}][Tf_2N]$ there are marginally higher values of c_s and $c_{0.2}$ associated with $[S_{221}][Tf_2N]$ for each carbon material. This may result from a denser packing of ions under polarisation at the electrode surface, or the utilisation of a greater fraction of pores by the smaller $[S_{221}]$ cation. The absence of a methyl group in $[S_{221}][Tf_2N]$ may also reduce the distance between the charge centre of the adsorbed cation and the electrode surface, thereby displacing a greater amount of charge in the electrode.

Resistances (R_i, ESR and R_{dc}) exhibited by cells using $[S_{221}][Tf_2N]$ are larger than those using $[S_{222}][Tf_2N]$ when paired with RC200, 250 and 300. This could be expected due to the lower conductivity of $[S_{221}][Tf_2N]$. Surprisingly when the sulfonium based ILs are paired with RC350, $[S_{221}][Tf_2N]$ displays substantially lower resistances despite exhibiting lower conductivity than $[S_{222}][Tf_2N]$. This is also evident when comparing the rate performance of the sulfonium based ILs paired with RC350 in Fig. 5(a) and (b). It may be the case that by further increasing the pore width of the carbon a similar drastic reduction in resistances with $[S_{222}][Tf_2N]$ would be observed.

This behaviour may arise from the smaller cation volume of $[S_{221}][Tf_2N]$ influencing the properties of the electrode–electrolyte interphase as it can be anticipated that this region will display a conductivity mechanism different to that in than the bulk.

3.5 EDLC performance – ammonium based ionic liquids

Moving on to consider the behaviour of cells using ammonium based cations, where the volume of the cation is greater than that of the anion, spectra for cells generated using each of the carbons with $[N_{2224}][Tf_2N]$ are presented in Fig. 6(a) and with $[N_{221(1O2)}][Tf_2N]$ in Fig. 6(b). The most ideally capacitive behaviour is seen with RC300 for both ILs and furthermore, the highest values of specific capacitance and lower values of R_i and ESR are also observed with this carbon material (Table 4).

In Fig. 6(a) RC200 and RC250 are seen to behave similarly when paired with $[N_{2224}][Tf_2N]$, however the increase in pore width to RC300 results in a marked decrease in R_i and increase in specific capacitance. This is also seen in the values of specific capacitance determined through galvanostatic discharge. However it is noted that there is a substantial decrease in the value of c_s when RC350 is used that is not seen in the values of $c_{0.2}$ and $c_{2.0}$. This may be due to the high viscosity of $[N_{221(1O2)}][Tf_2N]$ hindering the perturbation signal at low frequencies. Nevertheless, Fig. 7(a) shows that RC300 and RC350 behave almost identically with respect to rate and perform significantly better than RC200 and RC250.

$[N_{221(1O2)}][Tf_2N]$ could be expected to perform very differently to $[N_{2224}][Tf_2N]$ due to the lower cation size, lower viscosity and higher conductivity. An

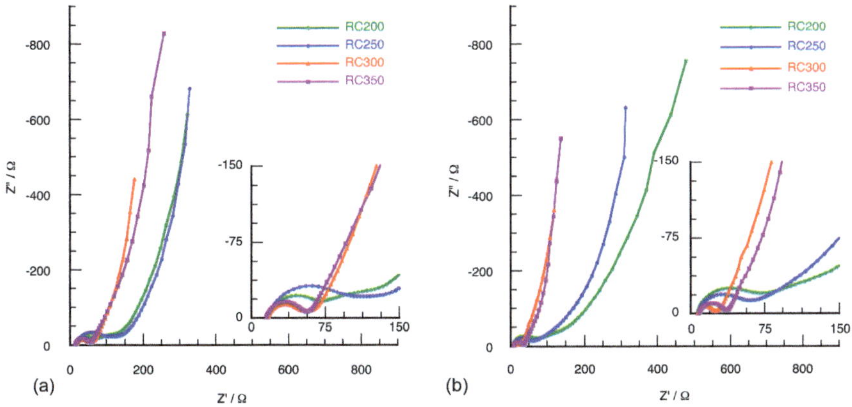

Fig. 6 Nyquist plots obtained using EIS for ammonium-based ILs with controlled porosity carbons including magnified high-frequency regions (inset). (a) $[N_{2224}][Tf_2N]$ and (b) $[N_{221(1O2)}][Tf_2N]$.

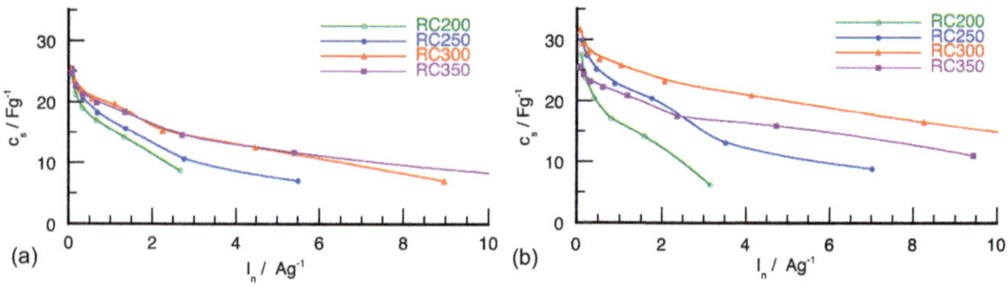

Fig. 7 Specific capacitance determined at different rates of constant current discharge between 0 and 3.0 V for cells using ammonium based ionic liquids and controlled porosity carbons. (a) $[N_{2224}][Tf_2N]$ and (b) $[N_{221(1O2)}][Tf_2N]$.

Table 4 Cell characteristics[a] for ammonium based ionic liquids combined with controlled porosity carbon materials

IL	RC	c_s [F g^{-1}]	R_s [Ω]	R_i [Ω]	ESR [Ω]	$c_{0.2}$ [F g^{-1}]	$c_{2.0}$ [F g^{-1}]	R_{dc} [Ω]
$[N_{2224}][Tf_2N]$	200	15.1	16.6	60.6	45.0	20.8	11.4	57.8
	250	16.6	17.2	84.0	48.8	22.1	13.3	63.8
	300	19.0	16.4	38.1	39.8	22.7	16.0	44.7
	350	14.1	15.0	42.8	37.8	22.3	16.4	39.4
$[N_{221(1O2)}][Tf_2N]$	200	16.3	7.4	61.3	42.1	23.6	12.0	46.5
	250	17.4	8.0	48.7	37.9	27.5	19.3	38.7
	300	22.7	7.4	16.0	21.6	28.4	23.2	19.4
	350	17.2	9.0	26.7	27.8	23.8	18.4	22.3

[a] c_s represents the specific capacitance (determined using EIS at 10 mHz), R_s the series resistance, R_i ionic resistance, ESR equivalent series resistance (determined at 1 kHz), $c_{0.2}$ specific capacitance determined at a current density of 0.2 A g^{-1} (and $c_{2.0}$ at 2.0 A g^{-1}) and R_{dc} represents the resistance determined using pulses of current at 2 mA cm^{-2}.

examination of Fig. 6(b) shows that there are subtle differences between the behaviour of the ammonium based ILs. At first glance there are clear similarities between the spectra obtained for RC200 and RC250 paired with each of the ILs, and there are significantly smaller values of R_i for RC300, increasing slightly as the pore width widens to RC350.

On closer inspection there is a noticeable difference in the values of R_s; $[N_{221(1O2)}][Tf_2N]$ typically exhibits values of roughly 8 Ω whereas $[N_{2224}][Tf_2N]$ consistently produces resistances around 16 Ω. Also, despite the values of R_i associated with RC200 being similar, there is a noticeable difference in the magnitude of R_i for the other carbon materials. This is a result of the difference in conductivity of the IL which may be attributed to the presence of the ether bond in $[N_{221(1O2)}][Tf_2N]$. In the case of RC200 (which exhibits the narrowest pore size), a large fraction the resistance originates from the presence of empty pores and electrolyte conductivity has little influence on the value of R_i.

At high frequencies there is little difference between RC300 and RC350 in the case of $[N_{2224}][Tf_2N]$ however, as with the sulfonium based ILs, there is a minimum R_i that coincides with a peak specific capacitance when RC300 is paired with $[N_{221(1O2)}][Tf_2N]$.

From Fig. 7(a) and (b) it can be seen that $[N_{221(1O2)}][Tf_2N]$ displays better performance than $[N_{2224}][Tf_2N]$ when paired with each of the carbons and that it is clear that the best performance is a combination of RC300 with $[N_{221(1O2)}][Tf_2N]$. Interestingly, Fig. 7(b) shows distinct kinks in the performance curves for the carbons possessing the narrower pore sizes (RC200 and 250), which are not present in the curves obtained using the other ILs. This may indicate the existence of a greater degree of ionic coordination resulting from the ether bond.

4 Conclusions

Ionic liquids were studied in relation to their performance as electrolytes in EDLCs with carbon electrodes possessing different pore characteristics. ILs consisting of sulfonium cations paired with the bis(trifluoromethanesulfonyl)-imide anion generally produced greater specific capacitances and lower resistances than the ammonium based ILs studied. This behaviour can be attributed to their smaller cation size, which strongly influences physical properties such as viscosity and ionic mobility. However it was seen that the ammonium based cations are associated with increased electrochemical stabilities, indicating that higher operating potentials are achievable with these electrolytes.

As the R/C ratio of the carbon precursor increased, mesopore volume and the average pore width increased, allowing for the comparison of EDLC performance with electrodes possessing slightly different pore characteristics. Despite displaying similar physical properties, the sulfonium based ILs performed remarkably differently with respect to pore width. The marginally smaller cation ($[S_{221}]$) performed better with the carbon exhibiting the largest pore width whereas peak performance for $[S_{222}][Tf_2N]$ was associated with a narrower pore size. A substantial reduction in resistance was observed at the larger pore width for $[S_{221}][Tf_2N]$ which is likely to be responsible for the improved performance.

Considerable differences between the performance of the ammonium based ILs were found and are related to their physical properties. In this case it is not only the smaller cation volume but also the inclusion of a methoxyethyl group in place of an ethyl side chain that affect physical properties and results in substantially improved electrochemical performance. This may be related to electron donation from the oxygen atom influencing the charge density of the cation and facilitating cation–cation interactions.

This study suggests that by altering the structure of the ions in the IL and by careful control of the pore characteristics of the electrode, the performance of EDLCs can be improved substantially. The development of more conductive and electrochemically stable ILs, coupled with tailored electrode materials, will enhance the energy density that can be attained by EDLCs and broaden the range of applications where they are employed.

Acknowledgements

The authors are indebted to the EPSRC for funding (EP/K021192/1).

References

1 B. E. Conway, *Electrochemical Supercapacitors – Scientific Fundamentals and Technological Applications*, Plenum, New York, 1999.
2 A. Burke, *J. Power Sources*, 2000, **91**, 37–50.

3 A. G. Pandolfo and A. F. Hollenkamp, *J. Power Sources*, 2006, **157**, 11–27.
4 R. Kötz and M. Carlen, *Electrochim. Acta*, 2000, **45**, 2483–2498.
5 P. J. Hall, M. Mirzaeian, S. I. Fletcher, F. B. Sillars, A. J. R. Rennie, G. O. Shitta-Bey, G. Wilson, A. Cruden and R. Carter, *Energy Environ. Sci.*, 2010, **3**, 1238.
6 J. R. Miller and A. F. Burke, *Electrochem. Soc. Interface*, 2008, 53–57.
7 P. Simon and Y. Gogotsi, *Nat. Mater.*, 2008, **7**, 845–854.
8 E. Frackowiak and F. Béguin, *Carbon*, 2001, **39**, 937–950.
9 J. R. Miller and P. Simon, *Science*, 2008, **321**, 651–652.
10 P. Simon and Y. Gogotsi, *Acc. Chem. Res.*, 2013, **46**, 1094–1103.
11 M. Winter and R. J. Brodd, *Chem. Rev.*, 2004, **104**, 4245–4269.
12 E. Frackowiak, Q. Abbas and F. Béguin, *J. Energy Chem.*, 2013, **22**, 226–240.
13 J. M. Blanes, R. Gutiérrez, A. Garrigós, J. L. Lizán and J. M. Cuadrado, *IEEE Trans. Power Electron.*, 2013, **28**, 5940–5948.
14 R. Carter, A. Cruden and P. J. Hall, *IEEE Trans. Veh. Technol.*, 2012, **61**, 1526–1533.
15 N. Omar, J. Van Mierlo, B. Verbrugge and P. Van den Bossche, *Electrochim. Acta*, 2010, **55**, 7524–7531.
16 R. A. Huggins, *Solid State Ionics*, 1996, **86–88**, 41–48.
17 A. Brandt, S. Pohlmann, A. Varzi, A. Balducci and S. Passerini, *MRS Bull.*, 2013, **38**, 554–559.
18 M. Armand, F. Endres, D. R. MacFarlane, H. Ohno and B. Scrosati, *Nat. Mater.*, 2009, **8**, 621–629.
19 M. C. Buzzeo, R. G. Evans and R. G. Compton, *ChemPhysChem*, 2004, **5**, 1106–1120.
20 M. Galiński, A. Lewandowski and I. Stępniak, *Electrochim. Acta*, 2006, **51**, 5567–5580.
21 T. Welton, *Chem. Rev.*, 1999, **99**, 2071–2083.
22 R. Mysyk, V. Ruiz, E. Raymundo-Piñero, R. Santamaria and F. Béguin, *Fuel Cells*, 2010, **10**, 834–839.
23 F. B. Sillars, S. I. Fletcher, M. Mirzaeian and P. J. Hall, *Phys. Chem. Chem. Phys.*, 2012, **14**, 6094–6100.
24 D. R. MacFarlane, J. Sun, J. Golding, P. Meakin and M. Forsyth, *Electrochim. Acta*, 2000, **45**, 1271–1278.
25 R.-S. Kühnel and A. Balducci, *J. Power Sources*, 2014, **249**, 163–171.
26 C. Arbizzani, M. Biso, D. Cericola, M. Lazzari, F. Soavi and M. Mastragostino, *J. Power Sources*, 2008, **185**, 1575–1579.
27 A. Balducci, R. Dugas, P. L. Taberna, P. Simon, D. Plée, M. Mastragostino and S. Passerini, *J. Power Sources*, 2007, **165**, 922–927.
28 M. Lazzari, F. Soavi and M. Mastragostino, *Fuel Cells*, 2010, **10**, 840–847.
29 A. B. McEwen, H. L. Ngo, K. LeCompte and J. L. Goldman, *J. Electrochem. Soc.*, 1999, **146**, 1687–1695.
30 T. Sato, G. Masuda and K. Takagi, *Electrochim. Acta*, 2004, **49**, 3603–3611.
31 H. Matsumoto, T. Matsuda and Y. Miyazaki, *Chem. Lett.*, 2000, 1430–1431.
32 D. Wei and T. W. Ng, *Electrochem. Commun.*, 2009, **11**, 1996–1999.
33 M. J. Monteiro, F. C. Camilo, M. C. C. Ribeiro and R. M. Torresi, *J. Phys. Chem. B*, 2010, **114**, 12488–12494.
34 A. J. R. Rennie, N. Sanchez-Ramirez, R. M. Torresi and P. J. Hall, *J. Phys. Chem. Lett.*, 2013, **4**, 2970–2974.

35 C. O. Ania, J. Pernak, F. Stefaniak, E. Raymundo-Piñero and F. Béguin, *Carbon*, 2006, **44**, 3126–3130.
36 C. Arbizzani, S. Beninati, M. Lazzari, F. Soavi and M. Mastragostino, *J. Power Sources*, 2007, **174**, 648–652.
37 S. Pohlmann, B. Lobato, T. A. Centeno and A. Balducci, *Phys. Chem. Chem. Phys.*, 2013, **15**, 17287–17294.
38 F. B. Sillars, S. I. Fletcher, M. Mirzaeian and P. J. Hall, *Energy Environ. Sci.*, 2011, **4**, 695–706.
39 A. J. R. Rennie and P. J. Hall, *Phys. Chem. Chem. Phys.*, 2013, **15**, 16774–16778.
40 Molinspiration Cheminformatics, http://www.molinspiration.com.
41 S. Fang, L. Yang, C. Wei, C. Peng, K. Tachibana and K. Kamijima, *Electrochem. Commun.*, 2007, **9**, 2696–2702.
42 H. Matsumoto, H. Sakaebe and K. Tatsumi, *J. Power Sources*, 2005, **146**, 45–50.
43 Z. Wang, Y. Cai, T. Dong, S. Chen and X. Lu, *Ionics*, 2013, **19**, 887–894.
44 T. Kondoh, A. Asano, J. Yang, K. Norizawa, K. Takahashi, M. Taguchi, R. Nagaishi, R. Katoh and Y. Yoshida, *Radiat. Phys. Chem.*, 2009, **78**, 1157–1160.
45 Z. B. Zhou, H. Matsumoto and K. Tatsumi, *Chem.–Eur. J.*, 2005, **11**, 752–766.
46 D. Weingarth, A. Foelske-Schmitz and R. Kötz, *J. Power Sources*, 2013, **225**, 84–88.
47 D. Weingarth, H. Noh, A. Foelske-Schmitz, A. Wokaun and R. Kötz, *Electrochim. Acta*, 2013, **103**, 119–124.
48 K. Xu, M. S. Ding and T. R. Jow, *Electrochim. Acta*, 2001, **46**, 1823–1827.
49 K. Xu, S. P. Ding and T. R. Jow, *J. Electrochem. Soc.*, 1999, **146**, 4172–4178.

Faraday Discussions

PAPER

Redox-active electrolyte for supercapacitor application

Elzbieta Frackowiak,* Mikolaj Meller, Jakub Menzel, Dominika Gastol and Krzysztof Fic

Received 25th March 2014, Accepted 6th May 2014

DOI: 10.1039/c4fd00052h

This paper reports the electrochemical behaviour of supercapacitor carbon electrodes operating in different aqueous solutions modified by various redox-active species (hydroxybenzenes, bromine derivatives and iodide). Three dihydroxybenzenes with varying stereochemistry, *i.e.*, –OH substitution, have been considered as electrolyte additives (0.38 mol L^{-1}) in acidic, alkaline and neutral solutions. High capacitance values have been obtained, especially for the acidic and alkaline solutions containing 1,4-dihydroxybenzene (hydroquinone). Bromine derivatives of dihydroxybenzenes were also considered as the additive in alkaline solution for use as a supercapacitor electrolyte, and a significant increase in capacitance value was observed. The redox couple investigated next was an iodide/iodine system, where 2 mol L^{-1} NaI aqueous electrolyte was utilized. In this case, the most promising faradaic contribution during capacitor operation was achieved. In particular, stable capacitance values from 300–400 F g^{-1} have been confirmed by long-term galvanostatic cycling (over 100 000 cycles), cycling voltammetry and floating. The mechanism of pseudocapacitance phenomena was discussed and supported by electrochemical and physicochemical measurements, *e.g.*, *in situ* Raman spectroscopy.

Introduction

Electrical double layer capacitors (EDLCs) have recently attracted great attention because of their high power rates and excellent cyclability. The total charge accumulated on an electrode/electrolyte interface is proportional to the electrochemically available surface area of the electrode, hence, only materials with well developed porosity, like activated carbons, can play a perfect role as an electrode for these devices.[1,2] Generally, the capacitance of the electrical double layer does not exceed 120–150 F g^{-1}, which results in a poor energy density, especially in an aqueous medium with a voltage window limited by water decomposition to *ca.* 1 V.[3] Obviously, in an aprotic organic medium or ionic liquid with a voltage window above 2.0 V, the energy density is significantly higher, but these

Poznan University of Technology, Institute of Chemistry and Technical Electrochemistry, Piotrowo 3, 60965 Poznan, Poland. E-mail: elzbieta.frackowiak@put.poznan.pl; Fax: +48 665 39 71; Tel: +48 665 36 32

electrolytes are rather environmentally unfriendly and require expensive assembling processes, which adversely affects their final price on the market. Apart from charge accumulation in the EDL, additional capacitance can be provided from faradaic reactions occurring on the electrode/electrolyte interface.[4] Transition metal oxides in symmetric and asymmetric configurations, nitrogen or oxygen enriched carbons with tailored functionalities or electrochemically conducting polymers are usually used for such capacitance enhancement.[5–11] On the other hand, pseudocapacitance from the electrode materials is always limited by slow diffusion and very often moderate electrolyte penetration. From this point of view, the pseudocapacitance originating from the redox active electrolyte solution[12–15] seems to be a more profitable one; the electroactivity of the iodine/iodide as well as bromine/bromide redox couples remains the origin of tremendous capacity values,[16–22] however, these systems still need to be optimized and unusual behaviour should be clearly elucidated. Hence, the I^-/I_2 system has been revisited here and investigated by us in detail. Electrolytes with quinone/hydroquinone[23–25] or with other benzene derivatives[26–30] can also contribute greatly to pseudocapacitance phenomena. Additionally, the molecular grafting of quinones cannot be excluded or neglected.[23,26,27,30–32] The abundance of dihydroxybenzene isomers encouraged us to study more precisely the influence of substituent location (*ortho*, *meta* or *para*) on their electrochemical activity and tentative application as redox active electrolytes for EDLCs in aqueous media. All these isomers will be called by their common names recommended by IUPAC; 1,4-dihydroxybenezene will be considered as hydroquinone, 1,3-dihydroxybenzene as resorcinol and 1,2-dihydroxybenzene as catechol. The practical application of redox active species such as hydroxybenzenes and their bromine derivatives as electrolyte additives, and iodides acting as the supercapacitor electrolyte, will be discussed.

Experimental

All electrochemical measurements were made under ambient conditions, *i.e.* at temperatures ranging from 21 °C to 25 °C. Two- and three-electrode capacitors were assembled in a Swagelok® system. For the acidic medium, gold current collectors were used. Stainless steel collectors served in alkaline and neutral media. Each electrode was composed of 85 wt% activated carbon, 10 wt% poly(vinylidene fluoride) (PVDF Kynar Flex 2801) and 5 wt% carbon black. Microporous carbon prepared by KOH activation[33] with a surface area of 1400 $m^2\ g^{-1}$ was used to form pressed pellets (9.8–10.3 mg) with a geometric surface area of 0.785 cm^2 per electrode and thickness of about 0.3 mm. Some electrochemical investigations were conducted on microporous carbon tissue (Kynol®) to eliminate the effect of the binder. 1 mol L^{-1} aqueous solution of H_2SO_4 with 1,2-, 1,3- or 1,4-dihydroxybenzene addition at a concentration of 0.38 mol L^{-1} served as the electrolyte. To study the effect of pH on the electrochemical activity of these species (especially 1,4-dihydroxybenzene), 6 mol L^{-1} KOH and 1 mol L^{-1} Li_2SO_4 have also been applied. The electrodes were soaked in the appropriate electrolytes and subjected to electrochemical measurements: cyclic voltammetry (CV) at various scan rates (1–20 mV s^{-1}), constant current charging–discharging (0.2–20 A g^{-1}), and electrochemical impedance spectroscopy (EIS) in the frequency range 100 kHz to 1 mHz with a voltage/potential amplitude of ±5 mV. To estimate capacitor ageing, polarization at a constant voltage, so called floating, was performed according to international standard IEC 62391-1. After every 2 hours of floating, galvanostatic testing (2 A g^{-1}) was done to measure the capacitance. Additionally, full electrochemical characterization (EIS, CV) was performed. For clarity, all values of the capacitance are expressed for one electrode, which means for the total system the capacitance is 25%. Moreover, physicochemical characterisation of the carbon materials before and after electrochemical performance testing was done in order to confirm the assumed grafting of the redox species. A potentiometric titration of the carbon materials was realized at ambient temperature (25 °C) using a Titrino 848 instrument (Metrohm®, Switzerland). The procedure for pK_a distribution and proton binding curve calculation was as follows: firstly, the pristine carbon material and carbon electrodes (*ca.* 10 mg) were stirred for 24 hours in 50 mL of 0.01 mol L^{-1} $NaNO_3$ solution with an ionic strength of 0.01. After stirring, the pH of the solution (containing carbon material) was measured and adjusted to 3 by controlled 0.1 mol L^{-1} HCl addition. After 15 minutes of pH stabilization, the solution was titrated with 0.1 mol L^{-1} NaOH until pH 11 was reached. The experimental data were transformed to proton binding isotherms Q(pH) representing the total number of protonated sites. Briefly, the proton binding isotherm is closely related to the pK_a distribution by the following equation:

$$Q(\mathrm{pH}) = \int q(\mathrm{pH}, \mathrm{p}K_\mathrm{a}) f(\mathrm{p}K_\mathrm{a}) \mathrm{dp}K_\mathrm{a}$$

A solution of this equation was obtained using the numerical procedure SAIEUS® (Solution of Adsorption Integral Equation Using Splines) by Jacek Jagiello (Micromeritics®, USA).

In order to compare and confirm the results obtained by the potentiometric titration, TGA experiments were done. All carbon materials were subjected to heating up to 950 °C with a 10 K min^{-1} heating rate using a 50 mL min^{-1} Ar protective flow in a Jupiter 209 thermomicrobalance (Netzsch®, Germany). The mass change was recorded simultaneously to mass spectra analysis of the decomposition products (done on an Aelos instrument by Netzsch®, Germany). Particularly, m/z = 18, 28 and 44 signals were studied, because one can assume that they might be related to H_2O, CO and CO_2 evolution. Furthermore, these compounds indicate the presence of oxygen functional groups, generated during the grafting process.

Apart from dihydroxybenzene, bromine derivatives, *i.e.*, dibromodihydroxybenzene, have also been studied as a source of redox reactions in the electrolyte. The amount of additive was 0.2 mol L^{-1} $C_6H_4Br_2O_2$ dissolved in different concentrations of KOH (1–6 mol L^{-1}).

To study the effect of halides on the capacitor performance, 2 mol L^{-1} NaI aqueous solution was used. This concentration was selected from previous studies.[18] For careful analysis of pseudocapacitive effects, *in situ* studies were performed using a DXR dispersive Raman microscope (Thermo Fischer Scientific, USA) equipped with a laser of 532 nm wavelength and 5 mW power, fluorescence and white light linearly corrected, with a medium cosmic radiation threshold detection.

Results and discussion

Detailed electrochemical characterization was realized in order to evaluate the electrochemical activity of the aforementioned redox active species by various methods.

Dihydroxybenzenes

Cyclic voltammetry (CV) performed at various scan rates (1 mV s^{-1} to 20 mV s^{-1}) for dihydroxybenzenes in 1 mol L^{-1} H_2SO_4 revealed the strong influence of hydroxyl group location on their electrochemical activity. CV curves at 10 mV s^{-1}, with measured current values recalculated to capacitance, are presented in Fig. 1.

A capacitor operating in pure 1 mol L^{-1} H_2SO_4 solution is also shown for comparison. The highest capacitance values were observed for the capacitor operating in 1 mol L^{-1} H_2SO_4 with 0.38 mol L^{-1} hydroquinone, *i.e.* 280 F g^{-1} in comparison to 124 F g^{-1} for the electrolyte without hydroquinone. For the capacitor operating in the electrolyte with catechol addition, the increase in capacitance is rather insignificant (140 F g^{-1}), and for the electrolyte with resorcinol, the capacitance is even worse than for unmodified 1 mol L^{-1} H_2SO_4 (95 F g^{-1}). This might be related to the poorer conductivity of the electrolyte solution, aggravated by resorcinol addition, as well as some pore blocking. More detailed information was obtained from a three-electrode cell. Cyclic voltammograms (5 mV s^{-1}) for the negative and positive electrodes separately are provided in Fig. 2 and 3.

Both electrodes have different electrochemical responses and they operate in a significantly shifted potential range, even if the capacitor works at a constant voltage (0.8 V) and the same pH. These results are in good agreement with the values obtained for the two-electrode cell. Typical redox behaviour with a sharp peak can be observed, especially for the electrolyte containing 0.38 mol L^{-1} hydroquinone. Apart from the characteristic reduction/oxidation of these species (immobilized on the positive electrode) disclosed by high capacitance values in a very narrow potential range (only 100 mV), hydroquinones can also be electrochemically "grafted" onto the carbon surface, which is an additional source of pseudocapacitance. Such electrochemical "grafting" is presented in Fig. 4, where

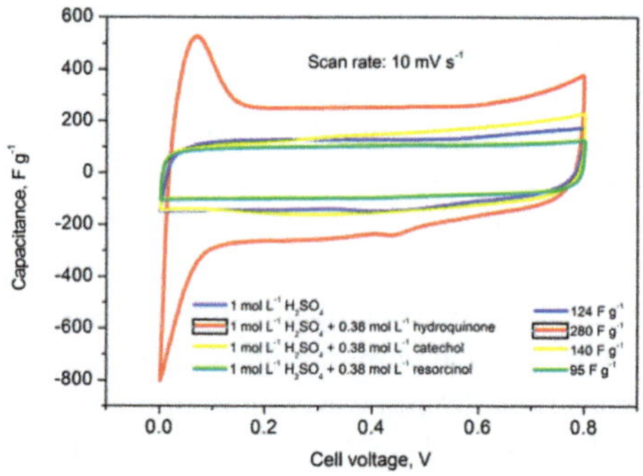

Fig. 1 Cyclic voltammograms for capacitors operating in 1 mol L^{-1} H_2SO_4 solution with different dihydroxybenzenes.

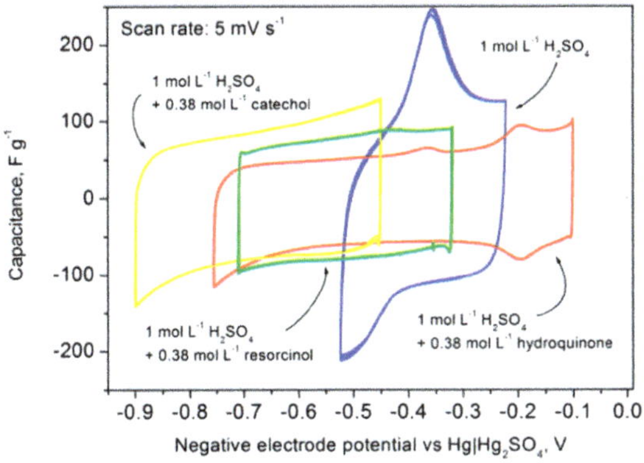

Fig. 2 Cyclic voltammograms for the negative electrode at a 5 mV s^{-1} scan rate operating in dihydroxybenzene-modified acidic electrolytes.

the total capacitance of a system operating in 1 mol L^{-1} H_2SO_4 with 0.38 mol L^{-1} hydroquinone increases with the cycle number. This confirms that quinone–hydroquinone functionalities can be "grafted" and/or strongly attracted through π bonding onto the carbon surface without aggravating the conductivity of the electrode. The capacitance originating from the electrostatic attraction of the ions in the sulphuric acid solution only was equal to *ca.* 125 F g^{-1} for the first cycle; after 550 cycles the total capacitance was 283 F g^{-1}. This clearly suggests a faradaic origin of the capacitance increase.

Unfortunately, it is impossible to determine precisely the contributions of the charge provided by the simple reduction and oxidation of the hydroquinone immobilized on the electrode/electrolyte interface, and the charge provided by the redox activity of quinone-type functionalities bonded to the surface.

A big discrepancy between the behaviour of the positive and negative electrodes is also observed well during the galvanostatic charge–discharge investigation at a current load of 0.5 A g^{-1} (Fig. 5). The positive electrode operates in a

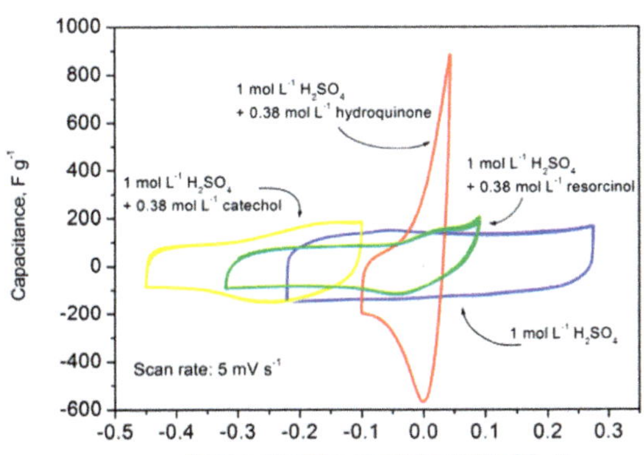

Fig. 3 Cyclic voltammograms for the positive electrode at a 5 mV s^{-1} scan rate operating in dihydroxybenzene-modified acidic electrolytes.

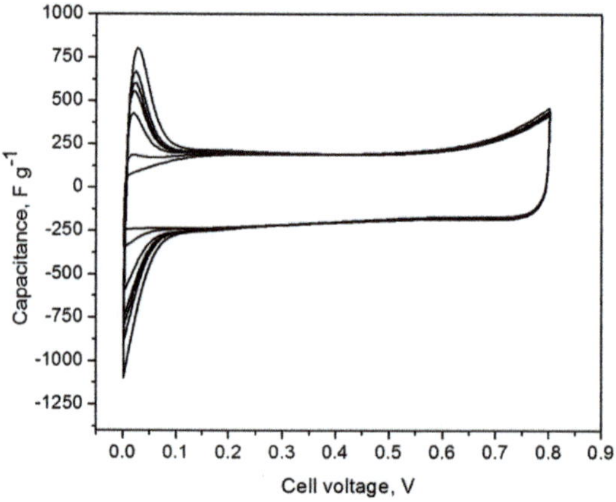

Fig. 4 The cyclic voltammogram (5 mV s^{-1}) for a capacitor operating in 1 mol L^{-1} H_2SO_4 with 0.38 mol L^{-1} hydroquinone additive.

0.1 V potential range. Unfortunately, the efficiency of this process is quite low (63%), which proves its high irreversibility, hence, practical application of this redox couple could be limited.

Electrochemical impedance spectroscopy results also follow the trend observed for the cyclic voltammetry and galvanostatic charging–discharging. Tremendous capacitance values revealed for the electrodes operating in 1 mol L^{-1} H_2SO_4 with 0.38 mol L^{-1} hydroquinone are assumed to be attributed to the redox process of the quinone/hydroquinone couple; high capacitance values (even *ca.* 1500 F g^{-1} at 1 mHz), recorded in low frequency regimes only, were accompanied by moderate charge propagation, which implies a diffusion-limited process contributing to a charge accumulation. Obviously, such high capacitance values will never have practical application. At high frequencies, no significant difference in electrochemical behaviour was observed; results in the form of Nyquist and Bode plots are presented in Fig. 6–8.

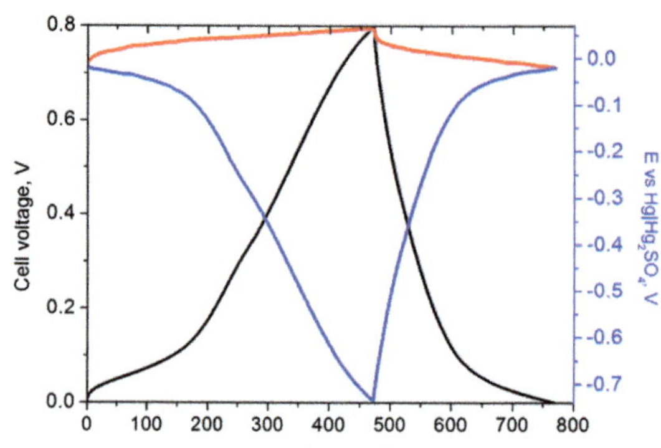

Fig. 5 The galvanostatic charging–discharging curve at a current load of 0.5 A g^{-1} for a capacitor operating in 1 mol L^{-1} H_2SO_4 with 0.38 mol L^{-1} hydroquinone additive.

For the positive electrode operating in 1 mol L^{-1} H_2SO_4 with 0.38 mol L^{-1} hydroquinone additive, a typical faradaic response was observed; this suggests that this electrode is mainly responsible for the redox contribution in the system (Fig. 7).

The change of phase angle (Fig. 8) also suggests different types of charge storage mechanisms, especially in the diffusion limited area, *i.e.*, in the low frequency region. For the negative electrode, the course is rather monotonic and confirms a rather electrostatic type of charge accumulation. On the other hand, the two humps observed for the positive electrode suggest the combination of typical faradaic and electrostatic (capacitive) processes.

The impedance spectra for the capacitors operating in various dihydroxybenzenes dissolved in 1 mol L^{-1} H_2SO_4 aqueous solution are presented in Fig. 9.

As it can be seen in Fig. 1, 2, 3 and 10, the electrochemical activity of the dihydroxybenzenes strongly depends on the hydroxyl group location (*para*, *ortho* or *meta*). In an acidic medium, hydroquinone with *para*-type substitution is the most active because of the distance between the hydroxyl groups on the benzene ring. The electrochemical activity of the quinone/hydroquinone redox couple is presented in Fig. 10. In the case of catechol, the hydroxyl groups can be oxidized even with aromatic structure deformation, but this process does not occur as easily as for hydroquinones. Resorcinol oxidation requires an additional activator in the reaction medium, because the oxidized form of this compound contains an unstable and thermodynamically non-preferable conjunction of *meta*-benzoquinoide bonds, so its oxidation was not observed.

Additionally, the substituent location (*para*, *ortho* or *meta*) strongly affects the electrode potential range. Significant shifts towards more cathodic values were recorded for catechol (−400 mV) and resorcinol (−200 mV). On the other hand, the potential range for positive electrodes operating in these electrolytes are narrower, which implies their faradaic activity, especially for the electrolyte with hydroquinone. Obviously, reduction/oxidation processes on activated carbon electrodes cannot show straightforwardly Nernstian behaviour with well-defined peaks because, on these electrodes, diffusion strongly affects the transport from

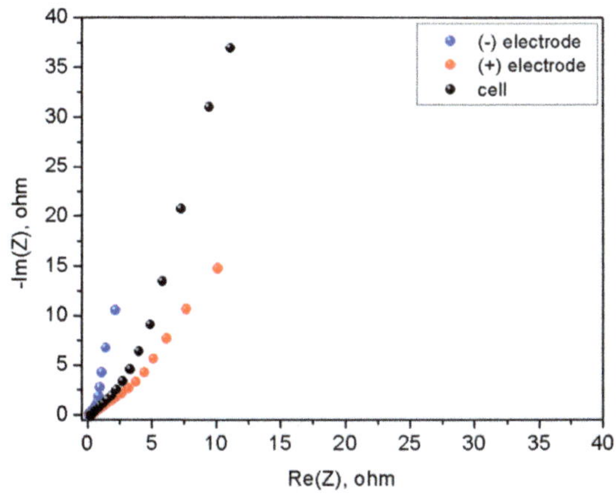

Fig. 6 The Nyquist plot for a capacitor operating in 1 mol L^{-1} H_2SO_4 with 0.38 mol L^{-1} hydroquinone additive.

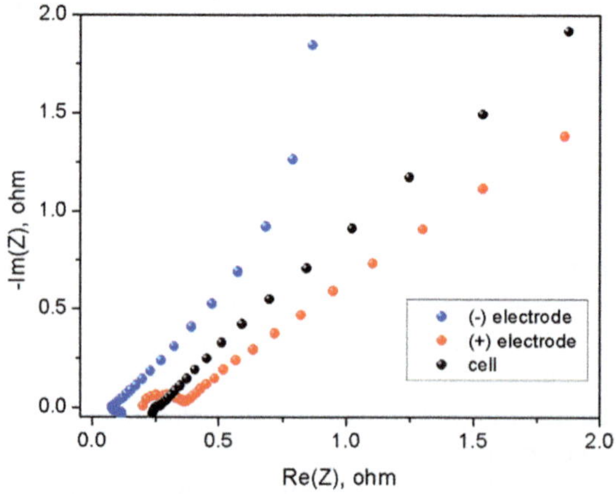

Fig. 7 A low frequency inset of the Nyquist plot for a capacitor operating in 1 mol L^{-1} H_2SO_4 with 0.38 mol L^{-1} hydroquinone additive.

the electrolyte bulk to the electrode surface, which results in an ill-defined potential wave. Moreover, considering the fact that the quinone reduction/oxidation is a proton consuming-process, the electrochemical activity should also depend strongly on the pH of the electrolyte. In order to investigate the influence of the pH, the three most often applied electrolytes, *i.e.*, 1 mol L^{-1} H_2SO_4, 1 mol L^{-1} Li_2SO_4 and 6 mol L^{-1} KOH, were selected and modified by a hydroquinone addition (0.38 mol L^{-1}). Cyclic voltammograms recorded at 10 mV s^{-1} scan rate are presented in Fig. 11.

The capacitance values are significantly higher for the capacitor operating in alkaline and acidic media (*i.e.*, 283 and 275 F g^{-1}, respectively) than for the neutral one (140 F g^{-1}). Additionally, obtained capacitance values are considerably higher than for the capacitors operating in unmodified electrolytes; for the

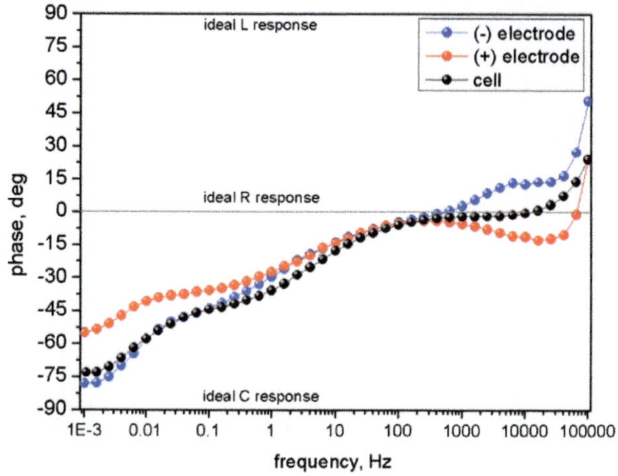

Fig. 8 The Bode plot for a capacitor operating in 1 mol L^{-1} H_2SO_4 with 0.38 mol L^{-1} hydroquinone additive.

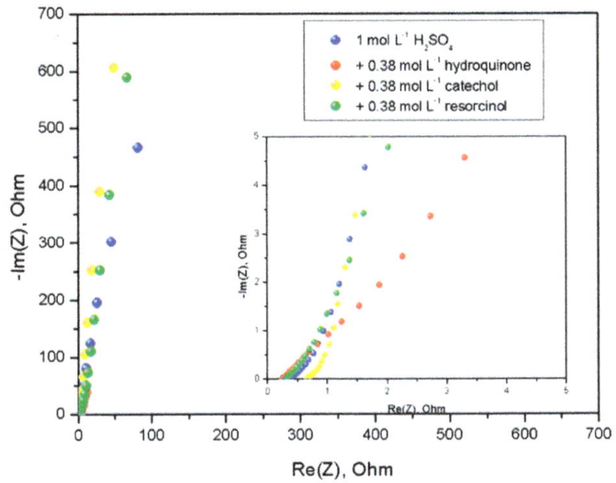

Fig. 9 Nyquist plots for capacitors operating in 1 mol L^{-1} H_2SO_4 with 0.38 mol L^{-1} dihydroxybenzene additives.

capacitor operating in 6 mol L^{-1} KOH electrolyte, the measured capacitance at the same scan rate was *ca.* 130 F g^{-1}, for 1 mol L^{-1} H_2SO_4 it was 123 F g^{-1}, and for the Li_2SO_4 solution, no significant difference in the capacitance was observed. Moreover, the different shapes of the cyclic voltammograms for the capacitors operating in acidic and alkaline media imply different quinone/hydroquinone reduction and oxidation mechanisms. In an acidic medium with readily accessible protons these processes occur quite easily along the applied voltage. In an alkaline medium, strong intermolecular hydrogen-bonding interactions between quinone and the hydroxyl and functional groups have a significant influence on the redox kinetics, therefore, the activity can be observed in a narrow voltage range, manifesting as small but reversible peaks. A three-electrode investigation confirmed that neither well-defined redox response is on the positive or negative electrode, however, an increase in their capacitance has been observed (Fig. 12).

Galvanostatic charge–discharge also confirms that in the alkaline medium no typical redox response is observed. A two- and three-electrode investigation at 0.5 A g^{-1} is presented in Fig. 13.

Finally, the redox activity of the quinone (Q)/hydroquinone (QH_2) redox couple can be described by the following reactions, in an acidic medium:

$$Q + e^- \leftrightarrow Q^{\bullet -} \tag{1}$$

$$Q^{\bullet -} + H^+ \leftrightarrow QH^{\bullet} \tag{2}$$

$$QH^{\bullet} + e^- \leftrightarrow QH^- \tag{3}$$

$$QH^- + H^+ \leftrightarrow QH_2 \tag{4}$$

giving finally:

$$Q + 2H^+ + 2e^- \leftrightarrow QH_2 \tag{5}$$

and without protons in an alkaline (HA) medium:

$$Q + e^- \leftrightarrow Q^{\bullet -} \quad (6)$$

$$Q^{\bullet -} + HA \leftrightarrow A^- + HQ^{\bullet} \quad (7)$$

$$HQ^{\bullet} + e^- \leftrightarrow HQ^- \quad (8)$$

$$HQ^- + HA \leftrightarrow H_2Q + A^- \quad (9)$$

giving finally:

$$Q + 2HA + 2e^- \leftrightarrow H_2Q + 2A^- \quad (10)$$

Additionally, self-protonation cannot be neglected.

Physicochemical measurements of both electrodes conducted separately after electrochemical polarization in the presence of H_2Q in an acidic medium are also in good accordance with the results obtained from the electrochemical

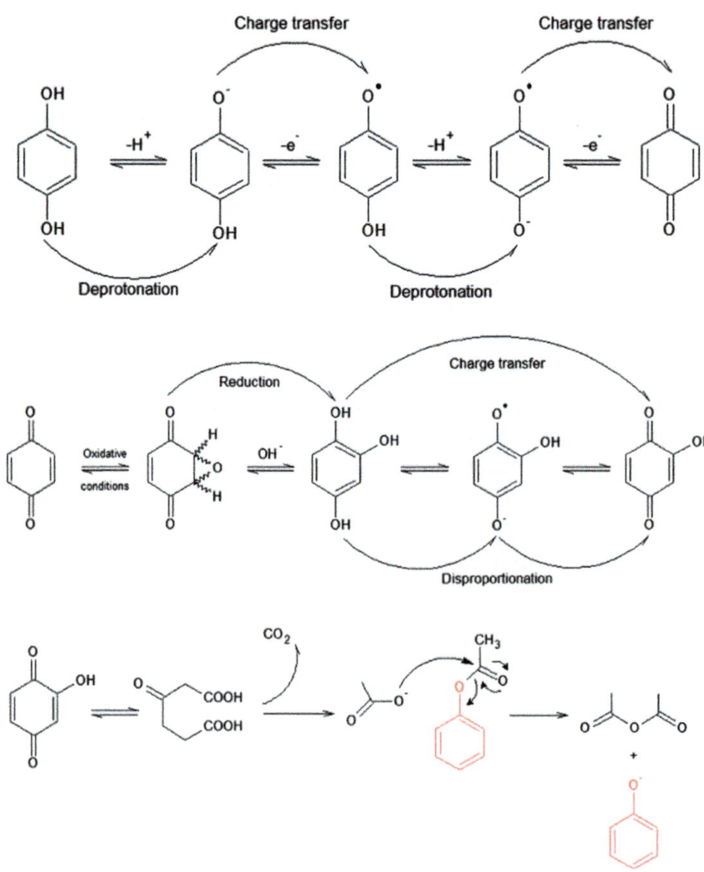

Fig. 10 The electrochemical activity of hydrobenzenes in an aqueous electrolytic medium on activated carbon electrodes.

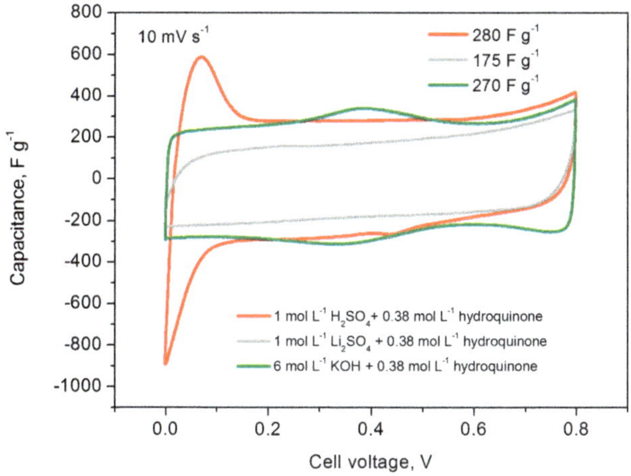

Fig. 11 Cyclic voltammograms for a capacitor operating in aqueous electrolytes of different pH with hydroquinone addition.

investigation. Proton binding curves (Fig. 14) revealed higher oxygen-based functionalities on the carbon electrodes, especially in the range of pK_a related to =CO functional groups (*i.e.* in the range 8–9) or for easily protonated ones with pK_a close to 5. Unfortunately, with the sensitivity of the method, one cannot determine precisely the types of functionalities, even if, in the literature, some assumptions are presented. Especially for the peaks in the pK_a range 8.5–10.0 one can observe some differences in response, however, they are too close to each other to differentiate and attribute them to typical functional groups. The most intense peaks observed at pK_a values of 3 and 11 should be related to the buffering effect of water.

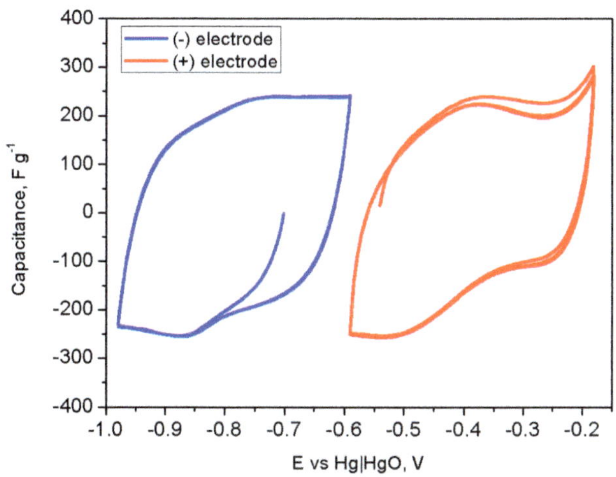

Fig. 12 Cyclic voltammograms for the positive and negative electrodes of a capacitor operating in 6 mol L^{-1} KOH with hydroquinone addition.

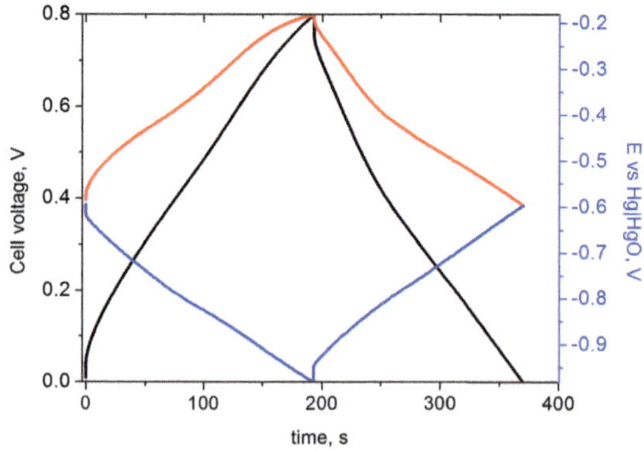

Fig. 13 The galvanostatic charging–discharging curve at current load of 0.5 A g^{-1} for a capacitor operating in 6 mol L^{-1} KOH with 0.38 mol L^{-1} hydroquinone additive.

TGA-MS experiments gave even more reliable evidence for the increase in oxygen-based functionalities. The mass change of the activated carbon, subjected to investigation and serving as a reference, was 7%. For the electrode after operation in unmodified 1 mol L^{-1} H_2SO_4, the mass change was 24%, with a significant decomposition signal of $m/z = 44$ at 200–300 °C, which suggests the evacuation of carboxyl-like functional groups. The mass change of the electrode after electrochemical operation in the electrolyte containing 0.38 mol L^{-1} of hydroquinone is 38%, which might be attributed to the decomposition of the generated oxygen-based species. Especially, the significant increase in the $m/z =$ 28 signal, related to =CO functional groups decomposing at higher temperatures, was observed. This suggests that, instead of carboxyl-like functionalities, some carbonyl species were generated. For the negative electrode of the capacitor operating in 1 mol L^{-1} H_2SO_4 containing 0.38 mol L^{-1} of hydroquinone, the recorded mass change was 26%. The kinetics of decomposition in this case suggests a major contribution from the fast evacuation of carboxyl-like

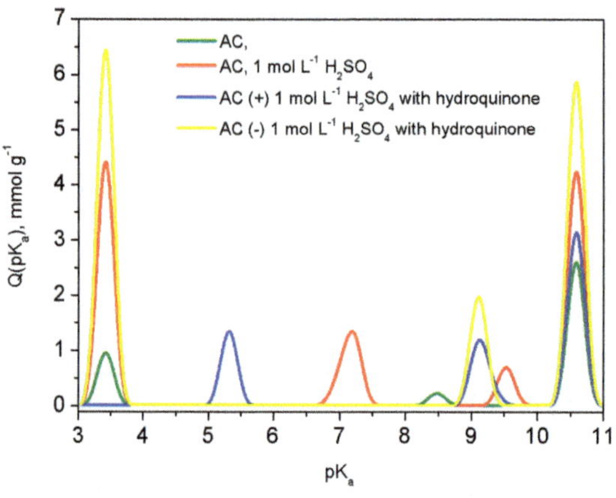

Fig. 14 Proton-binding curves for activated carbons.

functionalities, with some carbonyl-like species evacuating slowly at higher temperatures.

The redox activity of dibromodihydroxybenzene

Apart from pure dihydroxybenzenes as additives to the electrolyte, their bromine derivatives such as $C_6H_4Br_2O_2$ have been used as redox active species. The same microporous powdered carbon was used for electrode preparation in our investigations. $C_6H_4Br_2O_2$ compound was soluble only in the KOH medium. Therefore, different concentrations of the solvent (KOH), varying from 1 to 6 mol L^{-1}, were applied in order to determine the optimal operating condition of the system. It was found that a capacitor operating in 2 mol L^{-1} KOH resulted in the highest values of capacitance, *i.e.* about 270 F g^{-1}, from cyclic voltammetry and galvanostatic cycling (318 F g^{-1} from electrochemical impedance spectroscopy). It was also confirmed that the capacitor incorporating 0.2 mol L^{-1} $C_6H_4Br_2O_2$ + 2 mol L^{-1} KOH exhibited the highest capacitance, in comparison with 2 mol L^{-1} KOH and 0.2 mol L^{-1} $C_6H_6O_2$ + 2 mol L^{-1} KOH (Fig. 15 and 16).

The best capacitance performance was also noted in the case of the investigated compound in 2 mol L^{-1} KOH during galvanostatic charging–discharging (2 A g^{-1}), reaching a value of about 255 F g^{-1} and resulting in a 17% capacitance decay after 5000 cycles. The capacitance enhancement noted for the system with the bromide compound results from the substitution reactions between Br^- in the aromatic ring and the OH^- species present in the alkaline solution; these reactions might contribute additionally to the total capacitance obtained. Because of the strong alkaline pH, the reactions of Br^-/BrO_3^- play a rather minor role, but cannot be neglected. The presence of Br^- ions slightly shifts the potential of the quinone/hydroquinone redox couple (*ca.* 150 mV), hence, the capacitance is achieved at higher voltage values. Moreover, the substitution of Br^- by a OH^- group results in another source for pseudocapacitance, because *p*-dihydroxybenzene becomes trihydroxybenzene with two active configurations of the OH^- group able to undergo redox reaction within the hydroquinone/quinone couple: two in the *p*-position and two in the *o*-position.

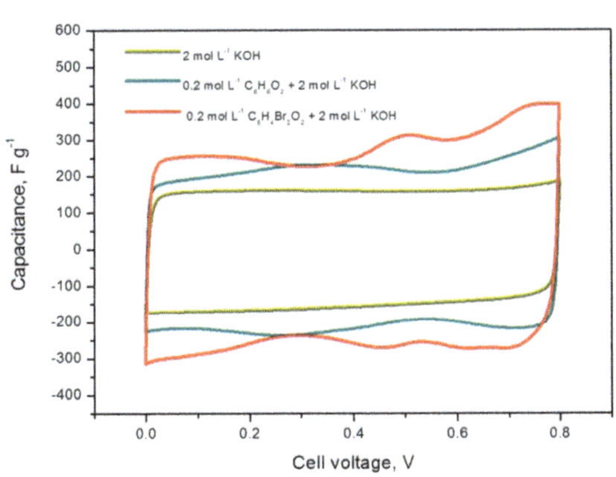

Fig. 15 Capacitance *vs.* voltage plots for the systems operating in 2 mol L^{-1} KOH with 0.2 mol L^{-1} $C_6H_6O_2$ and 0.2 mol L^{-1} $C_6H_4Br_2O_2$.

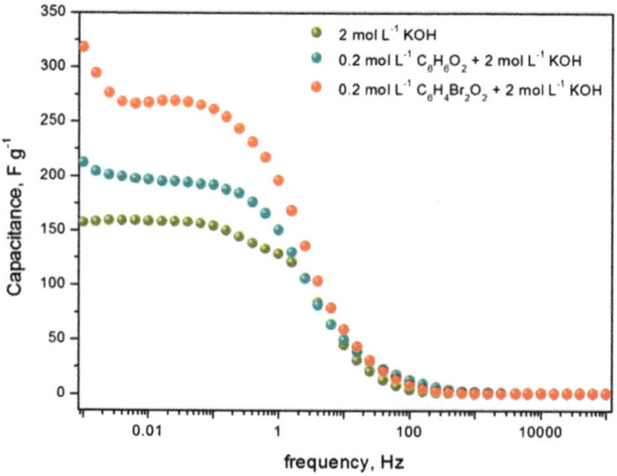

Fig. 16 Capacitance *vs.* frequency plots for the systems operating in 2 mol L^{-1} KOH with 0.2 mol L^{-1} $C_6H_6O_2$ and 0.2 mol L^{-1} $C_6H_4Br_2O_2$.

Iodide as redox active species

Our previous studies[16–18] on iodide for capacitor application have been focused on the electrochemical characterization and physicochemical properties of electrolytic solutions as well as the determination of the role of the cation. At present, special attention is devoted to the long-term cycling of sodium iodide aqueous solution as the electrolyte of the capacitor. 2 mol L^{-1} NaI solution exhibited a good conductivity value of 275 mS cm^{-1}. Floating was selected as the most destructive method for evaluation of the cycle life of the capacitor. The capacitor was constantly polarized at 1.2 V for 200 hours. During this time it was regularly tested by cyclic voltammetry, galvanostatic charge–discharge and electrochemical impedance spectroscopy to estimate its state of health. A plot of capacitance *vs.* floating time (Fig. 17) indicates an unusual capacitance enhancement with subsequent cycles. However, detailed analysis of this phenomenon shows how a careful but also critical evaluation of the data is necessary. For example, even if the galvanostatic method is considered as the best adapted for capacitance estimation, slightly varying values can be obtained from various formulas for C (considering the total discharge time or typical slope from the galvanostatic discharge curve).

Taking into account the full time of the capacitor discharging at a current load of 2 A g^{-1} (Fig. 18), higher capacitance values are obtained than those calculated from the slope of the galvanostatic discharge curve (blue points in Fig. 17). The capacitance calculated from the slope as a function of floating time (red points in Fig. 17) does not show a great tendency for growth, but it is extremely stable (varying from 250 to 280 F g^{-1} at 2 A g^{-1}).

This enormous capacitance increase during floating has also been monitored by cyclic voltammetry at 5 mV s^{-1} (Fig. 19) as well as by electrochemical impedance spectroscopy (Fig. 20).

The gradual capacitance increase from cyclic voltammetry in a narrow range of voltages (Fig. 19) matches very well with the charge transfer increase in Fig. 20

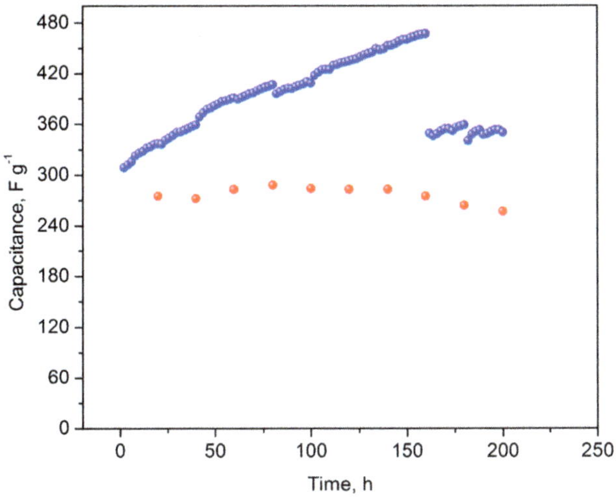

Fig. 17 Capacitance *vs.* floating time for a capacitor operating in 2 mol L^{-1} NaI solution calculated from two different formulas; blue points reflect capacitance calculated from full discharge time whereas red points concern capacitance estimated from the slope within 0.2–1.0 V.

that reflects the participation of redox reactions. It is noteworthy that the equivalent series resistance (ESR) is extremely small (0.3 ohm) and very stable during the entire floating time. Contrarily, the resistance R_{CT} gradually increases from negligible values to 6 ohm after floating.

Our experiments proved that 200 hours of floating at 1.2 V was equivalent to more than 110 000 galvanostatic cycles at 1 A g^{-1}. After this long-term galvanostatic cycling, 91% capacitance retention was still preserved.

Regardless of the testing method (floating or galvanostatic charge–discharge), no corrosion was observed on the stainless steel current collectors, which is an

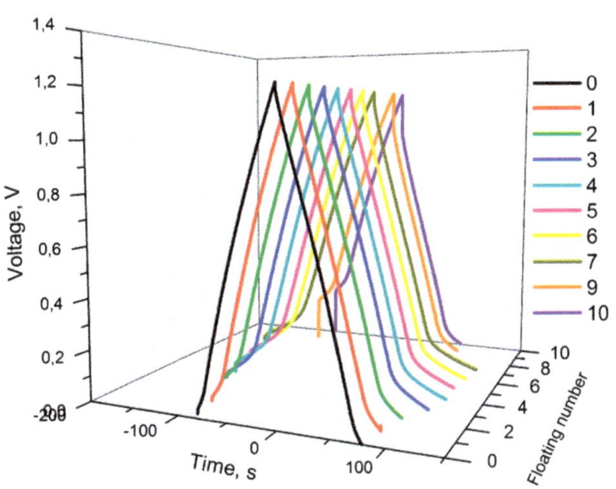

Fig. 18 Charge–discharge profiles for 2 A g^{-1} current load *vs.* floating time for 2 mol L^{-1} NaI solution.

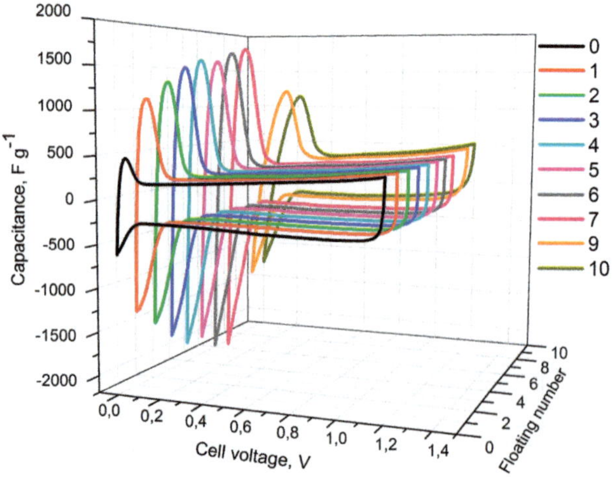

Fig. 19 Capacitance from cyclic voltammetry (5 mV s^{-1}) *vs.* floating time for 2 mol L^{-1} NaI solution.

additional benefit for practical application. Moreover, this halide is more environmentally friendly than bromine.

For elucidation of the charge storage mechanism of the iodide solutions, the *in situ* Raman spectroscopy method was used. Briefly, the D and G band ratio (Raman shifts of 1333 and 1597 cm^{-1}, respectively) for our activated carbon material was equal to 1.05, *i.e.*, typical for this type of disordered material. The shape of the ill-defined, broad D band confirms the expected distortions of the

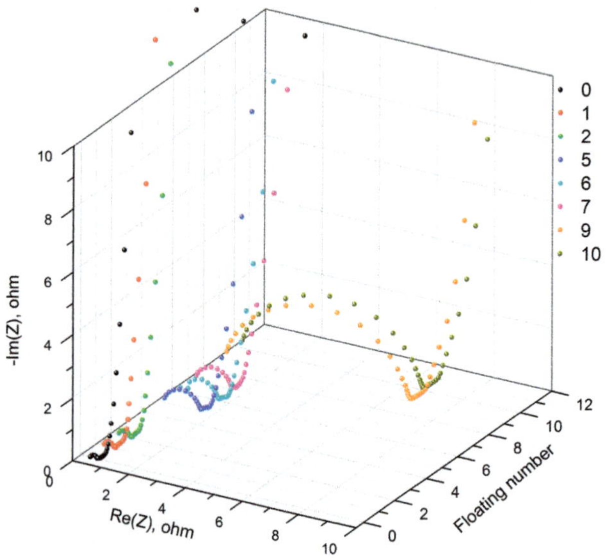

Fig. 20 Nyquist impedance spectra recorded after several floating tests for a capacitor in 2 mol L^{-1} NaI solution.

carbon network, being more benzenoid than graphitic (graphene-like). Moreover, the two-component G-band (at a Raman shift close to the E_{2g2} mode, ~1580 cm^{-1}) suggests various interactions between vibrational C–C covalent bonds within the aromatic plane with various lengths and strengths, which is considered as typical for disordered, activated carbons.

A carbon microelectrode (geometric area of 0.283 cm^2) made from AC powder as well as in the form of tissue placed on the gold support was polarized by cyclic voltammetry in the range from 0 to 750 mV *vs.* SCE and potentiostatically from open circuit potential (*ca.* 240 mV *vs.* SCE) to 750 mV *vs.* SCE with 50 mV potential steps. The cyclic voltammetry demonstrated a capacitive behaviour typical of the electrode in the potential range up to 550 mV *vs.* SCE and a significant increase in measured current above 600 mV *vs.* SCE, which is related to the iodide/iodine redox couple activity. It should be noted that spectra recorded in potentiostatic mode were collected after *ca.* 5 minutes of potential polarization, *i.e.* when the current response was stable. The current recorded in the EDL charging area was ~95 μA cm^{-2}, whereas in the redox activity region this value increased to ~330 μA cm^{-2}.

The D/G ratio and their Raman shifts did not change remarkably with the polarization of the carbon in the iodide solution, however, for the longer polarization at 750 mV *vs.* SCE, a small shift of 3 cm^{-1} was observed, especially for the D band. On the other hand, spectra in the low wavelength range were drastically modified by the polarization in the iodide electrolyte, with a remarkable increase of the I_3^- signal with polarization time. Raman spectra recorded in potentiostatic mode are shown in Fig. 21.

Considering the profiles of the recorded spectra, we propose the following mechanism. Firstly, I^- is oxidized to I_2, giving various broad bands in the Raman shift region of 190–170 cm^{-1} (for pure iodine 179 cm^{-1}). Iodine dissolved in the electrolyte solution might recombine with monoiodide I^- resulting in I_3^- species.[34,35] It is clearly seen that the I_3^- complex is formed from the $I_2 \cdot I^-$ complex, reinforcing the 168 cm^{-1} vibrational response. As expected, the Raman intensity increases with time, hence, one can assume that the concentration of I_3^- species increases and their presence quantitatively affects the profiles of the Raman spectra. Moreover, another peak at 108 cm^{-1} appears and can be attributed to the I_5^- complex. It should be noted that I_5^- consists of I_2 and I_3^- species, hence, their Raman activity is observed simultaneously. However, for a longer polarization time, the major contribution of I_3^- in the profile decreases, hence, a recombination of I_2 towards polyiodides such as I_5^- is observed. I_5^- starts to contribute to the spectra even after 1 h of polarization, with a small signal at 150 cm^{-1}, reinforced significantly after 6 h of polarisation and indicating bent C_{2v} modes of chained, discrete I_5^-. One can expect that the response at 368–270 cm^{-1} and 268–214 cm^{-1} can be ascribed to iodine in the molecular form, but also to C–I interactions. Especially, the signal recorded at 333 cm^{-1} and 327 cm^{-1} might be considered to be due to C–I_3^- and C–I_5^- bands, respectively, but their presence at lower Raman shifts (*e.g.* 165 and 115 cm^{-1}) cannot be directly confirmed because of the strong activity of the pure I_3^- and I_5^- bands. Generally, all forms of iodides present in the electrolyte solution and at the electrode/electrolyte interface reflect a complex Raman signal as a combination of major I_2 and I_n^- bands and an interaction between the bands of the I_2 molecule and related species.[35–37] Even if the *in situ* Raman experimental conditions are quite far from real capacitor

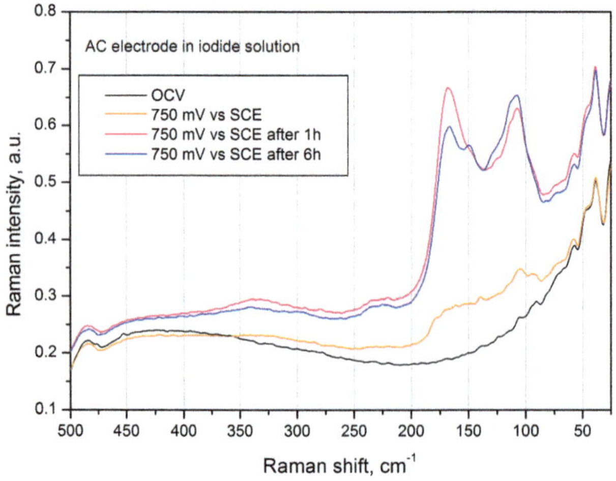

Fig. 21 Raman spectra recorded *in situ* for activated carbon fabric in 2 mol L^{-1} NaI during potentiostatic polarization.

operation, this technique supplies important information on this particular carbon/iodine interface.

Conclusions

The electrochemical behaviour of carbon-based electrochemical capacitors operating in different redox active solutions has been reported. Dihydroxybenzenes with varying –OH substitution (*ortho*, *meta*, *para*) as additives (0.38 mol L^{-1}) to various aqueous electrolytes served as the source of pseudocapacitance. The highest capacitance values (283 F g^{-1} at 5 mV s^{-1}) were observed for 0.38 mol L^{-1} hydroquinone dissolved in 1 mol L^{-1} H_2SO_4. Similar values were obtained for an alkaline electrolyte (6 mol L^{-1} KOH) with hydroquinone (275 F g^{-1} at 5 mV s^{-1}), which is considerably higher than for electrolytes without dihydroxybenzenes. Such intriguing behaviour is closely related to the electrochemical activity of these species, which depends strongly on their stereochemistry and the electrolyte pH. The physicochemical investigations performed in order to confirm the aforementioned assumptions are in good accordance with such conclusions. The significant increase in the mass change recorded during TGA experiments (more than 15%), as well as the change in proton binding curves, suggest that oxygen content in the electrode material increased. Some specific adsorption by π bonding of the dihydroxybenzenes onto the carbon network cannot be excluded, however, the electrochemical results encourage us to assume that some of these species were rather grafted onto the carbon surface. Unfortunately, the electrochemical reversibility of hydroquinones is not very high (*ca.* 60%), thus the practical application of this redox couple is rather doubtful. On the other hand, capacitance retention was good over 5000 cycles with a subsequent gradual decrease. An interesting capacitance behaviour has been shown for bromodihydroxybenzene used as the additive in an alkaline medium (2 mol L^{-1} solution). A large capacitance enhancement was observed because of the bromide substitution in the benzene ring. A significant increase in capacitance (50%) was found. Generally, a low regime was preferable for an efficient capacitor performance. The best results were obtained for alkali iodide, which played a dual role of electrolyte and redox active species simultaneously. The great advantage of this system is extraordinary reversibility due to the formation of iodine, and in turn, polyiodides (I_3^-, I_5^-). Even if the high reversibility of I_2/I_n^- was observed in a narrow range of potentials (below 200 mV), it greatly affected the total system. Electrochemical studies using floating, complemented by detailed capacitance measurements using other techniques, proved the exceptional performance of the capacitor based on iodides. The well documented enhancement of the capacitance during floating surely originates from the cumulative creation of I_n^- species, but the ultimate formation of iodates IO_3^- should not be excluded. Raman spectra recorded *in situ* during potentiostatic polarization provided good proof of the existence of I_n^- species as well as C–I interactions. The appeal of the iodide electrolytic solution for capacitor application is also due to the usage of a cheap stainless steel current collector.

Acknowledgements

The authors greatly acknowledge the Swiss–Polish project INGEC 107/2010.

Notes and references

1 E. Frackowiak and F. Béguin, *Carbon*, 2002, **40**, 1775.
2 *Carbons for Electrochemical Energy Storage and Conversion Systems*, ed. F. Béguin and E. Frackowiak, CRS Press Taylor & Francis Group, Boca Raton, 2010.
3 E. Frackowiak, *Phys. Chem. Chem. Phys.*, 2007, **9**, 1774.
4 *Supercapacitors: Materials, Systems and Applications*, ed. F. Béguin and E. Frackowiak, Wiley-VCH Verlag GmbH & Co., Weinheim, 2013.
5 J. Zhang and X. S. Zhao, *ChemSusChem*, 2012, **5**, 818.
6 S. Suematsu and K. Naoi, *J. Power Sources*, 2001, **97**, 816.
7 Y. J. Oh, J. J. Yoo, Y. I. Kim, J. K. Yoon, H. N. Yoon, J. H. Kim and S. B. Park, *Electrochim. Acta*, 2014, **116**, 118.
8 A. Yoshida, I. Tanahashi and A. Nishino, *Carbon*, 1990, **28**, 611.
9 C. T. Hsieh and H. Teng, *Carbon*, 2002, **40**, 667.
10 K. Okajima, K. Ohta and M. Sudoh, *Electrochim. Acta*, 2005, **50**, 2227.
11 H. A. Andreas and B. E. Conway, *Electrochim. Acta*, 2006, **51**, 6510.
12 E. Frackowiak, K. Fic, M. Meller and G. Lota, *ChemSusChem*, 2012, **5**, 1181.
13 K. Fic, E. Frackowiak and F. Béguin, *J. Mater. Chem.*, 2012, **22**, 24213.
14 S. T. Senthilkumar, R. K. Selvan, N. Ponpandian and J. S. Melo, *RSC Adv.*, 2012, **2**, 8937.
15 S. T. Senthilkumar, R. K. Selvan, Y. S. Lee and J. S. Melo, *J. Mater. Chem. A*, 2013, **1**, 1086.
16 G. Lota and E. Frackowiak, *Electrochem. Commun.*, 2009, **11**, 87.
17 G. Lota, K. Fic and E. Frackowiak, *Electrochem. Commun.*, 2011, **13**, 38.
18 J. Menzel, K. Fic, M. Meller and E. Frackowiak, *J. Appl. Electrochem.*, 2014, **44**, 439.

19 S. Yamazaki, T. Ito, M. Yamagata and M. Ishikawa, *Electrochim. Acta*, 2012, **86**, 294.

20 P. Barpanda, G. Fanchini and G. G. Amatucci, *Electrochim. Acta*, 2007, **52**, 7136.

21 P. Barpanda, G. Fanchini and G. G. Amatucci, *Carbon*, 2011, **49**, 2538.

22 P. Barpanda, Y. Li, F. Cosandey, S. Rangan, R. Bartynski and G. G. Amatucci, *J. Electrochem. Soc.*, 2009, **156**, A873.

23 S. Roldan, M. Granda, R. Menendez, R. Santamaria and C. Blanco, *J. Phys. Chem.*, 2011, **115**, 17606.

24 S. Roldan, C. Blanco, M. Granda, R. Menendez and R. Santamaria, *Angew. Chem.*, 2011, **123**, 1737.

25 S. Isikli and R. Diaz, *J. Power Sources*, 2012, **206**, 53.

26 S. Isikli, M. Lecea, M. Ribagorda, M. C. Carreno and R. Díaz, *Carbon*, 2014, **66**, 654.

27 G. Pognon, T. Brousse, L. Demarconnay and D. Bélanger, *J. Power Sources*, 2011, **196**, 4117.

28 G. Pognon, T. Brousse and D. Bélanger, *Carbon*, 2011, **49**, 1340.

29 D. M. Anjos, J. K. McDonough, E. Perre, G. M. Brown, S. H. Overbury, Y. Gogotsi and V. Presser, *Nano Energy*, 2013, **2**, 702.

30 Z. Algharaibeh and P. G. Pickup, *Electrochem. Commun.*, 2011, **13**, 147.

31 D. Bélanger and J. Pinson, *Chem. Soc. Rev.*, 2011, **40**, 3995.

32 P. S. Guin, S. Das and P. C. Mandal, *Int. J. Electrochem.*, 2011, 816202.

33 G. Lota, T. A. Centeno, E. Frackowiak and F. Stoeckli, *Electrochim. Acta*, 2008, **53**, 2210.

34 M. Mizuno, J. Tanaka and I. Harada, *J. Phys. Chem.*, 1981, **50**, 1789.

35 E. M. Nour, L. H. Chen and J. Laane, *J. Phys. Chem.*, 1986, **90**, 2841.

36 T. J. Marks and D. W. Kalina, *Extended Linear Chain Compounds*, ed. J. S. Miller, Plenum, New York, 1982, vol. 1, p. 197.

37 A. J. Blake, F. A. Devillanova, R. O. Gould, W. S. Li, V. Lippolis, S. Parsons, C. Radek and M. Schroder, *Chem. Soc. Rev.*, 1998, **27**, 195.

Faraday Discussions

ROYAL SOCIETY OF CHEMISTRY

PAPER

Sodium molybdate – an additive of choice for enhancing the performance of AC/AC electrochemical capacitors in a salt aqueous electrolyte

Q. Abbas, P. Ratajczak and F. Béguin*

Received 30th March 2014, Accepted 6th May 2014
DOI: 10.1039/c4fd00056k

Sodium molybdate (Na_2MoO_4) has been used as an additive to 1 mol L^{-1} lithium sulfate electrolyte for electrochemical capacitors based on activated carbon (AC) electrodes, in order to reduce the corrosion of stainless steel current collectors. We demonstrate that the MoO_4^{2-} anions improve the overall capacitance owing to pseudofaradaic processes. In a two-electrode cell, capacitance values of 121 F g^{-1} have been achieved up to 1.6 V using 1 mol L^{-1} Li_2SO_4 + 0.1 mol L^{-1} Na_2MoO_4, as compared to 103 F g^{-1} when 1 mol L^{-1} Li_2SO_4 is used. Further, by using a two-electrode setup equipped with a reference electrode, we could demonstrate that, at 1.6 V, the positive electrode potential reaches a value of 0.96 V *vs.* NHE in 1 mol L^{-1} Li_2SO_4, crossing the thermodynamic potential limit of oxygen evolution (E_{ox} = 0.846 V *vs.* NHE), and the pitting potential, E_{pit} = 0.95 V *vs.* NHE. By contrast, in 1 mol L^{-1} Li_2SO_4 + 0.1 mol L^{-1} Na_2MoO_4, the pseudofaradaic contribution occurring at −0.05 V *vs.* NHE due to MoO_4^{2-} anions drives the positive electrode to reach only 0.798 V *vs.* NHE. Hence, the oxidation of the AC and corrosion of the stainless steel current collector at the positive electrode are unlikely in Li_2SO_4 + Na_2MoO_4 when the capacitor operates at 1.6 V. During potentiostatic floating of the capacitor at 1.6 V for 120 hours in Li_2SO_4 + Na_2MoO_4, the capacitance and resistance remain constant at 125 F g^{-1} and ~1.0 Ω, respectively, while the resistance increases from 1.4 Ω to 3.1 Ω in Li_2SO_4. Overall, the addition of MoO_4^{2-} anions to Li_2SO_4 aqueous electrolyte allows the capacitance to be enhanced, corrosion of the positive stainless steel current collector to be inhibited and the AC/AC electrochemical capacitor to demonstrate stable performance up to 1.6 V.

1. Introduction

Electrochemical energy storage systems like electrical double-layer capacitors (EDLCs), also known as supercapacitors, have received considerable attention in recent years.[1–3] They are believed to be potential candidates to fill the energy gap

Institute of Chemistry and Technical Electrochemistry, Poznan University of Technology, Piotrowo 3, 60-965 Poznan, Poland. E-mail: francois.beguin@put.poznan.pl

between traditional accumulators and capacitors, and to fulfill the demand for delivering peak power in various applications.[4,5] One of the key research criteria for supercapacitors from an application point of view has been the enhancement of capacitance[6–8] which in turn improves the energy density, E (J), as given by eqn (1):

$$E = (1/2)\ CU^2 \tag{1}$$

where C represents the capacitance (F g^{-1}) and U the voltage (V). Other driving forces have been the use of low cost, safe, environment friendly and efficient construction materials.

Until recently, due to the use of acidic and basic electrolytes, it was believed that electrochemical capacitors based on activated carbon (AC) electrodes cannot work at voltages higher than 0.7–0.8 V in aqueous media.[5] However, by using alkali sulphate solutions as electrolytic media, is has been demonstrated that they can operate with good cycle life at voltage values up to 1.6–2 V.[9–11] Lithium sulphate (Li_2SO_4) stands out as the best salt among different alkali sulphates due to its high solubility and the good capacitance values of the systems. However, such good performance with these media has been achieved while using gold current collectors. For industrial viability, low cost collectors have to be used; considering the various possibilities, stainless steel seems to be the optimal choice.

Our recent studies on AC/AC capacitors in lithium sulfate with stainless steel current collectors have suggested that corrosion products accumulate between the AC electrode and the positive stainless steel current collector during potentiostatic floating at voltages higher than 1.5 V, and could be partially responsible for the blockage of AC porosity, reducing the power output and lifetime of the device.[12,13] Indeed, stainless steel may be corroded in an aqueous environment as a result of the production of iron oxides like Fe_2O_3 and Fe_3O_4. Such oxides are formed by gradual transformation of γFeOOH to acicular αFeOOH, and finally to sintered submicron powders of Fe_2O_3 and Fe_3O_4, resulting in the creation of new resistive elements in the equivalent electrical circuit.[14,15]

It has already been reported that the presence of $MoO_4{}^{2-}$ ions strengthens the hydrated iron oxide layer on the stainless steel surface in neutral aqueous solutions, inhibiting the corrosion provoked by aggressive anions like chlorides or sulfates.[16,17] It has been proposed that in neutral aqueous media, $MoO_4{}^{2-}$ interacts with Fe^{2+} to form ferrous-molybdate, which transforms in the presence of dissolved di-oxygen to insoluble protective ferric molybdate,[18] or that the non-oxidizing $MoO_4{}^{2-}$ forms a complex with iron which adsorbs on the metal surface.[19] Therefore, the purpose of this work is to use Na_2MoO_4 as an additive to the salt aqueous electrolyte in order to improve the performance of AC/AC supercapacitors with stainless steel current collectors. By comparison with systems containing only 1 mol L^{-1} Li_2SO_4, we show that the addition of Na_2MoO_4 improves the lifetime and performance of the supercapacitors by impeding the corrosion of the positive stainless steel current collector, as well as by giving higher capacitance values. A combination of electrochemical techniques, *i.e.* cyclic voltammetry, galvanostatic charge–discharge, chronoamperometry and electrochemical impedance spectroscopy has been used to understand the role of sodium molybdate and to analyze the performance of the supercapacitors.

2. Experimental

The current collectors (diameter 1.2 cm) were made of a low carbon content stainless steel 316L alloy consisting of the following major elements: Fe, C (0.02%), Cr (16%), Ni (10%) and Mo (2%). Their surface was cleaned with emery paper (P1000) before the investigations. The effect of corrosion was studied by cyclic voltammetry at 10 mV s^{-1}, both in 1 mol L^{-1} Li_2SO_4 and in a mixture of 1 mol L^{-1} Li_2SO_4 and 0.1 mol L^{-1} Na_2MoO_4, using a Teflon Swagelok® type three-electrode cell, with a stainless steel working electrode, gold counter electrode and Hg/Hg_2SO_4 (SME) reference electrode.

The carbon electrodes for supercapacitor construction were prepared by using capacitor grade DLC Supra 30 activated carbon from Norit (S_{BET} = 2071 m^2 g^{-1} evaluated by Micromeritics ASAP 2020), named as AC from now on in this manuscript. Pellets, with a diameter of 1.0 cm and mass of 7–10 mg, were prepared by mixing AC (80 wt%), carbon black SUPER C65 (Timcal, 10 wt%) and PTFE (Sigma Aldrich, 10 wt%) as binder in 10 mL of isopropanol. The dough obtained after mixing was rolled to a thickness of nearly 0.25 mm and was left to dry at 120 °C for ten hours under vacuum. Later, disk pellets were punched-out from the rolled sheet of electrode material.

Symmetric AC/AC systems were constructed using Teflon Swagelok® type two-electrode cells and stainless steel (316L) current collectors. A glassy microfiber filter (GF/A, Whatman™) of thickness 260 μm was used as separator. For investigating the electrochemical properties of the individual electrodes, we used a two-electrode setup equipped with a Hg/Hg_2SO_4 (SME) reference electrode; the same set-up was used for three-electrode investigations, one of the AC electrodes being selected as the working electrode. The experiments were performed in either 1 mol L^{-1} Li_2SO_4 (pH = 6.5, conductivity = 64 mS cm^{-1}), or in a 1 mol L^{-1} Li_2SO_4 + 0.1 mol L^{-1} sodium molybdate (Na_2MoO_4, from Sigma Aldrich) mixture (pH = 6.7, conductivity = 72 mS cm^{-1}).

The electrochemical properties of the capacitors were investigated by cyclic voltammetry (CV, 2 mV s^{-1}), galvanostatic cycling with potential limitation (GCPL, 200 mA g^{-1}) and impedance spectroscopy in the range of 1 mHz to 100 kHz, using a VMP3 multichannel potentiostat/galvanostat (Bio-Logic Instruments, France). The gravimetric capacitance C was calculated from the galvanostatic discharge of the two-electrode cells and expressed per electrode (F g^{-1}) using the formula (2):

$$C = (2)I/[(\mathrm{d}U/\mathrm{d}t)m_{\mathrm{am}}] \quad (2)$$

where I is the current (A), $\mathrm{d}U/\mathrm{d}t$ is the slope of the discharge curve (V s^{-1}), and m_{am} is the average mass of AC in one electrode (g). Accelerated ageing by potentiostatic floating has been applied to estimate the performance limits of the capacitors; in these experiments, the test cell is charged up to a certain voltage (in our experiments 1.6 V) and maintained at this value for periods of 2 hours at room temperature. Each period is followed by five galvanostatic charge–discharge cycles at 1 A g^{-1}, in order to estimate the discharge capacitance and the equivalent series resistance (ESR) from the voltage drop between the 5th charge and discharge when reversing the sign of the current. A total of 60 such sequences were applied, corresponding to a total floating time of 120 hours.

3. Results and discussion

3.1. Current collector corrosion

Fig. 1 represents the three-electrode cyclic voltammograms of stainless steel 316L in 1 mol L^{-1} Li_2SO_4 and in 1 mol L^{-1} Li_2SO_4 + 0.1 mol L^{-1} Na_2MoO_4. For both systems, the potential scan in the positive direction shows the active dissolution region from −1.2 V to −0.65 V *vs.* NHE, and an anodic peak followed by a passive region. The dominant species in the active dissolution region may be ferrous hydroxide, which precipitates on the electrode surface as its concentration exceeds the solubility product.[20] Further in the positive direction, Fe-hydroxide converts to different species between −0.65 V and 0.2 V *vs.* NHE, mainly Fe_2O_3 which is the major constituent of the passive layer.[21] The passive layer formed on the stainless steel surface in 1 mol L^{-1} Li_2SO_4 (black line) is due to oxides of iron and its alloying metals like Cr and Mo. This layer is thick in nature, protecting the stainless steel surface from corrosion caused by SO_4^{2-} anion attack.[22,23] Since the passive layer completely covers the electrode surface, it causes a fall in the anodic current density to a low value (j_{pass}), leading to the passive region between 0.2 V and 0.9 V *vs.* NHE. A further potential increase in the positive direction shows a sudden current leap associated with the starting of pitting (E_{pit}) at ~0.95 V *vs.* NHE. This region is called the transpassive region, where the current suddenly rises and a further potential increase results in di-oxygen evolution.[24]

Adding Na_2MoO_4 to 1 mol L^{-1} Li_2SO_4 results in a higher current for passive layer formation (red line) in the potential region between −0.65 V to 0.2 V *vs.* NHE; as a consequence the MoO_4^{2-} anions strengthen the passive layer and its thickness increases. This is further supported by the fact that the passivation current density (j_{pass}) in the case of Li_2SO_4 + Na_2MoO_4 is slightly lower than in the case of Li_2SO_4. Hence, the addition of MoO_4^{2-} to Li_2SO_4 improves the passivation of the stainless steel, protecting it against pitting corrosion caused by the attack of SO_4^{2-} anions and di-oxygen, which is produced due to water oxidation.

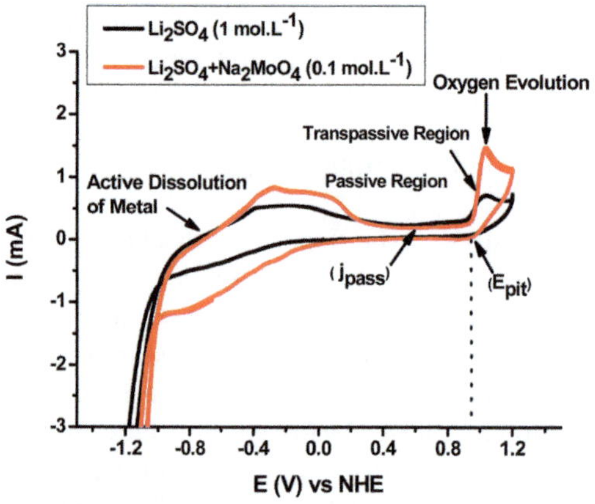

Fig. 1 Three-electrode cyclic voltammograms of stainless steel 316L at 25 °C (a scan rate of 10 mV s^{-1}) in 1 mol L^{-1} Li_2SO_4 and 1 mol L^{-1} Li_2SO_4 + 0.1 mol L^{-1} Na_2MoO_4. The dotted line (~0.95 V *vs.* NHE) shows the onset of pitting corrosion.

The explanation given above about the role played by Na_2MoO_4 in corrosion inhibition suggests that its addition to the 1 mol L^{-1} Li_2SO_4 solution could improve the performance of supercapacitors containing stainless steel current collectors and salt aqueous electrolytes.

3.2. Electrochemical properties of the AC electrode

Fig. 2 shows three-electrode cyclic voltammograms of AC with a gradual shift of the negative potential cut-off in 1 mol L^{-1} Li_2SO_4 and 1 mol L^{-1} Li_2SO_4 + 0.1 mol L^{-1} Na_2MoO_4. The vertical dotted lines at −0.383 V *vs.* NHE for Li_2SO_4 (pH = 6.5) and −0.395 V *vs.* NHE for Li_2SO_4 + Na_2MoO_4 (pH = 6.7) represent the thermodynamic potential values of water reduction in both media. In Li_2SO_4, at negative cut-off potentials higher than this value, the CVs display the typical rectangular shape related with charging the electrical double-layer. When the negative cut-off potential is lower than −0.383 V *vs.* NHE in Li_2SO_4, nascent hydrogen starts to be produced and is chemisorbed on the AC electrode until di-hydrogen evolution starts at approximately −0.8 V *vs.* NHE, as observed by the oscillations due to bubbling on the CVs (Fig. 2a).[11,25] When the potential scan is reversed to the positive direction, reversible hydrogen desorption starts at −0.2 V *vs.* NHE.

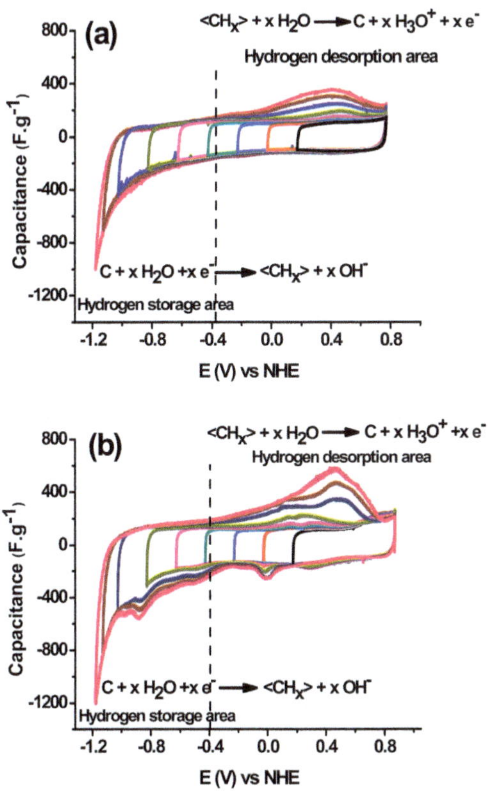

Fig. 2 Three-electrode cyclic voltammograms of the AC carbon electrode recorded at 2 mV s^{-1} in 1 mol L^{-1} Li_2SO_4 (a), and 1 mol L^{-1} Li_2SO_4 + 0.1 mol L^{-1} Na_2MoO_4 (b). The successive loops are obtained by a stepwise decrease of the negative cut-off potential. The dotted vertical lines correspond to the thermodynamic water reduction potential.

In Li_2SO_4 + Na_2MoO_4, the same sequence occurs with regard to electrical double-layer formation and reversible hydrogen storage, however new redox peaks are observed and, in the anodic region corresponding to hydrogen desorption (approximately −0.4 to 0.8 V *vs.* NHE), the hump is higher than in Li_2SO_4 (Fig. 2b). One also notices that hydrogen desorption is fully completed at 0.8 V in Li_2SO_4 + Na_2MoO_4, whereas at the same potential value in Li_2SO_4, the anodic redox current is still important. It suggests slightly different activation energies for the carbon–hydrogen bonds formed in the two electrolytic media.[26] Notwithstanding, in Li_2SO_4 + Na_2MoO_4 electrolyte, the observable oscillations due to di-hydrogen evolution appear at a slightly lower potential than in Li_2SO_4, *i.e.* around −1 V *vs.* NHE. This is supported by the different pH values measured at the electrode/electrolyte interface, *ca.* 10 and 8 in Li_2SO_4 + Na_2MoO_4 and Li_2SO_4, respectively. Hence, according to the Nernst law, in the case of Li_2SO_4 + Na_2MoO_4, the shift in di-hydrogen formation is explained by the lower local reduction potential of water. Due to this lower potential, the oxidation peak is more pronounced in the Li_2SO_4 + Na_2MoO_4 electrolyte, suggesting that more hydrogen is stored in the AC electrode using this medium.

The new peaks which appear in addition to the reversible hydrogen storage peaks in the Li_2SO_4 + Na_2MoO_4 electrolyte might be attributed to redox reactions involving the MoO_4^{2-} anion (Fig. 2b). The Pourbaix diagrams of molybdenum[27] show that the molybdate anions can be involved in the two redox equilibria (3) and (4):

$$MoO_4^{2-} + 4H^+ + 2e^- \leftrightarrow MoO_2 + 2H_2O \tag{3}$$

$$MoO_4^{2-} + 8H^+ + 6e^- \leftrightarrow Mo + 4H_2O \tag{4}$$

At pH = 6.7, the equilibrium potentials are −0.215 V and −0.384 V for equilibria (3) and (4), respectively. In Fig. 2b, peaks cannot be found strictly at these positions during the negative scan, probably because of the already mentioned pH change on the negatively polarized electrode surface. Moreover, the oxidation of water at the positive electrode of the cell may also contribute to a locally decreased pH, leading, at pH < 6, to the transformation of MoO_4^{2-} into $HMoO_4^-$ and further into MoO_3.[27] Even if the origin of the new redox peaks shown by the AC electrode in the Li_2SO_4 + Na_2MoO_4 medium is not perfectly elucidated, there is clearly a splitting and an overall increase of the anodic peak, revealing an additional pseudo-capacitive contribution to the electrical double-layer (EDL) capacitance. Apart from the differences in pseudocapacitive behavior, differences in the EDL capacitance are also observable between the two electrolytes in the region from *ca.* 0.2 to 0.8 V *vs.* NHE (black curves). In Li_2SO_4 + Na_2MoO_4, the capacitance reaches 120 F g^{-1} as compared to 98 F g^{-1} for Li_2SO_4. The higher value in Li_2SO_4 + Na_2MoO_4 than in Li_2SO_4 is essentially attributable to the higher conductivity of the former electrolyte as compared to the latter.

3.3. Electrochemical performance of the AC/AC two-electrode cells

The effects of adding Na_2MoO_4 to Li_2SO_4 were further investigated in a two electrode AC/AC cell equipped with a Hg/Hg_2SO_4 reference electrode. Fig. 3 shows, for the two electrolytic media, the potential limits of each electrode *vs.* NHE as function of cell voltage. In the case of Li_2SO_4 (pH = 6.5), the maximum

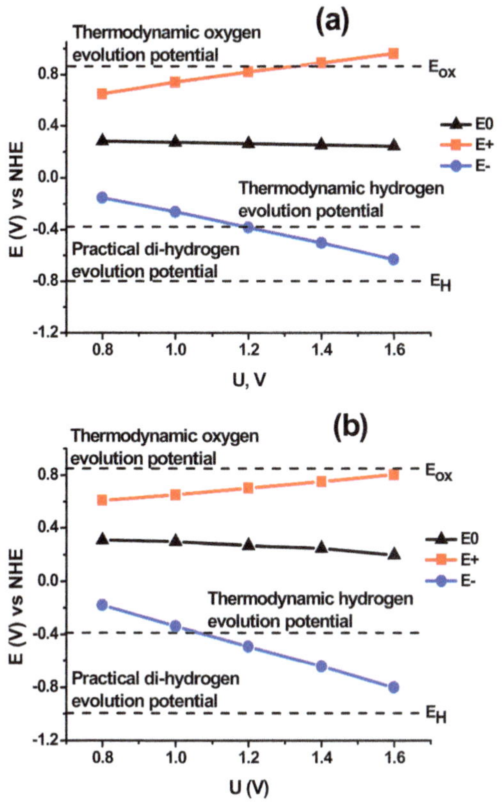

Fig. 3 Potential limits of the positive and negative AC electrodes during the galvanostatic (200 mA g^{-1}) charge–discharge of the AC/AC cells up to different values of the maximum voltage in 1 mol L^{-1} Li_2SO_4 (a), and 1 mol L^{-1} Li_2SO_4 + 0.1 mol L^{-1} Na_2MoO_4 (b) electrolytic media. E_+ and E_- represent the maximum and minimum potentials reached by the positive and negative electrodes, respectively. The E_0 values correspond to the rest potential of the electrodes after setting the cell voltage to 0 V.

potential of the positive electrode (E_+) is equal to the thermodynamic oxygen evolution potential for a voltage of 1.35 V (Fig. 3a). It means that, if the supercapacitor containing Li_2SO_4 as an electrolyte operates at 1.6 V, oxygen evolution will definitely occur at the positive electrode, causing corrosion of the stainless steel current collector and oxidation of the AC. By contrast, for the negative electrode in the Li_2SO_4 electrolyte, the minimum potential (E_-) is well above the practical di-hydrogen evolution potential of approximately −0.8 V *vs.* NHE established from Fig. 2a, even if the maximum voltage applied is 1.6 V.

Fig. 3(b) shows the potential limits in the presence of the Li_2SO_4 + Na_2MoO_4 electrolyte at various voltage values; at pH = 6.7 of this medium, the oxygen evolution potential is 0.84 V *vs.* NHE. Interestingly, for a voltage of 1.6 V, the maximum potential of the positive electrode (E_+) is at 0.798 V, and is shifted by approximately −0.162 V as compared to Li_2SO_4 (0.96 V *vs.* NHE), lying well below the oxygen evolution potential. This should practically prevent AC oxidation and stainless steel current collector corrosion at the positive electrode. Indeed, the dotted line in Fig. 1 shows that the pitting corrosion starts at $E_{pit} \sim 0.95$ V *vs.* NHE, which is far above $E_+ = 0.798$ V in the $Li_2SO_4 + Na_2MoO_4$ electrolyte at a voltage of 1.6 V. By contrast, at 1.6 V in Li_2SO_4, the potential of the positive electrode reaches 0.96 V *vs.* NHE and the latter is susceptible to pitting as it operates at a potential slightly higher than E_{pit}. Besides, the minimum potential E_- of the negative electrode is shifted to lower values. However, even at a voltage of 1.6 V, the E_- value is still higher than the practical di-hydrogen evolution potential of approximately −1.0 V *vs.* NHE which was estimated in Fig. 2b.

Table 1 shows the values of discharge capacitance for the positive and negative AC electrodes in 1 mol L^{-1} Li_2SO_4 and 1 mol L^{-1} Li_2SO_4 + 0.1 mol L^{-1} Na_2MoO_4. The capacitance values of the negative electrode are of the same order in the two electrolytic media; the marked increase at a voltage higher than 1.4 V is attributed to reversible hydrogen storage. The capacitance of the positive electrode is almost independent of voltage in Li_2SO_4, which is proof of typical EDL behavior. By contrast, for this electrode in the $Li_2SO_4 + Na_2MoO_4$ electrolyte, one can notice a remarkable increase of capacitance with voltage, due to pseudocapacitive contributions related to the presence of the molybdate anions.

Nyquist plots measured at 0.8 V on a two-electrode AC/AC cell equipped with a reference electrode in $Li_2SO_4 + Na_2MoO_4$ are presented in Fig. 4a. The nearly vertical lines at low frequency indicate the good capacitance properties of the electrodes and capacitor. The AC/AC capacitor exhibits a low equivalent series resistance (ESR) of 0.34 Ω; the comparable ESR values of 0.17 Ω and 0.22 Ω for the positive and negative electrodes, respectively, confirm the dominant contribution of electrolyte conductivity to this parameter. The equivalent distributed resistance of the capacitor (EDR = 3.75 Ω) is approximately the sum of the EDR values for positive (1.92 Ω) and negative (1.64 Ω) electrodes; such low values suggest a good propagation of the ionic species in the carbon pores. The charge transfer resistance R_{ct} of the positive (1.41 Ω) and negative (1.36 Ω) electrodes is nearly the same, and low, suggesting a good contact at the electrode/current collector interfaces; the R_{ct} measured for the capacitor (R_{ct} = 2.91 Ω) is approximately the sum of the values for the electrodes.

The low frequency part of the capacitance *vs.* frequency plot at 0.8 V (Fig. 4b) is in good agreement with the data presented in Table 1, showing a much higher capacitance of the positive electrode, due to contributions of the EDL and pseudo-faradic reactions associated with the molybdate anions. The capacitance of the negative electrode at 0.8 V is much lower than the positive one; as shown above, at this voltage, hydrogen storage is not involved, and only the EDL charging contributes to the capacitance value. Overall Fig. 4b shows that, as is predictable,

Table 1 Capacitance values (F g^{-1}) of the positive and negative AC electrodes during the galvanostatic (200 mA g^{-1}) charge–discharge of the AC/AC cells, up to different values of the maximum voltage in 1 mol L^{-1} Li_2SO_4 (a), and 1 mol L^{-1} Li_2SO_4 + 0.1 mol L^{-1} Na_2MoO_4 (b)

		0.8 V	1.0 V	1.2 V	1.4 V	1.6 V
$Li_2SO_4 + Na_2MoO_4$	C_+ (F g^{-1})	124	131	145	158	165
	C_- (F g^{-1})	76	83	90	96	107
Li_2SO_4	C_+ (F g^{-1})	96	98	99	104	102
	C_- (F g^{-1})	77	82	86	91	102

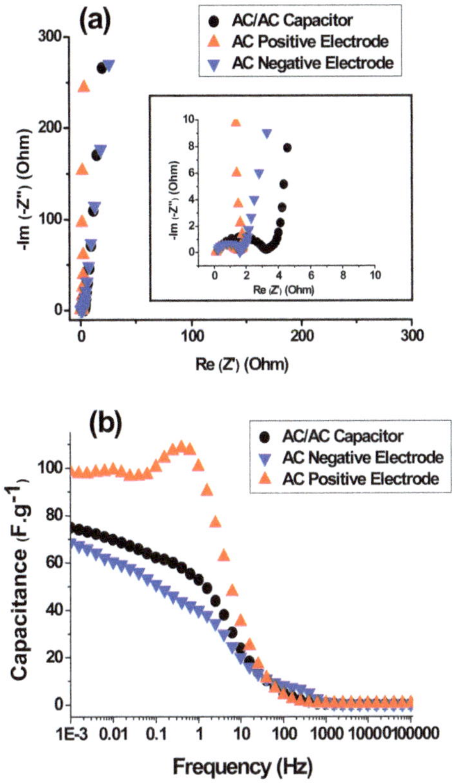

Fig. 4 Impedance spectroscopy data of an AC/AC cell in 1 mol L^{-1} Li_2SO_4 + 0.1 mol L^{-1} Na_2MoO_4 and of the individual electrodes at 0.8 V: (a) Nyquist plots; (b) capacitance *vs.* frequency.

the capacitance of the AC/AC capacitor is controlled mainly by the negative electrode.

The cyclic voltammograms of the AC/AC capacitors in 1 mol L^{-1} Li_2SO_4 and in 1 mol L^{-1} Li_2SO_4 + 0.1 mol L^{-1} Na_2MoO_4 are compared at 0.8 V and 1.6 V in Fig. 5. For both voltage ranges, the capacitance is higher in Li_2SO_4 + Na_2MoO_4 than in Li_2SO_4; the enhanced capacitance at 0.8 V seems solely due to the redox processes involving the molybdate ions, whereas, at 1.6 V, reversible hydrogen storage also contributes. The slightly worse charge propagation of the system in Li_2SO_4 + Na_2MoO_4 up to 0.8 V as compared to Li_2SO_4 is attributed to the pseudo-faradic contribution of the molybdate anions. Up to 1.6 V, the charge propagation is comparable in both electrolytic systems, due to the dominant effect of reversible hydrogen storage.[11]

Fig. 6 shows plots of capacitance and efficiency (discharge time/charge time) *vs.* voltage obtained from galvanostatic charge–discharge of the AC/AC capacitors in 1 mol L^{-1} Li_2SO_4 and 1 mol L^{-1} Li_2SO_4 + 0.1 mol L^{-1} Na_2MoO_4. Whatever the maximum voltage, the capacitance is higher in Li_2SO_4 + Na_2MoO_4 than in Li_2SO_4; the capacitance difference between the two systems is around 10 F g^{-1} at 0.8 V and it reaches 20 F g^{-1} at 1.6 V, confirming a larger pseudo-capacitive contribution related to the presence of molybdate anions at high voltage. Overall, at 1.6 V, the

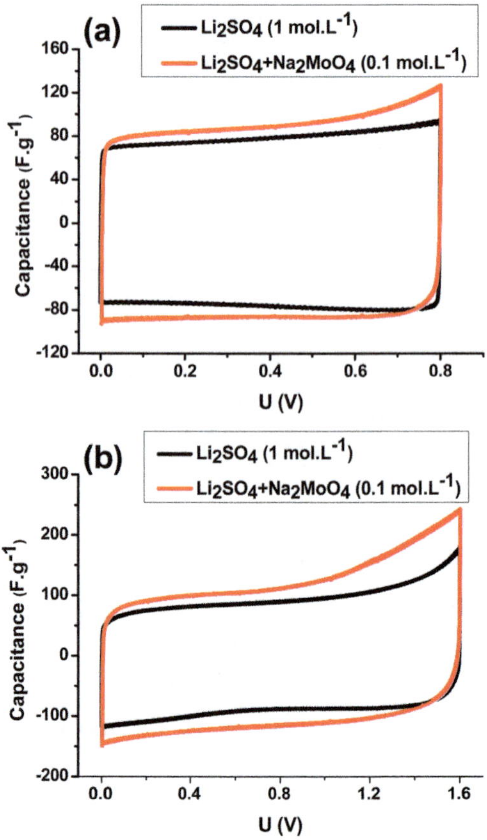

Fig. 5 Cyclic voltammograms (2 mV s^{-1}) of AC/AC capacitors in 1 mol L^{-1} Li_2SO_4 and 1 mol L^{-1} Li_2SO_4 + 0.1 mol L^{-1} Na_2MoO_4 electrolytes up to 0.8 V (a), and 1.6 V (b).

capacitor in $Li_2SO_4 + Na_2MoO_4$ displays a capacitance of 121 F g^{-1} as compared to 103 F g^{-1} in Li_2SO_4. An efficiency higher than 98% is observed over the entire voltage range for the AC/AC capacitor in Li_2SO_4, and only up to 1.2 V in the

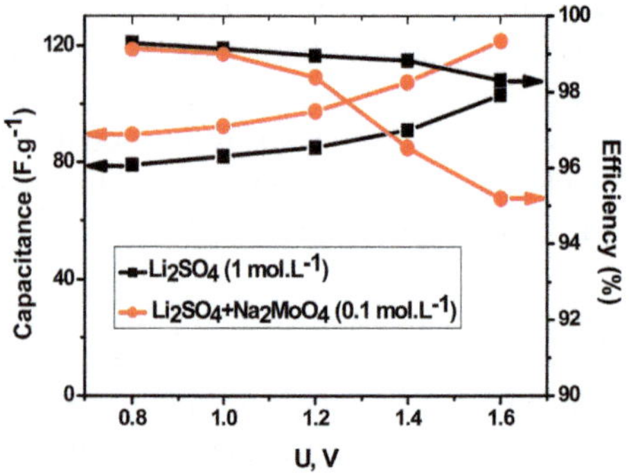

Fig. 6 Plots of capacitance and efficiency *vs.* maximum voltage from galvanostatic charge–discharge of the AC/AC capacitors in 1 mol L^{-1} Li_2SO_4 and 1 mol L^{-1} Li_2SO_4 + 0.1 mol L^{-1} Na_2MoO_4.

Li_2SO_4 + Na_2MoO_4 medium. At a voltage higher than 1.2 V, the efficiency in the later electrolyte decreases to reach ~95% at 1.6 V. This decay in efficiency could be attributed to irreversible redox processes occurring due to the presence of MoO_4^{2-} anions. Additionally, as shown in Fig. 2, the nature of the carbon–hydrogen bonds on the negative AC electrode slightly differs in both electrolytic media and this could also be responsible for the lower efficiency at 1.6 V.

The evolution of capacitance with the value of the charging–discharging current applied is presented in Fig. 7 for the AC/AC capacitors in the 1 mol L^{-1} Li_2SO_4 and 1 mol L^{-1} Li_2SO_4 + 0.1 mol L^{-1} Na_2MoO_4 electrolytes. In the two media, the capacitance slightly drops up to *ca.* 1 A g^{-1}, and then it is almost constant up to 10 A g^{-1}. Over the entire current range, the difference between the values for the two kinds of electrolyte is around 15–20 F g^{-1}. Above 1 A g^{-1}, it is likely that the charging of the electrical double-layer is the main contribution to the capacitance.

3.4. Accelerated ageing of the capacitors by floating at 1.6 V

The AC/AC capacitors in both types of electrolytes, without and with Na_2MoO_4, have been subjected to potentiostatic floating at 1.6 V over a total time of 120 hours. Every 2 hours, the values of capacitance and resistance were determined from a few galvanostatic cycles, and these are presented in Fig. 8. During the first 20 hours, the capacitance slightly increases for both electrolytic systems, and then it remains almost constant over the next 100 hours for the Li_2SO_4 + Na_2MoO_4 electrolyte, while it slightly decreases after a total floating time of 90 hours in the case of Li_2SO_4. The initial increase is attributed to the enhanced penetration of the ionic species into small micropores, provoked by the prolonged polarization at an elevated voltage.[12,13] A similar trend in floating conditions has been also observed in ref. 28 for a capacitor in [EMIM][BF_4] ionic liquid.

The beneficial effect of the Na_2MoO_4 inhibitor is more visible when considering the resistance (Fig. 8b). For the AC/AC capacitor in Li_2SO_4, the resistance first decreases at the beginning of the floating, is then constant, and dramatically increases after 90 hours of floating, which could be due to an increased electrode resistance, as described elsewhere.[29] Both the oxidation of the AC at the positive

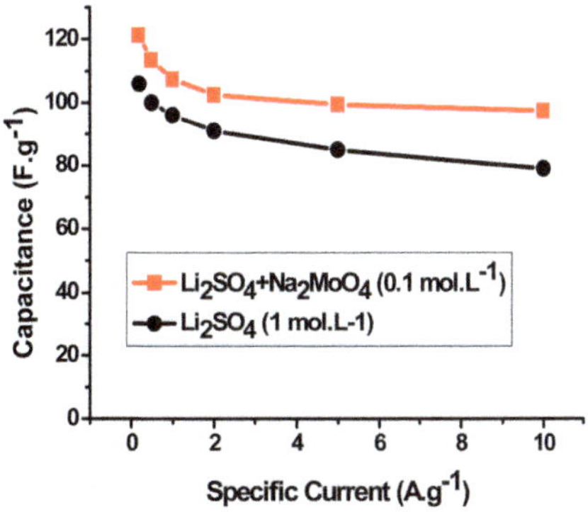

Fig. 7 Plots of capacitance *vs.* specific current (A g^{-1}) for the AC/AC capacitors in 1 mol L^{-1} Li_2SO_4 and 1 mol L^{-1} Li_2SO_4 + 0.1 mol L^{-1} Na_2MoO_4 charged up to 1.6 V.

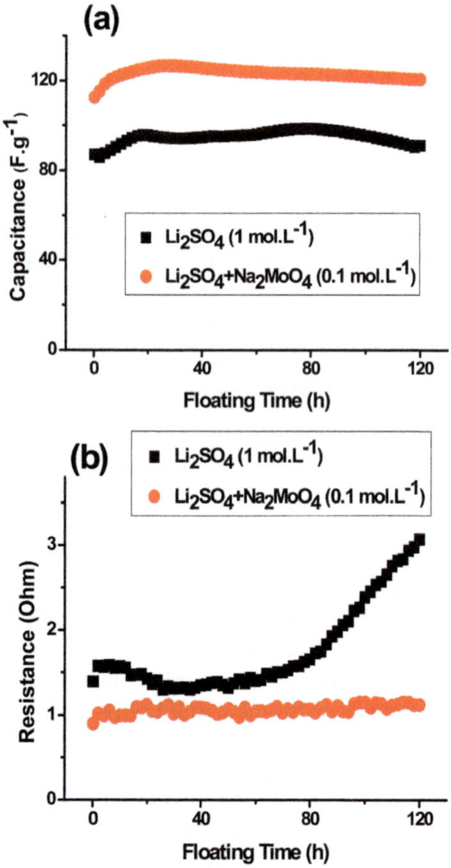

Fig. 8 The evolution of capacitance (a), and resistance (b), during the floating of the AC/AC supercapacitors in the 1 mol L^{-1} Li_2SO_4 and 1 mol L^{-1} Li_2SO_4 + 0.1 mol L^{-1} Na_2MoO_4 electrolytes at 1.6 V.

electrode and the accumulation of corrosion products between the AC disk pellet and the stainless steel current collector contribute to the increased resistance.[13] By contrast, in the Li_2SO_4 + Na_2MoO_4 electrolyte, the resistance is constant over the whole floating time and is always smaller than in Li_2SO_4. This confirms that the presence of sodium molybdate prevents stainless steel corrosion and the accumulation of corrosion products at the positive electrode/current collector interface. As shown in Fig. 3b, the maximum potential of the positive electrode is dramatically reduced by the addition of sodium molybdate and these undesirable side effects no longer occur when the AC/AC capacitor operates up to 1.6 V.

Cyclic voltammograms (CVs) have been recorded on the AC/AC capacitors in the two kinds of electrolytes after prolonged floating at 1.6 V, and are shown in Fig. 9. A comparison with the CV obtained before floating in Fig. 5b reveals, in the case of Li_2SO_4, a drop of capacitive current at voltages higher than 0.8–1 V. Such narrowing of the CVs has been already observed for capacitors in organic electrolytes, and it was demonstrated that this is due to porosity saturation by the ions when the carbon material has a relatively moderate pore volume.[30] As shown in ref. 12 and 13 and discussed above, when the capacitor is charged up to or above 1.6 V, the positive electrode potential is so high that the AC is oxidized, leading to a dramatic decrease of pore volume due to the blockage of pore entrances by oxygenated groups; at the same time, the stainless steel collector is corroded, which causes a low charge propagation resulting in a lowering of the capacitance. In the case of the Li_2SO_4 + Na_2MoO_4 electrolyte, the CV after prolonged floating (Fig. 9) nearly retains the same shape as for the fresh cell (see Fig. 5b), with good charge propagation up to 1.6 V. Once more, this is an indication that the capacitor with Li_2SO_4 + Na_2MoO_4 undergoes lesser ageing than the one with Li_2SO_4. As shown in Fig. 3b, at a voltage of 1.6 V in Li_2SO_4 + Na_2MoO_4, the potential of the positive electrode is well below the thermodynamic potential of water oxidation; consequently, under these conditions, the AC is not oxidized and the corrosion of the stainless steel does not occur.

Nyquist plots of the AC/AC capacitors in Li_2SO_4 and Li_2SO_4 + Na_2MoO_4, before and after 120 hours floating at 1.6 V, are shown in Fig. 10, and the equivalent series resistance (ESR), equivalent distributed resistance (EDR) and charge transfer resistance (R_{ct}) values calculated from these plots are given in Table 2. In Li_2SO_4, the three parameters dramatically increase after floating, whereas in Li_2SO_4 + Na_2MoO_4 they remain quasi-identical to the values before floating. The significant increase of R_{ct} resulting from the uneven current distribution after floating in Li_2SO_4 is due to the corrosion products generated on the positive stainless steel current collector and/or oxidation of the AC at the positive electrode. By contrast, no significant increase of R_{ct} is observed after floating in Li_2SO_4 + Na_2MoO_4, suggesting that the presence of molybdate anions prevents the formation of a low conductivity layer of corrosion products and the oxidation of the AC positive electrode.

The performance of the capacitor in the Li_2SO_4 + Na_2MoO_4 electrolyte remains almost unaltered after prolonged floating at 1.6 V. However, the capacitance *vs.* frequency plot (inset of Fig. 10b) displays a pseudocapacitive contribution at low frequency, probably due to (i) the generation of new functionalities on the AC surface during floating and/or (ii) the deposition of conductive MoO_2 or Mo on the negative AC surface, as given by equilibria (3) and (4).

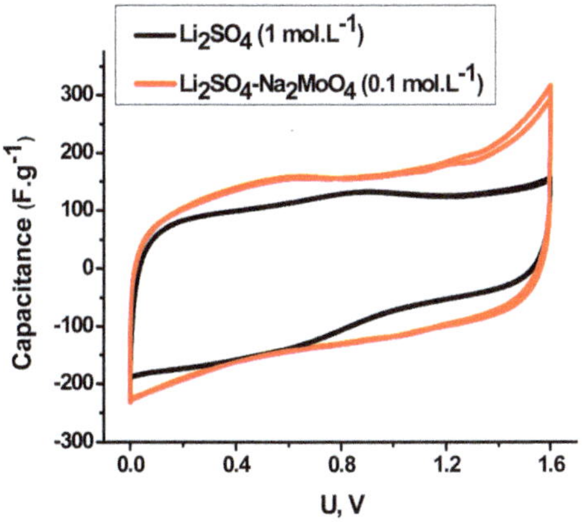

Fig. 9 Cyclic voltammograms (2 mV s^{-1}) of the AC/AC capacitors in 1 mol L^{-1} Li_2SO_4 and 1 mol L^{-1} Li_2SO_4 + 0.1 mol L^{-1} Na_2MoO_4 after floating at 1.6 V for 120 h.

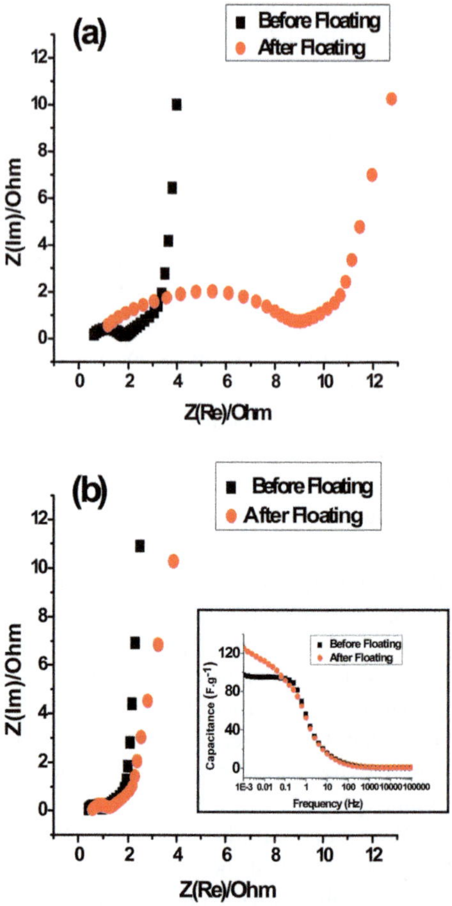

Fig. 10 Nyquist plots of the AC/AC capacitors at 0 V before and after floating at 1.6 V for 120 hours in 1 mol L^{-1} Li_2SO_4 (a), and 1 mol L^{-1} Li_2SO_4 + 0.1 mol L^{-1} Na_2MoO_4 (b). The inset in (b) shows the frequency dependence of capacitance before and after floating.

From the preceding discussion, it is now clearly proved that the resistance decreases and the capacitance increases when sodium molybdate is added to lithium sulfate in AC/AC capacitors. Additionally, the capacitor using Li_2SO_4 +

Table 2 ESR, EDR and R_{ct} values obtained from the Nyquist plots of the AC/AC capacitors in 1 mol L^{-1} Li_2SO_4 and 1 mol L^{-1} Li_2SO_4 + 0.1 mol L^{-1} Na_2MoO_4, before and after floating at 1.6 V for 120 h

		Li_2SO_4 + Na_2MoO_4	Li_2SO_4
ESR (Ω)	Before floating	0.43	0.45
	After floating	0.51	1.1
EDR (Ω)	Before floating	1.6	3.3
	After floating	1.7	10.2
R_{ct} (Ω)	Before floating	0.72	1.25
	After floating	0.77	8.02

Na_2MoO_4 has the ability to withstand potentiostatic floating at 1.6 V for 120 hours without observable changes in resistance and capacitance. The effects of floating on the AC and the nature of the corrosion products generated during the ageing of the capacitors in the Li_2SO_4 + Na_2MoO_4 electrolyte are currently being investigated.

4. Conclusions

Previous to this work, it was established that AC/AC electrochemical capacitors in alkali sulfate electrolytes operate with a good cycle life at voltages up to around 1.6–2 V. However, when stainless steel collectors were implemented in these systems, detrimental corrosion appeared on the positive collector, limiting the voltage to 1.5 V. In this paper, we have clearly demonstrated a dramatic improvement of cycle life at voltages up to 1.6 V when sodium molybdate (0.1 mol L^{-1}) is added to 1 mol L^{-1} Li_2SO_4 aqueous electrolyte. Besides their known corrosion inhibition role, the MoO_4^{2-} anions enhance the capacitance of the positive electrode through pseudo-faradic contributions. As a consequence, adding Na_2MoO_4 to 1 mol L^{-1} Li_2SO_4 results in shifting the maximum operating potential of the positive AC electrode from 0.96 V *vs.* NHE (in Li_2SO_4) to 0.798 V *vs.* NHE (in Li_2SO_4 + Na_2MoO_4), preventing it from oxidation and pitting corrosion of the stainless steel collector when the capacitor is polarized up to 1.6 V. Electrochemical investigations on AC/AC capacitors in Li_2SO_4 + Na_2MoO_4 subjected to potentiostatic floating at 1.6 V for 120 hours prove that the performance remains nearly unaffected with almost no change in R_{ct}. Besides, the capacitance of the AC/AC system reaches 121 F g^{-1} in Li_2SO_4 + Na_2MoO_4 as compared to 103 F g^{-1} for Li_2SO_4. Hence, sodium molybdate (Na_2MoO_4) is an additive of choice for supercapacitors using salt aqueous electrolytes and low cost stainless steel collectors. Further optimization of the electrolyte formulation allows energy density values comparable to those reached with organic electrolytes to be forecasted.

Acknowledgements

The authors are grateful to the Foundation for Polish Science for supporting the ECOLCAP project realized within the WELCOME program, co-financed from the European Union Regional Development Fund. Norit and Timcal are acknowledged for kindly providing the activated carbon and the carbon black, respectively.

Notes and references

1 B. E. Conway, *Electrochemical Supercapacitors: Scientific Fundamentals and Technological Applications*, Kluwer-Plenum, New York, 1999.
2 F. Béguin and E. Frackowiak, *Supercapacitors: Materials, Systems and Applications*, Wiley-VCH, Weinheim, 2013.
3 R. J. Miller and P. Simon, *Science*, 2008, **321**, 651.
4 E. Frackowiak and F. Béguin, *Carbon*, 2001, **39**, 937.
5 R. Kötz and M. Carlen, *Electrochim. Acta*, 2000, **45**, 2483.
6 M. Ramani, B. Haran, R. White, B. Popove and L. Arsov, *J. Power Sources*, 2001, **93**, 209.

7 E. Raymundo-Pinero, K. Kierzek, J. Machnikowski and F. Béguin, *Carbon*, 2006, **44**, 2498.
8 J. Chmiola, C. Largeot, P. Taberna, P. Simon and Y. Gogotsi, *Angew. Chem., Int. Ed.*, 2007, **47**, 1567.
9 L. Demarconnay, E. Raymundo-Pinero and F. Béguin, *Electrochem. Commun.*, 2010, **12**, 1275.
10 K. Fic, G. Lota, M. Meller and E. Frackowiak, *Energy Environ. Sci.*, 2012, **5**, 5842.
11 Q. Gao, L. Demarconnay, E. Raymundo-Pinero and F. Béguin, *Energy Environ. Sci.*, 2012, **5**, 9611.
12 P. Ratajczak, K. Jurewicz and F. Béguin, *J. Appl. Electrochem.*, 2014, **44**, 475.
13 P. Ratajczak, K. Jurewicz, P. Skowron, Q. Abbas and F. Béguin, *Electrochim. Acta*, 2014, **130**, 344.
14 K. Nomura and Y. Ujihira, *Hyperfine Interact.*, 1986, **29**, 1467.
15 J. H. Liu, C. Wen, M. Yu, S. M. Li and B. Wang, *Int. J. Electrochem. Sci.*, 2013, **8**, 4085.
16 A. M. Shams El Din and L. Wang, *Desalination*, 1996, **107**, 29.
17 T. Kodama and J. R. Ambrose, *Corrosion*, 1977, **33**, 155.
18 E. A. Lizlovs, *Corrosion*, 1976, **32**, 263.
19 Q. Qu, L. Li, S. Jiang, W. Bai and Z. Ding, *J. Appl. Electrochem.*, 2009, **39**, 569.
20 V. K. Gouda, in *Proc. 12th Int. Corrosion Cong.*, Houston, TX, USA, 1993, pp. 19–24.
21 Y. Cheng, J. Bullerwell and F. Steward, *Electrochim. Acta*, 2003, **48**, 1521.
22 B. G. Callaghan, *Corros. Sci.*, 1993, **35**, 1535.
23 C.-O. A. Olsson and D. Landolt, *J. Electrochem. Soc.*, 2001, **148**, B438.
24 S. Mischler, A. Vogel, H. J. Mathieu and D. Landolt, *Corros. Sci.*, 1991, **32**, 925.
25 K. Jurewicz, E. Frackowiak and F. Béguin, *J. Appl. Phys.*, 2004, **A78**, 981.
26 F. Béguin, M. Friebe, K. Jurewicz, C. Vix-Guterl, J. Dentzer and E. Frackowiak, *Carbon*, 2006, **44**, 2392.
27 M. Pourbaix, *Atlas d' Equilibres Electrochimiques*, Gauthier-Villars, Paris, 1963, p. 275.
28 D. Weingarth, A. Foelske-Schmitz and R. Kötz, *J. Power Sources*, 2013, **225**, 84.
29 S. Ishimoto, Y. Asakawa, M. Shinya and K. Noi, *J. Electrochem. Soc.*, 2009, **156**, A563.
30 R. Mysyk, E. Raymundo-Piñero and F. Béguin, *Electrochem. Commun.*, 2009, **11**, 554.

Faraday Discussions

PAPER

Reduced graphene oxide anchoring $CoFe_2O_4$ nanoparticles as an effective catalyst for non-aqueous lithium–oxygen batteries†

Yong Cao,‡ Sen-Rong Cai,‡ Shao-Cong Fan, Wei-Qiang Hu, Ming-Sen Zheng* and Quan-Feng Dong*

Received 21st April 2014, Accepted 7th May 2014
DOI: 10.1039/c4fd00075g

$CoFe_2O_4$ nanoparticles were uniformly anchored on reduced graphene oxide by a facile solvothermal method. The obtained $CoFe_2O_4$/reduced graphene oxide ($CoFe_2O_4$/rGO) hybrid was employed as catalyst for Li–O_2 batteries. It could effectively lower the ORR (oxygen reduction reaction) and OER (oxygen evolution reaction) overpotentials of the batteries and deliver a large capacity of 12 235 mA h g_{rGO}^{-1} (2116 mA h g_{hybrid}^{-1}). It also exhibited high cyclic stability.

Introduction

To meet the urgent energy demand of electric vehicles and large-scale energy storage, a new generation of lithium batteries, such as the Li–air (O_2) battery, has become a popular research topic for scientists.[1,2] As one of the metal–air batteries, the Li–O_2 battery has an extremely high theoretical energy density of ~11 500 W h kg^{-1}, excluding the mass of O_2 that can be obtained from the atmosphere at discharging, which is higher than that of any battery, and is equal to gasoline.[3]

In a typical Li–O_2 battery, the oxygen molecule is reduced to Li_2O_2 by combining with the lithium ions and electrons during the discharging process (the oxygen reduction reaction, ORR) with a thermodynamic potential of 2.96 V, and recovered during the subsequent charging process (the oxygen evolution reaction, OER).

However, large discharge and charge overpotentials exist due to the sluggish kinetics of the oxygen electrode, thus resulting in low energy efficiency. In view of

Department of Chemistry, College of Chemistry and Chemical Engineering, Xiamen University, State Key Laboratory of Physical Chemistry of Solid Surfaces, Xiamen University, Xiamen, Fujian, 361005, China. E-mail: qfdong@xmu.edu.cn; mszheng@xmu.edu.cn; Fax: +86-592-2185905

† Electronic supplementary information (ESI) available. See DOI: 10.1039/c4fd00075g

‡ These workers contributed equally to this work.

the poor rate capability and cyclic stability, current Li–O_2 batteries are still far from practical application.

Various electrocatalysts have been developed to improve the performance of the oxygen electrode, such as carbon materials, transition metal oxides and noble metals/oxides.[4–8] Graphene, an ideal two dimensional nanomaterial with unique physical and chemical properties, has been directly employed as a catalyst for the oxygen electrode of the Li–O_2 battery with a significantly improved ORR performance compared to other carbon materials.[9,10] It has also been used as the matrix for supporting transition metal oxide nanoparticles, such as MnO_2 and Co_3O_4, to form composite materials with synergistic effects for both ORR and OER.[11–13] Covalent hybrids of spinel $MnCo_2O_4$ and $NiCo_2O_4$ have higher electrochemical activities as ORR catalysts for Li–O_2 batteries than the single oxides.[14,15] $CoFe_2O_4$ has also been reported as a catalyst for the Li–O_2 battery when mechanically mixed with Super S carbon black and a binder and showed moderate activity and cyclic stability, but also a large discharge–charge overpotential.[16] We therefore speculate that graphene supported $CoFe_2O_4$ would improve the performance of $CoFe_2O_4$ as an oxygen electrode catalyst.

Herein, we report a $CoFe_2O_4$/reduced graphene oxide ($CoFe_2O_4$/rGO) hybrid which was prepared using a facile one-pot solvothermal procedure. The hybrid exhibited catalytic properties superior to those of pure rGO and its electrochemical performance as the oxygen electrode for Li–O_2 batteries was investigated in detail.

Experimental

Graphene oxide (GO) was prepared by the modified Hummers' method, while the $CoFe_2O_4$/rGO hybrid was synthesized by a solvothermal method. Detailed synthesis procedures are provided in the ESI.†

The crystallographic properties were investigated by powder X-ray diffraction (XRD, Philips X'pert Pro Super X-ray diffractometer using Cu Kα radiation with λ = 1.5418 Å) at a 2θ range of 10–80°. The mass content of $CoFe_2O_4$ in the hybrid was determined using a Diamond TG/DTA analyzer (Perkin Elmer) under a N_2–O_2 atmosphere at a heating rate of 5 °C min^{-1}. Detailed morphologies of the as-prepared samples were characterized by field emission scanning electron microscopy (FESEM, HITACHI S-4800) and transmission electron microscopy (TEM, JEOL-1400).

The cathode was prepared by casting the $CoFe_2O_4$/rGO hybrid and Kynar2801 (a copolymer based on PVDF) onto a nickel foam current collector with a mass ratio of 9 : 1. A Swagelok type electrochemical cell composed of electrolyte (1 M LiTFSI in TEGDME), a glass fiber membrane, a metallic lithium anode and the prepared cathode was used to test the electrochemical performance. The porous electrode was exposed to 1 atm O_2 in which the cell was tested. All parts of the cell were dried overnight in a vacuum oven at 80 °C before they were assembled into the Swagelok type cell in an argon-filled glove box. The electrochemical performance of the cell was tested at room temperature using a Neware battery test system (Neware Co., Ltd, Shenzhen, China).

Results and discussion

The X-ray diffraction (XRD) pattern of the as-prepared $CoFe_2O_4$/rGO hybrid is shown in Fig. 1a. All marked peaks can be indexed to spinel phase $CoFe_2O_4$

(JCPDS 00-022-1088) without any impurities. The calculated average particle size of the $CoFe_2O_4$ nanoparticle is about 13–15 nm based on the Scherrer equation. Note that the diffraction peak of the rGO cannot be observed in the pattern of the $CoFe_2O_4$/rGO hybrid, indicating the homogenous coating of $CoFe_2O_4$ nanoparticles on the rGO surface suppressing the stacking of graphene layers. The mass percentage of $CoFe_2O_4$ in the hybrid was measured by the TGA method as shown in Fig. 1b. The calculated mass content of $CoFe_2O_4$ is ~82.7%.

The morphology of the as-prepared rGO is shown in Fig. 2a, which displays typical interconnected sheets with wrinkles on the surface. SEM and TEM images of the $CoFe_2O_4$/rGO hybrid are shown in Fig. 2b and c. The rGO matrix can maintain a dispersive status after the loading of $CoFe_2O_4$ nanoparticles. The TEM image in Fig. 2c confirms a good dispersion of $CoFe_2O_4$ nanoparticles on the rGO surface. Energy dispersive X-ray spectrometry (EDS) mapping in Fig. 2d–g and S1 (ESI)† provides additional evidence of $CoFe_2O_4$ on the rGO surface from the distribution of Co, Fe, O and C elements.

N_2 adsorption–desorption isotherms and the pore-size distribution are shown in Fig. S2.† The N_2 sorption isotherm has a shape of irreversible type IV, which is characteristic of mesoporous materials. The large degree of hysteresis between the adsorption and desorption isotherms is attributed to macropores formed by

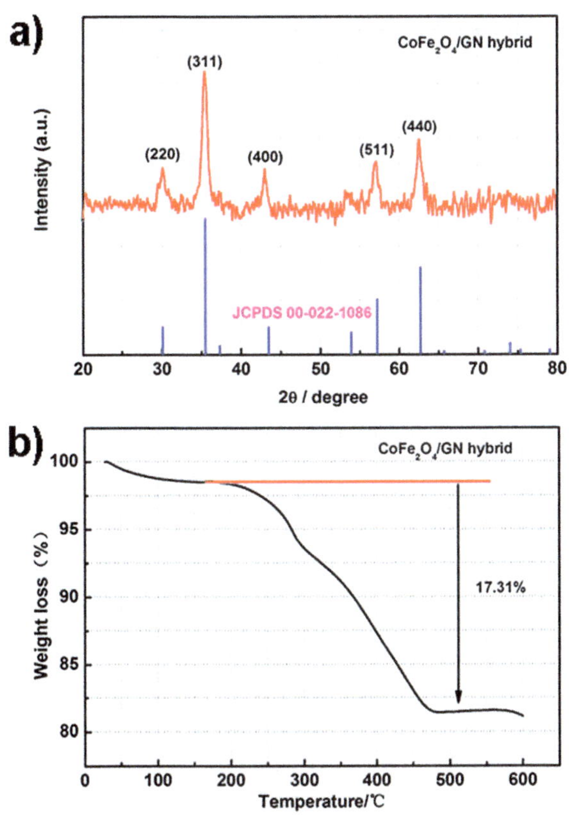

Fig. 1 (a) The X-ray diffraction pattern of the $CoFe_2O_4$/rGO hybrid. (b) The TG curve of the hybrid.

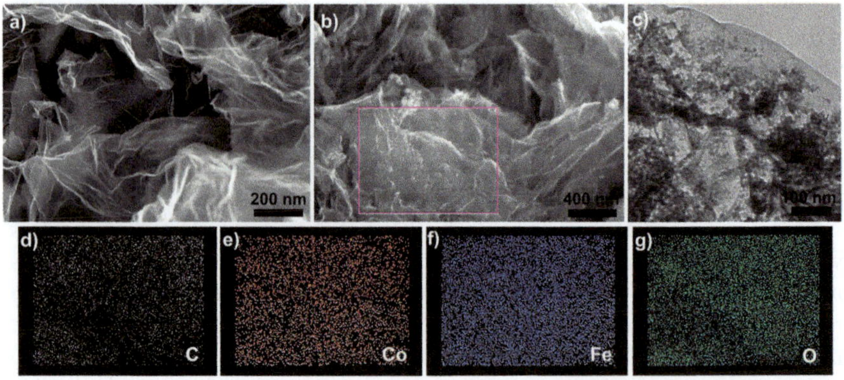

Fig. 2 (a) An SEM image of the rGO. (b) SEM and (c) TEM images of the $CoFe_2O_4$/rGO hybrid. (d)–(g) Element distribution of C, Co, Fe and O in the selected area of (b).

the graphene layers. The BJH pore-size-distribution shows that the major size of the pores falls into the range of 3–5 nm. The $CoFe_2O_4$/rGO hybrid has a BET specific surface area of 127 $m^2\ g^{-1}$ and an average pore volume of 0.432 $cm^3\ g^{-1}$. The large surface area ensures sufficient reaction sites and the large pore volume is favorable for the storage of the discharge product.

The electrochemical performance of the $CoFe_2O_4$/rGO hybrid as the cathode of Li–O_2 batteries was evaluated by a galvanostatic charge–discharge method. The voltage *vs.* capacity profile of a fully discharged/charged battery using the $CoFe_2O_4$/rGO hybrid cathode is displayed in Fig. 3. At a current density of 50 mA $g_{hybrid}{}^{-1}$, the battery delivered an extremely large discharge capacity of 12 235 mA h g^{-1} based on the mass of rGO in the hybrid, and a charge capacity of 11 850 mA h g^{-1} within the voltage window of 2.2–4.2 V. The calculated coulombic efficiency is as high as 96.9%. The discharge and charge capacities based on the total mass of the hybrid (rGO + $CoFe_2O_4$) are 2116 and 2050 mA h g^{-1}, respectively, corresponding to the top axis in Fig. 3. The discharge voltage platform is stable at around 2.7 V and the charge voltage platform is below 4.0 V. The energy efficiency

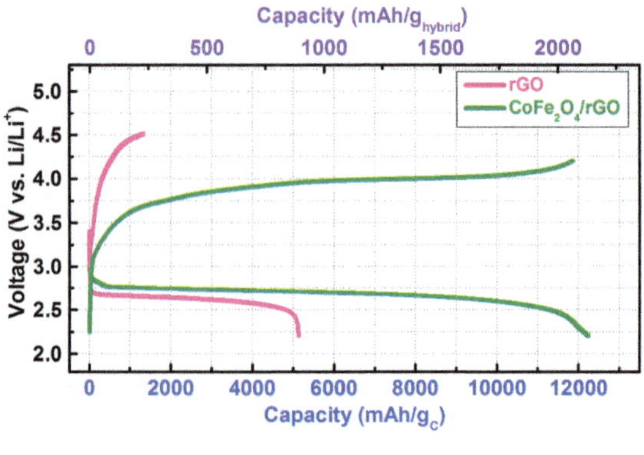

Fig. 3 A comparison of the voltage profiles of fully discharged Li–O_2 batteries with the $CoFe_2O_4$/rGO hybrid and pure rGO electrodes at a current density of 50 mA g^{-1}.

was about 68%, which is comparable with that of most reported transition metal oxides. For comparison, the performance of pure rGO was also tested under the same conditions. Although it could deliver a discharge capacity of 5140 mA h g^{-1}, the discharge overpotential was about 0.1 V larger than the $CoFe_2O_4$/rGO hybrid. Worse still, the charge capacity was only 1322 mA h g^{-1} even under the cut-off voltage of 4.5 V, and no charge platform was observed due to the large charge overpotential. Therefore, a synergistic effect was achieved after the hybridization of rGO and $CoFe_2O_4$ nanoparticles, which improved both the ORR and OER processes of the Li–O_2 batteries.

The cyclic capability of the $CoFe_2O_4$ was also evaluated at a current density of 50 mA g_{hybrid}^{-1} with a cut-off capacity of 1000 mA h g_C^{-1}. The battery had no capacity decay over 30 cycles, as shown in Fig. 4a. Selected voltage profiles of different cycles are shown in Fig. 4b. The battery exhibited excellent cycle stability over 20 cycles without pronounced vibration in both the discharge and charge profiles. The discharge median voltage and terminal charge voltage were also recorded, as shown in Fig. 4c. The initial discharge median voltage was 2.77 V which then slightly increased to 2.87 V in the subsequent five cycles. Although with a slight voltage degradation afterward, it was still above 2.6 V after 30 cycles. During charging, the terminal voltage of the initial ten cycles was about 4.1 V, but it then dropped and stabilized at 4.0 V in the subsequent cycles.

The improved performance of the $CoFe_2O_4$/rGO hybrid can be attributed to the effective hybridization of nanoscale $CoFe_2O_4$ particles and reduced graphene oxide. Firstly, rGO acts as the conductive matrix that is favorable for both electron-transport and electrolyte penetration into the electrode. The rGO matrix is highly dispersive and therefore benefits oxygen diffusion. Secondly, $CoFe_2O_4$ particles are uniformly anchored *in situ* on the rGO surface and act as an effective catalyst for both the ORR and OER. For one thing, the hybrid has a relatively high surface area and pore volume that can provide sufficient reaction sites and abundant space to store the discharge product Li_2O_2, promoting a large capacity delivery. For another, the $CoFe_2O_4$ nanoparticles with diameters of ~15 nm enable a fast reaction rate and short diffusion path length, resulting in fast kinetics of the oxygen electrode. Besides, it has been demonstrated that some carbon materials are not stable in the presence of Li_2O_2 and can be oxidized to Li_2CO_3, which is not easily decomposed during charging.[17] Also, defects on the carbon surface will accelerate the decomposition of the electrolyte, forming some by-products.[18] Some special electrode architectures in which the carbon surface (defects) was coated by catalyst particles could efficiently prevent the formation of by-products.[8,19] Therefore, although the rGO used here is effective for the ORR, the large charge overpotential directly causes battery degradation in the first charge process. While, after the anchoring of $CoFe_2O_4$ on the rGO surface, the ORR process was improved with a lower discharge overpotential and large capacity due to the synergistic effect of rGO and $CoFe_2O_4$. More importantly, the OER overpotential was significantly reduced, which is not only ascribed to the high catalytic activity of $CoFe_2O_4$, but also the efficient prevention of contact of rGO with Li_2O_2 by the $CoFe_2O_4$ particles.

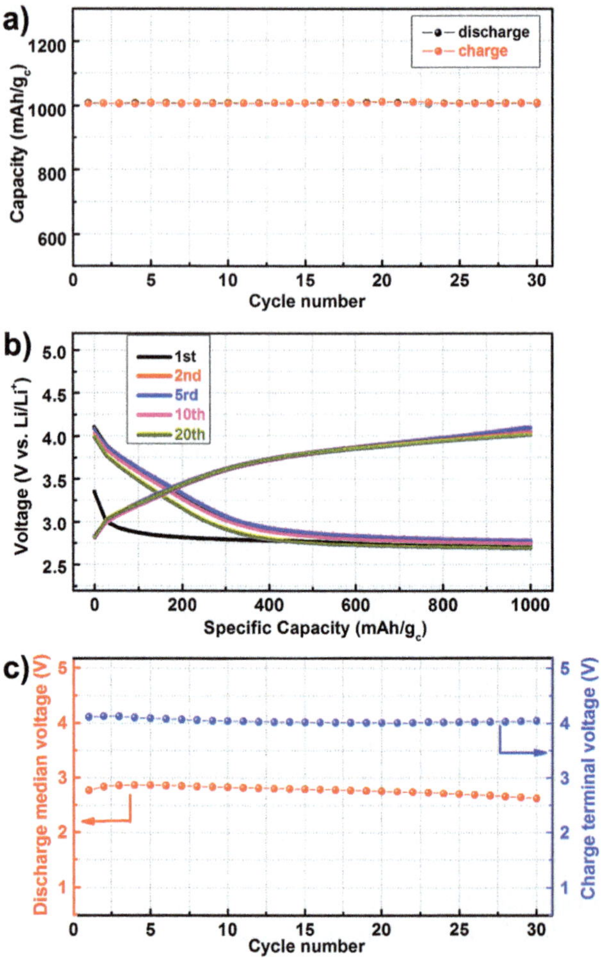

Fig. 4 (a) Cycling stability of the Li–O_2 battery at a current density of 50 mA g_{hybrid}^{-1}. (b) Discharging and charging profiles of selected cycles. (c) The variation of discharge median voltage and charge terminal voltage with cycle number.

Conclusions

In summary, a $CoFe_2O_4$/rGO hybrid was prepared through a facile solvothermal reaction. When employed as the oxygen cathode of Li–O_2 batteries, the $CoFe_2O_4$/rGO hybrid demonstrated bifunctional catalytic activity for both the ORR and OER. It showed a much higher specific capacity and energy efficiency than pure rGO. The improved performance can be attributed to the synergistic effect of the hybrid. The rGO acted as a conductive matrix that is favorable for both electron-transport and electrolyte penetration, while the anchored $CoFe_2O_4$ nanoparticles exhibited a catalytic role. Although some improvement has been achieved, the OER overpotential is still too large, so further study will be focused on the optimization of this type of catalyst.

Acknowledgements

The authors gratefully acknowledge the financial support from the Key Project of NSFC (U1305246, 21321062), and the tenth YUMIAO fund of Xiamen University (20120605).

Notes and references

1 M. Armand and J. M. Tarascon, *Nature*, 2008, **451**, 652–657.
2 R. Black, B. Adams and L. F. Nazar, *Adv. Energy Mater.*, 2012, **2**, 801–815.
3 P. G. Bruce, S. A. Freunberger, L. J. Hardwick and J.-M. Tarascon, *Nat. Mater.*, 2012, **11**, 19–29.
4 Y. Yang, Q. Sun, Y. S. Li, H. Li and Z. W. Fu, *J. Electrochem. Soc.*, 2011, **158**, B1211.
5 B. Sun, X. Huang, S. Chen, J. Zhang and G. Wang, *RSC Adv.*, 2014, **4**, 11115–11120.
6 Y. Qin, J. Lu, P. Du, Z. Chen, Y. Ren, T. Wu, J. T. Miller, J. Wen, D. J. Miller, Z. Zhang and K. Amine, *Energy Environ. Sci.*, 2013, **6**, 519–531.
7 B. Sun, P. Munroe and G. Wang, *Sci. Rep.*, 2013, **3**, DOI: 10.1038/srep02247.
8 Z. Jian, P. Liu, F. Li, P. He, X. Guo, M. Chen and H. Zhou, *Angew. Chem., Int. Ed.*, 2014, **53**, 442–446.
9 B. Sun, B. Wang, D. Su, L. Xiao, H. Ahn and G. Wang, *Carbon*, 2012, **50**, 727–733.
10 J. Xiao, D. Mei, X. Li, W. Xu, D. Wang, G. L. Graff, W. D. Bennett, Z. Nie, L. V. Saraf, I. A. Aksay, J. Liu and J.-G. Zhang, *Nano Lett.*, 2011, **11**, 5071–5078.
11 S. Liu, Y. Zhu, J. Xie, Y. Huo, H. Y. Yang, T. Zhu, G. Cao, X. Zhao and S. Zhang, *Adv. Energy Mater.*, 2014, DOI: 10.1002/aenm.201301960.
12 Q.-c. Liu, J.-j. Xu, Z.-w. Chang and X.-b. Zhang, *J. Mater. Chem. A*, 2014, **2**, 6081–6085.
13 Y. Cao, Z. Wei, J. He, J. Zang, Q. Zhang, M. Zheng and Q. Dong, *Energy Environ. Sci.*, 2012, **5**, 9765–9768.
14 H. Wang, Y. Yang, Y. Liang, G. Zheng, Y. Li, Y. Cui and H. Dai, *Energy Environ. Sci.*, 2012, **5**, 7931–7935.
15 B. Sun, J. Zhang, P. Munroe, H.-J. Ahn and G. Wang, *Electrochem. Commun.*, 2013, **31**, 88–91.
16 A. Debart, J. Bao, G. Armstrong and P. G. Bruce, *J. Power Sources*, 2007, **174**, 1177–1182.
17 M. M. Ottakam Thotiyl, S. A. Freunberger, Z. Peng and P. G. Bruce, *J. Am. Chem. Soc.*, 2012, **135**, 494–500.
18 B. D. McCloskey, A. Speidel, R. Scheffler, D. C. Miller, V. Viswanathan, J. S. Hummelshøj, J. K. Nørskov and A. C. Luntz, *J. Phys. Chem. Lett.*, 2012, **3**, 997–1001.
19 J. Lu, Y. Lei, K. C. Lau, X. Luo, P. Du, J. Wen, R. S. Assary, U. Das, D. J. Miller, J. W. Elam, H. M. Albishri, D. A. El-Hady, Y.-K. Sun, L. A. Curtiss and K. Amine, *Nat. Commun.*, 2013, **4**, DOI: 10.1038/ncomms3383.

Faraday Discussions

PAPER

In situ Raman study of lithium-ion intercalation into microcrystalline graphite

Christopher Sole, Nicholas E. Drewett and Laurence J. Hardwick*

Received 23rd April 2014, Accepted 8th May 2014
DOI: 10.1039/c4fd00079j

The first and second order Raman spectra of graphite during the first lithiation and delithiation have been investigated in a typical lithium-ion battery electrolyte. *In situ*, real-time Raman measurements under potential control enable the probing of the graphitic negative electrode surface region during ion insertion and extraction. The experimental results reveal the staging formation of a single particle within a free standing graphitic electrode. In particular, the *in situ* behaviour of the double resonance 2D band during the lithiation and delithiation of graphitic carbon has not been previously reported. The 2D band was observed to shift from 2681 to 2611 cm^{-1} and the band shape transformed into a single Lorentzian from 0.24 to 0.15 V *vs.* Li/Li^{+}, providing further information on the electronic structure and C–C bonding of stage 3 and 4 graphite intercalation compounds. The behaviour of the 2D band is in keeping with the Daumas–Hérold model of electrochemically derived intercalation, where the graphene layers are flexible and deform around domains of intercalating lithium ions.

1 Introduction

Currently, the major challenge in lithium battery research is to develop systems with even greater power and energy densities.[1–4] Improvements to battery properties will not only depend on the discovery of new materials, but also advances in the understanding of the surface and interfacial region of battery electrodes. Therefore gaining insight into the structure and chemical nature of these interfaces is of major importance.

Investigations into electrode reactions are considerably improved when electrode/solution interface spectra can be recorded as the electrochemical response is obtained. *In situ* Raman spectroscopy is a convenient and non-destructive tool for studying lithiation/delithiation processes within numerous battery electrodes, and it has been utilised to characterise a number of insertion materials over the past couple of decades, as summarised by two recent review articles.[5,6]

Stephenson Institute for Renewable Energy, Department of Chemistry, University of Liverpool, L69 7ZF, UK. E-mail: laurence.hardwick@liv.ac.uk

Highly crystalline graphitic materials are routinely used as the negative electrode in lithium-ion batteries. Their positive features include a high, reversible specific charge of up to the theoretical value of 372 A h kg^{-1} (of carbon) for the formation of the donor graphite intercalation compound (GIC) LiC_6, a good cycling stability (for portable electronic applications), a high electronic conductivity and low cost. However, problems persist with insufficient rate performance and electrode degradation over time. In addition, due to the low potential of intercalation (<0.2 V *vs.* Li/Li^+), lithium plating is a major safety risk at high charge rates.

The staging process of lithium intercalation into carbon was observed in a typical Li-ion battery electrolyte by Inaba *et al.*[7] during their pioneering *in situ* Raman microscopy studies. They observed that lithium intercalation proceeds through a series of staged graphite intercalation compounds (GICs), classified by a stage index, n, which represents the number of graphene layers separating the layers of intercalated ions. The Raman spectra for GICs with stage $n > 2$ are known to exhibit a doublet G band. The lower (E_{2g2}(i)) and upper (E_{2g2}(b)) frequency components are correspondingly associated with carbon-atom vibrations in interior graphite layers (not adjacent to the intercalate layer planes) and in bounding graphite layers (adjacent to the intercalate planes). The split in the E_{2g2} mode upon intercalation occurs primarily from changes in symmetry at the boundary layer, and secondarily from the electronic effects of the intercalate molecule. The E_{2g2}(i) band disappears for stage 1 and 2, where no graphite interior layer exists.

A quantitative measure of the intercalation stage index, n, can be derived from the relative intensities of the Raman doublet, R, by the following equation:[8]

$$R = \frac{I_i}{I_b} = \frac{\sigma_i}{\sigma_b}\frac{n-2}{2} \quad (n > 2) \tag{1}$$

where I_i and I_b represent the intensities of the interior E_{2g2}(i) and bounding E_{2g2}(b) layer modes respectively, and σ_i/σ_b is the ratio of the cross section for Raman scattering from the interior and bounding layers (a stage independent constant), which in the case of lithium intercalation is equal to 1.

The staging process during lithium insertion into graphitic carbons had been previously proposed by Dahn *et al.*[9] following detailed *in situ* powder X-ray diffraction (PXRD) studies. However, whilst PXRD probes the bulk electrode, Raman microscopy offers the ability to follow the processes of lithiation in individual carbon particles, or different areas of a single particle within the electrode.[10–12]

Improved understanding of electrochemical processes in functioning battery electrodes will require an insight into more localised lithiation mechanisms. Thus, a combination of surface and bulk characterisation is critical in realising the function of these materials under operating battery conditions.

Despite numerous *in situ* Raman studies of Li insertion into carbon,[7,10–20] the effect on the 2D band has yet to be reported. The 2D band is the overtone of the D band. Whilst the D band requires the presence of defects for activation, the overtone originates from a process whereby momentum can always be conserved by two phonons with opposite wave vectors,[21] hence the 2D band is always present. The 2D band (also referred to as G′ within the literature) has been used extensively in the research of sp^2 carbons. It allows determination of the number and orientation of graphene layers in few layered graphene samples (1–5 layers),[21] in addition to providing information on induced strain[22–24] and charging/doping effects.[25,26] In this study, we report on the behaviour of the 2D band during lithiation into microcrystalline graphite, in order to improve our understanding of both the model of Li^+ intercalation and the electronic structure of GICs during lithium intercalation and extraction.

2 Experimental

2.1 Free-standing electrode film preparation

Synthetic microcrystalline graphite (6 μm average particle size), poly(vinylidene fluoride-hexafluoropropylene) co-polymeric binder (Kynar-flex, Arkema) and dibutyl phthalate (DBP, Aldrich) were dispersed in acetone and then cast onto glass at a thickness of 60 μm. Once dry, the free-standing film was removed from the glass plate with the dibutyl phthalate plasticiser extracted using diethyl ether, leaving a porous film *ca.* 50 μm thick, which was cut into 6 mm diameter electrodes. These were dried under vacuum at 90 °C, weighed and then transferred to an argon filled glovebox (O_2, H_2O < 1 ppm) for Raman cell assembly. The loadings of the electrodes were 5 mg cm^{-2}, with a typical electrode mass being *ca.* 1.5 mg.

2.2 *In situ* Raman cell

The electrodes were used as active material within an *in situ* Raman cell (ECC-Opto-Std, El-Cell) configured as shown in Fig. 1, and similar to cells used in previous studies.[12,27] The cell is sealed to the atmosphere, enabling non-aqueous electrochemical processes to be investigated *in situ*. The aperture allows a Raman laser to be focussed upon the electrode's back surface during electrochemical cycling, so that Raman spectra may be obtained as a function of potential. The

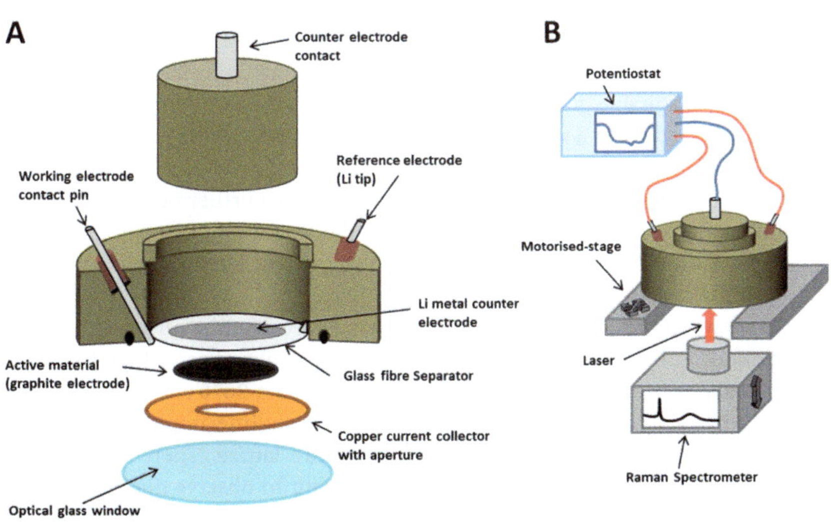

Fig. 1 (A) A schematic of the Raman cell assembly; (B) a schematic of the Raman cell in operation.

3-electrode configuration was prepared by placing a small piece of Li metal at the tip of the reference electrode contact pin. Once the cell was sealed, a negative pressure was applied with an electrolyte-filled syringe in order to wet the separator with 1 M $LiPF_6$ in EC–DMC electrolyte (BASF). The Raman cell was then hermetically sealed and removed from the glovebox, whereupon it was positioned atop the Raman microscope stage and connected to a potentiostat.

2.3 Raman spectroscopy

Raman spectra were recorded with a Raman microscope (Renishaw inVia), using a 633 nm wavelength laser focussed through an inverted microscope (Leica), *via* a 50× objective (Leica). The cell had an open circuit potential *ca.* 3.0 V (all quoted potentials are *vs.* Li/Li^+) and was galvanostatically cycled at 20 mA g^{-1} carbon (equivalent to a C-rate of C/19, where 1 C = the charge required to fully lithiate graphite to LiC_6 in 1 hour) between 5 mV and 1.5 V using a potentiostat (Biologic). Suitable filters were used to minimise the laser power at the surface to <1 mW cm^{-2}. Higher laser powers should be avoided so as to prevent both heating of the sample area and laser degradation of the surrounding electrolyte (which produces fluorescing species).

2.4 Scanning electron microscopy

The synthetic microcrystalline graphite particle and electrode film morphology were characterised using a scanning electron microscope (JEOL 7001F) operating at an accelerating voltage of 20.0 keV.

3 Results and discussion

A SEM image of the microcrystalline graphite is displayed in Fig. 2(A), showing that a distribution of *ca.* 6 μm sized platelet graphitic particles is present. The free standing electrode cast used for the *in situ* experiments is shown in Fig. 2(B), which displays the porosity generated within the film from the DBP extraction.

In order to acquire spectra with a good signal to noise ratio, the laser was focussed onto a suitable graphite particle as shown in Fig. 2(C), and the scattered light was collected from a volume of *ca.* 1–2 μm^3. The point for *in situ* measurements is carefully designated. On certain areas of the graphite electrode there is minimal detection of Raman peaks from the electrolyte. Furthermore, to allow for the refocussing of the laser throughout the experiment, and checking that the selected measurement spot has been maintained, a particle or area of the electrode with distinguishable features must be selected.

Fig. 3 shows the first and second order Raman spectra for the pristine synthetic graphitic carbon measured in this study. The main peaks observed are the D, G and 2D bands, appearing at 1336 cm^{-1}, 1580 cm^{-1} and *ca.* 2670 cm^{-1} respectively. The 1580 cm^{-1} peak (E_{2g2}), called the G band after crystalline graphite, is the only easily accessible Raman active mode of the infinite lattice. There is a second Raman active mode E_{2g1} at (41 cm^{-1}), but is difficult to observe due to its proximity to the Rayleigh line. The G mode is due to the relative motion of sp^2 carbon atoms in rings as well as chains. The peak at 1336 cm^{-1} is named the D band, from disordered graphite, and can be attributed to the breathing motion of sp^2

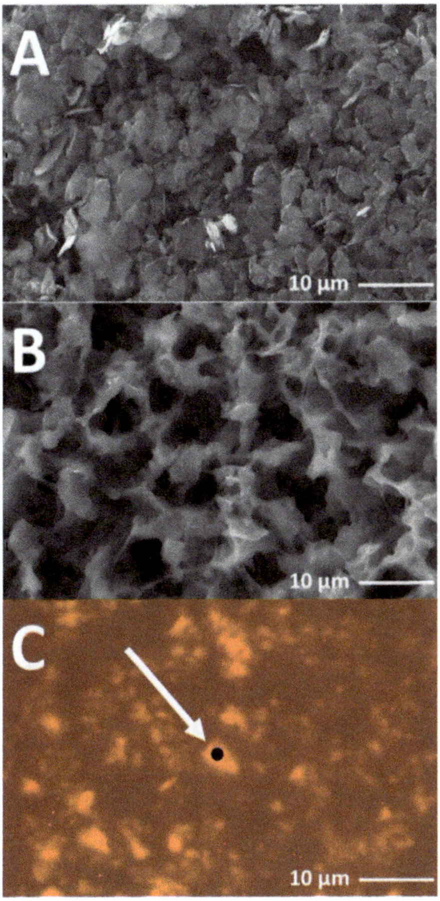

Fig. 2 (A) A SEM image of the microcrystalline graphite powder, (B) a SEM image of the free-standing electrode cast, (C) an optical microscope image of microcrystalline graphite particles from the Raman cell working electrode, the black spot indicates the measured graphite particle.

atoms in the rings at edge planes and defects in the graphene sheet. The origin of the D band has been discussed by Ferrari *et al.*[21]

Peak fitting of the 2D band gives two peaks in accordance with the literature[28,29] that have been designated 2D(1) and 2D(2), at 2649 cm^{-1} and 2686 cm^{-1} respectively. It should be noted that Cançado *et al.*[30] have proposed that the two-peak shape of the 2D band in bulk graphite actually arises from the convolution of an infinite number of peaks.

The load curve showing the first discharge and charge profile is shown in Fig. 4. As with other graphitic materials, the charge consumed for the formation of LiC_6 in the first cycle exceeds the theoretical maximum of 372 A h kg^{-1}. This is due to the partial reduction of electrolyte during the formation of a passivation film called the solid electrode interphase (SEI). The electrochemical performance of the *in situ* Raman cell matches the performance of similar graphitic materials previously reported in the literature.[31]

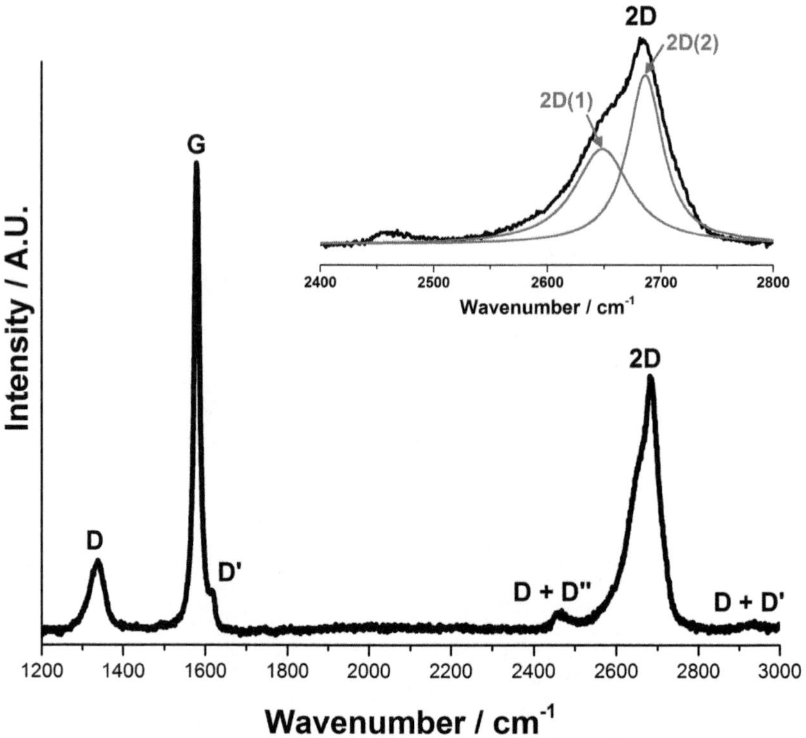

Fig. 3 Raman spectra of 6 μm graphitic carbon. The D, G and 2D bands are observed at 1336 cm^{-1}, 1580 cm^{-1} and 2670 cm^{-1} respectively. The 2D band can be peak fitted to give a sharp peak at 2686 cm^{-1} and a broad shoulder at 2649 cm^{-1}, labelled 2D(2) and 2D(1) respectively.

The results from the first lithium intercalation and de-intercalation cycle are shown in Fig. 5. The first order *in situ* Raman spectra compare well to the reported literature.[7,12,16] The potential (V) at which each spectrum was collected is displayed. All the spectra are base-line corrected and stacked arbitrarily up the *y*-axis to allow for clear visualisation. Bands due to the electrolyte are either not detected or negligible because of the employed confocal set-up, which allows a confocal resolution of 1–2 μm.[3]

At the open circuit potential (*ca.* 3.0 V *vs.* Li/Li^{+}), three main bands are observed in the region between 1000–3000 cm^{-1}: the G band at 1580 cm^{-1} and the 2D band at 2670 cm^{-1}. A weak D band is noted at 1330 cm^{-1}.

During lithiation, the series of first and second order band spectra may be divided into four stages of specific interest: the initial loss of D band intensity between potentials *ca.* 3.0–0.6 V; blue-shifting of the G band from 1580 cm^{-1} to 1590 cm^{-1}, accompanied by the gradual weakening (and eventual disappearance) of the 2D(1) band intensity between *ca.* 0.6–0.2 V; splitting of the G band into the E_{2g2}(i) (1575 cm^{-1}) and E_{2g2}(b) (1601 cm^{-1}) bands *ca.* 0.20–0.15 V, along with a major red-shift of the 2D(2) band; and finally the appearance of a weak peak around 1370 cm^{-1} and the gradual loss of all distinct Raman peaks below *ca.* 0.10 V, associated with the formation of highly conductive, low stage number GICs.

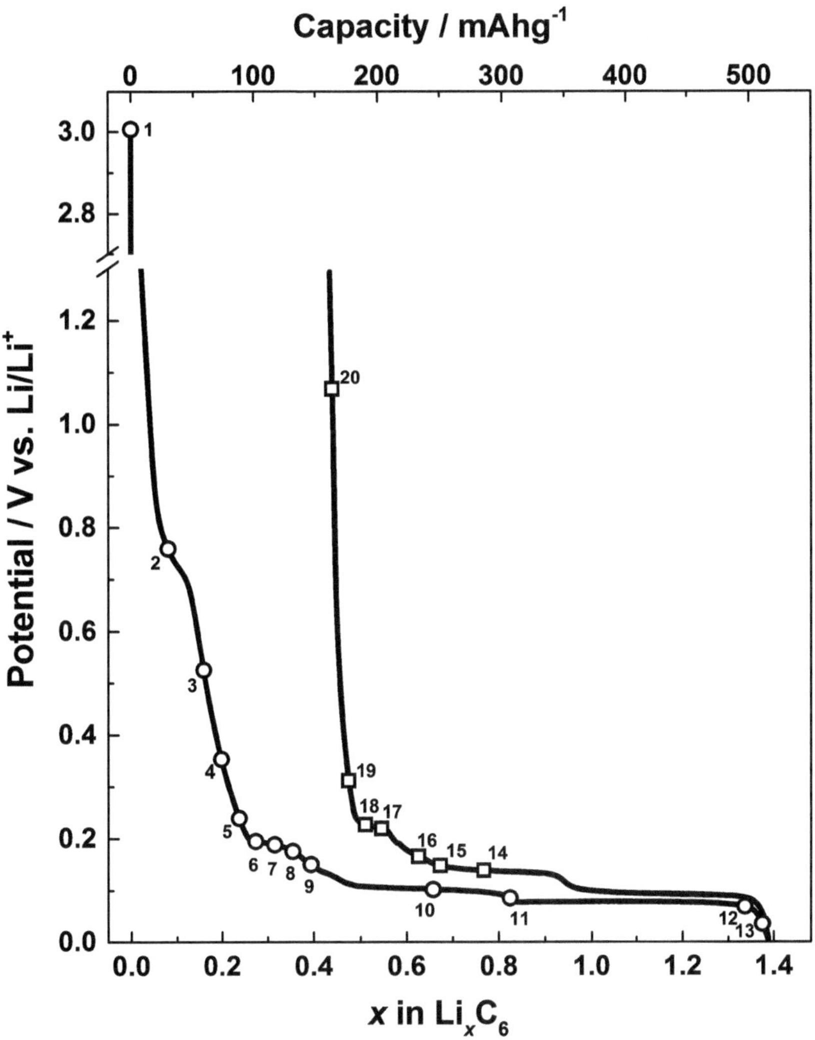

Fig. 4 A graph showing discharge and charge profiles for the microcrystalline graphite electrode in the *in situ* Raman cell against a metallic lithium counter/reference. Raman spectra acquisitions are marked with numbers corresponding to those shown in Fig. 5.

The reverse spectral series (lithium de-intercalation) may also be split into four stages of specific interest: the weak band at 1370 cm^{-1} reappears from *ca.* 0.14 V along with a broad G band at 1592 cm^{-1}; a weakening of the 1370 cm^{-1} band accompanies the growth, sharpening and blue-shift of the G band from 1592 to 1598 cm^{-1} to form the E_{2g2}(b) band of the stage 2 GIC at *ca.* 0.15 V; the return of the G band doublet, E_{2g2}(i) (1573 cm^{-1}) and E_{2g2}(b) (1601 cm^{-1}), as well as the returning 2D(2) band, between *ca.* 0.17–0.22 V; and eventually, the reappearance of a sharp singlet G band at 1586 cm^{-1} and the 2D(1) band, and subsequent band shifting back to the original OCP band positions between *ca.* 0.3–1.5 V.

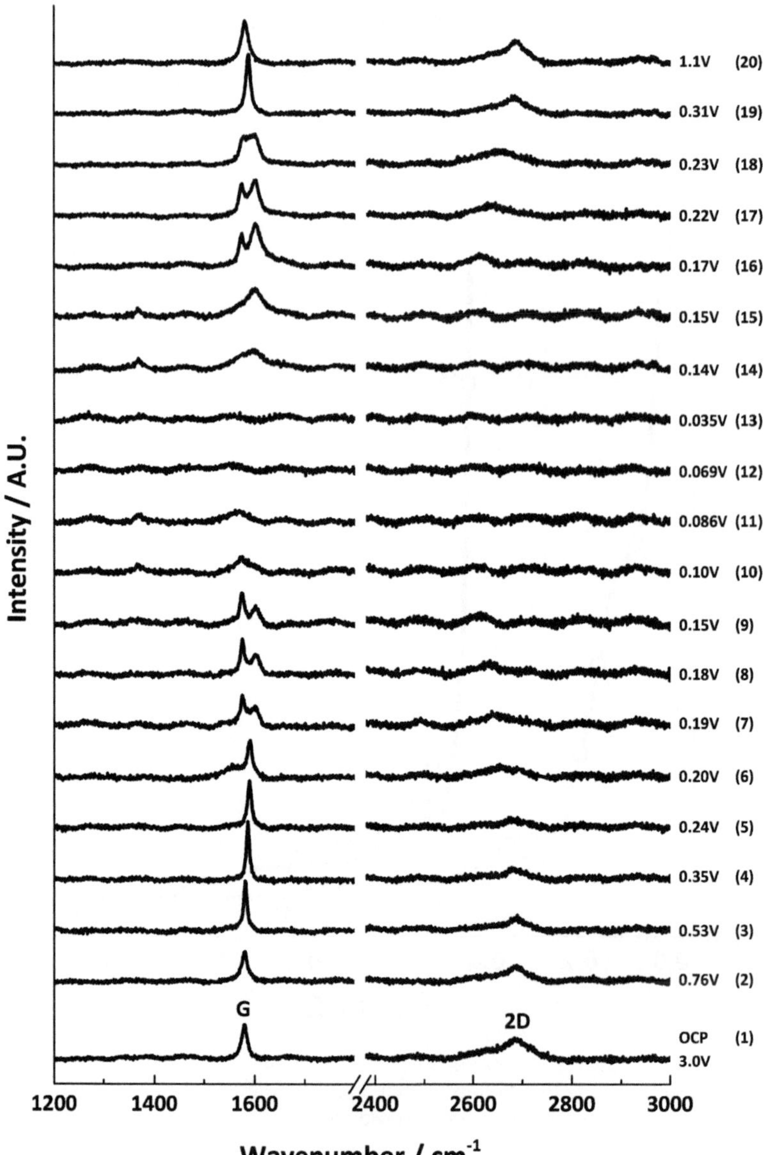

Fig. 5 *In situ* Raman spectral series for the first lithium insertion and extraction into 6 μm graphitic carbon; the potential at each spectral acquisition is labelled to the right.

3.1 The loss of the D band

During the galvanostatic first lithium intercalation, the first change observed in the Raman spectrum is the weakening and broadening of the D band. The spot measured in this graphitic material was highly ordered, resulting in a weak D band, which was consequently difficult to observe *in situ*. The loss of intensity of the D band is generally first observed to occur below 1.0 V,[12] which coincides with the formation of a SEI layer. It steadily decreases with potential until the band is lost into the signal noise below 0.6 V.

3.2 The shift and sharpening of the G band

The position of the G band is observed to shift from 1580 cm^{-1} to 1590 cm^{-1}. The frequency shift can be attributed to the increase of the force constants of the in-plane C–C bonds of the dilute stage 1 GICs. Within the region 0.6–0.2 V the G band shifted linearly with a potential slope of -28 ± 1 cm^{-1} V^{-1} (Fig. 6) which is close to the value of -29 cm^{-1} V^{-1} determined by Shi *et al.*[32] for graphite KS44. The peak shape first narrows and then widens over the potential range (Table 1). The narrowing and shift of the G band both begin below a potential of 0.6 V. This can be attributed to the beginning of lithium insertion at *ca.* 0.55 V.[14] The dilute stage 1 continues until 0.2 V, below which the formation of low stage number compounds begins. This causes the broadening of the G band as it also consists of weak signals from the emerging E_{2g2}(i) (1575 cm^{-1}) and E_{2g2}(b) (1601 cm^{-1}) bands.

3.3 The shift and shape change of the 2D band

The *in situ* behaviour of the double resonance 2D band during the lithiation and delithiation of graphitic carbon in a typical Li-ion battery electrolyte has not been previously reported. In contrast to the D band, which is widely reported to lose signal intensity between *ca.* 1–0.6 V,[12] the 2D band remains visible until below *ca.* 0.15 V, when the high conductivity of low stage number GICs makes band

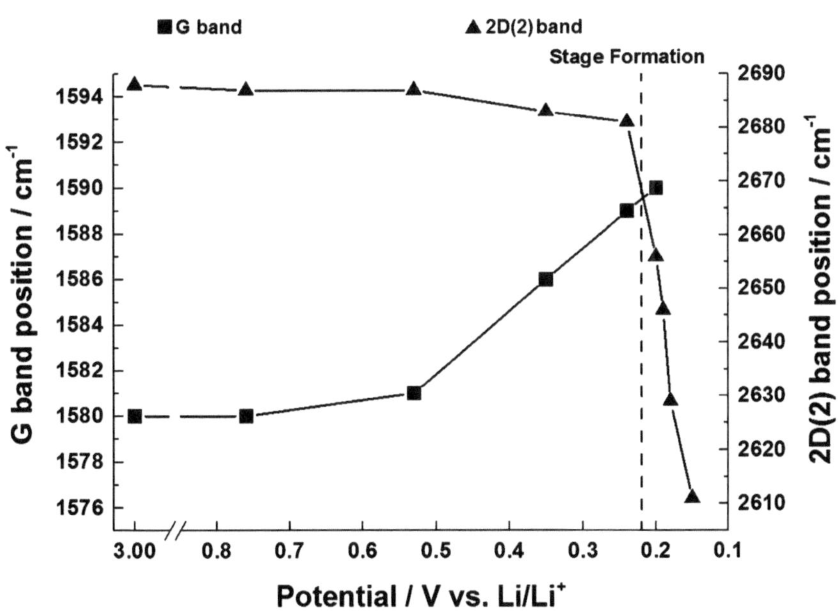

Fig. 6 The shift in G band and 2D(2) position within the potential range 3.0–0.15 V *vs.* Li/Li^+ for the first lithiation of graphite. Within the region 0.53–0.20 V the G band shifted linearly with potential (slope -27 cm^{-1} V^{-1}). Within the region 0.24–0.15 V the 2D(2) band also shifted linearly with potential (slope 802 cm^{-1} V^{-1}).

Table 1 *In situ* Raman peak positions and full width at half maximum (FWHM) of the first electrochemical lithium insertion and extraction into the microcrystalline graphite

	G		E_{2g2}(i)		E_{2g2}(b)		2D(1)		2D(2)	
E (V)	*ω* (cm^{-1})	FWHM (cm^{-1})	*ω* (cm^{-1})	FWHM (cm^{-1})	*ω* (cm^{-1})	FWHM (cm^{-1})	*ω* (cm^{-1})	FWHM (cm^{-1})	*ω* (cm^{-1})	FWHM (cm^{-1})
3.00	1580	12	—	—	—	—	2625	61	2688	53
0.76	1580	14	—	—	—	—	2625	57	2687	50
0.53	1581	9	—	—	—	—	2632	68	2687	36
0.35	1586	7	—	—	—	—	2622	30	2683	35
0.24	1589	9	—	—	—	—	2615	27	2681	45
0.2	1590	11	—	—	—	—	—	—	2656	75
0.19	—	—	1576	12	1599	15	—	—	2646	73
0.18	—	—	1575	9	1600	15	—	—	2629	59
0.15	—	—	1574	11	1601	15	—	—	2611	31
0.1	—	—	—	—	—	—	—	—	—	—
0.086	—	—	—	—	—	—	—	—	—	—
0.069	—	—	—	—	—	—	—	—	—	—
0.035	—	—	—	—	—	—	—	—	—	—
0.14	—	—	—	—	1592	57	—	—	—	—
0.15	—	—	—	—	1598	44	—	—	—	—
0.17	—	—	1573	8	1601	30	—	—	2614	51
0.22	—	—	1573	11	1598	22	—	—	2636	73
0.23	—	—	1577	16	1597	23	—	—	2649	90
0.31	1586	11	—	—	—	—	2630	48	2682	45
1.1	1579	15	—	—	—	—	2635	60	2685	39

observation difficult. As mentioned previously, no defects are required for activation of the 2D band and so it may be proposed that, whilst surface passivation/lithiation of defects may lead to the weakening and eventual disappearance of the D band, the 2D band remains visible. The 2D band position remains fairly constant until 0.53 V, including during the formation of the solid electrolyte interphase (below 1.0 V). Between 0.53 V and 0.24 V there is a gradual red-shift of the 2D(2) peak position of 21 ± 1 cm^{-1} V^{-1} accompanying the formation of the dilute stage 1 GIC signalled by the blue-shift of the G band. Between these potentials, the fitted 2D(1) peak intensity falls, making fitting difficult (due to the high background) and causing some uncertainty in the assigned position. Below 0.24 V the 2D(1) peak is no longer discernible.

At *ca.* 0.20 V, accompanying the onset of stage 4 GIC formation, the gradient of the 2D(2) peak red-shift increases significantly to a rate of 802 ± 87 cm^{-1} V^{-1} (Fig. 6, Table 1). A similar red-shift in the 2D position has previously been observed during the n-type doping of, and the intercalation of other metal ions into GICs.[24,26] This shift may be attributed to electronic doping and the increased in-plane (biaxial) lattice strain accompanying Li^+ insertion. The linear nature of this shift between 0.24–0.15 V, and the biaxial nature of the attributed lattice strain, implies that the graphene layers in the graphitic carbon are becoming increasingly distorted on lithium intercalation. This is in keeping with the Daumas–Hérold model of intercalation, where the graphene layers are flexible and deform around the intercalating lithium ions, and contrary to the Rüdorff model, which proposes a sequential filling up of alternating graphene interlayer spaces with no structural distortions induced within the individual graphene sheets, as seen in Fig. 7.

The apparent shape change of the 2D band towards a single Lorentzian peak suggests the graphene layers in the stage 4 GIC have become electronically similar to single layer graphene, most likely due to electronic decoupling resulting from the transfer of charge from the electron from the external circuit to the bounding layers.[24] The 2D band is no longer observable below 0.10 V. This suggests that stage 2 and stage 1 GIC are not visible in the Raman spectra as all graphene layers of the GIC are charged, and thus no longer give a Raman signal.

3.4 The splitting of the G band

As the potential continues to decrease below 0.20 V, the G band continues to broaden, with two side bands beginning to grow and become distinct bands. These peaks can be assigned to the intensities of the interior $E_{2g2}(i)$ (1575 cm^{-1}) and bounding $E_{2g2}(b)$ (1601 cm^{-1}) bands. These two bands quickly become prominent to the detriment of the dilute stage 1 band. This transformation points towards a phase transition from the dilute stage 1 to a staged phase with $n > 2$. Using eqn (1)[7] and the intensities of the bands from the peak fitting, the ratio R is calculated to be *ca.* 1.0, which signifies the establishment of a stage 4 lithium GIC at 0.19 V (spectrum number 7).

The phase transition is related to the change from a random distribution of lithium ions to a more compact distribution in the staged compound. The shift of the $E_{2g2}(b)$ band to 1601 cm^{-1} is due to a further increase of the C–C bond force constants.

3.5 The formation of the stage 2 GIC

The reduction and eventual disappearance of the $E_{2g2}(i)$ band is observed between 0.10 V and 0.069 V. However, the $E_{2g2}(b)$ band at 1601 cm^{-1} is also seen to disappear. Based on past reports of stage 2 GIC materials that have been produced

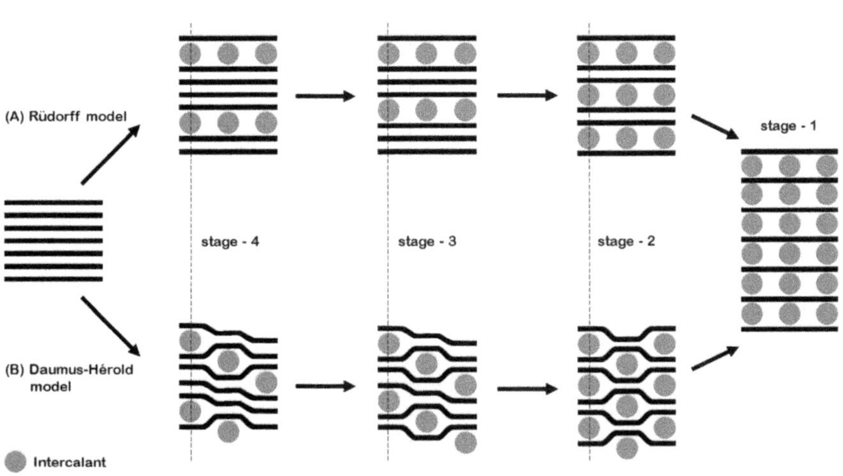

Fig. 7 A schematic of the (A) Rüdorff and (B) Daumas–Hérold models of ion intercalation into graphite.

from chemical synthesis, it is expected that the Raman spectrum should display a single band at *ca.* 1601 cm^{-1}.[33] This is clearly not the case and instead a broad band at 1550 cm^{-1} is seen. This is tentatively assigned as the blue-shifted E_{2g2}(b) band of a stage 2 liquid GIC, where there is no longer an ordered distribution of lithium within the planes. However, a recent article by Zabel *et al.*[34] on graphene sheets under strain shows a similar G-band shift for graphene sheets under *ca.* 1 bar of pressure, perhaps as a result of the contortion of graphene sheets seen from the Daumas–Hérold model of intercalation.

This shift can then be explained by charge transfer effects during intercalation of a donor species where electrons occupy the π^* orbital and thus weaken the C–C bond strength. Such an effect has been seen for the intercalation of donor species into less ordered carbon materials.[13,35,36] This broad band continues to blue-shift and broadens until it disappears into the noise at *ca.* 1540 cm^{-1}. Raman bands are thought to be no longer visible because of the increase of electrical conductivity of these low stage GICs, which leads to a reduction in optical skin depth, and results in a low Raman scattering intensity.[7]

3.6 The first de-lithiation

During the first de-lithiation, the reverse trend for the Raman spectra is seen, as shown in Fig. 5, as the potential is increased from 0.005 to 1.5 V. At 0.14 V, two bands are observed at 1370 cm^{-1} and 1592 cm^{-1}. The band at 1592 cm^{-1} is seen to shift and grow in intensity to 1601 cm^{-1} as the potential is increased. This indicates the formation of the E_{2g2}(b) band and thus the formation of a pure stage 2 GIC. The 1370 cm^{-1} band is absent by 0.17 V. At 0.17 V a second peak emerges at 1573 cm^{-1}, the E_{2g2}(i) band, which grows at the expense of the E_{2g2}(b) band. This marks the return to a stage 3 GIC, and then to a stage 4 by *ca.* 0.17 V.

From 0.22 to 0.24 V there is a mixture of stage 4 and dilute stage 1, and the doublet G band returns again to the single E_{2g2} band at 1586 cm^{-1}, which then blue-shifts back to 1579 cm^{-1}, showing the passage from dilute stage 1 to fully de-intercalated graphite. The shift in G band position is shown clearly between 0.31–1.1 V *vs.* Li/Li^+.

The Raman spectra before and after lithiation are shown in Fig. 8. Both show very similar features, and the relative intensities of the G and 2D bands remain the same. Both spectra display a weak D band, indicating that the area measured did not become further disordered during the first Li^+ insertion/extraction cycle. The lack of change in the spectra indicates that the measured graphite particle has experienced no significant structural change.

The reappearance of the 2D band is discernible at 0.17 V at 2614 cm^{-1} with a single Lorentzian fit that we have ascribed to the returning 2D(2) peak. This is a similar position to the last observable position during lithiation (2611 cm^{-1} at 0.15 V) suggesting the return of the stage 3 GIC. The 2D(2) peak red-shifts and broadens between *ca.* 0.17 V to 0.23 V until an eventual reforming of the low energy 2D(1) peak at 0.31 V. This splitting of the 2D band marks the renewed electronic coupling between the graphene layers and suggests the formation of the dilute stage 1 GIC. Both 2D(1) and 2D(2) peaks shift back towards their OCP positions.

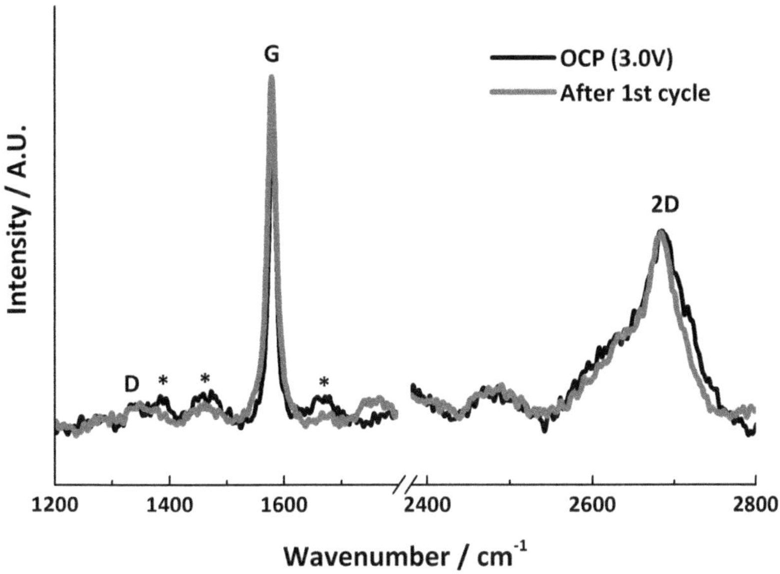

Fig. 8 *In situ* Raman spectra before and after lithiation, * indicates signals from electrolyte bands.

4 Summary and conclusions

Graphitic carbon remains the dominant active material used within lithium-ion negative electrodes. Understanding the mechanics of Li^+ intercalation into actual practical graphitic anodes is important in optimising this material in terms of performance, lifetime and safety. *In situ* Raman spectra of the first electrochemical lithiation and de-lithiation cycle at a single point on a graphite electrode were obtained with a good signal to noise ratio and minimal detection of electrolyte bands.

The obtained first order spectra agree with previous reports, in particular with the observation of the G-band splitting, indicating the formation of a staged GIC. The behaviour of the second order *in situ* Raman spectra was reported for the first time during the first lithium insertion and extraction cycle. The 2D band was observed to shift to lower wavenumbers (2681 to 2611 cm^{-1}) when stage 4 and 3 graphite intercalation compounds were formed. The shift of the 2D band supports the Daumas–Hérold model of electrochemically derived intercalation, where the graphene layers are flexible and deform around domains of ordered intercalated lithium ions. The opposite trends were observed in the Raman spectra during lithium extraction, indicating the lithium is removed in a reversible manner, with comparable initial and final spectra. This demonstrates that, at this particular point of the graphite electrode, no disordering of the structure resulted from a single cycle of lithium insertion and extraction.

Acknowledgements

We acknowledge the Engineering and Physical Sciences Research Council (EPSRC) for the funding of this research under grant number EP/K016954. The

Nanoinvestigation Centre at Liverpool (NiCaL) and the assistance with the SEM measurements from Stephen Hughes are also gratefully acknowledged.

References

1 A. S. Aricò, P. Bruce, B. Scrosati, J.-M. Tarascon and W. V. Schalkwijk, *Nat. Mater.*, 2005, **4**, 366.
2 P. G. Bruce, B. Scrosati and J.-M. Tarascon, *Angew. Chem., Int. Ed.*, 2008, **47**, 2930.
3 P. G. Bruce, S. A. Freunberger, L. J. Hardwick and J.-M. Tarascon, *Nat. Mater.*, 2012, **11**, 19.
4 J. B. Goodenough and K.-S. Park, *J. Am. Chem. Soc.*, 2013, **135**, 1167.
5 R. Baddour-Hadjean and J. P. Pereira-Ramos, *Chem. Rev.*, 2010, **110**, 1278.
6 V. Stancovski and S. Badilescu, *J. Appl. Electrochem.*, 2014, **44**, 23.
7 M. Inaba, H. Yoshida, Z. Ogumi, T. Abe, Y. Mizutani and M. Asano, *J. Electrochem. Soc.*, 1995, **142**, 20.
8 S. A. Solin, *Graphite Intercalation Compounds*, Springer-Verlag, Berlin, 1990.
9 R. Fong, U. Vonsacken and J. R. Dahn, *J. Electrochem. Soc.*, 1990, **137**, 2009.
10 K. Dokko, Q. F. Shi, I. C. Stefan and D. A. Scherson, *J. Phys. Chem. B*, 2003, **107**, 12549.
11 Y. Luo, W. B. Cai and D. A. Scherson, *J. Electrochem. Soc.*, 2002, **149**, A1100.
12 L. J. Hardwick, H. Buqa and P. Novak, *Solid State Ionics*, 2006, **177**, 2801.
13 M. Inaba, H. Yoshida and Z. Ogumi, *J. Electrochem. Soc.*, 1996, **143**, 2572.
14 W. W. Huang and R. Frech, *J. Electrochem. Soc.*, 1998, **145**, 765.
15 J. C. Panitz, F. Joho and P. Novak, *Appl. Spectrosc.*, 1999, **53**, 1188.
16 S. Migge, G. Sandmann, D. Rahner, H. Dietz and W. Plieth, *J. Solid State Electrochem.*, 2005, **9**, 132.
17 P. Novak, D. Goers, L. Hardwick, M. Holzapfel, W. Scheifele, J. Ufhiel and A. Wursig, *J. Power Sources*, 2005, **146**, 15.
18 L. J. Hardwick, H. Buqa, M. Holzapfel, W. Scheifele, F. Krumeich and P. Novak, *Electrochim. Acta*, 2007, **52**, 4884.
19 L. J. Hardwick, P. W. Ruch, M. Hahn, W. Scheifele, R. Kotz and P. Novak, *J. Phys. Chem. Solids*, 2008, **69**, 1232.
20 V. Baranchugov, E. Markevich, G. Salitra, D. Aurbach, G. Semrau and M. A. Schmidt, *J. Electrochem. Soc.*, 2008, **155**, A217.
21 A. C. Ferrari and D. M. Basko, *Nat. Nanotechnol.*, 2013, **8**, 235.
22 T. M. G. Mohiuddin, A. Lombardo, R. R. Nair, A. Bonetti, G. Savini, R. Jalil, N. Bonini, D. M. Basko, C. Galiotis and N. Marzari, *Phys. Rev. B: Condens. Matter Mater. Phys.*, 2009, **79**, 205433.
23 C. Metzger, S. Remi, M. Liu, S. V. Kusminskiy, A. H. Castro Neto, A. K. Swan and B. B. Goldberg, *Nano Lett.*, 2010, **10**, 6.
24 J. C. Chacón-Torres, L. Wirtz and T. Pichler, *ACS Nano*, 2013, **7**, 9249.
25 W. Zhao, P. H. Tan, J. Liu and A. C. Ferrari, *J. Am. Chem. Soc.*, 2011, **133**, 5941.
26 R. Parret, M. Paillet, J.-R. Huntzinger, D. Nakabayashi, T. Michel, A. Tiberj, J.-L. Sauvajol and A. A. Zahab, *ACS Nano*, 2013, **7**, 165.
27 L. J. Hardwick, M. Hahn, P. Ruch, M. Holzapfel, W. Scheifele, H. Buqa, F. Krumeich, P. Novák and R. Kötz, *Electrochim. Acta*, 2006, **52**, 675.
28 J. Nemanich and S. A. Solin, *Phys. Rev. B: Condens. Matter Mater. Phys.*, 1979, **20**, 392.

29 R. P. Vidano, O. B. Fischbach, L. J. Willis and T. M. Loehr, *Solid State Commun.*, 1981, **39**, 341.
30 L. G. Cançado, A. Reina, J. Kong and M. S. Dresselhaus, *Phys. Rev. B: Condens. Matter Mater. Phys.*, 2008, 77, 245408.
31 H. Buqa, A. Wursig, A. Goers, L. J. Hardwick, M. Holzapfel, P. Novak, F. Krumeich and M. E. Spahr, *J. Power Sources*, 2005, **146**, 134.
32 Q. F. Shi, K. Dokko and D. A. Scherson, *J. Phys. Chem. B*, 2004, **108**, 4789.
33 M. S. Dresselhaus and G. Dresselhaus, *Adv. Phys.*, 2002, **51**, 1.
34 J. Zabel, R. R. Nair, A. Ott, T. Georgiou, A. K. Geim, K. S. Novoselov and C. Casiraghi, *Nano Lett.*, 2012, **12**, 617.
35 M. Endo, C. Kim, T. Karaki, T. Fujino, M. J. Matthews, S. D. M. Brown and M. S. Dresselhaus, *Synth. Met.*, 1998, **98**, 17.
36 Z. X. Wang, X. J. Huang, R. J. Xue and L. Q. Chen, *Carbon*, 1999, **37**, 685.

Faraday Discussions

DISCUSSIONS

Carbon electrodes for energy storage: general discussion

Yury Gogotsi, Dirk Guldi, Richard McCreery, Chi-Chang Hu, Céline Merlet, François Béguin, Laurence Hardwick, Elzbieta Frackowiak, Julie Macpherson, Alexander Forse, George Zheng Chen, Katherine Holt, Robert Dryfe, Heisi Kurig, Surbhi Sharma, Patrick R. Unwin, Thomas Rabbow, Wanjing Yu, Fulian Qiu, Fernanda Juarez, Christopher Sole, Boris Dyatkin, Keith Stevenson, Yong Cao, Nico Cousens and Asa Noofeli

DOI: 10.1039/C4FD90026J

François Béguin opened the discussion of the paper by Yury Gogotsi: After vacuum treatment of the SiC-CDC at 700 °C, the specific surface area increases by around 1000 m^2 g^{-1}. You explain it by the removal of surface functional groups. However, the elemental analysis in table 1 shows almost the same composition for the initial CDC and 700 °C annealed. How can you explain such a high increase of surface area without any change of elemental composition?

Yury Gogotsi answered: We assume that the SSA increase is due to desorption of surface functional groups that take carbon atoms along.[1] However, adsorption of water, CO_2 and other species from the environment leads to increased oxygen content in the samples shown by X-ray energy dispersive spectroscopy.

1 B.J.M. Etzold *et al.*, *Chem. Mat.*, 2014, **26**, 3479–3484.

François Béguin continued: Thermal treatments of carbons either under neutral gas or under vacuum allow the removal of surface functional groups. However, they require a further 'passivation' of the surface, *e.g.* by dihydrogen, to neutralize the free radicals, often called dangling bonds. If not, the free radicals may lead to electrolyte decomposition, reducing the voltage window of the system.

Yury Gogotsi replied: I agree. We should do hydrogen treatment after vacuum annealing, in the same way as we do it after CDC synthesis.

Chi-Chang Hu asked: For the electrolytes containing large cations and anions, is the charge storage mechanism of carbide-derived carbons with sub-nanometer pores more reasonably attributed to an intercalation/de-intercalation process or specific adsorption of ions? How does the pore depth affect the charge storage/delivery rate?

Yury Gogotsi responded: I would call it electrosorption rather than intercalation. However, when the pore size is close to the ion size, expansion of the electrodes occurs[1] with the ion insertion process becoming similar to intercalation. The pore depth, as determined by the particle size or film thickness, controls the power to a large extent. However, above a certain critical pore length, the electrode cannot be fully charged and its capacitance becomes limited by the pore length, because some inner pores cannot be accessed by ions.[2]

1 M. M. Hantel *et al.*, *Electrochem. Comm.*, 2011, **13**, 1221–1224.
2 J. Chmiola *et al.*, *Science*, 2010, **328**, 480–483.

François Béguin remarked: One delegate suggested that ions could be intercalated. Pure intercalation would require an electron transfer. In other words, in EDLC based on porous carbon electrodes, it is considered that ions are purely electrosorbed. However, with some carbons, Koetz experiments demonstrated volume expansion of around 10%. Such high values suggest some constrained state of ions in narrow pores.

Yury Gogotsi answered: This is correct. We observed up to ~5% expansion for CDC.[1]

1 M. M. Hantel *et al.*, *Electrochem. Comm.*, 2011, **13**, 1221–1224.

Elzbieta Frackowiak enquired: Do you think that the pores of CDC carbons could be closed during heat treatment at temperatures over 1050 °C ? Such a phenomenon is very well known for various carbonaceous materials.

Yury Gogotsi replied: This depends on the carbon structure. During the CDC synthesis, graphitic ordering is seen above 1100 °C. CDCs produced at low temperatures (200–600 °C) can be restructured at temperatures below 1000 °C, but no structure collapse happens in most CDCs during annealing in vacuum up to about 1500 °C. They have a 3D covalent network, which is very stable, as it requires bond breaking for restructuring. Pore collapse and decrease in SSA happen at 1600 °C and above.

George Zheng Chen asked: What may be the reason for the different capacitance values of your carbon samples, apart from the pore sizes?

Yury Gogotsi answered: Surface chemistry and carbon structure (ordering), which determine the quantum capacitance, affect the electrode capacitance values at a given pore size.

Céline Merlet addressed Yury Gogotsi and George Zheng Chen: The models used in molecular simulations do not currently take into account the functionalities or defects present at the carbon surface in real systems. In simulations, the carbon is treated as a metallic electrode, which is probably close to reality for carbons synthesized at high temperature (which have a high electronic conductivity). Nevertheless, even with this representation, it is possible to observe capacitance variations for porous carbons with different pore sizes and local structures. A reason for this is the packing of the ions in the pores and the resulting charge separation that occurs. The larger/more efficient the charge separation, the better the charge storage efficiency. In large pores, or at planar surfaces, *e.g.* for a positive electrode, anions will always be partly surrounded by cations as there is space for them and because there are not enough carbon atoms to compensate the ionic charge. Hence, the charge stored on the carbon will be limited by the proximity of positive charges from the cations and the fact that the anion cannot affect a large number of carbon atoms. In small pores, the anion will lose its coordination shell from the cations as well as its solvation shell. The anion will be closer to a larger number of carbon atoms that it will affect. The avoidance of positive charges from the cations and the short distance between the anion and the carbon atoms will thus lead to a better charge storage efficiency. Another way of saying this is that for a given ionic charge, the charge developed on the carbon will be higher, thus with the same amount of ions, the charge storage will be higher.

George Zheng Chen responded: I appreciate this insightful explanation, although my original question was related to the origin of charge storage capacity on the carbon side or even inside the solid phase of carbon. The pore structure can affect the distribution and amount of charge balancing ions, and hence the overall capacitive performance. However, it is the response, but not the origin of the charge storage capacity of the carbon, particularly that beyond the predicted values from the specific surface areas.

Céline Merlet said: Following up on my previous comment, I think it is important that molecular dynamics simulations are extended to the representation of defects and functionalities at the surface of the carbon electrodes as these defects/functional groups lead to large changes in the capacitances and rate capabilities of supercapacitors. Simulation techniques are not ready for this yet as the electrode is currently represented as a metallic material, which is a good assumption for porous carbons synthesised at high temperatures but is not realistic for some types of carbon structures. When defects/functionalities are present, quantum capacitance can arise on top of the double layer capacitance. This has to be taken into account to realistically model different porous materials. There is thus a need to develop a representation of the electrode which includes the quantum capacitance as well as a suitable model for the interactions between the electrolyte and the functionalities. This is definitely a direction theoreticians should explore, keeping in mind that representing these effects is not trivial.

George Zheng Chen asked: What is inside the solid carbon phase (not the pores) that is responsible for charge storage?

Yury Gogotsi replied: It is disordered, primarily sp^2-bonded (graphitic) carbon. Ref. 1 provides information about the structure and composition of carbide-derived carbons. The ordering of the carbon increases and the content of functional groups on the surface decreases with increasing annealing temperature. The effect of vacuum annealing on CDC structure is described in ref. 2.

1 V. Presser, M. Heon and Y. Gogotsi, Carbide-Derived Carbons - From Porous Networks to Nanotubes and Graphene, *Adv. Funct. Mater.*, 2011, **21**(5), 810–833.
2 S. Osswald, J. Chmiola and Y. Gogotsi, Structural Evolution of Carbide-Derived Carbons upon Vacuum Annealing, *Carbon*, 2012, **50**(13), 4880–4886.

Céline Merlet addressed Yury Gogotsi and George Zheng Chen: My comment is that the electrode is actually already wet at zero applied potential as is usually seen in experiments[1] and was also observed in molecular dynamics simulations.[2] For the voltages studied in the molecular dynamics simulations, it was shown that the charge storage is not linked to ions entering the pores but to an exchange of ions between the bulk and the electrodes. The total volume of electrolyte in the pores was constant when changing the potential, but the ratio of anions with respect to cations was evolving such that the liquid inside the pores was not neutral. The charge is then compensated by the charge developed on the carbon electrode. The fact that ions are already adsorbed at zero potential doesn't mean that the electrode is charged, it just means that the numbers of anions and cations in the electrode are equal when there is no applied potential. Locally, small charges can be developed on the carbon, but the overall charge on the electrode is zero. To understand a little better the different capacitances observed for different carbon structures, molecular dynamics situations looking at solvation numbers and local charges on the carbon were conducted.[3] The main conclusion of this study was that the more confined the ions are, the more desolvated/decoordinated they are and the more efficient is the charge storage. The structure of the porous carbons allows for a good charge separation when a potential is applied between the electrodes, thus allowing an efficient energy storage.

1 For example by NMR, see A. C. Forse *et al.*, *J. Phys. Chem. C*, 2014, **118**, 7508.
2 C. Merlet *et al.*, *Nature Materials*, 2012, **11**, 306.
3 C. Merlet *et al.*, *Nature Comm.*, 2013, **4**, 2701.

George Zheng Chen answered: I agree with you about how the solution side of the carbon/electrolyte interface (double layer) affects the capacitance of the material, but I am more interested in knowing how the carbon side of the interface contributes to charge storage.

George Zheng Chen enquired: Your analysis and explanation about the effect of the pore structure of carbide derived carbon on the capacitive performance is very interesting and insightful. This consideration is obviously focused on the electrolyte side of the double layer. I would be grateful if you could explain if the changes in the porous structures of the carbon might have also brought about variation in the charge storage capacity on the carbon side, particularly when the carbon changes from amorphous to partially graphitized and fully graphitized structures.

Yury Gogotsi replied: Yes. Quantum capacitance, surface area and the pore size of the carbon affect the capacitance of the carbon-electrolyte system. They all change as the CDC structure changes from amorphous to graphitic.

Fernanda Juarez commented: My comment is about whether the ions inside of the pores are charged or not and the nature of the interaction between them. We have studied the insertion of cations in several nanotubes in a previous publication,[1] concluding that the main force driving this process is the image interaction between the ions and the nanotube walls. In addition to this, we have observed more interesting features in new simulations that are not yet published. The first one is that the same electrostatic force arises between halogen anions and carbon nanotubes. The second and more important observation is that there is a correlation between the ratio of the nanotube diameter to the ionic radius and the position of the ion in the tubes. When this ratio is large enough, the ion adsorbs near to the wall, not in the centre of the nanotube. In this case we have observed a change in the adsorption mechanism, increasing the covalent interaction and decreasing the electrostatic part. Nevertheless, even in this situation the ions (cations or anions) are still charged, and with an absolute value close to one. Therefore, after these observations, it is my opinion that it is very likely to find fully charged ions inside of the pores in the systems presented.

1 Goduljan *et al.*, *Electrochem. Comm.*, 2014, **45**, 48–51.

Yury Gogotsi added: I agree.

Thomas Rabbow said: The results from Gogotsi *et al.* show that the volume of subnanometer pores per mass increases strongly for SiC-CDC and slightly for TiC-CDC after annealing at 700 °C. Despite the strong increase of subnanometer pores for SiC-CDC, the capacitance decreases. In the case of TiC-CDC a slight increase of the capacitance is observed which corresponds to a slight gain of subnanometer pores. The pore model with pore sizes adequate for the applied ions looks confirmed for certain cases (as reported in the talk) but there are contradictions for the presented results on SiC-CDC and TiC-CDC, which limit the generalization. Therefore I would like to discuss a different view on this topic which does not fully contradict the pore model, but could extend it. It has been shown that basal and edge plane oriented graphite differs by more than an order of magnitude in differential capacitance (basal plane: 3 μF cm^{-2} *vs.* edge plane: 50–70 μF cm^{-2}) especially after annealing.[1]

The substrate investigated by Gogotsi *et al.* might be regarded as a mixture of amorphous and polycrystalline carbon. The crystalline phases can exhibit an edge plane or basal plane orientation at the surface of each particle. The estimation of the edge plane orientation and the number of defects could be done by oxygen adsorption–desorption (see *e.g.* active surface area concept by Laine *et al.*[2]). It would be interesting to investigate the active surface area after each annealing step and to correlate it with the capacitance and the volume of different pore sizes. It is an obvious assumption that inside a pore edge plane oriented crystal faces are predominant. Therefore it can be expected that a high number of (sub)-nanometer pores enhance the active surface area and thereby the capacitance. From this point of view the decrease of gravimetric capacitance (Fig. 3 of the paper) for both samples after high temperature annealing (T = 1400 °C) might be explained by fewer defects and a lower active surface area. The transitional increase of capacitance for TiC-CDC after annealing at 700 °C might result from the removal of surface groups as reported, without changing the active surface area (which needs higher temperatures). I would be very thankful for comments on this scheme or related literature.

1 See *e.g.* J.-P. Randin and E. Yeager, *J. Electrochem. Soc*, 1971, **118**, 711.
2 N. R. Laine *et al.*, *J. Phys. Chem.*, 1963, **67**, 2030.

Yury Gogotsi responded: I cannot fully agree that CDC can be regarded as a mixture of amorphous and polycrystalline carbon. We have done very extensive X-ray and neutron scattering studies[1] showing that as-produced CDCs have short range order but lack long-range order. Thus, they are amorphous materials.[2] Therefore, instead of edge *vs.* basal plane terminology, I'd discuss the presence of active sites, which can be distorted bonds, 5- or 7-member rings, ad-atoms, *etc.* They are probably responsible for differences in ion adsorption (capacitance) at the same pore size, and the number of active sites changes after annealing. Unfortunately, we still poorly understand which ones are important for achieving the highest capacitance.

1 *e.g.* J. C. Palmer *et al.*, in *Characterization of Porous Solids VIII*, RSC, 2009, Special Publication No. 318, pp. 56–63.
2 J. C. Palmer *et al.*, *Carbon*, 2010, **48**, 1116–1123.

Thomas Rabbow said: Independent from the very different crystal sizes of TiC-CDC and SiC-CDC, I could imagine that it makes sense to characterize the surface by the more specific oxygen adsorption–desorption (active surface area). Since oxygen gets chemisorbed at graphitic edge planes and defects it might be a more adequate measure for the surface size which contributes to the capacitance. If two carbons with a similar active surface area but different pore size distributions are compared one could divide between both effects.

Yury Gogotsi replied: This would be helpful. We performed thermal desorption analysis on CDC in the past,[1] but have not done it on the samples used in this study.

1 C. Portet *et al.*, *Thermochimica Acta*, 2010, **497**(1/2), 137–142.

François Béguin responded: It has been proved on different kinds of fibers (herringbone, platelet and planes parallel to the fiber axis) that both basal and edge planes contribute to capacitance, with slightly higher contribution for the edge planes. The measurement of active surface area is based on di-oxygen chemisorption; it means it measures the reactivity of carbon *vs.* di-oxygen. Therefore, it cannot be directly correlated with capacitance. By contrast, the ASA could probably be correlated with the cycleability of the system.

Boris Dyatkin added: We completed thermogravimetric analysis in an air environment and dynamic water vapour sorption (Fig. S3 and S4 in the ESI of the paper) to assess changes in the reactive sites on the pore walls. For water vapour sorption, we observed a direct correlation between increasing hydrophobicity and higher annealing temperatures for TiC-CDC and SiC-CDC materials. This was particularly reflected by the fact that water uptake occurred at higher relative humidity values for samples annealed at higher temperatures. Given that the overall pore size distribution did not drastically change during the process (*i.e.* no

Table 1 Analysis of oxidation onset and oxidation profile for air TGA SiC-CDC nanoparticles

Annealing Temperature	Onset of Oxidation	Key Oxidation Temperature	Full Width at Half Maxima of Oxidation Peak
Initial	424.1 °C	566.6 °C	73.6 °C
700 °C	453.4 °C	563.4 °C	78.7 °C
1200 °C	459.1 °C	574.3 °C	55.4 °C
1400 °C	445.8 °C	592.7 °C	48.5 °C
1800 °C	529.3 °C	627.9 °C	38.0 °C

significant mesoporosity was added), the most likely explanation was the shift of H_2O uptake mechanism from adsorption on functional groups to capillary action. This is expected to occur for materials with fewer functional groups and is in line with the oxygen composition data derived from TGA data collected under inert atmosphere, and XPS measurements. Furthermore, analysis of oxidation onset and oxidation profile (summarized in Tables 1 and 2) showcases a general trend of oxidation onset occurring at higher temperatures, combined with more narrow combustion profiles. This suggests a reduction in surface active sites (that would initiate oxidation at lower temperatures) and more homogeneous carbon compositions, especially after annealing at 1400–1800 °C. We are therefore likely reducing the active oxygen sites on pore walls after annealing, even at 700 °C, and begin to remove defects on pore wall surfaces only above a certain temperature treatment threshold.

Heisi Kurig asked: In the article, the chemical composition of functional groups was determined using XPS analysis. However, it is well known that X-rays do not penetrate materials more than 10 nm. How has it been verified that the results demonstrated can be expanded to the composition of inner pores also? Are there any other methods that support these results?

Yury Gogotsi answered: XPS is a surface technique. However, we did sputtering and analyzed subsurface layers as well. Also, we performed EDS analysis to determine the elemental composition. While both techniques give semi-quantitative results, they clearly show the trends in the composition change after vacuum annealing at different temperatures.

Table 2 Analysis of oxidation onset and oxidation profile for air TGA TiC-CDC microparticles

Annealing Temperature	Onset of Oxidation	Key Oxidation Temperature	Full Width at Half Maxima of Oxidation Peak
Initial	287.8 °C	440.4 °C	63.3 °C
700 °C	380.2 °C	470.8 °C	78.0 °C
1200 °C	409.9 °C	522.2 °C	48.3 °C
1400 °C	485.1 °C	587.9 °C	47.5 °C
1800 °C	528.9 °C	619.5 °C	43.8 °C

Dirk Guldi enquired: Has EPR spectroscopy been employed to characterise the free radical in the carbon materials, and if so has it been used to monitor saturation/interconversion into a stable material?

Yury Gogotsi replied: EPR has been used to study free radicals in carbons. For example, we recently used it to study surface graphitization of nanodiamond.[1] The increasing sensitivity of the dangling bond EPR line width to air content was used as indication of graphitization of the external layers of the nanodiamond. We have not used EPR to study bond saturation in CDC.

1 A. M. Panich *et al.*, *J. Physics: Cond. Matter*, 2013, **25**, 245303.

Keith Stevenson remarked: What is the defining length scale for defining the charge transfer process? It is difficult to rely on bulk characterisation techniques as they don't match the same length scale. We need more model systems that can precisely probe on a sub <2 nm scale.

Yury Gogotsi responded: The double layer thickness is within a few nanometers and the corresponding charge transfer processes occur at the nanometer scale. I agree that model systems (*e.g.* study of electrolytes at a single- or few-layer graphene surface[1]) are useful for understanding the mechanisms. However, they cannot provide the same confinement effect and I cannot suggest a good technique to probe individual ions confined in nanometer pores on a sub <2 nm scale. Therefore, we still have to rely on bulk characterization techniques when studying ions in porous carbons.

1 H. Zhou *et al.*, *ACS Nano*, 2012, **6**, 9818–9827.

Patrick Unwin said: There was an excellent poster from van Engers *et al.* on a surface force apparatus with graphene electrodes (poster P22). In principle, this would appear to be a particularly powerful means of examining surface

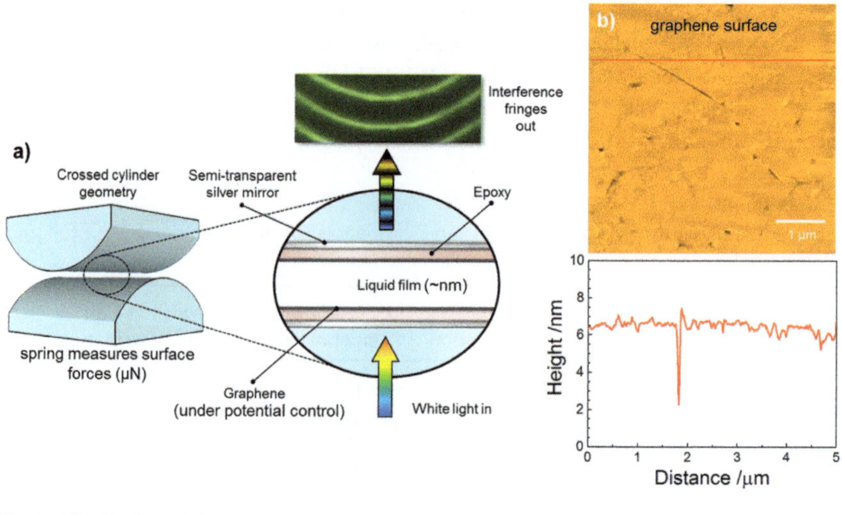

Fig. 1 The basics of the graphene surface force balance.

confinement effects, although I do not know if sufficiently close (nm and sub-nm) spacings can be achieved. I wonder if Nico Cousens, an author of the poster, could comment?

Yury Gogotsi added: I fully agree – this is a nice work and the technique is applicable to studying double-layer between carbon surfaces.

Nico Cousens answered: We have recently developed a surface force balance – also known as apparatus – with graphene surfaces (rather than conventional mica surfaces). We have successfully measured long range double layer forces in aqueous electrolytes at different graphene surface potentials. We are optimistic that we will soon be able to measure short range forces, such as those arising from layering of ionic liquids between 1–10 nm. The relevance of graphene to carbon electrodes, along with the ability to control the surface potential and confine ions to sub-nanometre gaps, will hopefully provide useful insights into surface confinement effects in (carbon based) electrochemical devices. A paper detailing the work has recently been accepted in Langmuir.[1] I have supplied a figure (Fig. 1) outlining the basics of the graphene surface force balance.

1 Britton *et al.*, *Langmuir*, 2014, **30**(38), 11485–11492.

Boris Dyatkin commented: The impact of quantum capacitance on total charge storage in porous carbons remains an important question. Biener *et al.*[1] showed how structural defects and surface groups affect the densities of state of planar graphene, and how that, in turn, affects both overall charge storage and screening of similarly charged ions. However, Kubuzono *et al.*[2] showed that quantum capacitance contributions remain significant only as long as the number of graphene layers remains below 4. We do not yet know whether planar graphene models can be applied to confined ions in cylindrical or slit pore configurations. Furthermore, while conventional assumption suggests that the number of graphene layers between two pores is 2 (the two pore walls), this is likely to change for different particle sizes with branched pore architectures. Furthermore, recent WAXS studies by Smarsly *et al.*[3] show some changes in graphene sheet stacking along the [002] direction for CDCs produced at higher temperatures. We expect that vacuum annealing likely also produces changes in the stacking, which further underscores the increased need for more extensive mathematical modeling of the structure and surfaces of porous carbon materials.

1 Biener *et al.*, *J. Phys. Chem. C.*, 2014, **118**(1), 4–15.
2 Kubuzono *et al.*, *Sci. Rep.*, 2014, **3**, 1595.
3 Smarsly *et al.*, *J. Phys. Chem. C*, 2014, **118**(29), 15705–15715.

Fulian Qiu asked: In the paper, 20 nm to 2.0 μm sized carbide-derived carbons with an average pore size of ~0.8 nm were employed as electrode materials. Areas measured using gas adsorption should be different from effective surface areas accessible by ions in the supercapacitors. When the particle size decreases, the internal pore surface to total surface area (internal pore surface plus inter-particle pore surface) ratio will decrease. For example, if one 2 μm particle is smashed into 20 nm particles, the total surface area of the 20 nm particles would be roughly 100 times bigger than that of the 2 μm particle. In this way, for a supercapacitor electrode, the inter-particle pore surface area will be much bigger for the case of 20 nm particles than for that of 2 μm particles, and it could be comparable to the internal pore surface area for the 20 nm carbon particle electrode. If the self discharge process is controlled by internal pores, how does the particle size influence the self discharge process?

Yury Gogotsi answered: The inter-particle pore surface area is bigger in the case of 20 nm particles than in that of 2 micron particles, but it is a small fraction of the internal pore surface area for the 20 nm carbon particle electrode (>1000 $m^2 g^{-1}$).

François Béguin opened the discussion of the paper by Asa Noofeli:† What is the meaning of ESR at 1 kHz? Similarly what is the difference between series resistance RS and ESR? Generally the ESR is defined as the intercept of a Nyquist plot with the X-axis at high frequency.

Asa Noofeli communicated in reply: ESR (equivalent series resistance) is the real component of the impedance measured at 1 kHz and is provided as a frame of reference as some instruments can only determine the impedance at this frequency. In this paper, RS represents the intercept of the Nyquist plot with the real axis.

George Zheng Chen commented: Regarding the use of 1 kHz for the resistance measurement of a supercapacitor, I think it is often sufficient for fast analysis, particularly if the interest is to monitor the trend of resistance change with a dynamic variation of voltage or pressure of the supercapacitor.

François Béguin asked: Ionic liquids display low electrical conductivity and high viscosity. For this reason, the authors have used carbons with a high amount of mesopores. Unfortunately, the porosity of such carbons is poorly developed and the density is extremely low. This leads to a very poor volumetric capacitance, while industry is just looking for the highest values of this parameter. The conductivity of ILs can be enhanced by increasing temperature. Unfortunately, this will lead to a decrease of stability window and consequently to losing the advantage expected of using ILs. I agree with one comment of the authors that, for stationary applications, the volume of the systems is not an important issue. If it is the case, why not to use aqueous electrolytes which are very cheap, safe and environment friendly and can be implemented with microporous carbons of high density?

Asa Noofeli communicated in response: Our aim of using carbons with various pore sizes was to study its effect on device performance. I agree that the chosen ionic liquids may not compare favourably with other electrolytes in terms of conductivity, nevertheless they have other advantages such as low volatility/flammability which may be useful. The relatively high viscosity could possibly be tackled by using higher operating temperatures. This is currently under investigation and unfortunately I don't have any results to present at the moment.

† Asa Noofeli's paper was presented by Peter Hall, University of Sheffield, Sheffield, UK.

However the operating window of aqueous electrolytes are limited and may not be suitable for all applications.

Alexander Forse enquired: Is it possible to modify your synthesis method to introduce some microporosity into the carbon? This may boost the energy density.

Asa Noofeli communicated in answer: Yes it is possible. We can modify the carbonization and activation processes to produce the required porosity. However, in the case of ionic liquid electrolytes, the presence of micropores tends to increase the overall resistance of the device.

Julie Macpherson asked: Water incorporation into ionic liquids is a potential problem for you, resulting in a reduction in the magnitude of the solvent window. How do you ensure against this?

Asa Noofeli communicated in reply: All ionic liquids are stored/handled inside an argon filled glovebox.

George Zheng Chen opened the discussion of the paper by Elzbieta Frackowiak: Your work on redox active electrolytes in supercapacitors is very interesting and stimulating as a new approach toward electrochemical energy storage. I would like to know how you have calculated the capacitance values for your two-electrode cell. My understanding is that for peak-shaped CVs and non-linear GCDs, the charge/voltage ratio is not the same as capacitance.

Elzbieta Frackowiak responded: Thank you for your encouraging comment. Capacitance values were verified in two different ways. Full time of discharging was considered for the first method, giving the higher value of capacitance, however, the energy extracted from this low voltage region is doubtful. For more reliable calculations, we considered only the slope of the linear part of the galvanostatic profile (0.2–1.0 V); such a method gave slightly lower values but we find them more appropriate. It is noteworthy that capacitors based on iodide aqueous solutions demonstrate excellent cyclability, especially with stainless steel current collectors.

George Zheng Chen asked: In Fig. 4 and 8 of the paper, it can be seen that the gain of charge storage is at low cell voltages, which means the storage is low in energy content.

Elzbieta Frackowiak answered: I agree that, in the low cell voltage region, energy storage is very moderate in the case of the hydroquinone additive (Fig. 4 and 5 of the paper) as well as in iodide solutions (Fig. 18 of the paper), however, integrating the total charge with a smart electrotechnical configuration might even efficiently exploit this low cell voltage region.

George Zheng Chen remarked: The diffusion or transport of the redox active species inside the activated carbon layer should have a strong influence on the charge storage behaviour. How thick and how porous was the activated carbon layer on the electrode, because the redox active species are not anchored on the electrode surface, and can diffuse back to the bulk electrolyte after charging or discharging, and then become irrecoverable? This may contribute to self-discharge. However, because the redox reactions occur at low voltages, such diffusion induced self-discharge may not be observed before the cell voltage decay from self-discharge of the charge stored in the double layer has come to a sufficiently low value.

Elzbieta Frackowiak replied: The thickness of the electrode was 300 μm. The specific surface area of the carbon electrode was 20% lower than the pristine activated carbon powder. The total pore volume of the electrode was 1.039 cm^3 g^{-1}, the micropore volume was 0.716 cm^3 g^{-1}, whereas the mesopore volume was equal to 0.269 cm^3 g^{-1}. The redox active species, depending on their type, can be partly grafted or immobilized into the carbon network (*e.g.* hydroquinone species) or specifically sorb/interact with the carbon structure in the case of iodide species, giving C-I binding. In the literature it has been found that iodine can play the role of a p-dopant of carbon and even increase its conductivity. Concerning the carbon texture, it has been shown by us that mesopores play a crucial role for the redox activity of the iodide/iodine couple, because they match very well with the size of polyiodides (the final product of redox reactions). The concentration of iodide ions in the electrolyte bulk is quite high in comparison to the charge utilised during the charging/discharging process, so even if some part of the process would be irreversible, we believe that the concentration of ions preserves the high efficiency of the charge/discharge process.

Regarding self-discharge – we measured this value over 24 h after full charging to 0.8 V and we observed 200 mV of self-discharge, with a significant drop of voltage during the first two hours.

Fulian Qiu remarked: In Fig. 17 of the paper, does the capacitance as a function of floating time show different trends from two different calculation procedures? In the two-electrode supercapacitor configuration, on the anode, iodide ions were oxidized to form iodine, and the iodine diffused away from the electrode surface, then combined with iodide to form complex species; this would depend on the time scale in the CV and charge-discharge process (scan rates and charge/discharge currents). On the cathode, it should be pure double-layer charge–discharge responses initially or on a shorter time scale. The device is an asymmetric supercapacitor. During the long duration floating experiments, complex species generated in the anode region could diffuse through the separator to the cathode surface, and electrochemical reduction processes occur on the electrode surface depending on the separator structure. In this way, the supercapacitor would be a battery. As asymmetric supercapacitors are different from pure electrochemical double-layer supercapacitors, charge–discharge processes are much slower in asymmetric supercapacitors, their discharge curves are not linear, and it is difficult to calculate capacitance using the slope accurately. The total discharge time method for the capacitance calculation would be more proper.

Elzbieta Frackowiak answered: We do agree that the redox activity of iodide species might suggest a battery type behaviour, but it is solely connected with the positive electrode. When separating the voltammetry responses for both electrodes, the positive electrode shows a very well defined peak in a narrow range of potential (according to iodine activity from the Pourbaix diagram). This high additional capacitance originates from iodide-based redox couples, formed during polarisation of the positive electrode. However, the total capacitor presents quite a rectangular shape of the cyclic voltammogram as well as a triangular shape of the galvanostatic profile, that allows us to consider such a system as a capacitive one with a significant contribution from pseudocapacitance. Typically, the slower the regime, the more significant the contribution from Faradaic reactions. Of course, we can assume that during long term cycling, the redox species diffuse between the electrodes creating a battery type response; but we would like to note that, even after several thousand of cycles, the major part of the galvanostatic profile is triangular (Fig. 18 of the paper), without any remarkable plateau induced by a 'pure' redox response. Hence, we decided to calculate the capacitance value using two possible methods typical for evaluation of capacitor performance (Fig. 17 of the paper). Remarkably, we noticed that after long-term cycling, the pH of the electrolytic solution changed from a neutral one to more alkaline (close to 8). Hence, the gradual aggravation of capacitor performance after durable cycling could be explained by pH change. We proved that iodates formed in a more alkaline medium show definitively worse performance (low capacitance values, poor solubility).

George Zheng Chen asked: Diffusion of the redox active species in the electrolyte should have played a role in the rate of charging and discharging. Fig. 4 of the paper shows the CVs recorded at 5 mV s^{-1}, what may happen if the scan rate is increased, say to 100 mV s^{-1}? On the other hand, the redox reaction only occurs in a particular and narrow range of electrode potentials or cell voltages. What cell voltage was applied for recording the impedance data presented in Fig. 6, 7, 8, 9, 16 and 20 of the paper?

Elzbieta Frackowiak responded: The redox response gradually disappears with an increase of scan rate; Faradaic reactions are typically limited by diffusion. At a scan rate of 100 mV s^{-1} we observed a more resistive behaviour of the CV profile, especially in the case of a hydroquinone additive in acidic solution (Fig. 4 of the paper). The impedance spectra in Fig. 6, 7, 8, 9, 16 and 20 of the paper were recorded at open circuit voltage (OCV).

Chi-Chang Hu queried: From a comparison of the CV curves shown in Fig. 1 in the text, how can the double-layer capacitive currents in the whole investigation potential region be significantly enlarged by adding dissolved electroactive species (*e.g.* H_2Q here)?

Elzbieta Frackowiak answered: The cyclic voltammogram presented in Fig. 1 of the paper concerning the activity of hydroquinone in an acidic medium has been recorded after 500 subsequent voltammetry cycles (similar to the profile shown in Fig. 4 of the paper). Capacitive current response combines electrostatic attraction of ions together with Faradaic reactions (Fig. 10 of the paper). Separation of the

EDL response and the Faradaic one might be possible by applying different scan rates (CV), current regimes (galvanostatic loading) or studying time constant values by electrochemical impedance spectroscopy. It has to be noted that the Faradaic response of porous electrodes does not reflect the typical peak profile observed for non-porous electrodes.

Dirk Guldi asked: Has the pH dependence of the different diols been investigated, especially with respect to the reversibility of the redox chemistry, that is, oxidation and reduction of the diole ?

Elzbieta Frackowiak responded: Various diols with different OH group substitutions have been investigated in neutral, alkaline and acidic media to prove a significant role of protons in their redox chemistry. Fig. 1 of the paper clearly shows the highest redox activity of 1,4-dihydroxybenzene (hydroquinone) in acidic and alkaline electrolytic solutions. Unfortunately, the reversibility of the redox activity is moderate, especially in an acidic medium (1 mol L^{-1} sulfuric acid).

Fulian Qiu opened the discussion of the paper by François Béguin by commenting: In Fig. 20 of the paper, does the charge-transfer resistance increase with floating time? In fact, there are two charge-transfer processes; two processes become more distinguishable with floating time. If the processes are from the anode electrode only, they originate from the current collector–carbon layer interface and electrochemical reaction on the carbon surface.

Yury Gogotsi asked: What electrochemical reaction occurs when sodium molybdate salt is added? Is there any deposition of molybdenum oxide or other reaction products on carbon?

François Béguin answered: Addition of sodium molybdate (Na_2MoO_4) to 1 mol L^{-1} Li_2SO_4 causes an increase in the conductivity of the electrolyte mixture resulting in enhanced EDL capacitance up to 0.8–1.0 V. When the voltage is increased up to 1.6 V, pseudofaradaic contributions due to the presence of MoO_4^{2-} (equilibria (3) and (4) in the manuscript) and reversible hydrogen storage come into play at the negative electrode. In post-mortem investigations of capacitors after floating at 1.6 V for 120 h, we could detect a MoO_2 deposit on the negative stainless steel current collector at the periphery of the AC electrode. Besides, no deposit was observable on the positive side due to the negative shift of potential as explained in the manuscript. Further investigations are in progress to analyze the carbon electrodes and especially to detect MoO_2 in the negative one.

Chi-Chang Hu remarked: The speaker demonstrated that the AC/AC system can achieve 2.2 V in aqueous media containing Li_2SO_4. However, after the floating test, it is not stable even at 1.6 V if the electrolyte does not contain Na_2MoO_4. What happens to the positive and/or negative electrodes? Why does Na_2MoO_4 improve the cycle stability of carbon? Was Na_2MoO_4 deposited onto the carbon electrodes?

François Béguin replied: In ref. 10–11 of the manuscript, it was demonstrated from galvanostatic charge–discharge cycling – where the cell voltage only stays temporarily at the specified value – that the AC/AC systems are able to operate up to ~2 V, while using gold current collectors. When using stainless steel current collectors and potentiostatic floating – which is to keep the cell at such a high voltage for a long period of time (120 h) – the maximum voltage in 1 mol L^{-1} Li_2SO_4 is 1.5 V. At voltages higher than 1.6 V, carbon is oxidized at the positive electrode and the current collector is corroded, which leads, after some floating time, to a drop of capacitance and an increase of resistance (detailed information is given in ref. 12–13 of the manuscript). When Na_2MoO_4 is added to the electrolyte, the maximum potential of the positive electrode is shifted to lower values, and consequently at a voltage of 1.6 V the positive electrode is not oxidized. A MoO_2 deposit was detected on the negative electrode, according to equation (3) of the manuscript.

Chi-Chang Hu continued: Due to the enlargement of capacitance as well as the increase in the iR drop of the resultant cell with floating test time, there should be a certain relationship between the formation of oxygen-containing functional groups and the floating test time. What is the role of Na_2MoO_4 during the floating test period with regard to protecting the oxidation/functionality of activated carbon on both electrodes?

François Béguin answered: In our previous works (ref. 12 and 13 of the paper), we demonstrated that under floating conditions at or beyond 1.6 V, oxygenated functionalities are developed on the positive electrode, resulting in an increase of resistance and decrease of capacitance. This happens mainly because the maximum potential of the positive AC electrode is above the water oxidation potential limit, and obviously the amount of oxygenated surface functionalities increases with the floating time. However, the resistance increase is also partly due to the generation of corrosion products on the positive stainless steel current collector which accumulate at the AC/collector interface. When MoO_4^{2-} is added to the electrolyte, the maximum potential of the positive AC electrode is lowered by about −0.162 V for an operating voltage of 1.6 V, *i.e.* below the oxidation limit of water, preventing the oxidation of carbon and corrosion of stainless steel.

Laurence Hardwick opened the discussion of the paper by Yong Cao: What is the mechanism of catalysis in your system, *i.e.* how does the catalyst work?

Yong Cao replied: Thank you for your question. The mechanism of $Li\text{-}O_2$ batteries is now unclear due to the use of different catalysts and electrolytes. In our previous work, we just searched for some effective catalysts and constructed the $Li\text{-}O_2$ battery test system. Now, we are proceeding with some deep study of the mechanism through *in situ* XRD and Raman and also XPS.

Laurence Hardwick continued: As you form a solid discharge product on the cathode surface, it is hard to imagine in a conventional sense how the catalyst works?

Yong Cao answered: This is a good comment. Some soluble catalysts have been reported with improved performance, which may be a research hot topic in the future.[1,2]

1 H.-D. Lim *et al.*, *Angew. Chem.*, 2014, **126**, 4007–4012.
2 Y. Chen *et al.*, *Nat. Chem.*, 2013, **5**, 489.

George Zheng Chen asked: In your work, what is defined as a catalyst? Can you give experimental demonstration of the catalysis as you have claimed?

Yong Cao responded: Thanks for your question. In Li-O_2 batteries, the discharge and charge reactions contain oxygen reduction and evolution reactions, but the sluggish kinetics of the oxygen electrode result in large discharge and charge overpotentials. So a catalyst is important to improve the kinetics of the oxygen electrode. In our work, when using the $CoFe_2O_4$–rGO composite as the cathode, the discharge overpotential, especially the charge overpotential, is significantly reduced compared with pristine rGO. Also, the capacity was improved.

George Zheng Chen continued: Would you please explain what results in your work show the catalytic activity as you have stated in the title of your paper? In other words, what electrode reactions have been catalysed and how?

Yong Cao replied: Both the discharge and charge overpotentials were significantly reduced, which can be attributed to the catalytic activity of the $CoFe_2O_4$–rGO composite, compared with the pristine rGO.

Chi-Chang Hu said: For Li-air batteries, the preferred product of the cathode is Li_2O_2. What is the preferred electron transfer number of the catalyst? From the preferred product, the preferred electron transfer number is 2. Can the authors obtain the preferred products and better performances if using a catalyst with the electron transfer number of the ORR equal to 2?

Yong Cao responded: Indeed, a catalyst with a 2-electron process would benefit the formation of the ideal product Li_2O_2. But in fact, the ORR process in the non-aqueous system is not similar to that in the aqueous system. The key issue in current Li-O_2 batteries is not the discharge process, but the charge process. Li_2O_2 has been confirmed as the discharge product in different electrolytes with different catalysts. However, some side-products form in the charging process due to either the oxidation of carbon in the presence of Li_2O_2 or the decomposition of electrolytes on the carbon surface. These side products, such as Li_2CO_3, are not easily electrodecomposed during charging, and thus result in a large overpotential and low round-trip efficiency.

Wanjing Yu said: I am just wondering if it is possible to calculate the specific capacity of $CoFe_2O_4$? The mass of rGO is only 17.3%. The $CoFe_2O_4$ might also contribute to the capacity. Is it reasonable to calculate the specific capacity based only on the mass of rGO?

Yong Cao replied: In early works, the capacity was calculated based on the mass of carbon in the electrode. We also provide the specific capacity based on the total mass of rGO + $CoFe_2O_4$ on the top axis of Fig. 3 of the paper. In fact, $CoFe_2O_4$ contributes no capacity in the voltage window of 2.2–4.2V.

Laurence Hardwick remarked: Did you use any other characterisation tools to confirm the composition of the discharge product (Li_2O_2)?

Yong Cao responded: Yes, we have also conducted XRD and Raman, but the signals were not so strong, which may be ascribed to the poor crystallinity of the Li_2O_2. But we are trying to improve it.

Laurence Hardwick continued: Does Li_2O_2 have a clear and distinct signal in XPS?

Yong Cao added: Yes, the binding energies of Li-O in Li_2O_2 and Li_2CO_3 are different from each other. You can refer to ref. 1.

1 S. Liu *et al.*, *Adv. Energ. Mater*, 2014, **4**, 1301960.

Robert Dryfe asked: In the experimental section you state that an atmosphere of oxygen is used to perform the experiments. However, can you be sure that there is no residual carbon dioxide in your sample? In other words, is it possible that the carbon deposits come from reduction of atmospheric CO_2, rather than reaction of the lithium salts with the reduced graphene oxide?

Yong Cao answered: We used a modified vacuum drier as the test device in which our batteries were placed. The vacuum drier was firstly evacuated and then filled with high purity oxygen. So we can exclude the influence of CO_2 from the atmosphere.

Surbhi Sharma enquired: Did you perform any XPS studies on your $CoFeO_4$–rGO samples to check for any residual impurities and the C:O ratio in GO and rGO?

Yong Cao replied: I am sorry to tell you that I did not do the work you mentioned.

Dirk Guldi opened the discussion of the paper by Laurence Hardwick: What is the rationale for the Raman shift ?

Laurence Hardwick answered: The rationale for the Raman shift of the 2D band is that it occurs from charge transfer onto the graphene sheets of the graphite intercalation compound, and from where the graphene layers are flexible and deform around domains of intercalating lithium ions in stage 3 and 4 GICs in particular. We will be investigating separating out the relative contributions in the near future.

Dirk Guldi continued: What is the local and overall concentration of lithium between the graphene sheets upon intercalation?

Laurence Hardwick responded: The stage 3 GIC, where we see the shift of the 2D band, has the formula LiC_{18}.

Christopher Sole commented: The large downshifting of the 2D Raman peak for our graphite electrodes upon stage 4 and stage 3 formation (0.2–0.15 V *vs.* Li) may be related to strain (or at least doping induced bond expansion) as suggested by Chacón-Torres *et al.*[1] in K^+ intercalated graphite intercalation compounds. This paper claims that the disappearance of the 2D peak for stage 2 and stage 1 compounds is due to excessive charge on Li bounding layers, thus the observed 2D peak for stage 4 and stage 3 comes from the nonbounding (interior) graphene layers which are much less charged. The split G peak at high stage number also suggests strain as the downshift of the interior G peak can be explained by bond expansion and not doping. Interestingly, this strain argument was briefly proposed by W. W. Huang and R. Frech in 1998[2] without looking at the 2D band.

1 Chacón-Torres *et al.*, Manifestation of Charged and Strained Graphene Layers in the Raman Response of Graphite Intercalation Compounds, *ACS Nano*, 2013, 7, 9249.
2 W. W. Huang and R. Frech, *In Situ* Raman Studies of Graphite Surface Structures during Lithium Electrochemical Intercalation, *J. Electrochem. Soc.*, 1998, **145**, 765.

François Béguin said: You conclude that your Raman data support the Daumas–Herold model of lithium intercalation. However, there is not a clear statement of the experimental information which allows you to draw such a conclusion. Could you explain which experimental data allow you to discriminate between the Rüdorff and Daumas–Herold models?

Laurence Hardwick replied: We hypothesise that the significant shift of the 2D band is the result of charge transfer and the deformation of graphene layers around domains of intercalating lithium ions. In the Rudorff model the ions enter into discrete layers – causing no deformation. In the Daumas–Herold model the graphene layers are required to slightly deform. In addition, as has been experimentally observed by us and before by M. Inaba, H. Yoshida and Z. Ogumi (ref. 13 in the paper), the G-band is seen to shift which can be explained by the presence of a dilute stage 1 graphite intercalation compound – again supporting the Daumas–Herold model in which intercalants are present within each layer.

Yury Gogotsi asked: How can you separate contributions from charge transfer and mechanical strain?

Laurence Hardwick answered: We are looking into experiments to do this – as suggested from this discussion. Certainly computational modelling, pdf and strain studies will be important to understand these contributions. A comment that we stated in the paper is that the graphene sheets deform around the lithium intercalated domains. We admit we have to be careful in calling this 'strain' as this is a larger scale property of the whole material.

Yury Gogotsi addressed Christopher Sole and Laurence Hardwick: Did you compare the Raman shift after intercalation to the Raman shift of graphite (graphene or carbon nanotubes) subjected to purely mechanical stress?

Laurence Hardwick responded: No, we have not done so. In the literature there are many investigations looking at mechanical stress on various forms of carbon – but I am yet unaware of any studies looking at the effect of mechanical stress on the Raman spectrum of graphite intercalation compounds (GICs). Certainly this type of study will be important in understanding the role of localised graphene sheet deformation in stage 2, 3, 4 GICs.

Christopher Sole added: We have not attempted such a measurement although it could be a good idea, the literature values for graphene depend a lot on substrate, preparation and laser used so it is difficult (not impossible!) to compare. The effect of uniaxial stress splits the 2D band so we can rule this out. Biaxial stress (as we propose) should have a similar downshifting effect, however, we need to take care to separate the effects of doping and strain. This may be possible using a technique similar to Lee *et al.*,[1] but they say this is limited to SLG and possibly difficult due to the dispersion of the 2D peak with laser wavelength (we use 633 nm in the paper).

1 Lee *et al.*, Optical separation of mechanical strain from charge doping in graphene, *Nat. Commun.*, 2012, **3**, 1024.

Yury Gogotsi continued: You may wish to conduct X-ray pair distribution function (PDF) analysis of intercalated samples to determine if there is a change in carbon–carbon bond length, and compare the observed strain to that measured by Raman spectroscopy.

Laurence Hardwick replied: That is an excellent suggestion, which we had not previously considered and would be a complementary measurement technique to probe structural changes at various stages in graphite intercalation compounds.

Keith Stevenson remarked: Does the 2D peak broaden as a function of lithiation potential? It is not clear how the intensity decreases and peak width changes (table 2 of the paper). They should be correlated.

Laurence Hardwick responded: As the intensity of the 2D(1) peak decreases so does the FWHM generally, although possibly the peak width may be obscured by background at low peak intensity The 2D(2) peak is relatively sharp and intense at the beginning and end. It broadens and increases in intensity as it shifts dramatically (at 0.2 V stage formation) although possibly this is due to incorporating leftover 2D(1) intensity. In short, it is challenging with current data (signal to noise ratio) to separate the contributions from the 2D(1) and 2D(2) peaks.

Keith Stevenson continued: It is not clear how the 2D overtone is reflective of compressive strain and buckling of the sheet, as the resolution of this band is low.

Laurence Hardwick answered: After our conservation I agree with your point we should avoid using the term strain. We hypothesise that what we observed is a localised distortion of the graphene sheet due to the distribution of lithium domains within the graphitic structure, and this is reflected by the shift in the 2D peak.

Keith Stevenson added: Do different particles show the same trends with lithiation/delithation?

Laurence Hardwick replied: This is a good point. Certainly, in past Raman studies[1] we observed variation in lithium concentration at low current densities during lithiation. We have not looked at multiple points in this instance yet for the 2D peak. This heterogeneity of lithium insertion into practical graphitic electrodes is an interesting observation that has been observed by many authors – such as Aurbach, Kostecki, Novak and Yazami.[2–5] However single point measurements on other graphitic carbons (as yet unreported by us) show similar trends for the 2D band to those reported within the Faraday paper.

1 L. J. Hardwick *et al.*, *Solid State Ion.*, 2006, **177**, 2801–2806.
2 E. Markevich, G. Salitra, M. D. Levi and D. Aurbach, Capacity fading of lithiated graphite electrodes studied by a combination of electroanalytical methods, Raman spectroscopy and SEM, *J. Power Sources*, 2005, **146**(1–2), 146–150.
3 R. Kostecki and F. McLarnon, Microprobe study of the effect of Li intercalation on the structure of graphite, *J. Power Sources*, 2003, **119–121**, 550–554
4 P. Maire, A. Evans, H. Kaiser, W. Scheifele and P. Novak, Colorimetric determination of lithium content in electrodes of lithium-ion batteries, 2008, *J. Electrochem. Soc.*, **155**(11), A862–A865.
5 Y. Reynier, R. Yazami and B. Fultz, XRD evidence of macroscopic composition inhomogeneities in the graphite–lithium electrode, *J. Power Sources*, 2007, **165**(2), 616–619.

Alexander Forse said: Have you performed any quantum chemistry calculations of the Raman frequencies for different intercalation stages? This could offer some insight into the mechanism and help distinguish between the different intercalation models mentioned in the paper.

Laurence Hardwick answered: This is an excellent suggestion – and no, we have not performed any calculations. We recognise that we will require these types of calculations to investigate whether the effect we see (*i.e.* the large downshift of the 2D band) can be wholly attributed to doping, or may also include a contribution from highly localised deformation of graphene layers around intercalated lithium domains.

Dirk Guldi asked: Are you considering Raman mapping to probe local disorder, stress, *etc.*?

Laurence Hardwick replied: Yes – we are considering mapping to probe multiple areas of the electrode to see if band shifts are uniform across the electrode surface.

Dirk Guldi continued: Have you changed the excitation in your Raman experiments to see if the effect on the 2D band shift and shape is wavelength dependent?

Laurence Hardwick responded: We have not done these measurements yet – and we plan to do so. We would expect to see a shift as the first order D-band has been shown to be excitation energy dependent. Looking at the data from Chancón-Torres *et al.*[1] on CxK compounds, they use a different wavelength laser and see a different wavenumber shift.

1 Chacón-Torres *et al.*, Manifestation of Charged and Strained Graphene Layers in the Raman Response of Graphite Intercalation Compounds, *ACS Nano*, 2013, 7, 9249.

Dirk Guldi asked: How do you intercalate lithium into graphite?

Laurence Hardwick replied: We are intercalating the lithium ions electrochemically, using a sealed *in situ* cell.

Katherine Holt addressed François Béguin and Laurence Hardwick: This is a question about the proposed models for Li ion intercalation into graphite. In the Rudorff model shown in Fig. 7 in the paper, Li is seen inserting between only some of the layers and not others. If this model is correct, what determines which layers (and how many) the Li will insert into? Is this statistical or is there a chemical preference for particular layers?

François Béguin answered: You asked exactly a good question about the Rudorff model: "why the reagent would select particular intervals and not others?". This model is perfect to account for the X-ray diffraction data but not to explain the reactivity of graphite with stage changes. This is the reason why the Daumas–Herold model has been introduced, in order to express that all intervals are equivalent; simply the reagent occupies a part of these intervals, being separated by pleats from the unoccupied regions.

Laurence Hardwick added: The Rudorff model was the earliest structural model to explain PXRD data of graphite intercalation compounds (GICs). All undergo an increase of the separation between the graphene sheets, the *d*-spacing (3.35 Angstroms). For lithium GICS this separation increases by 10% to 3.7 Angstroms. It could be argued that there is an additional energy barrier for the insertion of the first Li^+ for the separation increase – for it to squeeze in – which is then removed for the second Li^+ as the gap has already widened (unless of course the 1st Li^+ has diffused sufficiently into the bulk and the interface has relaxed back to the original separation). What I believe is that there is no chemical preference for a particular layer insertion (there is the added complication of solid electrolyte interphase formation – adding to the distribution of interfacial resistances in a porous electrode). I believe that in the initial stages of electrochemically driven intercalation there is a dilute stage; one GIC in which all layers have a diffuse amount of Li^+ inserted.

François Béguin continued: Does the 2D band show evidence of the dilute 1st stage?

Laurence Hardwick responded: The 2D band does shift moderately during the dilute stage 1 by *ca.* 22 $cm^{-1}\ V^{-1}$. The G-band shifts by *ca.* $-28\ cm^{-1}\ V^{-1}$ during the dilute stage 1 phase (*ca.* 0.55 to 0.22 V *vs.* Li/Li^+ (See Fig. 6 in the paper). This effect is likely to be from doping of the graphene layers.

Boris Dyatkin stated: It would be interesting to implement this *in situ* Raman spectroelectrochemistry approach to analyze anion intercalation into graphite cathodes. Read *et al.*[1] presented evidence that anion insertion likely produces stage 1 intercalated graphite at 5.2 V. Observing changes in the 2D band of the material during PF_6^- or BF_4^- intercalation would yield novel insights into changes in the graphite structure.

1 Read *et al.*, *Energy Environ. Sci.*, 2014, 7(2), 617–620.

Laurence Hardwick replied: Yes I would agree with this comment. We have in the past looked at the G and D bands during BF_4^- intercalation,[1] but we have not looked at the 2D band yet, but plan to do so.

1 L. J. Hardwick *et al.*, An *in situ* Raman study of the intercalation of supercapacitor-type electrolyte into microcrystalline graphite, 2006, *Electrochim. Acta*, **52**(2), 675–680.

Richard McCreery said: When you see the shift in the 2D band position during lithium ion intercalation, do you see a corresponding shift in the D band?

Laurence Hardwick answered: Unfortunately the D-band (when clearly visible – if we are measuring in a more 'disordered' area) weakens and then vanishes into the signal noise on initial Li^+ insertion, before any shift in the 2D band is observed, *ca.* by 0.8 V *vs.* Li/Li^+. The weakening of the D-band is shown in ref. 12 of the paper. We would be motivated to try to measure Li^+ intercalation into a more highly disordered carbon – to see if we could measure any D-band shift before the signal fully disappears.

Faraday Discussions

PAPER

Electrochemistry of well-defined graphene samples: role of contaminants

Hollie V. Patten,[a] Matěj Velický,[a] Nick Clark,[bc] Christopher A. Muryn,[ad] Ian A. Kinloch[b] and Robert A. W. Dryfe*[a]

Received 10th April 2014, Accepted 12th May 2014
DOI: 10.1039/c4fd00064a

We report the electrochemical characterisation of well-defined graphene samples, prepared by mechanical exfoliation. Mechanical exfoliation is the method of choice for high purity graphene samples, despite the inherent complexity of the approach and the small scale of the resultant flakes. However, one important, yet presently unclear area, is the role of adsorbates such as processing residue, on the properties of the graphene layer. We report high resolution microscopic and electrochemical characterisation of a variety of poly(methyl methacrylate) (PMMA) transferred graphene samples, with the explicit aim of investigating the relationship between electrochemical activity and sample purity.

Introduction

There has been enormous interest in the physical properties of graphene, the two-dimensional form of carbon, in recent years. One of the main proposed technological applications of graphene is in the area of electrochemical conversion and storage, with applications including supercapacitors,[1–3] photovoltaics,[4–6] fuel cells[6,7] and batteries.[8] However the "graphene" used in many of the studies for these applications is frequently prepared by chemical methods, *i.e.* graphite oxidation. The heterogeneous nature of the resultant reduced graphene oxide samples makes their structural characterisation inherently more difficult. A further problem with the electrochemical study of samples derived from graphene oxide is the presence of metallic impurities, such as the manganese frequently used as an oxidising agent, which have been reported to be responsible for catalytic processes initially attributed to the graphene sample itself.[9] Thus, to

[a]*School of Chemistry, University of Manchester, Oxford Road, Manchester M13 9PL, UK. E-mail: robert.dryfe@manchester.ac.uk; Fax: +44 (0)161 275 4598*
[b]*School of Materials, University of Manchester, Oxford Road, Manchester M13 9PL, UK*
[c]*Centre for Mesoscience and Nanotechnology, University of Manchester, Oxford Road, Manchester M13 9PL, UK*
[d]*Photon Science Institute, University of Manchester, Oxford Road, Manchester M13 9PL, UK*

fully understand the properties of graphene as an electrode material,[10,11] it is vital that studies of structurally well-defined graphene samples are performed.

Well-defined films of monolayer graphene can be prepared using the chemical vapour deposition (CVD) method.[12] However, the highest quality graphene samples, and thus those generally used for transport studies, for example, are those derived from mechanical exfoliation (ME), which involves the isolation of single graphene flakes on a suitable substrate.[13] Somewhat surprisingly, a relatively small fraction of the many electrochemical studies of graphene reported to date use samples prepared by CVD[14] and an even smaller number of papers are concerned with the electrochemical behaviour of graphene derived from ME,[15–17] which is probably related to the intrinsic difficulties (the need for a clean room environment to fabricate the electrode device, the inherent small size of the flakes) associated with such samples. Since the earliest reports of both ME and CVD derived graphene, polymer layers have been employed for transfer of the graphene samples from one substrate to another, with poly(methyl methacrylate) (PMMA) a popular transfer material. However, PMMA residues often persist on the graphene samples after (nominal) dissolution of the polymer film, and very recently a number of articles have begun to address the question of how these residues influence the electronic properties of the graphene samples.[18-21]

The reliance on such polymer transfer methods with ME graphene has meant that, despite this material having the highest intrinsic quality of graphene samples, the role of surface contaminants on the sample properties needs investigation. Although an increasing number of papers have been dedicated to cleaning the graphene surface, the effect of transfer residues on the electrochemical response is still unclear. As discussed in more detail below, a variety of procedures (solvent washing, vacuum annealing and exposure to controlled gaseous environments) have been used to remove polymer residue from high quality graphene samples. In the present study we investigate a range of cleaning procedures and the impact of these on the electrochemistry.

Experimental

Preparation of mechanically exfoliated (ME) graphene samples

The basic method of ME graphene preparation is described below. Additionally, five subsequent cleaning regimes were explored for the ME graphene monolayers, to assess the extent to which polymer and solvent residues associated with cleaning affected the electrochemical response.

Monolayer graphene samples were prepared by the ME of natural graphite (NGS Naturgraphit GmbH) onto Si/SiO_2 (90 nm oxide layer) substrates. Initially *ca.* 2 cm × 2 cm Si/SiO_2 (90 nm) substrates were cleaned with an O_2 plasma for 10 min (Moorfield Etcher, UK). Immediately after the substrates were removed from the plasma chamber, thin layers of natural graphite on tape were firmly placed onto them (graphite side down), ensuring all air bubbles were removed, and left on the wafers for at least 24 hours to enhance the adhesion between the graphite and SiO_2 wafer. To remove the tape and tape residues, samples were immersed in hot (80 °C) methyl iso-butyl ketone (MIBK, Sigma-Aldrich, UK) for *ca.* 15 min, followed by fresh hot MIBK for a further 10 min. The samples were then immersed in hot acetone (Sigma-Aldrich, UK) (5 min) before cooling to room temperature (5 min). Finally the samples were placed in iso-propyl alcohol (IPA,

Sigma-Aldrich, UK) for 10 min, dried using N_2 and baked on a hotplate at 110 °C (10 min). A final (low angle) tape peel was then performed on the sample.

Monolayer flakes of interest (>100 µm in diameter) were identified using optical microscopy (Nikon LV100-50iPol). PMMA (3% 950 K in anisole) (Micro-Chem Corp, MA, USA) was spin coated onto the samples (3000 RPM, 60 s) before heating on a hotplate at 120 °C (2 min). This process was repeated once. Tape windows, of between 0.5 mm and 1 mm diameter, made using a hole punch, were then placed over the flakes before immersion in KOH for a minimum of 4 hours. The KOH was used to etch the SiO_2 layer. The ME graphene samples were then transferred onto fresh Si/SiO_2 substrates, previously cleaned by sonication in acetone (10 min) and IPA (10 min), before drying with N_2. The excess PMMA was removed by immersion in acetone (10 min) and IPA (10 min). Fig. 1 shows a schematic for the preparation of the ME graphene samples and transfer process.

Additional cleaning procedures were investigated: the graphene sample prepared by the "standard" cleaning regime we have previously employed[15,22] for electrochemical experiments with ME graphene is denoted sample 1. In this case, the graphene is cleaned for 10 minutes each with acetone and IPA. Sample 2 was exposed to a wider range of solvents previously reported in the literature for PMMA removal:[18,19,23] successive washes in acetone for 16 h each and IPA (10 min); an overnight acetic acid wash with a 10 minute IPA rinse; and finally a 60 min chloroform wash with a 10 min IPA rinse, with Raman spectroscopy and atomic force microscopy (AFM) performed following each washing stage. As noted in the Introduction, vacuum annealing of graphene samples has been reported as a clean-up protocol:[20,23-27] vacuum baking at 215 °C for 16 hours at 10^{-5} mbar is denoted sample 3, whereas sample 4 was exposed to an ultra-high vacuum (10^{-9} mbar, 50 °C (2 h), 101 °C (3 days)). Sample 5 was also annealed, but in this case a H_2–Ar[28-30] (10% H_2) environment was used: the sample was heated from room temperature to 50 °C (ramping at 5 °C min^{-1}) held for 1 h, ramped to 110 °C (held for 1 h), then heated to 250 °C (held for 4 h), and subsequently 270 °C (for 7 h).

After the various cleaning procedures, electrical contacts were made to the graphene flakes using silver epoxy (RS components) to connect to copper wire (Advent, UK). All samples were stored in glass Petri dishes to avoid any contaminants previously reported to affect samples when stored in plastic sample boxes.[31]

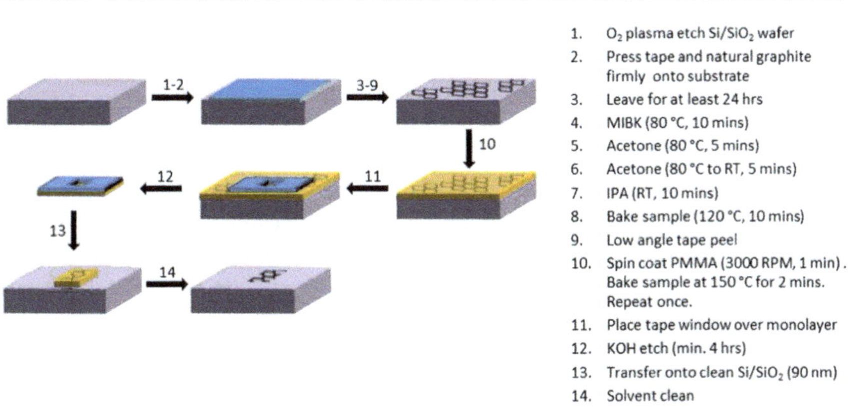

Fig. 1 A schematic to show the preparation of the ME graphene samples and transfer process.

Characterisation techniques

Optical microscopy. Optical microscopy (Nikon Eclipse LV100ND optical microscope with a DS-Fi2 U3 CCD camera, both Nikon Metrology, UK) was used to identify monolayer graphene flakes before transfer and to characterise them after transfer. Both bright-field and dark-field images were collected.

Atomic force microscopy. Peak force mode AFM measurements were performed in air using sharp silicon nitride lever (SNL) probes (Bruker, UK) with a Multimode 8 AFM. Images were analysed using Nanotec WSxM software.

Raman spectroscopy. Raman spectra of the samples were recorded using a 633 nm RL633 HeNe laser (RM System 1000 Mk1, Renishaw, UK). Extended scans were used to show peaks where the Raman shift was between 1000 cm^{-1} and 3000 cm^{-1} (15 s per spectrum with 5 accumulations using a 50× objective). The D, G and 2D peaks in the Raman spectra were analysed.

Electrochemistry. All aqueous solutions were prepared from Milli-Q reagent water (Millipore Corp.) with a resistivity of 18.2 MΩ cm at 25 °C. The redox active species are hexamine ruthenium chloride, $Ru(NH_3)_6^{3+}$ (99% Sigma-Aldrich, UK) and ferrocenylmethyltrimethylammonium hexafluorophosphate, $FcTMA^+$, both dissolved in 6 M LiCl (Sigma-Aldrich, UK) as a supporting electrolyte. The $FcTMA^+$ was prepared from ferrocenylmethyltrimethylammonium iodide (99% Strem Chemicals Ltd., UK) *via* metathesis with silver hexafluorophosphate (99.5% Strem Chemicals Ltd., UK).[32–34] The redox couples chosen are well-studied, outer-sphere systems, showing quasi-reversible ($Ru(NH_3)_6^{3+}$) and reversible ($FcTMA^+$)[35] kinetics on graphite substrates.

Cyclic voltammograms (CV) employed a PGSTAT302N potentiostat (Autolab, UK) using a three electrode set-up where the working electrode (WE) was graphene on Si/SiO_2 (samples 1–5), the reference electrode (RE) was a Ag/AgCl wire (Ag wire (Ag coated with PTFE 0.37 mm diameter, Advent, UK) previously anodised in a saturated potassium chloride (Sigma-Aldrich, UK) solution) and the counter electrode (CE) was a Pt wire (0.15 mm diameter, Advent, UK).

A localised electrochemical cell was formed using a droplet, which was held at the end of a pipette as previously reported.[15,36] Borosilicate glass capillaries (1.5 mm o.d. × 0.86 mm i.d., Intracel, UK) were pulled (Sutter puller P-97 Flaming/Brown) to a fine tip measuring *ca.* 1 μm in diameter and were back filled with a solution containing the redox active species and supporting electrolyte, using a syringe and micro-filler (World Precision Instruments, USA). The RE and CE were placed inside the pipette, which was then positioned close to the WE (the ME graphene) using a motorised manipulator (Siskiyou MX7630) before a droplet was formed and held between the WE and the pipette with the aid of a micro-injector (PV820 Pneumatic PicoPump, World Precision Instruments, USA). The sample was placed on the stage of the optical microscope mentioned above (GXML3030 Upright Materials Microscope, attached to a GXCAM-9 camera), hence the pipette was placed at a low (*ca.* 45°) angle with respect to the microscope stage.

Results & discussion

Sample characterisation

Representative optical micrographs of mechanically-exfoliated (ME) graphene flakes, transferred to a Si/SiO_2 wafer, are shown in Fig. 2. Exfoliation of natural graphite typically gives flakes with dimensions in the 0.1–1 mm range. Although the micrographs (Fig. 2(a) and (b)) suggest that the sample shown is continuous, dark-field imaging of the sample reveals wrinkles and defects, introduced during the transfer process. The samples are ultimately contacted and "spotted" with droplets of electrolyte solution containing one of the redox mediators (shown in Fig. 2(d)), which make local electrochemical cells (*vide infra*). Imperfections in the sample are more readily seen using AFM, as shown in Fig. 3, which shows micrographs and corresponding line profiles from a ME sample washed with different organic solvents, which have each been reported as useful for PMMA removal.[18-21] It is clear from Fig. 3(a) that the "default" acetone/IPA wash leaves considerable quantities of contaminant across the sample; this is attributed to PMMA residue that is not removed during the transfer process. The solvent treatments shown successively in Fig. 3 reveal improvements on the acetone/IPA wash: a prolonged acetone treatment, followed by an IPA rinse (Fig. 3(b)), an acetic acid wash for 16 hours followed by a 10 minute IPA rinse (Fig. 3(c)) and a 60 minute chloroform wash, followed by a 10 minute IPA rinse (Fig. 3(d)). The acetone/acetic acid/chloroform treatment in particular has removed much of the polymer residue. Raman spectra from the solvent treated samples are shown in Fig. 4. The 2D peak position (close to 2625 cm^{-1} in each case) confirms that the graphene is a monolayer, while the upshift of the G peak in the acetone and, particularly, the chloroform treated cases is indicative of sample doping (G peak positions of 1579.6 ± 2.4 cm^{-1} and 1583.9 ± 2.5 cm^{-1}, respectively). Vacuum annealing of a separate graphene sample, in contrast, gave unsatisfactory results (see Fig. 5). AFM indicated that the sample was largely free of debris, to a scale which was at least as good as the chloroform-washed sample of Fig. 3, however the Raman spectrum had degraded from the pristine monolayer response, with new bands at 1322 cm^{-1} and 2461 cm^{-1}, which indicate defect formation and sample contamination, respectively. The optical micrograph shown in Fig. 5(d) reveals that distinct fractures in the monolayer have formed on vacuum exposure, which we attribute to "degassing" of solvent residues or gases trapped between the monolayer and the substrate, possibly exacerbated by differential thermal

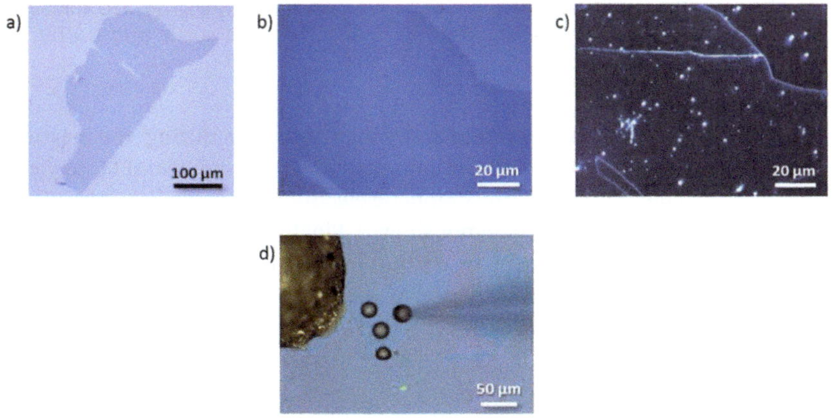

Fig. 2 Optical microscopy images of mechanically exfoliated (ME) graphene. Brightfield images show (a) a monolayer flake, (b) a selected area of the flake, (c) a darkfield image showing the same selected area, (d) the electrochemical set-up.

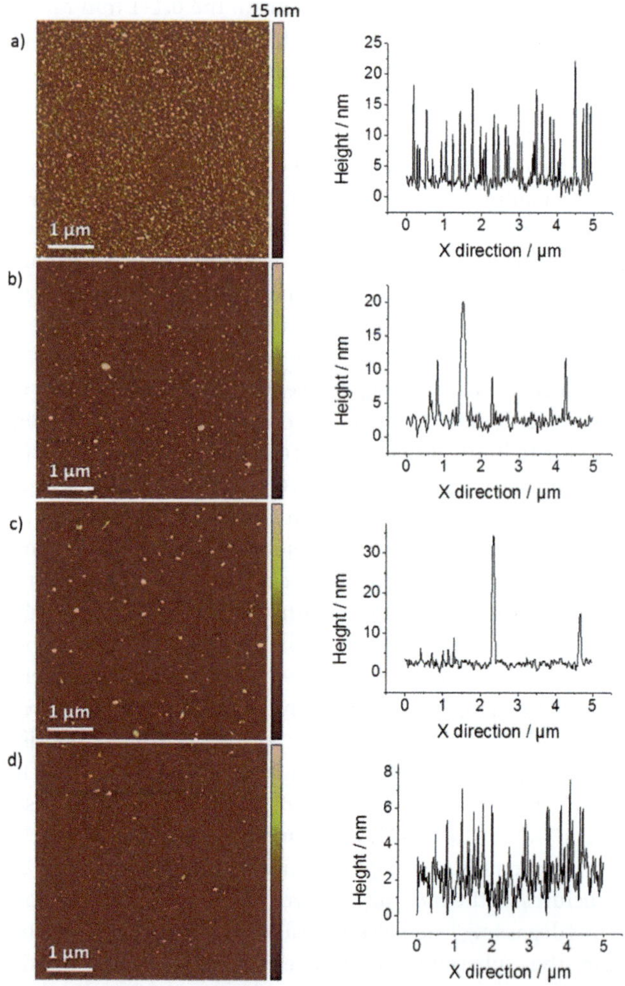

Fig. 3 AFM height images with a corresponding line profile (from centre) showing residue removal from different wet cleaning methods: (a) acetone (10 min), IPA (10 min), (b) successive washes in acetone, acetone (16 h), IPA (10 min), (c) acetic acid (16 h), (d) chloroform (60 min), IPA (10 min).

expansion of the ME graphene and the wafer substrate during the heating process. The ultra-high vacuum annealing method was found, *via* AFM, to yield higher levels of contaminant on the graphene electrode, most likely as a result of contaminants in the chamber (data not shown). Finally, the H_2–Ar annealing was found to give some improvement in residue in the AFM, although the Raman spectra (Fig. 6) again showed evidence of sample degradation, suggestive of amorphous carbon contamination or hydrogenation of either the PMMA residues or the graphene itself.[24,37]

Electrochemical response

The voltammetric response of each of the samples was recorded, using the microdroplet approach, with $Ru(NH_3)_6^{3+}$ and $FcTMA^+$, which are both quoted to

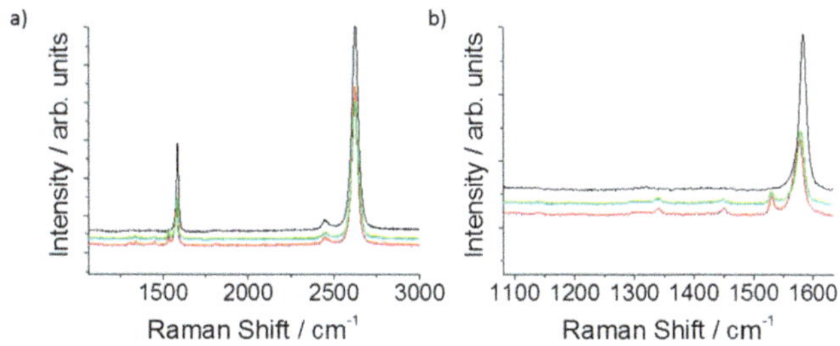

Fig. 4 (a) Raman spectra after different wet cleaning procedures: acetone (16 h), IPA (10 min) (red); acetic acid, IPA (10 min) (green); chloroform (60 min), IPA (10 min) (black). (b) Enlargement of the wavenumber region where the D and G peaks occur.

be model, outer-sphere redox couples. Droplets of 20–30 μm diameter were employed in this configuration, with the electrolyte contacted within the pipette containing the counter and reference electrodes. This size of droplet is small enough to contact the basal plane of the graphene, avoiding any cracks in the sample, see Fig. 2(d), but large enough to allow the response within the droplet to be approximated as a linear diffusive one, at least for short timescales. The sample prepared *via* the H_2–Ar annealing process (sample 5) was found to be inactive, which is attributed to problems with the contact made to the sample due to extensive cracking of the monolayer (detected using dark-field microscopy). Reproducible voltammetric responses were, however, obtained with both redox couples using the remaining four samples, *i.e.* numbers 1–4. The rate of electron transfer for both redox couples was found from the dependence of the forward and reverse peak separations over a range of scan rates, for microdroplets deposited at various locations on the graphene samples. When employing

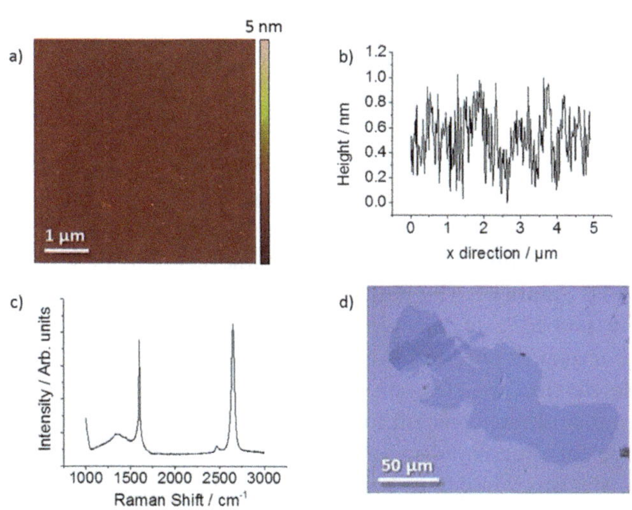

Fig. 5 (a) AFM height image with (b) corresponding line profile; (c) the Raman spectrum and (d) optical micrograph of a monolayer ME graphene flake after being vacuum annealed at *ca.* 10^{-5} mbar, 215 °C (16 h).

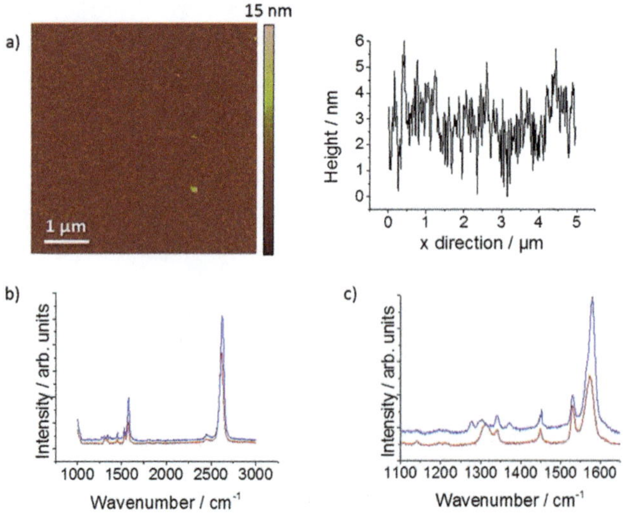

Fig. 6 (a) AFM height image with corresponding line profile of ME graphene after annealing in the presence of H_2–Ar at 270 °C (7 h), (b) Raman spectra after annealing at 250 °C (4 h) (red), and 270 °C (7 h) (blue).

$FcTMA^{+/2+}$ as the redox couple the Nicholson method of analysis was used to determine the electron transfer kinetics.[38,39] However, when employing $Ru(NH_3)_6{}^{3+/2+}$ as the redox couple the Klingler–Kochi method was preferred due to the large peak separations.[40] The kinetics were generally found to be close to the reversible limit for droplets of *ca.* 20–30 μm diameter (averaged data for several droplets for sample 1: 1.89 (±0.99) × 10^{-2} cm s^{-1}, sample 2: 2.76 (±1.59) × 10^{-2} cm s^{-1}, sample 3: 1.09 (±0.99) × 10^{-2} cm s^{-1}, sample 4: 1.24 (±1.52) × 10^{-2} cm s^{-1}) when employing $FcTMA^+$ as the redox couple, making it harder to discriminate between intrinsic variations in electrochemical activity between each sample. Note that the kinetics were close to the reversible limit measurable with this system, hence the relatively large error in the rate constants quoted. Representative voltammetry of the $FcTMA^+$ couple obtained from samples 1 and 3 is shown in Fig. 7.

The electrochemical behaviour observed for the $FcTMA^+$ couple, with fast electron transfer kinetics, is entirely consistent with that observed previously for this couple on carbon nanotubes.[33,41–43] To the best of our knowledge, there has only been one other study investigating this couple on graphene surfaces,[35] in this case using CVD graphene. The kinetics of $FcTMA^+$ electron transfer on the CVD sample were probed using an electrochemical microscopy technique with a higher mass transport rate, and therefore a higher upper bound to the range of measureable electron transfer rate constants. The authors of the earlier study reported a standard rate constant for the $FcTMA^+$ couple that is close to the one reported here for monolayer samples of CVD graphene, although the rate of electron transfer was found to increase with an increasing number of CVD layers. In a separate study, the oxidation of another water-soluble ferrocene derivative (ferrocenemethanol) has been reported to be fast, *i.e.* at the upper bound of the measurement techniques used, on both ME graphene, and on graphene samples

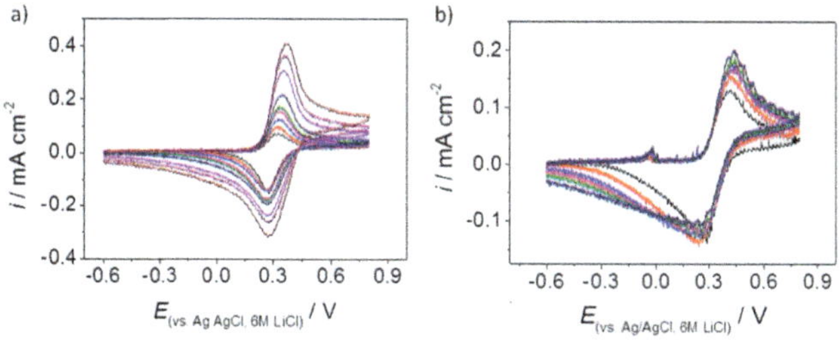

Fig. 7 Representative voltammetry obtained for the $FcTMA^+$ couple on (a) sample 1 (default solvent wash), and (b) sample 3 (vacuum annealed). Scan rates: 0.1, 0.15, 0.2, 0.25, 0.3, 0.4, 0.6, 0.8 and 1.0 V s^{-1}. Note the difference in current density for sample 3, which is half of that of sample 1. This is most likely due to a droplet being placed over a crack in the flake that is not visible using the optical microscope (crack size < 1 μm).

derived from chemical vapour deposition (CVD).[44] As an outer-sphere redox couple, the $FcTMA^+$ response would be expected to be independent of surface state, and only sensitive to the level of sample doping, however given the fast exchange kinetics for this couple, it seems that any changes in doping due to the different cleaning regimes adopted herein do not significantly affect the measured voltammetry for this molecule.

By contrast, the data in Fig. 8 shows that the kinetics of $Ru(NH_3)_6^{3+/2+}$ reduction/re-oxidation were substantially slower on the graphene samples (sample 1: 2.87 (±2.90) × 10^{-5} cm s^{-1}, sample 2: 1.46 (±0.69) × 10^{-7} cm s^{-1}, sample 3: 3.60 (±4.79) × 10^{-6} cm s^{-1}, sample 4: 4.42 (±3.07) × 10^{-7} cm s^{-1}); these values are somewhat slower than the behaviour reported previously for this couple on the basal plane of graphite. Quasi-reversible kinetics with a relatively slow standard electron transfer rate of 1 × 10^{-3} cm s^{-1}, have been reported for $Ru(NH_3)_6^{3+/2+}$ on basal plane highly oriented pyrolytic graphite.[45,46] The kinetics of this redox couple have also been investigated on electrodes modified with commercial CVD graphene samples, although it has been implicitly assumed that the basal plane of the graphene is inactive with respect to electron transfer in this case.[47] The slower electron transfer kinetics for $Ru(NH_3)_6^{3+/2+}$ relative to the reported values for basal plane graphite also, in contrast to the $FcTMA^+$ case, manifest themselves in a slight variation of the standard rate constant with sample pre-treatment, with sample 2 displaying slower kinetics than the other wet cleaned sample, sample 1.

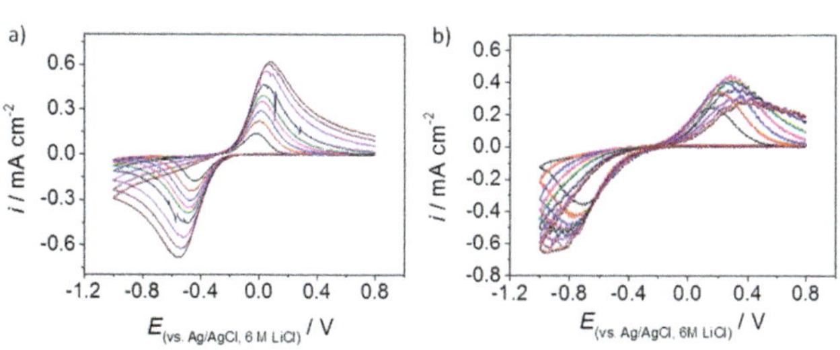

Fig. 8 Representative voltammetry obtained for the $Ru(NH_3)_6^{3+}$ couple on (a) sample 1 (default solvent wash), and (b) sample 3 (vacuum annealed). Scan rates: 0.1, 0.15, 0.2, 0.25, 0.3, 0.4, 0.6, 0.8 and 1.0 V s^{-1}.

One explanation for this is that the doping seen with sample 2 (see Fig. 4) has shifted the Fermi level for this sample into a region with a lower density of states, although why sample 2 should show slower kinetics for the $Ru(NH_3)_6^{3+/2+}$ case, and not for the $FcTMA^+$ case, is not immediately clear.

Conclusion

Measurable electron transfer kinetics are observed on graphene samples for both mediators in this study, which indicates that the basal plane of this material is electrochemically active. With regard to the specific aim of the work, the main finding is a rather limited sensitivity of the electron transfer kinetics to the specific pre-treatment regime of the graphene electrode. Whereas the presence of polymer and solvent residue, and the associated changes in doping, have been shown to have a substantial effect on the transport properties of graphene samples,[18] the electron transfer data appears to be relatively insensitive to the state of the sample surface, at least for the case of the most common contaminant encountered with exfoliated graphene, namely PMMA residue from the transfer process. As well as the variations in doping induced by the different sample treatments, one might expect the different residual levels of adsorbates to influence the rate of electron transfer, not least by blocking significant fractions of the sample surface (see Fig. 3, 5 and 6), given that the deleterious effects of surfactants on the electrochemical response of CVD graphene samples have been noted previously,[48] as has the effect of sample "ageing" on the response of graphite electrodes.[49] However such variations are not large for the sample treatments attempted here.

Acknowledgements

We thank the EPSRC (grants EP/I005145/1, EP/K039547/1 and EP/K016954/1) for financial support and Dr M Bissett for useful discussions.

References

1 Z. Bo, Z. Wen, H. Kim, G. Lu, K. Yu and J. Chen, *Carbon*, 2012, **50**, 4379.
2 M. F. El-Kady, V. Strong, S. Dubin and R. B. Kaner, *Science*, 2012, **335**, 1326.
3 M. M. Hantel, T. Kaspar, R. Nesper, A. Wokaun and R. Kötz, *Electrochem. Commun.*, 2011, **13**, 90.
4 J. Liu, M. Notarianni, G. Will, V. T. Tiong, H. Wang and N. Motta, *Langmuir*, 2013, **29**, 13307.
5 D. Chen, H. Zhang, Y. Liu and J. H. Li, *Energy Environ. Sci.*, 2013, **6**, 1362.
6 N. G. Sahoo, Y. Z. Pan, L. Li and S. H. Chan, *Adv. Mater.*, 2012, **24**, 4203.
7 H. J. Choi, S. M. Jung, J. M. Seo, D. W. Chang, L. M. Dai and J. B. Baek, *Nano Energy*, 2012, **1**, 534.
8 G. Kucinskis, G. Bajars and J. Kleperis, *J. Power Sources*, 2013, **240**, 66.
9 L. Wang, A. Ambrosi and M. Pumera, *Angew. Chem., Int. Ed.*, 2013, **52**, 13818.
10 J. Xia, F. Chen, J. Li and N. Tao, *Nat. Nanotechnol.*, 2009, **4**, 505.
11 M. D. Stoller, C. W. Magnuson, Y. Zhu, S. Murali, J. W. Suk, R. Piner and R. S. Ruoff, *Energy Environ. Sci.*, 2011, **4**, 4685.

12 X. Li, W. Cai, J. An, S. Kim, J. Nah, D. Yang, R. Piner, A. Velamakanni, I. Jung, E. Tutuc, S. K. Banerjee, L. Colombo and R. S. Ruoff, *Science*, 2009, **324**, 1312.

13 K. S. Novoselov, A. K. Geim, S. V. Morozov, D. Jiang, Y. Zhang, S. V. Dubonos, I. V. Grigorieva and A. A. Firsov, *Science*, 2004, **306**, 666.

14 D. A. C. Brownson and C. E. Banks, *Phys. Chem. Chem. Phys.*, 2012, **14**, 8264.

15 P. S. Toth, A. T. Valota, M. Velicky, I. A. Kinloch, K. S. Novoselov, E. W. Hill and R. A. W. Dryfe, *Chem. Sci.*, 2014, **5**, 582.

16 A. Das, S. Pisana, B. Chakraborty, S. Piscanec, S. K. Saha, U. V. Waghmare, K. S. Novoselov, H. R. Krishnamurthy, A. K. Geim, A. C. Ferrari and A. K. Sood, *Nat. Nanotechnol.*, 2008, **3**, 201.

17 F. Chen, Q. Qing, J. Xia, J. Li and N. Tao, *J. Am. Chem. Soc.*, 2009, **131**, 9908.

18 J. W. Suk, W. H. Lee, J. Lee, H. Chou, R. D. Piner, Y. Hao, D. Akinwande and R. S. Ruoff, *Nano Lett.*, 2013, **13**, 1462.

19 M. Her, R. Beams and L. Novotny, *Phys. Lett. A*, 2013, **377**, 1455.

20 A. Pirkle, J. Chan, A. Venugopal, D. Hinojos, C. W. Magnuson, S. McDonnell, L. Colombo, E. M. Vogel, R. S. Ruoff and R. M. Wallace, *Appl. Phys. Lett.*, 2011, **99**, 112108.

21 C.-J. Shih, Q. H. Wang, S. Lin, K.-C. Park, Z. Jin, M. S. Strano and D. Blankschtein, *Phys. Rev. Lett.*, 2012, **109**, 176101.

22 A. T. Valota, I. A. Kinloch, K. S. Novoselov, C. Casiraghi, A. Eckmann, E. W. Hill and R. A. W. Dryfe, *ACS Nano*, 2011, **5**, 8809.

23 Z. Cheng, Q. Zhou, C. Wang, Q. Li, C. Wang and Y. Fang, *Nano Lett.*, 2011, **11**, 767.

24 J. Hong, M. K. Park, E. J. Lee, D. Lee, D. S. Hwang and S. Ryu, *Sci. Rep.*, 2013, **3**, 2700.

25 R. Zan, C. Muryn, U. Bangert, P. Mattocks, P. Wincott, D. Vaughan, X. Li, L. Colombo, R. S. Ruoff, B. Hamilton and K. S. Novoselov, *Nanoscale*, 2012, **4**, 3065.

26 Z. H. Ni, H. M. Wang, Z. Q. Luo, Y. Y. Wang, T. Yu, Y. H. Wu and Z. X. Shen, *J. Raman Spectrosc.*, 2010, **41**, 479.

27 X. Wang, L. Jinjin, Z. Qing, Z. Yuan and Z. Mengke, *Key Eng. Mater.*, 2013, **562–565**, 85.

28 Y. Huang, X. Dong, Y. Shi, C. M. Li, L.-J. Li and P. Chen, *Nanoscale*, 2010, **2**, 1485.

29 A. M. Goossens, V. E. Calado, A. Barreiro, K. Watanabe, T. Taniguchi and L. M. K. Vandersypen, *Appl. Phys. Lett.*, 2012, **100**, 073110.

30 R. Raj, S. C. Maroo and E. N. Wang, *Nano Lett.*, 2013, **13**, 1509.

31 Z. Li, Y. Wang, A. Kozbial, G. Shenoy, F. Zhou, R. McGinley, P. Ireland, B. Morganstein, A. Kunkel, S. P. Surwade, L. Li and H. Liu, *Nat. Mater.*, 2013, **12**, 925.

32 J. L. Conyers and H. S. White, *Anal. Chem.*, 2000, **72**, 4441.

33 I. Dumitrescu, P. R. Unwin, N. R. Wilson and J. V. Macpherson, *Anal. Chem.*, 2008, **80**, 3598.

34 M. N. Szentirmay and C. R. Martin, *Anal. Chem.*, 1984, **56**, 1898.

35 A. G. Güell, N. Ebejer, M. E. Snowden, J. V. Macpherson and P. R. Unwin, *J. Am. Chem. Soc.*, 2012, **134**, 7258.

36 A. T. Valota, P. S. Toth, Y.-J. Kim, B. H. Hong, I. A. Kinloch, K. S. Novoselov, E. W. Hill and R. A. W. Dryfe, *Electrochim. Acta*, 2013, **110**, 9.

37 D. C. Elias, R. R. Nair, T. M. G. Mohiuddin, S. V. Morozov, P. Blake, M. P. Halsall, A. C. Ferrari, D. W. Boukhvalov, M. I. Katsnelson, A. K. Geim and K. S. Novoselov, *Science*, 2009, **323**, 610.

38 R. S. Nicholson, *Anal. Chem.*, 1965, **37**, 1351.

39 I. Lavagnini, R. Antiochia and F. Magno, *Electroanalysis*, 2004, **16**, 505.

40 R. J. Klingler and J. K. Kochi, *J. Phys. Chem.*, 1981, **85**, 1731.

41 J. Kim, H. Xiong, M. Hofmann, J. Kong and S. Amemiya, *Anal. Chem.*, 2010, **82**, 1605.

42 I. Heller, J. Kong, H. A. Heering, K. A. Williams, S. G. Lemay and C. Dekker, *Nano Lett.*, 2004, **5**, 137.

43 T. S. Miller, N. Ebejer, A. G. Guell, J. V. Macpherson and P. R. Unwin, *Chem. Commun.*, 2012, **48**, 7435.

44 W. Li, C. Tan, M. A. Lowe, H. D. Abruña and D. C. Ralph, *ACS Nano*, 2011, **5**, 2264.

45 K. R. Kneten and R. L. McCreery, *Anal. Chem.*, 1992, **64**, 2518.

46 K. K. Cline, M. T. McDermott and R. L. McCreery, *J. Phys. Chem.*, 1994, **98**, 5314.

47 D. A. C. Brownson, L. J. Munro, D. K. Kampouris and C. E. Banks, *RSC Adv.*, 2011, **1**, 978.

48 D. A. C. Brownson, J. P. Metters, D. K. Kampouris and C. E. Banks, *Electroanalysis*, 2011, **23**, 894.

49 A. N. Patel, M. G. Collignon, M. A. O'Connell, W. O. Y. Hung, K. McKelvey, J. V. Macpherson and P. R. Unwin, *J. Am. Chem. Soc.*, 2012, **134**, 20117.

Faraday Discussions

PAPER

Controlled covalent modification of epitaxial single layer graphene on 6H-SiC (0001) with aryliodonium salts using electrochemical methods†

K. J. Stevenson,*[a] P. A. Veneman,[a] R. I. Gearba,[a] K. M. Mueller,[a] B. J. Holliday,[a] T. Ohta[b] and C. K. Chan[b]

Received 10th March 2014, Accepted 6th May 2014
DOI: 10.1039/c4fd00038b

Controlled covalent modification of graphene *via* electrochemically assisted grafting of molecules is expected to be a robust method for tuning the doping levels and work function and therefore enabling the deployment of graphene in photovoltaic and battery applications. By using aryliodonium salts, in particular, bis(4-nitrophenyl) iodonium tetrafluoroborate, the grafting density can be adjusted from 4×10^{13} to 3×10^{14} molecules per cm^2. New insights on the grafting mechanism and the reactivity of the graphene are reported here. Clean basal planes were found to have increased reactivity compared to atomic-level point defects (single vacancies). High resolution scanning tunnelling microscopy (STM) shows that some of the grafts present three-fold symmetry, which may indicate that the grafts are pairs of molecules. The point of attachment of the second molecule is under investigation using computational work which includes simulations of the STM images. Elongated as well as extended grafts (larger than 4 nm) are also observed.

Introduction

Properties such as ballistic electron transport[1] with carrier mobilities as high as 10^4 cm^2 V^{-1} s^{-1} make graphene a forerunner candidate for electronic applications such as transistors. However, graphene is a semimetallic "gapless" material and many efforts in the scientific community have been focussed on finding ways to modify the graphene's band structure (*e.g.* create a bandgap) while maintaining the ballistic transport properties. The confinement of the carriers in one dimension (1D) by nanostructuring graphene into nanoribbons has been proposed to open a bandgap. Recently, Baringhaus *et al.*[2] showed that ballistic

[a]*Energy Frontier Research Center & Department of Chemistry, The University of Texas at Austin, Texas, USA. E-mail: stevenson@cm.utexas.edu*
[b]*Sandia National Laboratories, Albuquerque, New Mexico, USA*

† CCDC 990680. For crystallographic data in CIF or other electronic format see DOI: 10.1039/c4fd00038b

transport over micrometer distances in 40 nm wide graphene nanoribbons can occur accompanied by the opening of a transport bandgap which is important for fast switching. Tuning of the electronic properties of graphene also is important for non-transistor applications. In particular, owing to its broadband optical transparency,[3] graphene holds promise to replace the expensive transparent conductors such as indium-doped tin oxide electrodes in photovoltaics. Graphene is also emerging as a high performance electronic conductor for energy storage and conversion.[4] To this end the electronic properties and surface chemistry of graphene needs to be tuned, while maintaining its remarkable conductivity and transparency. Chemical modification of graphene by both non-covalent and covalent routes has been explored. The non-covalent modification of graphene usually involves the adsorption of charge transfer compounds such as organic molecules or atoms onto the surface.[5–7] This approach gives tunability over the bandgap and carrier concentration and has the advantage of maintaining the integrity of the graphene structure and, therefore, its high electrical conductivity. However, non-covalent interactions (van der Waals, electrostatic or π–π stacking) are weak leading to unstable systems, especially when other device processing steps follow as is the case for photovoltaic devices. On the other hand, covalent modifications disrupt the graphene structure and therefore can compromise, to some extent, the electronic conductivity. Judiciously controlled covalent modifications can, in principle, balance the induced local disruption of the conjugation. It should be noted that using such methods the work function will be modified only locally such that the high conductivity of intrinsic graphene will channel charge carriers (electron transport) down to these local hot spots.

Diazonium chemistry is one of the most commonly used methods to covalently functionalize graphene. The spontaneous dissociation[8–12] or electrochemical reduction of diazonium salts[13] is followed by the covalent bonding of the resulting radical with a carbon atom in the graphene lattice. The induced disorder in the graphene structure by conversion of the C-atoms from sp^2 to sp^3 hybridization naturally induces the formation of a bandgap. The grafting of nitrophenyl moieties has been shown to open a $\sim$380 meV optical bandgap[14] and calculations indicate that high coverages (equivalent to two molecules per 8C-atoms) have the potential to open a $\sim$2 eV gap. Farmer *et al.*[15] observed that diazonium salts are stable p-type dopants for graphene, while preserving the charge carrier transport in graphene ribbons. Moreover, the authors achieved controlled doping by varying the salt concentration. In a separate study, Paulus *et al.*[16] showed that graphene doping can be tuned by covalently attaching diazonium salts with different functionalities. Surprisingly, transmission electron microscopy studies showed that under specific reaction conditions the nitrophenyl grafting preserves the hexagonal symmetry of the graphene lattice. The lattice constant was doubled compared to the pristine graphene.[17] These results show the potential for precise tuning of the electronic properties of graphene by covalent attachment of molecules.

However much debate exists in the literature over the reaction mechanism[18] as well as the chemical reactivity of graphene.[19,16] It has been argued that single layer graphene (SLG) is more reactive than bilayer (double-layer) graphene (DLG) or multilayer graphene, which was explained by the existence of electron–hole puddles.[17,20] This finding was also supported by the work of Koehler *et al.*[21] who measured the electron transfer rate constants (k) for the reaction with diazonium salts to be 0.195 and 0.044 min^{-1} for SLG and DLG, respectively. The same authors have shown that the reactivity at the step edges is even higher ($k = 0.241$ min^{-1}). In addition, other parameters such as doping, band structure near the Fermi level, and defects[18] have been discussed as possible parameters that affect the graphene reactivity. In particular, there is an unsettled discussion over the reactivity of the basal plane *versus* step edges or defects in the basal plane. Studies of electron transfer (ET) kinetics performed on highly oriented pyrolytic graphite (HOPG) and glassy carbon (GC) by employing a variety of redox systems are used as a reference to understand graphene behaviour. The traditional view is that the basal plane of HOPG has very little to no activity[22–24] with the step edges presenting most of the electroactive sites.[25–28] However, recent studies show that the basal plane has significant activity.[29–31] The discrepancies in the results are explained on one hand by the use of tools with low spatial resolution to quantify the density of step edges and basal plane defects. On the other hand, the reactivity of the basal plane is shown to depreciate (foul) over time therefore questioning the validity of some of the studies.[29]

Herein we present a model single layer graphene system for examining the controlled covalent modification of graphene electrodes epitaxially grown on SiC. Functionalization is achieved using aryliodonium salts[32] which, by contrast with their diazonium counterparts, have a more negative reduction potential thereby reducing spontaneous functionalization.[33–35] The modified graphene samples are characterized using ensemble electrochemical techniques as well as highly localized methods such as high resolution Scanning Tunnelling Microscopy (STM).

Results and discussion

In our previous study[32] graphene functionalization was performed by electrochemical reduction of bis(4-trifluoromethyl-phenyl)iodonium tetrafluoroborate. In the present work, the trifluoromethyl functionality is replaced with a redox-active nitro group ($-NO_2$) that offers the advantage of being probed directly using electrochemical methods to quantify the grafting density. Prior to functionalizing the graphene samples, grafting studies have been performed on glassy carbon (GC) as a more assessable control system. Both symmetric bis(4-nitrophenyl)-iodonium tetrafluoroborate (**BNPI**) and asymmetric (4-nitrophenyl)phenyl-iodonium tetrafluoroborate (**NPPI**) salts are used in this study as an intimate probe to the radical attachment chemistry. The use of an asymmetric salt offers the possibility to graft two different molecules at the same time. Understanding the grafting mechanism at the molecular level is important for designing future systems involving grafting of both donor and acceptor molecules. The typical grafting profiles of these salts on GC electrodes are shown in Fig. 1. The first sweep of the cyclic voltammogram shows a faradaic response which is attributed to the electrochemical reduction of the iodonium salt and the subsequent grafting of the aryl radicals to the surface of the electrode. The reduction of **BNPI** occurs at a more positive potential compared to **NPPI**. This observation can be easily rationalized by considering the more electron deficient nature of the **BNPI** due to the second nitro functionality which facilitates electron transfer to electrogenerate a radical anion. Upon the second sweep, a lower current is observed as well as a shift of the reduction potential of the iodonium salts to more negative

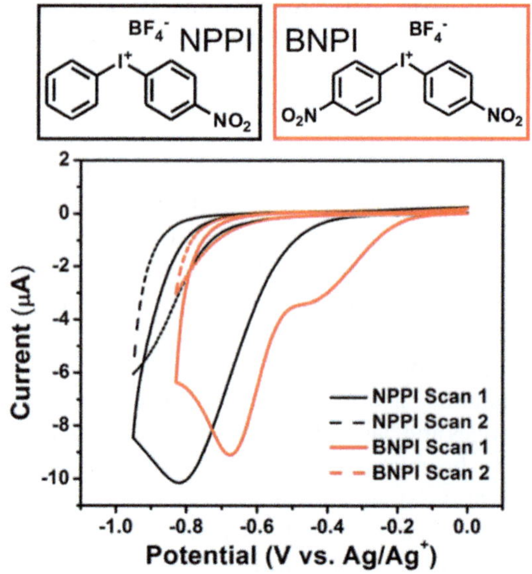

Fig. 1 Representative cyclic voltammograms for the grafting of **NPPI** (black) and **BNPI** (red) to a glassy carbon electrode from 0.5 mM solutions in CH_3CN.

potentials. This response is consistent with the typical electrografting profile of iodonium salts throughout the literature and the passivation of the surface of the electrode with an insulating film.[34,35] Following grafting, it is possible to estimate the surface coverage (grafting density) using Faraday's law since the attached *p*-nitrophenyl moiety exhibits a reversible one-electron redox response.

Fig. 2 shows the voltammetric response of the GC electrode after the electrochemically assisted grafting of **BNPI** and **NPPI**. Integration of the cathodic peak that occurs at *ca.* −1.5 V *vs.* Ag/Ag$^+$ allows for the determination of the total charge passed during the sweep which can be converted into surface coverage, Γ, (molecules per cm^2) by use of the following equation:

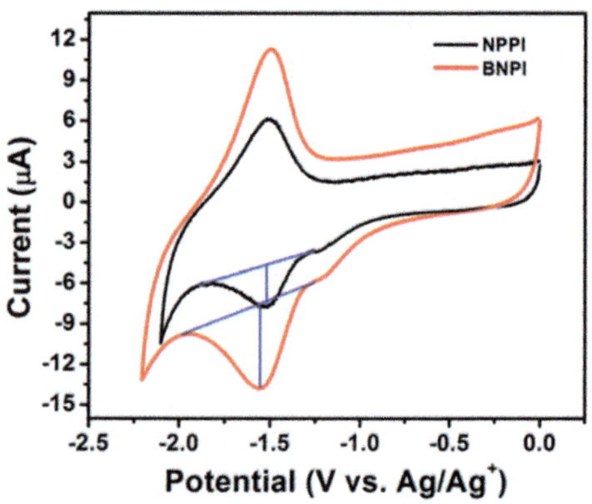

Fig. 2 Representative voltammetric responses for the reversible one electron transfer to the *p*-nitrophenyl surface functionalities on the surface of a glassy carbon electrode as the result of grafting from a 0.5 mM solution of **NPPI** (black) and 0.5 mM solution of **BNPI** (red).

$$\Gamma = \frac{QN_A}{n\,FA}$$

where Q is the total charge passed through the cell as determined by integration of the cathodic peak (reduction of the $-NO_2$), N_A is Avogadro's number, n is the number of electrons passed during the redox event ($n = 1$ in this case), F is Faraday's constant, and A is the surface area of the electrode expressed in cm^2. The geometric electrode area is determined using ferrocene by employing the Randles–Sevcik equation.[36]

Asymmetric diaryliodonium salts provide a unique opportunity to graft two different aryl species to the surface of an electrode starting from a single precursor. However, the ratio of the two aromatic moieties that are grafted to the electrode surface is an important detail that needs to be controlled in order to appropriately control the properties of the functionalized surface. Previous studies have examined into how the electronic nature of the substituent's on the aromatic systems influences the dissociation of the iodonium salts and their subsequent grafting.[34] Computational and experimental studies by Fontanesi *et al.*[37] have shown that the dissociation of alkynyl(aryl)iodonium salts is largely influenced by the electronic density of the aryl group substituents in such a way that electron withdrawing functionalities favour the grafting of the aryl ring *versus* the alkynyl group by destabilizing the C_{aryl}–I bond leading to the formation of the aryl radical bearing the withdrawing moiety.[37,38] A similar conclusion was reached for diaryliodonium salts by Vase *et al.* who showed that ~80% of the grafted species contain the nitro functionality when using the asymmetric (3-nitrophenyl)-phenyliodonium hydrogensulfate salt (compared to symmetrical bis(3-nitrophenyl)iodonium hydrogensulfate salt).[34]

Similar results are obtained under high functionalization conditions when using the **BNPI** and **NPPI** salts. Electrochemical grafting of **BNPI** and **NPPI** was accomplished as the result of 10 potential sweeps at 50 mV s^{-1} from 0 to −0.85 V and 0 to −0.95 V, respectively. In both cases, the concentration of the salt solution was 1 mM. These results, as determined by an average for five different glassy carbon electrodes for each salt, showed a coverage of $(3.0 \pm 0.6) \times 10^{14}$ molecules per cm^2 for **BNPI** while a coverage of $(2.7 \pm 0.4) \times 10^{14}$ molecules per cm^2 was found for **NPPI**. This equates to a grafting of ~90% of the nitro functionalized aryl ring when using the asymmetric **NPPI** salt. However, while exploring lower coverages, a reduced ratio of grafted *p*-nitrophenyl to phenyl groups was observed indicating effects other than the electronic nature of the diaryliodonium salts may be influencing the final composition of the organic film at high surface concentrations. Control over the degree of functionalization is obtained using cyclic voltammetry by varying the salt concentration and cycled the most negative potential (negative vertex) of the CV sweep.

At coverages less than 2.2×10^{14} molecules per cm^2 (as determined by integration of the cathodic reduction peak of the *p*-nitrophenyl moiety), linear control of the surface concentration is observed for **BNPI** by varying the concentrations of the iodonium salt in solution prior to grafting (see Fig. 3). A plateau effect was observed for coverages above 2.2×10^{14} molecules per cm^2 (corresponding to grafting solutions >0.80 mM). Similarly, the deposition of *p*-nitrophenyl groups from **NPPI** gave a linear increase in surface concentration by increasing the concentration of **NPPI** in solution before grafting. The decreased slope of the

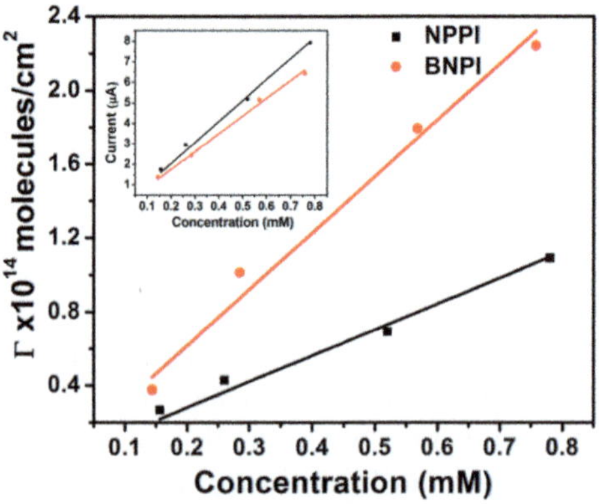

Fig. 3 Calibration curve showing the control of surface coverage using **NPPI** (black, slope $= 1.5 \times 10^{17}$ molecules per cm^2 per mol) and **BNPI** (red, slope $= 3.1 \times 10^{17}$ molecules per cm^2 per mol) by varying the concentration of the diaryliodonium grafting solution. Inset shows peak currents observed during grafting scans *versus* concentration of the iodonium salts.

NPPI calibration curve is expected given the competing pathway for phenyl grafting, however; the slope of the curve is nearly half that of the slope for the **BNPI** calibration curve (despite similar slopes for $i_{p,c}$ during grafting) indicating very little electronic influence on the grafting ratio as a result of the *p*-nitrophenyl moiety.

At first inspection, these results seem to contradict the ~90% grafting ratio observed under high grafting conditions when using electronic bond dissociation arguments alone as previously described for the alkynyl(aryl)iodonium salts. However, we have found this result is consistent with the results when considering the layer-by-layer deposition of the aryl functionalities on the surface of the electrode.

As depicted in Fig. 4, deposition of a monolayer of *p*-nitrophenyl groups from **BNPI** leaves the surface covered in aryl moieties that only contain electronically deactivated positions (denoted D) sterically open for the further deposition of aryl groups. This would hinder the formation of multilayers and decrease the total concentration of *p*-nitrophenyl groups on the surface of the electrode. However, for the asymmetric **NPPI** salt, the deposition of phenyl groups (alongside the *p*-nitrophenyl groups) leaves positions on the phenyl

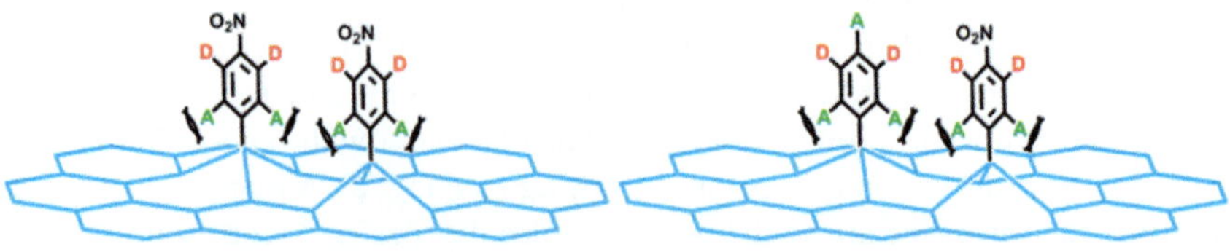

Fig. 4 Diagram showing the electronic and steric effects imposed on the first aryl layer when the glassy carbon electrode is modified with pure *p*-nitrophenyl functionalities from **BNPI** grafting (left) *versus* the mixed coverage obtained *via* grafting of the asymmetric **NPPI** (right). (A) Denotes activated positions while (D) stands for deactivated positions.

ring that are activated towards further grafting allowing for multilayer coverage to grow more efficiently which would lead to the appearance of a higher grafting ratio of *p*-nitrophenyl/phenyl. It is therefore suspected that the effect of the substituent on the cleavage of the iodonium salt is more pronounced for the cleavage of the alkynyl(aryl)iodonium salts than for the class of diaryliodonium salts studied here.

It should be noted that the observed ~50% grafting ratio of *p*-nitrophenyl to phenyl from the asymmetric **NPPI** salt could be a consequence of the more negative potential required for the reduction of the **NPPI** ($i_{p,c} = -820$ mV) as compared to **BNPI** ($i_{p,c} = -680$ mV). The more negative reduction potential increases the likelihood of reducing the aryl radicals further to their anionic form. The conversion of the aryl radicals to the corresponding anions should be more efficient for the electron deficient system containing the nitro functionality.

Therefore, should it happen that cleavage of the electron deficient ring is favoured, the observed decreased grafting efficiency may be attributed to the increased conversion of the *p*-nitrophenyl radical to the anionic form which is inactive towards grafting at the glassy carbon surface.

Further support that the electronic effects have less influence than previously thought can be found by inspection of the reported crystal structures for diaryliodonium salts. It is suspected that the electronic effects play a significant role in weakening the I–C_{aryl} bond (thus leaving it prone to cleavage upon reduction). Support for this hypothesis should be shown in the crystal structures as a lengthened bond between the iodine and the carbon from the aryl ring containing the electron withdrawing groups. Fig. 5a shows the distribution of the I–C_{aryl} bonds reported in the literature.[39]

The bonds fall within relatively narrow range between 2.018 Å and 2.174 Å and have an average bond length centered at 2.09 ± 0.02 Å. However, the distribution of bond lengths is much less meaningful than comparing the differences in I–C_{aryl} bond lengths within individual diaryliodonium salts. Fig. 5b shows a histogram displaying the variations of the differences in bond lengths (ΔBL) between two inequivalent I–C_{aryl} bonds within specific molecules calculated using the following equation:

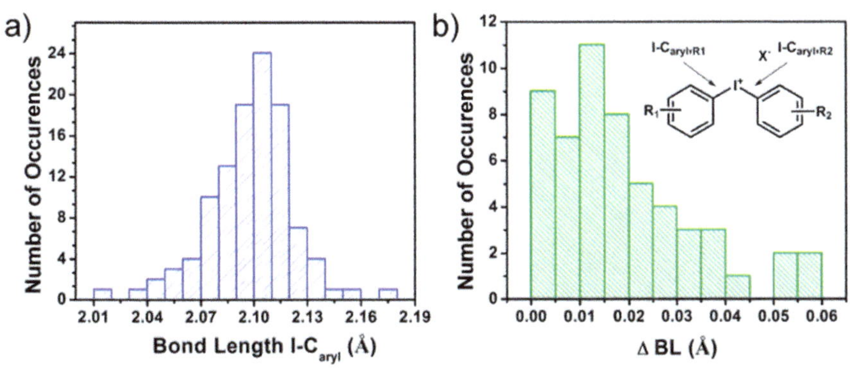

Fig. 5 (a) Histogram of bond lengths from reported crystal structures of diaryliodonium salts of the generic form shown in the inset to (b). The distribution of the differences of the I–C_{aryl,R_1} and I–C_{aryl,R_2} bond lengths of the iodonium salts are shown in (b).

$$\Delta BL = |(I\text{–}C_{aryl,R_1}) - (I\text{–}C_{aryl,R_2})|.$$

We find that 80% of the reported structures display a ΔBL of 0.03 Å or less, which indicates very little discrepancy between the different aryl groups. One would expect a much greater variation for significantly differing electronic functionalities with the rings containing electron withdrawing groups displaying longer bond lengths. However, examination of the crystal structures reveals the opposite trend. Crystal structures containing significantly stronger electron withdrawing groups on one aryl ring are observed to have equivalent or even shorter bond lengths than the opposing aryl ring indicating a stronger I–C_{aryl} bond for the electron deficient moiety.[40–43] However, the bond lengths still fall within a very narrow range which supports the idea that a synergistic effect is observed around the $I^{(III)}$ center where the total electronic nature of the salt is dispersed across both of the I–C_{aryl} bonds rather than a localized weakening.

In order to determine if these trends in the literature hold true for the asymmetric system presented here, X-ray quality crystals of **NPPI** were grown from a saturated solution in a DCM–MeOH–Et_2O (1 : 10 : 100) mixture. Additional details of X-ray structure determination and parameters may be found in the ESI.† The bent geometry of **NPPI** is consistent with that observed in the literature with a C1–I1–C7 bond angle of 94.15° (Fig. 6). Furthermore, the aryl ring containing the strongly electron withdrawing nitro functionality shows a bond length of I1–C7 of 2.111(5) Å, which is slightly shorter compared to the bond length of I1–C1 = 2.117(4) Å measured for the electronically neutral phenyl ring, consistent with the trends presented above. However, the overlap in the standard deviation of the two bond lengths makes them statistically equivalent, which is consistent with the likelihood of equal grafting ratio observed at low concentrations before any multilayer effects would come into play.

Graphene samples epitaxially grown on SiC were functionalized using the same method described for the GC electrodes. The graphene was epitaxially grown on 6H-SiC (0001) using previously described procedures.[44] Before

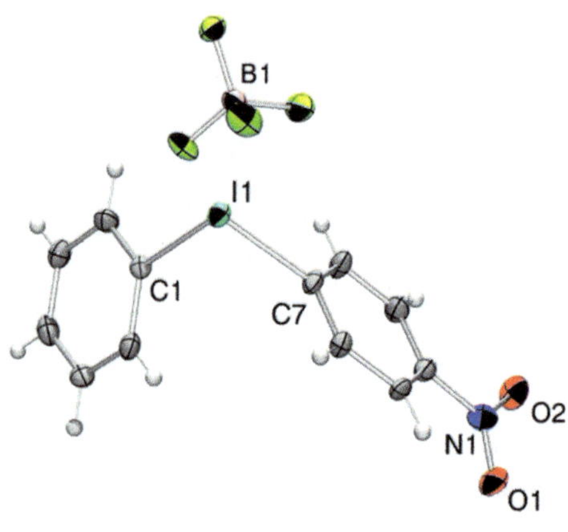

Fig. 6 ORTEP diagram of the single X-ray crystal structure of **NPPI**. Thermal ellipsoids are drawn at 50% probability. All hydrogens were modelled geometrically and refined using a riding model.

functionalization with **BNPI**, the different graphene samples were characterized using STM.

As reported previously graphene growth on SiC is mediated by the formation of a buffer layer (BL) or zero layer-graphene (Fig. 7a).[45] This layer consists of carbon atoms arranged in a graphene like lattice with 30% of the C-atoms covalently bound to the underlying SiC substrate. Dangling bonds in the SiC substrate negatively charge the BL and renders the graphene layer n-type doped. Although the graphene layer that grows on top of this BL is structurally nearly perfect, it is electronically coupled to the BL, which induces local variation in the density of states (Fig. 7c).[46] Fig. 7b shows a high resolution STM image of graphene on BL conductive 6H-SiC (0001). Coupling to the BL gives rise to an interference pattern characterized by the appearance of bright spots with six fold symmetry. It has been shown that decoupling of the graphene from the SiC substrate can be achieved by eliminating the Si–C bonds *via* H-intercalation.[45] Angle Resolved Photoelectron Spectroscopy measurements show that *n*-layer graphene transforms into (*n* + 1) graphene layers after H-intercalation. As predicted by theory[47]

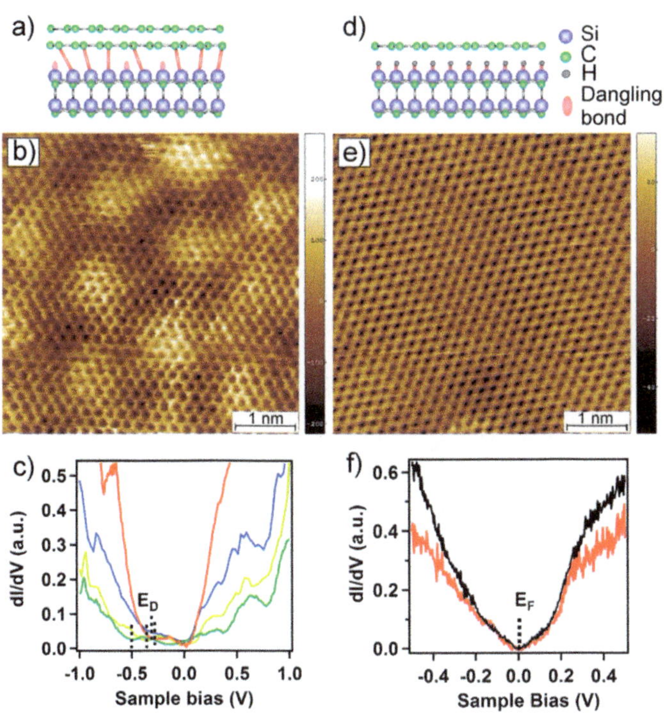

Fig. 7 (a) Schematic representation (side view) of graphene grown on SiC showing the presence of the buffer layer (BL). (b) STM image (V_{bias} = 25 mV) of the BL graphene on conductive SiC. The bright spots are the result of the electronic interference (charge transfer) with the BL. (c) d*I*/d*V* spectra collected in different locations on the BL graphene. The electronic heterogeneity introduced by the BL is reflected in the local variation of the Dirac point. Schematic representation of quasi-free standing graphene on insulating obtained by H-intercalation (d) and the corresponding STM image (e). (f) Averaged d*I*/d*V* spectra on BL-free graphene on SiC acquired in two different regions. No local variation in the density of states is observed.

and measured experimentally[45] the electronic structure of single layer graphene on the hydrogen-passivated SiC surface is minimally perturbed near the Dirac point. Fig. 7e shows the STM image of a quasi-free graphene layer on insulating SiC obtained by H-passivation.

The graphene film is extremely flat (80 pm). In contrast to the graphene on BL SiC, the local d*I*/d*V* measurements (Fig. 7f) show no variation in the density of states and confirm the intrinsic doping in the BL-free graphene. The BL-free samples are ideal for studying the electronic changes induced in graphene by molecular grafting.

Contrary to what is reported in the literature[48] atomic defects exist in graphene on H-intercalated SiC. Fig. 8a and b shows high resolution STM images of such atomic vacancies. By imaging the sample over a 7 × 7 nm grid, the average defect density (measured over approx. 30 images) was found to be $(2 \pm 0.5) \times 10^{12}$ vacancies per cm^2. This defect density is four orders of magnitude larger than the natural point defect density in graphite.[49] Note that such defects were not observed on single layer graphene formed on BL conductive SiC. The relatively low growth temperature of graphene on SiC of 1550 °C (ref. 44) (compared to the sublimation temperature of graphite at 3675 °C) facilitates the appearance of non-equilibrium defects such as vacancies. Therefore these defects are expected to be present to some extent in the BL graphene samples as well. It is unclear what causes the high defect density on the BL-free compared to BL graphene. One can

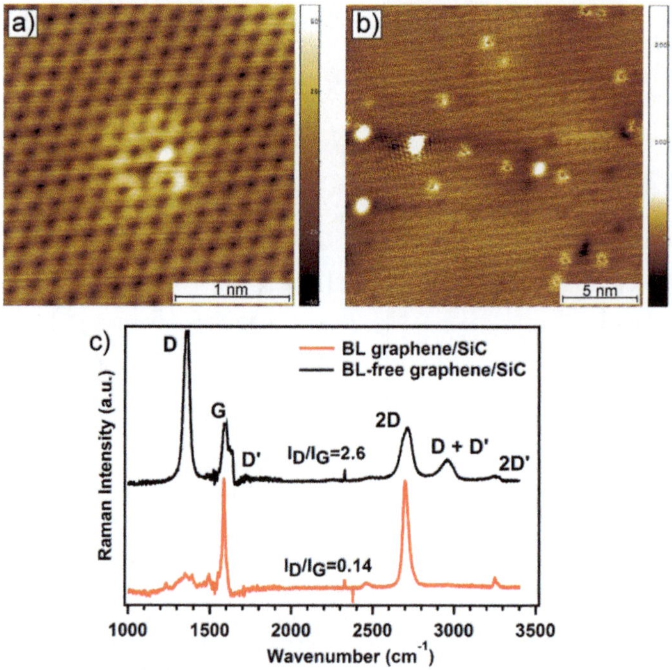

Fig. 8 High resolution (a) and larger scale (20 × 20 nm) (b) STM images (V_{bias} = 30 mV, $I_{setpoint}$ = 40 pA) of graphene on H-intercalated insulating SiC showing native atomic defects or single vacancies. (c) Raman spectra for BL and BL-free graphene samples measured at 514.5 nm. The black curve was shifted vertically for clarity.

speculate that these defects may have been introduced during the H-intercalation process. Furthermore, Raman spectroscopy was used to quantify the relative defect density corresponding to BL and BL-free graphene samples (Fig. 8c). Cançado *et al.*[50] showed that Raman spectroscopy can be used in principle to quantify the amount of point defects in graphene.

The presence of sp^3 hybridized C-atoms on the covalent graphene lattice (peak at ~1350 cm^{-1} routinely called D peak) is indicative of defects located both at the step edges or the basal plane. One can clearly see that the area of the D peak for the BL-free graphene is larger than for BL graphene. The samples used in this study are single layer graphene over approximately 80% of the surface area.[44] The rest is covered by multilayer (2–3 ML) graphene mostly at the SiC step edges. Since this is valid for both BL and BL-free one can conclude that the increased area of the D peak for the BL-free graphene is a clear indication of a significantly larger vacancy defect density. Single vacancies lead to the appearance of resonances at the Fermi level[51] accompanied by scattering of the electron waves.[51,52] One dangling bond always remains owing to geometrical reasons. Therefore one can expect higher reactivity at these point defects.

Graphene samples on BL-free conductive SiC were modified with **BNPI** using the procedure described previously for GC. Fig. 9 shows an excellent agreement between the grafting densities for **BNPI** on GC and graphene.

STM studies were performed on BL-free graphene samples on insulating SiC modified with 0.15 mM **BNPI** with a surface coverage of 4×10^{13} molecules per cm^2.

The large scale STM image in Fig. 10a shows the presence of several modifications identified as ~1 nm bright globular structures (green circles). The modifications appear as bright spots since the $-NO_2$ group is electron withdrawing. Different types of grafts have been observed. Fig. 10b shows a high resolution STM image (V_{gap} = 30 mV) of one of the most commonly observed modifications. The electronic structure of the graphene around the graft is perturbed as evidenced by the extensive scattering. The corresponding DOS map (Fig. 10c) recorded at $V_{gap} = -350$ mV two bright lobes which may indicate that the modification is actually a pair of molecules.

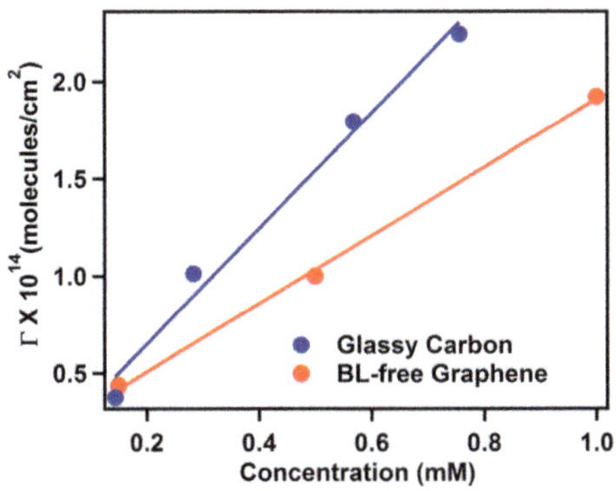

Fig. 9 Surface coverage using **BNPI** estimated from the integration of the reduction of the $-NO_2$ peak for GC and BL-free graphene.

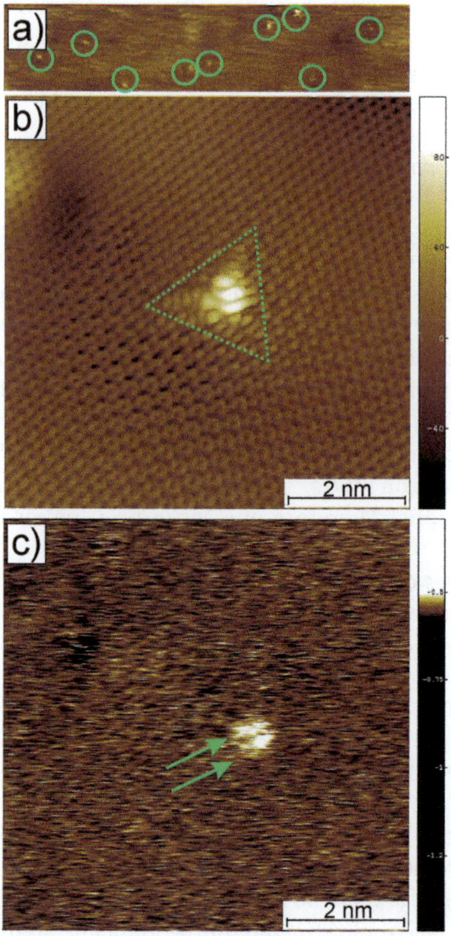

Fig. 10 (a) Large area (17 × 85 nm) STM image of **BNPI** modified BL-free graphene on insulating SiC. The modifications are identified with green circles. (b) High resolution STM image ($V_{bias} = 30$ mV, $I_{setpoint} = 40$ pA) showing a modification which displays three fold symmetry (the green triangle is a guide to the eye) and the corresponding density of states map ($V_{bias} = -350$ mV) (c).

The point of attachment of the succeeding aryl radical relative to the first is not clear. The graphene lattice has two carbon atoms per unit cell forming two sublattices denoted as A– and B–. Thermodynamic considerations[18] show that if the first attachment occurs on sublattice A– the second grafting will proceed with the higher probability in the *para* then *ortho* positions of the B sublattice in order to quench the spin. The modification appears in the STM with three-fold symmetry suggesting that the second molecule attaches on the same sublattice. It is often argued that if the reactions are kinetically driven, then grafting will occur on the same sublattice.[19] Ongoing computational work which include simulation of the STM images of the modified graphenes are expected to shed light on the attachment of the second molecule as well as on the structure of the different types of modifications. Occasionally extended modified regions (larger than 4 nm) have been observed. It is unclear what causes the appearance of such regions.

As discussed previously, the issue of whether the basal plane of graphene is more electrochemically active *versus* the plane edge and defect sites has been a pervasive question in the field of graphene modification. Where do redox-active molecules react and preferentially attach? Is it to the pristine graphene basal plane or do they preferentially decorate the defect points, step edges and various other structural flaws that might be present on the sample? As shown previously, our BL-free graphene samples have a vacancy density of the order of 10^{12} defects per cm^2. Based on the linear dependence of the grafting density with the

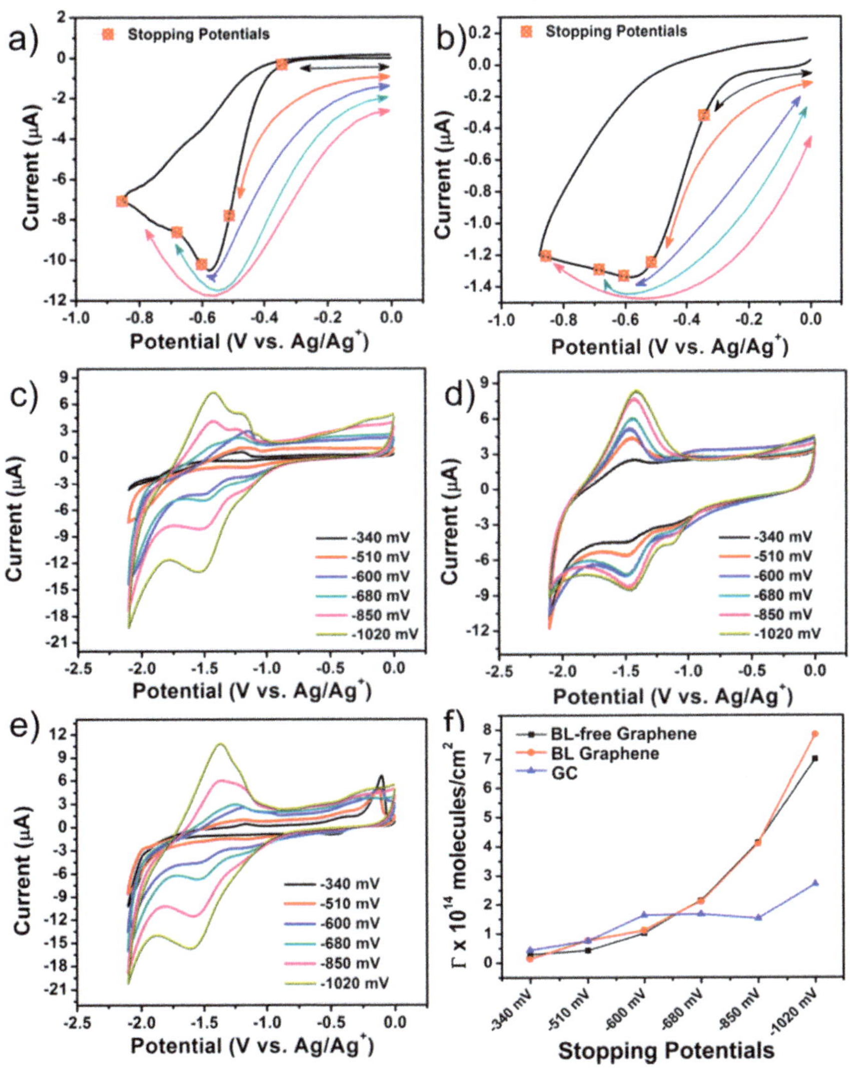

Fig. 11 Stopping potentials overlaid on the voltammogram for covalent functionalization with **BNPI** on BL-free graphene on insulating SiC (a) and glassy carbon (b). A final stopping potential at −1.02 V was also performed for each carbon substrate (point not shown in graphs (a) and (b)). The reduction of the *p*-nitrophenyl moiety after grafting to BL graphene on insulating SiC (c), GC (d) and BL-free graphene on insulating SiC (e). (f) The grafting coverage estimated for each of the stopping potentials.

concentration (Fig. 9, red curve) we argue that the vacancies are not preferentially targeted by the nitrophenyl radicals and that the reaction occurs rather at the undefective basal plane. Moreover, our STM studies on BL-free graphene clearly show regions with modifications surrounded by defects densities of the order of 10^{12} defects per cm^2.

To contribute further to answering this question, we analyzed the grafting behaviour for three samples: graphene on BL-free SiC, graphene on BL SiC and GC.

These specific samples were chosen because of the varying defect densities. As discussed previously the graphene samples have a very low number of step edges but the BL-free sample has a higher content of point defects (vacancies) as confirmed by Raman spectroscopy and STM. The GC electrode which consists of a combination of basal and edge planes should contain more defect sites than either graphene samples. The experiments were performed by sweeping twice to chosen potentials (referred to as stopping potentials). The resulting coverage was determined as shown previously by integration of the cathodic peak for nitrophenyl reduction. Grafting was performed using a 0.15 mM **BNPI** solution.

The stopping potentials for both GC and graphene samples are shown in Fig. 11a and b. An additional point was taken outside of the grafting window at −1.02 V.

Fig. 11c–e show the reduction scans for the nitrophenyl moieties grafted on BL graphene, GC and BL-free graphene and for the different stopping potentials (respectively). A strong agreement was found between the BL and BL-free graphene samples (Fig. 11f). Although more defective than any of the graphene samples the GC electrode shows similar coverages as well. These results demonstrate that although grafting at the defects site is thermodynamically favoured the reaction is most probably governed by kinetics.

Experimental

Materials

Dry solvents were prepared using an Innovative Technology, Pure Solv solvent system with double purifying column. 4-Nitrobenzeneboronic acid (Alfa Aesar), 1-iodo-4-nitrobenzene (Alfa Aesar), iodobenzene (Alfa Aesar) and *m*-chloroperoxybenzoic acid (*m*-CPBA) ≤77% (Sigma Aldrich) were used as received. The iodonium salts of bis(4-nitrophenyl)iodonium tetrafluoroborate (**BNPI**)[53] and (4-nitrophenyl)phenyliodonium tetrafluoroborate (**NPPI**)[54] were prepared using methods described by Olofsson *et al.* and confirmed by NMR using a Varian DirectDrive 400 (^{1}H 400 MHz, ^{13}C{^{1}H} 100 MHz, ^{19}F 376 MHz) referenced to residual solvent peaks (^{1}H and ^{13}C{^{1}H}) and $CFCl_3$ (^{19}F). Further confirmation was obtained by mass spectrometry using an Agilent 6530 quadrupole time of flight mass spectrometer introduced *via* electrospray ionization. The iodonium salts were stored under a nitrogen blanket, protected from light and kept at 0 °C.

Crystal structure determination for NPPI

X-ray experimental for **NPPI**: crystals of **NPPI** were grown by cooling a saturated solution of **NPPI** in a DCM–MeOH–Et_2O (1 : 10 : 100) to 0 C. The data were collected on a Rigaku AFC12 diffractometer with a Saturn 724 + CCD using a graphite monochromator with MoKα radiation (λ = 0.71073 Å). The data were collected at 153 K using an Oxford Cryostream low temperature device. Data reduction was performed using the Rigaku Americas Corporation's Crystal Clear version 1.40. The structure was solved by direct methods using SIR97 [54] and refined by full-matrix least-squares on F^2 with anisotropic displacement parameters for the non-H atoms using SHELXL-97.[56] Structure analysis was aided by use of the programs PLATON98 [57] and WinGX.[58,59] The hydrogen atoms on carbon were calculated in ideal positions with isotropic displacement parameters set to $1.2 \times U_{eq}$. of the attached atom ($1.5 \times U_{eq}$ for methyl hydrogen atoms). The hydrogen atoms bound to nitrogen were located in a ΔF map and refined with isotropic displacement parameters.

The data was checked for secondary extinction effects but no correction was necessary. Neutral atom scattering factors and values used to calculate the linear absorption coefficient are from the International Tables for X-ray Crystallography (1992).[60]

Methods

The electrochemistry experiments were performed in a nitrogen atmosphere glovebox using previously dried solvents. The supporting electrolyte, *n*-tetrabutylammonium hexafluorophosphate ($[(n\text{-Bu})_4N][PF_6]$) (Oakwood), was purified *via* recrystallization three times from boiling ethanol and then dried for 72 hours under active vacuum at 100 °C. Experiments were performed using a GPES system from Eco. Chemie. B. V. A three electrode cell was used to carry out the electrochemical studies using different carbon substrates (glassy carbon or graphene films on SiC) as the working electrodes, a platinum wire counter electrode and a $Ag/AgNO_3$ reference electrode (silver wire dipped in a 0.01 M silver nitrate solution in CH_3CN containing 0.1 M $[(n\text{-Bu})_4N][PF_6]$). Reference electrode was checked prior to use against the well-established ferrocene redox couple to ensure a good reference.

The covalent modification of the glassy carbon electrodes using **BNPI** and **NPPI** to generate the calibration curves were accomplished as follows. All glassy carbon electrodes were polished prior to grafting using 0.3 μm and 0.05 μm Al_2O_3 suspensions in distilled water and then rinsed with distilled water and acetone. The diameter of the glassy carbon button was 3.0 mm. Blank scans from 0 to −2.1 V were performed to ensure no signals from the *p*-nitrophenyl groups were observed indicating a clean electrode surface. Stock solutions of the desired iodonium salt were prepared in CH_3CN containing 0.1 M $[(n\text{-Bu})_4N][PF_6]$. All grafts were performed *via* cyclic voltammetry using two cathodic sweeps at 50 mV s^{-1} unless otherwise noted. The potential sweeps used to accomplish the grafting were from 0 to −0.85 V for **BNPI** and 0 to −0.95 V for **NPPI**. The more negative sweep width for **NPPI** is used due to its lower reduction potential. In general, the most cathodic potential used during grafting sweeps is ~150 mV past the peak reduction potential. Following grafting, the electrode is removed from the grafting solutions, washed with copious amounts of CH_3CN and transferred to a blank solution containing only electrolyte. To generate the data points for the calibration curve, the surface coverages were determined as follows. The *p*-nitrophenyl redox events are observed by two successive sweeps from 0 to −2.1 V at 1.00 V s^{-1}. The second scan is integrated to obtain the charge transferred during the redox event in order to achieve better reproducibility as the first scan yields undefined and inconsistent results. This is most likely the result of some remaining contaminants from the graft and rinse solutions remaining physisorbed to the surface despite rinsing with copious amounts of CH_3CN.

Potential controlled experiments on glassy carbon were performed as described above with the exception of the varying stopping potential used throughout the experiment during grafting. The same glassy carbon electrode was used for all stopping potentials.

The electrochemically assisted modification of the graphene samples during the potential control experiments was accomplished in a similar manner. To ensure good electrical contact to the graphene, the copper alligator clip was wrapped with gold foil prior to clamping onto the sample. The sample was rinsed with CH_3CN prior to grafting. No polishing procedures were used. The sample was then dipped into the prepared solution of **BNPI** and grafting was accomplished with two potential sweeps from 0 V to the predetermined stopping voltage. Care was taken to ensure that there was no contact between the clip and the electrolyte solution. The sample was then removed from the grafting solution, washed with copious amounts of CH_3CN, and allowed to dry. The sample was then dipped into a blank electrolyte solution and the redox event of the grafted *p*-nitrophenyl moieties was observed by two potential sweeps from 0 to −2.1 V at 100 mV s^{-1}. A slower scan rate was required for the graphene samples due to the high sheet resistance associated with a single layer of graphene. The second scan is then integrated to determine the surface coverage. The same process was repeated for each of the stopping potentials in the order of increasingly more negative potentials.

The graphene samples were grown on both insulating and conducting SiC.

The STM images were obtained using an Omicron system operated at 77 K using cut PtIr tips. The graphene samples were washed thoroughly with CH_3CN prior to introduction into the vacuum chamber where it was degassed at 425 K for a few hours prior to imaging. The DOS were recorded using a lock-in technique with a 20 mV wiggle voltage at 712 Hz.

The Raman data were recorded using a Renishaw inVia Raman microscope and a 514.5 nm argon laser with a 50× objective. To remove interference with strong Raman bands from the SiC substrate, the laser was defocused and spectra of the bulk SiC were obtained and subtracted from the graphene spectra.

Conclusions

Reproducible, controlled covalent grafting of graphene grown epitaxially on SiC was obtained using electrochemical methods. Functionalization is achieved using both symmetric and asymmetric aryliodonium salts, which contrary to diazonium salts do not exhibit spontaneous functionalization. It was found that the grafting density *p*-nitrophenyl groups using the symmetric salt (**BNPI**) can be precisely tuned between 4×10^{13} up to 3×10^{14} molecules per cm^2. Electron transfer kinetic studies have been performed using graphene samples with high (graphene on H-intercalated SiC) and low (graphene on buffer layer SiC) densities of single point defects (vacancies). It was found that for a given cycled potential and salt concentration the grafting densities are similar which implies that the electron transfer kinetics at the graphene basal plane is fast. There is no evidence for preferential attachment at the defects sites. High resolution STM imaging of the modified graphene samples show that some of the grafts present a three-fold symmetry which may indicate that they are pairs of molecules. The attachment of the second nitrophenyl group on the graphene lattice is not clear. Ongoing computational work will include simulation of STM images of different types of drafts are expected to shed light on the structure of these particular grafts. Elongated as well as extended grouping of clustered grafts have been also observed. Ongoing Angle Resolved Photoemission Spectroscopy will help shed light on the changes in the electronic structure (doping level, bandgap opening) induced by grafting.

Acknowledgements

This work was supported by an Energy Frontier Research Center funded by the U.S. Department of Energy, Office of Science, Office of Basic Energy Sciences on "Understanding Charge Separation and Transfer at Interfaces in Energy Materials" (EFRC:CST, Award Number DE-SC0001091). T.O. also acknowledges support from the U.S. Department of Energy, Office of Science, Office of Basic Energy Sciences, Division of Materials Science and Engineering. T.O. and C.K.C. acknowledge support from Sandia National Laboratories Laboratory Directed Research and Development. Sandia National Laboratories is a multiprogram laboratory managed and operated by Sandia Corporation, a wholly owned subsidiary of Lockheed Martin Corporation, for the U.S. Department of Energy's National Nuclear Security Administration under contract DE-AC04-94AL85000.

Notes and references

1 K. S. Novoselov, A. K. Geim, S. V. Morozov, D. Jiang, Y. Zhang, S. V. Dubonos, I. V. Grigorieva and A. A. Firsov, *Science*, 2004, **306**, 666–669.
2 J. Baringhaus, M. Ruan, F. Edler, A. Tejeda, M. Sicot, AminaTaleb-Ibrahimi, A.-P. Li, Z. Jiang, E. H. Conrad, C. Berger, C. Tegenkamp and W. A. de Heer, *Nature*, 2014, DOI: 10.1038/nature12952.
3 R. R. Nair, P. Blake, A. N. Grigorenko, K. S. Novoselov, T. J. Booth, T. Stauber, N. M. R. Peres and A. K. Geim, *Science*, 2008, **320**, 1308.
4 Y. Huang, J. Liang and Y. Chen, *Small*, 2012, **8**, 1805–1834.
5 C. Coletti, C. Riedl, D. S. Lee, B. Krauss, L. Patthey, K. von Klitzing, J. H. Smet and U. Starke, *Phys. Rev. B: Condens. Matter Mater. Phys.*, 2010, **81**, 235401.
6 Y.-H. Zhang, K.-G. Zhou, K.-F. Xie, J. Zeng, H.-L. Zhang and Y. Peng, *Nanotechnology*, 2010, **21**, 065201.
7 D. Wei, Y. Liu, Y. Wang, H. Zhang, L. Huang and G. Yu, *Nano Lett.*, 2009, **9**, 1752–1758.
8 E. Bekyarova, M. E. Itkis, P. Ramesh, C. Berger, M. Sprinkle, W. A. de Heer and R. C. Haddon, *J. Am. Chem. Soc.*, 2009, **131**, 1336–1337.
9 J. R. Lomeda, C. D. Doyle, D. V. Kosynkin, W.-F. Hwang and J. M. Tour, *J. Am. Chem. Soc.*, 2008, **130**, 16201–16206.
10 J. Liu, R. Wang, L. Cui, J. Tang, Z. Liu, Q. Kong, W. Yang and J. Gooding, *J. Phys. Chem. C*, 2012, **116**, 17939–17946.
11 M. Z. Hossain, M. A. Walsh and M. C. Hersam, *J. Am. Chem. Soc.*, 2010, **132**, 15399–15403.

12 H. Zhang, E. Bekyarova, J.-W. Huang, Z. Zhao, W. Bao, F. Wang, R. C. Haddon and C. N. Lau, *Nano Lett.*, 2011, **11**, 4047–4051.

13 L. Gan, D. Zhang and X. Guo, *Small*, 2012, **8**, 1326–1330.

14 S. Niyogi, E. Bekyarova, M. E. Itkis, H. Zhang, K. Shepperd, J. Hicks, M. Sprinkle, C. Berger, C. N. Lau, W. A. deHeer, E. H. Conrad and R. C. Haddon, *Nano Lett.*, 2010, **10**, 4061–4066.

15 D. B. Farmer, Y.-M. Lin, A. Afzali-Ardakani and P. Avouris, *Appl. Phys. Lett.*, 2009, **94**, 213106.

16 G. L. C. Paulus, Q. H. Wang and M. S. Strano, *Acc. Chem. Res.*, 2013, **46**, 160–170.

17 H. Zhu, P. Huang, L. Jing, T. Zuo, Y. Zhao and X. Gao, *J. Mater. Chem.*, 2012, **22**, 2063–2068.

18 D. Jiang, B. G. Sumpter and S. Dai, *J. Phys. Chem. B*, 2006, **110**, 23628–23632.

19 E. Bekyarova, S. Sarkar, S. Niyogi, M. E. Itkis and R. C. Haddon, *J. Phys. D: Appl. Phys.*, 2012, **45**, 154009.

20 R. Sharma, J. H. Baik, C. J. Perera and M. S. Strano, *Nano Lett.*, 2010, **10**, 398–405.

21 F. M. Koehler, A. Jacobsen, K. Ensslin, C. Stampfer and W. J. Stark, *Small*, 2010, **6**, 1125–1130.

22 M. T. McDermott, K. Kneten and R. L. McCreery, *J. Phys. Chem.*, 1992, **96**, 3124–3130.

23 R. J. Bowling, R. T. Packard and R. L. McCreery, *J. Am. Chem. Soc.*, 1989, **111**, 1217–1223.

24 X. Ji, C. E. Banks, A. Crossley and R. G. Compton, *ChemPhysChem*, 2006, **7**, 1337–1344.

25 T. J. Davies, R. R. Moore, C. E. Banks and R. G. Compton, *J. Electroanal. Chem.*, 2004, **574**, 123–152.

26 C.-Y. Lee and A. M. Bond, *Anal. Chem.*, 2009, **81**, 584–594.

27 N. L. Ritzert, J. Rodríguez-López, C. Tan and H. D. Abruña, *Langmuir*, 2013, **29**, 1683–1694.

28 R. L. McCreery and M. T. McDermott, *Anal. Chem.*, 2012, **84**, 2602–2605.

29 S. C. S. Lai, A. N. Patel, K. McKelvey and P. R. Unwin, *Angew. Chem., Int. Ed.*, 2012, **51**, 5405–5408.

30 A. N. Patel, M. G. Collignon, M. A. O'Connell, W. O. Y. Hung, K. McKelvey, J. V. Macpherson and P. R. Unwin, *J. Am. Chem. Soc.*, 2012, **134**, 20117–20130.

31 W. Yuan, Y. Zhou, Y. Li, C. Li, H. Peng, J. Zhang, Z. Liu, L. Dai and G. Shi, *Sci. Rep.*, 2013, **3**, 2248.

32 C. K. Chan, T. E. Beechem, T. Ohta, M. T. Brumbach, D. R. Wheeler and K. J. Stevenson, *J. Phys. Chem. C*, 2013, **117**, 12038–12044.

33 S. M. Dirk, S. Pylypenko, S. W. Howell, J. E. Fulghum and D. R. Wheeler, *Langmuir*, 2005, **21**, 10899–10901.

34 K. H. Vase, A. H. Holm, K. Norrman, S. U. Pedersen and K. Daasbjerg, *Langmuir*, 2007, **23**, 3786–3793.

35 M. Weissmann, S. Baranton and C. Coutanceau, *Langmuir*, 2010, **26**, 15002–15009.

36 N. G. Tsierkezos, *J. Solution Chem.*, 2007, **36**, 289–302.

37 C. Fontanesi, C. A. Bortolotti, D. Vanossi and M. Marcaccio, *J. Phys. Chem. A*, 2011, **115**, 11715–11722.

38 N. Florini, M. Michelazzi, F. Parenti, A. Mucci, M. Sola, C. Baratti, V. D. Renzi, K. Daasbjerg, S. U. Pedersen and C. Fontanesi, *J. Electroanal. Chem.*, 2013, **710**, 41–47.
39 I. J. Bruno, J. C. Cole, P. R. Edgington, M. Kessler, C. F. Macrae, P. McCabe, J. Pearson and R. Taylor, *Acta Crystallogr., Sect. B: Struct. Sci.*, 2002, **58**, 389–397.
40 F. Bailly, P. Barthen, H.-J. Frohn and M. Köckerling, *Z. Anorg. Allg. Chem.*, 2000, **626**, 2419–2427.
41 H.-J. Frohn, M. E. Hirschberg, R. Boese, D. Bläser and U. Flörke, *Z. Anorg. Allg. Chem.*, 2008, **634**, 2539–2550.
42 C. R. Hubbard, V. L. Himes, A. D. Mighell and S. W. Page, *Acta Crystallogr., Sect. B: Struct. Crystallogr. Cryst. Chem.*, 1980, **36**, 2819–2821.
43 S. Schäfer and T. Wirth, *Angew. Chem., Int. Ed.*, 2010, **49**, 2786–2789.
44 T. Ohta, N. C. Bartelt, S. Nie, K. Thürmer and G. L. Kellogg, *Phys. Rev. B: Condens. Matter Mater. Phys.*, 2010, **81**, 121411.
45 C. Riedl, C. Coletti, T. Iwasaki, A. A. Zakharov and U. Starke, *Phys. Rev. Lett.*, 2009, **103**, 246804.
46 V. W. Brar, Y. Zhang, Y. Yayon, T. Ohta, J. L. McChesney, A. Bostwick, E. Rotenberg, K. Horn and M. F. Crommie, *Appl. Phys. Lett.*, 2007, **91**, 122102.
47 I. Deretzis and A. La Magna, *Phys. Rev. B: Condens. Matter Mater. Phys.*, 2011, **84**, 235426.
48 S. Goler, C. Coletti, V. Piazza, P. Pingue, F. Colangelo, V. Pellegrini, K. V. Emtsev, S. Forti, U. Starke, F. Beltram and S. Heun, *Carbon*, 2013, **51**, 249–254.
49 H. Hiura, *J. Mater. Res.*, 2001, **16**, 1287–1292.
50 L. G. Cançado, A. Jorio, E. H. M. Ferreira, F. Stavale, C. A. Achete, R. B. Capaz, M. V. O. Moutinho, A. Lombardo, T. S. Kulmala and A. C. Ferrari, *Nano Lett.*, 2011, **11**, 3190–3196.
51 M. M. Ugeda, I. Brihuega, F. Guinea and J. M. Gómez-Rodríguez, *Phys. Rev. Lett.*, 2010, **104**, 096804.
52 J.-H. Chen, W. G. Cullen, C. Jang, M. S. Fuhrer and E. D. Williams, *Phys. Rev. Lett.*, 2009, **102**, 236805.
53 K. J. Stevenson, P. A. Veneman, R. I. Gearba, K. M. Mueller, B. J. Holliday, T. Ohta, C. K. Chan, M. Kim and J. R Chelikowsky, manuscript in preparation.
54 M. Bielawski, D. Aili and B. Olofsson, *J. Org. Chem.*, 2008, **73**, 4602–4607.
55 A. Altomare, M. C. Burla, M. Camalli, G. L. Cascarano, C. Giacovazzo, A. Guagliardi, A. G. G. Moliterni, G. Polidori and R. Spagna, *J. Appl. Crystallogr.*, 1999, **32**, 115–119.
56 G. M. Sheldrick, *SHELXL-97 Program for the Refinement of Crystal Structures*, University of Göttingen, Göttingen, Germany, 1997.
57 A. L. Spek, *PLATON, A Multipurpose Crystallographic Tool*, Utrecht University, The Netherlands, 1998.
58 L. J. Farrugia, *WinGX an Integrated System of Windows Programs for the Solution, Refinement and Analysis of Single Crystal X-Ray Diffraction Data, Version 1.64*, University of Glasgow, Glasgow, U.K., 1999.
59 L. J. Farrugia, *J. Appl. Crystallogr.*, 1999, **32**, 837–838.
60 *International Tables for X-ray Crystallography*, ed. A. J. C. Willson, Kluwer Academic Press, |Boston, 1992, vol. C, Tables 4.2.6.8 and 6.1.1.4.

Faraday Discussions

PAPER

Electrochemical characterisation of graphene nanoflakes with functionalised edges

Mailis M. Lounasvuori, Martin Rosillo-Lopez, Christoph G. Salzmann, Daren J. Caruana† and Katherine B. Holt*

Received 7th March 2014, Accepted 9th April 2014
DOI: 10.1039/c4fd00034j

Graphene nanoflakes (GNF) of diameter *ca.* 30 nm and edge-terminated with carboxylic acid (COOH) or amide functionalities were characterised electrochemically after drop-coating onto a boron-doped diamond (BDD) electrode. In the presence of the outer-sphere redox probe ferrocenemethanol there was no discernible difference in electrochemical response between the clean BDD and GNF-modified electrodes. When ferricyanide or hydroquinone were used as redox probes there was a marked difference in response at the electrode modified with COOH-terminated GNF in comparison to the unmodified BDD and amide-terminated GNF electrode. The response of the COOH-terminated GNF electrode was highly pH dependent, with the most dramatic differences in response noted at pH < 8. This pH range coincides with partial protonation of the carboxylic acid groups as determined by titration. The acid edge groups occupy a range of bonding environments and are observed to undergo deprotonation over a pH range *ca.* 3.7 to 8.3. The protonation state of the GNF influences the oxidation mechanism of hydroquinone and in particular the number of solution protons involved in the reaction mechanism. The voltammetric response of ferricyanide is very inhibited by the presence of COOH-terminated GNF at pH < 8, especially in low ionic strength solution. While the protonation state of the GNF is clearly a major factor in the observed response, the exact role of the acid group in the redox process has not been firmly established. It may be that the ferricyanide species is unstable in the solution environment surrounding the GNF, where dynamic protonation equilibria are at play, perhaps through disruption to ion pairing.

1 Introduction

This paper addresses the electrochemical behaviour of well-defined and characterised graphene nanoflakes (GNF) using cyclic voltammetry (CV) in the presence

Department of Chemistry, University College London, 20 Gordon St, London WC1H 0AJ, United Kingdom. E-mail: k.b.holt@ucl.ac.uk

† Presenting author. E-mail: d.j.caruana@ucl.ac.uk.

of solution redox species. Previous studies have used graphene from a variety of sources and with differing defect density and impurity content to construct or modify electrodes and to perform voltammetry with standard redox probes.[1–9] Electrodes with well-defined exposed surfaces fabricated from single-layer graphene sheets, produced by mechanical exfoliation or by chemical vapour deposition (CVD), have been used to carry out CV using ferrocene-methanol (FcMeOH)[3] and ferricyanide ($[Fe(CN)_6]^{3-}$).[4] Heterogeneous rate constants for FcMeOH oxidation were found to be one or two orders of magnitude greater than at basal plane graphite for mechanically exfoliated and CVD graphene respectively.[3] The enhanced kinetics were attributed to corrugations in the graphene sheet, giving rise to strain and curvature that may enhance reactivity, perhaps through the creation of midgap states.[10,11] Likewise, kinetics for the reduction of ferricyanide were found to be improved two-fold for defect-free single layer graphene over those obtained under the same conditions for multi-layer graphite.[4] Additionally it was found that defects, such as holes, in the graphene surface made little difference to the voltammetric response of the samples.

An alternate approach has been to drop-coat graphene platelets, obtained through reduction of graphene oxide (GO)[12–17] or through substrate-free gas-phase synthesis[18] onto carbon electrodes and to study the CV response of various redox species at the modified electrode. When graphene flakes of diameter *ca.* 500 nm were immobilised onto electrodes in this way it was found, in contrast to the studies above, that electron transfer kinetics of a number of solution analyte species (*e.g.* ferrocyanide, $[Ru(NH_3)_6]^{3+}$) were slower at the graphene-modified surfaces. This was attributed to the high proportion of graphene basal plane, where electron transfer kinetics are slower than at edge sites. In addition to these studies there exist many reports of the use of reduced GO materials in the form of composites or pastes for the detection of a range of analytically important species (for reviews see ref. 19–21). Although enhanced electrochemical response and analytical advantages have been cited, it is unclear whether the better performance of these electrodes is due to higher surface areas or the presence of redox active or catalytic impurities.[22] A lack of clarity as to purity of material, defect density and identity of oxygen content means in many cases it is difficult to truly assess the electrochemical response of pristine graphene.

Traditional consensus has been that electron transfer at graphitic materials is dominated by the edge plane[18,23,24] and it has been shown that the intentional generation of oxygen-containing defects increases reactivity.[23] However, recently it has been shown that the basal plane of highly oriented pyrolytic graphite (HOPG) can exhibit fast electron transfer for outer-sphere redox couples.[23,25–29] High resolution electrochemical surface imaging studies have shown that electron transfer at carbon nanotube walls and the graphene basal plane surfaces is fast and reversible and limited only by available density of states.[28,30–32] This contrasts with previous studies reporting exceedingly sluggish kinetics at the basal plane of graphitic materials. One possible explanation for these discrepancies is in preparation of the materials before electrochemical investigation. Adsorption of organic impurities onto freshly prepared graphene has been shown to take place within minutes on exposure to a typical laboratory atmosphere[33] and such a surface layer may lead to inhibition of electron transfer at carbon surfaces.

Notwithstanding the conflicting reports of basal plane activity, it is accepted that the edge plane of graphitic materials shows enhanced electrochemical activity due to the presence of high energy defects such as dangling bonds and oxygen functionalities. The interaction of various redox species with oxygen functionalities at carbon electrodes has been investigated extensively by McCreery and co-workers.[34–36] Common redox probes can be classified roughly into three categories: those which are insensitive to surface termination (FcMeOH, $[Ru(NH_3)_6]^{3+/2+}$); those which interact with specific oxygen functionalities (such as $Fe^{3+/2+}$ with C=O) and those which are surface sensitive but apparently do not interact with specific oxygen-containing groups ($[Fe(CN)_6]^{3-/4-}$).[37] Given that even the most carefully prepared graphene samples may have some oxygen content, it seems important to determine the influence of these functionalities on the electrochemical response. As graphene is increasingly being manufactured *via* reduction of graphene oxide, where an array of oxygen groups persist in the final product, the interaction of oxygen moieties with solution species will influence how well the material performs in electrochemical applications.

Our approach in this study is to use novel GNF materials with average diameter of just 30 nm.[38] The basal plane of the GNF is predominantly defect free and hence contains negligible oxygen content. In contrast, the edges of the flakes are decorated with carboxylic acid (COOH) functionalities. The high density of edge COOH groups makes this an ideal material with which to study the role of oxygen species on electrochemical response, as their influence is greatly amplified due to the small size of the flakes. Modification of the edge groups by transformation into amide groups also allows us to probe the influence of the edge group acidity and ability to hydrogen bond. A cartoon of the structures of the two types of flakes is shown in Fig. 1 (not to scale). In this paper we show how COOH-terminated GNF (c-GNF) and amide-terminated GNF (a-GNF) modified electrodes interact with the common redox probes FcMeOH, ferricyanide and hydroquinone.

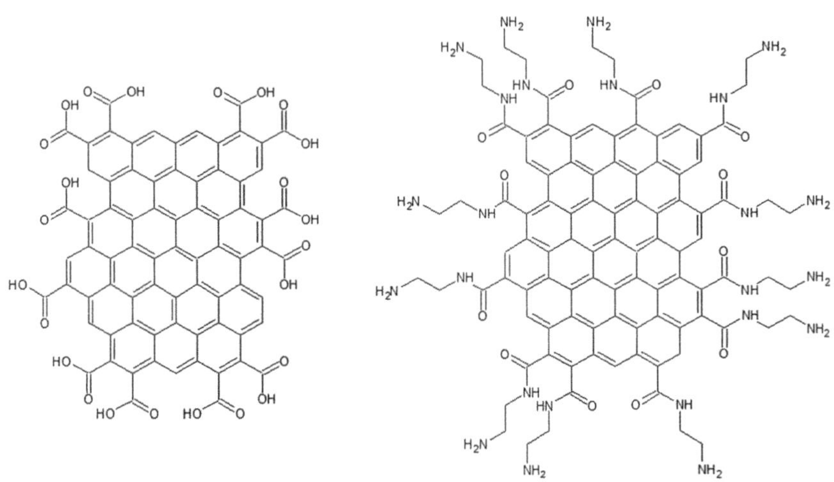

Fig. 1 Cartoon of the structure of the COOH-terminated GNF (c-GNF) and amide-terminated GNF (a-GNF) (not to scale).

2 Experimental procedures

2.1 Electrochemical measurements with GNF-modified electrodes

CV was carried out using a μ-Autolab potentiostat (Ecochemie, NL) coupled with GPES software. A boron-doped diamond (BDD) disk, 3 mm in diameter and sealed in PEEK (Windsor Scientific) was used as the working electrode, either unmodified or modified with a layer of adsorbed GNF. A platinum wire, coiled at the end to increase the surface area, served as a counter electrode. The reference electrode was Ag/AgCl in saturated KCl and all potentials are reported relative to it. The BDD electrode was polished using a 1.0 micron alumina suspension, rinsed with ultrapure water, and polished again using a 0.05 micron alumina suspension. Finally, the electrode was rinsed thoroughly with ultrapure water and dried using an ambient air flow. The GNF samples were drop-cast from aqueous suspensions of known concentration onto the freshly polished BDD electrode using a micropipette and allowed to dry under ambient conditions. After drying, the electrode was rinsed thoroughly with ultrapure water to remove any poorly adhered material from the surface and dried using an ambient air flow. The resulting amount of GNF on the electrode was estimated at 1.5 ± 0.5 μg in all experiments, and all CVs were recorded using a freshly modified electrode.

Electrochemical experiments were carried out in phosphate buffer solution (PBS) or KCl as background electrolyte (for concentrations refer to text). PBS of different pH in the range 4.5 to 9.2 was prepared by mixing different proportions of the constituent buffer salts KH_2PO_4 and K_2HPO_4. Redox probes hydroquinone (H_2Q), ferrocenemethanol (FcMeOH) and potassium ferricyanide ($K_3[Fe(CN)_6]$) were obtained from Sigma Aldrich and used as received. All solutions were prepared fresh daily with doubly deionised water, taken from a Milli-Q water purification system, with a resistivity of not less than 18.2 MΩ cm at 25 °C. For experiments in deoxygenated solutions and with air-sensitive chemicals such as H_2Q, high purity argon was bubbled through electrolyte solutions for 30 minutes to remove dissolved oxygen, and the gas flow was maintained over the surface of the solution during electrochemical experiments.

2.2 Synthesis and characterisation of c-GNF and a-GNF

The c-GNF were prepared from chemical oxidation of carbon nanotubes, as reported previously[38] with some modification to the method to allow for scale-up.[39] For the amidation reaction, the c-GNF were first treated with oxalylchloride in THF to prepare the acid chlorides, and then reacted with an excess of ethylenediamine in dry THF. The amidated product (a-GNF) was washed with water several times and dried in a vacuum. Characterisation of the flakes was carried out using X-ray photoelectron spectroscopy (XPS) with a Thermo Scientific K-Alpha XPS system with a monochromated Al K-alpha source ($E = 1486.6$ eV). XPS peak fitting was achieved using XPSPeak software. Fourier Transform Infra-Red (FTIR) spectroscopy was performed using a Bruker Tensor 27 spectrometer and the Opus data collection programme in Attenuated Total Reflectance (ATR) mode with a diamond crystal as the internal reflection element. Average diameter of the flakes was determined using Tapping Mode Atomic Force Microscopy, with the GNF spin-coated onto a smooth HOPG substrate.

An equivalence point and approximate pK_a for the c-GNF was obtained by titration of an aqueous suspension of dispersed c-GNF with NaOH. The hydrophilic nature of the COOH edge groups means the GNF disperse readily in water and other polar solvents. The NaOH solution was standardised prior to titration using potassium hydrogen phthalate (KHP). NaOH and KHP were placed in a desiccator for 12 hours prior to use. Water was either boiled or deoxygenated with argon before use. All solutions were kept under argon throughout the experiment. A micropipette was used to measure the volume of NaOH additions.

3 Results

3.1 Characterisation of COOH- and amide- terminated GNF by XPS and FTIR

The C1s region XPS spectra of the c-GNF used for this study were similar to the data shown in Fig. 2(b) in ref. 38. The dominant carbon bonding environments are consistent with sp^2 carbon (C=C, *ca.* 285 eV) and the presence of carboxylic acid groups (*ca.* 289 eV). The lack of a peak corresponding to sp^3 carbon–carbon bonding confirms the high aromaticity and low defect density of the GNF interior. Thus the XPS data confirm both the defect-free nature of the material and the uniformity and high density of the COOH edge groups. The a-GNF give rise to a C1s XPS spectrum consistent with full conversion of the COOH termination to $C(O)NHC_2H_2C_2H_2NH_2$ functionalities. FTIR spectra of c-GNF and a-GNF samples are consistent with the schematic structures shown in Fig. 1 and the C1s XPS data as well as FTIR spectra reported previously.[38] Carbonyl stretches at *ca.* 1720 cm^{-1} and 1650 cm^{-1} are observed for the COOH- and amide- terminated GNF respectively, confirming the presence of these functionalities. Given reports that GO materials experience chemical transformation between oxygen functionalities over time in water[40] we recorded the IR response of c-GNF suspended in water at regular intervals over six months. There was no change to the IR spectrum over this time period indicating that the COOH functionalities remained intact and stable almost indefinitely in water.

3.2 Determination of acidity of c-GNF using pH titration

Fig. 2 shows the titration curve for the addition of aliquots of $(236 \pm 6) \times 10^{-4}$ M NaOH to an aqueous suspension of c-GNF. As the XPS and IR characterisation show no detectable concentration of other acidic functionalities present in the flakes, the observed behaviour can be attributed solely to the COOH edge groups.

When the weak acid COOH edge groups of the c-GNF are exposed to water a dynamic equilibrium is established, where the acid groups become deprotonated:

$$\text{GNF-COOH} + H_2O \leftrightarrow \text{GNF-COO}^- + H_3O^+ \quad (1)$$

Addition of a small amount of strong base to the solution results in reaction of OH^- with the solution protons and hence the equilibrium is perturbed. Once the number of mols of OH^- added is equal to the number of weak acid groups (the equivalence point, found from the first derivative curve), further addition of base results in a rapid increase in pH. For c-GNF we observe more complex behaviour than would be expected for a single acid species dissolved in water. On addition of *ca.* 200 to 400 μl NaOH an increase in pH is observed, but this is not the sharp rise expected if all of the acid groups underwent deprotonation with the same pK_a. It is

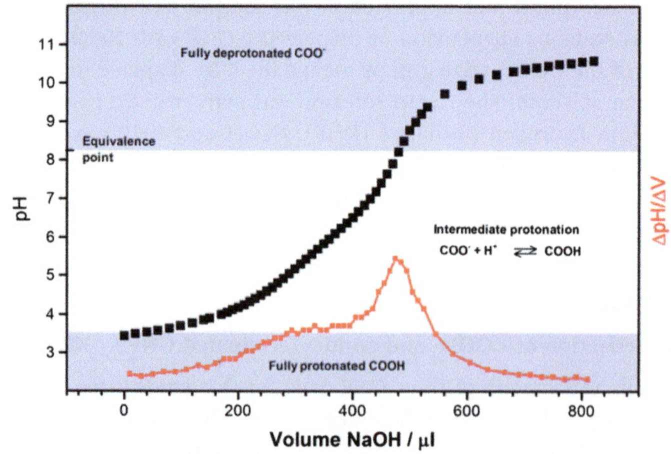

Fig. 2 Titration curve (black) and first derivative (red) for c-GNF.

probable that different bonding environments or electrostatic/hydrogen-bonding interactions between neighbouring groups result in a range of acid–base behaviours among the edge-group population. After addition of further NaOH an inflection point is observed in the first derivative curve, which we interpret as the point at which all COOH groups are fully deprotonated and which gives us an equivalence point of pH 8.3. This allows us to estimate the number of COOH groups as 7×10^{-3} mol of acid groups per gram of GNF material. The pK_a of a weak acid is defined as the pH at which half of the base required to reach the equivalence point is added and for the c-GNF pK_a is therefore estimated as 4.5. However given the wide pH range over which deprotonation is observed, the usefulness in reporting a single pK_a value for the c-GNF is questionable.

3.3 Electrochemical response of electrode-immobilised GNF in PBS

Fig. 3 shows the response of c-GNF, a-GNF and clean BDD in pH 4.6 and 9.2 solutions over the potential range −0.3 V to 1.0 V. In both solution conditions the response of the GNF layer can be observed over the background response of the BDD at potentials above *ca.* 0.4 V and below −0.2 V. Currents are larger for the a-GNF than c-GNF under the same conditions. No faradaic peaks are observed for the GNF modified electrodes in the range 0 to 0.4 V. As shown in the inset to Fig. 3, in deoxygenated solution the cathodic currents below −0.2 V are greatly diminished, showing the reduction currents in the main figure can be attributed to oxygen reduction. This indicates that both GNF samples can catalyse oxygen reduction better than the BDD, which is a poor electrocatalyst for this reaction in comparison to sp^2 carbon materials. The oxidation response is unaffected by the presence of oxygen in solution and is attributed to direct oxidation of the GNF.

These data show that the GNF adhere well to the surface of the BDD electrode for the duration of the experiment. At the present time it is unclear in what orientation the GNF are arranged on the electrode surface, as their small size and transparency makes the immobilised layer difficult to characterise. However as the coverage and flake orientation will clearly be important for fully

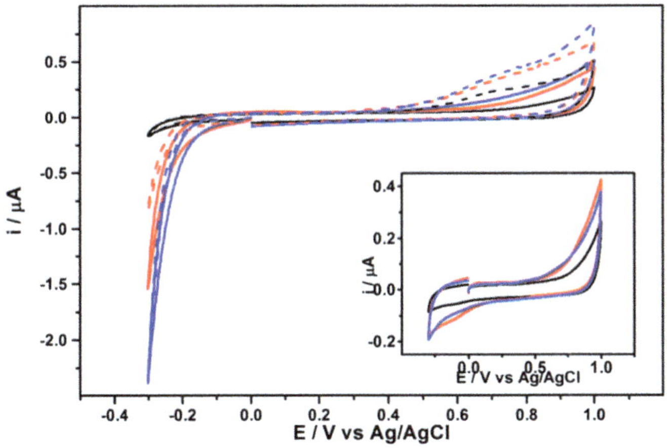

Fig. 3 Background response of BDD and GNF in pH 4.6 and 9.2. Black line: clean BDD. Red line: c-GNF. Blue line: a-GNF. Solid line: 0.1 M KH_2PO_4 (pH 4.6), dashed line: 0.1 M K_2HPO_4 (pH 9.2). Inset: 0.1 M KH_2PO_4 solution deoxygenated with argon. Scan rate 50 mV s^{-1}. First scans shown.

understanding the electrochemical response, work is ongoing to determine and control the morphology of the electrode layer.

3.4 Electrochemical response of c-GNF and a-GNF towards FcMeOH

The CV response of the c-GNF modified electrode towards 0.5 mM FcMeOH was compared to that obtained at clean BDD under the same conditions, using different pH PBS as the background electrolyte (Fig. 4). No difference in electrochemical response could be discerned between the BDD and the GNF-modified BDD under any conditions. For both electrode types the response was that of a reversible one electron transfer with a $\Delta E_p = |E_p{}^{ox} - E_p{}^{red}| = 60$ mV at a scan rate of 100 mV s^{-1}, close to the theoretical expected value of 59 mV for a one electron

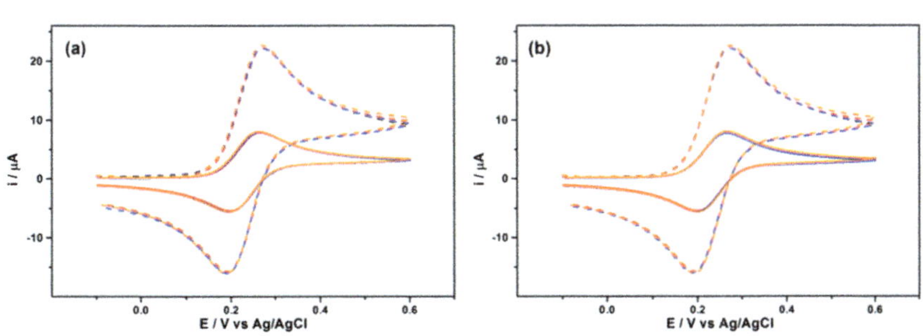

Fig. 4 CVs of 0.5 mM FcMeOH (a) at clean BDD in 0.1 M KH_2PO_4 (pH 4.6) (black line) and 0.1 M K_2HPO_4 (pH 9.2) (red line); at c-GNF modified BDD in 0.1 M KH_2PO_4 (pH 4.6) (blue line) and 0.1 M K_2HPO_4 (pH 9.2) (orange line); (b) at clean BDD in 0.1 M KH_2PO_4 (pH 4.6) (black line) and 0.1 M K_2HPO_4 (pH 9.2) (red line); at a-GNF modified BDD in 0.1 M KH_2PO_4 (pH 4.6) (blue line) and 0.1 M K_2HPO_4 (pH 9.2) (orange line). Scan rate 100 mV s^{-1} (solid line), 1 V s^{-1} (dashed line).

process. At scan rates up to 1 V s^{-1} the response remained identical at both clean BDD and the GNF modified electrode (ΔE_p *ca.* 80 mV at both electrodes). Electrode modification with a-GNF likewise showed no change in the CV response from that of clean BDD.

This result is not unexpected, as the FcMeOH/FcMeOH$^+$ redox couple is known to be relatively surface-insensitive and outer-sphere in nature. However, adsorption of this species to graphene[3] has been reported, indicating some surface interaction that could influence the electrochemical response. In this case we see no evidence of adsorption and also no indication that the protonation state of the c-GNF plays any role in the redox response of this probe. The GNF materials likewise show no electrode blocking effects that inhibit electrochemical response and no sign of limitation in electron transfer kinetics due to low density of states or lack of surface adsorption sites. On the other hand no enhancement in electron transfer kinetics is noted either, although the response at the underlying BDD is also close to reversible, so it would be difficult to determine any improvement.

3.5 Electrochemical response of c-GNF and a-GNF towards ferricyanide

In this GNF study we have two very clearly defined types of edge functionality with which to probe the interaction of ferricyanide with carbon electrode surfaces. The high density of COOH groups available on the c-GNF allows us to study both the electrostatic interaction between the ferricyanide and acid groups in different protonation states and the effect of acid–base equilibria on the redox response. The a-GNF allows us to probe the influence of the electronegative carbonyl moieties but in the absence of the deprotonation equilibria exhibited by the COOH groups.

Fig. 5a shows CVs for 0.5 mM ferricyanide at a clean BDD electrode in background electrolyte of 0.1 M pH 4.6 and 9.2 PBS. At the BDD electrode the peak separation ΔE_p remains constant at 65 ± 2 mV over the whole pH range examined (pH 4.6–9.2) indicating close to reversible electron transfer kinetics. The $E^{0'}$ of the couple, taken as ½($E_p^{ox} + E_p^{red}$), shifts towards higher values with increasing pH, being found at *ca.* 50 mV higher E at pH 9.2 than at pH 4.6. The peak currents for oxidation and reduction also decrease marginally over the same pH range. The ferri/ferrocyanide electron transfer process has been shown to be inhibited at oxygen-terminated BDD surfaces;[41] however we observe effectively reversible electron transfer kinetics at the (oxidised) BDD electrodes used in this study. Fig. 5a suggests a small degree of interaction of $[Fe(CN)_6]^{3-}$ with the BDD surface, however the effect is very small in comparison with some previous studies.

Fig. 5b shows the response of 0.5 mM ferricyanide at pH 4.6 and 9.2 at a c-GNF modified electrode. The response at this electrode is found to be very dependent on pH, particularly for pH < 8. Peak currents for both oxidation and reduction decrease and ΔE_p increases as the pH is lowered: at pH 7 $\Delta E_p = 109$ mV; pH 6 $\Delta E_p = 120$mV; pH 5 $\Delta E_p = 213$ mV and pH 4.6 $\Delta E_p = 250$ mV. This indicates that electron transfer becomes slower under these experimental conditions, which could be attributed to a change in the nature of the redox molecule, an unfavourable interaction with the electrode surface (or loss of a favourable interaction) or formation of an adsorbed inhibiting layer on the electrode. The response is not time or potential dependent, as it is observed immediately from the first CV scan and the response does not get worse with cycling (currents rather increase

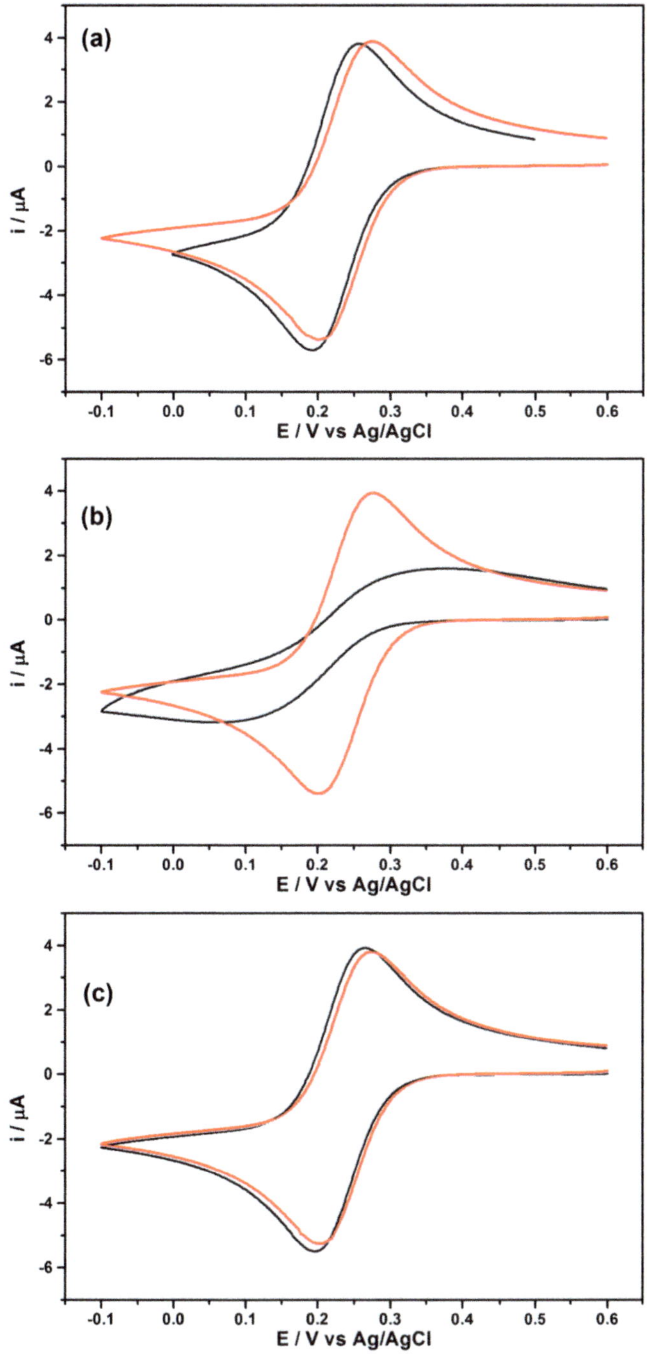

Fig. 5 The influence of pH on the ferri/ferrocyanide redox reaction at different electrodes. (a) Clean BDD; (b): BDD modified with c-GNF; (c): BDD modified with a-GNF. Supporting electrolyte: 0.1 M KH_2PO_4 (pH 4.6) (black line); 0.1 M K_2HPO_4 (pH 9.2) (red line). Potassium ferricyanide concentration 0.5×10^{-3} M. Scan rate: 50 mV s^{-1}.

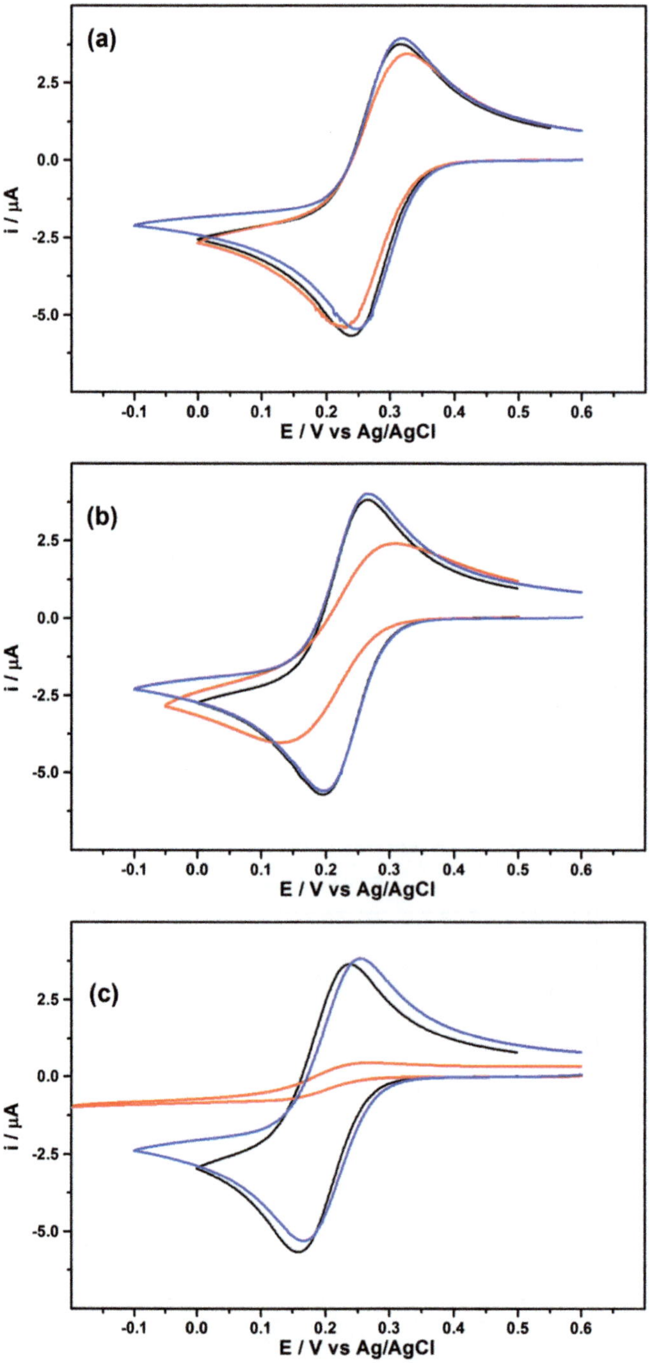

Fig. 6 The influence of supporting electrolyte concentration on the ferri/ferrocyanide redox reaction at different electrodes. Black line: clean BDD; red line: BDD modified with c-GNF. Blue line: BDD modified with a-GNF. Supporting electrolyte (a) 1.0 M KCl, (b) 0.1 M KCl, (c) 0.01 M KCl. Potassium ferricyanide concentration 0.5×10^{-3} M. Scan rate: 50 mV s^{-1}.

marginally with consecutive scans). This would indicate that the effect cannot be attributed to formation of a surface film that deposits as a function of time or applied potential. However it does not preclude the fast, spontaneous formation of an adsorbed layer, formed independently of applied potential.

Fig. 5c shows the response of the a-GNF modified electrode towards 0.5 mM ferricyanide in 0.1 M pH 4.6 and pH 9.2 PBS. For this electrode there is no pH dependence on the voltammetric response and the electron transfer kinetics appear only slightly less reversible than at clean BDD ($\Delta E_p = 70 \pm 1$ mV at 50 mV s^{-1}) over pH range 4.6–9.2. In fact the response is less pH dependent than at a clean BDD electrode. Thus it is apparent that carbonyl, amide or amine functionalities have little influence on the electrochemical response of ferri/ferrocyanide at GNF-modified electrodes. The observed inhibition of current at the c-GNF electrode can therefore be attributed specifically to the presence of the acid functionalities.

To investigate the c-GNF electrode further, CVs of ferricyanide were carried out in different ionic strength solution. In these experiments the background electrolyte was KCl so the solutions are not buffered, but are all in the pH range 5.9–6.2. Fig. 6 shows CVs at clean BDD, c-GNF and a- GNF with 0.5 mM ferricyanide in 1.0 M (a), 0.1 M (b) and 0.01 M KCl (c). Increasing the concentration of the supporting electrolyte causes a shift in $E^{0'}$ to more positive potentials for all electrodes studied here, confirming that the electrolyte plays a role in the redox equilibrium. At high ionic strength (1.0 M KCl) the CV response at all three electrodes is reversible, but at 0.1 M KCl currents at the c-GNF electrode are much reduced and ΔE_p is significantly increased. In 0.01 M supporting electrolyte the responses at clean BDD and amide-GNF modified electrodes are still reversible but the CV at the c-GNF electrode shows significant inhibition. The response appears sigmoidal, resembling the CV expected at an array of microelectrodes, or response through pinholes of an electrode partially covered in insulating material. If the experiment is repeated at similarly low ionic strength, but with the addition of KOH to bring the pH to 8.5, the CV returns to a more reversible form (albeit with $\Delta E_p = 231 \pm 21$ mV) indicating that it is solution acidity as well as ionic strength which is the important factor in this behaviour. Quantitatively similar behaviour is seen when the experiment is repeated in the same concentrations of NaCl and PBS electrolyte.

The ferri/ferrocyanide redox couple is often used as a standard probe, even though it has been shown to be inner sphere in nature and very sensitive to the electrode surface.[36,42–48] A systematic investigation into the influence of surface oxygen functionalities using glassy carbon electrodes showed that the ferricyanide CV response did not show a dependence on any specific surface oxygen groups, although it was sensitive to the presence of adsorbates.[36] However, other studies have shown pH dependence in the electron transfer kinetics of this couple at carbon electrodes, the process becoming slower as pH is increased.[42] This effect was attributed to the presence of surface carboxylic acid functionalities that become deprotonated and hence negatively charged in more alkaline solutions; therefore electrostatic repulsion results between the electrode surface and the negatively charged redox species. We estimate from titration of the c-GNF (Section 3.2) that the COOH edge groups are fully deprotonated at pH higher than 8, hence we might expect ferricyanide reduction to be inhibited at more alkaline pH. However we observe relatively reversible electrochemistry at pH 7 and above, and

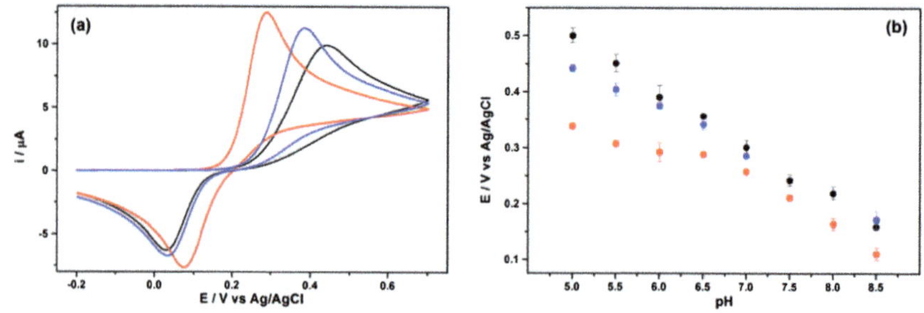

Fig. 7 (a) CV of 0.5 mM hydroquinone at clean BDD electrode (black line), c-GNF (red line) and a-GNF (blue line). Supporting electrolyte: 0.1 M pH 5.5 PBS. Scan rate: 50 mV s^{-1}. First scans shown. (b): Peak potential of hydroquinone oxidation as a function of pH at clean BDD electrode (black), c-GNF (red) and a-GNF (blue).

at pH 9.2 (where all of the COOH will be deprotonated and negatively charged) the response is identical to that at a clean BDD electrode. Therefore an electrostatic argument for the observed behaviour is clearly inappropriate in this case. In fact, we see slower kinetics at pH < 8, where according to the titration curve in Fig. 2 a wide range of COOH protonation states may be present. It may be that the dynamic acid–base equilibria in play at the edge of the GNF over this pH range plays a role in the observed electrochemical response.

Study of the ferri/ferrocyanide redox couple is made still more difficult due to its complex solution chemistry, in particular its preference for ion-pairing with solution cations[46–48] and propensity to lose ligands and form aggregates that are intermediates to Prussian Blue film deposition.[49–52] Additionally acid–base equilibria involving protonation of the nitrogen of the cyanide ligands becomes important over some pH ranges (pK_a of $H[Fe(CN)_6]^{3-}$ is *ca.* 4.2[53]). At low ionic strength the electrostatic interaction between the electrode and the redox probe will be enhanced as screening by solution ions in the double layer is less effective. These conditions seem to amplify the inhibiting effect of the COOH groups on the ferricyanide electrochemistry. Additionally the stability of the ferricyanide species may also be affected by the low ionic strength conditions as ion-pairing with K^+ will be less effective at the lower cation concentration. Possible factors influencing the observed electrochemistry are discussed further in Section 4 below.

3.6 Electrochemical response of c-GNF and a-GNF towards hydroquinone

The oxidation of hydroquinone (H_2Q) to benzoquinone (Q) was studied at clean BDD, c-GNF modified and a-GNF modified electrodes over the pH range 5.0 to 8.5. At BDD the response was very irreversible, as has been reported previously[54] with ΔE_p being *ca.* 250 mV at pH 8 and >450 mV at pH 5. Modification of the BDD with a layer of c-GNF resulted in a decrease in peak separation (ΔE_p *ca.* 200 mV at pH 8, 250 mV at pH 5). The CV response of the BDD electrode, the c-GNF electrode and the a-GNF electrode towards 0.5 mM H_2Q in 0.1 M pH 5.5 PBS is shown in Fig. 7a. A shift in both oxidation and reduction peaks towards reduced overpotential is observed at the c-GNF electrode; however the shift in oxidation peak potential is greater than that for the reduction peak. Indeed it was found that the anodic shift in the reduction peak position of *ca.* 40 mV compared to the peak at clean BDD was constant and independent of pH over the range tested. In contrast, the cathodic shift in the oxidation peak for c-GNF compared to BDD was pH dependent and was greater at lower pH. A small improvement in electron transfer kinetics is observed for the a-GNF electrode, with respect to the oxidation peak current, although little change to ΔE_p is observed on modifying the electrode.

The effect of pH on the H_2Q oxidation peak potential is plotted in Fig. 7b for BDD, c-GNF and a-GNF. A linear relationship is observed between peak position and pH for oxidation at the BDD electrode. Analysis of the gradient gives a 96 mV per pH unit relationship. The H_2Q/Q redox reaction is usually considered a $2e^-/2H^+$ process:

$$Q + 2e^- + 2H_3O^+ \leftrightarrow H_2Q + 2H_2O \quad (2)$$

Or in alkaline solution:

$$Q + 2e^- + 2H_2O \leftrightarrow H_2Q + 2OH^- \quad (3)$$

A 59 mV shift in peak position with pH is predicted for a Nernstian $2e^-/2H^+$ process. The deviation from this relationship observed at BDD may be related to the sluggish electrode kinetics for this reaction, which does not allow for such Nernstian analysis. Further elucidation of the reaction mechanism at BDD is beyond the scope of the current study and we confine further discussion to the observed differences at the c-GNF electrode. For the c-GNF electrode two distinct behaviours can be noted. At pH > 7 the relationship between peak position and pH is similar to that seen at BDD (93 mV per pH unit). However at pH < 7 a different gradient of 33 mV per pH unit can be fitted to the data. Clearly a change in reaction mechanism takes place at *ca.* pH 7, or alternatively the manner in which the H_2Q reactant or Q product interacts with the c-GNF changes in this pH range.

A 33 mV shift with pH unit is consistent with a $2e^-/1H^+$ process (predicted as 29.5 mV per pH unit):

$$Q + 2e^- + H_3O^+ \leftrightarrow HQ^- + H_2O \quad (4)$$

Such a reaction mechanism is unlikely in acidic, buffered solution as it requires H_2Q to be deprotonated, which is not possible given its pKa is 9.9.[55] A mechanism with a $2e^-/1H^+$ relationship would however be possible in the presence of an additional, non-solution, source of protons, such as the COOH-terminating groups offered by the c-GNF:

$$Q + 2e^- + \text{GNF-COOH} + H_3O^+ \leftrightarrow H_2Q + \text{GNF-COO}^- + H_2O \quad (5)$$

In the case of the H_2Q oxidation reaction it has been shown that modification of glassy carbon electrodes with phthalate bases (which contain two COOH groups) shifts the oxidation potential cathodically.[56] The proposed mechanism involves surface COO^- groups accepting the protons liberated in the oxidation of H_2Q and thus stabilising the reaction products. In effect the shift in oxidation potential is a thermodynamic consequence of the change in reaction mechanism rather than an improvement in electron transfer kinetics.

A similar process may be taking place in our system. Some improvement in electron transfer kinetics is observed on modifying the BDD electrode with GNF, as can be seen in the increase in oxidation peak currents for both a-GNF and c-GNF electrodes (Fig. 7a and b). The a-GNF electrode also shows a small cathodic shift in oxidation potential as pH is lowered, indicating improved electron transfer kinetics. H_2Q/Q can interact with the GNF *via* hydrogen bonding or electrostatic interactions with the edge groups or by hydrophobic or π–π interactions with the GNF basal plane. Such surface adsorption has been proposed to explain the improved electron transfer kinetics for this process (both oxidation and reduction) experienced at sp^2 carbon materials[57] in comparison to BDD, where limited surface adsorption is believed to take place.[41] However the marked change in proton concentration dependence noted at pH < 7 is unique to the c-GNF and strongly suggests the COOH groups play a role in the reaction mechanism, as shown in eqn (5). At pH > 7 the solution is sufficiently basic to allow the deprotonation accompanying H_2Q oxidation to proceed predominantly *via* a solution phase mechanism involving OH^- as the base (eqn (3)). However at pH < 7 there is a strong thermodynamic driving force for the COOH edge groups of the GNF to protonate, concomitant with conditions where there are fewer basic solution species. At this point a mechanism such as that shown in eqn (5) begins to dominate and is reflected by the change in proton concentration dependence of the oxidation peak position.

4 Discussion

The investigation described above has revealed important differences in the effect of COOH-termination and amide-termination of the GNF on electrochemical response towards ferricyanide and H_2Q. Edge termination was shown to have no influence on electron transfer for FcMeOH and indeed GNF-modified and clean BDD electrodes showed no difference in response towards this outer-sphere redox species. However, strong dependence on edge termination was noted for redox processes that are inner sphere or dependent on proton concentration. Some explanation of the role of the deprotonated COOH groups was discussed above with respect to the oxidation of the H_2Q species, where a direct function in the reaction mechanism was identified. However the role of the $COOH/COO^-$ functionalities in the electrochemical response of the ferri/ferrocyanide is less clear and requires further discussion.

Although it is difficult to provide a definitive explanation, it is known that ferri-/ferrocyanide can be unstable in solution, particularly at low ionic strength and low pH. Cyanide ligand loss and subsequent adsorption/decomposition of ferrocyano-species onto metal and carbon electrodes are well documented.[43,44,47,50–52,58] The $[Fe(CN)_6]^{4-/3-}$ redox reaction is believed to take place *via* activated ion-paired complexes such as $K_2[Fe(CN)_6]^{2-/1-}$. If these ion-pair complexes cannot form, for example at low ionic strength, then the electron transfer rate is much slower.[46] The electrochemical response of ferricyanide in the presence of c-GNF at pH < 7 suggests a lack of stability of the ion-paired redox species and hence sluggish electron transfer kinetics. At low ionic strength the response is consistent with a spontaneous deposition of blocking species on the electrode surface, indicating the real lack of stability of the redox molecule in these solution conditions.

In the pH range below 8 the c-GNF are clearly involved in dynamic protonation equilibria. Due to the differing bonding environments of the COOH groups and electrostatic and hydrogen-bonding interactions between neighbouring groups, protonation/deprotonation takes place over a wide pH range, as shown by the titration curve in Fig. 2. The instability of ferricyanide in these conditions may be due to the environment within the diffusion layer surrounding the GNF, where some of the carboxylic acid groups may be acidic enough to protonate the ferricyanide, promoting cyanide ligand loss in the form of HCN and allowing deposition of films similar in nature to Prussian Blue.[59] Solutions of ferricyanide at pH 3.6 have been reported to have different UV-Vis spectral features to those at higher pH (indicating protonation or ligand loss) and to develop blue precipitates on standing.[60] Although our solution pH values of 4.6–6 would not be considered acidic enough to cause decomposition of ferricyanide, a higher concentration of protons may be present close to the electrode surface due to the high density of carboxylic acid functionalities. The effect is exacerbated in low ionic strength solutions, as ferricyanide is considerably less stable in solution in the absence of ion-pairing to K^+.

Interestingly, when experiments with the c-GNF are repeated with the $[Ru(CN)_6]^{4-}$ as the redox couple the CV response is found to be independent of pH. The process appears reversible over the pH range 4.5–9.2 with no evidence of the inhibition and proposed surface film formation seen for $[Fe(CN)_6]^{3-}$. This would suggest that a mechanism requiring specific interaction between the COOH groups and the cyanide ligands of the redox species can be ruled out and it is more likely the complex solution chemistry of the ferricyanide molecule that results in the observed response.

5 Conclusion

The influence of edge group termination of GNF materials on electrochemical performance has been studied with respect to interaction with the common redox probes FcMeOH, ferricyanide and H_2Q. Neither a-GNF nor c-GNF show a detrimental effect on electron transfer rate with respect to the outer-sphere FcMeOH redox couple or indeed the more surface sensitive ferricyanide at pH > 8 (when c-GNF is fully deprotonated). These results are consistent with the observed fast electron transfer kinetics towards these species obtained using single layer graphene electrodes.[3,4] The high density of carboxylic acid or amide functionalities do not appear to perturb the electrochemical response under these reaction conditions.

The very high density of carboxylic acid groups of the c-GNF flakes and the absence of other oxygen-containing functionalities allow us to specifically investigate the effect of these highly charged and acidic groups on electrochemical response. Titration experiments reveal that in solution pH of *ca.* 4 to 8 the edge groups are present in a range of protonation states. In these intermediate states the c-GNF has a dramatic influence on the electrode response towards ferricyanide reduction. The response towards H_2Q reduction likewise shows deviation from the response at clean BDD at pH < 7 and in more acidic pH conditions the edge groups clearly play a role in the reaction mechanism. Although more investigations are required to fully understand the role of the partly protonated acid edge groups, it is clear that they have a significant influence on the observed electrochemistry.

Oxidised carbon electrodes will present a range of surface oxygen functionalities. Hence it can be difficult to isolate the interaction of redox probes with a specific surface moiety. The strength of this study are the well-characterised and uniform GNF that enable us to attribute changes in electrochemical response to specific functionalities. However the exact arrangement of the COOH edge groups on the outside of the GNF, the electrostatic interactions with neighbours and degree of hydrogen bonding still remain to be elucidated. Further studies are underway to understand the dynamic environment of the GNF edges in solution conditions.

Acknowledgements

We thank EPSRC/UCL Chemistry Department for studentship for Mailis Lounasvuori and Martin Rosillo-Lopez. Christoph Salzmann thanks the Royal Society for the award of a URF. The work was funded partly by EPSRC grant EP/J010006/1.

References

1 N. G. Shang, P. Papakonstantinou, M. McMullan, M. Chu, A. Stamboulis, A. Potenza, S. S. Dhesi and H. Marchetto, *Adv. Funct. Mater.*, 2008, **18**, 3506–3514.

2 D. A. C. Brownson and C. E. Banks, *Phys. Chem. Chem. Phys.*, 2011, **13**, 15825–15828.

3 W. Li, C. Tan, M. A. Lowe, H. D. Abruña and D. C. Ralph, *ACS Nano*, 2011, **5**, 2264–2270.

4 A. T. Valota, I. A. Kinloch, K. S. Novoselov, C. Casiraghi, A. Eckmann, E. W. Hill and R. A. W. Dryfe, *ACS Nano*, 2011, **5**, 8809–8815.

5 A. Bonanni, A. Ambrosi and M. Pumera, *Chem. - Eur. J.*, 2012, **18**, 4541–4548.

6 C. Tan, J. Rodríguez-López, J. J. Parks, N. L. Ritzert, D. C. Ralph and H. D. Abruña, *ACS Nano*, 2012, **6**, 3070–3079.

7 E. Kibena, M. Mooste, J. Kozlova, M. Marandi, V. Sammelselg and K. Tammeveski, *Electrochem. Commun.*, 2013, **35**, 26–29.

8 N. L. Ritzert, J. Rodríguez-López, C. Tan and H. D. Abruña, *Langmuir*, 2013, **29**, 1683–1694.

9 B. Zhang, L. Fan, H. Zhong, Y. Liu and S. Chen, *J. Am. Chem. Soc.*, 2013, **135**, 10073–10080.

10 K. Xu, P. Cao and J. R. Heath, *Nano Lett.*, 2009, **9**, 4446–4451.

11 F. Guinea, M. I. Katsnelson and M. A. H. Vozmediano, *Phys. Rev. B: Condens. Matter Mater. Phys.*, 2008, **77**, 075422.

12 A. Ambrosi, A. Bonanni, Z. Sofer, J. S. Cross and M. Pumera, *Chemistry (Weinheim an der Bergstrasse, Germany)*, 2011, **17**, 10763–10770.

13 C. Shan, H. Yang, J. Song, D. Han, A. Ivaska and L. Niu, *Anal. Chem.*, 2009, **81**, 2378–2382.

14 M. Zhu, C. Zeng and J. Ye, *Electroanalysis*, 2011, **23**, 907–914.

15 Y.-R. Kim, S. Bong, Y.-J. Kang, Y. Yang, R. K. Mahajan, J. S. Kim and H. Kim, *Biosens. Bioelectron.*, 2010, **25**, 2366–2369.

16 M. Mallesha, R. Manjunatha, C. Nethravathi, G. S. Suresh, M. Rajamathi, J. S. Melo and T. V. Venkatesha, *Bioelectrochemistry*, 2011, **81**, 104–108.
17 H. Du, J. Ye, J. Zhang, X. Huang and C. Yu, *J. Electroanal. Chem.*, 2011, **650**, 209–213.
18 D. A. C. Brownson, L. J. Munro, D. K. Kampouris and C. E. Banks, *RSC Adv.*, 2011, **1**, 978–988.
19 K. R. Ratinac, W. Yang, J. J. Gooding, P. Thordarson and F. Braet, *Electroanalysis*, 2011, **23**, 803–826.
20 Q. Y. He, S. X. Wu, Z. Y. Yin and H. Zhang, *Chem. Sci.*, 2012, **3**, 1764–1772.
21 S. X. Wu, Q. Y. He, C. L. Tan, Y. D. Wang and H. Zhang, *Small*, 2013, **9**, 1160–1172.
22 M. Pumera, A. Ambrosi and E. L. K. Chng, *Chem. Sci.*, 2012, **3**, 3347–3355.
23 C. X. Lim, H. Y. Hoh, P. K. Ang and K. P. Loh, *Anal. Chem.*, 2010, **82**, 7387–7393.
24 R. Sharma, J. H. Baik, C. J. Perera and M. S. Strano, *Nano Lett.*, 2010, **10**, 398–405.
25 M. A. Edwards, P. Bertoncello and P. R. Unwin, *J. Phys. Chem. C*, 2009, **113**, 9218–9223.
26 S. C. S. Lai, A. N. Patel, K. McKelvey and P. R. Unwin, *Angew. Chem., Int. Ed.*, 2012, **51**, 5405–5408.
27 A. N. Patel, M. G. Collignon, M. A. O'Connell, W. O. Y. Hung, K. McKelvey, J. V. Macpherson and P. R. Unwin, *J. Am. Chem. Soc.*, 2012, **134**, 20117–20130.
28 N. Ebejer, A. G. Gueell, S. C. S. Lai, K. McKelvey, M. E. Snowden and P. R. Unwin, *Annu. Rev. Anal. Chem.*, 2013, **6**, 329–351.
29 A. N. Patel, S.-y. Tan and P. R. Unwin, *Chem. Commun.*, 2013, **49**, 8776–8778.
30 A. G. Gueell, N. Ebejer, M. E. Snowden, J. V. Macpherson and P. R. Unwin, *J. Am. Chem. Soc.*, 2012, **134**, 7258–7261.
31 T. S. Miller, N. Ebejer, A. G. Gueell, J. V. Macpherson and P. R. Unwin, *Chem. Commun.*, 2012, **48**, 7435–7437.
32 A. G. Gueell, K. E. Meadows, P. V. Dudin, N. Ebejer, J. V. Macpherson and P. R. Unwin, *Nano Lett.*, 2014, **14**, 220–224.
33 Z. Li, Y. Wang, A. Kozbial, G. Shenoy, F. Zhou, R. McGinley, P. Ireland, B. Morganstein, A. Kunkel, S. P. Surwade, L. Li and H. Liu, *Nat. Mater.*, 2013, **12**, 925–931.
34 C. A. McDermott, K. R. Kneten and R. L. McCreery, *J. Electrochem. Soc.*, 1993, **140**, 2593–2599.
35 P. Chen, M. A. Fryling and R. L. McCreery, *Anal. Chem.*, 1995, **67**, 3115–3122.
36 P. Chen and R. L. McCreery, *Anal. Chem.*, 1996, **68**, 3958–3965.
37 R. L. McCreery, *Chem. Rev.*, 2008, **108**, 2646–2687.
38 C. G. Salzmann, V. Nicolosi and M. L. H. Green, *J. Mater. Chem.*, 2010, **20**, 314–319.
39 C. G. Salzmann and M. Rosillo-Lopez, in preparation.
40 A. M. Dimiev, L. B. Alemany and J. M. Tour, *ACS Nano*, 2013, **7**, 576–588.
41 M. C. Granger, M. Witek, J. Xu, J. Wang, M. Hupert, A. Hanks, M. D. Koppang, J. E. Butler, G. Lucazeau, M. Mermoux, J. W. Strojek and G. M. Swain, *Anal. Chem.*, 2000, **72**, 3793–3804.
42 M. R. Deakin, K. J. Stutts and R. M. Wightman, *J. Electroanal. Chem.*, 1985, **182**, 113–122.
43 C. Lee and F. C. Anson, *J. Electroanal. Chem.*, 1992, **323**, 381–389.
44 C. M. Pharr and P. R. Griffiths, *Anal. Chem.*, 1997, **69**, 4673–4679.

45 M. C. Granger and G. M. Swain, *J. Electrochem. Soc.*, 1999, **146**, 4551–4558.
46 L. M. Peter, W. Dürr, P. Bindra and H. Gerischer, *J. Electroanal. Chem.*, 1976, **71**, 31–50.
47 C. Beriet and D. Pletcher, *J. Electroanal. Chem.*, 1993, **361**, 93–101.
48 M. Noel and P. N. Anantharaman, *Analyst*, 1985, **110**, 1095–1103.
49 W. Huang and R. McCreery, *J. Electroanal. Chem.*, 1992, **326**, 1–12.
50 J. Kawiak, P. J. Kulesza and Z. Galus, *J. Electroanal. Chem.*, 1987, **226**, 305–314.
51 S. Pons, M. Datta, J. F. McAleer and A. S. Hinman, *J. Electroanal. Chem.*, 1984, **160**, 369–376.
52 A. Więckowski and M. Szklarzyk, *J. Electroanal. Chem.*, 1982, **142**, 157–170.
53 P. L. Domingo, B. Garcia and J. M. Leal, *Can. J. Chem.*, 1987, **65**, 583–589.
54 I. Duo, C. Levy-Clement, A. Fujishima and C. Comninellis, *J. Appl. Electrochem.*, 2004, **34**, 935–943.
55 J. H. Baxendale and H. R. Hardy, *Trans. Faraday Soc.*, 1953, **49**, 1140–1144.
56 J. Medina-Ramos, T. M. Alligrant, A. Clingenpeel and J. C. Alvarez, *J. Phys. Chem. C*, 2012, **116**, 20447–20457.
57 S. H. DuVall and R. L. McCreery, *Anal. Chem.*, 1999, **71**, 4594–4602.
58 A. A. Karyakin, *Electroanalysis*, 2001, **13**, 813–819.
59 P. L. Domingo, B. García and J. M. Leal, *Can. J. Chem.*, 1990, **68**, 228–235.
60 R. Yang, Z. Qian and J. Deng, *J. Electrochem. Soc.*, 1998, **145**, 2231–2236.

Faraday Discussions

PAPER

Probing the charging mechanisms of carbon nanomaterial polyelectrolytes†

Stephen A. Hodge,[a] Hui Huang Tay,[a] David B. Anthony,[a] Robert Menzel,[a] David J. Buckley,[b] Patrick L. Cullen,[b] Neal T. Skipper,[b] Christopher A. Howard[b] and Milo S. P. Shaffer*[a]

Received 17th March 2014, Accepted 30th April 2014
DOI: 10.1039/c4fd00043a

Chemical charging of single-walled carbon nanotubes (SWCNTs) and graphenes to generate soluble salts shows great promise as a processing route for electronic applications, but raises fundamental questions. The reduction potentials of highly-charged nanocarbon polyelectrolyte ions were investigated by considering their chemical reactivity towards metal salts/complexes in forming metal nanoparticles. The redox activity, degree of functionalisation and charge utilisation were quantified *via* the relative metal nanoparticle content, established using thermogravimetric analysis (TGA), inductively coupled plasma atomic emission spectroscopy (ICP-AES) and X-ray photoelectron spectroscopy (XPS). The fundamental relationship between the intrinsic nanocarbon electronic density of states and Coulombic effects during charging is highlighted as an important area for future research.

Introduction

Carbon nanomaterials, including both single-walled carbon nanotubes (SWCNTs) and graphene have impressive intrinsic mechanical, thermal and (opto)electronic properties.[1] Applications of these materials have been slow to emerge due to challenging synthesis and post-synthesis processing requirements. The desire for stable suspensions/solutions of monodispersed species to enable liquid-phase processing has led to the development of numerous aqueous-surfactant[2] and non-aqueous[3] dispersion procedures. Unfortunately, most routes rely on high-power ultrasonication to exfoliate and disperse the nanomaterials, which tends to degrade the desirable properties, for example *via* shortening and the introduction of structural defects.[4] Subsequently, lengthy ultracentrifugation steps are typically

[a]*Department of Chemistry, Imperial College London, London, SW7 2AZ, UK. E-mail: m.shaffer@imperial.ac.uk*
[b]*Department of Physics and Astronomy, University College London, London, WC1E 6BT, UK*

† Electronic Supplementary Information (ESI) available: Further characterisation of nanocarbon reactions (TGA, ICP-AES, XPS, UV-vis). Electrochemical measurements of metal compounds including the derivation of their reduction potentials. See DOI: 10.1039/c4fd00043a

required to remove non-exfoliated material, leading to limited overall process yields and scalability.

An attractive approach to dissolve carbon nanomaterials without damaging their structure is *via* charging protocols, as outlined in Fig. 1. Fullerene (C_{60}), SWCNT and graphene polyelectrolyte anions, known as fulleride, nanotubide and graphenide species, respectively, can be generated following reduction using alkali metal/liquid NH_3,[5–8] alkali metal naphthalide/tetrahydrofuran (THF),[9,10] or electrochemically in non-aqueous electrolytes.[11] In addition, electrochemical oxidation was recently demonstrated to generate nanotube cations, known as nanotubium species.[12] All reductive charging techniques insert electrons into the carbon π^* orbitals, shifting the Fermi energy of the nanocarbon density of states (DOS), resulting in increased reactivity. Subsequent reaction may take place *via* redox or electrophilic addition reactions that may involve a single-electron transfer (SET) (*i.e.* radical based reactions).[13–16,18,19]

The degree of functionalisation (d.o.f.), typically reported as a ratio of functional addends to carbon atoms on the nanocarbon species, is a quantitative measure of the reaction yield. However, reactivity depends strongly on the amount of charge per carbon atom,[20] its potential or chemical activity, the nature of the reactant added, and the reaction mechanism(s).

It follows that to enhance the d.o.f., the charge per carbon atom should be increased, and a reactant with a more favourable reduction or SET potential added. The degree of nanocarbon exfoliation is also a crucial factor in enhancing the d.o.f.; however, steric constraints also contribute and may become dominant, for fully exfoliated SWCNTs/graphenes, in which all atoms in principle lie on the 'surface'.[16,17] Ultimately, controlling the d.o.f. is critical: for example, too many functional addends can cause the loss of (opto)electronic and mechanical properties, too few and there may be no benefit to nanocarbon solubility or compatibility with a polymer matrix.[22] The first nanotube redox potentials have been probed using spectroelectrochemical methods,[23–25] and the redox potentials of reactants in aqueous systems are well studied. However, the exact potential of

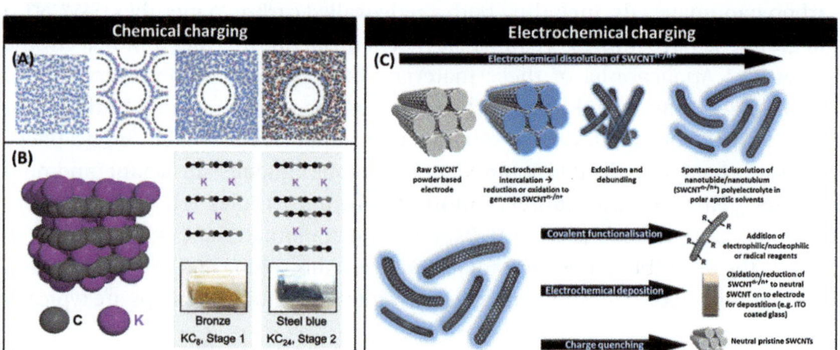

Fig. 1 Selected examples of charged carbon nanomaterials. (A) Nanotubide anions generated using alkali metal/liquid ammonia. Reproduced with permission from ref. 5. (B) Graphite intercalation compounds (GICs) generated by the vapour transport method[21] using potassium. (C) Electrochemical processing to generate nanotubide anions and nanotubium[12] cations. Modified with permission from ref 11. The charged materials shown in (A) and (B) are used in the current investigation.

charged nanocarbon polyelectrolytes and redox potentials of reactants in anhydrous, aprotic solvents are not well known.

The DOS of nanocarbon species are known to be complex but continuous, leading to the observation of bulk-continuum voltammetry.[11,12] The DOS for their ions has not yet been determined and is likely to vary with charge density. In this discussion paper, the potential/chemical activity of nanotube and graphene polyelectrolyte anions will be used to reduce a range of metal salts to generate zero oxidation state metal atoms/nanoparticles (NPs) on the carbon surface, as a route to probe the unclear relationship between added charge density and potential.

It has been observed that nanotubes/graphene act as reducing agents towards metal salts; even neutral SWCNTs can spontaneously reduce Au^{3+} and Pt^{2+} salt solutions to form their respective $M^{0}_{(metal)}$ NPs on the nanotube surface.[26] Nanotubides and graphenides can significantly increase the yield of this metal deposition due to the higher reducing potential of the carbon.[27] In principle, the use of metal salts should proceed by simple redox chemistry, avoiding electrophilic addition and radical reactive pathways. The initial hypothesis was that, for

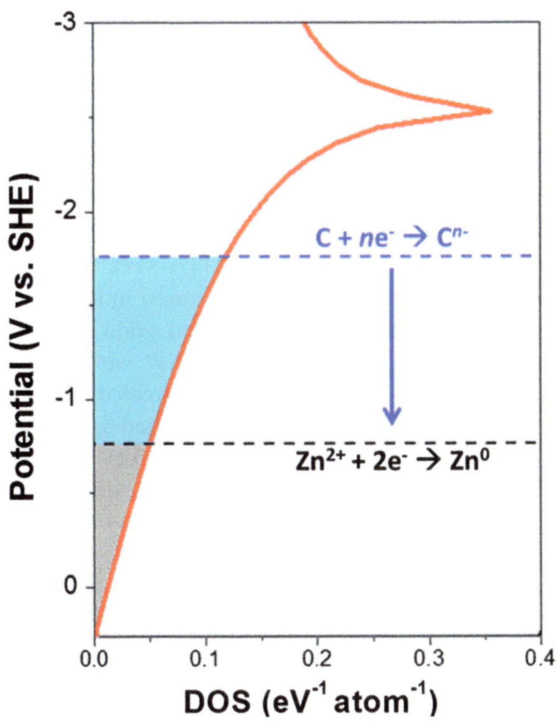

Fig. 2 DOS of a graphene crystal based on the tight-binding approximation (data values from ref. 30), neglecting Coulombic effects. The position of the graphene Fermi level (E_F) was previously determined[9] as +0.022 V *vs.* saturated calomel electrode (SCE) (~+0.263 V *vs.* SHE). The dashed black line represents the standard reduction potential for Zn^{2+} (−0.76 V *vs.* SHE), the dashed blue line represents a possible position of the nanocarbon reduction potential. Available electrons for reaction with Zn^{2+} are shaded blue, whereas the unavailable electrons are shaded grey. As the available charge is used, the potential of the nanocarbon will shift towards the Zn^{2+} reduction potential until the reaction reaches completion.

nanocarbon polyelectrolyte anion reactions with metal salts, the resulting metal to carbon (M : C) atomic ratio (*i.e.* the d.o.f.) should be defined by the initial charge transfer and the metal reduction potential. Based on the quantified d.o.f., the reduction potential of the nanocarbon polyelectrolyte should be predicted by integrating the areas of reacted electrons (charge utilisation yield) and unavailable charge (between the metal reduction potential and the undoped nanocarbon Fermi level) on the nanotube/graphene DOS plot, shown by the schematic diagram in Fig. 2.

Results and discussion

Sodium nanotubide experiments

Preliminary experiments used raw HiPco SWCNTs, reduced using solutions of sodium metal in liquid ammonia to give a NaC_{10} SWCNT salt following evaporation of the ammonia (note that a small amount of ammonia remains coordinated to the metal). This salt was left to spontaneously dissolve into anhydrous *N,N*-dimethylformamide (DMF) as in previous methodology.[5] The resulting dissolved SWCNT fraction used for reaction had a concentration of 0.6 mg mL^{-1} (80% yield of dissolution).

Two metal chloride salts ($MnCl_2$ and $ZnCl_2$), typically used as precursors to generate the respective $M^0{}_{(metal)}$ NPs using a wide range of reducing agents,[28,29] were added to the nanotubide dispersions in a three times excess to sodium. For these NPs, exposure to the atmosphere following reaction is likely to cause partial oxidation of the NPs. Following oxidative thermogravimetric analysis (TGA) of the washed, filtered and dried nanotube/metal samples, metal oxide residues remain (Electronic Supplementary Information (ESI) Fig. S1†). Raw HiPco contains ~30 wt% iron oxide originating from the catalyst used in its synthesis. Reaction with $MnCl_2$ showed only a subtle increase in metal oxide content from the raw material, however, the $ZnCl_2$ reaction yielded a ~7 wt% increase. Scanning electron microscopy (SEM) images of metal halide treated SWCNTs before TGA revealed NPs containing manganese and zinc, confirmed using energy-dispersive X-ray spectroscopy (EDX) (Fig. 3). These relative reactivities were expected, in-line with the reduction potentials of their respective metal salts (standard reduction potentials: $Zn^{2+} + 2e^- \rightarrow Zn^0$, −0.76 V, and $Mn^{2+} + 2e^- \rightarrow Mn^0$, −1.18 V *vs.* SHE (standard hydrogen electrode)).

Having successfully generated NPs for the manganese and zinc systems, the ability to extend this electrochemical series to a whole set of metal NPs exists. However, due to the presence of the (oxidised) iron catalyst, the possibility of competitive reduction reactions cannot be avoided. Due to the unknown charge per carbon atom in the dissolved fraction compared to the undissolved material, and the complex density of states (including mixtures of metallic and semi-conducting species with different band gaps and Fermi level positions) presented by nanotube samples, it is complex to confirm the exact reduction potential of the NaC_{10} SWCNT salt. Instead, a simpler charged carbon nanomaterial was studied. Graphite intercalation compounds (GICs) that can be exfoliated to monolayer graphenide species,[6] produced from various types of graphite are an ideal model system for the current study, offering high purity and an absence metal contaminants.

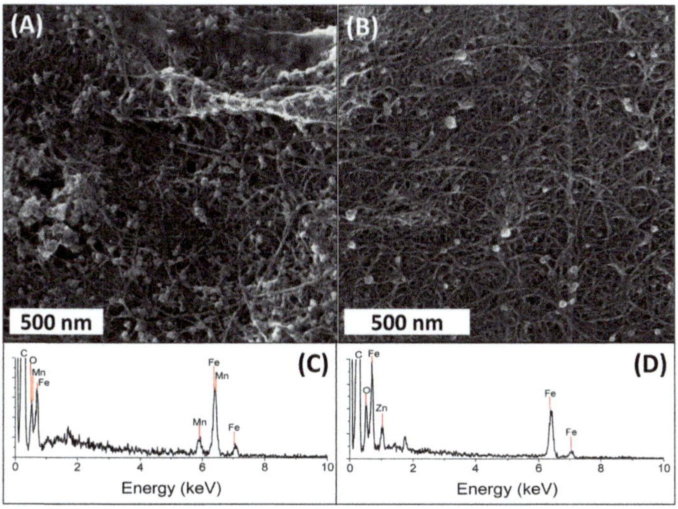

Fig. 3 SEM images of the metal NPs generated from reaction of NaC_{10} nanotubide with (A) $MnCl_2$ and (B) $ZnCl_2$, respectively. (C) and (D) are the complementary EDX spectra for Mn and Zn NPs, respectively.

Potassium graphenide experiments

GICs have been studied for many years.[31] The stage 1 compound, KC_8, as shown in Fig. 1(B), has been used directly as a strong reducing agent in many organic reactions.[32,33] KC_8 is more reducing than naphthalene,[34] generating the naphthalide radical anion in THF (−2.24 V *vs.* SHE), and has been exploited to prepare active metals including Zn, Sn, Ti, Fe, Pd and Ni (from reaction with the metal halides) highly dispersed on the graphite surface.[32] These metal-graphites demonstrated high catalytic activity for a wide range of synthetically useful reactions. A suggested mechanism for KC_8 reactivity was proposed by Ebert.[35] In this model, two potassium species, K^0 and K^+, exist between the graphite layers in order to account for the incomplete charge transfer to the graphite. In this way, KC_8 was proposed to exhibit a "hybrid" chemistry, in which this simple two species model may reflect a variable reduction potential as the charge is consumed. In the case of solubilised potassium graphenide, the charge transfer and consequent reduction potential of the KC_8 will be altered by solvation effects including the stabilisation of graphenide sheets and the degree of potassium ion dissociation.

Further insight can be gained from recent measurements of the electronic structure of bulk potassium GICs.[36] In crystals of KC_8 and KC_{24}, the Fermi energy relative to that of pure graphite is shifted as the electrons occupy the graphitic π^* band. The Fermi level shift has been measured using photoemission spectroscopy[36] to be −1.35 eV in KC_8 and ∼−0.75 eV in KC_{24}. It is important to note that, in these unexfoliated materials, there is only partial charge transfer (∼44%) from the potassium atoms to the π^* bands, with the rest thought to reside in an electronic band associated with the intercalant superlattice.[37]

In the experiments presented here, the potassium ions of solvated graphenide become dissociated from the charged graphene sheets to some extent (with some degree of counterion condensation[38]), presumably leading to an increase in charge transfer from the potassium to the graphene π^* band. The Fermi level shift in the limit of complete charge transfer from the potassium to the graphene can be approximated[39] using the relationship, $E_F = -\sqrt{n\pi}\cdot\hbar v_F$, where v_F is the Fermi velocity, $\hbar v_F = 5.52$ eV Å and n is the number of electrons per graphene unit cell area (5.24 Å^2) (KC_8, $n = 0.0477$, KC_{24}, $n = 0.0158$). Thus, Fermi level shifts of −2.14 eV (−1.88 V *vs.* SHE) and −1.23 eV (−0.997 V *vs.* SHE) are calculated for KC_8 and KC_{24}, respectively. Testing a range of metal salts with a range of reduction potentials should allow a comparative experimental reduction potential to be deduced. Ideally, the reactivities of KC_8 and KC_{24} should fall within these Fermi energy shifts for partial and complete charge transfer.

Potassium GICs were produced *via* the vapour transport method,[21] to give the corresponding characteristic bronze stage 1 KC_8 and steel blue stage 2 KC_{24} compounds (Fig. 1B). Graphenide dispersions were produced in dry *N*-methyl-pyrrolidone (NMP). Controlling the reaction stoichiometry as a route to produce smaller NPs, or even covalently bound metal atoms to the graphenides was investigated. Reactions were performed using the exact stoichiometry as number of charges available for reduction, *e.g.* for M^{2+} salts, $M^{2+}:K = 0.5:1$, M^+ salts, $M^+:K = 1:1$, *etc.*, with results shown in Table 1 and ESI Table S1.† Elemental analysis of the TGA residues with inductively coupled plasma atomic emission spectroscopy (ICP-AES) was used to accurately discriminate between the alkali metal oxide content and the generated metallic NPs. Since ICP-AES has limits of detection on the order of parts per billion (ppb), the combination of these techniques allows the unequivocal quantification of the carbon to metal atomic ratio.

As with the case of NaC_{10} nanotubide reduction, there is still a higher reactivity for highly charged graphenides (KC_8) towards zinc compared to manganese. KC_{24}, however, does not have sufficient reducing power to react with the manganese(II) salt.

Copper salts, with a reduction potential similar to that of uncharged graphene, would be expected to reveal the highest reactivity. In fact, a very low reactivity was observed for KC_8 with $CuCl_2$; this failure could be due to the one electron reduction of $CuCl_2$ to the intermediate CuCl.[40]

To enhance the reactivity, highly soluble mesityl copper(I) (CuMes) was used in order to avoid the two-step reduction.[41] The solution turned red/brown, indicating

Table 1 Reactivity of KC_8 and KC_{24} graphenide polyelectrolytes with metal salts/complexes

Metal cation (M^{n+})	Reduction potential (V *vs.* SHE)	KC_8 (M : C atomic ratio)	Charge utilisation yield[cd] (%)	KC_{24} (M : C atomic ratio)	Charge utilisation yield[cd] (%)
Mn^{2+} ($MnCl_2$)	−2.01[a], −1.18[b]	446	3.6	No reaction	–
Zn^{2+} ($ZnCl_2$)	−1.13[a], −0.76[b]	107	15.0	489	9.8
Cu^{2+} ($CuCl_2$)	−0.34[b]	887	1.8	–	–
Cu^+ (CuMes)	+0.44[a], +0.52[b]	10	78.5	–	–

[a] Derivation from cyclic voltammetry experiments in 0.1 M $KClO_4$/NMP. [b] Standard reduction potential. [c] TGA and ICP measurements. [d] Refer to Fig. 2 for charge utilisation yield definition. – not measured.

the formation of $Cu^0_{(metal)}$ in solution; subsequent UV-vis absorption spectroscopy (ESI Fig. S2) revealed the surface plasmon absorption (~570 nm) typical for metallic Cu NPs.[42] The resultant Cu NPs generated by KC_{24} were surprisingly air stable for approximately 1 week. Since the Cu NPs were free in solution and not bound to the graphene, filtration of the reaction mixture in air would have resulted in the oxidation of unreacted CuMes to $Cu^0_{(metal)}$. Therefore, KC_8 generated Cu NPs were precipitated with toluene and air-free centrifugation was performed to sediment the graphene/Cu NPs. The supernatant containing unreacted CuMes was removed and the graphene/Cu NPs were filtered and washed for subsequent characterisation. It was not possible to precipitate the KC_{24} generated NPs for reaction quantification. Since the reduction potential for Cu^+ lies below the Fermi level for undoped graphene,[9] a 100% charge utilisation yield would be expected. However, only a 78.5% yield was observed by TGA and ICP-AES, probably due to the extensive processing steps involved in extracting the KC_8-Cu NP sample.

The metal halide to potassium stoichiometry was also investigated for the KC_8-$ZnCl_2$ reaction, including a three times excess of $ZnCl_2$ as used for the previous nanotubide experiments (Table 2). SEM (Fig. 4) revealed nanoparticles formed at all stoichiometries, and the overall d.o.f. slightly increased with increased $ZnCl_2$ concentration. X-ray photoelectron spectroscopy (XPS) (ESI Fig. S3) showed the typical Zn $2p_{3/2}$ binding energy peak position in zinc oxide,[43] ZnO (~1022 eV). Due to the comparatively larger size of the NPs formed at excess zinc concentrations, the surface sensitive XPS measurements are likely to underestimate the true Zn : C atomic ratio. Combined TGA and ICP measurements probe the full bulk of the samples and, therefore, provide a more reliable determination of the d.o.f.

Deriving the reduction potentials of KC_8 and KC_{24}

In the present case, reactions were performed using 10^{-3} M concentrations in non-aqueous solutions. Consequently, the reduction potentials of the metal salts/complexes are likely to be significantly different to standard reduction potentials that are based on 1 M solute concentrations, 25°C. The reduction potential of the metal salt/complex is shifted dependent on its concentration as defined by the Nernst equation, $E_{M^{n+}/M} = E^0_{M^{n+}/M} + ((RT/nF)\ln[M^{n+}])$, where E^0 is the standard reduction potential, and n is the number of electrons involved. It will be important to adjust the reduction potentials accordingly, in order to determine the extent of reaction.

For M^{2+} reactions (M^{2+} : K = 0.5 : 1): KC_8 (2.6 mM M^{2+}), $E = E^0 + (-0.076$ V), KC_{24} (0.87 mM M^{2+}), $E = E^0 + (-0.091$ V). For M^+ reactions (M^+ : K = 1 : 1): KC_8 (5.2 mM M^+), $E = E^0 + (-0.135$ V), KC_{24} (1.74 mM M^+), $E = E^0 + (-0.163$ V). A higher

Table 2 Comparison between KC_8-$ZnCl_2$ reactivity with varying Zn^{2+} : K stoichiometry

Stoichiometry Zn^{2+} : K	KC_8 (Zn : C atomic ratio) TGA/ICP	KC_8 (Zn : C atomic ratio) XPS
0.5 : 1	107	161
3 : 1	92	205
10 : 1	88	259

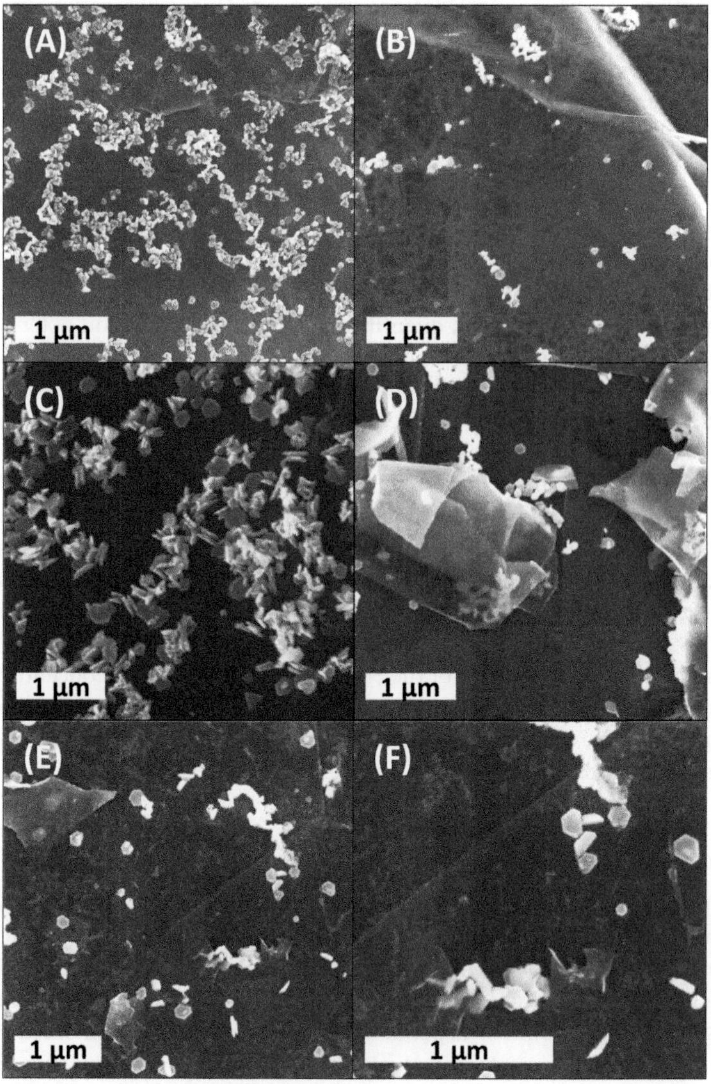

Fig. 4 SEM images of the generated Zn NPs from reaction of KC_8 with $ZnCl_2$ at varying stoichiometry. Two different regions are shown for each case, (A),(B) Zn^{2+} : K 0.5 : 1; (C),(D) Zn^{2+} : K 3 : 1; (E),(F) Zn^{2+} : K 10 : 1.

concentration of Zn^{2+} ions promotes a more favourable reduction to zinc (at Zn^{2+} : K 3 : 1, 15.6 mM, $E = E^0 + (-0.053\text{ V})$), and Zn^{2+} : K 10 : 1, 52 mM, $E = E^0 + (-0.038\text{ V})$), leading to the observed higher degree of functionalisation, Zn : C (by TGA/ICP-AES), and relatively larger nanoparticle formation, refer to Table 2 and Fig. 4.

The reduction potentials for $MnCl_2$, $ZnCl_2$ and CuMes at typical reaction concentrations were deduced from cyclic voltammetry experiments in 0.1 M $KClO_4$/NMP electrolyte (Fig. 5). The Ag/$AgNO_3$ non-aqueous reference electrode was calibrated against ferrocene/ferrocenium (Fc/Fc^+) (+0.08 V *vs.* Ag/$AgNO_3$) and then adjusted to the SHE (Fc/Fc^+ = +0.64 V *vs.* SHE). The derivation of metal salt/

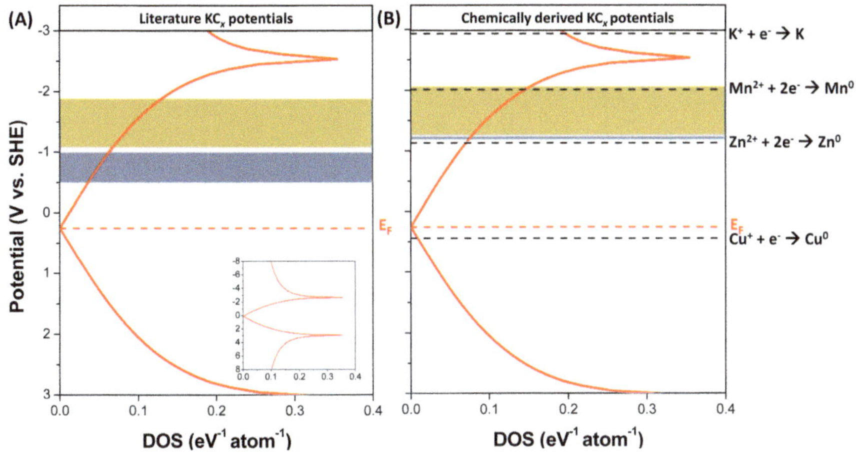

Fig. 5 DOS of a graphene crystal based on the tight-binding approximation (data values from ref. 30). Black dashed lines represent the reduction potentials for the metal compounds derived in this study in 0.1 M $KClO_4$/NMP. In (A), the bronze and blue shaded bands represent the approximate levels of doping for KC_8 and KC_{24}, respectively, based on experimental data for bulk GIC crystals[36] (partial charge transfer) and approximations[39] for complete charge transfer. Inset shows the perfect graphene DOS over a wider potential range. In (B), the bronze band represents the derived reduction potentials for KC_8 based on charge utilisation yields in reactions with $MnCl_2$, $ZnCl_2$ and CuMes. The blue line represents the derived reduction potential for KC_{24} based on the charge utilisation yield during the reaction with $ZnCl_2$.

complex reduction potentials from their cyclic voltammetric behaviour is discussed in ESI Fig. S4. The experimentally determined metal reduction potentials *vs.* SHE used for deriving the KC_8 and KC_{24} reduction potentials are: $MnCl_2$ (−2.01 V), $ZnCl_2$ (−1.13 V), CuMes (+0.44 V). Fig. 5 shows a schematic view of all the above reduction potentials with respect to the graphene DOS.

As described in Fig. 2, integrating the DOS between the reduction potential of the metal salt and the zero density Fermi level of undoped graphene affords an area of unavailable electron density for reaction. Integration of the respective “charge utilisation yield” (Table 1) above the metal salt potential provides an estimate of the KC_x reduction potential. The KC_{24} reduction potential could only be derived from its reactivity with $ZnCl_2$; integration of the DOS to 9.8% charge utilisation above the Zn potential gives −1.21 V *vs.* SHE symbolised by the blue coloured line in Fig. 5B.

For KC_8, the $MnCl_2$ reaction (3.6% charge utilisation) affords a reduction potential of −2.04 V *vs.* SHE, however, KC_8 reactivity with $ZnCl_2$ (15% charge utilisation) suggests a reduction potential of −1.25 V *vs.* SHE. Even at a 20-fold increase of Zn^{2+} : K stoichiometry (10 : 1), the reduction potential is relatively unchanged (−1.27 V *vs.* SHE). The reaction of KC_8 with CuMes suggests a reduction potential of −1.70 V, falling within the same region. This range of potentials is symbolised by the bronze coloured band in Fig. 5B. Based on the observed reactivity with $MnCl_2$, it would be reasonable to suggest that the KC_8 potential should lie towards the upper region of this band. The derived reduction potentials closely resemble the band corresponding to literature potentials discussed above (Fig. 5A). The significant difference between the KC_8 reduction

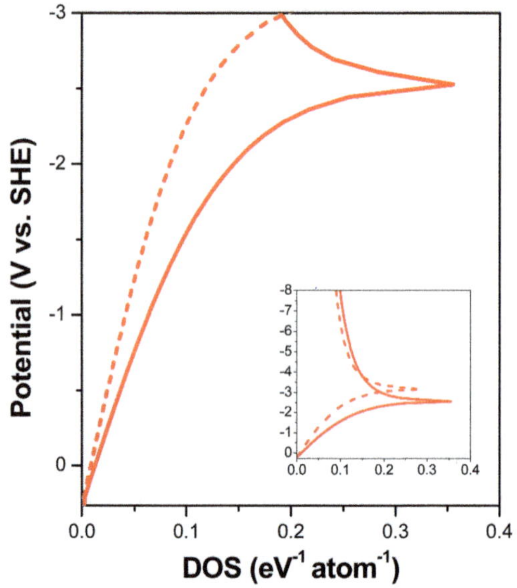

Fig. 6 Semi-quantitative estimate of Coulombic effects on the graphene DOS. The solid curve represents the tight-binding DOS as used above. The dashed and dotted curves represent the possible effect of Coulombic interactions due to the additional electron density in doped graphene. Inset shows the modified graphene DOS over a wider potential range.

potentials derived from the manganese and zinc reactivity might be due to various factors; most importantly, the DOS presented here is for a perfect graphene system and may not accurately represent the true DOS of highly doped graphene in solution.

Coulombic interactions that increase with added electron density are likely to affect the DOS such that there is a non-linear "stretching" of the density of states. This effect is semi-quantitatively represented in Fig. 6. Here, the potentials in the DOS plot are shifted with the addition of successive electron density, by the equation[44] $\Delta V = ne/C$, where ne is the added electron density and C is the capacitance. Qualitatively, with increased Coulombic repulsion, a slightly more reducing KC_8 potential would be obtained relative to the potentials derived above. This effect may explain the shift in range between Fig. 5A and 5B, and in particular the deposition of manganese. In addition, the change in shape of the DOS would tend to bring the estimates based on the different metals closer together. However, further theoretical simulations that incorporate both Coulombic and solvation effects are required to complement the proposed theory.

Experimental

Nanocarbon polyelectrolyte preparation and subsequent reactions were performed by SAH inside a nitrogen-filled Lab Master SP glove box (mBraun, Germany) containing <1 ppm O_2, H_2O at all times unless stated.

Preparation of charged carbon polyelectrolytes

HiPco nanotubide salt (NaC_{10}) was produced by DJB using the sodium/liquid ammonia reduction method[5] using raw HiPco SWCNT powder (Batch: R2-172, NanoIntegris, USA). After removing the ammonia, the remaining nanotubide salt was added to DMF (anhydrous grade, 99.8% further dried by 3 Å molecular sieves, both purchased from Sigma-Aldrich, UK) and left to spontaneously dissolve for at least 7 days. Fractions of the supernatant were then taken for subsequent reactions.

KC_8 and KC_{24} were made by CAH and PLC, respectively, using the vapour transport method[21] from graphite powder (4965-113.4G, Batch: 02426DA, Sigma-Aldrich, UK) and were confirmed to be phase pure by X-ray diffraction (within the limits of the measurement). For each experiment, 20 mL, NMP (anhydrous grade, 99.5%, further dried by 4 Å molecular sieves, both purchased from Sigma-Aldrich, UK), was added to 10 mg KC_x in a 100 mL Young's tap Schlenk tube. The sample was removed from the glove box, mildly sonicated for 30 min (ultrasonic cleaner, VWR, UK) and returned to the glove box for subsequent reactions.

Metal salts/complexes

$MnCl_2$ (anhydrous, beads, ~10 mesh, 99.999% trace metals basis), $ZnCl_2$ (anhydrous, beads, amorphous, ~10 mesh, 99.99% trace metals basis) and $CuCl_2$ (99.999% trace metals basis) were purchased from Sigma-Aldrich, UK, and used as- received. CuMes was purchased from Strem Chemicals, UK, and used as-received.

Nanotubide polyelectrolyte reactions

MCl_2 salt (M = Mn, Zn) was added to 3 mL DMF and added to 2 mL nanotubide dispersion (~1.2 mg nanotubide). An overall stoichiometry of 3 : 1 M : Na was used. The reaction was left to stir for 24 h before being removed from the glove box, washed and filtered under vacuum with 100 mL each of DMF, water, chloroform and ethanol (reagent grade, from Sigma-Aldrich, UK, and VWR, UK). Products were characterised by TGA and SEM.

Graphenide polyelectrolyte reactions

0.1 M stock solutions of MX_2 (M = Mn, Zn and Cu) salts and CuMes were made in NMP. Aliquots were added to the KC_x dispersion to give the desired stoichiometry of metal to potassium. The reaction was left to stir for 72 h before being removed from the glove box, washed and filtered under vacuum with 100 mL each of NMP, water, chloroform and ethanol. Products were dried and characterised as detailed in main article. The KC_8-CuMes sample was precipitated with toluene and centrifuged before filtration of the sediment. Air-free centrifugation was performed using a Sigma 2–16K centrifuge with Sigma 12139-H rotor at 15,000 g, 30 min. FEP Oak Ridge centrifuge tubes (Thermo Scientific, UK)were used and sealed with PTFE tape to maintain an inert environment. The sediment was filtered under vacuum with 100 mL each of toluene, THF, NMP, water, chloroform and ethanol (reagent grade, from Sigma-Aldrich, UK, and VWR, UK).

Characterisation techniques

Scanning electron microscopy (SEM). SEM was carried out by DBA and HHT on a high resolution field emission gun scanning electron microscope (FEGSEM) Leo Gemini 1525, with a built-in energy dispersive and wavelength dispersive X-ray spectrometer (EDX) (INCA, using INCA suite software V4.15, 2009, Oxford Instruments Plc., UK). Images were taken with an aperture of 30 μm typically at 5 keV. The gun voltage was increased to 15 keV for EDX spectra.

SEM/EDX samples were prepared on Al stubs using either silver dag, Leit adhesive carbon tabs or solution drop cast onto Al foil covered stubs. All SEM preparation products purchased from Agar Scientific, UK with Al foil supplied by VWR, UK. A representative EDX blank spectra was taken at regions away from the observed nanoparticles to ensure spectra were representative.

Thermogravimetric analysis (TGA). TGA experiments were carried out by SAH. TGA for the analysis of nanotubide reactions was performed using a Perkin Elmer Pyris 1 TGA. Samples were heated from 30 to 100 °C at 10 °C min^{-1}, and then held isothermally at 100 °C for 30 min in an inert atmosphere (N_2, 60 mL min^{-1}) to remove residual solvents. The temperature was then ramped to 850 °C at 10 °C min^{-1} in an oxidative atmosphere (air, 60 mL min^{-1}).

TGA for the analysis of graphenide reactions was performed using a Mettler Toledo TGA/DSC 1 with a GC200 flow controller. The TGA was coupled to a mass spectrometer (Hiden MS fitted with a 200 a.u. quadrupole sensor). Samples were heated from 30 to 100 °C at 35 °C min^{-1}, and then held isothermally at 100 °C for 30 min in an inert atmosphere (N_2, 60 mL min^{-1}) to remove residual solvents. The temperature was then ramped to 850 °C at 10 °C min^{-1} with a switch from N_2 to air at 600 °C. An inert gas flow was used between 100–600 °C to ensure there was no weight loss from trapped solvents, that would change the overall calculated carbon mass.

Inductively coupled plasma atomic emission spectroscopy (ICP-AES). ICP-AES was performed by SAH using a Perkin Elmer ICP 2000 DV OES (dual view optical emission spectrometer) in axial mode. Data was recorded using WinLab32 software. TGA residues (0.1–0.3 mg) were dissolved in 10 mL 10% HCl in water and further diluted by a factor of 10 for sampling. Calibration was performed using 1, 5 and 25 ppm solutions of ICP multi-element standard solution IV (Merck, Germany). A blank 10% HCl in water sample was carried out immediately following calibration, as a reference point for samples.

X-ray photoelectron spectroscopy (XPS). XPS spectra of powdered graphite samples were recorded by RM using a Thermo Scientific K-Alpha instrument using focused (400 μm spot) monochromatic Al-Kα radiation at a pass energy of 40 eV. The binding energies were referenced to the sp^2 C 1s peak of graphite at 284 eV.

UV-vis absorption spectroscopy. UV-vis spectroscopy was performed by SAH using a Perkin Elmer Lambda 950 spectrometer with samples measured in a sealed 4 mm path length optical glass cuvette.

Electrochemical characterisation. A Solartron 1287 Potentiostat (Solartron Analytical, UK) was used for the cyclic voltammetric (CV) measurements using CorrWare 2 software. CV experiments were performed inside a glove box. Experiments and analysis were performed by SAH and HHT. A 10 mL three electrode electrochemical cell was used, with platinum wire working and counter electrodes

(Laboratory Reagent, 0.5 × 100 mm) purchased from Fisher Scientific, UK. The reference electrode, Ag/Ag^+, containing 0.01 M silver nitrate and 0.1 M tetrabutylammonium perchlorate in acetonitrile, was purchased from IJ Cambria Scientific, UK. Anhydrous ($KClO_4$) was purchased from Sigma-Aldrich and used as-received. Electrolytic solutions of 0.1 M $KClO_4$ in NMP were further dried by 3 Å molecular sieves and stored in a glove box.

Conclusions

In this paper, nanocarbon polyelectrolyte anions have been used to reduce a series of metal salts/complexes, to attempt to quantify the reduction potential of these charged solutions. In characterising the degree of functionalisation and charge utilisation yields, combined TGA and ICP-AES techniques have provided accurate quantification by overcoming issues arising from the presence of alkali metal hydr(oxides) in TGA residues and the surface sensitivity of techniques such as XPS.

While it is difficult to confirm the exact electrochemical potentials of these dissolved, charged nanocarbons, the results obtained for potassium graphenide systems closely match previous experimental and theoretical predictions for partial and complete charge transfer from potassium to carbon. The reactivity appears to occur across a continuum of redox potentials, dominated by the filling of the electronic DOS of these nanocarbons, benefitting from the solvation and subsequent dissociation of graphenide and K^+ ions.

However, it is still not fully understood how the DOS may change as a function of doping, especially since electron affinities and Coulombic interactions will play a major role at increased charge densities. Particularly with carbon nanotubes, the distribution of charge will be determined by the specific nanotube physical properties (diameter, length), electronic type (metallic/semi-conducting), dielectric properties, and purity (metal catalyst/defective and amorphous carbons). The alkali metal type and its presence as the metallic form also have an important role. Other alkali metal GICs including LiC_x may demonstrate higher reactivity than KC_8, due to the higher reduction potential of Li and the ability to transfer more charge (LiC_6). However, solvation effects will play a key role in determining the GIC chemical activity.

To summarise, future experiments investigating a complete set of metal salts/complexes may allow the accurate determination of nanocarbon reduction potentials. Understanding these potentials, the nature of the electron transfer, and the extent of remaining unused charge, will help to improve the functionalisation of these technological relevant advanced carbons. As an aside, the observed production of metal nanoparticles is an interesting phenomenon, and may be applicable for a whole range of materials, from catalysts to nanoelectronic devices. Further nanoparticle stability and microscopy studies are required to probe the possible size distribution control that can be achieved with different stoichiometry.

Acknowledgements

The authors would like to thank EPSRC EP/G007314/1 and EP/L001896/1 for funding. HHT would like to thank the National Research Foundation, Energy Innovations Programme Office, Singapore.

Notes and references

1 M. S. Dresselhaus, G. Dresselhaus and P. Avouris, *Carbon nanotubes: Synthesis, structure, properties, and applications*, Springer, New York, 2001.
2 V. C. Moore, M. S. Strano, E. H. Haroz, R. H. Hauge, R. E. Smalley, J. Schmidt and Y. Talmon, *Nano Lett.*, 2003, **3**, 1379–1382.
3 S. D. Bergin, Z. Sun, P. Streich, J. Hamilton and J. N. Coleman, *J. Phys. Chem. C*, 2010, **114**, 231–237.
4 T.-J. Park, S. Banerjee, T. Hemraj-Benny and S. S. Wong, *J. Mater. Chem.*, 2006, **16**, 141–154.
5 S. Fogden, C. A. Howard, R. K. Heenan, N. T. Skipper and M. S. P. Shaffer, *ACS Nano*, 2012, **6**, 54–62.
6 E. M. Milner, N. T. Skipper, C. A. Howard, M. S. P. Shaffer, D. J. Buckley, K. A. Rahnejat, P. L. Cullen, R. K. Heenan, P. Lindner and R. Schweins, *J. Am. Chem. Soc.*, 2012, **134**, 8302–8305.
7 C. A. Howard, H. Thompson, J. C. Wasse and N. T. Skipper, *J. Am. Chem. Soc.*, 2004, **126**, 13228–13229.
8 C. A. Howard, J. C. Wasse, N. T. Skipper, H. Thompson and A. K. Soper, *J. Phys. Chem. C*, 2007, **111**, 5640–5647.
9 A. Catheline, C. Valles, C. Drummond, L. Ortolani, V. Morandi, M. Marcaccio, M. Iurlo, F. Paolucci and A. Penicaud, *Chem. Commun.*, 2011, **47**, 5470–5472.
10 A. Pénicaud, P. Poulin, A. Derré, E. Anglaret and P. Petit, *J. Am. Chem. Soc.*, 2005, **127**, 8–9.
11 S. A. Hodge, S. Fogden, C. A. Howard, N. T. Skipper and M. S. P. Shaffer, *ACS Nano*, 2013, **7**, 1769–1778.
12 S. A. Hodge, M. K. Bayazit, H. H. Tay and M. S. P. Shaffer, *Nat Commun*, 2013, 4.
13 Y. Chen, R. C. Haddon, S. Fang, A. M. Rao, P. C. Eklund, W. H. Lee, E. C. Dickey, E. A. Grulke, J. C. Pendergrass, A. Chavan, B. E. Haley and R. E. Smalley, *J. Mater. Res.*, 1998, **13**, 2423–2431.
14 M. K. Bayazit, A. Suri and K. S. Coleman, *Carbon*, 2010, **48**, 3412–3419.
15 R. Graupner, J. Abraham, D. Wunderlich, A. Vencelová, P. Lauffer, J. Röhrl, M. Hundhausen, L. Ley and A. Hirsch, *J. Am. Chem. Soc.*, 2006, **128**, 6683–6689.
16 D. Wunderlich, F. Hauke and A. Hirsch, *Chem. - Eur. J.*, 2008, **14**, 1607–1614.
17 T. Morishita, A. J. Clancy and M. S. P. Shaffer, *J. Mater. Chem. A*, 2014, **2**, 15022.
18 A. Mukherjee, R. Combs, J. Chattopadhyay, D. W. Abmayr, P. S. Engel and W. E. Billups, *Chem. Mater.*, 2008, **20**, 7339–7343.
19 J. Chattopadhyay, S. Chakraborty, A. Mukherjee, R. Wang, P. S. Engel and W. E. Billups, *J. Phys. Chem. C*, 2007, **111**, 17928–17932.
20 F. Hof, S. Bosch, S. Eigler, F. Hauke and A. Hirsch, *J. Am. Chem. Soc.*, 2013, **135**, 18385–18395.
21 C. A. Howard, M. P. M. Dean and F. Withers, *Phys. Rev. B: Condens. Matter Mater. Phys.*, 2011, **84**, 241404.
22 S. A. Hodge, M. K. Bayazit, K. S. Coleman and M. S. P. Shaffer, *Chem. Soc. Rev.*, 2012, **41**, 4409–4429.
23 Y. Tanaka, Y. Hirana, Y. Niidome, K. Kato, S. Saito and N. Nakashima, *Angew. Chem., Int. Ed.*, 2009, **48**, 7655–7659.
24 Y. Hirana, Y. Tanaka, Y. Niidome and N. Nakashima, *J. Am. Chem. Soc.*, 2010, **132**, 13072–13077.

25 D. Paolucci, M. M. Franco, M. Iurlo, M. Marcaccio, M. Prato, F. Zerbetto, A. Pénicaud and F. Paolucci, *J. Am. Chem. Soc.*, 2008, **130**, 7393–7399.
26 H. C. Choi, M. Shim, S. Bangsaruntip and H. Dai, *J. Am. Chem. Soc.*, 2002, **124**, 9058–9059.
27 M. K. Bayazit, S. A. Hodge, A. R. J. Clancy, R. Menzel, S. Chen and M. S. P. Shaffer, unpublished work.
28 J. F. Bondi, K. D. Oyler, X. Ke, P. Schiffer and R. E. Schaak, *J. Am. Chem. Soc.*, 2009, **131**, 9144–9145.
29 S. R. Ghanta, M. H. Rao and K. Muralidharan, *Dalton Trans.*, 2013, **42**, 8420–8425.
30 P. Hadley, *Density of states of a graphene crystal in the tight-binding approximation*, http://lamp.tu-graz.ac.at/~hadley/ss1/bands/tbtable/2d_graphene_dos.html.
31 M. S. Dresselhaus and G. Dresselhaus, *Adv. Phys.*, 1981, **30**, 139–326.
32 D. Savoia, C. Trombini and A. Umanironchi, *Pure Appl. Chem.*, 1985, **57**, 1887–1896.
33 A. Weitz and M. Rabinovitz, *Synth. Met.*, 1995, **74**, 201–205.
34 L. B. Ebert, D. R. Mills and J. C. Scanlon, *Mater. Res. Bull.*, 1982, **17**, 1319–1328.
35 L. B. Ebert, *Carbon*, 1985, **23**, 585–587.
36 Z.-H. Pan, J. Camacho, M. H. Upton, A. V. Fedorov, C. A. Howard, M. Ellerby and T. Valla, *Phys. Rev. Lett.*, 2011, **106**, 4.
37 S. A. Solin and H. Zabel, *Adv. Phys.*, 1988, **37**, 87–254.
38 D. Voiry, C. Drummond and A. Penicaud, *Soft Matter*, 2011, **7**, 7998–8001.
39 S. Pisana, M. Lazzeri, C. Casiraghi, K. S. Novoselov, A. K. Geim, A. C. Ferrari and F. Mauri, *Nat. Mater.*, 2007, **6**, 198–201.
40 M. Takahashi, J. Kamada, K. Iwata, K. Goto, H. Watanabe and S. Tamai, *Bull. Chem. Soc. Jpn.*, 2008, **81**, 168–170.
41 C. Barriere, K. Piettre, V. Latour, O. Margeat, C.-O. Turrin, B. Chaudret and P. Fau, *J. Mater. Chem.*, 2012, **22**, 2279–2285.
42 M. Singh, I. Sinha, M. Premkumar, A. K. Singh and R. K. Mandal, *Colloids Surf., A*, 2010, **359**, 88–94.
43 N. T. Mai, T. T. Thuy, D. M. Mott and S. Maenosono, *CrystEngComm*, 2013, **15**, 6606–6610.
44 R. W. Murray, *Chem. Rev.*, 2008, **108**, 2688–2720.

Faraday Discussions

PAPER

Spontaneous formation of metallic nanostructures on highly oriented pyrolytic graphite (HOPG): an *ab initio* and experimental study

Maria F. Juarez,*[a] Silvina Fuentes,[bc] Germán J. Soldano,[a] Lucia Avalle[bc] and Elizabeth Santos[ac]

Received 21st March 2014, Accepted 28th May 2014
DOI: 10.1039/c4fd00047a

We have investigated the decoration of step-edges of HOPG by Ag, Au and Pt using experimental and theoretical approaches. Metallic nanowires can be formed on bare or functionalized step-edges. Energy dispersion analysis indicates the presence of oxygenated groups. The experiments showed that nanowires can be obtained with the three metals along the step-edges, but the shapes and morphologies are very different. We have found that the interaction between the metal wires and the carbon follows the sequence: Pt > Au > Ag. The electronic redistribution between the atoms participating in the bond between the metallic nanowire and the step-edges shows a complicated pattern. The density of electronic states projected on the different atoms indicates that there are different orbitals participating in the bonds.

Introduction

Nanostructures are promising materials because of the capability to tune their properties by changing their morphology. One of the most interesting methods for their synthesis is the deposition on highly oriented pyrolytic graphite (HOPG). The topology of nanostructures resulting from metal deposition strongly depends on the interactions between metal–substrate, metal–metal and on the chemical environment. In the case of metallic substrates, pseudomorphic layer-by-layer deposits are usually obtained because of the strong metal–substrate interactions. On the contrary, metals interact very weakly with the coordinately saturated graphite basal plane surfaces; therefore the deposits are rarely epitaxially oriented on HOPG. In this case, the deposition process occurs mostly by a Volmer–Weber

[a]*Institute of Theoretical Chemistry, Ulm University, D-89069 Ulm, Germany. E-mail: fernanda.juarez@uni-ulm.de*
[b]*Facultad de Ciencias Exactas y Naturales, Universidad Nacional de Catamarca, Catamarca, Argentina*
[c]*Instituto de Física Enrique Gaviola (IFEG-CONICET), Facultad de Matemática, Astronomía y Física, FaMAF-UNC, 5000 Córdoba, Argentina*

mechanism,[1] *i.e.*, an instantaneous nucleation and subsequent three-dimensional growth mode of deposition. A variety of shapes and sizes can be obtained depending on the pre-treatment and environment used for the deposition.

The group of Penner has performed a systematic investigation to find appropriate conditions in order to obtain either narrowly disperse in size homogeneously distributed metal nanoparticles on atomically smooth HOPG,[2–5] and also dimensionally uniform nanowires (NW) on step-defects of HOPG.[6–10] Others groups were also very active in this field.[11–21] Metallic nanostructures on HOPG have been prepared by various methods such as chemical vapor deposition (CVD) using organometallic precursors,[19] physical vapor deposition (PVD) under controlled temperature of the substrate,[10] spontaneous deposition at open circuit potential (SP-ocp) immerging the HOPG in a solution containing ions of the metal to be deposited,[3,15,16,20] electrochemical deposition applying potential pulses (ED),[2–6,11,13,14,21] electrochemical step-edge decoration (ESED)[7,8] in order to obtain NW structures, galvanic displacement (GD) powered by the concurrent oxidation of insoluble derivatives.[9]

The group of Arvía[12,21] have obtained palladium mesoislands on HOPG with different shapes depending on the selected potential window for the electrodeposition. This growth mode change was attributed to the influence of chloride adsorbed on gold islands. Both these nanostructures show markedly different catalytic activity for the oxygen reduction reaction. There is evidence from energy dispersive X-ray analysis (EDX) that chloride is also present in spontaneously deposited platinum nanoparticles.[15] Shen *et al.* explained this phenomenon of the presence of chloride groups and internal charges in the nanoparticle by an incomplete reduction in the spontaneous deposition. They proposed that the existent of incompletely oxidized functionalities at the step edges, other defects of the delaminated graphite layer and the hydrogen atoms formed on HOPG surface could be part of the driving force for the electroless Pt deposition.

ED produces highly monodispersed nanoparticles (NP) homogeneously distributed on the whole surface, while SP-ocp preferentially conduces to deposit on defects forming NW structures. Cross *et al.*[10] used ensembles of 2–15 nm diameter gold nanoparticles obtained by PVD as nucleation templates for the preparation by ESED of gold nanowires.

Pre-oxidizing HOPG by anodic current at high potentials can eliminate spontaneous deposition.[3,15,16] In some cases, a chemical,[22] physical or electrochemical[7] pre-treatment of the HOPG is performed to induce defects, which are more active for the deposition. We can distinguish between two different domains on the surface: terraces where the carbon atoms are sp^2 bonded, and step-edges that have chemically active dangling bonds.[18] On one hand, by applying a positive potential the defects can react to form carbonyls, ethers, hydroxyls, and other oxygen containing functionalities.[7,19] On the other hand, if the applied potential is high enough, a complete oxidation of the defects takes place and spontaneous reduction of the metal cannot occur. A major task in the state of the art of this topic is to control the structure and functionalization of step-edges.

In previous works, we have performed *ab initio* investigations of both free-standing metal NW[23] and supported on step-edges.[24,25] More recently, we have found that in O_2/H_2 atmosphere, the graphene edges are mainly oxidized.[26] On the armchair edges, the terminal carbons are carboxyl groups, but on the zigzag edges, we have found stable ketones and aldehyde groups.

Now in this work, we focus on NW structures preferentially formed on functionalized step-edges by spontaneous deposition. Our goal is to understand the driving forces, which produce such structures and disentangle the metal–metal and metal–substrate interactions on perfectly hydrogenated and oxidized edges. We shall compare the stability of nanowires and small clusters on the HOPG edges.

Experimental

Highly oriented pyrolytic graphite (HOPG) provided by SPI Supplies (SPI-3, 7 × 7 × 1 mm) was used as a substrate for spontaneous deposition of different metals (Au, Pt and Ag). Previous to the experiments, the surfaces were prepared by cleaving the HOPG substrates and peeling off the top layers using an adhesive tape (macroporous Hipoalergic Clear Ind. Arg). The freshly cleaved surface obtained by means of this procedure[27] shows defects consisting of atomic steps, 0.2–0.3 nm and steps of several or dozens of atomic layers.[28]

The spontaneous metal deposition was carried out at open circuit potential (ocp) using different deposition times (1 s, 10 s, 90 s, 180 s). Immersion solutions of 10 mM H_2PtCl_6 (E. Merck, Schurchardt) in 0.1 M HCl, pH 2.6 were used to obtain nanostructures of Pt. In the case of Au and Ag nanostructures, solutions of 10 mM $HAuCl_4$ (Sigma Aldrich) and 10 mM Ag_2SO_4 (Tetrahedron) in 0.1 M H_2SO_4 (Baker analyzed-ACS. Reagent) were employed, respectively.

After the formation of the nanostructures the electrode was removed from the deposition solution and was characterized by scanning electron microscopy (SEM) and energy dispersive microanalysis (EDS) using a FEG SEM/CARL ZEIZZ-Sigma instrument.

The electrocatalytic activity for the hydrogen evolution reaction was qualitatively evaluated by cyclic voltammetry at a scan rate of 0.010 V s^{-1} and 0.050 V s^{-1} in a 0.5 M solution of H_2SO_4 (prepared from 98.2% purity Baker analyzed-ACS. Reagent). The electrochemical experiments were performed with a potentiostat – Metrohm – Autolab Galvanostat, Model PGSTAT302/302N at 25 ± 1°C.

Computational details

Periodic density functional theory (DFT) calculations based on plane waves have been performed as implemented in the DACAPO code.[29] The electron–ion interactions were accounted through ultrasoft pseudopotentials,[30] while the valence electrons were treated within the generalized gradient approximation (GGA) in the version of Perdew, Burke and Ernzerhof (PBE).[31] An energy cutoff of 450 eV was used with a (12 × 1 × 1) *k*-points Monkhorst–Pack grid[32] for zigzag and armchair unit cell. For the relaxations the convergence criterion was achieved when the total forces were less than 10 meV $Å^{-1}$.

Atomic charges were obtained using the Bader analysis method, which is an intuitive method to calculate the charge enclosed around an atom.[33,34] The software used was written by Arnaldsson *et al.* and was obtained from ref. 35–37.

All the optimized structures and electronic density differences were plotted with XCrySDen software.[38]

Modeling

We modeled the graphite steps using graphene nanoribbons (GNRs), *i.e.* graphene sheets which are periodic in one direction (x axis) and truncated by two edges (in the y axis). The edges correspond to the so-called armchair and zigzag configurations. One side of the nanoribbon is closed with hydrogen and the other one is used to study the adsorption of the metal wires. The vacuum between the periodic edges, and between graphene planes is greater than 12 Å. The width of the nanoribbon (from edge to edge) was optimized previously (ref. 26), and a distance of 9.27 Å for zigzag edges and of 8.03 Å for armchair edges was used. The carbon–carbon distance was taken from the optimized graphene lattice constant: 2.46 Å which is in excellent agreement with experimental values.[39] The two bottom layers of GNRs and the hydrogen adsorbed atoms were kept fixed and the rest was fully relaxed, *i.e.* the metal atoms, the graphite step and the corresponding functionalization of the ending.

Results and discussion

1. Morphological characterization

Fig. 1 shows SEM images obtained after spontaneous deposition of Au at different times. The decoration of step-edges with NPs of homogeneous size is evident from these pictures. The NPs coalesce into continuous NWs at large deposition times. NWs of about 2 μm length and 250 nm width were obtained after 3 minutes.

After peeling the HOPG, besides carbon the energy dispersive microanalysis of the fresh surface also shows traces of oxygen (Fig. 2, upper part). After immerging the HOPG in the deposition solution containing $HAuCl_4$, the ED spectrum evidences the presence of Au and an increase of oxygen (Fig. 2 bottom). It is possible that during the metal growth, functionalization of the defects simultaneously takes place. Therefore, carbonyls, ethers or hydroxyl groups can be formed, supplying the necessary reducing equivalents to deposit the metal.

Fig. 3 shows SEM images of SD-ocp of Pt at different times. Similar to Au, Pt mainly nucleates on the step-edge defects of the HOPG. However, at longer deposition times, the NPs coalesce stronger than in the case of Au. They form almost spherical conglomerates and the NWs are shorter than for Au. There is also the tendency to form circular structures, as observed previously by Zoval *et al.*[3] The EDS shows the presence of Pt and small amounts of chloride. A similar situation has been observed by Gimeno *et al.*[21] for Pd deposition on HOPG. They explained that clusters initially formed acquire a compact round disk shape, because the adsorbed chloride assists the isotropic inter-terrace surface diffusion of palladium adatoms. In the present case, the enhancement of Pt adatom surface diffusion promoted by chloride ions can also explain the appearance of compact rounded islands. Shen *et al.*[15] propose that the reduction of $PtCl_6^{2-}$ to Pt is incomplete in the spontaneous deposition process. Therefore, there is still the possibility that chloride remains attached to Pt NPs.

In the case of Ag deposited spontaneously on HOPG, the nucleation also starts on the step-edges, but the growth can also occur in directions perpendicular to defects, as can be observed in Fig. 4. Although NWs are also formed at longer deposition times, their structure is irregular and porous-like. They seem to be formed by cubic nanocrystals linked by the vertices, such as can be observed from

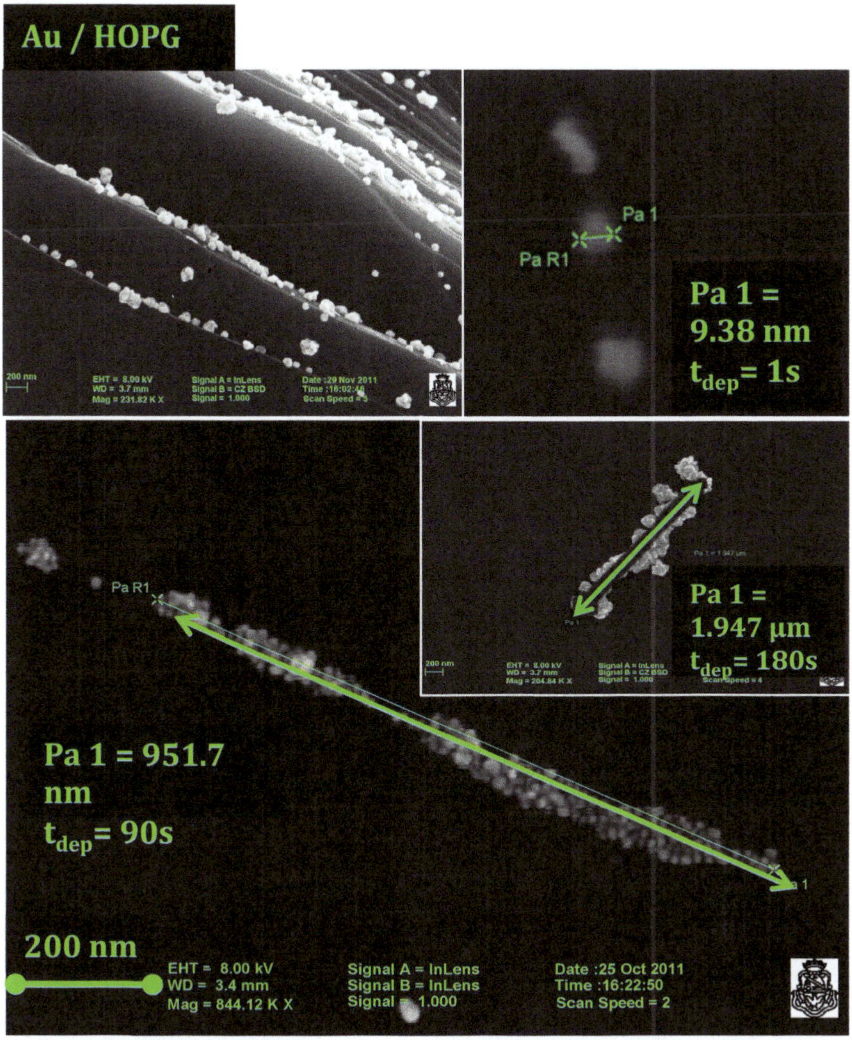

Fig. 1 SEM images of Au deposited on HOPG at different times.

the zoom of the image shown in the inset of Fig. 4. The cubic shape indicates that the nanocrystals expose predominantly (100) facets. EDS analysis shows besides Ag, the presence of traces of sulphur, probably coming from the sulphate electrolyte used for the deposition and an incomplete reduction. Although Ag displays a unique oxidation state, it can occur a similar phenomenon as in the case of Pt. Traces of Ag_2SO_4 salt could be also contained in the metallic NPs.

2. Substrate: stability of the graphene endings

Among the various possible atomic arrangements at step-edges of HOPG, the armchair and the zigzag are particularly stable configurations (Fig. 5, left).

We have considered three different endings for the zigzag and armchair steps: bare, hydrogenated, and oxidized terminations. The optimized structures of the steps used in this work are shown in Fig. 5.

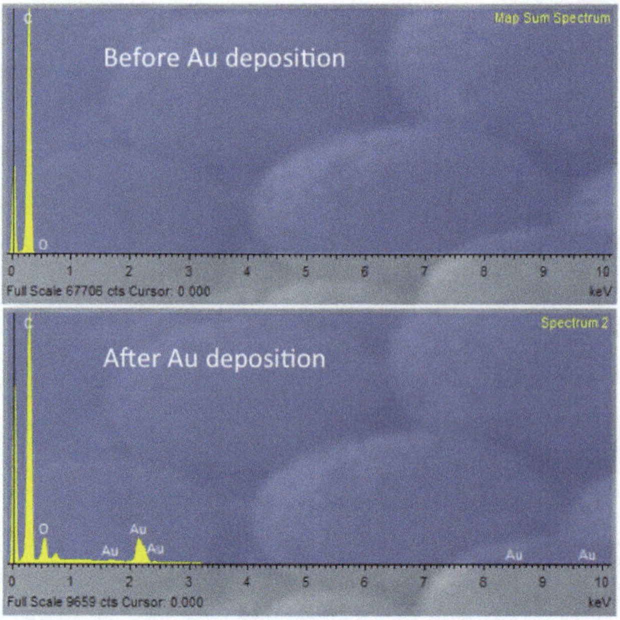

Fig. 2 Energy dispersive spectra (EDS) obtained immediately after delamination of the HOPG plane (upper part), and after electroless Au deposition by immersion of the HOPG in a solution containing $HAuCl_4$ (bottom part).

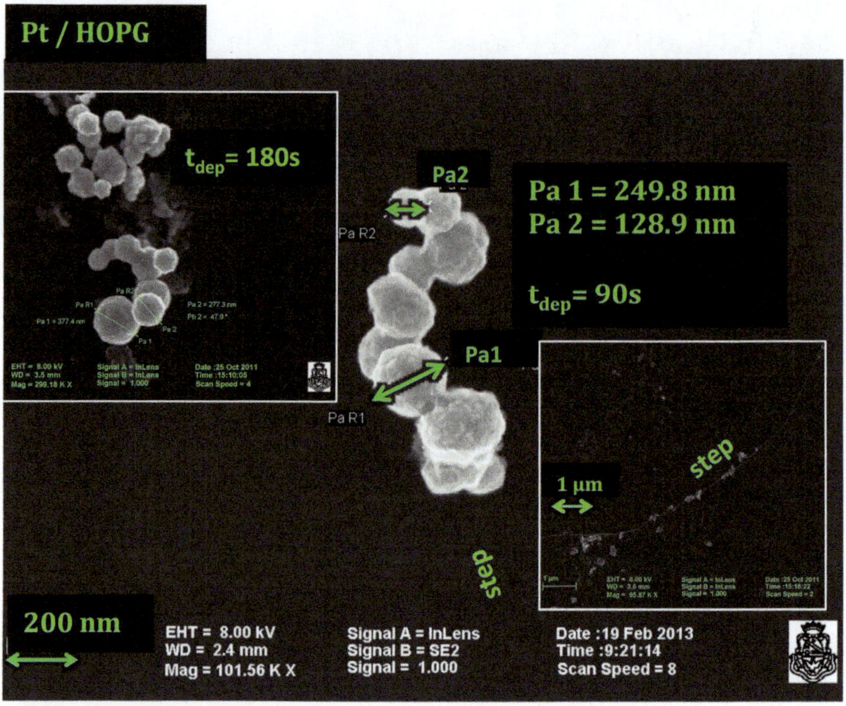

Fig. 3 SEM images of Pt deposited on HOPG at different times.

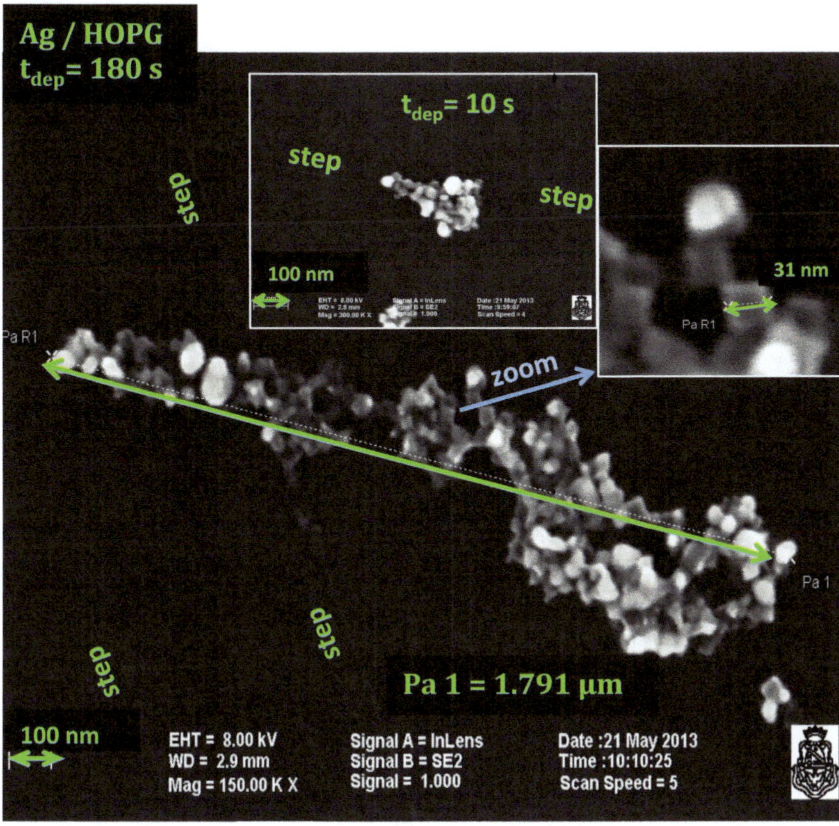

Fig. 4 SEM images of Ag deposited on HOPG at different times.

In the zigzag steps, the bare ending consists of dangling sp^3 orbitals, centered in the carbon atoms. The lonely electrons in such orbitals are very reactive, making the system unstable in real conditions. However, it is interesting to consider the simplest termination in order to understand the nature of the interaction between the metal wires and the graphene steps.

In a previous work[26] we have studied the thermodynamic stability of several functionalizations of the graphene endings. We have only considered functional groups containing C, O and H atoms, but we have found that in a wide range of temperature and pressure, two carboxylic groups placed opposite are the most stable ending on the zigzag steps. On the armchair endings, two distant lactone rings are the preferential termination.

The stability of the structures has been evaluated by calculating the energy for the formation reaction:

$$\mathrm{C_{graphene}} + \frac{1}{2}n\mathrm{H_2} + \frac{1}{2}m\mathrm{O_2} \rightarrow \mathrm{C_{step}H}_n\mathrm{O}_m \tag{R1}$$

$$\Delta E_{\mathrm{form}} = \frac{E\left(\mathrm{C_{step}H}_n\mathrm{O}_m\right) - E\left(\mathrm{C_{graphene}}\right) + \frac{1}{2}nE(\mathrm{H_2}) + \frac{1}{2}mE(\mathrm{O_2})}{L_{\mathrm{x}}} \tag{1}$$

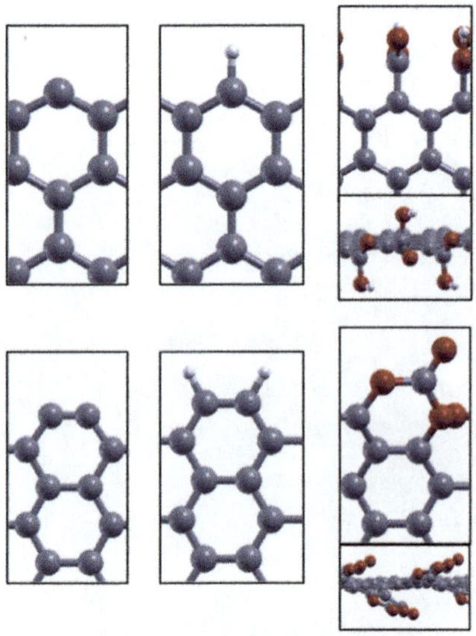

Fig. 5 Top views of the optimized structures of the zigzag (upper part) and armchair (bottom) steps: bare (left), hydrogenated (middle), and oxidized (right) terminations. A side view has been included for the oxidized steps in order to show the three-dimensional arrangement of the functional groups.

where $C_{step}H_nO_m$ is the functionalized graphene step, $C_{graphene}$ is an infinite layer of graphene, and H_2 and O_2 are the hydrogen and oxygen molecules in the vacuum, respectively. According to this definition, a positive formation energy means that energy is required in order to create a functionalized step. For the sake of comparison, all the energy values have been normalized by the length L_x of the corresponding step, making it possible to compare the results obtained in zigzag and armchair steps. The formation energies of systems shown in Fig. 5 are collected in Table 1.

The steps with unpaired electrons (the bare steps) are the most unstable, whereas the oxidized steps are the most stable ones. The armchair steps are slightly more stable than zigzag steps in all the studied cases. However, the differences enlarge if we consider the length per step atom. The separation between carbon atoms in zigzag steps is larger than in armchair systems (2.46 Å *vs.* 2.13 Å), making the formation energy per atom bigger in the zigzag steps.

Table 1 Formation energies of the functionalized graphene steps shown in Fig. 5

Step		ΔE_{form} (eV Å^{-1})
Zigzag	Bare	1.08
	Hydrogenated	0.088
	Oxidized	−0.95
Armchair	Bare	0.98
	Hydrogenated	0.038
	Oxidized	−1.14

The formation energies of the hydrogenated endings are very close to zero, denoting that energy required to cleavage the graphene layer is compensate with newly formed C–H bonds.

3. Metal nanowires on graphene steps: energetics, and structural analysis

In this part, we shall consider separately different aspects of the adsorption of gold, silver, and platinum metal to form wires on the steps. We shall start with the geometric analysis of the optimized structures and we shall continue with energetics of the adsorption process.

On the bare steps, the geometrical arrangement of the atoms was the same for the three considered metals. A schematic plot of the geometries is shown in Fig. 6. Table 2 presents the most important bond lengths and angles of the metal wires adsorbed on different graphene steps.

On the zigzag steps, every metal atom is bound to a carbon atom with a distance between them that is shorter than metal–metal bond distances in gold, platinum and silver free-standing wires. Therefore, and in order to improve the bond between the metal atoms, the wire loses its lineal geometry. As we can observe in Table 2, the larger distortion angles θ_{wire} correspond to the metal wires supported on armchair steps, where the metal–metal bond length is even larger than in zigzag steps. The carbon–metal atom distance decreases in the sequence: Ag > Au > Pt, in zigzag and armchair steps.

The adsorption of metal wires was also studied in the hydrogenated steps. The optimized systems are plotted in Fig. 7. It is straightforward to notice that the final geometry depends on the nature of metal atoms.

On the hydrogenated armchair steps, the final structure is similar for the three metals. In these systems, one of the hydrogen atoms is bound to the two metal atoms. This interaction is stronger in platinum than in silver, and it increases the metal–metal bond length in the former case, and makes the opposite in the later case. We have also found a correlation between the increase in the metal–metal bond length and the reduction of angle θ_{wire}.

Another investigated system is the platinum nanowire supported on oxidized steps. Different views are plotted in Fig. 8, in order to show the special geometries of adsorption. It is important to point out that on the zigzag steps, the platinum atoms break the carboxylic groups, take away the OH and produce new bonds with the oxygen atoms. On the armchair steps, only one metal atom (of a total of four) is bound to one lactone cycle. In this case, there are two bonds that attach the

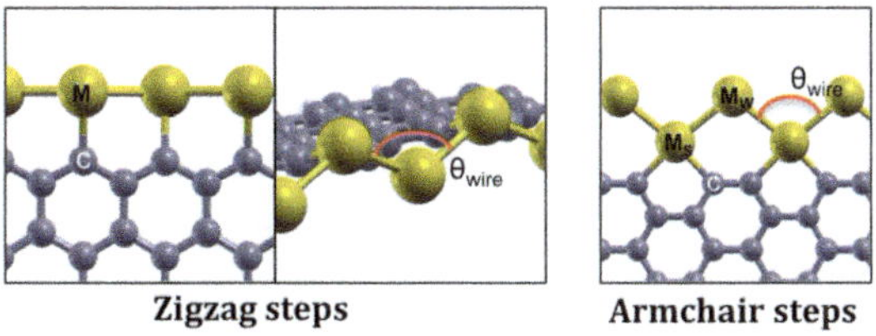

Fig. 6 Top view of the optimized metal wires adsorbed on bare zigzag (left) and armchair (right) steps. A side view of the metal wire in the zigzag step has been included in order to show the lateral structure of the wire.

Table 2 Geometrical parameters of the metal nanowires supported on bare, hydrogenated and oxidized graphene steps. The corresponding values for free standing wires are shown for sake of comparison

(1) Ag

	Free	Bare		Hydrogenated	
		Zigzag	Armchair	Zigzag	Armchair
d_{MM}/Å	2.70	2.75	2.79	2.88	2.72
θ_{wire}	180	127	99	117	103
d_{CM}/Å	—	2.16	2.17	2.16	2.17

(2) Au

	Free	Bare		Hydrogenated	
		Zigzag	Armchair	Zigzag	Armchair
d_{MM}/Å	2.66	2.74	2.74	2.60	2.84
θ_{wire}	180	128	102	142	97
d_{CM}/Å	—	2.09	2.12	2.13	2.13

(3) Pt

	Free	Bare		Hydrogenated		Oxidized	
		Zigzag	Armchair	Zigzag	Armchair	Zigzag	Armchair
d_{MM}/Å	2.41	2.55	2.60	2.55	2.66	2.76	2.49[a]
θ_{wire}	180	150	110	148	106	126	151[b]
d_{CM}/Å	—	1.94	2.07	1.97	2.09	2.01	2.08

[a] The value corresponds to the average of values obtained for the four distinguishable Pt atoms. [b] The value corresponds to the angle between the atoms in the ends and the middle of the wire.

wire, one with oxygen atom outside of the ring, and the other one with most oxidized carbon atom in the lactone.

Platinum wires adsorbed on oxidized steps exhibit two different behaviours. On the zigzag steps, the nanowire is strongly attached to the step, increasing the metal–metal bond largely. On the other hand, the platinum wire supported on the armchair step is weakly adsorbed, and the final geometry of the nanowire is quite similar to the one in vacuum.

In order to analyse the stability of the previous systems, we have calculated several formation reactions considering the functionalization of the steps.

Adsorption on bare steps. The adsorption of a nanowire in a bare step can be described using the main reaction in Scheme 1. The energy of such adsorption is given by the equation

$$\Delta E_{\text{ads}} = \frac{E(\text{C}_{\text{step}}\text{M}_m) - E(\text{C}_{\text{step}}) + mE(\text{M}_{\text{wire}})}{L_{\text{x}}} \quad (2)$$

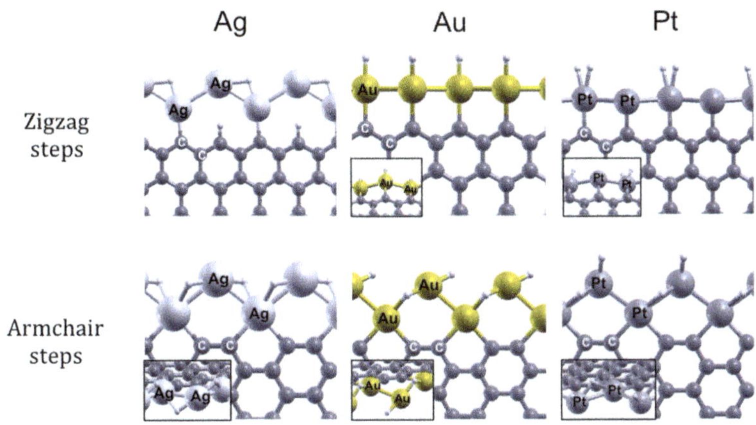

Fig. 7 Top view of the optimized metal wires adsorbed on hydrogenated zigzag (upper part) and armchair (bottom) steps. Tilted and side views of the metal wire in the zigzag and armchair step have been included in order to show the lateral distortion of the nanowire structure.

where $E(C_{step}M_m)$, $E(C_{step})$, and $E(M_{wire})$ are the energies of adsorbed nanowire, bare step, and free standing wire, respectively. In order to decompose the adsorption energy ΔE_{ads} in its fundamental parts, we have used a thermodynamic cycle adding three reactions: (1) the distortion of a relaxed nanowire, (2) the distortion of a relaxed step, and (3) the binding between the two unrelaxed parts (wire + step).

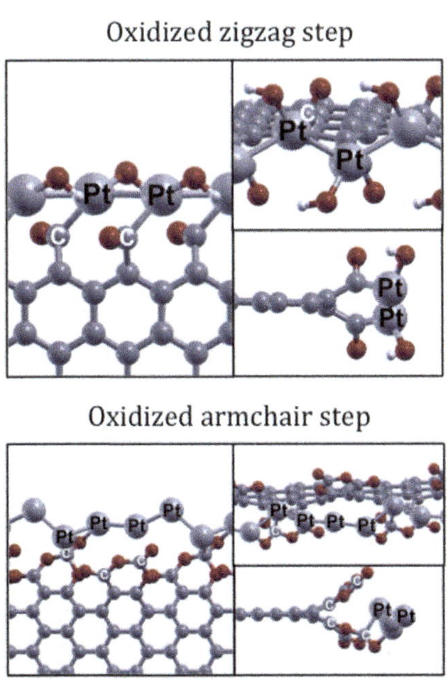

Fig. 8 Top and side views of the optimized Pt metal wires adsorbed on oxidized zigzag (upper part) and armchair (bottom) steps.

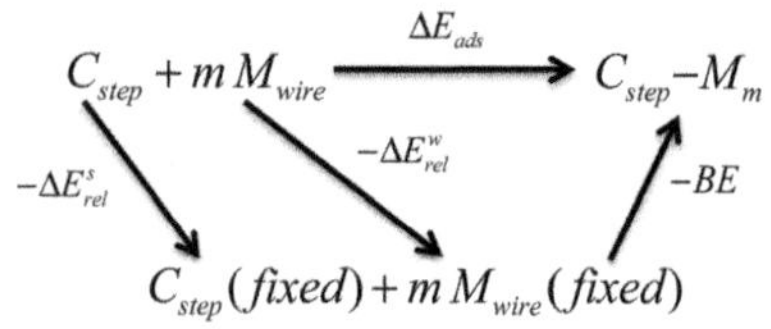

Scheme 1 Adsorption reaction of a nanowire in a bare step. The thermodynamic cycle decomposes the adsorption process in three different reactions: (1) distortion of the nanowire, (2) distortion of the substrate (graphene step), and (3) union of the two parts (step + wire).

The energy required to produce the distortion of the nanowires or the steps is the opposite of the relaxation energy of them, and can be described using the equations:

$$\Delta E_{\mathrm{rel}}^{\mathrm{s}} = \frac{E(\mathrm{C_{step}}) - E\left(\mathrm{C_{step}^{fixed}}\right)}{L_{\mathrm{x}}} \tag{3}$$

$$\Delta E_{\mathrm{rel}}^{\mathrm{w}} = \frac{E(M_{\mathrm{wire}}) - E\left(M_{\mathrm{wire}}^{\mathrm{fixed}}\right)}{L_{\mathrm{x}}} \tag{4}$$

where $E(\mathrm{M_{wire}^{fixed}})$ and $E(\mathrm{C_{step}^{fixed}})$ are the energies of the graphene step and the metal nanowire with the same geometry as in the adsorbed system (wire + step). According to these definitions, a negative value means that the energy of the relaxed (free) substrate (or adsorbate) is lower than energy of the fixed system. Therefore, in this case, it is necessary to give energy to the system in order to modify the structure of the substrate or the adsorbate.

The final contribution to the adsorption is the binding energy between the adsorbate (nanowire) and the substrate (graphene step):

$$\mathrm{BE} = \frac{E\left(\mathrm{C_{step}^{fixed}}\right) + mE\left(\mathrm{M_{wire}^{fixed}}\right) - E(\mathrm{C_{step}M}_m)}{L_{\mathrm{x}}} \tag{5}$$

It is important to notice that either the nanowire or the step, must have exactly the same geometry as in the final system, so the binding energy is only the result of the chemical interaction between the two of them. As it follows from eqn (5), a positive value corresponds to an attractive interaction between substrate and adsorbate. From the thermodynamic cycle, the adsorption can be reconstructed as is shown in the equation:

$$\Delta E_{\mathrm{ads}} = -\mathrm{BE} - \Delta E_{\mathrm{rel}}^{\mathrm{s}} - \Delta E_{\mathrm{rel}}^{\mathrm{w}}$$

The energies are summarized in Table 3.

The values obtained in Table 3 lead to several conclusions.

The first conclusions are about the distortion in the geometries of the wires and steps. We have found that the relaxation energies for the nanowires are small in all the cases and they are within the DFT error (0.1 eV). The relaxation energies

Table 3 Adsorption energies for nanowires supported on bare zigzag and armchair steps. The relaxation and binding energies have been included to decompose the adsorption energy, according to the thermodynamic cycle shown in Scheme 1

Step	M	ΔE_{ads} (eV Å^{-1})	BE (eV Å^{-1})	ΔE^{w}_{rel} (eV Å^{-1})	ΔE^{s}_{rel} (eV Å^{-1})
Zigzag	Ag	−0.42	0.44	0.023	−0.042
	Au	−0.50	0.58	0.056	−0.133
	Pt	−1.17	1.24	−0.002	−0.064
Armchair	Ag	−0.37	0.47	0.029	−0.130
	Au	−0.63	0.76	0.042	−0.170
	Pt	−0.80	1.04	0.012	−0.254

for the steps are larger in the armchair than in the zigzag case, the platinum wire being the one that produces the biggest distortion.

Secondly and according to the analysis of the binding energy, we have found that the interaction between the metal wires follows the sequence: Pt > Au > Ag. We want to point out here that there is a correlation between the increase in the binding energies and the reduction in the bond length d_{CM}, that follows the enhancement in the bond between the substrate and the metal nanowire.

Finally we have observed that, due to the fact the relaxation energies are mainly negligible, the adsorption energies can be estimated from the binding energies. The only exception is the platinum wire on armchair steps, where the distortion energy is significant.

So far we have only considered the adsorption of nanowires from bare graphene steps. However, it is also possible to produce the same structures using hydrogenated or oxidized steps as starting surfaces, and removing the functional groups:

$$C_{step}H_n + mM_{wire} \rightarrow C_{step} - M_m + \frac{1}{2}nH_2 \quad \text{(R2)}$$

$$C_{step}H_nO_k + mM_{wire} \rightarrow C_{step} - M_m + \frac{1}{2}nH_2 + \frac{1}{2}kO_2 \quad \text{(R3)}$$

The energies of the reactions (R2) and (R3) are named ΔE_{ads2}, and ΔE_{ads3}, respectively. The values obtained for the three metal wires are provided in Table 4. According to these results, the adsorption of the nanowires needs external energy in order to completely remove the oxide from the step. A similar situation occurs with the hydrogenated steps, the only exception being the platinum wires on the zigzag steps.

Adsorption on hydrogenated steps. Several reactions can be defined for the formation of the same system, but we shall consider only the following ones:

$$C_{step}H_n + mM_{wire} \rightarrow C_{step} - M_mH_n \quad \text{(R4)}$$

$$C_{step} - M_m + \tfrac{1}{2}nH_2 \rightarrow C_{step} - M_mH_n \quad \text{(R5)}$$

While both reactions have the same final product, the reactants are different and their meanings too. Reaction (R4) is the adsorption of a metal wire with partial or total displacement of hydrogen from the step. The first case

Table 4 Adsorption energies for nanowires supported on bare steps as they are defined by reactions (R2) and (R3). All the energies have been normalized by the length L_x of the unit cell

Step	M	ΔE_{ads2} (eV Å^{-1})	ΔE_{ads3} (eV Å^{-1})
Zigzag	Ag	0.58	1.60
	Au	0.50	1.53
	Pt	−0.17	0.85
Armchair	Ag	0.57	1.76
	Au	0.31	1.50
	Pt	0.14	1.33

corresponds to silver on zigzag steps, and in all the cases, hydrogen atoms make new bonds with the metal atoms.

Reaction (R5) does not represent the adsorption of any metal atoms but of hydrogen atoms, with a chance of partial remotion of metal atoms. Again, the last case corresponds to silver on zigzag steps. The adsorption of hydrogen atoms is an important reaction that has been widely used to study the reactivity of an extensive list of systems (surfaces, modified surfaces, nanostructures).

The energies ΔE_{hydr1} (reaction (R4)) in Table 5 indicate that the formation of these systems requires energy for gold and silver nanowires. Platinum has not only a strong interaction with the graphene steps but also with the hydrogen atoms. Therefore, the adsorption of Pt atoms compensates the breaking of the C–H bonds.

As the results for the energies ΔE_{hydr2} show in Table 5, the adsorption of hydrogen on the supported metal wires improves the stability of the systems, lowering the final energy and gaining energy after the reaction. This effect is more remarkable for silver on zigzag steps and gold on armchair endings.

Adsorption of Pt on oxidized steps. The last reaction studied is the adsorption of a Pt wire on an oxidized step:

$$\mathrm{C_{step}H}_n\mathrm{O}_k + m\mathrm{M_{wire}} \rightarrow \mathrm{C_{step}} - \mathrm{M}_m\mathrm{H}_n\mathrm{O}_k \qquad \text{(R6)}$$

The energies are briefly summarized in Table 6. We have found that it is still possible to adsorb platinum on the oxidized step, but the energy obtained is smaller than in bare or hydrogenated steps. On zigzag steps, a possible explanation for this result is the large amount of energy invested on the breaking of OH

Table 5 Energies of the reactions (R4) and (R5), involving nanowires supported on hydrogenated steps. All the energies have been normalized by the length L_x of the unit cell

Step	M	ΔE_{hydr1} (eV Å^{-1})	ΔE_{hydr2} (eV Å^{-1})
Zigzag	Ag	0.22	−0.36
	Au	0.45	−0.05
	Pt	−0.24	−0.08
Armchair	Ag	0.64	−0.15
	Au	0.31	−0.39
	Pt	−0.18	−0.10

Table 6 Energies of the reaction (R6), between platinum nanowires and oxidized steps. All the energies have been normalized by the length L_x of the unit cell

Step	M	ΔE_{oxid} (eV Å^{-1})
Zigzag	Pt	−0.14
Armchair	Pt	−0.07

group from the carboxylic ending. The situation is quite different on the armchair step, where the nanowire is attached to a lower number of atoms from the step.

4. Metal nanowires on graphene steps: density of states (DOS), charges and electronic density differences

In this final part, we shall describe with a certain detail the nature of the bond between the metal wires and the bare zigzag and armchair steps. We shall use the results obtained through Bader analysis, electronic density difference and atomic projected DOS (pDOS). And we shall finish with a short description of the platinum wires on three functionalized graphene steps.

The atomic charges obtained through the Bader analysis are summarized in Table 7. On the zigzag steps, the charge on the metal atoms depends on the nature of the metal. The largest value in this case was obtained for the silver nanowire. We have also found that the positive charges on the metal atoms correlate with the negative charge on the carbon atoms. These results suggest that there is a small charge transfer between the metal and carbon atoms.

On the armchair steps, we have two different metal atoms: M_s (closer to the step) and M_w (far from the step). On one hand, the positive charges on the atom M_s are larger than in the zigzag steps, but they are independent of the nature of the atom. But on the other hand, the charges on the atom M_w are negative and depend on the metal. As in the zigzag steps, the carbon atoms are negatively charged too.

In order to get a better description, we have also used the electronic density difference defined by the equation:

$$\Delta\rho = \rho(\mathrm{C_{step}M}_m) - \rho(\mathrm{C_{step}^{fixed}}) + \rho(\mathrm{M}_m^{\mathrm{fixed}}) \tag{6}$$

where $\rho(\mathrm{C_{step}M}_m)$, $\rho(\mathrm{C_{step}^{fixed}})$, and $\rho(\mathrm{M}_m^{\mathrm{fixed}})$ are the electronic densities of the adsorbed nanowire, the fixed step and the fixed wire, both fixed at the same geometry that the adsorbed nanowire, respectively. According to this definition, a positive value corresponds to an accumulation of electrons.

Table 7 Bader charges for the carbon and metal atoms in the metal wires supported on bare graphene steps

	Zigzag steps				Armchair steps		
Charge	Ag	Au	Pt	Charge	Ag	Au	Pt
q_M	+0.19	+0.09	+0.06	$q_{\mathrm{M_s}}$	+0.35	+0.36	+0.35
q_C	−0.14	−0.10	−0.07	$q_{\mathrm{M_w}}$	−0.05	−0.17	−0.13
				q_C	−0.18	−0.13	−0.16

Fig. 9 shows the density difference obtained for the selected systems. Besides the fact that the obtained plots are complex, we can observe some trends. In all the systems studied there is an accumulation of electrons between the carbon and metal atoms. In the zigzag steps, there is also an increase of density between the metal atoms in the wire.

On the armchair we can observe that the density in the atom M_s remains constant in the three metal, but it is rather different for the atom M_w. This observation corroborates the result obtained with the Bader analysis.

It is important to notice that, even when the atomic charges calculated with the Bader method are small, the local distribution of electrons can be rather complex and full of redistributions between different atoms and orbitals in the same atom.

Finally, we want to point out that the complex structure of the pictures indicates that there are more that one kind of orbitals participating in the bonds, and changing their populations during the adsorption.

To complete the analysis of the nature of the wire/step bond, we have projected the DOS in the sp and d bands of the metal atoms, and the sp band of the carbon atoms. Fig. 10 shows the results for the nanowires supported on the zigzag and armchair steps.

On the zigzag steps, the d and sp bands of the metal atoms contribute equally to the bond with the step, whereas on the armchair steps, the sp band has a minor role.

After adsorption of the nanowire, we can observe a shift in the d bands to lower energies, with the only exception of silver on zigzag steps. On the armchair steps, the atom M_s has a larger shift than the atom M_w.

Finally it is interesting to analyse the case of the silver nanowire adsorbed on the zigzag step. Atomic charges have shown that the silver atom is positively charged, with a small charge of +0.19. However, the density difference indicates that the situation is more complicated, and there is also an accumulation of

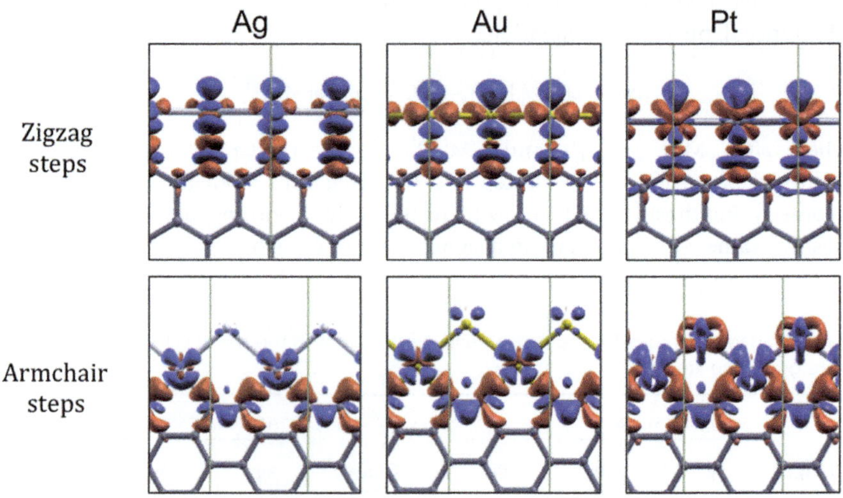

Fig. 9 Top view of the electronic density difference obtained using eqn (6). Negative (positive) values are plotted in blue (red). The isocontours correspond to a value of 0.006 electron per Å^3.

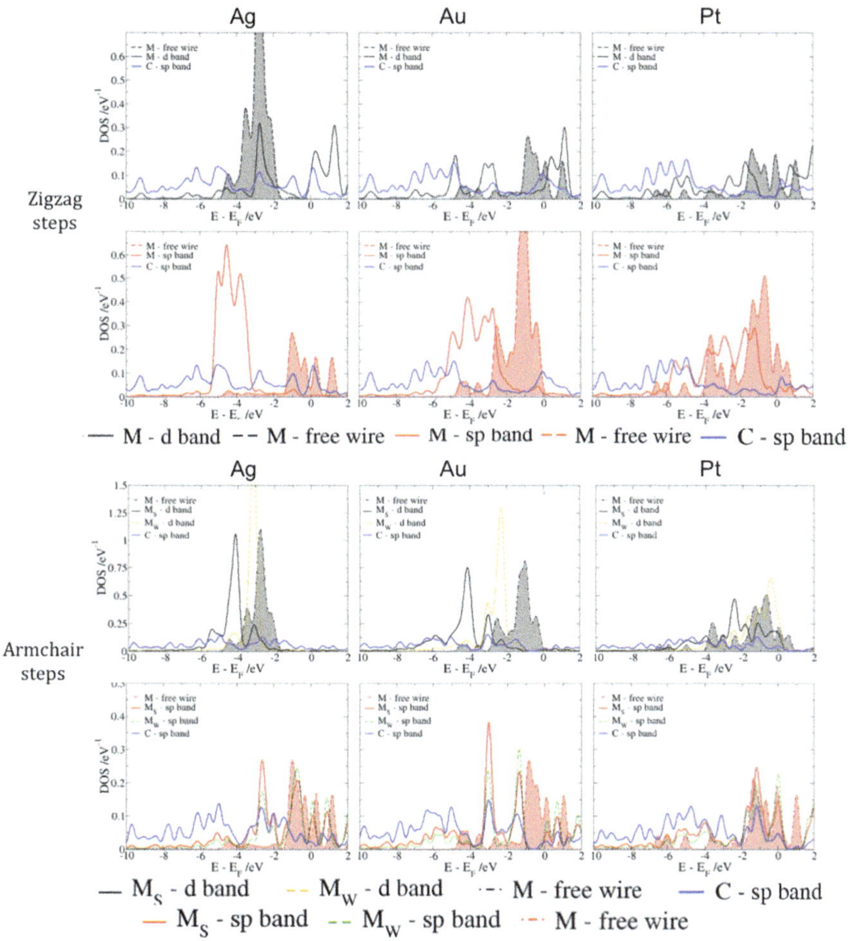

Fig. 10 Atomic projected DOS of metal nanowires adsorbed on bare graphene steps. The projections are plotted in different colors as follows: blue lines for the sp band of C atom; black and red lines for the metal d and sp band, respectively. The filled curves have been added for the sake of comparison, and correspond to the orbitals in the free standing wire.

electrons in the silver atom. The pDOS supports this observation and gives more information. We can observe in Fig. 10 that the occupation of the d band is smaller on the supported wire than on the free-standing wire. In Fig. 10, we can also notice that the loss of electrons in the d band is compensated by an increase

Table 8 Bader charges for the carbon and Pt atoms in the platinum nanowires supported on bare, hydrogenated, and oxidized graphene steps

	Zigzag steps				Armchair steps		
Charge	Bare	Hydrogenated	Oxidized	Charge	Bare	Hydrogenated	Oxidized
q_M	+0.06	+0.15	+0.54	q_{M_s}	+0.35	+0.26	+0.21[a]
q_C	−0.07	−0.15	+1.6	q_{M_w}	−0.13	+0.05	−0.06[b]
				q_C	−0.16	−0.21	+2.9[c]

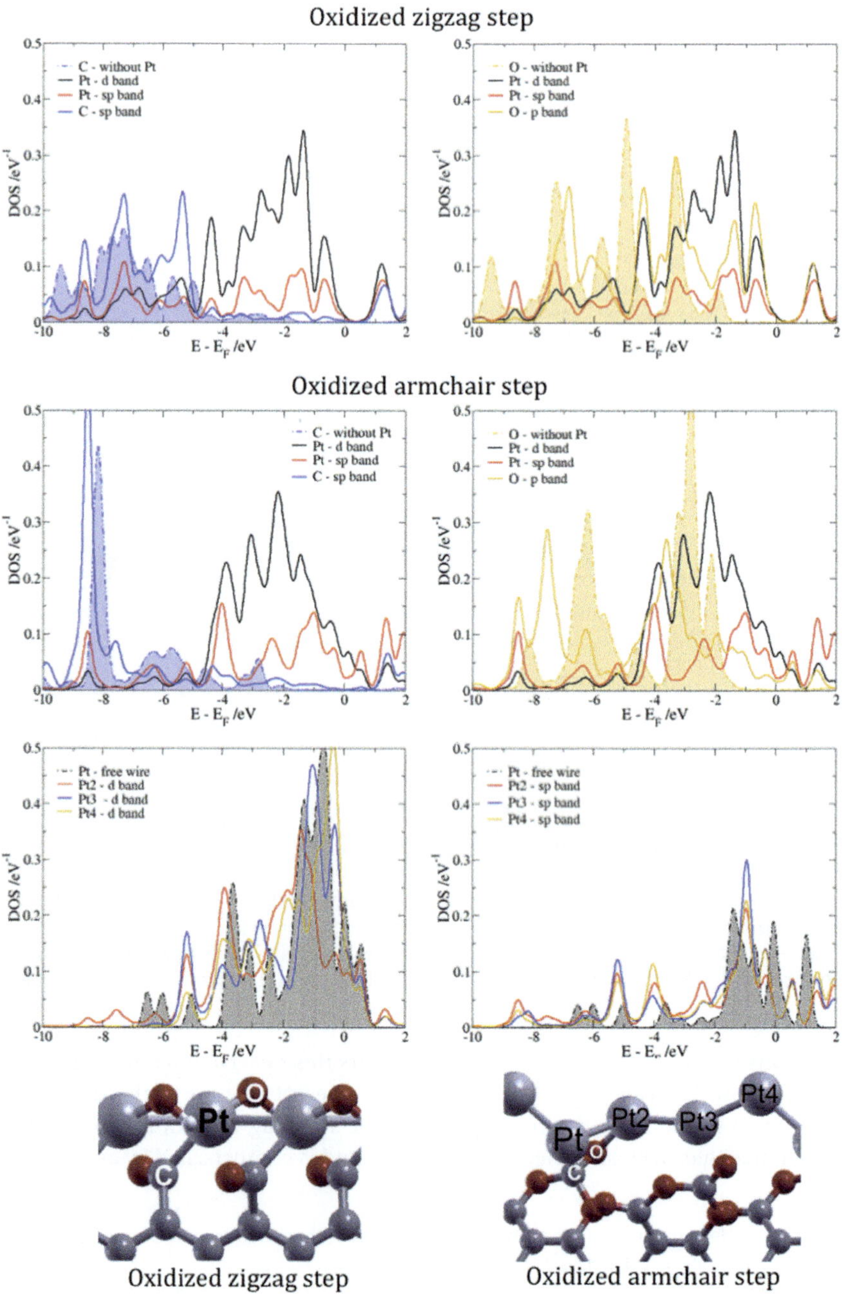

Fig. 11 Atomic projected DOS of Pt nanowires adsorbed on oxidized graphene steps. The projections are plotted in different colors as follows: blue lines for the sp band of C atom; black and red lines for the metal d and sp band, respectively. The filled curves have been added for the sake of comparison, and correspond to the orbitals in the free standing wire.

in the DOS of the sp band. Therefore, the net charge in the silver is closer to zero but the electron difference shows that there is an important redistribution of the electrons.

We shall conclude this section with the analysis of platinum wires attached to different graphene edges. The Bader charges for the platinum and carbon atoms in these systems are shown in Table 8.

The calculated charges of platinum atoms present two interesting cases. On the oxidized steps we found the most positive charge of the selected systems, whereas on the oxidized armchair step, the platinum shows a significant lowering of the positive charge. In the former case, the large charge is due to the bond between the platinum atom and an oxygen atom (from an OH group). However, in the second case the nanowire is weakly adsorbed through a bond between the platinum atom and the oxidized carbon atom.

The DOS of the platinum nanowires adsorbed on the oxidized steps is shown in Fig. 11. There, the DOS has been projected in the atomic orbitals as follows: d and sp band of platinum atoms, sp band of carbon, p band of oxygen. We have only considered the carbon and oxygen atoms directly bound to the platinum atoms. In order to improve the analysis, we have also included pDOS of the carbon and oxygen atoms in the relaxed, oxidized step (without metal atoms), and of the platinum atom in the free standing wire.

The comparison between the bands of carbon and oxygen atoms with and without the metal atoms, allow us to conclude that the oxygen atom has a fundamental role in the formation of the new bonds. The new orbitals can be found at energies closer to the Fermi level. On the armchair steps, we can also see an important new state below 4 eV of the Fermi level.

The orbitals of the platinum atoms in the nanowire supported in the armchair step present small changes in comparison with the orbitals of the free standing wires. This result is in line with our previous observation that the nanowire is weakly attached to edge and the structure of the wire is slightly modified.

Conclusions

We have contrasted two totally different approaches: an experimental overview of spontaneous deposition of Ag, Au and Pt and theoretical investigations of the initial stages of the decoration of step-edges of HOPG with those metals.

The experiments showed that nanowires can be obtained with the three metals along the step-edges, but the shapes and morphologies are very different. While Au can form very long and thin nanowires by merging small nanoparticles, in the case of Pt there is a tendency of the nanoparticles to coalesce in much larger and almost spherical clusters, probably facilitated by the presence of chloride anions. The decoration with Ag is less favourable and frequently the growth of the nanowires occurs perpendicular to the step-edges with nanocrystals showing preferentially (100) facets. Energy dispersive microanalysis showed the presence of oxygen, indicating that functionalities such as carboxyls, ethers, aldehydes or hydroxyls, could be present at defects.

Theoretical calculations of nanowire formation on bare and functionalized step-edges have been performed for different morphologies and chemical groups. We have found that the interaction between the metal wires and the carbon follows the sequence: Pt > Au > Ag. In the case of hydrogenated edges, the formation of nanowires requires energy for gold and silver nanowires. Platinum has not only a strong interaction with the step-edges but also with the hydrogen atoms. Therefore, the adsorption of Pt atoms compensates the breaking of the C–H bonds. We have found that it is still possible to adsorb platinum on the oxidized step, but the energy obtained is smaller than in bare or hydrogenated steps. The old idea that the driving force for electroless metal deposition derives from the supply of electrons by incompletely oxidized functionalities is partially confirmed from the charge and charge difference calculations. However, the landscape is much more complicated. It is important to notice that, even when the atomic charges calculated with the Bader method are small, the local distribution of electrons can be rather complex and full of redistributions between different atoms and orbitals in the same atom. Finally, we want to point out that the complex structure of the pictures indicates that there are more that one kind of orbitals participating in the bonds, and changing their populations during the adsorption, as confirmed from the density of states projected on the different atoms.

In this contribution, we have set the starting points at the two extremes of a bridge between experiments and theory to understand the metal decoration of step-edges on HOPG. On one hand, an analysis under systematic control of the experimental conditions is required. On the other hand, the inclusion of a more realistic environment such as the presence of chloride or solvent molecules is necessary in the theoretical description.

Acknowledgements

This work is part of the research network FOR1376 financed by the Deutsche Forschungsgemeinschaft. We thank PIP-CONICET 112-201001-00411, PICT-2008-00088 and PICT- 2012-2324 (Agencia Nacional de Promoción Científica y Tecnológica, FONCYT, préstamo BID) for support. We are also grateful for support from FaMAF for the funding through the Segundo Plan de Inversiones para Equipamientos 2010–2011 and a generous grant of computing time from the Baden-Württemberg.

References

1 E. Budevski, G. Staikov and W. J. Lorenz, *Electrochemical Phase Formation and Growth*, VCH Verlagsgesellschaft GmbH, Weinheim, 1996.
2 J. V. Zoval, R. M. Stiger, P. R. Biernacki and R. M. Penner, *J. Phys. Chem.*, 1996, **100**, 837.
3 J. V. Zoval, J. Lee, S. Gorer and R. M. Penner, *J. Phys. Chem. B*, 1998, **102**, 1166.
4 H. Liu and R. M. Penner, *J. Phys. Chem. B*, 2000, **104**, 9131.
5 H. Liu, F. Favier, K. Ng, M. P. Zach and R. M. Penner, *Electrochim. Acta*, 2001, **47**, 671.
6 R. M. Penner, *J. Phys. Chem. B*, 2002, **106**, 3339.
7 E. C. Walter, B. J. Murray, F. Favier, G. Kaltenpoth, M. Grunze and R. M. Penner, *J. Phys. Chem. B*, 2002, **106**, 11407.
8 E. C. Walter, M. P. Zach, F. Favier, B. J. Murray, K. Inazu, J. C. Hemminger and R. M. Penner, *ChemPhysChem*, 2003, **4**, 131.
9 R. A. W. Dryfe, E. C. Walter and R. M. Penner, *ChemPhysChem*, 2004, **5**, 1879.

10 C. E. Cross, J. C. Hemminger and R. M. Penner, *Langmuir*, 2007, **23**, 10372.
11 M. S. El-Deab, T. Sotomura and T. Ohsaka, *J. Electrochem. Soc.*, 2005, **152**, C730.
12 A. Hernández Creus, Y. Gimeno, P. Díaz, L. Vázquez, S. González, R. C. Salvarezza and A. J. Arvía, *J. Phys. Chem. B*, 2004, **108**, 10785.
13 T. Brülle and U. Stimming, *J. Electroanal. Chem.*, 2009, **636**, 10.
14 T. Brülle, W. Ju, Ph. Niedermayr, A. Denisenko, O. Paschos, O. Schneider and U. Stimming, *Molecules*, 2011, **16**, 10059.
15 P. Shen, N. Chi, K.-Y. Chan and D. L. Phillips, *Appl. Surf. Sci.*, 2001, **172**, 159.
16 G. Lu and G. Zangari, *Electrochim. Acta*, 2006, **51**, 2531.
17 O. V. Cherstiouk, P. A. Simonov and E. R. Savinova, *Electrochim. Acta*, 2003, **48**, 3851.
18 H.-B.-R. Lee, S. H. Baeck, T. F. Jaramillo and S. F. Bent, *Nano Lett.*, 2013, **13**, 457.
19 M. Aktary, C. E. Lee, Y. Xing, S. H. Bergens and M. T. McDermott, *Langmuir*, 2000, **16**, 5837.
20 P. M. Quaino, M. R. Gennero de Chialvo, M. E. Vela and R. C. Salvarezza, *J. Argent. Chem. Soc.*, 2005, **93**, 215.
21 Y. Gimeno, A. Hernández Creus, P. Carro, S. González, R. C. Salvarezza and A. J. Arvía, *J. Phys. Chem. B*, 2002, **106**, 4232.
22 W. Xia, C. Jin, S. Kundu and M. Muhler, *Carbon*, 2009, **47**, 919.
23 E. Santos, P. Quaino, G. Soldano and W. Schmickler, *Electrochem. Commun.*, 2009, 1764.
24 G. J. Soldano, P. Quaino, E. Santos and W. Schmickler, *J. Phys. Chem. C*, 2013, **117**, 19239.
25 G. J. Soldano, P. Quaino, E. Santos and W. Schmickler, *ChemPhysChem*, 2010, **11**, 2361.
26 G. Soldano, F. Juarez, B. Teo and E. Santos, *Carbon*, 2014, **78**, 181.
27 H. Chang and A. J. Bard, *Langmuir*, 1991, **7**, 1143.
28 N. Patel, M. G. Collignon, M. A. O'Connell, W. O. Hung, K. McKelvey, J. V. Macpherson and P. R. Unwin, *J. Am. Chem. Soc.*, 2012, **134**, 20117.
29 http://www.fysik.dtu.dk/campos, accessed October 2007.
30 D. Vanderbilt, *Phys. Rev. B: Condens. Matter*, 1990, **41**, 7892.
31 J. P. Perdew, K. Burke and M. Ernzerhof, *Phys. Rev. Lett.*, 1996, **77**, 3865.
32 J. H. Monkhorst and D. J. Pack, *Phys. Rev. B: Solid State*, 1976, **13**, 5188.
33 R. F. W. Bader, W. H. Henneker and P. E. Cade, *J. Chem. Phys.*, 1967, **46**, 3341.
34 R. F. W. Bader, P. E. Cade and P. M. Beddall, *J. Am. Chem. Soc.*, 1971, **93**, 3095.
35 G. Henkelman, A. Arnaldsson and H. Jónsson, *Comput. Mater. Sci.*, 2006, **36**, 354.
36 S. Gudmundsdóttir, W. Tang, G. Henkelman, H. Jónsson and E. Skúlason, *J. Chem. Phys.*, 2012, **137**, 164705.
37 http://theory.cm.utexas.edu/bader/, accessed July 2011.
38 A. Kokalj, *Comp. Mater. Sci.*, 2003, **155**, 28, code available from http://www.xcrysden.org/.
39 C. Kittel, *"Introduction to Solid State Physics"*, Wiley, New York, 3rd edn, 1967.

PAPER

Nanodiamond surface redox chemistry: influence of physicochemical properties on catalytic processes

Thomas S. Varley,* Meetal Hirani, George Harrison and Katherine B. Holt*

Received 17th March 2014, Accepted 18th March 2014
DOI: 10.1039/c4fd00041b

Modification of an electrode with an immobilised layer of nanodiamond is found to significantly enhance the recorded currents for reversible oxidation of ferrocene methanol (FcMeOH). Current enhancement is related to nanodiamond diameter, with enhancement increasing in the order 1000 nm < 250 nm < 100 nm < 10 nm < 5 nm. We attribute the current enhancement to two catalytic processes: i) electron transfer between the solution redox species and redox-active groups on the nanodiamond surface; ii) electron transfer mediated by $FcMeOH^+$ adsorbed onto the nanodiamond surface. The first process is pH dependent as it depends on nanodiamond surface functionalities for which electron transfer is coupled to proton transfer. The adsorption-mediated process is observed most readily at slow scan rates and is due to self-exchange between adsorbed $FcMeOH^+$ and FcMeOH in solution. $FcMeOH^+$ has a strong electrostatic affinity for the nanodiamond surface, as confirmed by *in situ* infrared (IR) experiments.

1. Introduction

In recent years electrochemical experiments have been undertaken using electrodes composed of, or modified with, undoped nanodiamond materials.[1] Given that bulk diamond is intrinsically insulating, with a band gap of 5.47 eV, it may seem unlikely that undoped diamond can perform as an electrode material. However, unsaturated bonding at the diamond surface and the presence of oxygen functionalities can give rise to surprisingly rich redox chemistry.[2,3] For diamond nanomaterials the large surface atom to bulk atom ratio allows surface properties to dominate, hence nanodiamond is able to participate in electrochemical processes despite its insulating nature. Different types of nanodiamond with varying surface termination have been investigated for electrochemical activity.[1–5] Most commonly used are detonation nanodiamond powders that have

Department of Chemistry, University College London, Christopher Ingold Building, London, WC1H 0AJ, UK. E-mail: t.varley@ucl.ac.uk; Tel: +44 (0)20 7679 4623

been used to construct cavity electrodes,[5–7] sintered into pellet electrodes[8] or drop-coated from solution onto electrodes.[2,3,9] Detonation nanodiamond is synthesised by detonation of carbon-based explosives in the absence of oxygen, resulting in a carbonaceous soot from which the nanodiamond powder can be extracted and purified.[10] The resulting powder is composed of relatively monodisperse diamond nanoparticles, with average diameters of 5 to 10 nm, as shown in the TEM image in Fig. 1a. Detonation nanodiamond exhibits an array of surface carbon-oxygen functionalities (including ketones, esters, alcohols and acids) that have been characterised using FTIR[11,12] and XPS.[13,14]

Electrodes modified with an immobilised surface layer of detonation nanodiamond show an enhanced, catalytic electrochemical response towards various analytes such as nitrite,[7] oxygen[15] and haemoglobin,[16] as well as common redox probes like $Ru(NH_3)_6^{3+/2+}$ and $Fe(CN)_6^{3-/4-}$.[2,3,9]

It is proposed that surface functionalities, defect sites and unsaturated bonding give rise to surface electronic states with energies within the band gap of undoped diamond.[3] It has been demonstrated that these surface states can act as both electron donors and acceptors and can support catalytic redox processes in the presence of specific redox-active molecules *via* a feedback mechanism.[1–3,9] This mechanism is illustrated schematically in Fig. 1b. Briefly, a species such as FcMeOH can undergo oxidation at an electrode to produce $FcMeOH^+$. In the presence of nanodiamond, the $FcMeOH^+$ may then be reduced by redox active functionalities on the nanodiamond surface, resulting in regeneration of FcMeOH. This continual regeneration of FcMeOH in the electrode diffusion layer results in an enhanced catalytic current. *In situ* IR studies of 5 nm diameter nanodiamond in the presence of redox probes have implicated surface quinone- and hydroquinone-like moieties in this redox mediation.[11] Catalytic current enhancements at nanodiamond-modified electrodes, *via* the mechanism shown in Fig. 1b, have previously been reported for charged species such as $Ru(NH_3)_6^{3+/2+}$ and $Fe(CN)_6^{3-/4-}$.[3,9] At nanodiamond-modified electrodes, peak currents for these species increase 2–4 fold over diffusion-controlled currents at the unmodified electrode.[1–3,9] However, as the charge on the nanodiamond surface changes with the protonation state of the surface functionalities,

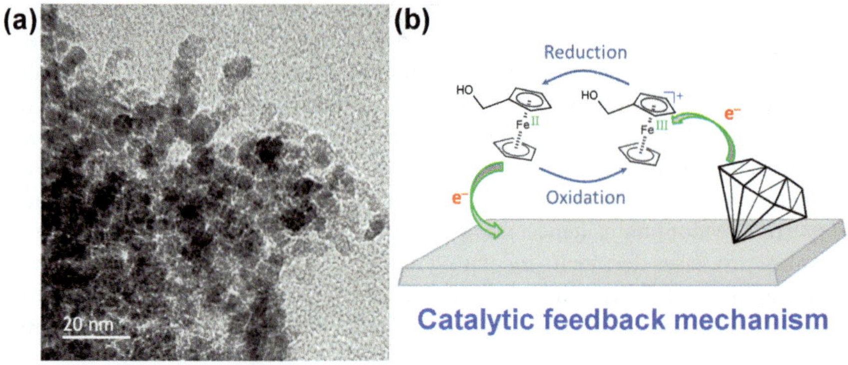

Fig. 1 (a) TEM micrograph of 10 nm diameter nanodiamond. (b) Illustration of the nanodiamond feedback mechanism. The green arrows depict the direction of electron transfer.

electrostatic interactions between the surface and redox probe must be taken into account when carrying out analysis.

In this paper we focus on nanodiamond materials of different size and morphology and their interaction with the outer-sphere and uncharged redox probe FcMeOH. We find that electrodes modified with nanodiamond exhibit significantly enhanced catalytic currents for the reversible oxidation of this neutral species compared to charged redox species. We investigate the electrochemical response when changing the diameter of the nanodiamond particles used and the pH and ionic strength of the electrolyte solution. Analysing the effect of these physicochemical parameters, alongside factors such as redox species concentration and voltammetric scan rate, provides further insight into the mechanisms of current enhancement and nanodiamond surface properties.

2. Experimental

2.1. Equipment

All electrochemical experiments were carried out using a μAutolab PGSTAT potentiostat (Eco Chemie, Utrecht, Netherlands) running GPES (v4.9), or a CHI910B Scanning Electrochemical Microscope/bipotentiostat (CH Instruments, Inc. Austin, Texas, USA) running CHI (v6.28). Electrochemistry was performed using a three electrode setup. The working electrode was a boron-doped diamond disk (d = 3.0 mm), press fitted into a PEEK body (Windsor Scientific, Slough, United Kingdom). The counter electrode was fabricated in-house from platinum wire and sheet, cut into flags of approximately 4 cm^2. A stable reference potential was achieved by using an Ag/AgCl (sat. KCl) RE-5B reference half-cell (BASi, West Lafayette, America).

A Carbolite MTF 1200 horizontal tube furnace was used to pre-treat and clean all nanodiamond samples. The temperature control unit and associated electronics were designed and built in house. Mid-infrared spectra were recorded (100 scans) in attenuated total reflectance (ATR) mode using a Bruker Tensor 27 Fourier transform infrared (FTIR) spectrophotometer (Bruker, Coventry, United Kingdom) fitted with a room temperature DLaTGS detector at 4 cm^{-1} resolution. The Platinum ATR accessory consisted of a diamond prism operating with one reflection. Background spectra were collected prior to each experiment (100 scans). Transmission electron micrograph (TEM) images were recorded using a Jeol JEM 2100 TEM with a 200 kV accelerating voltage using a LaB6 filament. All nanoparticles were deposited from the same solutions used for modifying electrode surfaces. Carbon coated copper TEM grids were used as the nanoparticle support.

2.2. Chemicals and solutions

Nanodiamond (10 nm mean particle size, measured by TEM analysis, >97%), potassium chloride (puriss grade, >99.5%), potassium dihydrogen phosphate (>98%), potassium phosphate dibasic (puriss, >99%), potassium hexacyanoferrate trihydrate (98.5–102.0%) and ferrocene methanol (97%) were all purchased from Sigma-Aldrich (Dorset, England). Other nanodiamond samples were gifted by a collaborator. All solutions were prepared using water from a Millipore Milli-Q Gradient A10 water filter (18.2 MΩ cm). Electrodes were cleaned by mechanical polishing with MasterPrep alumina polishing suspension, 0.05 μm (Buehler, Düsseldorf, Germany), and deionized water using 2-7/8″ microcloth polishing pads (Buehler, Düsseldorf, Germany). Electrochemical measurements were carried out in either potassium chloride (0.1 M) with a recorded pH of 5.88, or a mixture of the phosphate buffer solutions potassium dihydrogen phosphate (0.1 M) and potassium phosphate dibasic (0.1 M) mixed in different proportions to achieve the required pH.

All nanodiamond samples were treated in a furnace at 425 °C for 4 h (including temperature ramp times, which were set at 10 °C min^{-1}) in air. This was to remove any sp^2 graphitic carbon from the surface and to maximise the number of oxygen terminating functional groups.[17] Nanodiamond suspensions were prepared by adding nanodiamond to purified water, centrifuging twice (13 200 rpm for 10 min) and briefly sonicating to ensure large aggregates were broken up. The mass per volume concentrations of the various nanodiamond solutions were: 5 nm ND 1.50 mg ml^{-1}; 10 nm ND 1.50 mg ml^{-1}; 100 nm ND 1.55 mg ml^{-1}; 250 nm ND 1.60 mg ml^{-1} and 1000 nm ND 1.7 mg ml^{-1}.

2.3. Preparation of nanodiamond-modified electrodes

Nanodiamond films were drop coated onto clean, dry electrodes by pipetting the various nanodiamond suspensions (2.5 μl) onto the electrode surface. The droplet was positioned and manoeuvred to ensure complete coverage of the electrode's surface. The nanodiamond coated electrode was left to air dry for 20 min and checked under an optical microscope to ensure even and complete coverage.

3. Results

3.1. Characterisation of nanodiamond materials using FTIR and TEM

Fig. 2 shows representative FTIR spectra and TEM micrographs for two of the nanodiamond samples used in this study: 5 nm detonation nanodiamond and 100 nm microcrystalline nanodiamond (produced by mechanical grinding of synthetic diamond). Two different detonation nanodiamond materials were studied and examination with TEM (Fig. 1a, Fig. 2b) revealed the powder to be composed of discrete spherical and relatively monodisperse particles; one sample with mean particle diameter 5 nm and another with a slightly larger particle diameter closer to 10 nm. In both samples the individual nanoparticles aggregate together to form larger agglomerates. The surface of all of the nanodiamond samples used is highly oxidised due to the pretreatment procedure of heating in air, described in section 2.2 above. The FTIR spectrum of the 5 nm nanodiamond is complex, with a great number of different functionalities contributing to the vibrational spectrum. These functionalities have been characterised and discussed in detail previously,[11,12] and include alcohols and carbonyl species: 3393 cm^{-1} (νOH), 2926 cm^{-1} (νCH_2), 2847 cm^{-1} (νCH), 1729 cm^{-1} (νC=O) and 1634 cm^{-1} ($\nu OH_{Abs.}$). The convoluted set of peaks between 1000–1500 cm^{-1} contains a mix of bending and stretching modes of various C–C, C–H and C–O containing functionalities. Of most interest to the present study is the presence of redox-active functionalities such as unsaturated ketone or 'quinone-like' moieties that can participate in redox cycling.[11]

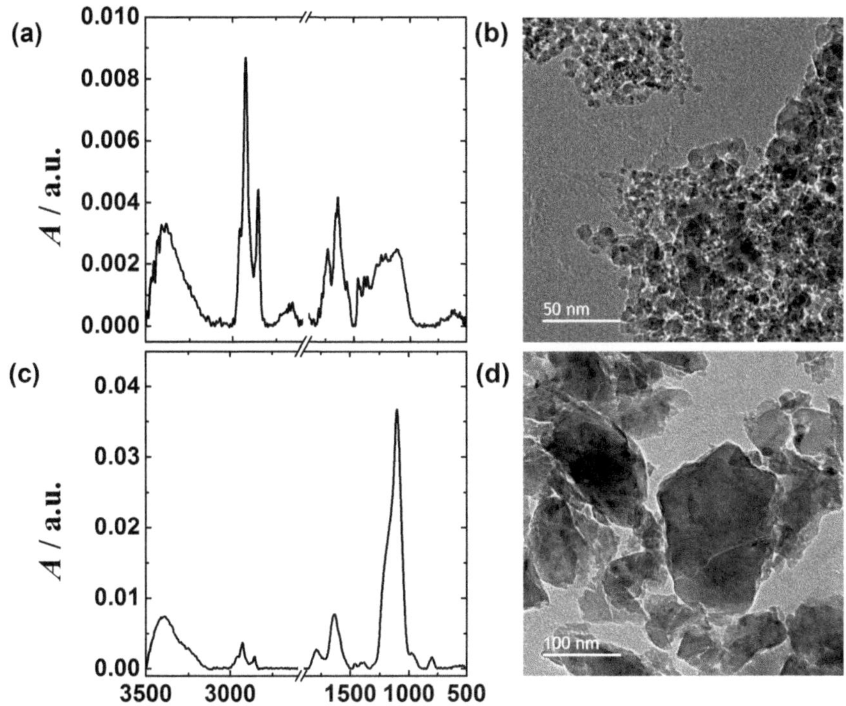

Fig. 2 ATR-FTIR spectra and TEM micrographs of 5 nm, (a) and (b), and 100 nm, (c) and (d), nanodiamond drop-cast films.

The 100, 250 and 1000 nm samples have pronounced crystal facets, consistent with nanodiamond produced through mechanical grinding of High Temperature High Pressure synthesis diamond.[18] Analysis of the size distribution of the 100 nm nanodiamond sample gives a mean particle size of 98 nm with a standard deviation of 35 nm; the modal average size is *ca.* 100–120 nm.

The crystalline nanodiamond samples all show a large absorption peak at *ca.* 1100 cm^{-1} in the IR spectra, attributable to nitrogen centres at defect sites within the bulk material.[19] The IR spectra of these larger nanodiamonds also show absorption bands attributed to surface functionalities at 3394 cm^{-1} (νOH), 2923 cm^{-1} (νCH_2), 2855 cm^{-1} (νCH), 1790 cm^{-1} (νC=O), 1631 cm^{-1} ($\nu OH_{Abs.}$) and 1103 cm^{-1} (νN-V^-), demonstrating the presence of similar surface functionalities to those on the detonation nanodiamond particles. Although the nanodiamond samples were produced by different synthetic methods, making direct comparison difficult, the oxidative pre-treatment results in similar surface oxygen functionalities, allowing us to anticipate similar electrochemical properties.

3.2. Cyclic voltammetry (CV) investigations of nanodiamond-modified electrodes

3.2.1 The effect of nanodiamond particle diameter on the voltammetric response of FcMeOH. Fig. 3a shows the cyclic voltammograms (CV) of 1 μM FcMeOH at a clean boron doped diamond (BDD) electrode (black line) and a BDD electrode modified with 100 nm nanodiamond particles (red/blue lines). Both the

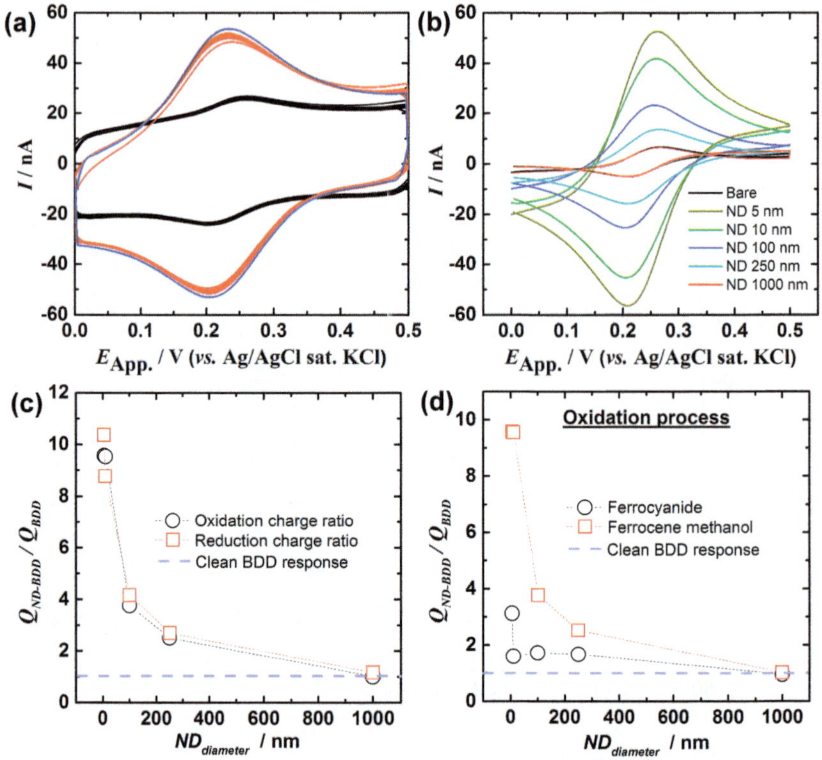

Fig. 3 (a) Voltammograms of FcMeOH (1 μM) in KCl electrolyte solution (0.1 M, pH 5.88) at a BDD (black) and 100 nm nanodiamond-modified BDD working electrode (red: first 10 scans, blue: final 10 scans, 100 scans total), $\nu = 10$ mV s^{-1}. (b) Voltammograms (5th of 5 scans) of FcMeOH (1 μM) in KCl electrolyte solution (0.1 M, pH 5.88) on a BDD (black) and various different nanodiamond-modified BDD working electrodes (coloured lines, $d = 5$–1000 nm), $\nu = 20$ mV s^{-1}, with background subtraction. (c) Charge ratio of the FcMeOH peak enhancement ($Q_{ND\text{-}BDD}$ / Q_{BDD}) *vs.* nanodiamond size, calculated from data presented in (b). (d) Charge ratio of the FcMeOH/$Fe(CN)_6^{4-}$ peak enhancement ($Q_{ND\text{-}BDD}$: Q_{BDD}) *vs.* nanodiamond size calculated from data presented in (b) for FcMeOH and identical experiments carried out using $K_4Fe(CN)_6$. The blue dashed line highlights the charge ratio response for clean, un-modified BDD.

oxidation and reduction currents at the nanodiamond modified electrode show similar enhancements of *ca.* four times that of the clean BDD surface. This current enhancement is stable to repeated cycling over multiple scans. The peak separation $\Delta E_p = |E_p^{ox} - E_p^{red}|$ is ~30 mV at the nanodiamond-modified electrode, in contrast to the theoretically predicted value of 59 mV for a diffusion-controlled one electron transfer. Fig. 3b shows CVs of FcMeOH at electrodes modified with nanodiamonds of different diameters.

The 1000 nm nanodiamond has a negligible effect on the electrochemical response, the CV being essentially identical to that at the clean BDD. Enhanced currents are observed at the other electrodes, with peak currents increasing in the order 250 nm < 100 nm < 10 nm < 5 nm. Oxidation and reduction peaks are enhanced equally in each case. The degree of current enhancement at the 10 nm and 5 nm nanodiamond-modified electrodes is quite dramatic. This has been quantified in Fig. 3c, where a 'charge enhancement ratio' has been calculated for each nanodiamond diameter. This is defined as the charge passed during oxidation or reduction at the nanodiamond-modified electrode divided by that passed at the clean BDD electrode under the same conditions, *i.e.* $Q_{ND\text{-}BDD}/Q_{BDD}$. Charge passed during the oxidation of FcMeOH or reduction of $FcMeOH^+$ is increased by a factor of ten when the electrode is modified with a layer of 5 nm or 10 nm nanodiamond. A four-fold enhancement is noted for the 100 nm particles, falling to two-fold for the 250 nm nanodiamond, and no enhancement is seen for an electrode modified with 1000 nm particles.

In comparison to previous studies using the $Fe(CN)_6^{3-/4-}$ redox probe,[9] the current enhancements for FcMeOH are considerably higher. This is shown in Fig. 3d, where the charge enhancement ratio for $Fe(CN)_6^{4-}$ oxidation is compared to that for FcMeOH oxidation at electrodes modified with different diameter nanodiamond. The data shows the same relationship between charge enhancement and nanodiamond diameter for both redox species, however, the magnitude of enhancement shows a strong dependence on the chemical nature of the redox species, with FcMeOH currents being significantly more enhanced.

3.2.2. The effect of the solution pH on the CV response of FcMeOH at nanodiamond-modified electrodes. CVs for 1 μM FcMeOH at clean BDD and 100 nm nanodiamond-modified BDD electrodes in different solutions of pH 5–9 are shown in Fig. 4a. At the BDD electrode the CV response of FcMeOH is independent of pH, producing voltammograms with ΔE_p *ca.* 60 mV, consistent with one-electron, reversible, diffusion-controlled electrochemistry. The currents are

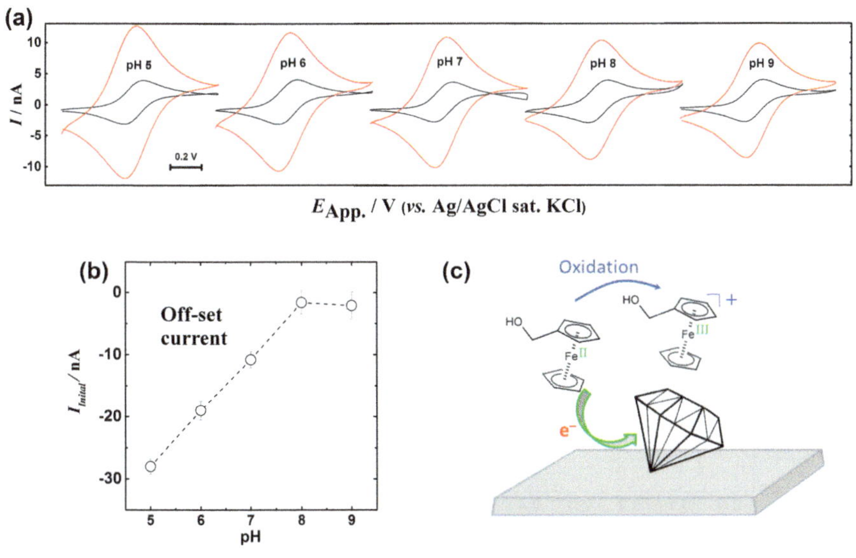

Fig. 4 (a) Cyclic voltammograms (1st scan) of FcMeOH (1 μM) in aqueous solutions of K_2HPO_4–KH_2PO_4 (0.1 M) in various ratios to produce pH 5–9 electrolytes, at the clean BDD electrode (black) and nanodiamond modified BDD electrode (red), recorded at 5 mV s^{-1}. (b) Initial off-set currents measured with different electrolytes of pH 5–9 (data collected from the 1st scan of the experiments displayed in (a)). (c) Illustration of FcMeOH oxidation performed by the nanodiamond. The green arrow depicts the direction of electron transfer.

enhanced at the 100 nm nanodiamond-modified electrodes over all pH values, but the enhancement is pH dependent with I_p increasing as the pH is lowered. As for Fig. 3a, ΔE_p for the nanodiamond-modified electrode is ~30 mV over the whole pH range. I_p *vs.* ν analysis suggests mixed diffusion controlled and adsorption behaviour.

Each of the CVs at the nanodiamond-modified electrodes exhibits a non-zero 'off-set' reduction current at 0 V. At this potential there should be no species in the solution available to be reduced (only FcMeOH is present and this can only be oxidised). We have reported this phenomenon previously for 5 nm nanodiamond-modified electrodes in solutions containing $Fe(CN)_6^{4-}$.[9] It is attributed in the present case to reduction of $FcMeOH^+$ species within the diffusion layer at the electrode surface. $FcMeOH^+$ is proposed to be generated spontaneously by the oxidation of FcMeOH at the nanodiamond surface (Fig. 4c). This oxidation takes place through electron exchange between nanodiamond surface functional groups, which are reduced concomitantly with the generation of $FcMeOH^+$. The magnitude of the off-set current is pH dependent as shown in Fig. 4b. At pH 5, significant initial reduction currents are observed, but these become negligible at pH 8 and above.

3.2.3. The effect of solution ionic strength on the CV response of FcMeOH at nanodiamond-modified electrodes. The CVs for 1 μM FcMeOH at a 100 nm nanodiamond-modified BDD electrode in electrolytes of varying ionic strength are shown in Fig. 5. The ionic strength was varied by altering the concentration of the KCl electrolyte across five orders of magnitude. As the ionic strength is decreased, a pre-peak emerges from the oxidation wave, which shifts to more negative potentials with decreasing ionic strength (increasing the pre-peak to main-peak separation). The height of the main oxidation wave is decreased with decreasing ionic strength, while that of the reduction peak increases. In CV experiments the emergence of a pre-peak to an oxidation peak is usually taken as evidence of outer-sphere adsorption of the oxidation product at the electrode

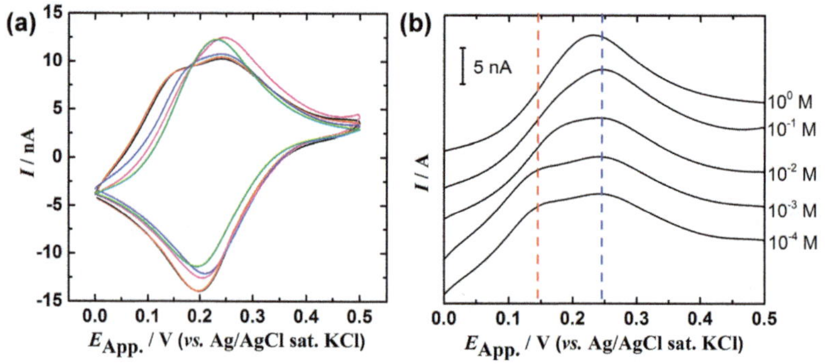

Fig. 5 (a) Cyclic voltammograms (5th scan) of FcMeOH (1 μM) in aqueous solutions containing various concentrations of KCl (black = 10^{-4} M, red = 10^{-3} M, blue = 10^{-2} M, magenta = 10^{-1} M, green = 10^{0} M, pH = 5.06–6.12) at a nanodiamond modified (100 nm) BDD working electrode, recorded at 5 mV s^{-1}. (b) Stacked plot of the oxidation process only. The blue dashed line identifies the standard FcMeOH diffusion controlled peak (as seen for the clean BDD electrode), whereas the red dashed line highlights the emerging pre-peak.

surface. In this case, the oxidation product $FcMeOH^+$ is likely to experience a positive electrostatic interaction with the negatively charged nanodiamond surface.[20,21] Decreasing the ionic strength of the medium will increase the extent of adsorption, as the length to which the electric field of a surface can be felt is inversely proportional to the concentration of electrolyte (as described by DLVO theory).[22,23] This is consistent with the observations in Fig. 5b, where the adsorption pre-peak can be seen to become more prominent as ionic strength is decreased and hence more $FcMeOH^+$ adsorbs onto the nanodiamond surface.

3.2.4. The effect of CV scan rate on the CV response of FcMeOH at nanodiamond-modified electrodes. The studies described in this section were carried out using 5 nm nanodiamond modified electrodes, as these showed the most extreme current enhancements. CVs were carried out in 1 μM and 10 μM FcMeOH over scan rates of 1–500 mV s^{-1} (shown in Fig. 6, where currents have been normalised for clarity). Below scan rates of 10 mV s^{-1}, two oxidation waves are clearly observed; one wave is positioned at the same E_p as the diffusion-controlled oxidation peak at the clean BDD electrode, and the other is a pre-peak to this main wave. A pre-peak to the reduction wave is also observed. As the scan rate is increased the two waves merge to form one broad peak located midway between the pre-peak and main peak positions. At scan rates above *ca.* 100 mV s^{-1} only one peak is observed, at the position expected for a simple diffusion-controlled mechanism. However, at very low concentrations of FcMeOH (1 μM), at the faster scan rates, this peak becomes a steady state current response, indicating that oxidation is no longer taking place under diffusion-control.

3.3. *In situ* ATR-IR spectroscopy of 5 nm nanodiamond in the presence of FcMeOH

The electrochemical data is strongly suggestive of adsorption, of the oxidation product $FcMeOH^+$ in particular, at the nanodiamond surface. *In situ* ATR-FTIR spectroscopy experiments were therefore performed to monitor changes in the IR

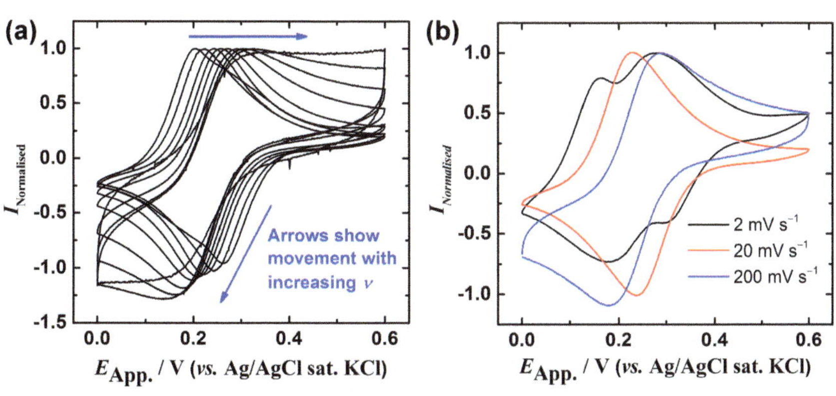

Fig. 6 (a) Normalised current (by max.) voltammograms of FcMeOH (1 μM) for scan rates of 1, 2, 5 10, 20, 50, 100, 200 and 500 mV s^{-1}. (b) Voltammograms of FcMeOH (10 μM) for scan rates of 2, 20 and 200 mV s^{-1}. Experiments were performed with a 300 s pre-experiment equilibration time (no applied potential) in KCl electrolyte solution (0.1 M, pH 5.88) at a nanodiamond-modified (5 nm) BDD electrode.

absorption at the nanodiamond/solution interface in solutions containing FcMeOH or $FcMeOH^+$. The experiments were performed by drop-casting a 5 nm nanodiamond film onto the ATR crystal. An enclosed liquid cell was clamped over the nanodiamond modified ATR crystal and gently filled with water to solvate the nanodiamond layer. After allowing time to equilibrate, a small volume of FcMeOH or $FcMeOH^+$ solution was gently introduced to the cell, and the FTIR spectrum between 3500–500 cm^{-1} was recorded over 120 min. Fig. 7 shows the IR difference spectra obtained after 120 min of nanodiamond film immersion in the FcMeOH or $FcMeOH^+$ solution, along with a spectrum of the solid FcMeOH for comparison.

In the absence of nanodiamond, there is negligible change over 120 min in the IR spectra of the ATR crystal in FcMeOH solution. With the nanodiamond film, the presence of FcMeOH produces increases in infrared absorption at *ca.* 3200–3300 cm^{-1} (νOH) and 1650 cm^{-1} (δOH), *i.e.* predominantly in regions associated with OH vibrational modes. There are a greater number of changes observed for the nanodiamond film exposed to $FcMeOH^+$ solution. These occur in regions associated with OH, C–H and C–O stretches and vibrations: 2925 cm^{-1} (νCH_2, asym.), 2845 cm^{-1} (νCH_2, asym.), 1648 cm^{-1} (δOH), 1576–1393 cm^{-1} ($\delta/\omega CH_2$) and 1044 (νC–C–O), showing many features in common with the solid FcMeOH spectrum.[24]

4. Discussion

4.1. Mechanism of the catalytic current enhancement

With the exception of 1000 nm nanodiamond, modification of the BDD electrode with an immobilised layer of nanodiamond results in enhanced CV currents for the reversible oxidation of FcMeOH. Catalytic current enhancements have previously been reported for the redox probes $Fe(CN)_6^{3-/4-}$, $IrCl_6^{2-/3-}$ and $Ru(NH_3)_6^{2+/3+}$, among others;[1–3,6,9] however, currents achieved here for the neutral FcMeOH species are considerably higher than those observed for charged redox molecules. The current enhancement is partly attributed to a catalytic feedback mechanism, as shown in Fig. 1b. In this scheme, FcMeOH undergoes oxidation at

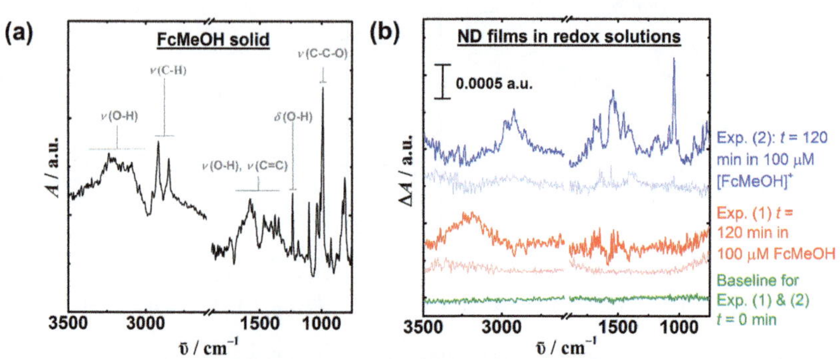

Fig. 7 (a) ATR-FTIR spectrum of FcMeOH (solid powder, black line), and (b) ATR-FTIR difference spectra (coloured lines) recorded for solutions of FcMeOH (red) and $FcMeOH^+$ (blue) on a clean ATR prism (transparent line), and in the presence of a nanodiamond drop-cast film (solid line). All aqueous solutions contained KCl (0.1 M, pH 5.88).

the BDD electrode to produce $FcMeOH^+$. FcMeOH is then regenerated rapidly by the reduction of $FcMeOH^+$ at the nanodiamond interface. On the CV return scan the opposite process is proposed to take place; FcMeOH generated at the electrode is immediately oxidised back to $FcMeOH^+$ after donating an electron to the nanodiamond surface. For this mechanism to be feasible, the oxidation and reduction of nanodiamond surface functionalities must be at similar potentials to the standard potential of the solution redox probe. Electrode-immobilised 5 nm detonation nanodiamond has been shown to exhibit reversible redox peaks in this potential range, which shift potential with changing solution pH.[9] In addition, *in situ* IR studies have provided evidence that 'quinone-like' surface moieties on the nanodiamond are involved in the redox cycling.[11] Although the 100 nm and 250 nm diamond particles used in this study are synthesised by different methods to the 5 nm nanodiamond and possess different morphologies, the observed current enhancements suggest similar surface species on each that can contribute to a catalytic reaction mechanism. All of the samples were oxidised prior to use and among the oxygen functionalities present it is likely that 'quinone-like' and 'hydroquinone-like' species are present on each to support the catalytic cycle in Fig. 1b.

4.2. Relationship between the particle size and the current enhancement

In this study an inverse relationship between particle size and observed current enhancement was observed for both the oxidation of FcMeOH and reduction of $FcMeOH^+$ in the presence of 5, 10, 100 and 250 nm diameter nanodiamonds. The extent of the enhancement observed, quantified as the 'charge enhancement ratio', depends on the availability of nanodiamond surface functionalities to support the feedback mechanism. The concentration of these functionalities will depend on the nanodiamond particle size as well as the surface density of these groups. This latter factor will vary with oxidation pre-treatment, as well as the type of exposed crystal facets, edge sites and the defect density. A log–log plot of charge enhancement ratio with nanodiamond diameter produces a straight line, suggesting that a monomial power law relates these two properties. A straightforward relationship between the surface area to volume ratio of the particles and the charge enhancement ratio (σ) should result from a $\sigma = 3r^{-1}$ fit to the data (where r is the particle radius). However, a more complex relationship is found. Fractal scaling could be one explanation for the unusual relationship obtained, and it should also be noted that the 100 nm and 250 nm samples are not spherical, like the 5 and 10 nm samples. Most likely, the density of redox active sites on the different diamond samples differs. The highly active detonation nanodiamond surface is very defective and has much in common with amorphous carbons; it exhibits an incredibly diverse and dense coverage of functionalities. In contrast, the 100 nm and 250 nm samples are more 'diamond-like' in their character and thus the surface defect density and degree of sp^2 bonding is much lower, resulting in fewer active sites. The 1000 nm diamond does not support the catalytic current enhancement (although neither does it block electrode activity). It is proposed that there are insufficient surface groups present on this material to produce a detectable catalytic current. The same mechanism is likely to take place at the surface of the 1000 nm diamond, however, the surface atom to bulk atom ratio is much smaller, so currents generated *via* this mechanism will be negligible in comparison to the diffusion controlled currents.

4.3. pH dependence of the current enhancement

Current enhancement with 100 nm nanodiamond-modified electrodes was found to increase in the order pH 9 < pH 8 < pH 7 < pH 6 < pH 5. The redox response of FcMeOH at the BDD electrode is independent of pH, hence the ability of the nanodiamond to contribute to the catalytic cyclic must be the proton dependent step. 5 nm nanodiamond exhibits reversible redox peaks in voltammetry that shift by *ca.* 60 mV to more negative potentials with each increase in pH unit. This is typical of a coupled reaction of one proton per electron transferred and consistent with a quinone/hydroquinone redox transformation, which in aqueous solution is a $2H^+/2e^-$ reaction.[9] In the present study, we propose that similar surface species are present on the 100 nm nanodiamond and that the redox potentials of these functionalities are similarly pH dependent. The pH 5 response suggests the potentials of the surface species are coincident with the potential of the FcMeOH/$FcMeOH^+$ couple at this pH. This match allows the most efficient electron transfer to take place between the nanodiamond and the solution species and hence the current enhancement is greater. As the pH is increased, the potential of the surface functionalities moves towards more negative potentials (according to Nernstian analysis of the proton dependence of the process), while that of the FcMeOH/$FcMeOH^+$ couple is unchanged. As the potentials of the solution and surface species are not so well matched, the driving force for the catalytic process is decreased and the current enhancement is not so pronounced. Hence smaller current enhancements are noted at pH 9.

4.4. pH dependence of the magnitude of the off-set current

Further evidence for pH dependent electron transfer between the nanodiamond surface groups and the FcMeOH species comes from inspection of the magnitude of the off-set reduction current at 0 V in Fig. 4. As described above, we attribute this current to the reduction of $FcMeOH^+$ present in the diffusion layer. The $FcMeOH^+$ is proposed to be generated by the spontaneous oxidation of FcMeOH at the nanodiamond (ND) surface in the absence of an applied potential:

$$\mathrm{FcMeOH} + \mathrm{ND}_{\mathrm{ox.}} \rightarrow \mathrm{FcMeOH}^+ + \mathrm{ND}_{\mathrm{red.}}$$

For such a process to be thermodynamically feasible, the nanodiamond surface functionalities must undergo reduction at a higher potential than the potential of the FcMeOH/$FcMeOH^+$ couple. The magnitude of the reduction current for the $FcMeOH^+$ species suggests this is the case at pH < 8 with the largest driving force for the process occurring at the lower pH values. At pH 8 and above the reduction currents are very small, indicating that electron transfer between the FcMeOH and the nanodiamond is negligible. This is consistent with the proton dependence of the potential of the surface functionalities. As pH increases, the potential shifts to more negative values and by pH 8 is proposed to be found below that of the FcMeOH/$FcMeOH^+$ couple, hence spontaneous electron transfer does not take place.

4.5. Ionic strength dependence of the voltammetric response

At low ionic strengths, a pre-peak to both the main oxidation and reduction waves becomes evident. This, along with the observed increased width of the voltammetric peaks in the presence of nanodiamond, as well as the decreased peak to peak separation (*ca.* 30 mV), is diagnostic of adsorption of both FcMeOH and $FcMeOH^+$ at the electrode surface. As no evidence for adsorption is observed at the clean BDD electrode it is assumed that there is a strong affinity of the FcMeOH and $FcMeOH^+$ species for the nanodiamond surface. The positively charged $FcMeOH^+$ can clearly interact *via* electrostatic interactions with the negatively charged surface oxygen functionalities. Some current enhancement for the $Ru(NH_3)_6^{3+}$ species at the 5 nm nanodiamond modified electrodes has previously been attributed to such an interaction. The strong adsorption of the neutral FcMeOH is more surprising as the nanodiamond surface is considered quite hydrophilic and less likely to interact with a largely hydrophobic molecule. However, the adsorption of hydrophobic species to oxidised nanodiamond has previously been reported and in fact exploited for the delivery of water-insoluble drugs.[25] The sp^2 bonding present at the nanodiamond surface provides ample opportunity for hydrophobic and π–π stacking interactions with the aromatic pentacyclo- moieties of the FcMeOH molecule.

4.6. Contribution of the adsorption processes to the enhanced current response

The currents for FcMeOH oxidation at the nanodiamond-modified electrodes are much greater than observed previously for the $Fe(CN)_6^{4-}$ species. As the potential of the two species is similar, the same redox interactions with the nanodiamond surface groups are proposed. The difference in current enhancement is attributed to the extent to which the redox molecule can approach closely to the surface of the nanodiamond and the resulting efficiency of electron tunnelling between the two. The very negatively charged $Fe(CN)_6^{4-}$ is unlikely to be able to approach closely to the surface due to electrostatic repulsion, hence in the absence of any other factors it is not surprising that FcMeOH shows a higher current response that $Fe(CN)_6^{4-}$. However, the adsorption behaviour of FcMeOH and $FcMeOH^+$ is clearly a contributor to the very significant and unexpectedly large current enhancements we observe.

As stated above, FcMeOH undergoes spontaneous oxidation at the nanodiamond surface to generate $FcMeOH^+$. This positively charged species will interact very strongly with the surface and become adsorbed. The strong interaction is illustrated by the affinity of the $FcMeOH^+$ for the nanodiamond surface, as found in the IR studies in Fig. 7. The contribution of the Gibbs energy of adsorption can shift the redox potentials of an adsorbed species relative to the solution species and results in the observation of pre-peaks in the CV.[26,27] Here we propose that the surface adsorbed $FcMeOH^+$ also participates in the catalytic feedback mechanism, as shown in Fig. 8. Solution FcMeOH undergoes electron exchange with adsorbed $FcMeOH^+$ to generate solution $FcMeOH^+$ and adsorbed FcMeOH. The adsorbed FcMeOH can be re-oxidised to adsorbed $FcMeOH^+$ by the underlying BDD electrode, or by a redox process with the nanodiamond surface groups. Thus, in addition to the catalytic currents generated by the mechanism shown in Fig. 1b, the adsorbed species also mediate a catalytic process. The accumulative influence of both processes explains why currents are so significantly enhanced for this solution species compared to others previously studied.

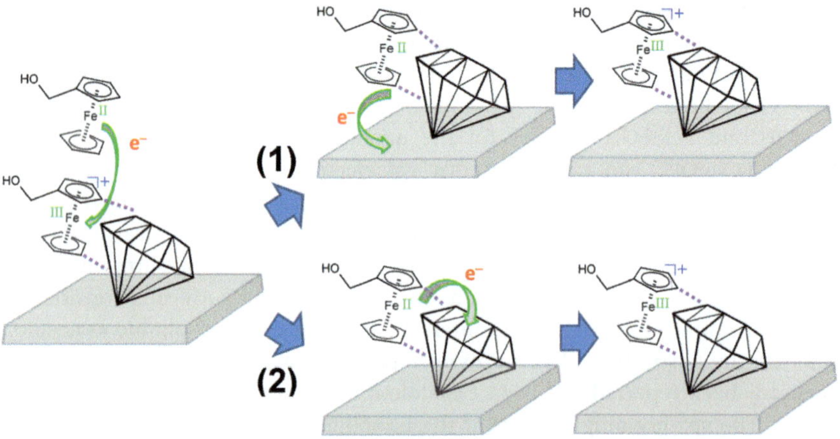

Fig. 8 Illustration of nanodiamond surface adsorbed $FcMeOH^+$ and its associated redox reaction. Re-oxidation of the FcMeOH product can then occur by either (1) the underlying electrode (when at sufficient potential) or (2) reaction with nanodiamond surface functionalities. The green arrows depict the direction of electron transfer.

4.7. Scan rate dependence of the voltammetric response

The scan rate dependence of the response illustrates further the different processes contributing to the catalytic currents. At slow scan rates, the pre-peak for the adsorption mediated process is very distinct from the main peak during the oxidation process. The two catalytic processes take place at distinct potentials; first *via* the adsorbed $FcMeOH^+$ species and then *via* the nanodiamond surface groups, indicating that adsorbed $FcMeOH^+$ is easier to reduce than the nanodiamond surface under these conditions. The factor determining the rate of the adsorption-mediated mechanism is the availability of adsorbed $FcMeOH^+$ to oxidise the solution FcMeOH. Assuming the rate of regeneration of adsorbed $FcMeOH^+$ by reduction at the BDD electrode or nanodiamond surface is fast enough, the limit to this reaction is simply how much $FcMeOH^+$ is adsorbed at the nanodiamond interface. The adsorption-mediated process is so distinctly observed at slower scan rates because the flux of FcMeOH is smaller under these conditions, and hence mass transport of reactant can keep up with the rate of regeneration of the available $FcMeOH^+$. At faster scan rates, the higher flux of solution FcMeOH leads to diffusion-controlled currents overwhelming the adsorption response and the pre-peak merges with the main diffusion-controlled peak. At higher scan rates and low concentrations a steady-state current begins to emerge (Fig. 6a). In this regime, we assume that catalysis takes place largely by the nanodiamond-mediated mechanism shown in Fig. 1b. The steady state current shows that the rate of $FcMeOH^+$ generation at the BDD electrode is exactly matched by the rate of $FcMeOH^+$ reduction by the nanodiamond surface functionalities. At slow scan rates a peak response is obtained for this process, as a thick diffusion layer containing $FcMeOH^+$ builds up and there are insufficient nanodiamond surface functionalities to maintain the catalytic cycle. At faster scan rates the diffusion layer is much thinner, so there is less $FcMeOH^+$ for the nanodiamond to reduce, hence the rate of electron transfer from the nanodiamond can keep up with the rate of $FcMeOH^+$ generation, leading to a steady state current response.

5. Conclusions

Modification of a BDD electrode with an immobilised layer of nanodiamond was found to enhance currents for the reversible oxidation of FcMeOH. Current enhancement is inversely related to nanodiamond diameter, with enhancement increasing in the order 1000 nm < 250 nm < 100 nm < 10 nm < 5 nm. In comparison to previous studies using negatively charged redox probes, we find that currents are much larger for the neutral species FcMeOH under the same experimental conditions. We attribute the current enhancement to two catalytic processes: i) electron transfer between the solution redox species and redox active groups on the nanodiamond surface; ii) electron transfer mediated by $FcMeOH^+$ adsorbed onto the nanodiamond surface. The first process is pH dependent as it is believed that 'quinone-like' and 'hydroquinone-like' species on the nanodiamond surface mediate the redox process and thus it is coupled to proton transfer. The $FcMeOH/FcMeOH^+$ redox cycle is more efficiently catalysed than the $Fe(CN)_6{}^{4-/3-}$ transformation reported previously, as the neutral and positive redox species are attracted rather than repelled by the negatively charged nanodiamond surface. The adsorption mediated process is observed most readily at slow scan rates and is due to electron self-exchange between adsorbed $FcMeOH^+$ and FcMeOH in solution. $FcMeOH^+$ has a strong affinity for the nanodiamond surface, as confirmed by *in situ* ATR-FTIR experiments.

This study provides insight into the interaction of solution species with diamond and carbon surfaces under different physicochemical conditions. It also reveals mechanistic details of catalytic processes associated with the surfaces of nanodiamond materials. Catalytic currents within electrochemistry are of interest as they can allow energy intensive reactions to proceed at lower overpotentials (or allow a greater rate of reaction at the same overpotential). Such catalysts have applications in important sectors such as synthesis, energy conversion and molecular detection. Due to the particular mechanism of the nanodiamond electrocatalysis, nanodiamonds could find use within sensing devices, for example allowing detection of analytes at very low concentrations.

Acknowledgements

The authors would like to thank the EPSRC for funding this research (Grant No. EP/J010006/1). We also wish to extend our gratitude to Christoph Salzmann and Martin Rosillo-Lopez for use of their furnace and Steve Firth for assistance with TEM analysis.

References

1 For a review see: K. B. Holt, Electrochemistry of Nanodiamond Particles in *Nanodiamond*, ed. Oliver A. Williams, Royal Society of Chemistry, 2014, pp. 128–150.

2 K. B. Holt, C. Ziegler, D. J. Caruana, J. Zang, E. J. Millán-Barrios, J. Hu and J. S. Foord, *Phys. Chem. Chem. Phys.*, 2008, **10**, 303.

3 K. B. Holt, *Phys. Chem. Chem. Phys.*, 2010, **12**, 2048.
4 W. Hongthani and D. J. Fermin, *Diamond Relat. Mater.*, 2010, **19**, 680.
5 J. B. Zang, Y. H. Wang, L. Y. Bian, J. H. Zhang, F. W. Meng, Y. L. Zhao, S. B. Ren and X. H. Qu, *Electrochim. Acta*, 2012, **72**, 68.
6 J. B. Zang, Y. H. Wang, S. Z. Zhao, L. Y. Bian and J. Lu, *Diamond Relat. Mater.*, 2007, **16**, 16.
7 L. H. Chen, J. B. Zang, Y. H. Wang and L. Y. Bian, *Electrochim. Acta*, 2008, **53**, 3442.
8 I. A. Novoselova, E. N. Fedoryshena, E. V. Panov, A. A. Bochechka and L. A. Romanko, *Phys. Solid State*, 2004, **46**, 748.
9 K. B. Holt, D. J. Caruana and E. J. Millán-Barrios, *J. Am. Chem. Soc.*, 2009, **131**, 11272.
10 V. N. Mochalin, O. Shenderova, D. Ho and Y. Gogotsi, *Nat. Nanotechnol.*, 2012, **7**, 11.
11 J.-S. Tu, E. Perevedentseva, P.-H. Chung and C.-L. Cheng, *J. Chem. Phys.*, 2006, **125**, 174713.
12 J. Scholz, A. J. McQuillan and K. B. Holt, *Chem. Commun.*, 2011, **47**, 12140.
13 O. Shenderova, A. M. Panich, S. Moseenkov, S. C. Hens, V. Kuznetsov and H.-M. Vieth, *J. Phys. Chem. C*, 2011, **115**, 19005.
14 A. Dementjev, K. Maslakov, I. Kulakova, V. Korolkov and V. Dolmatov, *Diamond Relat. Mater.*, 2006, **15**, 1813.
15 W. Zhao, J.-J. Xu, Q.-Q. Qiu and H.-Y. Chen, *Biosens. Bioelectron.*, 2006, **22**, 649.
16 J.-T. Zhu, C.-G. Shi, J.-J. Xu and H.-Y. Chen, *Bioelectrochemistry*, 2007, **71**, 243.
17 S. Osswald, G. Yushin, V. Mochalin, S. O. Kucheyev and Y. Gogotsi, *J. Am. Chem. Soc.*, 2006, **128**, 11635.
18 A. Krueger, *Chem.-Eur. J.*, 2008, **14**, 1382.
19 I. Kiflawi, A. E. Mayer, P. M. Spear, J. A. Van Wyk and G. S. Woods, *Philos. Mag. B*, 1994, **69**, 1141.
20 O. A. Williams, J. Hees, C. Dieker, W. Jäger, L. Kirste and C. E. Nebel, *ACS Nano*, 2010, **4**, 4824.
21 M. Ozawa, M. Inaguma, M. Takahashi, F. Kataoka, A. Kruger and E. Osawa, *Adv. Mater.*, 2007, **19**, 1201.
22 J. P. Hsu and Y. C. Kuo, *J. Colloid Interface Sci.*, 1997, **185**, 530.
23 P. Debye and E. Hückel, *Physikal. Z.*, 1923, **24**, 185.
24 G. Socrates, *Infrared characteristic group frequencies – Tables and charts*, Wileyand Sons, 2nd edn, 1994, pp. 94–156.
25 M. Chen, E. D. Pierstorff, R. Lam, S-Y. Li, H. Huang, E. Osawa and D. Ho, *ACS Nano*, 2009, **3**, 2016.
26 R. H. Wopshall and I. Shain, *Anal. Chem.*, 1967, **39**, 1514.
27 R. H. Wopshall and I. Shain, *Anal. Chem.*, 1967, **39**, 1535.

DISCUSSIONS

Role of surface contaminants, functionalities, defects and electronic structure: general discussion

Richard McCreery, Chi-Chang Hu, Julie Macpherson, Matěj Velický, John Foord, Matteo Duca, Katherine Holt, Manuel Alvarez-Guerra, Robert Dryfe, Surbhi Sharma, Patrick R. Unwin, Jingping Hu, George Zheng Chen, Fulian Qiu, Robert Hamers, Mark Newton, Andrew N. J. Rodgers, Philip A. Ash, Deborah Lomax, Keith Stevenson, Aleix Güell, Thomas Varley, Stephen Hodge and Fernanda Juarez

DOI: 10.1039/C4FD90027H

Richard McCreery opened the discussion of the paper by Robert Dryfe: You stated that Ru(hexamine) shows a depression of the outer sphere ET rate, on graphene, possibly due to the proximity of its redox potential to the Dirac point. If so, then other redox systems with different E^os should show an increasing kinetic trend as the potential moves from the Dirac point. Has anyone observed such a trend?

Robert Dryfe answered: This has not been explicitly investigated, to the best of my knowledge. We have plotted this data for the limited number of redox couples (four) we have investigated, *i.e.* the FcTMA and $Ru(NH_3)_6{}^{3+}$ couples reported in this manuscript, as well as ferricyanide and hexachloroiridate; the latter pair from data that is currently unpublished. Plotting the decadic logarithm of the standard rate constant on monolayer graphene (using the 'sample 1' preparation) as a function of the standard reduction potential yields the graph shown (Fig. 1), which does give an approximate correlation. This does support the density of states argument advocated by the questioner – given that the Dirac point of graphene is reported[1] to be close to the reduction potential of the $Ru(NH_3)_6{}^{3+}$ complex, the complex displaying the slowest kinetics. However, given the limited dataset, we have also taken the liberty of attempting this kinetic correlation with redox potential for a wider set of data, namely the rate constants reported on CVD-prepared monolayer graphene by Prof. Abrũna and co-workers.[2] This plot revealed no strong correlation, so the best we can say is that the question is still open.

1 Q. H. Wang *et al.*, *Nat. Chem.*, 2012, **4**, 724.
2 N. Ritzert *et al.*, *Langmuir*, 2013, **29**, 1683.

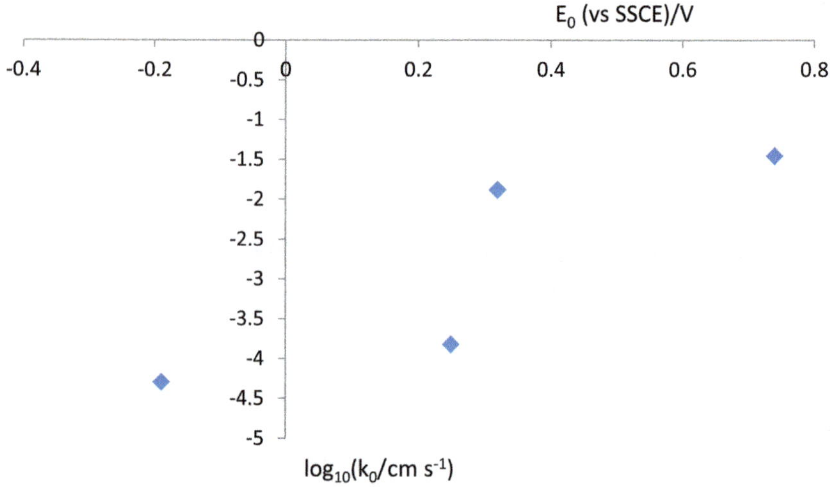

Fig. 1 Decadic logarithm of the standard rate constant on monolayer graphene as a function of the standard reduction potential.

Matěj Velický addressed Richard McCreery and Robert Dryfe: Rob Dryfe mentioned that the low electrode kinetics observed for the hexaammineruthenium (HAR) redox mediator are possibly related to the fact that the standard redox potential of HAR could be close to the Dirac point in graphene – where the density of states (DOS) is low – which would result in slower kinetics. This is indeed one of the early explanations we adopted, however, our recent work on graphene *vs.* bulk graphite electrodes (unpublished) indicates that there might be another explanation for this behavior. Monolayer graphene does generally exhibit kinetics slower than bulk graphite, however, the absolute value of HAR kinetics is still relatively slow on bulk graphite (~10^{-5} to 10^{-4} cm s^{-1}), which does not support the Dirac point proximity explanation. Furthermore, unpublished results suggest that the variation in kinetics is most prominent for thin graphene samples (*i.e.* below 10–20 layers), which makes sense as the DOS in these flakes will be affected much easier than in bulk graphite. Also it is important to add that the work presented here has been carried out using natural graphite, in contrast with the vast majority of electrochemical studies on HOPG, so any comparison with other literature data is very limited.

Julie Macpherson asked Robert Dryfe: Is acetone and chloroform doping in graphene p-type or n-type and by how much does the Fermi level shift? Where would the E^{o} for ruthenium hexamine ($Ru(NH_3)_6^{3+}$) sit on the band diagram for doped graphene?

Robert Dryfe replied: Chloroform is reported to intercalate between graphene and the oxide-covered silicon wafer support[1] leading to significant (by *ca.* 0.5 eV) p-doping of the graphene: we see slightly slower kinetics for the chloroform-treated sample. The same reference reports that acetone leads to much lower doping than chloroform.

1 H. H. Kim *et al.*, *ACS Nano*, 2013, 7, 1155.

Patrick Unwin enquired: Studies from our group (*e.g.* ref. 1 and 2) have shown that there are apparently significant changes in the HOPG surface and resulting electrochemical response over time, and that one needs to make measurements as soon as possible after cleavage to capture the behaviour of a freshly cleaved surface. How do you deal with the time issue for studies of exfoliated graphene and what happens to the electrochemical response over time?

1 S. C. S. Lai, A. N. Patel, K. McKelvey and P. R. Unwin, *Angew. Chem. Int. Ed.*, 2012, **51**(22), 5405–5408.
2 A. N. Patel *et al.*, *J. Am. Chem. Soc.*, 2012, **134**(49), 20117–20130.

Robert Dryfe responded: This is another good point and my answer relates to the one given to Julie Macpherson, in response to her question about adsorbates in graphene. The time issue is difficult to control in our mechanically exfoliated samples since we generally identify and transfer the flake first, and make electrical contact afterwards, meaning that there is a long atmospheric exposure time (typically 24 h). Dr Velický has, however, used our pipette approach to perform an '*in situ*' cleavage of graphite, that is to use the pipette to exfoliate fresh material which is immediately exposed to the solution and not the atmosphere (currently unpublished data). We find much more reversible kinetics for the slowest of the redox couples, *i.e.* the ruthenium hexamine species, for the freshly exfoliated samples. This is not, admittedly, done for thin 'graphene-like' samples, as it would be difficult to perform such a cleave without breaking a thin sample, but – at least for graphite – our findings certainly agree with the work reported in the two papers you cite from your laboratory.

Julie Macpherson asked: Given the issues surrounding graphene contamination during transfer and solvent washes, with even the length of time the material is exposed to the atmosphere or solution of interest potentially playing a role, do you foresee an application role for graphene electrodes in the future?

Robert Dryfe answered: This is a good question; the short answer is yes, I do, but it is essential that we do further work to reveal the intrinsic electron transfer response of 'true' graphene, *i.e.* one free of accidental contamination. I think that, as well as contamination due to polymer residues and solvents, adsorption of organic material from the atmosphere is a significant factor behind much of the variability seen in graphene electrochemistry. We need to develop protocols to isolate and transfer graphene samples without exposure to such adventitious carbon. I don't think this has been achieved yet.

Aleix Güell remarked: It is quite well known, the influence of the substrate on graphene's electronic properties, and electron transfer in particular.[1] This is the reason for the search focusing on screening methods such as h-BN, silane monolayers, *etc.* My concern is that, working with such big graphene flakes, and considering its known impermeability,[2] the chances to trap air-humidity between graphene and the substrate are high. How would this affect macroscopic experiments? Have you observed variation within a flake in electrochemical activity when using microscopic droplet experiments?

1 Q. H. Wang *et al.*, *Nature Chemistry*, 2012, **4**, 724.
2 J. S. Bunch *et al.*, *Nano Lett.*, 2008, **8**, 2458.

Robert Dryfe replied: We do see localised variation of electrochemical activity within single flakes. I accept the questioner's suggestion that trapped solvent (or other) residues underneath the flake could be responsible for this variation, however, similarly local adsorbates on top of the flake (as alluded to above) could equally be responsible, so for the moment we cannot readily distinguish between these two possibilities.

George Zheng Chen opened the discussion of the paper by Keith Stevenson: I was very impressed by your insightful talk on single layer graphenes. How large are these single layer graphenes in terms of area? Will the size of the graphene layer affect the double layer capacitance, or will the graphene layer itself be charged?

Keith Stevenson responded: Thanks for your comments. The original nature materials paper[1] that describes the growth of single layer graphene on SiC indicates that the SiC terraces are about 3×50 μm and that the graphene forms single domains over the individual terraces. Our Sandia researcher (Calvin Chan) says that the physical grains of graphene are actually much larger than that, on the order of 100s of microns to millimeters. They typically cut the SiC wafer to 1 cm^2 sizes and grown graphene on SiC of this size. At Sandia they have performed LEEM studies and don't see multiple graphene domains at once, as the microscope's field of view is smaller than 100 micrometers. The question is, what do you really refer to as a grain of single layer graphene, because even though a single graphene crystallite is millimeters large, whenever it goes over a terrace edge on SiC it will be interacting with another layer of graphene in that region. These samples are approximately 20% bilayer graphene. We haven't looked specifically at the double layer capacitance in aqueous electrolytes, only in ACN TBAPF6 electrolytes. To our knowledge the single layer graphene on hydrogen intercalated SiC has no residual surface charge. The single and double atom vacancies are possibly passivated by hydrogen or other heteroatoms, but we don't see anything in the STM images to suggest that the defects are trapping charge at these sites.

1 K. V. Emtsev *et al.*, *Nat. Mat.*, 2009, **8**, 203.

Robert Dryfe said: Fig. 11, in particular Fig. 11(e), of your paper shows the comparison of surface coverage between your graphene samples and glassy carbon. Given the debate about the relative reactivity of graphene *vs.* graphite, and about the activity of the graphene basal plane, could you comment on the relative reactivity of your graphene samples *vs.* HOPG?

Keith Stevenson replied: We did not perform any studies on HOPG. The problem with HOPG is that the quality of the substrate is strongly dependent on the provider, therefore it is hard to establish a clear correlation between the local properties (*e.g.* the density of step edges and basal plane defects) of the HOPG and the reactivity. We plan to obtain some high quality HOPG from McCreery and perform complimentary experiments to compare the relative reactivity of single layer graphene *vs.* HOPG.

Patrick Unwin remarked: We have recently studied the diazonium functionalisation of HOPG surfaces at both the macroscale and microscale.[1] Our data suggest that step edge defects play little role in the electrochemical generation of diazonium radicals or the subsequent modification (resulting in passivation of the electrode surface). At the macroscale, we studied 2 different grades of HOPG whose step edge density differed by more than 2 orders of magnitude. The cyclic voltammetric response during repetitive sweeps, during which the HOPG surfaces became increasingly passivated, was independent of the HOPG grade, with both materials showing similar electrochemical responses (Fig. 2). At the microscale, we have used scanning electrochemical cell microscopy (SECCM)[1] to pattern high quality HOPG at a series of spots, through meniscus contact from the SECCM pipette, under chronoamperometric conditions for defined time periods (1–17 s). As shown in the AFM images in Fig. 3, local functionalization creates well-defined spots on the surface and there is no noticeable change in the spots in regions with step edges compared to those without. Modification of the sp^2 graphite surface was confirmed by Raman microscopy.[1]

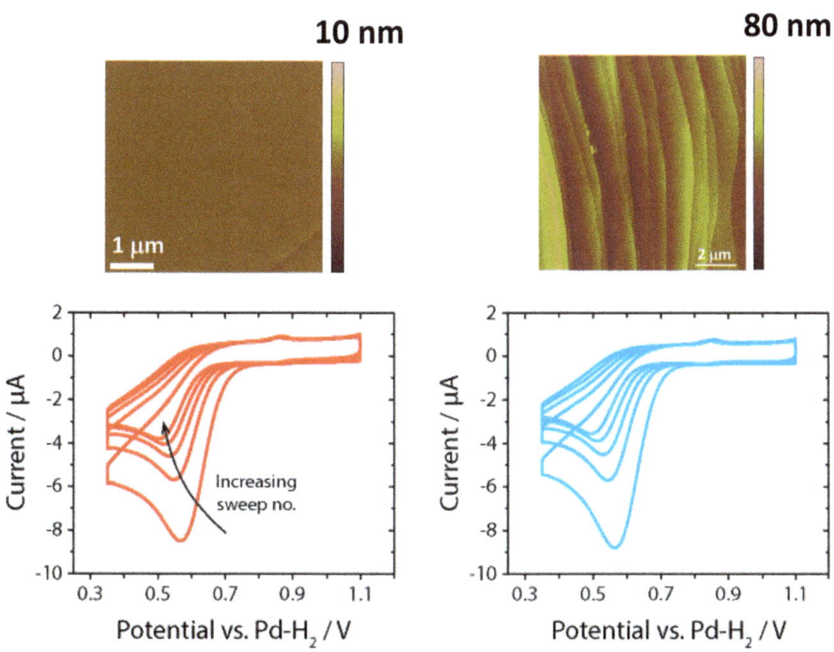

Fig. 2 CVs (0.2 V s^{-1}) for the reduction of 4-carboxybenzenediazonium (0.1 mM in 50 mM aqueous H_2SO_4) as a function of increasing sweep number for 2 grades of HOPG with widely different step edge densities. Left: ungraded, high quality HOPG, originating from Dr Arthur Moore and kindly provided by Prof. Rick McCreery (University of Alberta). Right: SPI-3 grade HOPG. Voltammetric data from ref. 1.

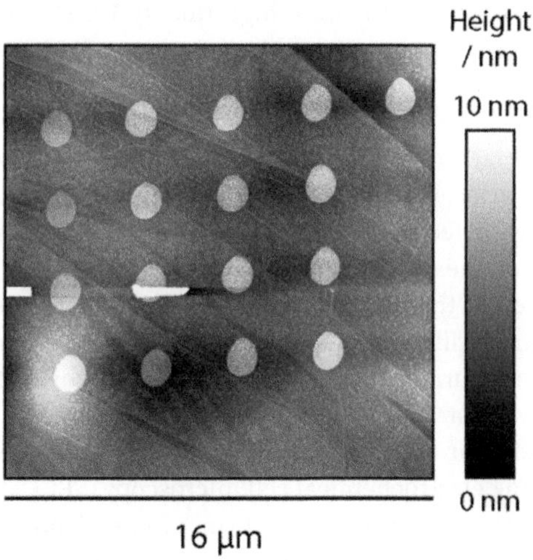

Fig. 3 Array of diazonium-functionalised spots on high quality HOPG, visualised by ex-situ AFM, following modification by SECCM for defined times from 0.5 s (top left) to 8.5 s (top right) at 0.5 s intervals, through the reduction of 4-carboxybenzenediazonium (0.1 mM in 25 mM aqueous H_2SO_4). The substrate potential was set at a value corresponding to the peak current of initial voltammetric scans on HOPG. Data taken from ref. 1.

1 P. M. Kirkman, A. G. Güell, A. S. Cuharuc and P. R. Unwin, *J. Am. Chem. Soc.*, 2014, **136**(1), 36–39.

Richard McCreery asked: Kariuki and McDermott reported images of 'mushrooms' formed by nucleation of diazonium-derived radicals attaching to edge plane sites on HOPG. Doesn't this indicate faster reactivity at edge sites?

Patrick Unwin answered: The CV responses (above and in ref. 1) at 2 grades of HOPG with a difference in step edge density of >2 orders of magnitude suggest that step edges are not important in controlling the functionalization of HOPG surfaces by diazonium-derived radicals and oligomers. Additionally, in the same study, the detailed AFM analysis we carried out on portions of the HOPG surface modified locally by SECCM showed fairly uniform deposits, with no detectable difference between the pristine basal surface and regions intersected by step edges. We did not see 'mushroom' features on the surface.

1 P. M. Kirkman, A. G. Güell, A. S. Cuharuc and P. R. Unwin, *J. Am. Chem. Soc.*, 2014, **136**(1), 36–39.

Richard McCreery commented: A general comment based on organic chemistry regarding the radical attack of HOPG: in order to attach to the basal plane the system would lose aromaticity in three rings, to form a sp^3 center. To attack an edge site and add to a double bond would affect aromaticity in only one ring. So basal attack would seem to be energetically more difficult.

Keith Stevenson responded: We agree that from a thermodynamic standpoint the energy required to break an sp^2 bond to form a sp^3 center at the basal plane of

HOPG should be higher than reacting at an edge site. We have performed preliminary theoretical investigations that support this model. However, since we do not find these single site modifications located close to single or double carbon vacancies as imaged by STM, this suggests that the reaction mechanism is kinetically controlled (*i.e.* the rate limiting step is not carbon–carbon bond breaking but rather a concerted reaction involving possibly the LUMO of graphene and the radical of the iodonium salt. Since we are performing electrochemical reactions we are far from thermodynamic equilibrium, so a thermodynamic argument does not seem to hold when trying to understand the reactivity at basal planes *vs.* defects and step edges. In the presented studies we find that the reactivity is rather controlled by kinetics as suggested by data presented in the paper (see Fig. 11f).

Richard McCreery remarked: Breton *et al.*[1] reported that diphenylpicrylhydrazil (DPPH), which is a radical scavenger, limited many diazonium modifications to real monolayers, by trapping additional phenyl radicals in solution formed after the initial monolayer. We haven't tried it, but others have said it works well.

1 Breton *et al.*, *Chem. Mater.*, 2013, **25**, 2905.

Keith Stevenson answered: This is an excellent suggestion. We are currently conducting more experiments to see if this changes the surface coverage.

Julie Macpherson asked: How do you control the reaction conditions in the functionalization of graphene so that you do not disturb the hexagonal symmetry?

Keith Stevenson replied: Wherever the covalent attachment occurs, it does break the hexagonal symmetry (A-B symmetry) which is accompanied by strong alteration of the grephene's local electronic properties. This is evidenced by the pronounced scattering observed in the STM images around the modifications (Fig. 10b of the paper).

Philip A. Ash opened the discussion of the paper by Daren Caruana:† Recently there have been several demonstrations of photodriven catalysis using immobilised light harvesting moieties. For example, light-triggered H_2 production by hydrogenases bound to CdTe quantum dots[1] or CdS nanorods,[2] or graphene quantum dots as photosensitizers for solar cell applications.[3] With regards to your graphene nanoflakes, is it possible that a systematic study of the effect of edge termination on the rate of electron transfer could provide insight into desirable linker characteristics for covalent attachment of quantum dots to electron acceptors such as semiconducting surfaces or electrocatalysts? Could you use the nanoflakes to inject electrons photochemically into directly attached redox-active species?

1 Greene *et al.*, *J. Am. Chem. Soc.* 2012, **134**, 11108–11111.
2 Wilker *et al.*, *J. Am. Chem. Soc.* 2014, **136**, 4316–4324.

† Daren Caruana's paper was presented by Katherine Holt, University College London, London, UK.

3 Roy-Mayhew *et al.*, *Chem. Rev.* 2014, **114**, 6323–6348.

Katherine Holt responded: This is a very exciting idea for an application! As the edge groups of our GNF are so well-defined, it make systematic study and controlled modification a real possibility. Also we should be able to control the numbers and distribution of the attached species on the flakes. However, the photochemical stability of the GNF needs to be studied to ensure they could be used in photo-driven catalysis.

Matteo Duca asked: Have you tried adding crown ethers to the electrolyte to sequester K^+ ions, thus highlighting their role in ion-pair formation with ferricyanide?

Katherine Holt answered: This is a good idea and definitely something to explore. I am not aware of any previous studies using ferricyanide where this has been attempted.

Matteo Duca said: Concerning the oxidation of hydroquinone at c-GNF (subsection 3.6 of your paper), experiments in unbuffered media should confirm or rule out the occurrence of a concerted proton-coupled electron transfer (CPET). If CPET is operative, the oxidation wave is expected to shift to a higher potential with respect to a buffered media having the same pH, while its peak potential should be independent of the pH for pH < pK_a of the phenolic OH. Additionally, the use of D_2O instead of H_2O can highlight a kinetic isotope effect, thus corroborating the CPET mechanism. Similar experiments played a pivotal role in the mechanistic study of CPET during the oxidation of phenol at a glassy carbon electrode in aqueous media.[1]

1 C. Costentin *et al.*, *PNAS*, 2009, **106**, 18143–18148.

Katherine Holt replied: We have some additional experimental data that can be used to answer this question. Experiments were carried out in unbuffered media in the pH range 5.5 to 8.2. The oxidation peak potentials are plotted in Fig. 4b together with data from experiments in buffered media. The data show that in unbuffered media the E_p is shifted to higher potentials compared to buffered solutions and, crucially, the E_p values are independent of pH over this pH range. The behaviour of hydroquinone in D_2O was also studied. Fig. 4a shows the oxidation wave at clean BDD and c-GNF modified BDD in H_2O and D_2O. The pD of the D_2O solution, based on the smaller dissociation constant of D_2O, is estimated to be 0.4 units higher than the pH of a corresponding H_2O solution and therefore about 6.6. When D_2O is used as the solvent, the main oxidation peak potential shifts to higher values compared to H_2O of similar pH.

A second set of peaks at more negative potentials appear in unbuffered solutions after the initial oxidation wave. Fig. 5a gives a comparison of the CVs at clean BDD and c-GNF modified BDD at pH 8.21. The ratio of peak I to peak II is plotted in Fig. 5b as a function of pH. The peak separation of *ca.* 60 mV is indicative of reversible kinetics, and the peak potentials are independent of pH over the pH range examined. The fact that the peak height of this reversible redox wave decreases with decreasing pH suggests it is due to a stepwise PET as

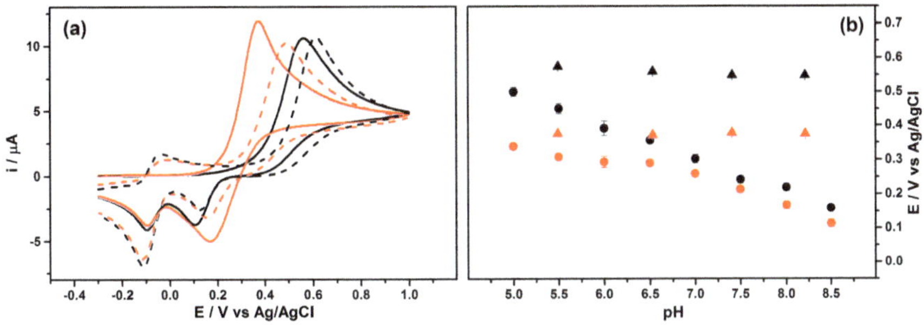

Fig. 4 (a) CV of 0.5 mM hydroquinone at clean BDD electrode (black line) and c-GNF modified electrode (red line) in H_2O (solid line) and D_2O (dashed line). Supporting electrolyte: 0.1 M KCl. The pH of the H_2O electrolyte solution was adjusted to 6.5 with KOH. Scan rate: 50 mV s^{-1}. First scans shown. (b) Peak potential of hydroquinone oxidation as a function of pH at a clean BDD electrode (black) and c-GNF modified electrode (red). Circles represent buffered media and triangles unbuffered media.

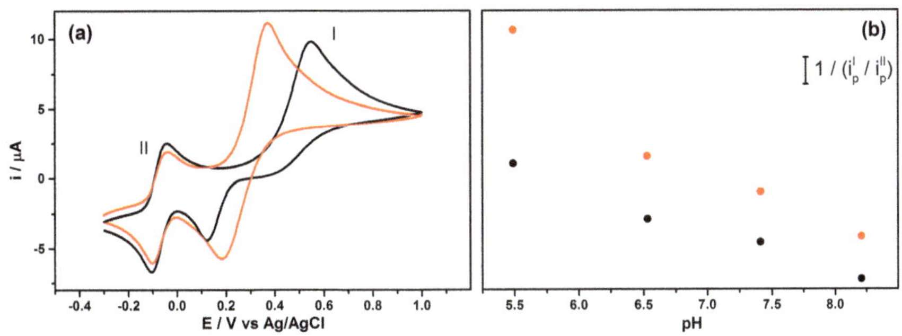

Fig. 5 (a) CV of 0.5 mM hydroquinone at clean BDD electrode (black line) and c-GNF modified electrode (red line) in unbuffered H_2O. The pH of the H_2O electrolyte solution was adjusted to 8.21 with KOH. Supporting electrolyte: 0.1 M KCl. Scan rate: 50 mV s^{-1}. Second scans shown. (b) Ratio of peak heights of hydroquinone oxidation as a function of pH at clean BDD electrode (black) and c-GNF modified electrode (red).

described by Costentin *et al.*[1] The reversible kinetics support the assignment of this mechanism. Upon modification of the electrode with c-GNF, there is no shift in E_p, but the peak height is smaller. These observations are in agreement with the conclusions made in Section 3.6 of our paper regarding the role played by COOH groups in the reaction mechanism. When c-GNF are present on the electrode surface they can act as a proton source and sink, allowing the reaction to proceed to a greater extent *via* the concerted pathway. The results presented here are in agreement with those reported by Costentin *et al.*[1] in the above-mentioned paper and strongly suggest that the oxidation of hydroquinone at both BDD and c-GNF modified BDD follows the CPET mechanism in buffered media. In unbuffered media, two competing pathways are operative: CPET, which dominates in acidic solutions, and PET in basic conditions.

1 C. Costentin *et al.*, *J. Am. Chem. Soc.*, 2008, **130**, 15817–15819.

Keith Stevenson asked: The pH titration slide that you showed is very intriguing as the extrapolated pK_a for carboxylic functionalities is more basic than in solution. Is this similar to that seen for SAMs? What fraction of functionalised sheets contain these surface functionalities? How do you know that you don't have mixtures of sheets with varying functionalities and surface coverages?

Katherine Holt answered: We have been quite cautious about extrapolating a pK_a value from this titration curve, as while we can identify an equivalence point the curve is not the form we would expect for one single solution species. However, it does seem that being bound to the nanoflakes does influence the pK_a in a manner that might be observed for other surface bound acid groups, for example in a SAM. Although we are reasonably sure that carboxylic acid groups are the only acidic oxygen functionality present on the flakes, it is difficult to know their distribution. For example, we could have a population of flakes with most of the edge functionalised and another population that is less functionalised. So far we have assumed an average edge coverage for all of the flakes, but identifying different fractions in the sample would be useful. We think most flakes must be terminated as they are dispersible in water. Size may also be a factor – the flakes are a range of diameters, so of course the smaller flakes may experience more interactions between neigbouring groups than larger flakes.

Patrick Unwin enquired: In your paper you mentioned that you were carrying out studies to determine the film morphology and architecture, which will clearly be important in understanding the electrode responses. Have you made any progress with this?

Katherine Holt replied: This work is still ongoing as it is very difficult to characterise these small, optically transparent flakes and the BDD electrodes are likely too rough to allow AFM imaging of the flake distribution on the surface. TEM imaging of drop-cast films (on a TEM grid) shows some evidence of stacking parallel to the surface and we could assume that the flakes align in a similar manner on the electrode surface. In order to control the film morphology better we are exploring two approaches: (i) adding divalent cations to the GNF suspension to bring about their complexation and precipitation; the structures formed can then be characterised *ex situ* before being drop-coated onto the electrode; (ii) modifying a working electrode (*e.g.* SAM on gold) and then assembling the flakes on this layer in defined orientation to explore electron transfer across the flake from edge to edge.

George Zheng Chen remarked: In Fig. 3 of your paper, why is there almost no change in the double layer charging current?

Katherine Holt responded: It is true there is only a small increase in double layer charging for the modified electrode (*e.g.* over the potential range 0–0.4 V). We haven't investigated this aspect of the response in great detail, but perhaps this suggests only a very thin and relatively non-porous layer of GNF on the surface and hence no significant increase in surface area.

John Foord asked: How porous are the graphene films? Are you sampling only the outer layers or the entire film in the electrochemical experiments?

Katherine Holt answered: It is difficult to characterise the films on the electrode as they are so thin and optically transparent, but we suspect they are a few layers thick. If you add more graphene to the electrode surface it simply falls off as it is so hydrophilic and hence dispersible in solution. TEM imaging of flakes drop-coated onto the TEM prism show some evidence of stacking, so we are probably studying electron transfer through multiple layers (and therefore contact with the redox probe may only be at the outer layer) but there will be some porosity allowing some ingress of the redox probe to the inner layers. We need to have more control and understanding of how the GNF order onto the electrode to have better understanding of their behaviour.

Julie Macpherson remarked: In your diamond nanoparticle study[1] you report evidence of FcMeOH adsorption on the 'sp^2' or defected surface of the nano-diamond and you refer to work by Abrũna (reference 3 in your paper) who also reports FcMeOH adsorption on graphene surfaces. You don't see any evidence for FcMeOH adsorption on your graphene nanoflakes. Do you think we should be concerned about the possibility of FcMeOH adsorption on sp^2 surfaces? By tradition FcMeOH is considered outer sphere and therefore non-interacting with the surface.

1 T. S. Varley, M. Hirani, G. Harrison and K. B. Holt, *Faraday Discuss.*, 2014, DOI: 10.1039/C4FD00041B.

Katherine Holt replied: In the diamond nanoparticle study[1] we see evidence from cyclic voltammetry and IR spectroscopy that FcMeOH and $FcMeOH^+$ adsorb onto the diamond surface. However the adsorption is quite weak, as removing the modified electrode from the solution and rinsing it seems to remove the adsorbed species. We also see no evidence of the FcMeOH species on the rinsed diamond powders in subsequent XPS analysis. In the case of the diamond surfaces it also seems that adsorption of $FcMeOH^+$ is stronger than adsorption of neutral FcMeOH, which seems to indicate an electrostatic interaction with the negatively charged, oxidised diamond, as well as some pi-pi interactions. In Abruña's work[2] he reports electrochemical evidence for adsorption of FcMeOH to a graphene surface. Much is weakly adsorbed and is removed with repeated cycling. However, about 1% of a monolayer appears to be very stable. He suggests this adsorption occurs "predominantly at local defect sites". He also states "This value is comparable to what we observe on bulk graphite surfaces (~2% monolayer coverage)". Additionally he cites a reference for adsorption of ferrocene on multiwalled carbon nanotubes[3] which makes interesting reading. In the latter publication they see ferrocene adsorption onto the walls of nanotubes in different orientations (though it should be noted that surfactant may be present and play a role). They suggest that certain orientations are preferred and adsorption is increased when oxygen functionalities are present on the nanotubes, and that interaction is perhaps preferably with the Fc^+ species in this case. So to answer your question, a variety of sources seem to suggest that FcMeOH (and Fc) can interact with sp^2 surfaces either *via* pi-pi interactions (weak) with the pristine surface, or perhaps more importantly *via* some electrostatic interaction with oxygen groups or defect sites (stronger). However we do not see any obvious signs of adsorption in the case of our graphene nanoflakes. However your question raises an interesting point about the outer sphere nature of these probes. I think electron transfer can still be called as outer sphere because interaction with the surface is not necessary for fast electron transfer, but that is not to say that some interaction cannot take place.

1 T. S. Varley, M. Hirani, G. Harrison and K. B. Holt, *Faraday Discuss.*, 2014, DOI: 10.1039/C4FD00041B.
2 W. Li *et al.*, *ACS Nano*, 2011, **3**, 2264.
3 D. Zheng *et al.*, *Thin Solid Films*, 2008, **516**, 2151.

Surbhi Sharma said: You prepare carboxylated graphene sheets by unzipping CNTs. How do you confirm or establish that the carboxylation process does not leave any other oxygen related defects (particularly –OH groups that are almost unavoidable) on the basal and edge planes? Do you perform any XPS analysis?

Katherine Holt responded: We have used two different types of c-GNF in our studies. One sample was identical to that reported in ref. 1, where the carbon 1s XPS peak could be fit with constituent peaks at 284.7 eV, 286.7 eV and 288.4 eV. The peak positions are consistent with C–C sp^2 bonding, C–O bonding (alcohol, epoxide, ketone *etc.*) and C(O)O bonding (carboxylic acid, anhydride, lactone) respectively. The second sample was used for all studies in this paper and the C 1s data could be fit very satisfactorily with 2 peaks – at 284.7 eV and 288.4 eV. In other words the C–O content was very small, if present at all. Although only data for the second sample is reported here, we see broadly similar behaviour for the first sample in the pH dependence of the responses for the different redox probes. The responses are however more enhanced for the flakes used in this study, we believe because the concentration of carboxylic acid groups is higher.

1 C. G. Salzmann, V. Nicolosi and M. L. H. Green, *J. Mater. Chem.*, 2010, **20**, 314.

Fulian Qiu asked: In Fig. 5 and 6 of the paper, the ferri/ferrocyanide redox reaction showed slower responses on the –COOH modified graphene. If the –COOH group on the modified graphene is reduced electrochemically, and the electrode becomes a graphene electrode, what do you expect from the ferri/ferrocyanide redox reaction?

Katherine Holt answered: When the –COOH GNF is modified to give an amide termination we find that the electrochemical response to ferri/ferrocyanide becomes reversible. This suggests that it is the acidic group which is causing the slower response. I would imagine if the –COOH groups were also reduced electrochemically (*e.g.* to –OH or even –H termination) the response towards the redox probe would also be reversible as the acid groups have been removed.

Philip A. Ash remarked: You have studied the effect of ionic strength and pH on the ferri/ferrocyanide redox reaction at graphene nanoflake electrodes. Have you also tried adding a small positively charged redox mediator such as methyl viologen? High concentrations of KCl are known to solubilise Prussian blue, and oligomeric hexacyanoferrate moieties have been proposed as possible 'blocking' species at electrode surfaces. Such species will lead to a build-up of negative charge at the nanoflake edges that could lead to a greater electrostatic repulsion than the $COOHCOO^-$ termination alone. Is it possible that a positively charged mediator would restore reversibility of the ferri/ferrocyanide redox reaction if this were the case?

Katherine Holt replied: This is something we have not tried but would be a simple experiment to test the inhibiting mechanism.

Philip A. Ash continued: Could you use infrared or Raman spectroscopy to identify the 'blocking' species responsible for the irreversibility of the ferri/ferrocyanide redox couple with the COOH-terminated nanoflakes? Spectroscopic techniques should give insight into protonation/deprotonation equilibria and reactions occurring at the nanoflake edges as a function of potential, pH *etc.* Do you think that direct spectroelectrochemical measurements of the electrode-immobilised nanoflakes would be helpful in this respect, as you should observe oxidation and reduction of an adsorbed species even in the case where it 'blocks' the solution response?

Katherine Holt responded: We are currently exploring using *in situ* IR spectroelectrochemistry to investigate the stability of ferri/ferrocyanide in the presence of the c-GNF. We have also looked at the protonation state of the flakes in solutions of different pH and these results are being prepared for publication. Direct spectroelectrochemistry would be useful because we can observe changes to the GNF termination, as well as changes to the redox species itself. We'd like to use Raman spectroscopy as well, however for some reason our flakes fluoresce so we have been unable to get good low enough backgrounds to see the Raman response.

Philip A. Ash said: If cyanide ligand loss occurs due to dynamic protonation equilibria at the COOH-terminated nanoflake edges below pH 8, you would expect to see evolution of HCN in a time-resolved infrared measurement, and subsequent appearance of peaks due to the 'blocking' species. Such a measurement would be complicated, however, by peaks due to mixed-valence species containing bridging cyanide (such as Prussian blue, Berlin green *etc.*) that all have peaks in the 2040–2100 cm^{-1} region, *i.e.* between those observed for ferri- and ferrocyanide. It is also unclear whether the extent of adsorption would be limited to monolayer formation or whether the edge of the nanoflakes could nucleate growth of a thicker hexacyanoferrate film.

Katherine Holt answered: Yes, our preliminary IR data has suggested formation of species which absorb strongly in this region when the c-GNF are present. It is very difficult to assign the bands to specific species as many reports in the literature are not too precise about their assignments! However it is clear that precursor molecules to film formation are formed over time in the presence of the c-GNF. We are presently studying the time dependence of this repsonse under different conditions, including in the presence of D_2O rather than water.

Julie Macpherson commented: In section 3.5, in regard to the bare BDD surface, the authors comment "The ferri/ferrocyanide electron transfer process has been shown to be inhibited at oxygen terminated surfaces, however we observe effectively reversible electron transfer kinetics at the (oxidised) BDD electrodes used in this study". We also see reversible electron transfer kinetics at our oxidised BDD electrodes (but for the reverse process *i.e.* ferro/ferricyanide) in 0.1 M KNO_3. The surface is alumina polished prior to use and we used XPS to characterise the surface functional groups on the oxygenated surface; the surface was found to contain C–O–C, C–OH and C=O groups. The electrode was polycrystalline, doped above the metallic threshold and thick enough to be removed from the wafer so it was freestanding. It also was shown to contain negligible sp^2 carbon. It was only when we anodically pretreated the electrode in 0.1 M H_2SO_4 using applied potentials *e.g.* 3 V for 60 s, that the electron transfer kinetics for this couple slowed considerably. XPS now showed the surface to contain groups such as COOH on the surface. Hence we thought that in some way these groups were responsible for the change in electron transfer kinetics. The authors note in their paper that the response of ferricyanide is very inhibited by the presence of carboxylic acid terminated GNF at pH < 8. We describe our work in ref. 1.

1 Hutton *et al.*, *Anal. Chem.*, 2013, **85**, 7230–7240.

Katherine Holt responded: Very interesting that the same functionalities as diamond have been shown to inhibit the ferri/ferrocyanide redox couple! We find in general that all of our BDD electrodes, if carefully polished and cleaned by electrochemical cycling, show reversible responses towards this redox couple. Its clear that we are using good quality BDD (little sp^2 content) that is not over-oxidised.

Julie Macpherson opened the discussion of the paper by Stephen Hodge: If you were able to use metallic nanotubes of a defined chirality where all metal catalyst had been removed, do you think this material would show greater reductive capabilities than the KC_8 and KC_{24} materials you worked with, given the different band structures and DOS?

Stephen Hodge answered: At high charging stoichiometries, the differences between metallic and semi-conducting SWCNTs tend to disappear, as the Fermi level will shift beyond the band gap. At low doping levels (below 1 charge to 100 C atoms) there will be obvious differences in the electron affinity of metallic *vs.* semi-conducting species. The total amount of charge that might be successfully transferred to SWCNTs before the appearance of excess alkali metal in equilibrium, is unknown; the threshold may be slightly higher for SWCNTs than graphite (based on some of our own preliminary data). However, investigating the maximum levels of doping for SWCNTs is more challenging than graphite, due to the heterogeneity of the samples; ARPES which generates excellent data for doped graphites is not readily accessible for SWCNTs.

For the same M : C charging ratios, the tendency for charge dissociation is likely to differ between SWCNTs and graphene salts due partly to the density of states, but probably more significantly due to changes in geometry and solvation energies/entropies. The capacitance of SWCNTs is also likely to be different from graphene, creating a different Coulombic shift. The dependence of effective dielectric constant or polarisability on nanotube type, as a function of doping level, also remains to be determined. Overall, since SWCNTs can probably be charged to a higher M:C ratio than graphite, one would assume that they have a lower reducing power at an equivalent M:C ratio.

Julie Macpherson continued: Do you think you will be able to reduce Al^{3+} using your graphene intercalated materials? If not, would it be possible using the nanotube system proposed in the previous question?

Stephen Hodge replied: The reactivity with manganese chloride was only slight (~3% yield). Since the standard reduction potential for Al^{3+} is a further 0.5 V, it is unlikely that Al^{3+} could be reduced with KC_8. Perhaps lithium intercalated graphite *e.g.* LiC_6, or more highly charged carbon nanotubides could provide the necessary higher reduction potentials for Al deposition.

Robert Dryfe enquired: Can you say something about the product that you form? Is monolayer graphene produced?

Stephen Hodge responded: We did not do any microscopy studies in this paper, but there are published papers showing the formation of monolayer graphene sheets following charging methods such as the route deployed here.[1] Simple discharging causes restacking, unless preventative measures are taken. The degree of exfoliation may be a factor in limiting the reduction potential of the graphenide.

1 Milner *et al.*, *J. Am Chem. Soc.*, 2012, **134**, 8302–8305.

Robert Dryfe continued: Your method is based on the exfoliation and stabilisation of graphene through charging of the graphene sheets. However, the data presented suggests that there is significant re-aggregation of the graphene: is it possible to use the metal deposition to prevent re-aggregation?

Stephen Hodge answered: The reduction of the metal salts to metal nanoparticles discharges the nanocarbons causing re-aggregation and a high level of restacking. Graphenide reactions that deposit less mobile, more localised species can help to prevent restacking or even enhance solubility.[1] It may be possible to produce a similar effect with metal or metal-containing species. Alternatively, to produce monolayer graphene decorated with metal nanoparticles, it may be simpler to deposit or assemble the monolayer graphene first and then submerse the film in a solution containing the metal salts.

1 Morishita *et al.*, *J. Mater. Chem. A*, 2014, **2**, 15022–15028.

Keith Stevenson remarked: What is the degree of reduction of the potassium? Is it metallic?

Stephen Hodge replied: In the paper we discussed the model proposed by Ebert[1] for the GIC KC_8, which proposes the presence of a mixture of metallic K^0

and K^+ ions between the graphite layers, relating to incomplete charge transfer to the graphite. For our solubilised graphenide species, and based on our experimental results, the potassium is likely to have transferred all of its charge to the carbon giving purely K^+ species. The central hypothesis of our paper is that the redox potential of the transferred electron depends on the level of doping, in a continuous fashion depending on the density of states and Coulombic effects. At the most charged level, the redox potential approaches that of metallic potassium.

1 L. B. Ebert, *Carbon*, 1985, **23**, 585–587.

Keith Stevenson continued: Have you measured the open circuit potential in solution? Can it be used to understand if you have a more reduced form of carbon?

Stephen Hodge responded: In fact, the original basis of this paper was to study the redox reactions on carbon nanotubide solutions using open circuit potential (OCP) measurements. In principle, this method would have allowed us to study the kinetics of the reactions and the OCP should have plateaued at the reduction potential of the metal salt. Unfortunately, there appeared to be a slow but significant self-discharge process occurring independently of any metal salt addition, and we could not obtain reliable results. The delay between experiment setup and stable OCP measurement was enough to have significantly lowered the carbon reduction potential. Therefore, we went down the route of performing the reactions as shown in this paper, but it is an experiment we should return to at a later date.

John Foord said: Can you vary the size of the metal nanoparticles by changing reaction conditions, and how adherent are the particles to the supports?

Stephen Hodge answered: In the case of $ZnCl_2$ reactions, we demonstrated that with increased metal salt to charge ratio the nanoparticle size was increased from 10s of nanometres to ~100 nm. There is most likely a balance between thermodynamics (nucleation) *vs.* kinetics (growth), depending on the redox potential of the metal, the degree of carbon charging, and the salt concentration. Further experiments might explore these effects more explicitly to generate smaller metal nanoparticles which might, for example, be more catalytically active.

Andrew Rodgers asked: Are the negatively charged carbon nanomaterials soluble in aqueous solutions?

Stephen Hodge replied: Unfortunately not, as the charge required to dissolve nanotubes/graphene is beyond the water and oxygen reduction potentials. All experiments must be performed in a glove box; subsequent exposure of these charged nanocarbons to the atmosphere results in their rapid discharge and possible covalent functionalisation *via* protonation and hydroxylation (see ref. 1).

1 Hof *et al.*, *J. Am. Chem. Soc.* 2013, **135**, 18385.

John Foord enquired: You mention in your paper that one of the motivations is to develop more practical processing conditions for carbon nanomaterials. But the work you describe seems to focus on rather air-sensitive chemistry. Is this practical?

Stephen Hodge responded: We use reductive charging methods as a route to non-destructive processing, avoiding the concentrated acid treatments, ultrasonication and ultracentrifugation typically employed. We have a number of publications in this area and typically use the increased reactivity for a variety of practical processing steps, including producing solutions, purification, or covalent modification reactions.[1,2] The focus of this Faraday paper was to use mildly dispersed materials to explore some fundamental issues, rather than to develop the process chemistry further. Although air-sensitive chemistry is required, these methods are actually highly scalable, and can be applied to bulk volumes (unlike, for example, ultracentrifugation). As an example, the nanotubide chemistry has recently been commercialised.[3] In fact, air sensitive chemistry can become easier as volumes increase since adventitious moisture becomes less significant.

1 Fogden *et al.*, *ACS Nano*, 2012, **6**, 54.
2 Hodge *et al.*, *ACS Nano*, 2013, 7, 1769.
3 http://www.linde-gas.com/en/products_and_supply/electronic_gases_and_chemicals/carbon_nanotubes/index.html.

John Foord continued: It appears from the work that XPS gave you rather unreliable results. I was wondering why this was – it normally works quite well for nanocomposite materials provided they are homogeneously prepared?

Stephen Hodge answered: The resulting hybrid metal nanoparticle–nanocarbon materials are not particularly homogeneous, and metal nanoparticle sizes differ (10–100 nm) with varying metal salt stoichiometry. Since even the smallest metal nanoparticle sizes are around the critical range of the XPS depth sensitivity, significant changes in apparent composition can be anticipated due to sampling artefacts.

Katherine Holt opened the discussion of the paper by Fernanda Juarez: In your paper you describe that modelling shows the most thermodynamically stable arrangement for oxidised zig-zag edges of graphene are carboxylic acid groups. For armchair edges lactone rings are preferred. In the paper we presented earlier[1] we found that our graphene nanoflakes are almost exclusively carboxylic acid terminated. Does your data therefore suggest that our flakes must have zig-zag edges rather than armchairs?

1 M. M. Lounasvuori *et al.*, *Faraday Discuss.*, 2014, DOI: 10.1039/c4fd00034j.

Fernanda Juarez replied: It is rather difficult to say how a real (experimental) system looks like just using computational simulation results, but it is not impossible. In a previous work,[1] we studied in detail the thermodynamic stability of more than 80 possible endings for zigzag and armchair steps. We obtained 8 zigzag and 5 armchair structures that are thermodynamically stable in different H_2/O_2 environments, and 4 of these endings (2 zigzag and 2 armchair) have carboxylic groups. Therefore, it is possible to have armchair steps with carboxylic groups. Nevertheless, more information is necessary to decide whether the endings are zigzag or armchair carboxylic groups. One way to do it is by calculating infrared frequencies of the stable systems, which are very sensitive to changes in the surrounding structure.

1 Soldano *et al.*, *Carbon*, 2014, **78**, 181.

Katherine Holt continued: How do the carboxylic acid groups stack around the edges?

Fernanda Juarez responded: In Fig. 5 of our manuscript there is a picture showing the distribution of the carboxylic groups on the zigzag edges. To describe this better we can say that the carboxylic groups are perpendicular to the graphene surface, face-to-face, but with the –OH groups on opposites sides. One important observation is that there are no hydrogen bonds in this arrangement.

Julie Macpherson asked: In the introduction to your paper you state "... if the applied potential is high enough, a complete oxidation of the defects takes place and spontaneous reduction of the metal cannot occur". Please can you clarify this statement, in particular what do you mean by complete oxidation and why is spontaneous reduction not possible?

Fernanda Juarez answered: Typical pre-treatment of the HOPG to avoid spontaneous deposition is anodic polarization. Several authors claim that this process completely eliminates all the reducing functionalities (ref. 3, 15 and 16 in our manuscript). However, both the nature of such species and the mechanism of this pre-oxidation treatment in order to avoid spontaneous deposition are unclear. It was suggested by Penner (ref. 3 in our manuscript) that the incompletely oxidized functionalities act as electron donors during the spontaneous reduction of the metals.

Julie Macpherson continued: For the work on the adsorption of Pt on oxidised steps, you say that the energy obtained on the oxidised step is smaller than on the bare or hydrogenated steps. However I thought that Penner always advocated that in order to obtain preferential electrodeposition along step edges and not the basal plane on HOPG, it was necessary to apply potential pulses in order to selectively oxidise the step. This seems to contradict your findings?

Fernanda Juarez replied: This contradiction is actually apparent and there are two aspects to be considered. First, as we have already explained in the previous question, the nature of the functionalities is unknown. We can only claim that the oxidized systems, which we have investigated, are less favorable for the deposition than bare steps. It is also expected that the process on the former should not be spontaneous and additional energy is required so the reaction can occur. Second, the procedure followed by Penner involves the application of an external potential pulse in order to form the nucleus (non-spontaneous deposition), not contradicting our findings.

Richard McCreery enquired: Your computer model considers mainly thermodynamic stability of metals on various graphene sites. Does it take into account the effect of hydration of the metal ions from which the metals are deposited?

Fernanda Juarez responded: Our model does not consider the hydration in any explicit or implicit manner. In DFT there are some models to add solvent implicitly, like using the polarizable continuum method (PCM). Unfortunately, the PCM can only be used in systems without periodic boundary conditions, like we did. Besides the fact that we did not use solvent, we believe that its role can be important in the kinetics of the reaction. However, in our manuscript we were focussed on disentangling the nature of the interactions between the carbon edges and the metal wires. A second stage of this work should be adding the solvent.

Manuel Alvarez-Guerra said: I was wondering about the computational time required for dealing with such complex structures in your *ab initio* approach. What size of sample or number of molecules can you include to obtain results in reasonable time?

Fernanda Juarez replied: In the calculation of those systems by DFT, there are a lot of steps that are not published but that require a considerable effort. These steps usually involve running several tests in order to find both the optimal computational conditions (functionals, pseudopotentials, basis sets) and the best software (in some complicated cases). However, if we do not consider that stage, then it usually takes a week to optimize the structure of a system with 40 atoms (where 20 of them are free to move), using at least 32 processors. Properties calculations can take one day per system, but the post-processing of the information takes much more time.

Deborah Lomax asked: The study aims to pair experimental with computational methods to investigate metal deposition at step edges on HOPG. With HOPG, the deposition environment involves step edges – several layers in height – on a terrace, whilst the modelling considers only the edge of a single graphene layer. Is it valid to compare results from these different approaches given the different substrate geometries, and how can the modelling be improved to reflect this?

Fernanda Juarez answered: As we already said in our manuscript: "In this contribution, we have set the starting points at the two extremes of a bridge between experiments and theory to understand the metal decoration of step-edges on HOPG". We truly believe that the real system is much more complex than our extremely simple simulation, and we know the limits of the simplifications we made. However, the study of real systems must be done in several steps, disentangling the nature of different interactions at each stage. We also think that this is a helpful first step in the right direction. A more realistic description that includes several layers of graphene, should describe correctly both metal/graphene and graphene/graphene interactions. This is not so easy from a computational point of view. The problem is that the interactions between the graphene layers are only well described by the functionals (and pseudopotentials) including Van-der-Waals corrections. It has not been systematically tested yet, how accurate these functionals reproduce the metal/graphene and metal/metal interactions.

Deborah Lomax continued: Has the influence of relativistic effects on the metal-step edge interactions been investigated?

Fernanda Juarez responded: As is usual, we have included the relativistic effects in the pseudopotentials, for all our calculations. There is a very interesting review[1] showing that relativistic effects are well represented in the valence electrons by the pseudopotential approximation and are successfully transferred to molecules or to the solid state. In particular, and as far as we know, there are no studies about this effect in the metal/graphene step interaction. However, in a recent publication,[2] it has been shown that the relativistic effects can strengthen the Au–C bond and they must be included in order to observe the adsorption of CO in metal clusters. A similar effect has been observed for platinum.[3]

1 P. Schwerdtfeger, *ChemPhysChem*, 2011, **12**, 3143.
2 Kuang *et al.*, *Physica E*, 2012, **44**, 2132.
3 Heinemann *et al.*, *J. Chem. Phys.*, 1996, **104**, 4642.

John Foord remarked: The calculations you describe are based on thermodynamic predictions. But many growth processes are driven by kinetic effects. How computationally expensive would it be to introduce kinetic barriers into the theory?

Fernanda Juarez replied: This is an interesting question because there is more than one possible answer. In studying chemical reactions, it has been a major problem to develop an accurate theory to calculate reaction rates. One simply way to do this is by obtaining the kinetic barriers of the most relevant reactions, through the nudge elastic band (NEB) method. However, the main difficulty it is to find the 'relevant' reactions, and there is always the chance to 'forget' the reaction with the lowest barrier. In any case, and just talking about the computational cost, this procedure can be applied to any of the studied systems in reasonable time.

On the other hand, there are some important points that are not going to be represented using the NEB method: the charge transfer and the role of the solvent. In theory, solvent molecules can be added explicitly during NEB calculations, but it is not feasible to obtain results in a rational time. In our group, Schmickler and Santos have developed a theory that allows kinetic barriers to be obtained, considering the solvent effect and electron transfer.[1] Nevertheless, using this theory requires extended work to obtain the 'relevant' reactions. In this sense, our manuscript is setting up the foundations to find those reactions. Another way to obtain reaction rates is through molecular dynamics. Here again, a previous work has to be done to obtain a wide range of chemical potentials to represent the interactions between the graphene (edges and terraces) and the metal atoms. This is the step that is always done with the very accurate tool of DFT calculations. I want to remark that the objective of this work is to understand the chemical nature of the metal–carbon bond in the graphene edges, and it is only a preliminary step to unravel the chemistry of the very complicated experimental system.

1 Santos *et al.*, *Phys. Rev. B*, 2009, **79**, 235436.

John Foord continued: I was interested to see that the silver particles deposited are (100) terminated cubes rather than (111). Is this a manifestation of the role of the solution in interacting with the surfaces involved?

Fernanda Juarez communicated in reply: During the meeting, I did not have an answer for that question, and we have not realized the importance of that result. Now, we have found in another publication[1] that silver nanocubes with (100) facets are difficult to prepare in solution, and a capping agent must be used. In that case, it was found that the coordination of the silver nanoparticles was made through the O and N atoms of the pyrollidone group. Our hypothesis is that in our experimental system, the sulphate/bisulphate anions could stabilize the (100) facets over the (111), which are usually more stable. However, a deeper experimental and theoretical study is required.

1 Sun *et al.*, *Science*, 2002, **298**, 2176.

Julie Macpherson opened the discussion of the paper by Thomas Varley: In the experimental section you say that all nanodiamond surfaces were treated in a furnace to remove any graphitic sp^2 carbon from the surface and to maximise the number of oxygen terminating groups. However, your explanation for what is going on assumes redox groups on the surfaces associated with sp^2 carbon. Can you describe for me what the surfaces of the two different nanodiamonds (detonation and HPHT) look like after the experimental treatment process?

Thomas Varley replied: The two different types of nanodiamond have very similar surface functionality (as shown by FTIR), which is due to the pre-treatment method employed. We heated the samples to 425 °C for 4 h as outlined in the referenced paper by Osswald *et al.*[1] This falls within the narrow temperature range that removes amorphous and sp^2 carbon material by oxidation, leaving the sp^3 crystalline core intact. Depending on the experimental conditions, Osswald and co-workers found that the surface functionalities could be controlled. By heating in air, carbonyl, acid and alcohol functionalities were created. Our attempts at using this method agree with the author's findings, although, the exact mechanism of how the oxygen containing groups form is still an on-going debate in the literature. It is likely that the outer shell (a few atomic layers) of the detonation nanodiamond particles is closer in structure to tetrahedral amorphous carbon than diamond and this is why oxygen functionalities that have sp^2 character are formed on the surface.

1 Osswald *et al.*, *J. Am. Chem. Soc.*, 2006, **128**, 11635–11642.

Julie Macpherson continued: How does the surface of the HPHT diamond change after mechanical grinding (this also relates to the previous question)?

What is the source of the HPHT? Is it in microcrystalline form, single crystal, and how pure is the material to start with?

Thomas Varley responded: We obtained the HPHT diamond used in this study from a collaborator and as such do not have information on the surface chemistry of diamond prior to the grinding. However, we have characterised our as-received material *via* FTIR and XPS and no evidence of metallic impurities was present. The heat treatment (oxidation) we employed was carried out in order to standardise our samples, in addition to providing the oxygen containing surface functionalities. As we don't have much information about the origin of the HPHT diamond samples used in this study, we have started instead to use similar diamonds from a different commercial source (Microdiamant), from whom we receive quite detailed technical information about their origin. The characterisation we have carried out so far suggests similar surface chemistry for the samples from different origins.

Julie Macpherson added: How does the density of sp^2 or 'defect' sites vary between detonation diamond and mechanically ground HPHT?

Thomas Varley answered: Detonation nanodiamond has a surface that contains large amounts of tetrahedral amorphous carbon,[1] meaning it is very defective compared to the surfaces formed through mechanical grinding. Therefore, the density of defect sites is much greater for the nanodiamonds fabricated by detonation synthesis as opposed to those made by mechanical grinding.[2]

1 V. N. Mochalin, O. Shenderova, D. Ho and Y. Gogotsi, *Nature Nanotechnology*, 2012, 7, 11–23.
2 K. B. Holt, *Phys. Chem. Chem. Phys.*, 2010, **12**, 2048–2058.

Robert Hamers asked: Could you explain the catalytic feedback process a bit more? I think most of us are familiar with the catalytic feedback process in an AFM where there are two electrodes and so the product at one electrode can diffuse and become a reactant at the other electrode. In your experiments, with only one electrode, it is less clear how a catalytic enhancement could occur. Could you please explain in more detail?

Thomas Varley replied: In the case presented in this paper, there are two mechanisms for catalytic enhancement: 1) *via* surface states on the nanodiamond, and 2) mediated by adsorbed FcMeOH/$FcMeOH^+$. The first mechanism has been described by us previously for the $[Fe(CN)_6]^{3-/4-}$ redox couple[1] and generally results in current enhancements of up to 2 times the diffusion-limited current compared to the non-modified electrode. The second mechanism seems more specific to FcMeOH and is a consequence of its ability to adsorb on the nanodiamond surface. This mechanism is largely responsible for the much larger enhancements observed for this redox species (*ca.* 10 times the diffusion-limited current). To describe the first mechanism in more detail: at sufficient potentials FcMeOH can undergo oxidation at the surface of the BDD electrode to produce $FcMeOH^+$. In the presence of nanodiamond, $FcMeOH^+$ can be reduced by redox active surface functionalities on the nanodiamond, resulting in the regeneration of FcMeOH. The regeneration of FcMeOH in the electrode diffusion layer results in an enhanced catalytic current. On the reverse scan, the opposite process is proposed to take place; FcMeOH generated at the electrode is immediately oxidised back to $FcMeOH^+$ after donating an electron to the nanodiamond surface. For this mechanism to be feasible, the oxidation and reduction of nanodiamond surface functionalities must occur at similar potentials to the standard potential of the solution redox probe, *i.e.* FcMeOH. I guess the heart of your question is how this mechanism is sustained? It's different from feedback between two electrodes of course, because we need to explain how the nanodiamond can continuously regenerate the redox species, *i.e.* do the nanodiamond surface states not eventually run out of electrons (or holes) to allow the catalytic process to continue? The answer to this is that under some conditions the catalytic feedback cannot be sustained. It can be sustained as long as there are available surface nanodiamond functionalities within the diffusion layer. This is a function of redox probe concentration (low concentrations allow the mechanism to be sustained), scan rate (too slow and the nanodiamond states get depleted due to the thick diffusion layer; too fast and the large flux of the redox probe causes diffusion-controlled currents to dominate over the catalytic response – there is an optimum intermediate scan rate to observe the effect), surface chemistry of the nanodiamond (*i.e.* density of surface groups), electrode coverage of nanodiamond and potential of the redox probe relative to the available nanodiamond surface states. Under some conditions, for some probes, the enhancement we observe is decreased significantly after the first CV cycle; in other cases it can be maintained with continuous cycling. There are also a few routes to 'recharging' of the nanodiamond surface states, allowing them to continue contributing to the redox cycle. Within tunnelling distance of the electrode, those surface groups with reversible redox chemistry can be directly oxidised and reduced at the electrode;[1] as we cycle the potential during the experiments they are able to exchange electrons with the electrode and become 'recharged'. Further from the electrode, depending on their potential relative to the redox probe, the surface groups can exchange electrons with the redox probe and also become 'recharged'. Finally, due to the complex nature of the nanodiamond surface, electronic communication, *via* conjugation of pi systems, may result in electrons being able to conduct from one side of a nanodiamond particle to the other, again providing electrons to maintain the redox cycle. In addition to this mechanism, is the one mediated by adsorbed FcMeOH, as described in the paper. This results in much larger currents than those we have observed previously, and the currents are more easily sustained over repeated cycling. The positively charged $FcMeOH^+$ interacts very strongly with the negatively charged nanodiamond surface and becomes adsorbed (this strong interaction is illustrated in the IR studies presented in Fig. 7 of the paper). The contribution of the Gibbs energy of adsorption can shift the redox potentials of an adsorbed species relative to the solution species and results in the observation of pre-peaks in the voltammogram. The surface adsorbed $FcMeOH^+$ also participates in the catalytic feedback mechanism (as shown in Fig. 8 of the paper), by undergoing electron exchange with solution FcMeOH, to generate solution $FcMeOH^+$ and adsorbed FcMeOH. The adsorbed FcMeOH can be re-oxidised to adsorbed $FcMeOH^+$ by the underlying BDD electrode or by a redox process with nanodiamond surface groups. FcMeOH has a very fast rate constant for electron self-exchange (classic Marcus theory) and also rapidly undergoes outer sphere electron transfer with the underlying BDD electrode. It is actually the accumulative influence of the nanodiamond surface and adsorbed $FcMeOH^+$ processes that explains why currents are so significantly enhanced for the $FcMeOH/FcMeOH^+$ redox couple compared to others previously studied.

1 *e.g.* K. B. Holt, D. J. Caruana and E. J. Millán-Barrios, *J. Am. Chem. Soc.*, 2009, **131**, 11272.

Matteo Duca remarked: With respect to the observed 'off-set' reduction current and the spontaneous oxidation of FcMeOH at nanodiamond, has the open circuit potential been measured upon immersion of the working electrode in the FeMeOH-containing solution? This is a key parameter for comparable processes, such as 'electroless' deposition.

Thomas Varley responded: This is something we have planned to measure, but as yet have not. I completely agree that it is an important parameter and would provide useful information, especially when measured in different solutions (*i.e.* varying the electrolyte concentration, redox species concentration, nanodiamond coverage, pH, *etc.*).

Matteo Duca continued: With respect to the off-set current, can it be avoided by immersing the electrode into the electrolyte under potential control (in particular, by keeping the potential negative enough)?

Thomas Varley replied: The off-set current is a result of the nanodiamond oxidising the FcMeOH species in solution upon contact; this generates $FeMeOH^+$ at the electrode interface and it is the reduction of this species that gives rise to the off-set current. The only way you could prevent the off-set current is to ensure all the redox-active surface sites are fully reduced prior to introducing the nanodiamond to the FcMeOH. It is unlikely this can be achieved electrochemically as not all the surface sites that can carry out the oxidation will be within tunnelling distance of the electrode. Making the potential more negative will only increase the driving force for the reduction of the spontaneously formed $FcMeOH^+$ at the electrode. However, we have shown in other data that if you hold a potential of 0.0 V (*vs.* Ag/AgCl sat. KCl) for 300 s prior to starting the experiment, you no longer observe the off-set current, as this is sufficient time to reduce the $FcMeOH^+$ spontaneously generated by the nanodiamond at the electrode surface (the off-set current steadily decreases as the equilibration time approaches 300 s).

John Foord asked: Could the signal enhancements you describe be explained by adsorption/absorption/preconcentration effects, as commonly seen in porous layers, rather than the catalytic electron transfer cycle?

Thomas Varley answered: As shown in the paper we do suggest that the ferrocene methanol species is adsorbing, and this is contributing towards the overall current. However, on the time periods of our experiments it is unlikely, but cannot be ruled out completely, that pre-concentration is occurring. In addition, you would expect surface adsorbed species to exhibit a current which is proportional to scan rate. Thus at higher scan rates, the associated peak would dominate the voltammetric response as the diffusional component will only scale with the square root of scan rate; this is not what we observe. Finally, other spectroscopic data presented within the paper and previous work[1,2] shows that interaction with surface redox groups on the nanodiamond is occurring; these interactions can explain the observed enhancements. For example, if the current increases were solely due to absorption of redox probe into the porous nanodiamond layer, we would expect oxidation and reduction peaks to be equally enhanced in all cases – this is not observed for many redox probes.[3] However, the contributions towards the current enhancements are quite complex and likely include adsorption, absorption, catalytic effects and perhaps some mass transport effect. Which process dominates will depend on the redox probe used and solution conditions.

1 K. B. Holt, D. J. Caruana and E. J. Millán-Barrios, *J. Am. Chem. Soc.*, 2009, **131**, 11272.
2 J. Scholz, A. J. McQuillan and K. B. Holt, *Chem. Commun.*, 2011, **47**, 12140–12142.
3 K. B. Holt, *Phys. Chem. Chem. Phys.*, 2010, **12**, 2048–2058.

Matteo Duca enquired: A possible way to establish the nature of the adsorbed species would involve generating the adsorbate at the electrode, which is then transferred to clean electrolyte for a voltammetric measurement. Has such an experiment been tried and, if so, has it yielded any significant results?

Thomas Varley replied: This experiment has been tried but the adsorbed FcMeOH or $FcMeOH^+$ species are not bound strongly enough to detect when transferred to clean electrolyte. Similar experiments were carried out with XPS analysis of the nanodiamond modified electrode surface and nanodiamond powders that had been previously exposed to FcMeOH. The lack of any signal at the binding energy for iron and the lack of electrochemical response suggests the adsorption is non-specific.

Robert Hamers said: Have you investigated the dependence of the cyclic voltammogram peak height as a function of scan rate? That is a common way of distinguishing between reactions in solution (which become diffusion-limited) *vs.* reactions that involve surface species (and are therefore limited by available surface sites).

Thomas Varley responded: We have. Log-log plots of peak height *vs.* scan rate gave slopes between 0.6–0.8. A slope of 0.5 is expected for diffusion controlled reactions, whereas a slope of 1.0 is expected for surface confined reactions. Our values suggest a mixed response, which is what our mechanism suggests.

Philip A. Ash remarked: The FTIR data you report is very seductive, as it clearly implies preferential adsorption of the oxidised $FcMeOH^+$ species, suggesting that electron transfer could indeed be mediated by redox-active groups at the nanodiamond surface as you suggest. I wonder if it would be possible to take this experiment further, by adding a cocktail of solution redox mediators to change the solution potential and initiate or inhibit this electron transfer step in a stopped-flow potentiometric experiment? In principle the high surface area nanodiamond film could enhance the sensitivity of your spectroscopic measurement sufficiently to allow you to observe intermediates in the course of reaction, either through the use of isotope exchange or through the inclusion of a ferrocene derivative with a suitable redox-dependent infrared active marker. It is doubtful that this approach would be successful at a planar diamond electrode, however, as the overall absorbance of your adsorbed films is quite low.

Thomas Varley answered: This is an excellent idea and something we should consider. However, careful consideration of which are the redox species to be used in such a solution would be required. Regarding the preferential adsorption of $FcMeOH^+$, it is likely that both species are adsorbed through pi-pi stacking. However, the positively charged $FcMeOH^+$ will also be attracted to the negatively charged nanodiamond surface, although this interaction is pH dependant.

George Zheng Chen asked: Are the nanodiamond particles not insulators? If they are, how can electron transfer reactions proceed on them or be affected by them? Could it be the boundary region between the nanodiamond particle and the electrode substrate that is responsible for your observed catalytic effect?

Thomas Varley replied: Diamond is an insulator with an approximate band gap of 5.5 eV. However, it has previously been shown[1] that surface sites have redox activity and thus must have energies that are within diamond's band gap. It would not be correct for us to rule out effects due to the boundary region, however, spectroscopic evidence of changes in surface functionality after reactions with specific redox probes[2] coupled with the observation that the enhancement varies depending on the properties of the redox probe, provides strong evidence towards our proposed feedback mechanism.

1 K. B. Holt, D. J. Caruana and E. J. Millán-Barrios, *J. Am. Chem. Soc.*, 2009, **131**, 11272.
2 J. Scholz, A. J. McQuillan and K. B. Holt, *Chem. Commun.*, 2011, **47**, 12140–12142.

George Zheng Chen continued: Fig. 3b of the paper shows CVs obtained from electrodes loaded with nanodiamond particles of very different sizes, and it is clear that very similar kinetics are governing the electrode reactions. If so, how can the catalytic activity be affected by the particle sizes? Or, was your observed catalytic effect simply due to the sum of lengths of the boundaries between the particles and the electrode substrate increasing with decreasing particle size?

Thomas Varley responded: The kinetics of the catalytic reaction are not affected by the size of the nanodiamonds. This is because it is the same functionalities (but perhaps not identical species) that are responsible for the enhancement. The size of the nanodiamonds affects the density of the surface sites, thus the magnitude of the enhanced current.

George Zheng Chen asked: How can the nanodiamond particles attach to the electrode surface? What is the nature of the interaction?

Thomas Varley answered: We have not explored in any great detail the mechanism that holds the drop-coated nanodiamond particles onto the electrodes surface, but they are physisorbed quite strongly. We have checked that the material isn't affected by immersion into a solution using before and after optical images obtained *via* a microscope and the nanodiamond layer remains intact unless removed *via* rough physical treatment (*e.g.* polishing).

George Zheng Chen contined: Have you compared the current change with the number and size of the nanoparticles?

Thomas Varley replied: We have shown in this paper that the size of the nanodiamonds has a dramatic effect on the magnitude of the enhanced current. Smaller particles give larger enhancements due to the increased surface area. Previous work carried out in the laboratory has shown that the amount of nanodiamond present also affects the enhanced current.

George Zheng Chen added: Would it be possible that the electrochemistry proceeded at the boundary between the nanodiamond particles and the electrode substrate, instead of the insulating nanodiamond particles? Also, there must be a strong interaction to make the nanodiamond particles remain stable on the electrode surface, what could this interaction be?

Thomas Varley responded: The nanodiamond core is extremely resistive, and a good example of an electrical insulator as diamond is known to have a large band gap of *ca.* 5.5 eV. Previous work in our group[1] shows that nanodiamond has redox active surface sites that have orbital energies within diamond's band gap. Boundary region effects cannot be completely ignored, however spectroscopic evidence of changes in surface functionality after exposing nanodiamond with hexachloroiridate(IV), coupled with the observation that the enhancement varies depending on the properties of the redox probe,[2] reinforces our proposed feedback mechanism. Indeed there must be a strong interaction holding the nanodiamond particles onto the electrode surface. We have checked that the nanodiamond layer(s) aren't affected by immersion and removal from a solution (*via* optical microscopy). The nanodiamond layer remains constant unless removed *via* rough physical treatment (*e.g.* polishing). Nanodiamond is known to aggregate into small (*ca.* 50 nm) clusters, due to electrostatic and even covalent linkage of the particles. These clusters can again aggregate forming secondary clusters, but these are known to be less stable. Since we are using a carbon electrode it is highly feasible that similar interactions are occurring. Also, all of the experiments presented in this paper were conducted at positive potentials, which would provide strong attraction forces for the negatively charged nanodiamonds.

1 K. B. Holt, D. J. Caruana and E. J. Millán-Barrios, *J. Am. Chem. Soc.*, 2009, **131**, 11272.
2 J. Scholz, A. J. McQuillan and K. B. Holt, *Chem. Commun.*, 2011, **47**, 12140–12142.

Mark Newton said: Synthetic diamond is well know as an ideal material for ATR, particularly when studying very hard solids. It is well know that with hydrogen terminated intrinsic diamond, the surface conductivity[1] is sufficient to permit electrochemical investigations over a limited range of potentials and conditions. Is it possible to contemplate combining FTIR and electrochemistry on a diamond ATR prism?

1 F. Maier, M. Riedel, B. Mantel, J. Ristein, and L. Ley, Origin of Surface Conductivity in Diamond, *Phys. Rev. Lett.*, 2000, **85**, 3472.

Thomas Varley answered: This is an excellent idea. The experimental requirements, however, would be quite tricky although collaboration with a researcher who has the facilities to deposit such diamond films would make this type of experiment more feasible. That said, the requirement for the diamond to be hydrogen terminated, to permit the surface conductivity, would mean losing the redox-active surface sites (this would be a good experiment to prove they are required). It might be possible to incorporate nanodiamond particles into the surface of the film, in which case the *in situ* FTIR spectroelectrochemical measurements could be performed.

Julie Macpherson enquired: Have you observed any redox couples which don't show a response with the nanodiamonds?

Thomas Varley replied: Currently, we have tested $[Fe(CN)_6]^{3-/4-}$, $[IrCl_6]^{3-/2-}$, $[Ru(NH_3)_6]^{3+}$, $[Ru(CN)_6]^{3-/4-}$, ferrocene methanol, ferrocene di-methanol, ferrocene carboxylic acid and ferrocene di-carboxylic acid. All of these species showed a catalytic enhancement but to varying extents. Testing species that have E^{o} values further from the nanodiamond values are likely to exhibit no enhancement, but as yet none have been tested. It is difficult to identify a potential where the nanodiamond surface does not have surface active sites from the current potential windows investigated.[1] If the nanodiamonds are hydrogen terminated, then the catalytic enhancement has been shown to disappear for some redox probes, *e.g.* $[Ru(NH_3)_6]^{3+}$ (ref. 2). This is suggested to be because it is the oxygen content of the surface which gives rise to the redox activity. Once the oxygen groups are removed then the enhancement can no longer take place.

1 K. B. Holt, D. J. Caruana and E. J. Millán-Barrios, *J. Am. Chem. Soc.*, 2009, **131**, 11272.
2 K. B. Holt *et al.*, *Phys. Chem. Chem. Phys.*, 2008, **10**, 303–310.

Julie Macpherson continued: Do you know what the E^{o} is for the proposed redox active groups on the nanodiamond surface?

Thomas Varley responded: The redox active groups on the nanodiamond surface vary with pH and span a range of potentials between 0.1 V (pH 4) to 1.0 V (pH 9), *vs.* Ag/Ag/Cl. The potentials of these surface groups/states were obtained using differential pulse voltammetry of a layer of nanodiamond immobilised on an electrode[1]. Distinct peaks at discrete potentials were observed for the nanodiamond layer, some of which suggested reversibility in the redox response. The response is actually quite complex and varies with pH and with previous electrochemical cycling.

1 K. B. Holt, D. J. Caruana and E. J. Millán-Barrios, *J. Am. Chem. Soc.*, 2009, **131**, 11272.

Keith Stevenson asked: The CVs remind me of those of Royce Murray's charging of soluble gold clusters, where they exhibit a potential dependent charging response similar to charging an ultrasmall capacitor with different charge states. Is it possible that the nanodiamonds have an inverse charge storage structure? Have you done the experiment with soluble nanodiamonds?

Katherine Holt communicated in reply: Yes – an interesting observation! And something I noted when I carried out the original experiments. The evenly spaced peaks do seem to suggest the type of capacitive charging seen for gold nanoparticles terminated with an insulting SAM layer. In our case it would be an insulating core, coated with a conducting coating. It is certainly an interesting and exciting possibility that this could be taking place – however we have suggested that the peaks are due to redox chemistry of surface functionalities on the diamond surface, that can undergo oxidation and reduction at specific potentials. The reason for this is the pH dependence of the response and the fact that the peaks shift *ca.* 60 mV with each change in pH unit, suggesting a proton coupled electron transfer. A possible origin of this redox chemistry is quinone-type species on the surface of the nanodiamond.

Chi-Chang Hu queried: Due to the possible diffusion effect from the microelectrode, what is the coverage of the nanodiamond?

Thomas Varley replied: Experiments such as varying the surface coverage contributed towards us eliminating any microelectrode effects. Varying the coverage changes the enhancement factor, but not the voltammetric shape that would change if the nanodiamond were creating an array of microelectrodes on the BDD surface. An example of the effect that varying the amount of nanodiamond has on the voltammetry is shown in ref. 1, Fig. 1. Other evidence against a microelectrode effect includes the dependence shown of current on the identity of the redox species and solution pH. If the current enhancement were simply attributed to mass transport effects this should be independent of these factors. The coverage of the nanodiamond in the present work was measured by optical imaging, and estimated at *ca.* 40%. However, due to the size difference between the particles and the electrode, this technique is susceptible to errors. Therefore, coverage was controlled by the amount of material deposited, which was set at 4 μg per coating.

1 K. B. Holt, D. J. Caruana and E. J. Millán-Barrios, *J. Am. Chem. Soc.*, 2009, **131**, 11272.

Chi-Chang Hu continued: Can the author try to change the mass loading of nano-diamonds (*e.g.* 2 sizes) to evaluate the microelectrode effect?

Thomas Varley responded: This is something we have explored and reported previously.[1] Such experiments are important as they allow us to observe the effect of nanodiamond loading on the voltammetric wave shape and the extent of the Faradaic current enhancement to be explored. Further analysis of our data should subsequently help contribute towards determining the number of active sites per nanodiamond.

1 K. B. Holt, D. J. Caruana and E. J. Millán-Barrios, *J. Am. Chem. Soc.*, 2009, **131**, 11272.

Fulian Qiu commented: Inert nanodiamond immobilised on a electrode surface showed two catalytic processes for the oxidation of FcMeOH. You could disperse nanodiamonds into an electrolyte solution rather than immobilisation. If the mechanism of the two catalytic processes is singular, cyclic voltammograms should show the same behaviors as those recorded for the immobilised nanodiamond electrodes regarding the pHs and sizes of nanodimonds. This will rule out possible surface diffusion processes.

Jingping Hu asked: Nanodiamonds are known to tend to aggregate together easily. Does the aggregation issue have any big influence on your results? Have you considered using boron doped nanodiamonds in future experiments?

Thomas Varley answered: This is a good point. We do not expect the aggregates to affect our observations, other than dictating the amount of accessible surface area. Using boron doped nanodiamonds is not something we have considered, but is an excellent and interesting idea, as the conductivity of these nanoparticles would no doubt lead to an interesting electrochemical response.

Robert Hamers said: In your experiments, do you correct for differences in the microscopic surface area of your samples?

Thomas Varley replied: No, the reason for this being we are interested in determining the effect of the surface area on the catalytic current. We have shown that the enhancement is inversely proportional to particle size, thus, for the same mass of diamond, smaller particles (and so more surface area) produce a higher enhancement. However, this isn't a standard relationship, and further experiments are required to fully understand this observation.

Robert Hamers continued: Is it possible that the differences you see could be accounted for through differences in microscopic surface area? It seems possible that perhaps the detonation nanodiamond may not be completely insulating, and so you could be seeing effects due to changes in accessible microscopic surface area. Is that possible ?

Thomas Varley responded: The measured relationship between the effect of particle size and current enhancement does not agree with this hypothesis. For example, peak current should be directly proportional to the increase in area, but this is not what we observe. Of course, it is difficult to know exactly what the accessible surface is, due to aggregation for the smaller particles and orientation for the larger particles. Also, the fact that the enhancement is different for different redox species means that the current is related to specific interactions between the redox probe and the nanodiamond, not changes in the apparent electrode area. We cannot completely rule out surface conduction, although it has previously been reported that conductivity is only achieved for nanodiamond that is hydrogen terminated, or through the presence of defects within the extended diamond structure (*e.g.* boron doping). Surface defects and impurities become more important as the size of the diamond decreases. This has a huge effect on the observed properties of the nanodiamond.[1]

1 K. B. Holt *et al.*, *J. Phys. Chem. C*, 2009, **113**, 2761–2770.

John Foord enquired: How do the electrochemical peak currents you see at the nanodiamond treated electrode compare with those calculated from planar diffusion – are they bigger?

Thomas Varley answered: The peak currents are significantly bigger than those calculated from planar diffusion. In the paper, the enhancement is *ca.* 10 times greater than the expected current.

Faraday Discussions

PAPER

Photoemission from diamond films and substrates into water: dynamics of solvated electrons and implications for diamond photoelectrochemistry

R. J. Hamers, J. A. Bandy, D. Zhu and L. Zhang

Received 11th March 2014, Accepted 26th March 2014
DOI: 10.1039/c4fd00039k

Illumination of diamond with above-bandgap light results in emission of electrons into water and formation of solvated electrons. Here we characterize the materials factors that affect that dynamics of the solvated electrons produced by illumination of niobium substrates and of diamond thin films grown on niobium substrates using transient absorption spectroscopy, and we relate the solvated electron dynamics to the ability to reduce N_2 to NH_3. For diamond films grown on niobium substrates for different lengths of time, the initial yield of electrons is similar for the different samples, but the lifetime of the solvated electrons increases approximately 10-fold as the film grows. The time-averaged solvated electron concentration and the yield of NH_3 produced from N_2 both show maxima for films grown for 1–2 hours, with thicknesses of 100–200 nm. Measurements at different values of pH on boron-doped diamond films show that the instantaneous electron emission is nearly independent of pH, but the solvated electron lifetime becomes longer as the pH is increased from pH = 2 to pH = 5. Finally, we also illustrate an important caveat arising from the fact that charge neutrality requires that light-induced emission of electrons from diamond must be accompanied by corresponding oxidation reactions. In situations where the valence band holes cannot readily induce solution-phase oxidation reactions, the diamond itself can be etched by reacting with water to produce CO. Implications for other reactions such as photocatalytic CO_2 reduction are discussed, along with strategies for mitigating the potential photo-etching phenomena.

A. Introduction

Diamond is well known for its outstanding chemical and physical properties.[1] Included among these are a wide window of stability as a function of electrochemical potential and pH, and a very high overpotential for reduction of H^+ to H_2.[1,2] While the electrochemistry of diamond has been heavily exploited,[1,3–7] fewer studies have investigated its photoelectrochemistry.[8–11] One particularly

Dept. of Chemistry, University of Wisconsin-Madison, 1101 University Avenue, Madison, W 53706, USA

important property of diamond is that its very high conduction band energy leads to negative electron affinity (NEA).[12–16] In NEA, the conduction band lies above the free energy of a stationary electron in vacuum, often referred to as the "vacuum level". Termination of the surface with H atoms assures that the C–H surface dipole enables electron emission in a nearly barrier-free manner.[12] One consequence of negative electron affinity is that illumination of diamond with above-bandgap light leads to facile emission of electrons. The emission of electrons into vacuum has been widely studied because of its potential applications such as in display technologies.[17–19] The emission of electrons is not limited to vacuum, and in principle NEA can lead to emission of electrons into gases and even liquids. These environments are interesting chemically because they provide pathways toward new types of non-thermal chemistry.[20,21]

Recently we performed the first studies showing the ability of diamond to photo-emit electrons into water, with direct detection of solvated electrons enabled by transient absorption spectroscopy.[20]

Fig. 1 shows an energy level diagram for an H-terminated diamond surface in contact with an aqueous solution. Also depicted are some reduction reactions relevant to the diamond's unique photoemissive abilities, along with the reduction potentials at pH = 7. These potentials were obtained from tabulated standard potentials[22,23] using the Nernst equation to calculate values at pH = 7, leaving other reactants and products in their standard states. Of particular interest are diamond's ability to produce solvated electrons, atomic hydrogen, and the one-electron reduction of CO_2 to the CO_2 anion. These reactions occur at strong negative potentials that cannot be reached by any other common semiconductor electrode conduction band. Of these reactions, the formation of solvated electrons is the first step, since other reactants (such as CO_2 or H^+) are usually present at much lower concentrations compared with the ~55 Molar concentration of water molecules in pure water.

The reactions depicted closer to the valence band are important because, as described later, photoemission of electrons must be accompanied by oxidation

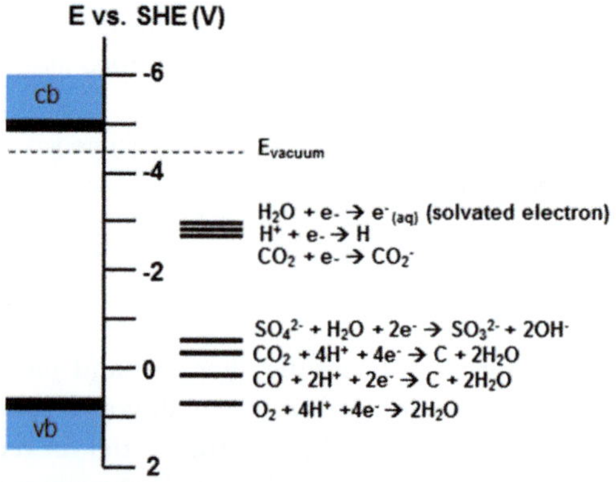

Fig. 1 H-terminated diamond in contact with aqueous electrolyte; electrochemical reduction potentials shown are for pH = 7 with other reactants and products at their thermodynamic standard states. Valence band (vb) and conduction band (cb) energies are typical for H-terminated diamond.

reactions. Fig. 1 depicts some potentially relevant reactions (written as reduction reactions, to maintain consistency with standard electrochemical literature) that can occur at the diamond surface under appropriate conditions.

Here we describe recent experiments investigating some of the properties of diamond and the adjacent electrolyte that influence the emission of electrons into water and some caveats that are important in understanding the effective use of diamond to induce new chemistry.

B. Results

B1. Solvated electrons from diamond thin films

In previous work,[20,21] we demonstrated that solvated electrons could be produced in water by excitation of diamond using above-bandgap light and that smaller, but non-zero, emission could also be detected at slightly longer wavelengths. Since diamond can be easily grown on a variety of different substrates, we aimed to understand whether diamond thin films on metallic substrates would provide similar performance compared with bulk diamond samples and, if so, to what extent the electron emission would depend on the thickness of the films.

We explored the emission of electrons from diamond thin films grown on niobium. Niobium substrates are widely used as substrates for diamond films used in water purification,[24,25] in part because niobium is chemically robust in water. Fig. 2 illustrates the apparatus used for transient absorption measurement of solvated electrons produced by diamond. Fig. 3 shows the change in transmission of the 705 nm beam after the sample is excited with above-bandgap light (213 nm) with data from a fresh Nb surface and also diamond thin films grown on Nb. These diamond thin films were grown *via* microwave plasma-enhanced chemical vapour deposition for different lengths of time ranging from 10 minutes to 4 hours. In each case the diamond thin film are immersed in a solution of deionized water. The decreases in intensity at $t = 0$ arise from the emission of solvated electrons, which have a broad absorption peak centered near 692 nm at room temperature.[26,27] Fig. 3 shows that solvated electrons are observed being emitted from both the starting Nb surface and from the diamond films on Nb

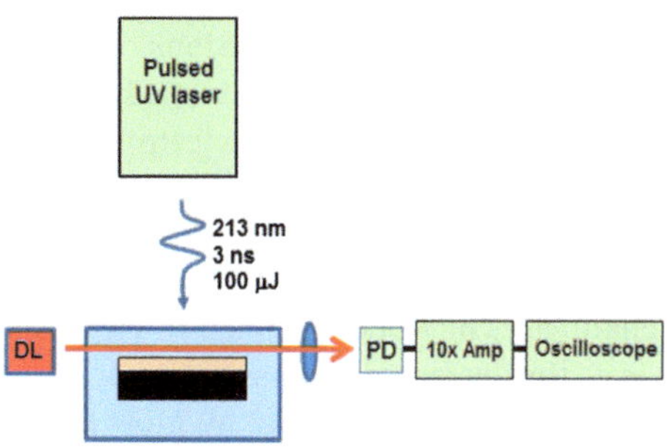

Fig. 2 Illustration of transient absorption apparatus for measuring solvated electron emission from diamond. The sample is a diamond thin film on Nb substrate, immersed in a cuvette filled with water. A fast photodiode (PD), amplifier and oscilloscope record the change in intensity of light from a 705 nm diode laser (DL) after the sample is excited by a pulsed UV laser.

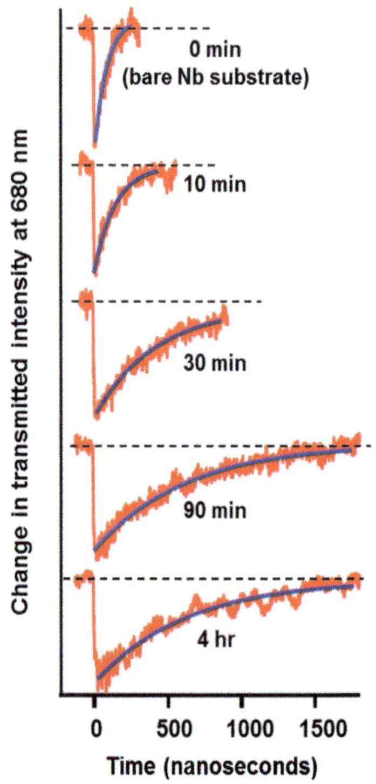

Fig. 3 Transient absorption measurements showing transient decrease in the transmission of the 705 nm laser. Measurements are shown on the starting bare Nb substrate and on diamond films grown on Nb for the indicated lengths of time up to 4 hours.

substrates. In additional control experiments we observe that a part of the solvated electron signal arises from two-photon excitation of water by the intense laser pulse. Thus, careful checking and control of laser intensity is required.

Due to the complex geometry of the 705 nm detection beam relative to the sample, the absolute amplitude of the transmitted intensity cannot be easily related to an electron concentration. However, as all samples use the same geometry, the amplitudes can be compared between different samples. The data in Fig. 3 were fitted to simple exponential functions of the form $\Delta I = A_0\exp^{-t/\tau}$. The resulting fits are shown in Fig. 3, and the extracted values of electron lifetime τ and the maximum amplitude of the solvated electron signal A_0 are graphed in Fig. 4.

The amplitude of the initial solvated electron signal at time = 0 is a direct measure of the efficiency of electron emission, while the lifetime is a measure of the available decay channels. The values graphed in Fig. 4 show that the instantaneous amplitudes (*e.g.*, the decrease in transmission at $t = 0$) are similar for the bare Nb surface and for the diamond films at all growth times, although a very small decrease in A_0 is observed (Fig. 4b) as the film thickness increases. Much more significant is that the *lifetime* of the solvated electrons increases from 77 ns on the bare Nb surface to a constant value of ~600 ns on samples grown for 2 hours or more (Fig. 4a).

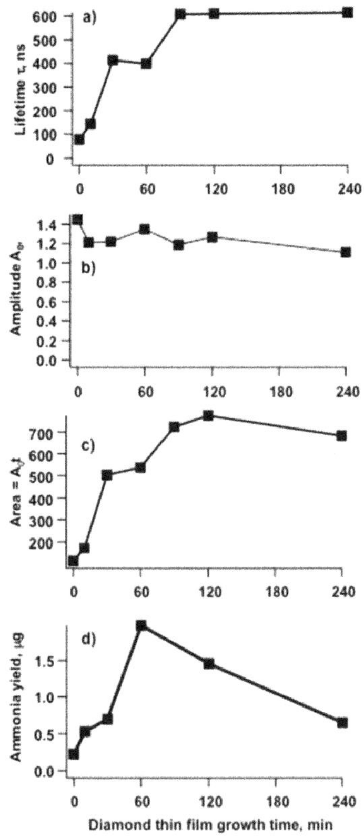

Fig. 4 Analysis of the data from Fig. 3, showing how growth time affects specific solvated electron properties and the resulting catalytic performance. a) electron lifetime, b) maximum amplitude of transient , c) total area under electron decay curve, and d) ammonia yield.

In our previous work[20,21] we showed that solvated electrons were able to induce photocatalytic reduction of N_2 to NH_3. This specific reaction is a stringent test of the ability to initiate new chemistry as this reaction has one of the largest known reaction barriers[28] and therefore cannot be performed efficiently with any known photocatalyst.[29] The critical step in the reaction is the formation of $N_2 + e^- + H^+ \rightarrow N_2H$, which requires approximately 3 eV of energy.[28] In photocatalytic applications, the overall yield of the reactions is expected to vary depending on the time-averaged concentrations of electrons. While electron-induced reactions, such as N_2 reduction, are complex and their overall kinetics may not be a simple function of the electron concentration, we calculated the overall area of the exponential curves in Fig. 4c using Area $= A_0\tau$, because the data are well fit by simple exponential decays. Since the maximum electron yield A_0 has a small decrease with increasing growth time, the area (Area $= A_0\tau$) has a slight maximum, which occurs after a time of approximately 2 hours.

Fig. 4d shows experimental data depicting the yield of ammonia (NH_3) produced when the samples of Fig. 4 were immersed in N_2-satured water with 10 mM Na_3SO_4, and illuminated continuously with light from a 450 W HgXe lamp.

These catalytic yield data show that there is a clear maximum in the ammonia yield on films grown for approximately 1 hour. While the diamond film growth time that yields the maximum electron yield with pulsed excitation (~2 hours) is slightly different than the growth time yielding the best photocatalytic activity with continuous illumination (~1 hour), both appear to show best performance using films that have been grown for approximately 1–2 hours. Electron microscopy images of the films (not shown) reveal that the diamond growth is continuous within ~30 minutes of growth and that the film thickness after 1–2 hours of growth is approximately 100–200 nm as measured by ellipsometry.

We also used scanning electron microscopy to characterize the changes in the diamond film growth. Fig. 5 shows a scanning electron microscope image of the diamond film on Nb after growth for 1 hour. At this point the surface shows diamond crystallites that are approximately 50–150 nm in size. Measurements of the film thickness show an average thickness of approximately 150 nm. Since the grain size and ellipsometric thickness are nearly identical, this indicates that the film is just on the threshold of being continuous. Images at longer times show larger grain sizes and a continued evolution of the morphology.

B2. Influence of solution pH on electron emission

While the data in Fig. 1–4 were obtained at pH = 7, the fate of solvated electrons is strongly dependent on the pH. It is well known that solvated electrons rapidly react with H^+ to form neutral atomic hydrogen atoms ($H^{\bullet}$), *via* $H^+ + e^- \rightarrow H^{\bullet}$. Hydrogen atoms are highly reactive, and in our recent computational study of N_2 reduction to NH_3[21] we found that this reaction plays a central role in the overall catalytic reduction efficiency. To test the dependence of electron emission on pH, we conducted studies using free-standing, boron-doped diamond substrates. These substrates were chosen because they can easily be purchased and act as a great standard. (The corrosion is only an issue if you bias the substrate positively, which we do not want to really do anyway. The acid does not react at a noticeable rate, even with copper, in case you did not know.)

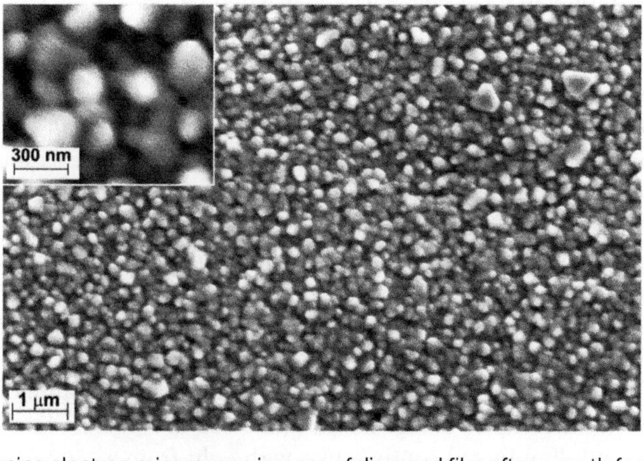

Fig. 5 Scanning electron microscope images of diamond film after growth for 1 hour. The inset shows a higher-magnification of a region from the same sample.

Fig. 6a shows transient absorption measurements of electron emission from a boron-doped diamond sample into solutions of varying pH, all measured using 213 nm excitation. In each case the pH was controlled using a NaH_2PO_4–Na_2HPO_4 buffer (1 mM of combined concentration) for pH = 5 and pH = 7, and using H_3PO_4 for pH = 2 and pH = 3. For pH = 7 and pH = 5 relatively long solvated electron lifetimes of 255 and 310 nanoseconds are observed; similar results are also observed at pH = 9. However, as the pH is decreased to pH = 3 the solvated electron lifetime decreases to 37 ns, and for pH = 2 the lifetime is <10 ns and limited by the time resolution of the detection electronics.

Fig. 6b shows the corresponding ammonia yield data as a function of pH, measured in N_2-saturated water. The overall ammonia yield reaches a maximum near pH = 7.

Our data indicate that for pH < 5 the solvated electron kinetics are controlled by the reaction $H^+ + e^- \rightarrow H^\bullet$. For pH > 5, the lifetimes are likely dominated by other reactions and/or possibly by recombination with the diamond substrate. The lifetimes observed with highly boron-doped diamond are slightly shorter than those observed on the diamond thin films grown on Nb. This suggests that there may be some impact of the diamond itself on the solvated electron lifetime,

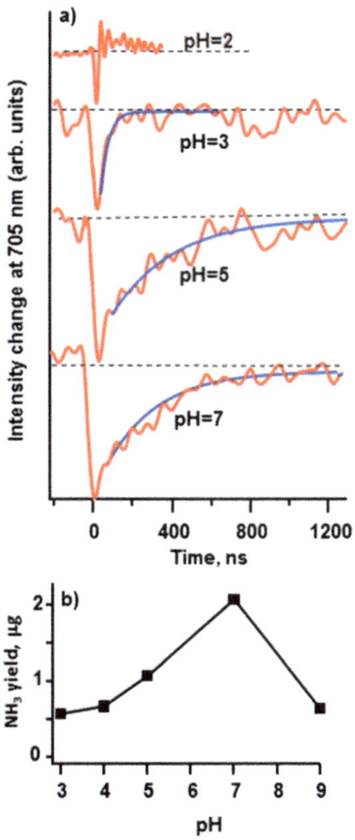

Fig. 6 Influence of pH on solvated electron dynamics and subsequent chemistry. a) Transient absorption spectra from boron-doped diamond at different values of pH. b) Ammonia yield as a function of pH.

possibly *via* image–charge effects or by material-dependent recombination pathways. Computational modelling studies of the overall N_2 to NH_3 reaction pathway show that the only feasible reaction sequence involves

$$\text{Diamond} + H_2O + \text{light} \rightarrow e^-_{(aq)}\ (\text{solvated electron}) \tag{1}$$

$$e^-_{(aq)} + H^+ \rightarrow H^\bullet \tag{2}$$

$$H^\bullet + N_2 \rightarrow N_2H \tag{3}$$

$$N_2H + H^\bullet \rightarrow N_2H_2. \tag{4}$$

N_2H_2 is a very stable molecule that can be rapidly reduced to NH_3.[28] In this reaction sequence, atomic hydrogen produced by reaction of solvated electrons plays a key role in two distinct steps. While it might be expected that the highest ammonia yield would be observed at the lowest pH (where the solvated electrons react most rapidly and presumably produce the highest concentration of H), the reaction yield is actually highest at intermediate pH. Detailed rate modeling is able to reproduce this trend by including a wider range of side-reactions.[21] One reaction of particularly great importance is the bimolecular reaction $H^\bullet + H^\bullet \rightarrow H_2$ (gas), which becomes a particularly important alternative reaction pathway at low pH. While much remains to be learned about the reaction pathways of such high-energy species, it is clear that the ability to produce solvated electrons by photoexcitation of diamond enables chemistry to be investigated in an entirely new regime not accessible with conventional electrochemistry or other semiconductor photoelectrochemistry.

B3. Photo-induced etching of diamond: caveats and implications for photocatalytic reduction of CO_2

The emission of electrons into water must always be accompanied by oxidation processes in order for the diamond to maintain charge neutrality. Consequently, when investigating diamond for catalytic reduction reactions, it is equally important to consider the oxidation reactions induced by the valence band holes.

In water, holes in diamond's valence band have several oxidation pathways. As shown in Fig. 1, for diamond samples with electron affinity of approximately −0.8 V, the 5.5 eV bandgap places the valence band approximately 4.7 eV below the vacuum level or approximately −0.26 V relative to the standard hydrogen electrode (SHE). Water oxidation at pH = 0 has a standard redox reduction potential ($O_2 + 4H^+ + 4e^- \rightarrow 2H_2O$) of 1.23 V *vs.* SHE, but at pH = 7 the reduction potential is approximately 0.81 V *vs.* SHE. Thus, if the valence band lies more positive than +0.81 eV (or equivalently, more than ~5.2 V below vacuum) then it is thermodynamically capable of oxidizing water to produce $O_2 + H^+$. However, diamond and other carbon-based materials have an additional pathway, which is through oxidation of the substrate itself. Many prior studies have shown that under acidic conditions carbon can be oxidized at potentials of ~0.8 V to form CO_2 and CO.[30–33] There are two important oxidation reactions of diamond itself,

$$C + H_2O \rightarrow CO_2 + 4H^+ + 4\,e^-$$

and

$$C + H_2O \rightarrow CO + 2H^+ + 2e^-.$$

The reduction potentials associated with the corresponding reduction reactions are[23,30]

$$CO_2 + 4H^+ + 4e^- \rightarrow C + 2H_2O\ E^0 = 0.207\ V\ (E = -0.206\ \text{at pH} = 7)$$

$$CO + 2H^+ + 2e^- \rightarrow C + H_2O\ E^0 = 0.518\ V\ (E = 0.105\ V\ \text{at pH} = 7)$$

At pH = 7, oxidation of carbon to form CO becomes possible for potentials more positive than 0.105 V, while oxidation to form CO_2 is possible for potentials more positive than −0.206 V (in both cases assuming gaseous products produced at 1 atmosphere pressure). In either case, at circumneutral pH, oxidation of the diamond substrate to form CO and/or CO_2 is both thermodynamically favored over oxidation of water.

Because complex, multi-electron reactions such as the above reactions for oxidation of water and of carbon are often strongly kinetically hindered, it is difficult to predict what valence band reactions will dominate. However, a key consequence of the above thermodynamic analysis is that unless steps are taken to provide alternative pathways for the valence-band holes, photo-induced etching of the diamond to produce CO and CO_2 as products is thermodynamically allowed. To illustrate this, we conducted experiments using H-terminated detonation nanodiamond in aqueous 0.1 M KCl solution. In these studies, nanodiamond powder was cleaned and then H-terminated by heating in H_2 at 750 °C. The H-terminated powder was then stirred into an aqueous solution of 0.1 M KCl. The solutions were then spared with argon, CO_2 or isotopically labeled $^{13}CO_2$ for 30 minutes and illuminated with UV light. Since natural-abundance CO_2 is ~99% $^{12}CO_2$, we refer to this as $^{12}CO_2$. After illuminating for 30 minutes, the gas-phase products were analyzed by Fourier-transform infrared spectroscopy.

Fig. 7 shows the resulting experimental spectra of the gaseous products in the primary CO absorption region, along with simulations of the expected spectra of ^{13}CO and ^{12}CO using the HITRAN spectral database.[34] These data show several important facts. First, ^{12}CO production is observed in the solutions saturated with $^{12}CO_2$, $^{13}CO_2$, and argon. In each case, the resulting spectrum matches almost perfectly the expected spectrum of ^{12}CO. Even when the solution was purged using isotopically pure $^{13}CO_2$, the major isotope of CO produced is ^{12}CO. This isotope distribution shows that the vast majority of the CO produced does not come from the gas-phase CO_2 in the solution, but must instead come from the diamond sample itself. This conclusion is confirmed by the fact that CO is also observed when the solution is extensively purged with ultra-pure argon. In this case the only significant source of carbon in the system is the diamond powder itself.

As we will report elsewhere, the problem of photo-induced oxidation of diamond can be alleviated, but great care must always be taken to ensure that the reactions are truly catalytic. One approach is to use conductive diamond samples,

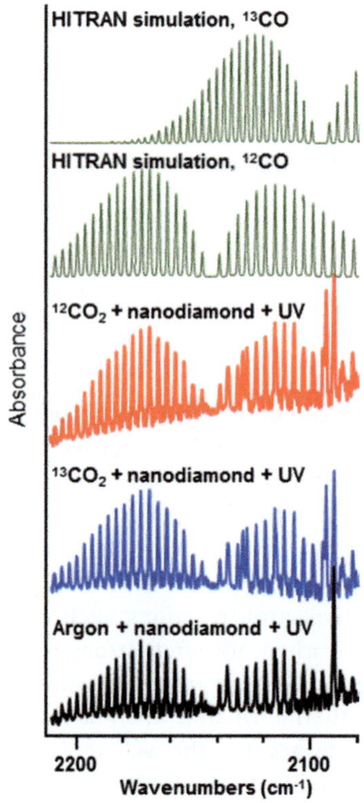

Fig. 7 Infrared measurements of CO produced from H-terminated nanodiamond illuminated in water with different dissolved gases, along with the computer simulations of spectra for ^{12}CO and ^{13}CO.

and to connect the diamond to another electrode, such as Pt, that has faster oxidation–reduction kinetics. The intentional addition of species that are readily oxidized by the Pt and/or by the valence band holes can also reduce the problem. This general approach was recognized more than 50 years ago in the seminal work by Fujishima and Honda using TiO_2 to photocatalyze the oxidation reaction.[35] In that work they used illuminated TiO_2 to oxidize organic species, but connected the TiO_2 to a Pt electrode to achieve reduction reactions that are otherwise very slow on TiO_2 because of its shallow conduction band energy.[35] On diamond the situation is reversed, with the electron emission being facile but corresponding oxidation reactions being sluggish and problematic.

The problem of photo-etching is most severe with nano-diamond powder, since in this case the overall low crystal quality and presence of many under-coordinated surface sites makes the surfaces more reactive. To avoid these problems, several modifications to the experimental procedures are necessary, with the goal of ensuring that the valence-band holes cannot oxidize diamond to CO or CO_2. These modifications, as explained in the experimental section, primarily involve using a Pt electrode in contact with a redox couple having facile oxidation kinetics. More detailed procedures and the enablement of selective CO_2 reduction without diamond etching will be described elsewhere.[36]

C. Discussion

While photoemission into water has some relation to photoemission into vacuum, there are also significant differences, such as the high electric field that can be present in the electrical double-layer, and the fact that electrons can be trapped near interfaces of a solid-liquid interface. Benderskii and co-workers[37] investigated photoemission from metals into liquids and noted that there are two possible mechanisms of photoemission. One, a direct surface effect that is controlled by the electric field component perpendicular to the surface, and two, a volumetric effect that is not dependent on the optical polarization. Under the conditions of our experiments (excitation directed normal to the surface) only the volumetric effect should be possible since with normal incidence there is no component of electric field perpendicular to the interface.[37]

The presence of solvated electrons was inferred as early as the 1970s from steady-state photocurrent measurements of photocurrents induced in water containing various electron acceptors and illuminated with ultraviolet light.[37–39] Yet, to the best of our knowledge the present work is the first direct measurement of solvated electrons from a metal by transient absorption spectroscopy. Our data show, somewhat surprisingly, that the major change observed in growing diamond films onto Nb is a dramatic increase in the *lifetime* of the solvated electrons. This can be understood on the basis of a simple model depicted in Fig. 8.

Metal surfaces contain a continuum of states, and therefore when an electron is injected into the solution and becomes solvated, it has a large number of degenerate states associated with the continuum of empty states of the metal. Thus, the electron can easily recombine and thermalize within the metal. In contrast, when an electron is ejected from a diamond-coated surface, the emission is essentially irreversible because the solvated electron energy is well below the conduction-band minimum. There are no (or few) electron states in diamond at the energy of the solvated electron, and therefore no possibility for the electron to recombine without simultaneously being able to lose large amounts ($\sim$3 eV) of energy. Thus, on metals it is likely that the lifetime is controlled by recombination in the metallic states, while in the case of a diamond-coated electrode the lifetime is controlled by chemical processes such as reduction of protons, described earlier. While much remains to be learned about the dynamics of electron emission and its implications for chemical processes, the ability to easily create solvated electrons from a solid-state material brings with it a range of new opportunities for high-energy chemistry.

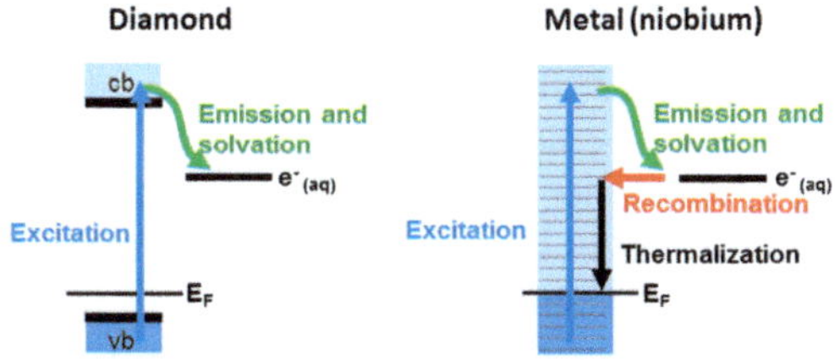

Fig. 8 Schematic depiction of processes controlling solvated electron lifetime on diamond *vs.* metal surfaces.

D. Experimental

Materials and methods

Growth of diamond thin films. Niobium substrates were seeded by polishing with 0–2 micron and 0–0.1 micron sized diamond particles. To remove any of the larger sized particles and aggregates, the polishing was followed by sonication in methanol followed by water. Diamond growth was accomplished using a microwave plasma-enhanced chemical vapor deposition system using a mixture of hydrogen and methane gas. Growth consisted of an initial 10 minute nucleation phase using H_2 and CH_4 flow rates of 200 standard cubic centimeters per minute (sccm) and 2 sccm, respectively, with a microwave power of 1000 W. The nucleation phase was followed by a growth phase using 200 sccm and 1 sccm flow rates for H_2 and methane, respectively, and a microwave power of 800 watts. In the growth phase, the substrate temperature is approximately 750° C as measured by an optical pyrometer. Finally, the growth phase was followed by a 10 min etching in hydrogen plasma only to remove/reduce sp^2 hybridized carbon on the surface. In all phases, the pressure in the chamber is maintained at 48 torr.

Measurement of solvated electrons. Solvated electrons were measured using transient absorption spectroscopy. An ultraviolet pulsed laser (NT340, EKSPLA, Inc., Vilnius, Lithuania), was used to excite the samples at 213 nm, with the pulse energy adjusted to 1.5 mJ using a polarizing prism. A 705 nm diode laser (Newport LP705-SF15) with a F240FC-780 fiber collimator passed immediately in front of the sample. The transmitted light was detected using a 1.2 GHz silicon detector, Newport model 818-BB-21, amplified using FEMTO amplifier, DHPCA-100 (100 V/A transimpedance, 20 MHz bandwidth) and recorded using a 4 GHz digital oscilloscope (Agilent Model DSO9404A). To reduce radio-frequency noise from the pulsed laser, a background subtraction is typically performed. After each sample is measured under the desired conditions, the solution is then acidified with H_2SO_4, which reacts rapidly with the solvated electrons but leaves all other properties unchanged; this background spectrum is recorded and subtracted from the original transient.

Measurement of ammonia yield. UV illumination used a Newport Model 6278 UV-enhanced Xenon lamp, 450 Watts electrical power. The light was passed through a water-cell to remove infrared light, and then collimated onto the sample. Ammonia yield experiments were performed in a 2-compartment cell with a separating glass frit. The sample was immersed in a solution of 10 mM Na_2SO_4; the reference electrode consisted of a Pt wire immersed in a I^-/I_3^- solution with an I^-/I_3^- concentration ratio of 250 : 1. In experiments reported here, the sample was also biased at −0.3 V relative to the reference electrode in order to minimize the possibility of the valence band holes inducing etching or other oxidation reactions. Control experiments without bias showed this small bias to have no significant effect on ammonia yield. Ammonia yield was measured using the indophenol blue method. In this method, 0.100 ml of a 1 M NaOH solution containing 5% salicylic acid and 5% sodium citrate (by weight) was added to 2 ml of sample solution, followed by addition of 0.02 ml of 0.05 M NaClO and 0.02 ml of an aqueous solution of 1% (by weight) $Na[Fe(NO)(CN)_5]$. The absorption spectrum was taken with a Shimadzu 2401PC Ultraviolet-Visible Spectrophotometer after allowing the mixture to stand for 1 hour. The formation of indophenol blue was determined by reading the absorbance at a wavelength of 700 nm. Appropriate standards were run to generate a working curve for quantitative analysis.

Diamond photo-etching. Detonation nano-diamond (4–6 nm diameter) was first hydrogen-terminated at 750 °C in a tube furnace at atmospheric pressure under a 50 sccm H_2 flow for 3.5 hours. It was then dispersed in 0.1 M KCl aqueous solution (nanodiamond powder concentration 1 g L^{-1}). 18.2 MΩ NanoPure (Barnstead) water was used to minimize any possible organic contamination. The solution was then purged with CO_2 or argon for 30 min under ambient pressure and exposed to ultraviolet light for 16 hours with constant stirring. The reaction products were passed through a valve directly into a custom-made infrared gas cell for analysis of the gas-phase products by Fourier-transform infrared spectroscopy (Bruker Vertex 70 with a liquid nitrogen-cooled HgCdTe detector).

E. Conclusions

Our results show that the dynamics of the solvated electrons play a crucial role in the overall efficiency of photocatalytic reactions at diamond surfaces. While both metals (niobium) and diamond are able to create solvated electrons upon excitation with ultraviolet light, the electrons produced at niobium surfaces exhibit a much shorter lifetime. The much longer lifetime for electrons produced by diamond likely arises from the absence of any electronic states in the diamond bandgap that are energetically degenerate with the solvated electrons. Therefore the bandgap of diamond causes electron emission to be an *irreversible* process: once the electron is emitted, its primary pathway involves chemical reduction reactions. In contrast, on metals the presence of degenerate states leads to shorter lifetimes for the solvated electrons. The longer electron lifetimes of the solvated electrons produced at diamond surfaces provides greater opportunity for the electrons to initiate new chemistry. New high-energy reactions such as N_2 reduction and, in principle, CO_2 reduction *via* a simple 1-electron reduction (instead of competing proton-coupled electron transfer processes) are therefore possible.

To realize the effective utilization of diamond for photocatalysis, it is essential to also control the valence-band holes to ensure the diamond surface is not corroded. These strategies can be chemical in nature by providing sacrificial reactants and/or by using alternative materials such as Pt to carry out the oxidation reactions necessary to maintain charge neutrality.

Acknowledgements

Studies of diamond thin film growth, diamond photoemission into aqueous media, and the resulting solvated electron dynamics were supported by the National Science Foundation Solid State and Materials Chemistry Program, Grant DMR-1207281. Studies of diamond-based CO_2 reduction were supported in part by the Air Force Office of Scientific Research under AFOSR Award FA9550-12-1-0063.

Notes and references

1 G. M. Swain and R. Ramesham, *Anal. Chem.*, 1993, **65**, 345.
2 H. B. Martin, A. Argoitia, U. Landau, A. B. Anderson and J. C. Angus, *J. Electrochem. Soc.*, 1996, **143**, L133.
3 G. M. Swain and R. Ramesham, *Anal. Chem.*, 1993, **65**, 345.
4 A. Hartl, *et al.*, *Nat. Mater.*, 2004, **3**, 736.
5 J. Rubio-Retama, *et al.*, *Langmuir*, 2006, **22**, 5837.
6 J. Hees, *et al.*, *ACS Nano*, 2011, **5**, 3339.
7 J. Hees, R. Hoffmann, N. J. Yang and C. E. Nebel, *Chem. - Eur. J.*, 2013, **19**, 11287.
8 Y. V. Pleskov, A. Y. Sakharova, M. D. Krotova, L. L. Bouilov and B. V. Spitsyn, *J. Electroanal. Chem.*, 1987, **228**, 19.
9 Y. V. Pleskov, A. Y. Sakharova, E. V. Kasatkin and V. A. Shepelin, *J. Electroanal. Chem.*, 1993, **344**, 401.
10 A. Y. Sakharova, *et al.*, *Russ J Electrochem*, 1995, **31**, 169.
11 Y. V. Pleskov, A. Y. Sakharova, A. V. Churikov, V. P. Varnin and I. G. Teremetskaya, *Russ J Electrochem*, 1996, **32**, 1075.
12 F. J. Himpsel, J. A. Knapp, J. A. Vanvechten and D. E. Eastman, *Phys. Rev. B*, 1979, **20**, 624.
13 L. Diederich, O. M. Kuttel, E. Schaller and L. Schlapbach, *Surf. Sci.*, 1996, **349**, 176.
14 D. Takeuchi, H. Kato, G. S. Ri, T. Yamada, P. R. Vinod, D. Hwang, C. E. Nebel, H. Okushi and S. Yamasaki, *Appl. Phys. Lett.*, 2005, **86**, 152103.
15 L. Diederich, O. Kuttel, P. Aebi and L. Schlapbach, *Surf. Sci.*, 1998, **418**, 219.
16 J. Cui, J. Ristein and L. Ley, *Phys. Rev. B: Condens. Matter*, 1999, **60**, 16135.
17 W. Zhu, G. P. Kochanski and S. Jin, *Science*, 1998, **282**, 1471.
18 K. Okano, K. Hoshina, S. Koizumi and K. Nishimura, *Diamond Relat. Mater.*, 1996, **5**, 19.
19 K. Okano, S. Koizumi, S. R. P. Silva and G. A. J. Amaratunga, *Nature*, 1996, **381**, 140.
20 D. Zhu, L. H. Zhang, R. E. Ruther and R. J. Hamers, *Nat. Mater.*, 2013, **12**, 836.
21 J. R. Christianson, D. Zhu, R. J. Hamers and J. R. Schmidt, *J. Phys. Chem. B*, 2014, **118**, 195.
22 H. A. Schwarz, *J. Phys. Chem.*, 1991, **95**, 6697.
23 S. M. Bratsch, *J. Phys. Chem. Ref. Data*, 1989, **18**, 1.
24 M. Fryda, *et al.*, *New Diamond Frontier Carbon Technol.*, 1999, **9**, 229.
25 H. Zanin, *et al.*, *J. Appl. Electrochem.*, 2013, **43**, 323.
26 F. Y. Jou and G. R. Freeman, *J. Phys. Chem.*, 1979, **83**, 2383.
27 F.-Y. Jou and G. R. Freeman, *J. Phys. Chem.*, 1977, **81**, 909.
28 T. Bazhenova and A. Shilov, *Coord. Chem. Rev.*, 1995, **144**, 69.
29 K. Hoshino, R. Kuchii and T. Ogawa, *Appl. Catal., B*, 2008, **79**, 81.
30 H. S. Choo, *et al.*, *J. Electrochem. Soc.*, 2007, **154**, B1017.
31 K. G. Gallagher, G. Yushin and T. F. Fuller, *J. Electrochem. Soc.*, 2010, **157**, B820.
32 H. S. Choo, *et al.*, *J. Power Sources*, 2008, **185**, 740.
33 J. O. Besenhard and H. P. Fritz, *Angew. Chem., Int. Ed. Engl.*, 1983, **22**, 950.
34 L. S. Rothman, *et al.*, *J. Quant. Spectrosc. Radiat. Transfer*, 2010, **111**, 1568.
35 A. Fujishima and K. Honda, *Nature*, 1972, **238**, 37.

36 L. H. Zhang, R. J. Hamers, manuscript in preparation.
37 V. Benderskii, S. D. Babenko, Y. M. Zolotovitskii, A. G. Krivenko and T. S. Rudenko, *J. Electroanal. Chem.*, 1974, **56**, 325.
38 A. V. Benderskii, V. A. Benderskii and A. G. Krivenko, *J. Electroanal. Chem.*, 1995, **380**, 7.
39 S. D. Babenko, V. A. Benderskii, Y. M. Zolotovitskii and A. G. Krivenko, *J. Electroanal. Chem.*, 1977, **76**, 347.

Faraday Discussions

PAPER

Synthesis of iodobiaryls and dibenzofurans by direct coupling at BDD anodes†

B. Elsler, D. Schollmeyer and S. R. Waldvogel*

Received 24th March 2014, Accepted 7th April 2014
DOI: 10.1039/c4fd00049h

The first direct oxidative phenol–arene cross-coupling reactions of an iodine-containing guaiacol derivative and the possible over-oxidation products of electron-rich phenols are described. Hereby, a "green" and targeted synthesis for dibenzofurans was developed.

Introduction

Dibenzofurans are important structural motifs in natural products[1–10] and materials science.[11,12] The broad range of their biological activity makes their synthesis interesting for pharmaceutical purposes. A final C–C or C–O coupling reaction of two aryls is often performed by using transition metal catalysts and requiring substrates equipped with leaving groups. Classic synthetic routes are *O*-arylations based on Cu[13,14] or Pd[10,15–17] catalysts (Fig. 1a). However, a greater significance is still attributed to C–C coupling for obtaining the desired dibenzofurans (Fig. 1b).

Because of the toxicity of the employed organometallic catalysts and formed by-products, direct C–C coupling reactions are of great interest. More recent

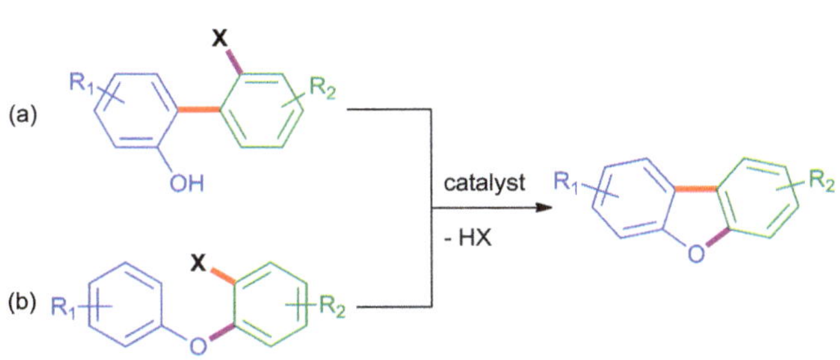

Fig. 1 Common pathways for the synthesis of dibenzofurans.

Institut für Organische Chemie, Johannes Gutenberg-Universität, Mainz, Duesbergweg 10–14, 55128 Mainz, Germany. E-mail: waldvogel@uni-mainz.de; Web: http//www.chemie.uni-mainz.de/OC/AK-Waldvogel/

† Electronic supplementary information (ESI) available. CCDC reference number 993358. For ESI and crystallographic data in CIF or other electronic format see DOI: 10.1039/c4fd00049h

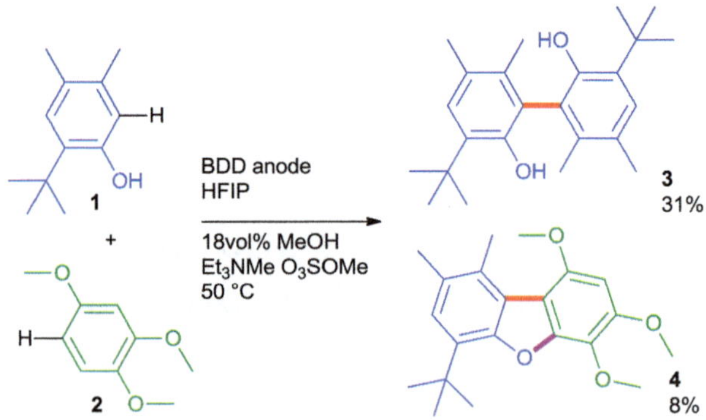

Fig. 2 Electrolysis of 1 and 2. Electrolysis conditions: 50 °C, constant current ($j = 2.8$ mA cm^{-2}), BDD anode, nickel-net cathode, undivided cell, $Q = 2$ F (phenol 1), supporting electrolyte: 0.09 M Et_3NMe O_3SOMe. Homo-coupling products of 2 were not detected.

methods exploit a direct Pd-catalyzed coupling of biaryl ethers requiring a single leaving functionality like a halogen,[18–23] N_2^+,[24] CO_2,[25,26] or CO.[27] Despite the clear need for development of mild reaction conditions, there are only a few examples of the direct cyclization of biaryl ethers without leaving groups described in the literature.[28–34] In the past decade, electrochemical methods have gained increasing attention for use in C–C coupling reactions. Since only electrons serve as the reagent and neither catalysts nor leaving functionalities are required, such transformations can be considered as "green" on all levels.[35] Direct electrochemical C–C bond formation was effectively realized *via* the oxidative dehydrodimerization of phenols[36–38] or arenes[39] and could be expanded to metal and reagent free cross-coupling reactions.[40–43] For anodic treatment, phenols are challenging substrates. In particular, simple phenols have an explicit tendency to form polycyclic scaffolds, creating a plethora of products.[44–46]

Here we report an electrochemical approach for the synthesis of dibenzofurans and their iodinated precursors. An electro-organic pathway, including the

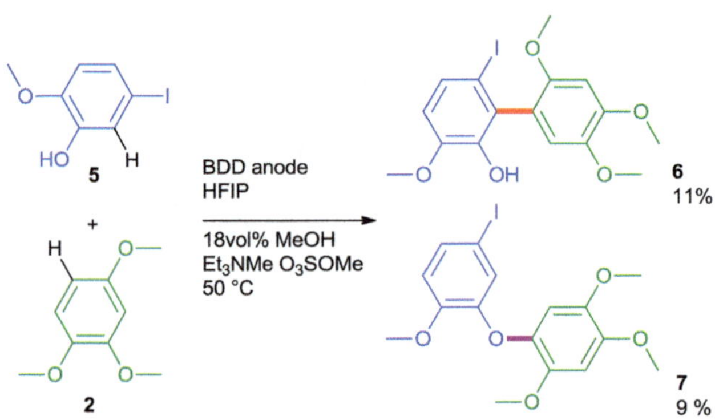

Fig. 3 Electrolysis of 5 and 2. Electrolysis conditions: 50 °C, constant current ($j = 2.8$ mA cm^{-2}), BDD anode and cathode, undivided cell, $Q = 2$ F (phenol 5), supporting electrolyte: 0.09 M Et_3NMe O_3SOMe. The ratio of cross-coupling products to homo-coupling product (determined by GC) = 10 : 1.

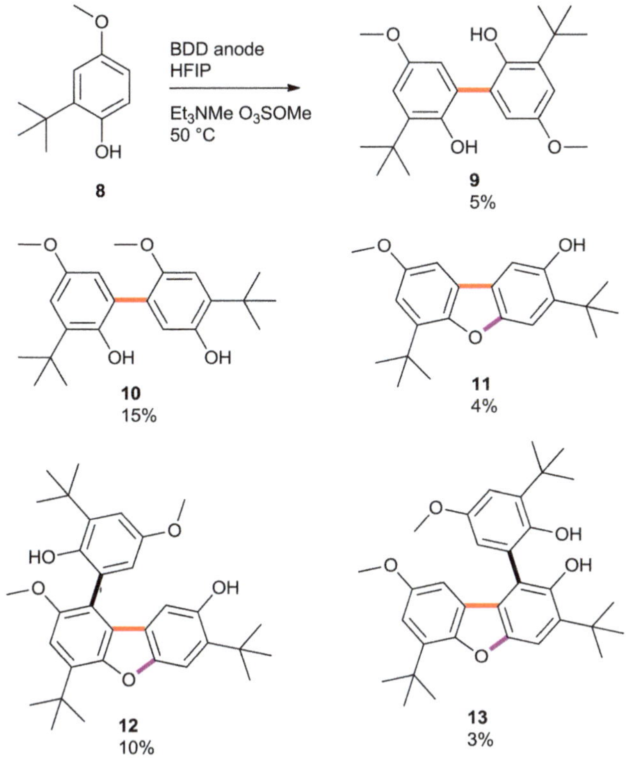

Fig. 4 Major components produced by electrolysis of **8**. Electrolysis conditions: 50 °C, constant current ($j = 4.7$ mA cm^{-2}), BDD anode, nickel-net cathode, undivided cell, $Q = 1$ F (phenol **8**), supporting electrolyte: 0.09 M $Bu_3NMe\ O_3SOMe$.

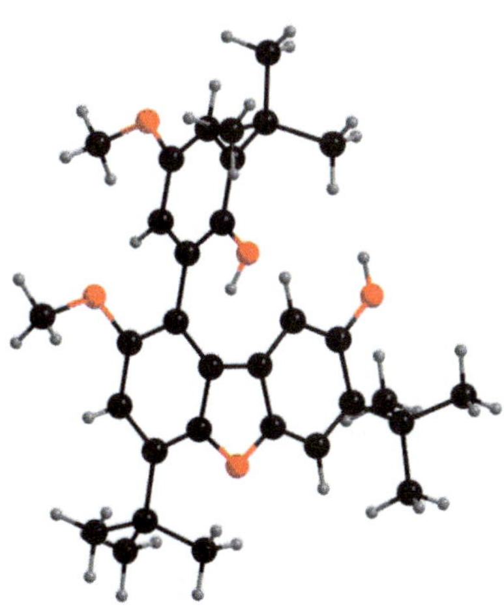

Fig. 5 Molecular structure of **12** from X-ray diffraction analysis of a suitable single crystal.

subsequent oxidation of cross- and homo-coupled aryls for this structural motif, is elucidated.

Results and discussion

Sterically demanding and electron-rich phenol components are suitable for cross-coupling reactions at boron doped diamond (BDD) electrodes. In some cases, if steric demand and the electron density of the substrates employed is too high, formation of homo-couplings and over-oxidation of the desired biaryls are observed. By electrolyzing 4,5-dimethyl-2-(1,1-dimethylethyl)phenol **1** and 1,2,4-trimethoxybenzene **2** (Fig. 2) the desired cross-coupling product is not detected. The major product is represented by the symmetrical biphenol **3** in 31% yield. Comparable syntheses require electrolysis of a derivative of **1** templated by a tetraphenoxy borate. The overall yield in this case is 19%.[47,48] Small amounts of dibenzofuran **4** indicate the subsequent oxidation of a non-symmetrical biaryl intermediate, which leads to *O*-arylation.

We were pleased to see that the iodine containing phenol **5** can be cross-coupled with **2**, which leads to biaryl **6** and the mixed biaryl ether **7** without any dehalogenation upon electrolysis at the BDD electrode (Fig. 3). With the previously mentioned reaction conditions, these structural motifs can easily be transformed into the corresponding dibenzofurans. This two-step synthesis allows very easy, environmentally benign and low-cost access to highly substituted, non-symmetrical dibenzofuran moieties.

The more electron rich 2-(1,1-dimethylethyl)-4-methoxyphenol **8** has a strong tendency to form over-oxidized products when subjected to homo-coupling (Fig. 4). Electrolysis in pure 1,1,1,3,3,3-hexafluoropropan-2-ol (HFIP) provides a broad structural variety of di- and trimeric structures. Despite the structural diversity, all substances can be purified by column chromatography and are interesting motifs for biological activity studies. Because of the spatial proximity of two hydroxy functions in products **12** (Fig. 5) and **13**, these products are also interesting precursors for ligands in transition metal catalysis. Using an anode material other than BDD, *e.g.* glassy carbon, the formation of **10** and **13** strongly decreases (see the ESI†).

Conclusions

Boron doped diamond electrodes exhibit unique synthetic potential which is not observed for other types of anode material. The anodic conversion of iodophenols with arenes leads to iodinated mixed biaryls and diaryl ethers without loss of the valuable iodo functionality. When no cross-coupling is carried out, the biphenols are anodically formed, followed by over-oxidation to the dibenzofurans. Although the reported products are formed in low yield, the simple starting materials and easy to perform protocol makes this pathway synthetically interesting.

Experimental

General procedure for anodic coupling reactions

A solution of a phenolic component (0.005 mol), an arene component (0.015 mol) and *N*-methyl-*N*,*N*,*N*-triethylammonium methylsulfate (0.68 g, 0.003 mol) in a

33 mL 1,1,1,3,3,3-hexafluoropropan-2-ol–methanol (0–18% v/v) mixture was transferred into an undivided L-type electrolysis cell equipped with a BDD anode and a nickel-net cathode.[49] A constant current electrolysis with a current density of 2.8 mA cm^{-2}–4.7 mA cm^{-2} was performed at 50 °C. After application of 965 C (2 F per mol phenol) the electrolysis was stopped and the solvent mixture was recovered *in vacuo* (50 °C, 200–70 mbar). Non-converted starting materials were recovered by short-path or bulb-to-bulb distillation (100 °C, 10^{-3} mbar) and the crude coupling products were purified by column chromatography (SiO_2, ethyl acetate–cyclohexane). If necessary, products were recrystallized by dissolving in dichloromethane and solvent diffusion with cyclohexane.

Synthesis of 2,2′-dihydroxy-3,3′-bis-(1,1-dimethylethyl)-5,5′,6,6′-tetramethylbiphenyl (3) and 8,9-dimethyl-4-(1,1-dimethylethyl)-1,3,4-trimethoxy-dibenzo[*b*,*d*]furan (4)

3 and **4** were synthesized by the general procedure with 0.89 g 4,5-dimethyl-2-(dimethylethyl)phenol and 2.55 g 1,2,4-trimethoxybenzene in a mixture of 27 mL HFIP and 6 mL MeOH. After evaporation of the electrolyte and unconverted material, the products were purified by column chromatography (cyclohexane : ethyl acetate = 9 : 1) and yielded **3** as a colorless solid (272 mg, 0,77 mmol, 31%) and **4** as yellow oil (148 mg, 0.43 mmol, 9%).

Analytical data of 3. R_f 0.90 (Cy:EA = 9 : 1); ^{1}H NMR (300 MHz, $CDCl_3$): 1.38 (s, 18H), 1.80 (s, 6H), 2.24 (s, 6H), 4.78 (bs, 2H), 7.11 (s, 2H); ^{13}C NMR (75 MHz, $CDCl_3$): 15.91 (C-7), 19.99 (C-8), 29.54 (C-10), 34.46 (C-9), 120.94, 128.03, 128.73, 133.34, 134.11, 150.35. MS (EI, 70 eV): *m*/*z* (%): 354.2 (64) $[M]^{\bullet+}$, 339.2 (36) $[M\text{-}CH_3^{\bullet}]^+$, 283.2 (100) $[M\text{-}CH_3^{\bullet}\text{-}C_4H_8]^+$; HRMS for $C_{24}H_{34}O_2$: (ESI) [M + Na]: calc.: 354.2559, found: 354.2559.

Analytical data of 4. R_f 0.51 (Cy:EA = 9 : 1); ^{1}H NMR (400 MHz, $CDCl_3$): 1.53 (s, 9H), 2.43 (s, 3H), 2.84 (s, 3H), 3.93 (s, 3H), 3.97 (s, 3H), 4.01 (s, 3H), 6.96 (s, 1H), 7.12 (s, 1H); ^{13}C NMR (75 MHz, $CDCl_3$): 17.27, 20.74, 29.87, 34.01, 56.19, 61.23, 61.45, 91.36, 110.95, 124.12, 124.86, 128.96, 130.49, 130.81, 138.57, 148.19, 152.77, 152.93, 153.25. MS (EI, 70 eV): *m*/*z* (%): 342.2 (100) $[M]^{\bullet+}$, 327.1 (100) $[M\text{-}CH_3^{\bullet}]^+$; HRMS for $C_{21}H_{27}O_4$: (ESI^+) $[M + H^+]$: calc.: 343.1904, found: 343.1907.

Synthesis of 2-hydroxy-6-iodo-2′,3,4′,5′-tetramethoxybiphenyl (6) and 5′-iodo-2,2′,4,5-tetramethoxybiphenylether (7)

6 and **7** were synthesized by the general procedure with 1.26 g 5-iodoguaiacol and 2.55 g 1,2,4-trimethoxybenzene in a mixture of 27 mL HFIP and 6 mL MeOH. After evaporation of the electrolyte and unconverted material, the products were purified by column chromatography (cyclohexane:ethyl acetate = 4 : 1) and yielded **6** as yellow oil (238 mg, 0.56 mmol, 11%) and **7** as an off-white solid (188 mg, 0,45 mmol, 9%).

Analytical data of 6. R_f 0.31 (Cy:EA = 4 : 1); ^{1}H NMR (400 MHz, $CDCl_3$): 3.76 (s, 3H), 3.84 (s, 3H), 3.89 (s, 3H), 3.94 (s, 3H), 5.65 (brs, 1H), 6.61–6.68 (m, 3H), 7.43 (d, J = 8.6 Hz, 1H); ^{13}C NMR (75 MHz, $CDCl_3$): 55.87, 56.02, 56.35, 56.90, 77.16, 91.32, 98.10, 111.71, 114.53, 119.74, 129.34, 129,34 143.05, 143.71, 146.90, 149.69, 150.88. MS (EI, 70 eV): *m*/*z* (%): 416 (100) $[M]^{\bullet+}$, 401 (30) $[M\text{-}CH_3^{\bullet}]^+$; HRMS for $C_{16}H_{17}IO_5$: (ESI) [M + Na]: calc.: 439.0018, found: 439.0021. Elemental analysis for $C_{16}H_{17}IO_5$: calc.: 46.17% C, 4.12% H, found: 46.29% C, 4.11% H.

Analytical data of 7. R_f 0.03 (Cy:EA = 4 : 1); ^{1}H NMR (400 MHz, CDCl$_3$): 3.77 (s, 3H), 3.78 (s, 3H), 3.89 (s, 3H), 3.91 (s, 3H), 6.64 (d, J = 2.1 Hz, 2H), 6.69 (d, J = 8.5 Hz, 1H), 6.82 (d, J = 2.1 Hz, 1H), 7.28 (dd, J = 8.5 Hz, 2.1 Hz, 1H); ^{13}C NMR (75 MHz, CDCl$_3$): 56.19, 56.57, 56.65, 57.25, 77.16, 82.12, 100.04, 107.03, 114.10, 123.99, 131.32, 136.49, 143.41, 145.26, 146.54, 148.73, 149.43. MS (EI, 70 eV): m/z (%): 416 (100) $[M]^{\cdot+}$, 401 (20) $[M\text{-}CH_3^{\cdot}]^+$; HRMS for $C_{16}H_{17}IO_5$: (EI) $[M^{\cdot+}]$: calc.: 416.01207, found: 416.01198.

Synthesis of 2,2′-dihydroxy-5,5′-dimethoxy-3,3′-bis-(1,1-dimethylethyl)biphenyl (9), 2,3′-dihydroxy-2′,5-dimethoxy-3,4′-bis-(1,1-dimethylethyl)biphenyl (10), 3,6-bis-(1,1-dimethylethyl)-2-hydroxy-8-methoxydibenzo[*b*,*d*]furan (11), 3,6-bis-(1,1-dimethylethyl)-9-(3-(1,1-dimethylethyl)-2-hydroxy-5-methoxyphenyl)-2-hydroxy-8-methoxydibenzo[*b*,*d*]furan (12) and 3,6-bis-(1,1-dimethylethyl)-1-(3-(1,1-dimethylethyl)-2-hydroxy-5-methoxyphenyl)-2-hydroxy-8-methoxydibenzo[*b*,*d*] furan (13)

Compounds **9** to **13** were synthesized by the general procedure with 3.61 g (20 mmol) 2-(dimethylethyl)-4-methoxyphenol in 30 mL HFIP. After evaporation of the electrolyte and unconverted material, the products were purified by column chromatography (cyclohexane:ethyl acetate = 99.5 : 0.5 + 1% AcOH) and yielded **9** as a brownish solid and the other products as colorless solids. **9:** 351 mg (0.98 mmol, 5%), **10:** 1.06 g (2.96 mmol, 15%), **11:** 124 mg (0.38 mmol, 4%), **12:** 335 mg (0.66 mmol, 10%), **13:** 101 mg (0.20 mmol, 3%).

Analytical data of 9. R_f 0.55 (Cy:EA = 9 : 1); ^{1}H NMR (400 MHz, CDCl$_3$): 1.43 (s, 18H) 3.78 (s, 6H), 5.01 (brs, 2H), 6.63 (d, J = 3.1 Hz, 2H), 6.96 (d, J = 3.1 Hz, 2H). ^{13}C NMR (101 MHz, CDCl$_3$): 29.2, 35.0, 55.6, 111.9, 115.0, 123.6, 138.8, 145.82, 153.3. Data according to the literature.[48]

Analytical data of 10. R_f 0.59 (Cy:EA = 9 : 1); ^{1}H NMR (400 MHz, CDCl$_3$): 1.43–1.49 (m, 18H), 3.80 (s, 3H), 3.83 (s, 3H), 4.81 (bs, 2H), 6.50 (s, 1H), 6.62–6.67 (m, 2H), 6.94 (d, J = 3.0 Hz, 1H), 6.98 (s, 1H). ^{13}C NMR (70 MHz, CDCl$_3$): 29.68, 29.74, 29.87, 35.08, 35.32, 55.87, 111.82, 113.97, 114.98, 115.56, 120.22, 123.98, 139.14, 141.10, 146.00, 146.36, 153.28, 153.31. MS (EI, 70 eV): m/z (%): 358 (80) $[M]^{\cdot+}$, 302 (50) $[M\text{-}C_4H_8]^{\cdot+}$.

Analytical data of 11. R_f 0.42 (Cy:EA = 9 : 1); ^{1}H NMR (400 MHz, CDCl$_3$): 1.51 (s, 9H), 1.53 (s, 9H), 3.89 (s, 3H), 4,93 (bs, 1H), 6.95 (d, J = 2.6 Hz, 1H), 7.12 (d, J = 2.6 Hz, 1H), 7.15 (s, 1H), 7.52 (s, 1H); ^{13}C NMR (70 MHz, CDCl$_3$): 29.80, 29.89, 34.67, 35.41, 56.03, 100.25, 106.76, 109.98, 112.80, 122.46, 124.71, 136.37, 136.93, 150.06, 150.30, 151.55, 155.39. MS (EI, 70 eV): m/z (%): 326 (50) $[M]^{\cdot+}$, 311 (100) $[M\text{-}CH_3^{\cdot}]^+$.

Analytical data of 12. R_f 0.48 (Cy:EA = 9 : 1); ^{1}H NMR (400 MHz, CDCl$_3$): 1.46 (m, 18H), 1.60 (s, 9H), 3.74 (s, 3H), 3.86 (s, 3H), 4.86 (bs, 1H), 5.25 (bs, 1H), 6.38 (s, 1H), 6.75 (d, J = 3.1 Hz, 1H), 7.04 (d, J = 3.1 Hz, 1H), 7.08 (s, 1H), 7.49 (s, 1H); ^{13}C NMR (101 MHz, CDCl$_3$): 29.70, 29.88, 29.88, 34.98, 35.26, 35.44, 55.78, 57.57, 108.20, 109.35, 109.65, 111.75, 114.99, 116.97, 121.65, 123.23, 124.66, 135.75, 137.57, 138.87, 146.75, 150.46, 150.49, 151.72, 152.08, 152.25. MS (EI, 70 eV): m/z (%): 489 (50) $[M\text{-}CH_3^{\cdot}]^+$; HRMS for $C_{32}H_{40}O_5$: (ESI) [M + Na]: calc.: 527.2773, found: 527.2769. Crystal structure determination of **12**: $C_{32}H_{40}O_5$, M_r = 504.66; colorless block-like crystals (0.28 × 0.30 × 0.48 mm^3), T = 173 K, λ(Mo-Kα) = 0.71073 Å, monoclinic space group $C2/c$, a = 27.352(1) Å, α = 90°, b = 10.7150(5) Å, β = 121.95(1)°, c = 24.122(1) Å, γ = 90°, V = 5999.2(8) Å^3, Z = 4, ρ(calc) = 1.212 g

cm^{-3}, $2\theta_{max} = 56.0°$, $\mu = 0.17$ mm^{-1}, $F(000) = 2344$, 16749 reflections, 7142 unique reflections ($R_{int} = 0.0263$), $w = 1/[\sigma^2(F_0^2) + (0.061*P)^2 + 4.97*P]$ while $P = (Max(F_0^2,0) + 2*F_c^2)/3$. GOF = 1.039, $R_1 = 0.0526$ [$I > 2\sigma(I)$], $R_1 = 0.0770$ [all data], $wR_2 = 0.1452$. In the solid state the biaryl is twisted by 94.985(5)°.

Analytical data of 13. R_f 0.46 (Cy:EA = 9 : 1); ^1H NMR (400 MHz, $CDCl_3$): 1.46 (m, 18H), 1.60 (s, 9H), 3.74 (s, 3H), 3.86 (s, 3H), 4.86 (bs, 1H), 5.25 (bs, 1H), 6.38 (s, 1H), 6.75 (d, J = 3.1 Hz, 1H), 7.04 (d, J = 3.1 Hz, 1H), 7.08 (s, 1H), 7.49 (s, 1H); ^{13}C NMR (101 MHz, $CDCl_3$): 29.36, 29.70, 29.83, 34.51, 35.04, 35.56, 55.02, 57.22, 100.82, 109.22, 109.95, 112.23, 112.48, 118.74, 120.67, 121.32, 124.65, 135.83, 137.84, 138.03, 147.51, 148.90, 149.95, 150.09, 151.12, 154.56. MS (EI, 70 eV): m/z (%): 489 (80) $[M-CH_3^{\bullet}]^+$; HRMS for $C_{32}H_{40}O_5$: (ESI) [M + Na]: calc.: 527.2773, found: 527.2775.

References

1 Y. Q. Ye, C. Negishi, Y. Hongo, H. Koshino, J.-i. Onose, N. Abe and S. Takahashi, *Bioorg. Med. Chem.*, 2014, **22**(8), 2442.
2 H. Huang, X. Feng, Z. Xiao, L. Liu, H. Li, L. Ma, Y. Lu, J. Ju, Z. She and Y. Lin, *J. Nat. Prod.*, 2011, **74**(5), 997.
3 C.-H. Lin, H.-S. Chang, C.-H. Liao, T.-H. Ou, I.-S. Chen and I.-L. Tsai, *J. Nat. Prod.*, 2010, **73**(10), 1628.
4 W. K. P. Shiu and S. Gibbons, *Phytochemistry*, 2009, **70**(3), 403.
5 Y. Feng, A. R. Carroll, R. Addepalli, G. A. Fechner, V. M. Avery and R. J. Quinn, *J. Nat. Prod.*, 2007, **70**(11), 1790.
6 J. Qu, C. Xie, H. Guo, W. Yu and H. Lou, *Phytochemistry*, 2007, **68**(13), 1767.
7 K. Kaniwa, T. Ohtsuki, Y. Yamamoto and M. Ishibashi, *Tetrahedron Lett.*, 2006, **47**(10), 1505.
8 S. Manniche, K. Sprogøe, P. W. Dalsgaard, C. Christophersen and T. O. Larsen, *J. Nat. Prod.*, 2004, **67**(12), 2111.
9 J. R. Carney, J. M. Krenisky, R. T. Williamson and J. Luo, *J. Nat. Prod.*, 2002, **65**(2), 203.
10 H. Teng, G. A. Thakur and A. Makriyannis, *Bioorg. Med. Chem. Lett.*, 2011, **21**(19), 5999.
11 M. Asakawa, P. R. Ashton, C. L. Brown, M. C. T. Fyfe, S. Menzer, D. Pasini, C. Scheuer, N. Spencer, J. F. Stoddart, A. J. P. White and D. J. Williams, *Chem. Eur. J.*, 1997, **3**(7), 1136.
12 K. Y. Tsang, H. Diaz, N. Graciani and J. W. Kelly, *J. Am. Chem. Soc.*, 1994, **116**(9), 3988.
13 J. Zhao, Y. Wang and Q. Zhu, *Synthesis*, 2012, **44**(10), 1551.
14 J. Zhao, Q. Zhang, L. Liu, Y. He, J. Li, J. Li and Q. Zhu, *Org. Lett.*, 2012, **14**(20), 5362.
15 B. Xiao, T.-J. Gong, Z.-J. Liu, J.-H. Liu, D.-F. Luo, J. Xu and L. Liu, *J. Am. Chem. Soc.*, 2011, **133**(24), 9250.
16 J. Liu, A. E. Fitzgerald and N. S. Mani, *J. Org. Chem.*, 2008, **73**(7), 2951.
17 K. Kawaguchi, K. Nakano and K. Nozaki, *J. Org. Chem.*, 2007, **72**(14), 5119.
18 L.-C. Campeau, M. Parisien, A. Jean and K. Fagnou, *J. Am. Chem. Soc.*, 2006, **128**(2), 581.
19 M. Parisien, D. Valette and K. Fagnou, *J. Org. Chem.*, 2005, **70**(19), 7578.
20 Z. Liu and R. C. Larock, *Org. Lett.*, 2004, **6**(21), 3739.

21 Z. Liu and R. C. Larock, *Tetrahedron*, 2007, **63**(2), 347.

22 J. M. Gajera, B. Gopalan, P. S. Yadav, S. D. Patil and L. A. Gharat, *J. Heterocycl. Chem.*, 2008, **45**(3), 797.

23 D. E. Ames and A. Opalko, *Synthesis*, 1983, (03), 234.

24 Z. Du, J. Zhou, C. Si and W. Ma, *Synlett*, 2011, (20), 3023.

25 C. Wang, I. Piel and F. Glorius, *J. Am. Chem. Soc.*, 2009, **131**(12), 4194.

26 A. Voutchkova, A. Coplin, N. E. Leadbeater and R. H. Crabtree, *Chem. Commun.*, 2008, **47**, 6312.

27 S. Maetani, T. Fukuyama and I. Ryu, *Org. Lett.*, 2013, **15**(11), 2754.

28 T. Ishida, R. Tsunoda, Z. Zhang, A. Hamasaki, T. Honma, H. Ohashi, T. Yokoyama and M. Tokunaga, *Appl. Catal., B*, 2014, **150–151**, 523.

29 S. Trosien, P. Böttger and S. R. Waldvogel, *Org. Lett.*, 2014, **16**(2), 402.

30 B. Liégault, D. Lee, M. P. Huestis, D. R. Stuart and K. Fagnou, *J. Org. Chem.*, 2008, **73**(13), 5022.

31 J. Griffiths, A. E. H. Machado, A. M. A. G. Oliveira, A. M. F. Oliveira-Campos and M. M. M. Raposo, *Helv. Chim. Acta*, 2003, **86**(8), 2900.

32 H. Hagelin, J. D. Oslob and B. Åkermark, *Chem. Eur. J.*, 1999, **5**(8), 2413.

33 B. Akermark, L. Eberson, E. Jonsson and E. Pettersson, *J. Org. Chem.*, 1975, **40**(9), 1365.

34 K.-P. Zeller and H. Petersen, *Synthesis*, 1975, **08**, 532.

35 P. Anastas and N. Eghbali, *Chem. Soc. Rev.*, 2009, **39**(1), 301.

36 A. Kirste, S. Hayashi, G. Schnakenburg, I. M. Malkowsky, F. Stecker, A. Fischer, T. Fuchigami and S. R. Waldvogel, *Chem. Eur. J.*, 2011, **17**(50), 14164.

37 A. Kirste, G. Schnakenburg and S. R. Waldvogel, *Org. Lett.*, 2011, **13**(12), 3126.

38 A. Kirste, M. Nieger, I. M. Malkowsky, F. Stecker, A. Fischer and S. R. Waldvogel, *Chem. Eur. J.*, 2009, **15**(10), 2273.

39 T. Morofuji, A. Shimizu and J.-i. Yoshida, *Angew. Chem. Int. Ed.*, 2012, **51**(29), 7259.

40 B. Elsler, D. Schollmeyer, K. M. Dyballa, R. Franke and S. R. Waldvogel, *Angew. Chem. Int. Ed.*, 2014, **53**(20), 5210.

41 T. Kashiwagi, B. Elsler, S. R. Waldvogel, T. Fuchigami and M. Atobe, *J. Electrochem. Soc.*, 2013, **160**(7), G3058.

42 A. Kirste, B. Elsler, G. Schnakenburg and S. R. Waldvogel, *J. Am. Chem. Soc.*, 2012, **134**(7), 3571.

43 A. Kirste, G. Schnakenburg, F. Stecker, A. Fischer and S. R. Waldvogel, *Angew. Chem. Int. Ed.*, 2010, **49**(5), 971.

44 J. Barjau, G. Schnakenburg and S. R. Waldvogel, *Angew. Chem. Int. Ed.*, 2011, **50**(6), 1415.

45 J. Barjau, G. Schnakenburg and S. Waldvogel, *Synthesis*, 2011, (13), 2054.

46 J. Barjau, P. Koenigs, O. Kataeva and S. R. Waldvogel, *Synlett*, 2008, (15), 2309.

47 I. M. Malkowsky, R. Fröhlich, U. Griesbach, H. Pütter and S. R. Waldvogel, *Eur. J. Inorg. Chem.*, 2006, (8), 1690.

48 I. M. Malkowsky, C. E. Rommel, R. Fröhlich, U. Griesbach, H. Pütter and S. R. Waldvogel, *Chem. Eur. J.*, 2006, **12**(28), 7482.

49 U. Griesbach, H. Pütter, S. R. Waldvogel, I. M. Malkowsky, U. Griesbach, H. Pütter and S. R. Waldvogel, *Eur. J. Org. Chem.*, 2006, (20), 4569.

Faraday Discussions

PAPER

Laser heated boron doped diamond electrodes: effect of temperature on outer sphere electron transfer processes

Lingcong Meng,†[a] James G. Iacobini,†[a] Maxim B. Joseph,[a] Julie V. Macpherson*[a] and Mark E. Newton*[b]

Received 20th March 2014, Accepted 4th April 2014
DOI: 10.1039/c4fd00044g

Thermoelectrochemical experiments can reveal significant information about electrochemical processes compared to ambient only measurements. Typical thermoelectrochemistry is performed using resistively heated wires or laser heated electrodes, both of which can suffer drawbacks associated with the electrode material employed. Boron doped diamond (BDD) is ideal for thermoelectrochemical investigations due to its extremely high thermal conductivity and diffusivity, extreme resistance to thermal ablation (can withstand laser power densities, P_d, of GW cm^{-2} for nanosecond pulses) and excellent electrochemical properties (low background currents and wide potential window). In this paper we describe the use of a pulsed laser technique to heat the rear of a 1 mm diameter conducting BDD disc electrode, which drives electrochemical solution reactions at the front face. Maximum electrode temperatures of 90.0 °C were recorded experimentally and confirmed by finite element modelling (FEM). The effect of laser pulsed heating (maximum 3.8 kW cm^{-2}; 10 ms on and 90 ms off) on the cyclic voltammetric response of two fast (reversible) outer sphere electron transfer redox mediators ($Ru(NH_3)_6^{3+/2+}$ and $IrCl_6^{2-/3-}$) are investigated. In particular, we observe pulsed increases in the current, which increase with increasing P_d. The potential of the peak current is shifted positively for the $Ru(NH_3)_6^{3+/2+}$ couple (in accordance with a positive temperature coefficient, β, +0.68 mV K^{-1}) and negatively for the $IrCl_6^{3-/2-}$ couple ($\beta = -0.48$ mV K^{-1}). Scanning backwards, in contrast to that observed for a macrodisc electrode in ambient solution, a cathodic peak is again observed for $Ru(NH_3)_6^{3+/2+}$ and an anodic peak for $IrCl_6^{3-/2-}$ couple. We attribute this response to the entropy of the redox reaction and the time-dependant change in mass transport due to the induced thermal gradients at the electrode/electrolyte interface. The observed responses are in qualitative agreement with FEM simulations.

[a]*Department of Chemistry, University of Warwick, Coventry, UK. E-mail: j.macpherson@warwick.ac.uk; Fax: +02476524112; Tel: +02476 573886*

[b]*Department of Physics, University of Warwick, Coventry, UK. E-mail: m.e.newton@warwick.ac.uk; Fax: +02476150897; Tel: +02476 150799*

† Both authors contributed equally in this work.

1 Introduction

The effect of temperature, T, in electrochemistry is of great importance[1,2] yet most electrochemical experiments are performed under ambient conditions. This is surprising since T plays a key role in determining the thermodynamics and kinetics of electron transfer processes,[3,4] affects adsorption/desorption processes and double layer phenomena and is intrinsically linked to the mass transport properties of the system. It is also often the case that model systems proposed for real processes operating at elevated temperatures *e.g.* oxygen reduction reaction associated with fuel cells, are carried out under ambient conditions.[5,6]

Thermoelectrochemistry[7] is the branch of electrochemistry where temperature is a controlled variable in the electrochemical experiment. The simplest means of controlling temperature is to use an isothermal approach, heating or cooling the system to the required temperature as appropriate. However, mass transport is often ill-defined due to uncontrolled thermal gradients in the solution, and systematically varying T is a slow process.[8] By moving to non-isothermal methodologies where heat is applied locally to the working electrode, a well-defined temperature gradient now exists at the electrode–electrolyte interface, meaning mass transport can be quantified in the system. It is also possible to heat and change the temperature much faster by utilising short thermal pulses. Finally, the temperature at the electrode/electrolyte interface can be driven above the boiling point. As the electrode only sits at these elevated temperatures for brief periods of time, bubbling is avoided.

Different approaches to non-isothermal heating have been adopted[3] although the most popular in the literature are Joule heating and laser heating. In the former, the electrode can either be heated directly or indirectly, although the direct approach gives rise to the fastest thermal changes and is most widely used. Here the electrode is typically heated *via* application of an alternating current or radio frequency waves.[3] Thermal modulation by laser heating was proposed in 1975 by Barker and Gardner.[9] Since then laser heating using a pulsed laser has become a popular method of choice.[10–14] The lasers employed typically range in power densities, P_d, (mW cm^{-2} – GW cm^{-2}), wavelength (355 nm–1064 nm), and pulse width (~10's of ps – 10's of ms) and have been used to examine a variety of different temperature dependant electrochemical phenomena, ranging from electrode kinetics,[13–16] double layer structure elucidation[17] and surface cleaning.[10,18] Typically the laser is either focused onto the back face of the electrode[13,14,18] or focused directly onto the front face, passing through solution[10–12,15] where energy will be lost, the amount dependant on the path length and absorption coefficient of solution. Local temperature increases typically vary from a few K to 500 K dependant on laser operation conditions.[12]

An important consideration in all thermoelectrochemical experiments is the geometrical arrangement and electrode material employed. Non-isothermal, Joule heated microwire electrodes have proven popular due to decreased Ohmic drop and higher rates of mass transport, compared to macroelectrodes.[3] In general, metal electrodes such as Pt and Au are employed due to their high electrical and thermal conductivities. However, the electrode materials have limitations. For example, it has been shown that even after mild laser treatments (P_d > 1.5 W cm^{-2}) the metal electrodes roughen[11] and at high temperatures oxides form, making the material less inert, and electrochemical interpretation more complicated.[19] Furthermore, the usable potential region becomes narrower when the temperature increases, which for electrocatalytic materials such as Pt is especially a problem.[20] Glassy carbon electrodes have also been employed in both laser heated[21] and Joule heated experiments[22] but were not robust enough for high temperature electrochemistry. In some cases the electrode was even found to combust.[22]

Boron doped diamond (BDD) has emerged as an important electrode material due to its unique properties such as wide potential window, chemical and mechanical robustness and low background currents.[23] These arise from the electrochemical stability of the surface brought about by the sp^3 bonding configuration of this non-metallic material. Furthermore, large area free-standing polycrystalline BDD (pBDD) can now be grown using microwave plasma chemical vapour deposition (MW-CVD) techniques[24] making applications more widespread. As the boron doping level in diamond is increased the resistivity decreases. The material becomes metal-like at B dopant concentrations $>1 \times 10^{20}$ B atoms cm^{-3}.[23,25] Crucially, BDD also possesses a very high thermal conductivity, *ca.* 600 W m^{-1} K^{-1} at 300 K,[26] for comparison the thermal conductivity of copper is *ca.* 400 W m^{-1} K^{-1}.[27] The thermal diffusivity of metallic pBDD is approximately 2–3 times higher than that of a metal; hence the entire electrode should rapidly attain a relatively uniform temperature, even when heated from the rear. Being mechanically robust, pBDD electrodes can be made very thin, and hence have a very low thermal mass. Thus high electrode temperatures can be achieved with modest input powers. Diamond oxidises in air only if heated over *ca.* 700 °C.[28] Previous studies investigating laser damage to diamond have shown the laser damage threshold for intrinsic polished polycrystalline diamond to be typically in excess of 1.5 GW cm^{-2} for nanosecond (ns) pulses.[29]

In this discussion paper, given the notably favourable properties of conducting diamond, we examine the use of pBDD as an electrode material for thermo-electrochemical studies. In particular using pulsed laser heating methodology we examine the redox behaviour of two different fast electron transfer outer sphere redox species at a macrodisk pBDD electrode, in widely different potential ranges, under heated conditions.

2 Experimental

Solutions

All solutions were prepared using Milli-Q water (resistivity 18.2 MΩ cm at 25 °C) and all reagents were used as received. The supporting electrolyte employed was 0.1 M potassium nitrate (KNO_3, Sigma-Aldrich, 99%) unless otherwise stated. Redox solutions contained either 1 mM ruthenium (III) hexaamine chloride ($Ru(NH_3)_6^{3+}$, Acros Organics, 98%) or 1 mM potassium hexachloroiridate (III) ($IrCl_6^{3-}$, Sigma-Aldrich, 98%). For open circuit potential measurements, potassium ferricyanide (III) ($Fe(CN_6)^{3-}$, Sigma-Aldrich, 99%), potassium hexacyanoferrate (II) trihydrate ($Fe(CN_6)^{4-}$, Sigma-Aldrich, ≥98.5%), ruthenium (II) hexaamine chloride ($Ru(NH_3)_6^{2+}$, Acros Organics, >99%) and potassium hexachloroiridate (IV) ($IrCl_6^{2-}$, Sigma Aldrich, 98%) solutions were also employed.

Pulsed laser heating set-up

A 9 mm diameter disk of freestanding conducting pBDD synthesised by MW-CVD (boron dopant level $\sim 3 \times 10^{20}$ B atoms cm^{-3}), was grown by Element Six (Harwell, Oxford). The pBDD was removed from the growth substrate and polished (lapped) on both sides to give a surface finish of 1.25 ± 0.05 nm (measured over typical areas of 20 μm × 20 μm). The resulting thickness of the electrode was 200 μm. The electrode was ohmically back contacted by sputtering (Edwards E606 sputterer) Ti (20 nm)/Au (400 nm) and annealing in a tube furnace for 5 hr at 450 °C. In order to perform electrochemical measurements during pulsed heating, a custom-built electrochemical cell was designed and fabricated in-house. The cell comprised two parts, a Teflon® container and a Perspex® window, held together with nylon screws. The Perspex® window contained a hole in the centre, across which the black pBDD disk was mounted (inside face) and held in place with adhesive Kapton® tape (100 μm thickness, stable up to 1000 °C, R.S. Components Ltd.). The electrode area exposed to solution was reduced, also using Kapton® tape, which contained a laser micromachined circular hole, 1 mm in diameter. A conductive track was made from the annealed Ti/Au contact on the back face of the pBDD electrode (which is not in contact with solution) using silver conductive paint (R.S. Components Ltd.) to the top of the Perspex® window, where electrical contact was made with a crocodile clip. The experimental set-up is shown in Fig. 1.

For the majority of experiments, cyclic voltammetry (CV) was performed simultaneously with pulsed laser heating *i.e.* temperature pulse voltammetry (TPV). All electrochemical experiments utilised a three electrode setup controlled by a potentiostat; (CompactStat, Ivium Technologies, The Netherlands) connected to a laptop computer. A thick coiled Pt wire was used as the counter electrode and a commercial saturated calomel electrode (SCE: CHI150, CH Instruments Inc. TX) served as the reference electrode. A diode laser (LM-D0296, LIMO) 914.7 nm, 30 W, was used for all pulsed temperature experiments, controlled by a laser diode

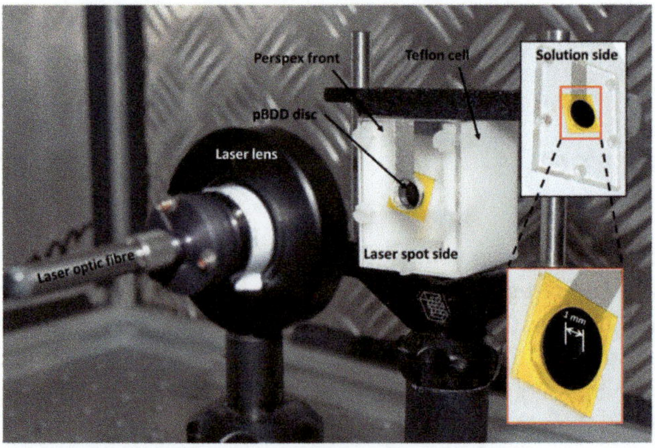

Fig. 1 Photograph of the pulsed laser heating experimental set-up. The laser beam is focused by the laser lens onto the rear of the pBDD electrode held in the Perspex® cell. Insets show the solution facing side of the electrochemical cell, detailing the black pBDD held in place with Kapton® tape which also defines the active electrode area (1 mm disc).

controller (LDC1000, Laser Electronics Ltd Lincolnshire, England). The delivery lens focused the laser to a 1 mm diameter central circular spot on the back face of the BDD electrode. A dual-channel function generator (AFG3022B, Tektronix) was used as a master controller for TPV experiments to ensure that the potentiostat and laser were synchronised when manually triggered. For the majority of experiments the CV potential scan rate was fixed at 100 mV s^{-1} (0.25 mV step potential). P_d was varied between 0.3 kW cm^{-2} and 3.8 kW cm^{-2} in conjunction with a laser duty cycle (ratio of laser on to off time) of 10 ms on (smallest pulse width that could be effectively employed) and 90 ms off.

Open Circuit Potential (OCP) measurements

OCP measurements were carried out using an isothermal approach to determine the temperature coefficient, $\beta = (\delta OCP/\delta T)$, for the different redox couples employed. A heated water bath attached to a jacketed glass cell was used to control the temperature of the electrochemical system. For OCP measurements three separate solutions were employed containing either (1) 0.5 mM $Fe(CN)_6^{4-}$ and 0.5 mM $Fe(CN)_6^{3-}$; (2) 0.5 mM $Ru(NH_3)_6^{3+}$ and 0.5 mM $Ru(NH_3)_6^{2+}$ or (3) 0.5 mM $IrCl_6^{3-}$ and 0.5 mM $IrCl_6^{2-}$. In all cases, one of the solutions was held at an elevated temperature using the jacketed cell, in the range 23 °C–63 °C, whilst the other was maintained at ambient temperature (23 °C). The two solutions were connected with a saturated KNO_3 salt bridge. Two conducting BDD, glass-sealed 1 mm diameter macrodisk electrodes, were constructed as described previously,[30] placed in the separate solutions and connected to a high impedance digital voltmeter. OCPs were recorded as a function of the temperature difference, ΔT, between the two solutions.

Atomic Force Microscopy (AFM)

AFM measurements were recorded in Tapping Mode™ with an Enviroscope equipped with Nanoscope IV controller (Bruker, UK).

COMSOL simulations

A model for simulating the induced temperature profile upon application of the laser pulse as well as the effect of the fluctuating temperature on convection, diffusion and thermodynamics was carried out using COMSOL Multiphysics 4.3b (COMSOL, Sweden) finite element modelling (FEM) software. A *2D* geometry, shown in Fig. 2 (not to scale) was built to describe the *2D* axisymmetric electrochemical cell. The model consists of the pBDD electrode, thickness 200 μm and 4.5 mm radius (domain a), mounted in Perspex®, (domain d), insulated with Kapton® tape, thickness 100 μm, (domain b), which also defines the radius of the electrode (0.5 mm) in contact with solution (domain c). The radius and height of the modelled system is 12.5 mm. Each domain is occupied by material with a defined density (ρ), heat capacity (C_p), thermal conductivity (κ) and dynamic viscosity (μ), as detailed in Table 1. The pressure constraint point (P_{con}), which is necessary for the solved equations to have a unique solution, in Fig. 2 was set to 1 atm. The location selected for this constraint was such that the solution was not significantly perturbed. The boundary conditions are summarised in Table 2. In Table 2, P (W) and A (cm^2) represent the total boundary power and the active area and $\boldsymbol{n}$ is the unit normal vector. A no heat flux boundary condition is applied to

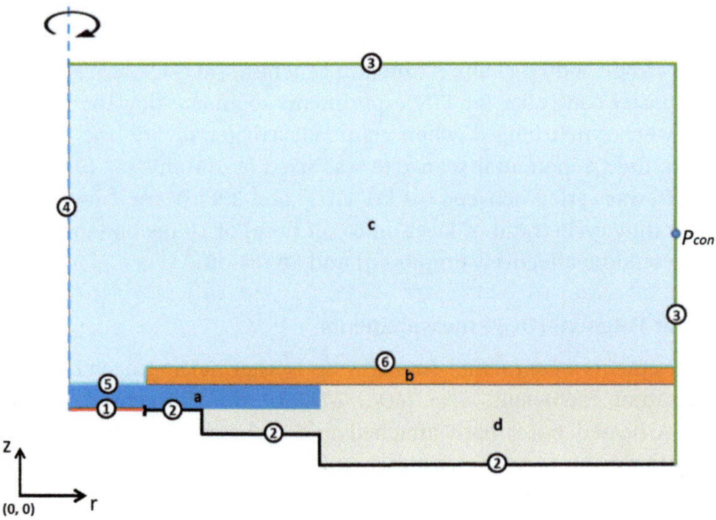

Fig. 2 Schematic of the simulated geometry (not to scale) of the electrochemical cell consisting of pBDD (a), Kapton® (b), water (c) and Perspex® (d).

Table 1 Summary of the properties for the different domains used for the simulation of temperature at the diamond electrode/electrolyte interface

	Diamond[a]	Kapton®[b]	Perspex®[c]	Water[d]
Heat Capacity C_p (J kg^{-1} K^{-1})	$C_p = 3.2T - 473.6$ (275 K < T < 500 K)	1090	1500	$12010 - 80.4 \times T + 0.310 \times T^2 - 5.38 \times 10^{-4} \times T^3 + 3.63 \times 10^{-7} \times T^4$ (273.15 K < T < 553.75K)
Density ρ (kg m^{-3})	3500	1420	1160	$838 + 1.40 \times T - 3.01 \times 10^{-3} \times T^2 + 3.72 \times 10^{-7} \times T^3$ (273.15 K < T < 553.75K)
Thermal Conductivity κ (W m^{-1} K^{-1})	$-1.25T + 1041.25$ (275 K < T < 500 K)	0.12	50	$-0.869+8.95\times10^{-3}\times T-1.58\times10^{-5}\times T^2+7.98\times10^{-9}\times T^3$ (273.15 K<T<553.75K)
Dynamic Viscosity μ (Pa s)	N/A	N/A	N/A	$1.38-2.12 \times 10^{-3} \times T + 1.36 \times 10^{-4} \times T^2 - 4.65 \times 10^{-7} \times T^3 + 8.90 \times 10^{-10} \times T^4 + 9.08 \times 10^{-13} \times T^5 + 3.85 \times 10^{-16} \times T^6$ (273.15 K < T < 413.15K)

[a] Manufacturer's data sheet (Element 6 Ltd.) [b] Manufacturer's data sheet (R. S. Components Ltd.) [c] Manufacturer's data sheet (Gilbert Curry Ltd.) [d] Expression taken from COMSOL 4.3b software.

boundary 2 (thermal insulator). No slip conditions were applied for boundaries 5 and 6. Gravity has an effect on the flow pattern of the fluid and thus a volume force (f_z) acting only in the z-dimension defined as

$$f_z = -\rho g \tag{1}$$

Table 2 Summary of boundary conditions used for the simulation of temperature at the diamond electrode/electrolyte interface

Boundary	Boundary type	Equation
1	Boundary heat source	$-n\cdot(-\kappa\nabla T) = P_d \; P_d = \frac{P}{A}$
2	Thermal insulation	$-n\cdot(-\kappa\nabla T) = 0$
3	Ambient temperature	293.15 °C
4	Axis of symmetry	$\nabla T\cdot n = 0$
5, 6	No slip fluid boundary	$\boldsymbol{v} = 0$
5	Butler Volmer kinetics	$k_f = k^0 e^{\frac{\alpha nF(E-(E^0+\beta\Delta T))}{RT}}$ $k_b = k^0 e^{\frac{(1-\alpha)nF(E-(E^0+\beta\Delta T))}{RT}}$ $j_{elec} = k_f C_{ox,elec} - k_b C_{red,elec}$

where g is the gravitational constant (~9.81 m s^{-2}), was employed.

Heat transfer involves various mechanisms, such as thermal diffusion, thermal convection, thermal radiation, *etc.* These mechanisms often occur simultaneously in the system. Thermal diffusion of heat from a region of high temperature to low temperature is described by eqn 2, for both solids and liquids, where α_D, the thermal diffusivity (m^2 s^{-1}) is given by eqn 3,

$$\frac{\partial T}{\partial t} = \alpha_D \nabla^2 T \tag{2}$$

$$\alpha_D = \frac{\kappa}{\rho C_p} \tag{3}$$

where t (s) represents time, and ∇^2 is the Laplace operator.

Convective flow of water in the system is treated as an incompressible flow of a Newtonian fluid which follows the Navier–Stokes equation, eqn 4,

$$\rho\left(\frac{\partial \boldsymbol{v}}{\partial t} + \boldsymbol{v}\cdot\nabla\boldsymbol{v}\right) = -\nabla p + \mu\nabla^2\boldsymbol{v} + f_z \tag{4}$$

where $\boldsymbol{v}$ (m s^{-1}) is the flow velocity, p (pa) is the pressure and ∇ is the grad operator.

A boundary heat source (boundary 1), reflecting P_d values in the range 0.3 kW cm^{-2} to 3.8 kW cm^{-2}, was employed to model the laser spot on the back of the diamond electrode. At $t = 0$ s, a laser pulse of 10 ms is applied to the centre of the electrode (1 mm in diameter), followed by 90 ms cooling time. The modelled temperature at the electrode–electrolyte interface is the average temperature across the 1 mm diameter electrode surface.

Simulation of the effect of laser heating on TPV was performed by considering the effect of the temperature pulses and corresponding transient perturbations in T on the mass transport properties, kinetics and thermodynamics of the system, as described by eqn 5–11. Transport of dilute species in solution depends on convection, migration and diffusion. Migration was ignored due to the presence of excess supporting electrolyte (0.1 M KNO_3). In the fluid phase, the contribution of diffusion and convection is given by:

$$D_i \nabla C_i + \boldsymbol{v} \nabla C_i = 0 \quad (5)$$

where D_i and C_i are the diffusion coefficient and concentration of the species respectively. The effect of changing T on D_i was calculated using the Stokes–Einstein equation:[31]

$$D_i = \frac{k_B T}{6\pi\mu r} \quad (6)$$

where k_B is the Boltzmann constant and r is the radius of the molecule, assumed to be temperature independent. r for $Ru(NH_3)_6{}^{3+}$ was determined as 0.28 nm using eqn 6, and a literature value[32] for D at 298 K of 8.8×10^{-6} cm^2 s^{-1}. The T induced change in μ is accounted for in the water material characteristics, as described in Table 1.

Simple outer sphere electron transfer was modelled at the electrode–solution interface (boundary 5) using Butler–Volmer (BV) kinetics. The current at the electrode, i_{elec}, is determined by:

$$i_{elec} = nAFj_{elec} \quad (7)$$

where n is the number of electrons transferred per redox species, A is the area of the electrode, F is Faraday's constant and j_{elec} is the flux of species at the electrode surface, defined by:

$$j_{ele\ c} = k_f C_{ox,elec} - k_b C_{red,elec} \quad (8)$$

where k_f and k_b are the reaction rates of the forward and back reactions respectively, described by:

$$k_f = k^0 e^{\frac{\alpha nF\left(E-\left(E^0+\beta\Delta T\right)\right)}{RT}} \quad (9)$$

$$k_b = k^0 e^{\frac{(1-\alpha)nF\left(E-\left(E^0+\beta\Delta T\right)\right)}{RT}} \quad (10)$$

where α is the charge transfer coefficient, here assumed to be 0.5, E is the electrode potential, E^0 is the standard potential (−0.18 V *vs.* SCE), R is the gas constant and ΔT is the change in temperature at the electrode/electrolyte interface. The standard rate constant (k^0) is temperature dependent, described by the Arrhenius equation,

$$k^0 = Ap\, e^{\frac{-E_a}{RT}} \quad (11)$$

where E_a is the activation energy and A_p is the pre-exponential factor, which is considered to be temperature independent. For $Ru(NH_3)_6{}^{3+/2+}$ A_p is calculated as 1026 cm s^{-1} assuming a literature value of 22.5 kJ mol^{-1} (ref. 33) for E_a and $k^0 =$ 0.1 cm s^{-1} (ref. 34) at 293.15 K (eqn 11).

3 Results and discussion

For all TPV experiments the basic premise is imposition of a pulsed laser to heat the surface of the electrode during CV. In CV, for a chosen scan rate and potential step, the number of data points are defined by scan rate/potential step and the time between potential steps is given by (number of data points)$^{-1}$. For the CV experiments described herein (using commercial software) the current is sampled immediately after each potential step. In TPV, the temperature pulse is superimposed on top of the CV and the current recorded, as shown schematically in Fig. 3. For the majority of experiments the laser is switched on and the electrode heated for the shortest time possible, 10 ms, and then switched off for 90 ms, with the cycle continuously repeated. For typical experimental parameters of 100 mV s^{-1} and 0.25 mV potential step, current values are collected every 2.5 ms, with a total of 400 data points per second.

Verification of the time-dependant average temperature at the front face of the BDD electrode during the laser pulse experiments was made from both measurement of the OCP (Fig. 4a) and thermal modelling (Fig. 4b), described in Section 2. β was calculated as -1.56 mV K^{-1} for $Fe(CN)_6^{3-}/Fe(CN)_6^{4-}$, from a plot of OCP *versus* ΔT ($R^2 = 0.98$) over the ΔT range of 40 °C. This value is in very good agreement with that determined previously, -1.6 mV K^{-1}.[35] With knowledge of β and the measured OCP, during laser heating, the corresponding interfacial electrode ΔT, *versus* time, t, behaviour can be determined; Fig. 4a shows the OCP and ΔT data for a P_d of 0.6 kW cm^{-2}, 1.2 kW cm^{-2} and 3.2 kW cm^{-2} over a period of 90 ms (the laser is on for the first 10 ms). The data shows that during the laser

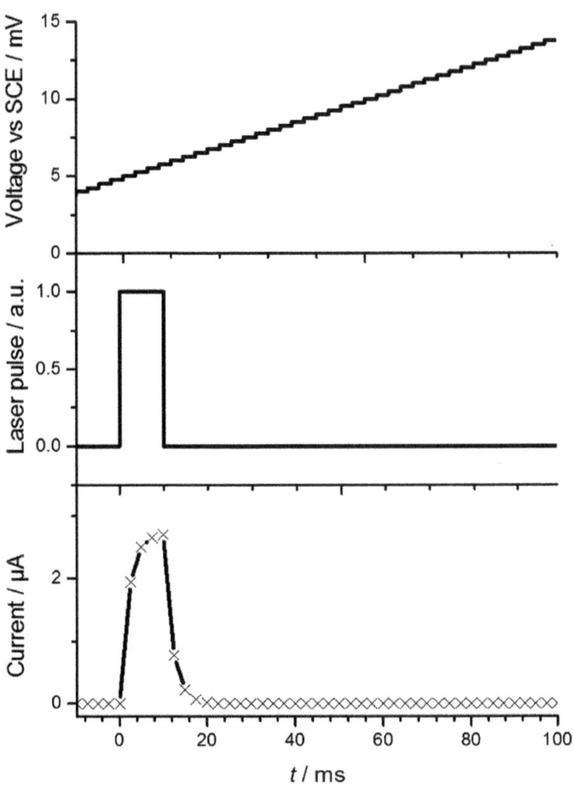

Fig. 3 Schematic graphs showing the time dependence of (a) voltage steps taken by the digital potentiostat, (b) applied laser pulses (10 ms on and 90 ms off) and (c) change in current measured as a result of laser heating.

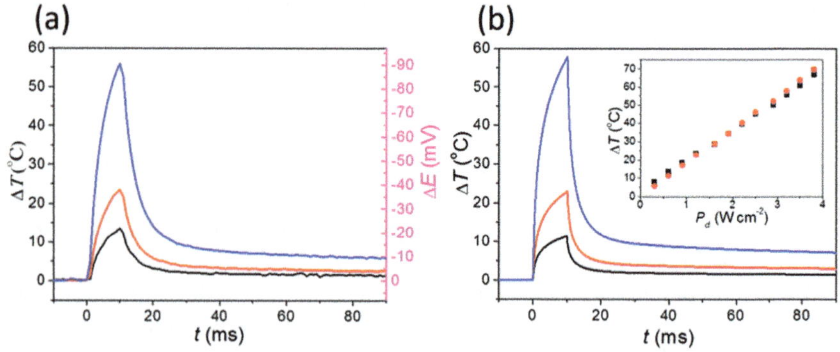

Fig. 4 (a) OCP and associated temperature rise (ΔT) *versus* time for a 90 ms time period, recorded between a laser heated 1 mm diameter pBDD disc electrode and a 1 mm diameter pBDD disc electrode at ambient temperature (23 °C). Solution contained 0.5 mM $Fe(CN)_6^{4-}$ and 0.5 mM $Fe(CN)_6^{3-}$ in 0.1 M KNO_3. The laser was pulsed at 0.6 kW cm^{-2} (black), 1.9 kW cm^{-2} (red) and 3.2 kW cm^{-2} (blue) for a 10 ms period and then switched off. The OCP data is converted to ΔT using the $Fe(CN)_6^{4-}/Fe(CN)_6^{3-}$ temperature coefficient $= -1.62$ mV K^{-1}. (b) Finite element modelled ΔT (at the electrode surface) of the laser heated pBDD electrode at $P_d = 0.6$ kW cm^{-2} (black), 1.9 kW cm^{-2} (red) and 3.2 kW cm^{-2} (blue). Insert shows maximum ΔT as a function of P_d for both experimental (black ■) and simulation (red ●).

pulse period, the temperature rises rapidly at the electrode surface. After the laser is turned off, the temperature decreases at a slower rate due to the lower value of thermal diffusivity for water, hindering heat transfer to bulk solution. For different P_d values in the range 0.3 kW cm^{-2} to 3.8 kW cm^{-2}, maximum T values at t_{10ms} at the electrode/electrolyte interface in the range 31.0 to 90.0 °C were obtained. Specifically, 31.0 °C (0.3 kW cm^{-2}); 36.5 °C (0.6 kW cm^{-2}); 41.5 °C (0.9 kW cm^{-2}); 46.4 °C (1.2 kW cm^{-2}); 51.6 °C (1.6 kW cm^{-2}); 57.3 °C (1.9 kW cm^{-2}); 62.9 °C (2.2 kW cm^{-2}); 68.4 °C (2.5 kW cm^{-2}); 73.1 °C (2.9 kW cm^{-2}); 78.9 °C (3.2 kW cm^{-2}); 84.1 °C (3.5 kW cm^{-2}) and 90.0 °C (3.8 kW cm^{-2}). Fig. 4b shows the ΔT profile determined from FEM simulations for a P_d of 0.6 kW cm^{-2}, 1.2 kW cm^{-2} and 3.2 kW cm^{-2} over 90 ms (the laser is on for the first 10 ms). Also given in the inset is a comparison of the maximum recorded ΔT for the two methodologies; OCP (■) and FEM modelling (●). As can be seen there is very good agreement between both methods of temperature determination. The employment of a higher P_d would allow even greater ΔT values to be reached.

To visualise propagation of the thermal gradient through the pBDD electrode and into solution during the laser pulse as a function of time, thermal modelling was performed. Fig. 5a shows the 2D temperature profile at $t = 10$ ms (*i.e.* at the end of the laser pulse) and Fig. 5b the z axial temperature profile (about the central axis of symmetry, $r = 0$) as a function of time, $t = 0$, 5, 10 and 20 ms for a P_d of (i) 1.2 kW cm^{-2} and (ii) 3.2 kW cm^{-2}, respectively. As Fig. 5a shows, for both P_d values, heat diffuses out from the laser heated zone hemispherically, diffusing much faster through the 200 μm thick pBDD than the surrounding solution, due to the higher thermal diffusivity of the conducting diamond. At both $t = 5$ ms and $t = 10$ ms (the end of the laser pulse), there is a small gradient in temperature through the BDD electrode which increases in absolute value as both P_d and t

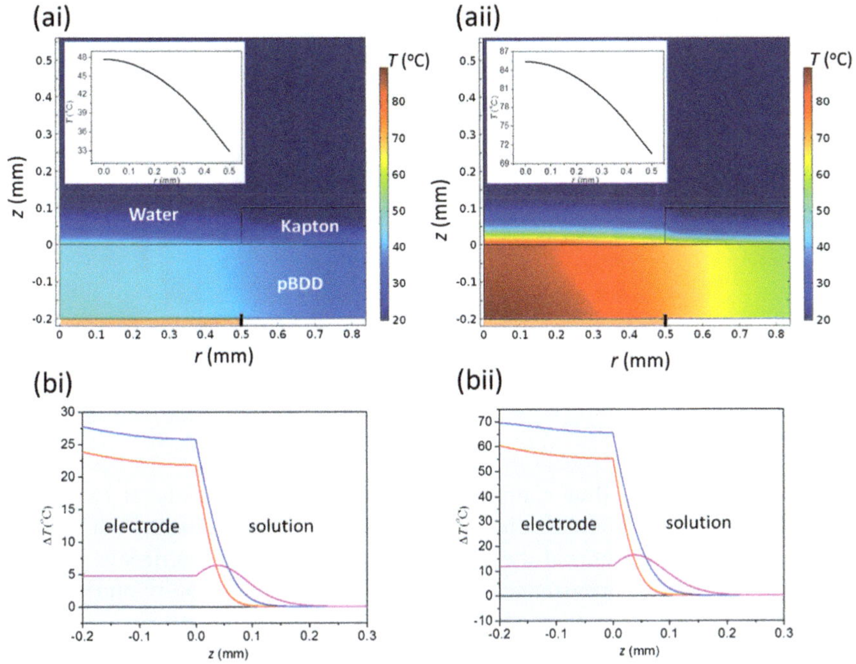

Fig. 5 (a) 2D temperature profile at $t = 10$ ms and (b) z axial temperature profile through pBDD to solution at $r = 0$ mm for times of 0 ms (black), 5 ms (red), 10 ms (blue) and 20 ms (purple) at P_d of 1.2 kW cm^{-2} (i) and 3.2 kW cm^{-2} (ii). The insets to (a) show radial temperature profiles at $z = 0$ mm (electrode surface) at $t = 10$ ms. The laser heated boundary in (a) is marked by an orange band and occurs at $z = -0.2$ and $0 < r < 0.5$.

increases. The thermal gradient extends *ca.* 200 μm (1.2 kW cm^{-2}) and 230 μm (3.2 kW cm^{-2}) from the electrode surface into solution. At $t = 20$ ms, after the laser has been switched off for 10 ms, ΔT has decreased significantly in the electrode (with a uniform T obtained vertically through the pBDD). The water, however, as it retains heat more effectively, due to a higher C_p, has a higher T than that of the diamond (at a defined distance from the electrode) as the pBDD is able to cool itself more rapidly.

With optical heating it is sensible to consider the laser damage threshold of the electrode material employed, especially as metallic pBDD absorbs very strongly at the wavelength (914 nm) of the diode laser employed herein. Previous studies indicate a laser damage threshold for intrinsic polycrystalline diamond typically to be in excess of 1.5 GW cm^{-2} for ns pulses,[29] which is far higher than the conditions employed here. This is verified in Fig. 6 which shows typical AFM images recorded in tapping mode at a scan rate of 20 μm s^{-1} before (a) and after (b) applying a P_d of 1.2 kW cm^{-2} (10 ms on, 90 ms off) for 120 cycles, typical to conditions employed herein. It can be seen from Fig. 6a that the typical surface roughness of the pBDD electrode is 1.25 ± 0.05 nm before laser heating, compared to 1.3 ± 0.1 nm after. The AFM images in Fig. 6 thus indicate no change in surface morphology at these modest fluxes. Note, this is in stark contrast to Pt electrodes which showed surface roughening with the electrode exposed to a P_d three orders of magnitude lower, for a similar timescale of exposure, but with ns

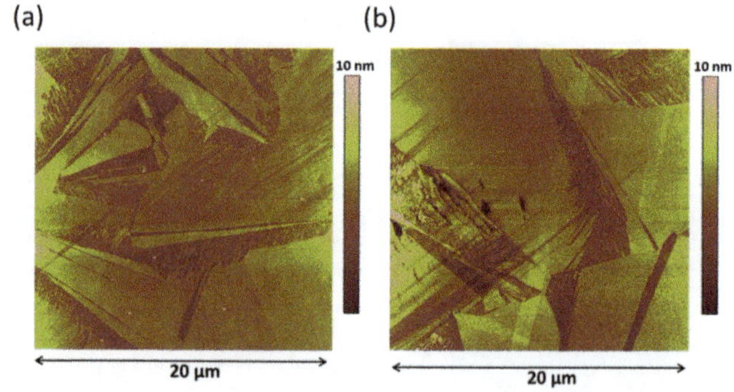

Fig. 6 Typical tapping mode AFM images recorded at a scan rate of 20 μm s^{-1} before (a) and after (b) applying P_d of 1.2 kW cm^{-2} (10 ms on, 90 ms off) for 120 cycles.

laser pulses.[11] We are thus confident that much shorter and higher P_d pulses could be used in conjunction with BDD electrodes in order to establish significantly higher temperatures at the pBDD electrode/electrolyte interface.

Initial experiments investigated the effect of pulsed temperature on the non-faradaic response of the electrode as a function of P_d (0, 0.6 kW cm^{-2}, 1.2 kW cm^{-2} and 3.2 kW cm^{-2}). Fig. 7 shows the current–time response in 0.1 M KNO_3 with the electrode held at a constant potential of 0 V *vs.* SCE for a P_d of 0.6 kW cm^{-2}, 1.2 kW cm^{-2} and 3.2 kW cm^{-2}, with an increased data sampling frequency of 1 kHz. The laser was pulsed on for 10 ms and off for 90 ms. Also shown is the current–time response (black line) under ambient conditions. It can be seen in Fig. 7 that higher P_d and hence larger temperature changes produce bigger

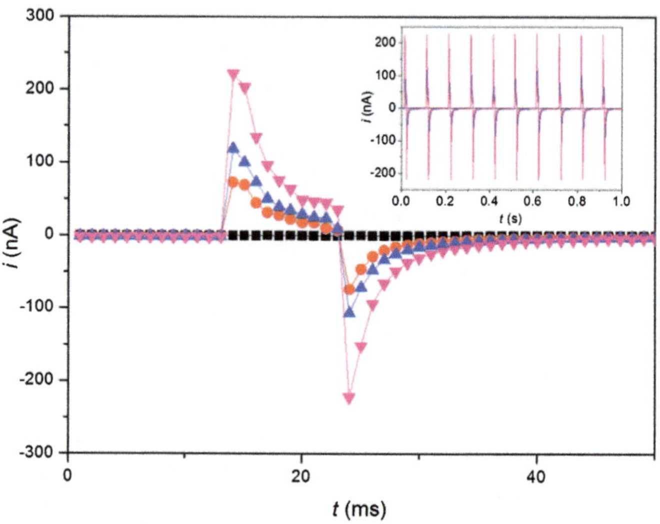

Fig. 7 Temperature pulsed chronoamperometry of pBDD in 0.1 M KNO_3 at P_d of 0 W cm^{-2} (black ■), 0.6 kW cm^{-2} (red ●), 1.2 kW cm^{-2} (blue ▲) and 3.2 kW cm^{-2} (purple ▼) at an applied potential of 0 V *vs* SCE. Insert shows time-dependent current at the potential of 0 V *vs.* SCE for a duration of 1 s.

current transients. The inset shows a zoomed out current–time trace recorded over 1 s. Each trace has a peak, which is associated with the laser pulse turning on and a reverse peak associated with the laser pulse turning off. The largest change in both cases occurs during the first ms and then the current transient decays.

Previous studies of the double layer using ns laser pulse techniques have shown that thermal heating leads to a virtually instantaneous restructuring of the double layer. The time constant of the data capture system employed herein is insufficient to probe such fast events, but still shows an effect. In particular, the current rise is associated with perturbation of the double layer resulting in a charging current, which decays with time in a similar fashion to a potential step experiment carried out under isothermal conditions. When the laser is turned off the temperature at the interface decreases to its original value, and due to restoration of the double layer, a current transient in the opposite direction is observed. The charge associated with the positive and negative current transients is approximately equal. For a P_d of 0.6 kW cm^{-2}, 1.2 kW cm^{-2} and 3.2 kW cm^{-2}, the charge transferred for both positive and negative transients is 0.31 nC and 0.30 nC, 0.49 nC and 0.48 nC, 0.95 nC and 0.91 nC, respectively.

Experiments investigating the effect of TPV on faradic processes were conducted by using a fast (reversible) one electron transfer outer sphere redox mediator, 1 mM $Ru(NH_3)_6^{3+}$ in 0.1 M KNO_3. The experimental TPV data is shown in Fig. 8a and Fig. 8b for the forward scan and backward scan respectively, under temperature pulsed (P_d of 1.2 kW cm^{-2}, pulse on 10 ms, off 90 ms) conditions (red line). The ambient CV (black line) for $Ru(NH_3)_6^{3+}$ reduction and subsequent oxidation shows a peak-to-peak separation of 65 mV and a reductive peak current of 2.18 μA in accordance with that predicted using the Randles–Sevick equation, assuming an electrode diameter of 1 mm and diffusion coefficient of 8.8×10^{-6} cm^2 s^{-1}.[32]

During TPV current spikes are obtained during laser heating, which decay during cooling, in a similar way to chronoamperometric transients produced during potential pulsing. A scan rate of 100 mV s^{-1} with a potential step of 0.25 mV is employed (sampling rate = 400 s^{-1}), resulting in 4 data points per laser pulse, where the last point records the largest current reading. Raw data is collected by the potentiostat (Fig. 8a and b, red line) and then sampled by taking every 40th point, resulting in the last sampling point during the laser pulse being plotted against potential. Fig. 8c shows "data sampled" TPV responses at P_d of 1.2 kW cm^{-2} (red line) and 2.5 kW cm^{-2} (green line).

Three striking features are observed in the "data sampled" laser pulsed TPV, compared to the CV recorded under ambient conditions; (1) the cathodic current for the forward scan increases as P_d increases. Specifically, the peak current has increased compared to that under ambient conditions by ×3.5 for 1.2 kW cm^{-2} and ×8.1 for 2.5 kW cm^{-2}. (2) The potential for the forward peak current shifts positively with increasing P_d *i.e.* −0.168 V for 1.2 kW cm^{-2} and −0.148 V for 2.5 kW cm^{-2} and; (3) a cathodic current is observed in magnitude for the backward scan, which increases in magnitude as the P_d is increased. β for the $Ru(NH_3)_6^{3+/2+}$ system was calculated as +0.68 mV K^{-1} from a plot of OCP *versus* ΔT ($R^2 = 0.99$) over a ΔT range of 40 °C, in close agreement with literature values.[11] From knowledge of β it is possible to calculate the change in entropy of the system, ΔS, through

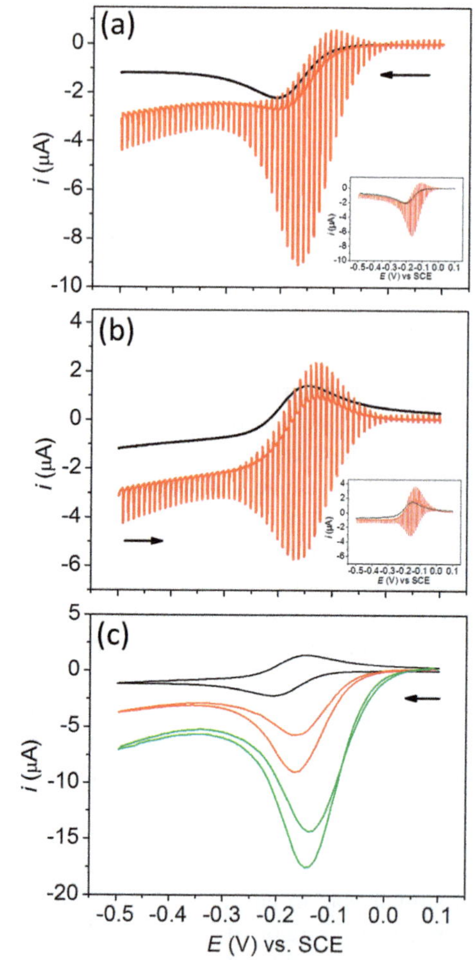

Fig. 8 TPV (red) at P_d of 1.2 kW cm^{-2} and ambient CV (black) recorded at a pBDD electrode in a solution containing 1 mM $Ru(NH_3)_6^{3+}$ in 0.1 M KNO_3 (a) forward scan and (b) backward scan. Inserts in (a) and (b) show both the simulated TPV (red) at P_d of 1.2 kW cm^{-2} and ambient CV. (c) "Data sampled" experimental TPVs at a P_d of 1.2 kW cm^{-2} (green), and 2.5 kW cm^{-2} (red), every 40th data point sampled. The CV recorded under ambient conditions (black) is not data sampled. Scan rate: 100 mV s^{-1}.

$$\Delta S = nF\beta = 65.6 \text{ J K}^{-1} \text{ mol}^{-1} \tag{12}$$

given $n = 1$.

The increase in cathodic (negative) current in the forward scan and observation of a cathodic current in the backward scan is due primarily to a combination of two factors, the entropy of the redox reaction and a change in mass transport due to the induced thermal gradient at the electrode/electrolyte interface. In these experiments, the temperature at the electrode/electrolyte interface is pulsed from ambient to hot repeatedly. Entropically, under steady-state conditions, a positive β indicates an additional cathodic current contribution to the ambient response (for both redox processes), near to the equilibrium potential.[36] Furthermore, the transient nature of the temperature increase also results in time-dependant increases in mass transport, due to both diffusion and convective effects, which serves to enhance the current relative to ambient conditions. By modelling all the contributions described in Section 2, qualitatively similar TPVs can be obtained, as shown in the inserts to Fig. 8a and b. The small discrepancy between model and experiment is likely to be due to employment of a 2D axisymmetric model. This model does not account for the fact the laser strikes the electrode perpendicular to the direction of gravity, instead a computationally less time consuming parallel arrangement was assumed, which is likely to underestimate convective water flow effects. More sophisticated 3D simulations are currently in progress.

In order to further explore the effect of TPV on redox electrochemistry, another fast electron transfer (reversible) outer sphere redox mediator, $IrCl_6^{3-}$, was employed. Isothermal experiments gave a β value of -0.48 mV K^{-1} ($R^2 = 0.99$) for $IrCl_6^{3-/2-}$ ($\Delta S = -46.3$ J K^{-1} mol^{-1}) recorded over a ΔT range of 40 °C. Fig. 9 shows a "data sampled" TPV (data collected as described for Fig. 8c) for a pBDD electrode in a solution containing 1 mM $IrCl_6^{3-}$ in 0.1 M KNO_3, at a scan rate of 100 mV s^{-1} for P_d of 0.6 kW cm^{-2}, 1.2 kW cm^{-2}, 1.9 kW cm^{-2} and 2.5 kW cm^{-2}. Under ambient conditions the CV for $IrCl_6^{3-}$ oxidation and subsequent reduction shows a peak-to-peak separation of 70 mV and an oxidative peak current of 1.96 μA in agreement with Randles–Sevick predictions, assuming an electrode diameter of 1 mm and diffusion coefficient of 7.5×10^{-6} cm^2 s^{-1}.[37] It can be seen in Fig. 9 that (i) the anodic peak current for the forward scan increases as P_d increases; (ii) the potential for the forward peak current shifts negatively with increasing P_d, as expected for a negative β value (0.731 V for 0.6 kW cm^{-2}; 0.711 V for 1.2 kW cm^{-2}; 0.691 V for 1.9 kW cm^{-2}; and 0.681 V for 2.5 kW cm^{-2}) and; (iii) as P_d increases the "data sampled" back scan TPV shows a very small cathodic current at 0.6 kW cm^{-2}, which disappears at higher P_d, resulting in the

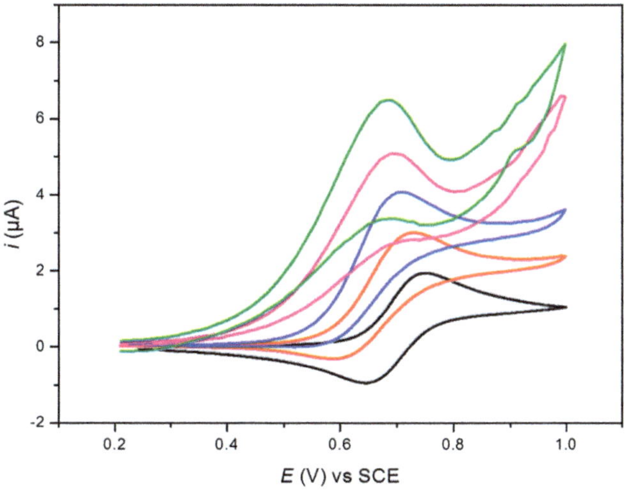

Fig. 9 "Data sampled" TPV using a pBDD electrode in a solution containing 1 mM $IrCl_6^{3-}$ in 0.1 M KNO_3, at P_d of 0.6 kW cm^{-2} (red), 1.2 kW cm^{-2} (blue), 1.9 kW cm^{-2} (purple) and 2.5 kW cm^{-2} (green). Every 40th data point is sampled. The CV recorded under ambient conditions (black) is not data sampled or subject to a laser pulse. Scan rate: 100 mV s^{-1}.

observation of double peaked anodic currents at the two highest P_d values. The data shown is thus in qualitative agreement with that seen for the $Ru(NH_3)_6^{3+}$ TPV system (Fig. 8).

4 Conclusions

An electrochemical system combined with a pulsed laser setup is employed to examine the effect of temperature on the electrochemical response of pBDD electrodes. Conducting diamond is an ideal material for thermoelectrochemical studies due to its wide potential window, low background currents, very high thermal conductivity and thermal diffusivity, mechanical robustness and ability to withstand P_d up to GW cm^{-2} without damage, for ns pulses. In these measurements, laser pulses of 10 ms duration (90 ms off) were delivered using a focused diode laser (maximum P_d 3.8 kW cm^{-2}) onto a ~ 1 mm spot on the rear side of a 200 μm thick pBDD electrode. The temperature attained at the pBDD electrode/electrolyte interface during this time period was assessed using both OCP measurements and FEM simulations; with both methods providing good agreement. A maximum temperature of 90.0 °C was achieved at the electrode/electrolyte interface using a P_d of 3.8 kW cm^{-2} for a heating time of 10 ms. Due to the high thermal diffusivity of pBDD, heat propagates into the electrode rapidly, without damage to the electrode surface, as evidenced by AFM. As diamond can withstand significantly greatly power densities even higher temperatures can be attained quickly by moving to significantly higher power lasers.

TPV measurements in supporting electrolyte were found to perturb the double layer, resulting in a positively peak-shaped charging current when the laser was on and restoration of the double layer, accompanied with a negative charging current, when the laser was turned off. The total charge transferred during the switch on and off periods was found to be approximately equal. In the presence of the fast electron transfer (reversible) outer sphere redox mediators, $Ru(NH_3)_6^{3+}$ (β = +0.68 mV K^{-1}) and $IrCl_6^{3-}$ (β = −0.48 mV K^{-1}), TPV resulted in chronoamperometric-like current increases during the CV which enhanced the forward peak current, increasing with increasing P_d. The peak potentials were shifted positively for $Ru(NH_3)_6^{3+/2+}$ and negatively for $IrCl_6^{3-/2-}$, compared to the CV recorded under ambient conditions, in accordance with their entropic positive and negative β values, respectively. Scanning backwards cathodic ($Ru(NH_3)_6^{3+/2+}$) and anodic ($IrCl_6^{3-/2-}$) peaked current response were also observed (leading to the observation of a double peak). FEM showed striking qualitative agreement with the experimental data obtained and differences are most likely due to constraints imposed due to technical limitations of the approach used. pBDD is thus a promising electrode material for high temperature thermoelectrochemical measurements. We plan to extend these studies by using higher power lasers to further increase the surface temperature and examine the effect of temperature on industrially relevant electrochemical processes.

Acknowledgements

JI acknowledges support from the EPSRC Analytical Fund. LM and MJ would like to thank the University of Warwick Chancellors International Scholarship Scheme and Element Six respectively for financial support. We acknowledge Prof. Patrick

Unwin for helpful discussions. We all thank Mr. Jon Newland (Warwick Chemistry) for photographing the experimental setup and Element Six for providing the pBDD used in these studies.

References

1 C. G. Ubah and E. Asselin, *ECS Trans.*, 2009, **19**, 3–20.
2 J. Zhang, Z. Xie, J. Zhang, Y. Tang, C. Song, T. Navessin, Z. Shi, D. Song, H. Wang and D. P. Wilkinson, *J. Power Sources*, 2006, **160**, 872–891.
3 P. Gründler, A. Kirbs and L. Dunsch, *ChemPhysChem*, 2009, **10**, 1722–1746.
4 C. L. Forryan and R. G. Compton, *Phys. Chem. Chem. Phys.*, 2003, **5**, 4226–4230.
5 S. H. Joo, S. J. Choi, I. Oh, J. Kwak, Z. Liu, O. Terasaki and R. Ryoo, *Nature*, 2001, **412**, 169–172.
6 K. Gong, F. Du, Z. Xia, M. Durstock and L. Dai, *Science*, 2009, **323**, 760–764.
7 P. V. Rysselberghe, *Electrochemical Affinity: Studies in Electrochemical Thermodynamics and Kinetics*, Paris, 1955.
8 G. G. Wildgoose, D. Giovanelli, N. S. Lawrence and R. G. Compton, *Electroanalysis*, 2004, **16**, 421–433.
9 G. Barker and A. Gardner, *J. Electroanal. Chem.*, 1975, **65**, 95–100.
10 E. Hershenhart, R. L. McCreery and R. D. Knight, *Anal. Chem.*, 1984, **56**, 2256–2257.
11 R. P. Akkermans, M. F. Suárez, S. L. Roberts, Q. Fulian and R. G. Compton, *Electroanalysis*, 1999, **11**, 1191–1202.
12 J. L. Brennan and R. J. Forster, *J. Phys. Chem. B*, 2003, **107**, 9344–9350.
13 J. F. Smalley, C. Krishnan, M. Goldman, S. W. Feldberg and I. Ruzic, *J. Electroanal. Chem.*, 1988, **248**, 255–282.
14 J. F. Smalley, S. W. Feldberg, C. E. Chidsey, M. R. Linford, M. D. Newton and Y. P. Liu, *J. Phys. Chem.*, 1995, **99**, 13141–13149.
15 A. Yamakata, T. Uchida, J. Kubota and M. Osawa, *J. Phys. Chem. B*, 2006, **110**, 6423–6427.
16 T. Hinoue, R. Harui, T. Izumi, I. Watanabe and H. Watarai, *Anal. Sci.*, 1995, **11**, 1–8.
17 V. Climent, B. A. Coles and R. G. Compton, *J. Phys. Chem. B*, 2002, **106**, 5988–5996.
18 T. Hinoue, N. Kuwamoto and I. Watanabe, *J. Electroanal. Chem.*, 1999, **466**, 31–37.
19 M. R. Mahoney and R. P. Cooney, *J. Phys. Chem.*, 1983, **87**, 4589–4591.
20 P. Gründler, A. Kirbs and T. Zerihun, *Analyst*, 1996, **121**, 1805–1810.
21 F. Qiu, R. G. Compton, F. Marken, S. J. Wilkins, C. H. Goeting and J. S. Foord, *Anal. Chem.*, 2000, **72**, 2362–2370.
22 P. Gründler, T. Zerihun, A. Möller and A. Kirbs, *J. Electroanal. Chem.*, 1993, **360**, 309–314.
23 L. A. Hutton, J. G. Iacobini, E. Bitziou, R. B. Channon, M. E. Newton and J. V. Macpherson, *Anal. Chem.*, 2013, **85**, 7230–7240.
24 R. Balmer, J. Brandon, S. Clewes, H. Dhillon, J. Dodson, I. Friel, P. Inglis, T. Madgwick, M. Markham and T. Mollart, *J. Phys.: Condens. Matter*, 2009, **21**, 3–5.
25 J. P. Lagrange, A. Deneuville and E. Gheeraert, *Diamond Relat. Mater.*, 1998, **7**, 1390–1393.

26 J. E. Field, *The properties of natural and synthetic diamond*, Academic Press London, 1992.
27 K. Yoshida and H. Morigami, *Microelectron. Reliab.*, 2004, **44**, 303–308.
28 P. John, N. Polwart, C. Troupe and J. Wilson, *Diamond Relat. Mater.*, 2002, **11**, 861–866.
29 R. Sussmann, G. Scarsbrook, C. Wort and R. Wood, *Diamond Relat. Mater.*, 1994, **3**, 1173–1177.
30 L. Hutton, M. E. Newton, P. R. Unwin and J. V. Macpherson, *Anal. Chem.*, 2009, **81**, 1023–1032.
31 E. L. Cussler, *Diffusion: mass transfer in fluid systems*, Cambridge, 2009.
32 J. V. Macpherson, D. O'Hare, P. R. Unwin and C. P. Winlove, *Biophys. J.*, 1997, **73**, 2771–2781.
33 F. Qiu, R. G. Compton, B. A. Coles and F. Marken, *J. Electroanal. Chem.*, 2000, **492**, 150–155.
34 H. V. Patten, K. E. Meadows, L. A. Hutton, J. G. Iacobini, D. Battistel, K. McKelvey, A. W. Colburn, M. E. Newton, J. V. Macpherson and P. R. Unwin, *Angew. Chem., Int. Ed.*, 2012, **51**, 7002–7006.
35 A. Olivier, E. Merienne, J. Chopart and O. Aaboubi, *Electrochim. Acta*, 1992, **37**, 1945–1950.
36 J. Valdes and B. Miller, *J. Phys. Chem.*, 1989, **93**, 7275–7280.
37 J. V. Macpherson, C. E. Jones and P. R. Unwin, *J. Phys. Chem. B*, 1998, **102**, 9891–9897.

Faraday Discussions

PAPER

Selection, characterisation and mapping of complex electrochemical processes at individual single-walled carbon nanotubes: the case of serotonin oxidation

Aleix G. Güell,[a] Katherine E. Meadows,[ab] Petr V. Dudin,[a] Neil Ebejer,[a] Joshua C. Byers,[a] Julie V. Macpherson[a] and Patrick R. Unwin*[a]

Received 26th March 2014, Accepted 4th April 2014
DOI: 10.1039/c4fd00054d

The electrochemical (EC) oxidation of the neurotransmitter, serotonin, at individual single-walled carbon nanotubes (SWNTs) is investigated at high resolution using a novel platform that combines flow-aligned SWNTs with atomic force microscopy, Raman microscopy, electronic conductance measurements, individual SWNT electrochemistry and high-resolution scanning electrochemical cell microscopy (SECCM). SECCM has been used to visualise the EC activity along side-wall sections of metallic SWNTs to assess the extent to which side-walls promote the electrochemistry of this complex multi-step process. Uniform and high EC activity is observed that is consistent with significant reaction at the side-wall, rather than electrochemistry being driven by defects alone. By scanning forward and reverse (trace and retrace) over the same region of a SWNT, it is also possible to assess any blocking of EC activity by serotonin oxidation reaction products. At a physiologically relevant concentration (5 μM), there is no detectable blocking of SWNTs, which can be attributed, at least in part, to the high diffusion rate to an individual, isolated SWNT in the SECCM format. At higher serotonin concentration (2 mM), oligomer formation from oxidation products is much more significant and major blocking of the EC process is observed from line profiles recorded as the SECCM meniscus moves over an SWNT. The SECCM line profile morphology is shown to be highly diagnostic of whether blocking occurs during EC processes. The studies herein add to a growing body of evidence that various EC processes at SWNTs, from simple outer sphere redox reactions to complex multi-step processes, occur readily at pristine SWNTs. The platform described is of general applicability to various types of nanostructures and nanowires.

[a]*Department of Chemistry, University of Warwick, Gibbet Hill Road, Coventry, CV4 7AL, United Kingdom. E-mail: p.r.unwin@warwick.ac.uk; Tel: +44 (0) 2476 523264*
[b]*Molecular Organisation and Assembly in Cells Doctoral Training Centre, University of Warwick, Gibbet Hill Road, Coventry, CV4 7AL, United Kingdom*

Introduction

The use of carbon nanotubes (CNTs) as electroactive materials in sensors and biosensors is of increasing interest, especially as a consequence of their chemical stability, high electrical conductivity, low price, biocompatibility and wide potential window.[1] Yet, the rational integration of CNTs into electrochemical sensors has been limited, until recently, by a lack of appropriate experimental tools with which to develop and assess fundamental models for electrochemistry at CNTs. The majority of experimental studies, focused on macroscopic measurements of large ensembles of CNTs,[1,2] have claimed that, even for simple outer sphere redox couples, heterogeneous electron transfer (HET) only takes place at edge–plane defects in multi-walled nanotubes,[3] or at opened oxygenated ends of single-walled nanotubes (SWNTs).[4,5] The side-wall – which constitutes the main area of a SWNT – has been proposed to be inert towards HET.[2–5] On the other hand, microscopic measurements at individual SWNTs have suggested highly active side-walls for electrochemistry,[6–9] in agreement with the response observed at other types of sp^2 carbon electrodes such as the basal surface of highly oriented pyrolytic graphite (HOPG)[10–18] or graphene,[19] which also exhibit high activity to HET.

Measurements at individual SWNTs are evidently much less ambiguous than those on macroscopic samples which typically average over millions of different CNTs. The first attempts to isolate and study the electrochemistry of an individual SWNT involved defining an electrochemical cell over a single nanotube by lithographic means.[6] Recently, we proposed an alternative approach, using scanning electrochemical cell microscopy (SECCM),[20,21] a pipette-based technique developed in our group, to confine the electrochemical (EC) measurement to small regions of a SWNT.[8,9] The convenience of this approach for the study of individual SWNTs resides in the EC cell being defined by a liquid meniscus at the tapered end of a double-barrelled pipette used as a probe, the size of which is easily tuneable from microns to tens of nanometres. Not only does this confer high spatial resolution, but also minimises the need to manipulate or prepare the sample, hence allowing it to be studied in close to its pristine state. In addition, the scanning capability of the moveable EC cell allows maps of EC activity to be built up, facilitating the determination of structure–activity correlations.[10,19,22] Thus, SECCM has been used to study HET on 2-D networks of pristine SWNTs grown by catalysed chemical vapour deposition (cCVD) on inert Si/SiO_2 supports.[8] By tuning the dimensions of the SECCM probe, it was possible to access SWNT side-walls, ends and cross junctions and assess the electrochemistry of two different outer sphere redox mediators[8] without any influence from (or need of) a support electrode.

Very recently, we have developed a new platform, combining flow-aligned SWNTs on Si/SiO_2 substrates with SECCM,[9] allowing the EC response to be mapped, together with a complete characterisation of the SWNT,[6,9] including the electronic properties, which can be either semiconducting or conducting.[23,24] This is a key aspect which tends to be ignored in macroscopic measurements of SWNT ensembles.

The aim of this paper is to build significantly on this new approach by considering more complex EC processes, such as multi-step processes involving

HET and coupled chemical steps, as in the detection of neurotransmitters, which is the focus herein. CNTs have generally been demonstrated as a suitable carbon-based electrode material for the detection of neurotransmitters, such as serotonin[25] (which plays a key role in numerous physiological processes, such as depression[26]) and dopamine.[27,28] Very low limits of detection are possible when SWNT network ensembles are used,[27] however the extent to which such complex processes are controlled by particular features on SWNTs is not well understood. Moreover, particularly in biosensing applications, carbon electrodes may suffer from fouling by reaction intermediates or products,[25,29,30] which is a key problem to assess when considering the future applications of SWNTs as *in-vivo* or *in-vitro* sensors or in long-term implants.

In this paper, we have used the individual SWNT platform, combined with SECCM,[20,21] to investigate the oxidation of serotonin. The high mass transport achievable with this platform[8,9,31] has allowed us to study this complex multi-step reaction[32,33] and determine the activity of SWNT side-walls and the impact of fouling by side reactions. The platform described allows an individual SWNT electrode to be selected and fully characterised with complementary techniques such as Raman spectroscopy, AFM and conductance measurements, paving the way to understand how surface structure and the electronic nature of a SWNT impact EC activity.

Experimental

Sample preparation

Flow-aligned SWNTs were grown on Si/SiO_2 (IDB Technologies Ltd.; n-type Si, 525 μm thick with 300 nm of thermally grown SiO_2) by cCVD, using Fe catalyst nanoparticles.[8] Each substrate was partially immersed in an aqueous solution of horse spleen ferritin protein (Sigma Aldrich) diluted 1 : 200 from the original concentration for 20 minutes, resulting in a band of catalyst nanoparticles along one side of the substrate (Fig. 1A). The substrates were dried in a nitrogen stream

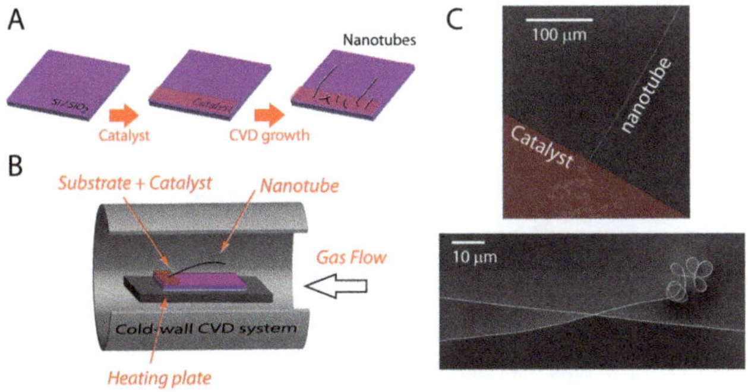

Fig. 1 (A) Si/SiO_2 wafers were used as substrates for carbon nanotube growth. The catalyst (ferritin) was deposited to only one edge of the sample, from where SWNTs were grown by cCVD over the region that is catalyst free. (B) Schematic of a cold-wall cCVD system, where a heating element heats the sample. The sample was placed so that the area with catalyst was away from the flow since the SWNTs grew against the flow. (C) FE-SEM images of flow aligned SWNTs. Straight long nanotubes (top) were synthesized, that were terminated in a tortuous shape (bottom).

and ashed in an oxygen plasma for 2 minutes at 100 W (K1050X plasma system, Emitech, UK) before being placed into a cold wall cCVD system (nanoCVD, Moorfields, UK), in which the samples were positioned directly onto the heating element, while the surrounding wall remained at a much lower temperature (<100 °C). The samples were oriented with the band of catalyst nanoparticles perpendicular to the direction of flow since nanotubes grow along the direction of gas flow,[34,35] Fig. 1B. Samples were heated to 950 °C under a flow of 150 sccm H_2 and 250 sccm Ar for 5 minutes. The growth period was initiated upon the introduction of ethanol to the system *via* a flow of 250 sccm Ar, bubbled through ethanol at 0 °C, whilst maintaining the flow of H_2 at 150 sccm. The duration of this growth phase was set between 5 and 10 minutes, depending on the desired nanotube length. After this period, the flow of Ar–ethanol and H_2 was terminated and the system was left to cool under a flow of 1000 sccm of Ar.

Metal contacts

The macroscopic electrical contact to the SWNTs was provided by a thermally evaporated (Minibox, Moorfields, UK) Pd band (~70 nm thick) with an adhesion layer of 2 nm of Cr. The area to be covered was defined by a stencil shadow mask, ensuring that the SWNTs remained pristine in the region of interest (ROI).

Silver electrodeposition

To confine the electrodeposition, a single-barrel pipette was used to define the EC cell, as described in Fig. 2A.[34] Briefly, a single channel glass pipette of approximately 200 μm tip opening was produced by laser-pulling (P-2000 Sutter Instruments) followed by polishing.[34] The pipette was filled with an aqueous solution (made up in Millipore Milli-Q water for all solutions in this study) of silver nitrate (10 mM) and supporting electrolyte (50 mM KNO_3), together with a silver wire electrode that acted as a quasi-reference counter-electrode (QRCE). The pipette was then approached to the sample by means of a micropositioner *x*-*y*-*z* stage

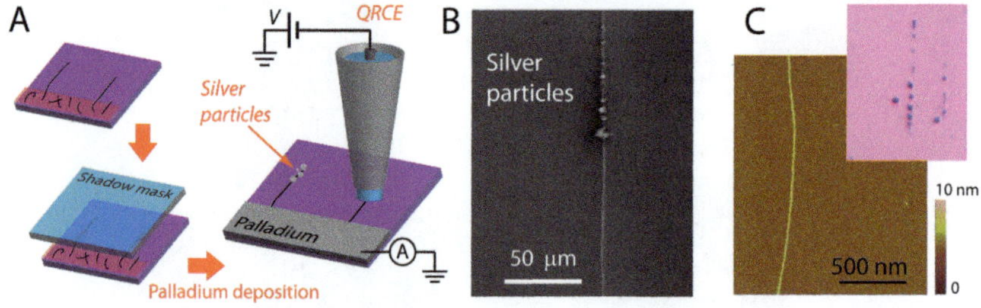

Fig. 2 (A) A shadow mask was used to select an area to be covered with a Pd thin film coincident with catalyst region. Once electrical contact had been made to the SWNTs, silver decoration of SWNTs was carried out by means of a single-barrel pipette filled with $AgNO_3$ and KNO_3 and a silver wire acting as a QRCE. A pristine length of SWNT between the macroscopic metal thin film and the silver nanoparticles was interrogated by complementary techniques. (B) FE-SEM image of a SWNT showing a silver decorated section and a pristine segment. (C) The silver particles can be easily visualised by optical microscopy (top) and the pristine area of the nanotube interrogated by AFM (bottom) is clean and free from residues and catalyst.

(Newport, USA) until the formation of a meniscus between the pipette and sample, defining an electrochemical cell. This was then moved laterally across the substrate, parallel to the thin metal contact, while a potential of −0.3 V was applied to the sample with respect to the QRCE in the pipette. Under these conditions, a change in the current was observed when a section of a SWNT (serving as a working electrode) was wetted by the meniscus, with the reduction of silver ions to silver particles taking place at the surface of the SWNT. Fig. 2B shows a field-emission scanning electron microscopy (FE-SEM) (Zeiss Supra-55VP) image of an aligned SWNT decorated by silver particles.

Atomic Force Microscopy (AFM)

Tapping-mode AFM imaging (Enviroscope, Bruker-Veeco, USA) was performed over pristine regions of the SWNTs under investigation (Fig. 2C, bottom).

Hg-ultramicroelectrode (Hg-UME) probe fabrication

A liquid mercury hemisphere was electrodeposited at the tip of a disc-shaped glass-sealed and polished platinum ultramicroelectrode (UME), of 12.5 μm radius. In Fig. 3A a schematic and an optical image of a Hg-UME probe (Hg hemisphere of 12.5 μm radius) is presented. The electrodeposition was carried out by means of a potentiostat (CH Instruments, 730A, TX, USA) in a two-electrode EC cell using the UME as a working electrode and silver wire as a QRCE in an aqueous solution of 10 mM mercurous nitrate (Acros organics) in 50 mM nitric acid (Fisher Scientific).[36] The Hg-UME probe was mounted on a manual *x*-*y*-*z* micropositioner that was used to approach and scan the probe across the sample. The silver particles facilitated the rough localisation of the SWNT and minimised the area of the sample that had to be scanned to seek the SWNT of interest. A potential bias of −0.5 V was applied between the Hg-UME probe and ground,

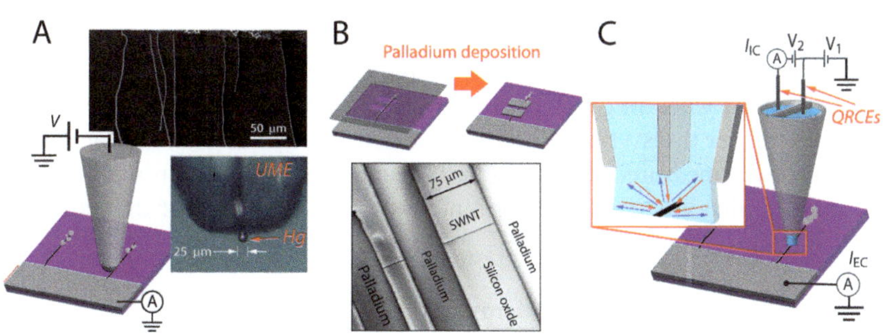

Fig. 3 (A) FE-SEM image of an array of flow-aligned SWNTs. The density required the use of a Hg-UME probe to electrically contact each SWNT individually. Schematic of the experimental setup, with a UME with a Hg drop (optical image). (B) Stencil shadow masks were employed in devices with one SWNT per sample. The silver particles facilitated the localization of the SWNT and the correct placement of the shadow mask. FE-SEM image of a device with three Pd contacts along one single SWNT. (C) Schematic of the SECCM operation: double-barrel pipette filled with the solution of interest and one chloridized silver wire in each barrel acting as QRCEs. SECCM was deployed in the pristine region of the SWNT. The combination of SECCM and SWNT provides high mass transport towards and away from the SWNT.

while the sample (SWNT) at ground (Fig. 3A). This ensured that an electronic current would flow through both metallic and semiconductor SWNTs (*vide infra*). Once an electrical contact between the protruding Hg drop and the SWNT was made, evidenced by a current flow, the probe position was fixed and current–voltage curves were measured (CH Instruments, 730A, TX, USA). At this point, the relative potential difference between the Hg-UME and the substrate was scanned (at 100 mV s^{-1}) between +3 V and −3 V, producing current–voltage curves that were used to determine the electronic nature of the SWNT.

Multiple contact SWNT device

For very low densities (one SWNT per cm^2 of sample), multiple metal contacts were applied over the SWNT employing a shadow mask with two (or more) windows to delimit the area of the thermal deposition of metal film (Pd). The visualisation of the silver electrodeposit aided the alignment of the mask (perpendicular to the SWNT) and ensured that there was only one SWNT bridging the metals contacts. An example of a typical device with multiple contacts (separated by different distances) is presented in Fig. 3B.

Microcapillary electrochemical method measurements

A single-barrel pipette was laser-pulled (P-2000, Sutter Instruments) to a sharp end and polished flat until a tip opening of 50 μm in diameter was obtained.[34] Two one-electron outer sphere redox couples were employed: the oxidation of ferrocenylmethyl-(trimethylammonium) ($FcTMA^+$), as the hexafluorophosphate salt synthesised in-house[37] and the reduction of ruthenium(III) hexamine ($Ru(NH_3)_6^{3+}$) (as the chloride, Strem Chemicals, Newbury Port, USA). The pipette was filled with 2 mM $FcTMA^+$ salt and 2 mM $Ru(NH_3)_6^{3+}$ salt and supporting electrolyte (100 mM KCl), together with a chloridized Ag wire as a QRCE. In a similar manner as for the silver decoration of SWNTs (see above), the probe was scanned across the sample, perpendicular to pristine sections of the SWNTs, in a region between the Pd contact and Ag decoration, while a potential for $FcTMA^+$ oxidation of 0.3 V *vs.* Ag/AgCl QRCE was applied *via* a potentiostat (CH Instruments, 730A, TX, USA), giving an EC current for both metallic and semiconducting SWNTs when in contact with the solution.[34] Once the moveable EC cell was centred over the SWNT of interest, a wide potential range cyclic voltammogram (CV) was measured to determine the activity (or not) towards $Ru(NH_3)_6^{3+}$ reduction, which was used to determine the electronic nature of the nanotube (*vide infra*).

Scanning Electrochemical Cell Microscopy

A complete description of SECCM can be found elsewhere.[9,10,15,19–21,31] SECCM employed double-barrel borosilicate glass pipettes as probes. These were pulled to a sharp taper by means of a laser puller (see above), yielding an opening of a desired size. In the present study, probes between 400 to 1000 nm were used, the dimensions of which were precisely determined by FE-SEM. To ensure a confined and stable meniscus contact during experiments, the outer walls of the pipette were silanised using dimethyltrichlorosilane (Aldrich). Then, both barrels of the pipette were filled with solution (5 μM–2 mM serotonin hydrochloride (Sigma-Aldrich) in 10 mM HEPES buffer and 0.1 M NaCl), and a chloridized silver wire was inserted into each barrel, which acted as a QRCE. A potential bias (V_2 in Fig. 3C) was applied between the QRCEs (0.5 V in the present study) to induce an ion conductance current (I_{IC}) across the thin film layer of solution protruding from the tapered end of the probe that connected both barrels. The probe was then mounted on a highly controllable *x-y-z* piezoelectric positioning instrument (for *z*-axis P-753.31C, *x-y*-axis P-622.1CD, Physik Instrumente), with the sample positioned vertically below the probe. The probe was oscillated (typical frequency 266 Hz, peak-to-peak oscillation approximately 50 nm) in the *z*-direction (perpendicular to the sample surface) to generate an alternating current (AC) component of the ion conductance current (I_{IC}) upon meniscus contact with the surface. As described in detail elsewhere,[20,31,38] this is due to the periodic motion of the meniscus changing the dimensions of the conductance cell. The AC component (magnitude) was extracted with a lock-in amplifier (SR830, Stanford Research Systems) and used to maintain a constant tip-sample separation during imaging due to the strong dependence of the AC amplitude on the tip-sample separation with meniscus contact.[20,31]

With the meniscus at the end of the pipette in contact with the substrate, the moveable EC cell was translated laterally across the surface at a speed of 0.5 μm s^{-1}, while measuring the surface EC current (I_{EC}).[21] A schematic of the whole SECCM set up is presented in Fig. 3C.

To drive electrochemical reactions at the SWNT, a bias V_1 (see Fig. 3C) was applied so that the effective potential applied to the SWNT was $-(V_1 + V_2/2)$ with respect to the QRCE. The sample was mounted so that the main scan axis (*x*-axis) was essentially perpendicular to the SWNT of interest. The scan pattern was designed so that the meniscus moved forward and reversed over the same line before moving to a new line, allowing for the identification and validation of EC features, but also enabling any changes in surface activity to be identified.

Raman spectroscopy

A 514 nm laser micro-Raman spectroscopy set up (Argon laser, 10 mW, 50× lens, NA 0.85, inVia micro-Raman, Renishaw, UK) was employed for the acquisition of Raman spectra over pristine sections of the SWNTs, roughly localised by the silver deposit.

Results and discussion

SWNT synthesis and characterisation

A key feature of the platform employed herein is that SWNTs were investigated in their pristine state and fully characterised. To achieve this, it was essential that steps such as the establishment of electrical contacts with the SWNTs and marking of parts of SWNTs *via* silver particle decoration to enable rough localisation (*vide infra*), were done with minimal manipulation or perturbation of the SWNTs. Millimetre-long straight SWNTs were ideal for this purpose: their long length enabled the use of metal thin film contacts, deposited with stencil shadow masks, avoiding photolithography or other techniques that could leave residues on the parts of the SWNT to be investigated electrochemically. In addition, the employment of long SWNTs allowed the establishment of a large pristine ROI set far from the catalyst (contact) region, ensuring that the EC response was free from any contribution of the catalyst particles from which the SWNTs were grown.

Various methods based on cCVD have been developed for growing horizontally aligned ultra-long SWNTs in the last few years.[35] Gas-flow-guided growth[39–41] or the use of a quartz single crystal substrate for lattice-induced growth[42,43] are some of the strategies used to achieve aligned arrays of SWNTs of large length. For the studies herein, flow alignment was used, as this provided flexibility in the substrate to be used and offered the production of SWNT arrays of lower density, facilitating access to individual SWNTs by SECCM and other techniques.

To obtain flow-directed horizontally-aligned ultra-long SWNTs, physical aspects of the process to maintain the SWNT floating, or lifted, above the substrate during growth needed to be considered, such as fast-heating[41,44,45] or ultralow feeding gas.[34,46] We used a cold-wall cCVD system[9] (Fig. 1B), in which only the heating plate where the sample was positioned – located at the centre of the chamber – reached the growth temperature. This system allowed fast-heating[41,44,45] (from room temperature to 950 °C in 7 min), generating local turbulences that are believed to facilitate the "take off" of the nanotube-catalyst and maintain it lifted above the substrate during growth. As a catalyst, Fe nanoparticles derived from ferritin solution were used,[8,9] rather than very active and commonly used Cu,[47] so as to achieve a low density of SWNTs on the sample,[40,47] thereby minimising SWNT overlap and the formation of bundles during flow growth.

Using the growth conditions described in detail in the experimental section, straight flow-aligned millimetre long SWNTs were grown on Si/SiO_2 surfaces. Interestingly, the growth direction was towards the gas flow, in contrast with the common conception in the literature that it is with the flow.[35,40,48,49] This suggests a tip-growth mechanism,[40] where the catalytic particle is at the flying end of the SWNT. Once the particle-nanotube reached a certain height, where there is a stable laminar gas flow regime,[50] the SWNT grows fairly straight for long distances, and its growth terminates when the flying SWNT falls to the substrate, possibly bouncing and rolling, in a turbulent regime, introducing to the end of the SWNT a tortuous twisting (Fig. 1C). Different growth times were evaluated, and it was concluded that maximum growth occurred within the first 5–10 min; extending the growth for longer periods neither increased the SWNT density nor the length of SWNTs.

Visualisation of SWNTs by local silver electrodeposition

To enable ready visualisation of the SWNTs by optical microscopy, silver particles were electrodeposited on a section of the SWNTs (see Experimental section), creating an ensemble of particles that marked one end of the SWNT (away from the Pd contact). Ensembles of particles were visible under FE-SEM (Fig. 2B) and optical microscopy (Fig. 2C, top), allowing the rough localisation of the SWNT and, in particular, an indication of the location of the pristine segment of the SWNT. Tapping-mode AFM of the nanotubes (Fig. 2C) was carried out, confirming the cleanliness of the sample and the area of interrogation (lack of residues and catalytic particles), and to determine the diameter of the nanotube. In the study herein, SWNTs were around ~1.2 nm in diameter, in agreement with the single-walled nature and the type of cCVD growth.[8,9]

Electronic characteristics of individual SWNTs

We consider and assess three different methods to identify the electronic character of individual SWNTs. First, conductance measurements were made by recording current (*I*) – voltage (*V*) curves across a defined portion of an SWNT. For metallic SWNTs, which could approach quantised conductance values at room temperature, there is often significant additional resistance at the metal contacts[51] or due to SWNT defects, giving rise to ohmic response with variable resistance values.[52] In contrast, for semiconductor SWNTs, that are considered p-type semiconductors,[53] the overlap of the metal–SWNT electronic states results in the presence of a Schottky barrier at the metal contact–semiconductor interface for hole or electron injection.[52] The value of this barrier depends on the band alignment at the metal–SWNT interface, and assuming the absence of interface states,[54] is determined by the work function of the metal contact, the work function of the SWNT (approx. 4.9 eV) and the band gap of the SWNT.[54] Thus, a metal with large work function such as Pd (5.1 eV), which we selected specifically for the studies herein, tends to form an ohmic contact for hole injection[55] because the Fermi level of the metal lies within the valence band of the SWNT. For the SWNTs used in this work this is particularly applicable because they are of relatively large diameter, which results in a small band gap.[24] It is worth noting that high work functions alone might not be sufficient to ensure an ohmic contact, as observed for Pt, which is due to the fact that poor wetting with the SWNT introduces additional tunnelling barriers.[56] Under these conditions, the response to hole injection is ohmic, whilst a large barrier exists to electron injection, leading to an asymmetry in the conductance *I*–*V* curves, with the device behaving as a rectifier or Schottky diode.[52,57] This is evident in the resulting *I*–*V* curves that were used to assess the electronic nature of SWNTs.[58] In Fig. 4A, two representative *I*–*V* curves for metallic (top) and semiconductor (bottom) SWNTs are presented, the former exhibiting a defined ohmic response, while the latter displays a rectifying behaviour, with small leakage current.[9]

An alternative technique to elucidate the electronic nature of CNTs is Raman spectroscopy.[59] Besides the ability to distinguish between metallic and

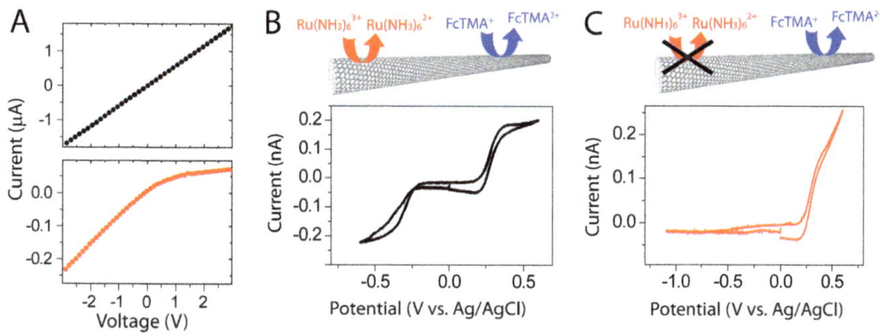

Fig. 4 (A) Representative current–voltage traces for a metallic (top) and semiconductor (bottom) SWNT. Voltage applied to the mobile mercury contact with respect to the sample. Interrogation of the EC activity on pristine individual SWNTs using a single-barrel pipette (~50 μm diameter) filled with two mediators, 2 mM $FcTMA^+$ and 2 mM $Ru(NH_3)_6^{3+}$, and supporting electrolyte (0.1 M KCl), and a Ag/AgCl wire as a QRCE. Cyclic voltammograms (100 mV s^{-1}) on a metallic (B) and semiconductor (C) SWNT.

semiconducting SWNTs,[60] a vast amount of information about CNTs can be obtained,[59] but for our purposes, we focused on two of the most important signals in the Raman spectra of SWNTs: the radial breathing mode (RBM) region, between 100 and 300 cm^{-1}, corresponding to the coherent vibration in the radial direction; and the G-band region found around 1580 cm^{-1}. The presence of peaks in the RBM region confirmed the single-walled nature of the CNTs under investigation, and the peak frequency was used to estimate the SWNT diameter,[61] resulting values in agreement with the ones obtained with AFM (~1.2 nm).

Fig. 5 shows how metallic and semiconducting SWNTs can be distinguished by focusing on the G-band region. Usually composed by two peaks, G+ (feature associated with carbon atom vibrations along the SWNT axis) and G− (feature associated with vibrations of carbon atoms in the circumferential direction), the metallic SWNTs G-band is composed by two components, one peak at 1590 cm^{-1} (denoted G+) and another broader and asymmetric peak with the Breit–Wigner–Fano lineshape, at lower frequency (labelled G−).[62] Semiconducting SWNTs, however, exhibit two intense Lorentzian lineshapes in the G-band region.[60] In addition, the high quality and essentially defect-free nature of the SWNTs synthesized herein was confirmed by the lack of any peak in the D-band around ~1350 cm^{-1}.

Recently, we have reported the use of SECCM[8,9,63] to determine the EC activity of SWNTs for two well-known and widely studied outer-sphere redox processes, the one-electron oxidation of $FcTMA^+$ and the one-electron reduction of $Ru(NH_3)_6^{3+}$, demonstrating the high EC activity of pristine SWNT side-walls. While both metallic and semiconductor SWNTs exhibited very high kinetic rate constants towards the oxidation of $FcTMA^+$, a huge difference between the EC current at metallic and semiconducting SWNTs was observed for the reduction of $Ru(NH_3)_6^{3+}$. The EC current was consistent with high activity at metallic SWNTs, but for semiconductor SWNTs the current was almost undetectable (except at high driving potentials where a small current was seen),[9] due to the redox potential of the $Ru(NH_3)_6^{3+/2+}$ couple lying in the charge depletion region of semiconducting SWNTs.[36] The simultaneous employment of these two redox couples in solution can thus be used to distinguish easily between metallic and semiconductor SWNTs.

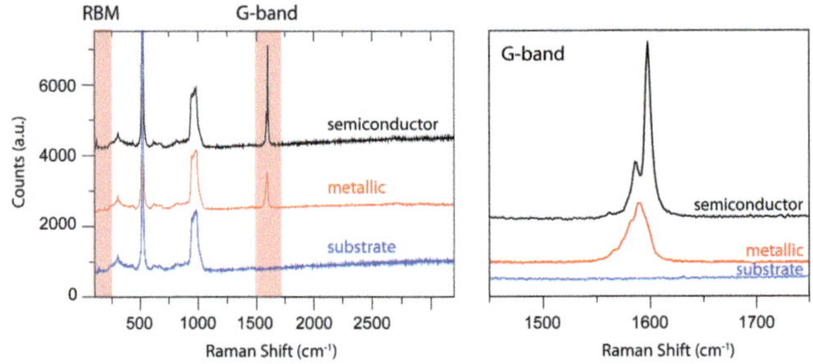

Fig. 5 Comparison of Raman spectra acquired on metallic (red line) and semiconducting (black line) SWNTs, as well as the supporting substrate (Si/SiO_2, blue line). The RBM and G-band regions are indicated with a light red box. The major difference in the Raman spectra between metallic and semiconducting SWNTs is found in the G-band region, both in shape and intensity, as shown in right-hand-side plot.

Fig. 4B and C show two typical CVs obtained with a 50 μm diameter single-barrel pipette filled with both redox couples (2 mM each of $FcTMA^+$ and $Ru(NH_3)_6^{3+}$), one carried out on a metallic SWNT (B) and the other on a semiconducting SWNT (C). Both responses show a clear sigmoidal signature for the $FcTMA^{+/2+}$ couple, but exhibit a strong difference in the cathodic region, with a voltammetric signature for $Ru(NH_3)_6^{3+}$ reduction only seen for the metallic SWNT. This confirms the suitability of this experiment to distinguish between metallic and semiconducting SWNTs.

Oxidation of serotonin

We used the platform outlined to study the EC response for serotonin EC oxidation at individual metallic SWNTs. We selected metallic SWNTs because these are most useful for general EC analysis, as highlighted above.

A typical SECCM image for the oxidation of 5 μM serotonin (in 10 mM HEPES and 0.1 M NaCl) is presented in Fig. 6A. The map of EC activity, acquired at an effective potential of 0.55 V *vs.* Ag/AgCl QRCE (just below the diffusion-limited potential, corresponding to *ca.* 80–90% of the limiting current value), shows a clear and uniform response along the entire length of the SWNT. Considering the type of SWNT employed in this work and the lack of D-peak in the Raman spectrum (Figure 5), the measured current cannot be attributed uniquely to EC oxidation at defect sites, which are found at a very low density (one defect per 4 μm on average).[64] Thus, the SWNT side-walls promote the EC oxidation process, as seen for other neurotransmitters on sp^2 carbon electrodes generally.[12–14,16,25] Indeed, the studies herein, together with other reports,[6–9] indicate clearly that the side-walls of SWNTs can easily support various redox reactions spanning simple outer-sphere couples and more complex multi-step processes.

The orientation of the moveable SECCM probe with respect to the SWNT during imaging means that scan profiles can be analysed to extract further

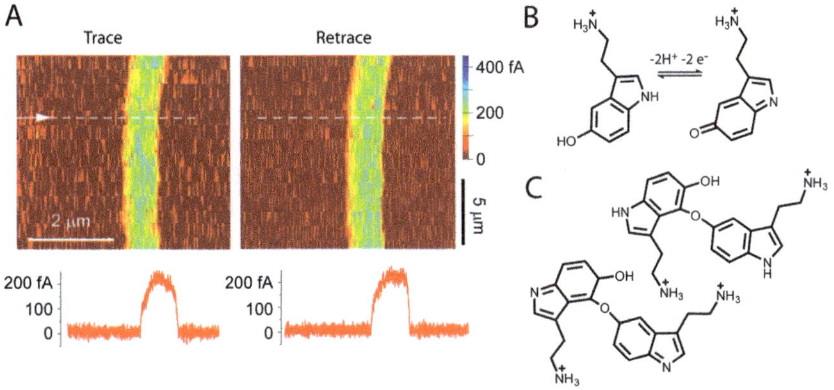

Fig. 6 (A) SECCM maps of EC current (trace – left and retrace – right) of the oxidation of serotonin (5 μM in 10 mM HEPES buffer and 0.1 M NaCl) acquired at 0.55 V *vs.* Ag/AgCl QRCE at a pipette scan velocity of 0.5 μm s^{-1}. Corresponding scan profiles (line plotted marked with an arrow) are also shown. (B) Scheme for the two-electron two-proton oxidation of serotonin. (C) Two examples of the major dimers formed in side reactions suggested to lead to insoluble polymer-like by-products.[33]

spatially resolved EC information, since the redox current is collected at 1 nm intervals along each line scanned by the pipette meniscus. Comparing the trace and retrace EC maps and considering the corresponding line profiles (Fig. 6A), highly consistent behaviour is observed. As the meniscus moves from the inert substrate (Si/SiO_2) to the active SWNT, the EC current gradually increases as the length of the nanotube accessed by the SECCM meniscus increases, with the current reaching a peak value when the SWNT lies at the centre of the meniscus from the pipette. The consistency of the trace and retrace scan images and the morphology of the line profiles (as expected for an active surface without any parasitic blocking process[9]) indicates that on this timescale (*ca.* 1.8 s total contact time of the meniscus with the SWNT in each pass for this particularly experiment) and serotonin concentration (5 μM) the SWNT remains highly active for electrochemistry.

It is important to point out that although the concentration of serotonin is low, the electrode is submitted to very high mass transport[8,9,31] yielding high peak current densities (~5.5 mA cm^{-2} considering that the entire area of the SWNTs was fully accessible), yet there is minimal (undetectable) fouling on the timescale of the experiment. The absence of fouling effects may be attributed to the extremely high mass transport to and from the SWNT electrode surface,[8,9,31] which facilitates the diffusion of the fouling products away from the electrode in this configuration before blocking can occur.[9] The serotonin oxidation process (Fig. 6B) also involves a side reaction (not shown) leading to oligomer formation (Fig. 6C).[33] The high mass transport rates of the SECCM/SWNT configuration means that this side reaction can be outrun at low serotonin concentration. In fact, these results highlight the considerable potential of individual SWNTs as an electrode material of high activity that is relatively immune to fouling in the detection of neurotransmitters. Fouling by reaction products can be minimised by simply enhancing mass transport of the species generated rather than needing to use preconditioning potentials or alternative materials (or surface functionalities). It is interesting to note that the configuration of a moving meniscus and individual SWNT also opens up the detection of trace amounts of material. The profiles in Fig. 6A correspond to the detection of just ~320 fC of charge. For the ($2e^-$) oxidation of serotonin (Fig. 6B) this equates to attomol detection levels, and even lower quantities are open to measurement based on the excellent signal to noise observed.

In order to investigate the impact of film formation on the resulting EC properties of SWNTs, a deliberately higher concentration of serotonin (2 mM) was employed in order to accelerate blocking film formation. In Fig. 7A, EC activity maps (trace and retrace) of serotonin EC oxidation at an individual metallic SWNT are shown, acquired at the same potential as for Fig. 6A (effective potential 0.55 V *vs.* Ag/AgCl QRCE in 10 mM HEPES buffer and 0.1 M NaCl). EC activity is again seen along the entire length of the SWNT, but due to the higher concentration, we observe different maps and line profiles from those seen in Fig. 6A. These are consistent with rapid fouling of the SWNT as soon as meniscus contact is made. Based on the data in Fig. 6A, one would predict a peak current of 80 pA, but – in fact – after increasing rapidly upon meniscus contact with the SWNT (left hand edge of the image and profile in Fig. 7A), the current decreases rapidly with further meniscus contact, and with the meniscus directly centred over the SWNT, the current is less than 10% of the value expected. Thus, once the meniscus makes

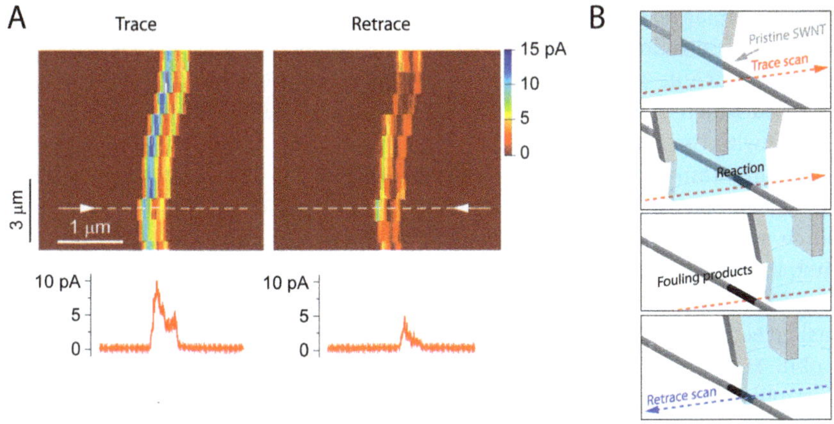

Fig. 7 (A) SECCM maps of EC current (trace – left and retrace – right) of the oxidation of serotonin (2 mM in 10 mM HEPES buffer and 0.1 M NaCl) at 0.55 V *vs.* Ag/AgCl QRCE on an individual SWNT. The asymmetry in the scan profiles is proof of the fouling of the electrode. Scan profiles plotted are marked with an arrow. (B) Sequence of processes during line profile measurements: the fouling products generated in the trace scan remain on the SWNT and on the retrace the SECCM meniscus scans over a fouled electrode.

contact with the SWNT and serotonin oxidation begins, the current decays due to fouling of the surface, despite the increasing area of the SWNT exposed to solution during imaging. This leads to a large asymmetry in the SECCM line profile shape.

The impact of the fouling products on the EC process at the SWNT is further confirmed with the retrace scan profiles, which do not exhibit the initial current spike (seen on the trace scan when the meniscus made initial contact) since the SWNT is already extensively fouled (Fig. 7B). The initial current in the retrace scan corresponds closely to that measured at the end of the trace linescan.

Taken together, the data in Fig. 6A and 7A highlight how SECCM can be used to assess complex EC processes. Moreover, the platform opens up a new route to trace EC detection of neurotransmitters at physiologically relevant concentrations, without any blocking effects which impact other EC techniques and carbon electrodes. Indeed, voltammetric measurements at conventional carbon electrodes struggle to achieve detection at the μM level,[29] yet an individual SWNT opens up this possibility (Fig. 6A). If, over large time periods, the SWNTs were to become contaminated, the pipette could simply be moved to another part of the SWNT or to another SWNT, suggesting SWNT arrays coupled with droplet cells as a new approach to trace electroanalysis.

Conclusions

SECCM imaging of individual metallic single-walled carbon nanotubes (SWNTs) has revealed that the side-walls are highly electrochemically active for the oxidation of serotonin, a complex, multi-stage process involving coupled heterogeneous electron transfer and chemical steps. No signs of electrode fouling were observed when physiologically relevant concentrations of serotonin were used (5 μM), although there was a deleterious impact of the EC process at much higher concentrations (2 mM) where the formation of blocking oligomers became more significant. This was readily seen in the morphology of the EC images recorded which indicated rapid fouling as soon as the SECCM meniscus made contact with the SWNT.

An important feature of the platform described is that a particular type of SWNT can be selected and investigated. We have presented 3 different ways in which metallic and semiconducting SWNTs can be identified and distinguished. First, the platform described enables micro-Raman spectroscopy to be applied to individual SWNTs, with the spectra recorded allowing the SWNT diameter and basic electronic characteristic to be determined. Second, we have shown how a mobile Hg-contact can be used as one terminal in a 2-terminal device in which electronic conductance *I–V* curves can be measured, from which a SWNT can be assigned as semiconducting or conducting. Third, the outer-sphere redox couple $Ru(NH_3)_6^{3+/2+}$ is very sensitive to the electronic character of a SWNT. In all cases, the SWNT can be located by these techniques and, by SECCM, because one end of a SWNT can be marked by Ag electrodeposition for detection by optical microscopy.

The field of electrochemistry at SWNTs has been held back for some time by the difficulty of studying individual well-characterised SWNTs, a general inability to link electrochemistry to the electronic properties of the SWNT, and a lack of suitable high resolution EC methodologies to probe electrochemistry at a length scale that is capable of resolving any role of defects, or other features of SWNTs, in the EC behaviour. The platform described allows these issues to be addressed and overcome, thereby greatly advancing understanding of the electrochemistry of SWNTs. Moreover, the platform should be applicable to other types of nanostructures which are of growing interest for EC technologies and fundamental charge-transfer studies.

Acknowledgements

This project was supported by the European Research Council through Project ERC-2009-AdG 247143-QUANTIF and EPSRC (EP/H023909/1 and MOAC CDT studentship), with additional support from the National Physical Laboratory. JCB acknowledges a Marie Curie International Incoming Fellowship under project number 329953 "POLYMAP". Equipment used in this research was obtained through Science City (AM2), with support from Advantage West Midlands and partial funding by the European Regional Development Fund.

References

1 I. Dumitrescu, P. R. Unwin and J. V. Macpherson, *Chem. Commun.*, 2009, **45**, 6886–6901.
2 R. L. McCreery, *Chem. Rev.*, 2008, **108**, 2646–2687.
3 C. E. Banks, T. J. Davies, G. G. Wildgoose and R. G. Compton, *Chem. Commun.*, 2005, 829–841.
4 A. Chou, T. Bocking, N. K. Singh and J. J. Gooding, *Chem. Commun.*, 2005.
5 C. E. Banks, R. R. Moore, T. J. Davies and R. G. Compton, *Chem. Commun.*, 2004, 1804–1805.

6 I. Heller, J. Kong, H. A. Heering, K. A. Williams, S. G. Lemay and C. Dekker, *Nano Lett.*, 2005, **5**, 137–142.

7 J. Kim, H. Xiong, M. Hofmann, J. Kong and S. Amemiya, *Anal. Chem.*, 2010, **82**, 1605–1607.

8 A. G. Güell, N. Ebejer, M. E. Snowden, K. McKelvey, J. V. Macpherson and P. R. Unwin, *Proc. Natl. Acad. Sci. U. S. A.*, 2012, **109**, 11487–11492.

9 A. G. Güell, K. E. Meadows, P. V. Dudin, N. Ebejer, J. V. Macpherson and P. R. Unwin, *Nano Lett.*, 2014, **14**, 220–224.

10 S. C. S. Lai, A. N. Patel, K. McKelvey and P. R. Unwin, *Angew. Chem., Int. Ed.*, 2012, **51**, 5405–5408.

11 A. N. Patel, M. G. Collignon, M. A. O'Connell, W. O. Y. Hung, K. McKelvey, J. V. Macpherson and P. R. Unwin, *J. Am. Chem. Soc.*, 2012, **134**, 20117–20130.

12 A. N. Patel, K. McKelvey and P. R. Unwin, *J. Am. Chem. Soc.*, 2012, **134**, 20246–20249.

13 A. N. Patel, S.-y. Tan, T. S. Miller, J. V. Macpherson and P. R. Unwin, *Anal. Chem.*, 2013, **85**, 11755–11764.

14 A. N. Patel, S.-y. Tan and P. R. Unwin, *Chem. Commun.*, 2013, **49**, 8776–8778.

15 P. M. Kirkman, A. G. Guell, A. S. Cuharuc and P. R. Unwin, *J. Am. Chem. Soc.*, 2014, **136**, 36–39.

16 S. Lhenry, Y. R. Leroux and P. Hapiot, *Anal. Chem.*, 2012, **84**, 7518–7524.

17 A. Anne, M. A. Bahri, A. Chovin, C. Demaille and C. Taofifenua, *Phys. Chem. Chem. Phys.*, 2014, **16**, 4642–4652.

18 P. L. T. M. Frederix, P. D. Bosshart, T. Akiyama, M. Chami, M. R. Gullo, J. J. Blackstock, K. Dooleweerdt, N. F. d. Rooij, U. Staufer and A. Engel, *Nanotechnology*, 2008, **19**, 384004.

19 A. G. Güell, N. Ebejer, M. E. Snowden, J. V. Macpherson and P. R. Unwin, *J. Am. Chem. Soc.*, 2012, **134**, 7258–7261.

20 N. Ebejer, M. Schnippering, A. W. Colburn, M. A. Edwards and P. R. Unwin, *Anal. Chem.*, 2010, **82**, 9141–9145.

21 N. Ebejer, A. G. Güell, S. C. S. Lai, K. McKelvey, M. E. Snowden and P. R. Unwin, *Annu. Rev. Anal. Chem.*, 2013, **6**, 329–351.

22 B. D. B. Aaronson, C.-H. Chen, H. Li, M. T. M. Koper, S. C. S. Lai and P. R. Unwin, *J. Am. Chem. Soc.*, 2013, **135**, 3873–3880.

23 I. Heller, J. Kong, K. A. Williams, C. Dekker and S. G. Lemay, *J. Am. Chem. Soc.*, 2006, **128**, 7353–7359.

24 J. W. G. Wilder, L. C. Venema, A. G. Rinzler, R. E. Smalley and C. Dekker, *Nature*, 1998, **391**, 59–62.

25 A. G. Güell, K. E. Meadows, P. R. Unwin and J. V. Macpherson, *Phys. Chem. Chem. Phys.*, 2010, **12**, 10108–10114.

26 M. Berger, J. A. Gray and B. L. Roth, *Annu. Rev. Med.*, 2009, **60**, 355–366.

27 P. Bertoncello, J. P. Edgeworth, J. V. Macpherson and P. R. Unwin, *J. Am. Chem. Soc.*, 2007, **129**, 10982–10983.

28 S. Sansuk, E. Bitziou, M. B. Joseph, J. A. Covington, M. G. Boutelle, P. R. Unwin and J. V. Macpherson, *Anal. Chem.*, 2012, **85**, 163–169.

29 A. N. Patel, P. R. Unwin and J. V. Macpherson, *Phys. Chem. Chem. Phys.*, 2013, **15**, 18085–18092.

30 W. Harreither, R. Trouillon, P. Poulin, W. Neri, A. G. Ewing and G. Safina, *Anal. Chem.*, 2013, **85**, 7447–7453.

31 M. E. Snowden, A. G. Güell, S. C. S. Lai, K. McKelvey, N. Ebejer, M. A. O'Connell, A. W. Colburn and P. R. Unwin, *Anal. Chem.*, 2012, **84**, 2483–2491.
32 P. Hashemi, E. C. Dankoski, J. Petrovic, R. B. Keithley and R. M. Wightman, *Anal. Chem.*, 2009, **81**, 9462–9471.
33 M. Z. Wrona and G. Dryhurst, *Bioorg. Chem.*, 1990, **18**, 291–317.
34 P. V. Dudin, M. E. Snowden, J. V. Macpherson and P. R. Unwin, *ACS Nano*, 2011, **5**, 10017–10025.
35 L. Huang, Z. Jia and S. O'Brien, *J. Mater. Chem.*, 2007, **17**, 3863–3874.
36 T. M. Day, N. R. Wilson and J. V. Macpherson, *J. Am. Chem. Soc.*, 2004, **126**, 16724–16725.
37 M. N. Szentirmay and C. R. Martin, *Anal. Chem.*, 1984, **56**, 1898–1902.
38 K. T. Rodolfa, A. Bruckbauer, D. J. Zhou, A. I. Schevchuk, Y. E. Korchev and D. Klenerman, *Nano Lett.*, 2006, **6**, 252–257.
39 Y. Wang, M. J. Kim, H. Shan, C. Kittrell, H. Fan, L. M. Ericson, W.-F. Hwang, S. Arepalli, R. H. Hauge and R. E. Smalley, *Nano Lett.*, 2005, **5**, 997–1002.
40 A. Reina, M. Hofmann, D. Zhu and J. Kong, *J. Phys. Chem. C*, 2007, **111**, 7292–7297.
41 S. Huang, X. Cai and J. Liu, *J. Am. Chem. Soc.*, 2003, **125**, 5636–5637.
42 S. J. Kang, C. Kocabas, T. Ozel, M. Shim, N. Pimparkar, M. A. Alam, S. V. Rotkin and J. A. Rogers, *Nat.Nanotechnol.*, 2007, **2**, 230–236.
43 A. Rutkowska, D. Walker, S. Gorfman, P. A. Thomas and J. V. Macpherson, *J. Phys. Chem. C*, 2009, **113**, 17087–17096.
44 S. Huang, M. Woodson, R. Smalley and J. Liu, *Nano Lett.*, 2004, **4**, 1025–1028.
45 S. Huang, B. Maynor, X. Cai and J. Liu, *Adv. Mater.*, 2003, **15**, 1651–1655.
46 Z. Jin, H. Chu, J. Wang, J. Hong, W. Tan and Y. Li, *Nano Lett.*, 2007, **7**, 2073–2079.
47 R. Cui, Y. Zhang, J. Wang, W. Zhou and Y. Li, *J. Phys. Chem. C*, 2010, **114**, 15547–15552.
48 X. Wang, Q. Li, J. Xie, Z. Jin, J. Wang, Y. Li, K. Jiang and S. Fan, *Nano Lett.*, 2009, **9**, 3137–3141.
49 S. Li, Z. Yu, C. Rutherglen and P. J. Burke, *Nano Lett.*, 2004, **4**, 2003–2007.
50 B. H. Hong, J. Y. Lee, T. Beetz, Y. Zhu, P. Kim and K. S. Kim, *J. Am. Chem. Soc.*, 2005, **127**, 15336–15337.
51 A. Bezryadin, A. R. M. Verschueren, S. J. Tans and C. Dekker, *Phys. Rev. Lett.*, 1998, **80**, 4036–4039.
52 S. Heinze, J. Tersoff, R. Martel, V. Derycke, J. Appenzeller and P. Avouris, *Phys. Rev. Lett.*, 2002, **89**, 106801.
53 J. Kong, N. R. Franklin, C. Zhou, M. G. Chapline, S. Peng, K. Cho and H. Dai, *Science*, 2000, **287**, 622–625.
54 F. Léonard and J. Tersoff, *Phys. Rev. Lett.*, 2000, **84**, 4693–4696.
55 Z. Chen, J. Appenzeller, J. Knoch, Y.-m. Lin and P. Avouris, *Nano Lett.*, 2005, **5**, 1497–1502.
56 A. Javey, J. Guo, Q. Wang, M. Lundstrom and H. Dai, *Nature*, 2003, **424**, 654–657.
57 H. M. Manohara, E. W. Wong, E. Schlecht, B. D. Hunt and P. H. Siegel, *Nano Lett.*, 2005, **5**, 1469–1474.
58 M. Biercuk, S. Ilani, C. Marcus and P. McEuen, in *Carbon Nanotubes*, Springer Berlin Heidelberg, 2008, vol. 111, pp. 455–493.

59 M. S. Dresselhaus, G. Dresselhaus, R. Saito and A. Jorio, *Phys. Rep.*, 2005, **409**, 47–99.
60 A. Jorio, A. G. Souza Filho, G. Dresselhaus, M. S. Dresselhaus, A. K. Swan, M. S. Ünlü, B. B. Goldberg, M. A. Pimenta, J. H. Hafner, C. M. Lieber and R. Saito, *Phys. Rev. B: Condens. Matter*, 2002, **65**, 155412.
61 A. Jorio, R. Saito, J. H. Hafner, C. M. Lieber, M. Hunter, T. McClure, G. Dresselhaus and M. S. Dresselhaus, *Phys. Rev. Lett.*, 2001, **86**, 1118–1121.
62 S. D. M. Brown, A. Jorio, P. Corio, M. S. Dresselhaus, G. Dresselhaus, R. Saito and K. Kneipp, *Phys. Rev. B: Condens. Matter*, 2001, **63**, 155414.
63 T. S. Miller, N. Ebejer, A. G. Güell, J. V. Macpherson and P. R. Unwin, *Chem. Commun.*, 2012, **48**, 7435–7437.
64 Y. Fan, B. R. Goldsmith and P. G. Collins, *Nat. Mater.*, 2005, **4**, 906–911.

Faraday Discussions

PAPER

Glutamate biosensors based on diamond and graphene platforms

Jingping Hu,[ab] Sirikarn Wisetsuwannaphum[b] and John S. Foord*[b]

Received 5th March 2014, Accepted 24th March 2014
DOI: 10.1039/c4fd00032c

L-Glutamate is one of the most important neurotransmitters in the mammalian central nervous system, playing a vital role in many physiological processes and implicated in several neurological disorders, for which monitoring of dynamic levels of extracellular glutamate in the living brain tissues may contribute to medical understanding and treatments. Electrochemical sensing of glutamate has been developed recently mainly using platinum, carbon fibre and carbon nanotube electrodes. In the present work, we explore the fabrication and properties of electrochemical glutamate sensors fabricated on doped chemical vapour deposition diamond electrodes and graphene nanoplatelet structures. The sensors incorporate platinum nanoparticles to catalyse the electrooxidation of hydrogen peroxide, glutamate oxidase to oxidise glutamate, and a layer of poly-phenylenediamine to impart selectivity. The performance of the devices was compared to a similar sensor fabricated on glassy carbon. Both the diamond and the graphene sensor showed very competitive performance compared to the majority of existing electrochemical sensors. The graphene based sensor showed the best performance of the three investigated in terms of sensitivity, linear dynamic range and long term stability, whereas it was found that the diamond device showed the best limit of detection.

1 Introduction

The integration of biological tissues with electronic devices, to enable electrical stimulation of cells and electrical transduction of cellular signals, is at present an active topic in research to improve understanding of neuronal function.[1,2] Whilst electrical stimulation and measurements of action potential can be achieved using microprobe electrodes, a more recent trend is the growth of cellular tissues at the interface with electrical devices such as microelectrode or open gated micro-transistor arrays to improve the quality of the data recorded and the study of cell assemblies.[3–5] A significant problem here is the design of

[a]Huazhong University of Science and Technology, School of Environmental Science and Engineering, Wuhan, P.R. China 430074. E-mail: hujp@hust.edu.cn; Tel: +86-27-87793948
[b]Chemistry Research Laboratory, Department of Chemistry, University of Oxford, Mansfield Rd, Oxford, OX1 3TA, UK. E-mail: john.foord@chem.ox.ac.uk; Fax: +44 1865 275410; Tel: +44 1865 275967

cell-semiconductor interfaces which are sufficiently stable in physiological media. For example, conventional Si chip technology undergoes drift and long term degradation because of the penetration of electrolyte ions into the transistor structure. The same problem is encountered with electrode and microelectrode assemblies for the electrochemical amperometric detection of the quantal release of neurotransmitters such as dopamine and norepinephrine. Here materials such as carbon fibre electrodes are often employed, but again degrade in the physiological media through oxidation and electrode fouling, resulting in poor sensitivity and stability.[6,7]

An interesting advance to the problem of finding materials that are stable in physiological media for these applications, is the use of diamond grown by chemical vapour deposition (CVD).[8,9] CVD diamond is now commercially readily available at modest price and when doped with boron becomes highly conductive, displaying excellent electrochemical properties. In particular the high phase stability, resistance to fouling, low background capacitive current and wide range of operating potentials makes it ideal for use in hostile environments.[10] The material also lends itself to unique, highly sensitive transistor designs which overcome the materials issues inherent to Si chip technologies, for detection of cell electrical signals.[3] The use of a single platform material both for electrochemical and transistor-based electrical measurements offer unique possibilities for the future design of bioelectronic devices.

One step towards achieving this is the development of diamond-based electrode interfaces for the selective electrochemical detection of the neurotransmitter species of interest. In this context previous studies have focussed on the use of boron-doped diamond (BDD) for detection of species such as dopamine, norepinephrine, epinephrine, melatonin, catecholamine and adenosine *in vitro*.[7,11–14] BDD electrodes have also been used to monitor serotonin *in vivo*, and demonstrated advantageous properties over carbon fibre electrodes including a very low background current and a high level of response stability even without the use of a protective polymer coating.[12] The application of BDD electrodes in the biosensing of glutamate has not been studied in any detail. Among the neurotransmitters detected by biosensors, L-glutamate is one of the most important in the mammalian central nervous system, playing a vital role in many physiological processes.[15–17] Its excess release is implicated in several neurological disorders, such as strokes, epileptic seizures and Parkinson's disease.[18–20] L-Glutamate is most widely monitored *in vivo* by microdialysis methods, which are limited by poor spatial and slow time resolution.[18] An alternative approach, using electrochemical sensing methods, has been developed recently, using platinum,[21,22] glassy carbon,[20,22] and carbon nanotube electrodes,[23] which provide an approach to an implantable amperometric sensor for continuous long term sensing, with a high spatial and temporal resolution.

Apart from the conventional solid carbon electrodes such as graphite and glassy carbon, over the past decade research in electrochemistry has focused increasingly on the use of dispersed forms of carbon of which perhaps the most important has been carbon nanotubes because of their high surface area, ease of chemical functionalisation, suitability for enzyme immobilisation, and enhanced electrochemical activity through edge–plane like defects.[24] Difficulties in purification however can give rise to undesired properties.[25] Graphene platelets are another form of dispersed carbon material,[26] which is now evoking high interest within electrochemistry. Attractive characteristics include high surface area (2630 m^2 g^{-1}, twice as much as that of SWCNTs),[27] low cost, ease of processing,[28] unique thermal and mechanical properties and high electrical conductivity. Graphene has therefore become an attractive and promising novel nanomaterial in various applications. For example, in fuel cell development, a graphene-based electrode was shown to outperform CNT-based ones in terms of electrocatalytic activity[29] and provided better macroscopic scale conductivity in a biosensing application.[30] The material has been widely employed for bioanalytical applications such as in detection of glucose,[31,32] dopamine,[33] NADH,[34,35] H_2O_2[36] and glutamate based on field effect transistor configuration using CVD grown graphene.[37]

The glutamate biosensor is based on the oxidation of glutamate in the presence of glutamate oxidase.

$$Glutamate + O_2 + H_2O \rightarrow \alpha - ketoglutarate + NH_3 + H_2O_2 \quad (1)$$

The H_2O_2 produced in this reaction is electroactive at electrodes such as Pt thus providing a method to infer the concentration of glutamate, although it is inactive at many typical carbon-based electrodes which are often used, necessitating addition of various electrocatalytic materials such as Pt nanoparticles,[21] hydrous iridium oxide,[38] Prussian blue[39] or peroxidase enzymes including horseradish peroxidase,[40] cytochrome c peroxidase,[41] or catalase enzyme,[42] or heme peptide such as hemoglobin.[43]

In the present work we thus investigate the extent to which electrochemical detection of glutamate can be achieved using conductive diamond and graphene as the electrode materials, activated with Pt electrocatalytic species that can catalyse the oxidisation of hydrogen peroxide.[21,22] However, even in the presence of platinum electrocatalyst, a high positive potential (>0.5 V vs Ag/AgCl) is expected for the oxidation of H_2O_2. At this potential, some endogenous electroactive species, such as ascorbate, urate, dopamine, acetaminophen, and catecholamines, are also oxidized. This undermines the selectivity of the biosensor[39] and requires screening strategies to eliminate the problem. The polymer polyphenylenediamine is used in this regard in the present work.

The performance of BDD and graphene based glutamate sensors formed in this way is explored in this article. In particular the sensitivity, selectivity and stability of the biosensors are characterised and compared with those based on the more established conventional glassy carbon based electrodes which we fabricate using similar methods.

2 Experimental

2.1 Reagents and solutions

The enzyme glutamate oxidase (GluOx, EC 1.4.3.11) from *Streptomyces sp.* was supplied from Sigma Aldrich, as was L-glutamic acid (Glu, >99.5%), *m*-phenylenediamine (PPD, >99%), glutaraldehyde (GA, 50% in water), hydrogen peroxide (50 wt% in water), L-ascorbic acid (AA), uric acid (UA, Sigma) and 4-acetamidophenol (AP, Aldrich). Dihydrogen hexachloroplatinate (IV) hexahydrate ($H_2PtCl_6 \cdot 6H_2O$, 99.9%) was obtained from Alfa Aesar. $NaBH_4$ (Fluka), Poly(diallyldimethylammonium chloride) solution 20 wt% in water (Aldrich) and sodium chloride (Fluka) were in analytical grade and used as supplied. Graphene oxide 0.5 wt% in water with a thickness of around 3 nm and 550 nm in size was purchased from Angstron Materials.

The supporting electrolyte used was phosphate buffered saline solution (PBS) consisting of 0.137 M NaCl (Aldrich, >99%), 0.0027 M KCl (Fluka, >99.5%), 0.01 M Na_2HPO_4 (Sigma-Aldrich, >99%) and 0.002 M KH_2PO_4 (Sigma, >99%), and adjusted to pH 7.4 with sodium hydroxide (Fisher, 98.7%). All solutions were prepared with distilled Milli-Q water (>18 MΩcm).

2.2 Instrumentation

Polished smooth (average roughness Ra < 30 nm) BDD and glassy carbon discs of 3 mm diameter and 0.5 mm thickness obtained from Element Six Ltd (Ascot, UK) were used as rotating disk electrodes (RDE). Experiments were conducted at 20 ± 2 °C using a CHI900B electrochemical workstation in a standard three-electrode setup with a Ag/AgCl reference electrode, against which all potentials quoted in this paper are measured with respect to, and a platinum counter electrode. Prior to use, the BDD electrode was cleaned by potential cycling between −1.6 and 2.5 V in 0.5 M sulphuric acid for 20 cycles to remove contaminants from the surface. The glassy carbon electrode was cleaned by polishing with alumina powder to a mirror finish and ultrasonicating in ethanol and water respectively. The electrode was stored in PBS in a fridge (4 °C) after GluOx functionalization.

TEM images were taken using JEOL JEM-2000FX equipment with an accelerating voltage of 200 kV. Tapping-mode AFM was conducted with a Digital Instruments Multimode Nanoscope Scanning Probe Microscope (Veeco Metrology Group). The XPS measurements were conducted using a hemispherical analyser spectrometer with Al K_α radiation ($h\nu = 1486.6$eV). Zeta potential was measured using a Zetasizer NanoZS (Malvern Instruments) at 25 °C and all the samples were diluted to 0.1 mg mL^{-1}.

3 Results and discussions

3.1 Preparation and characterisation of platinum nanoparticle decorated BDD electrodes and graphene nanoplatelets

The first step in sensor fabrication was the preparation of electrode materials decorated with Pt nanoparticles. Electrodeposition is a well-established approach which creates adherent metal nanoparticles on diamond electrodes, whereas coatings of pre-formed nanoparticles tend to be easily lost from the very inert diamond surface. Platinum particles were therefore electrodeposited on the BDD electrode from a solution of 2 mM H_2PtCl_6 dissolved in 0.5 M H_2SO_4, using a fixed electrode potential of −0.2 V for 10 minutes as we have described previously.[44] The exact procedure used leads to the formation of electrode surfaces decorated with stable, roughly 150 ± 30 nm diameter porous Pt clusters with a coverage of 3×10^8 particles cm^{-2}, as illustrated by the secondary electron microscope image in Fig. 1(a). The total Pt electroactive area is around 13 times bigger than the physical area of the electrode, for which details are given elsewhere.[44]

In contrast a wet chemical approach could instead be used to prepare Pt functionalised graphene, which has the ability to form more dispersed materials. To achieve this, graphene oxide was first functionalised with poly(diallyldimethyl

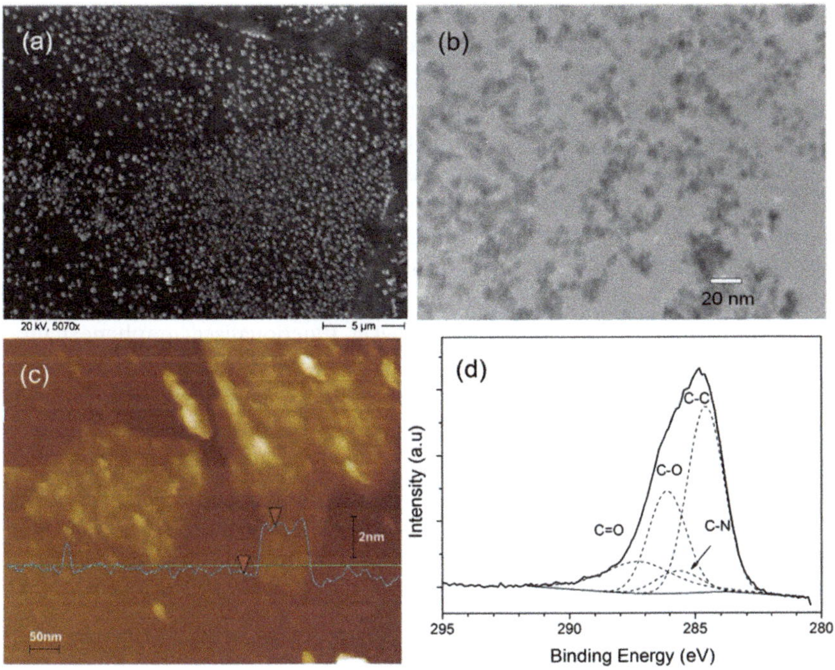

Fig. 1 SEM image of Pt modified diamond electrode showing the presence of Pt clusters on the diamond surface (a), TEM (b) and AFM (c) images of 50 wt% Pt on PDDA graphene nanoplatelets showing uniform distribution of Pt nanoparticles on graphene surface and the thickness of graphene sheet respectively and XPS spectrum of carbon 1s of the same sample (d).

-ammonium chloride) (PDDA) by mixing 50 mg graphene oxide with 0.5 wt% PDDA and 1 wt% NaCl in aqueous solution which was then ultrasonicated to obtain a stable suspension. Graphene oxide carries a negative charge, so the purpose of the PDDA which is a cationic polyelectrolyte is to change this to a positive charge, thus attracting the $[PtCl_6]^{2-}$ to the graphene surface. This was confirmed with zeta potential measurements; graphene oxide possesses a negative zeta potential of −26.7 mV and this switched to a positive zeta potential of 63.7 mV after functionalisation with PDDA, confirming the successful adsorption of PDDA on to the surface of graphene oxide. Due to this, PDDA functionalised graphene oxide becomes more susceptible for reaction with the $[PtCl_6]^{2-}$ complex and the amount of Pt on the graphene can be easily tuned by simply adjusting the quantity of the added platinum precursor. The mixture was filtered and washed several times to remove excess PDDA, and dried in air for 24 hours. Then 5mg of PDDA functionalised graphene oxide was dispersed in 10 mL water by ultrasonication followed by adding 10.5 mg H_2PtCl_6 and 50 mM HCl and was ultrasonicated for 15 minutes. 0.5 mL 200 mM $NaBH_4$ (in large excess) in 200 mM NaOH was added drop wise to this mixture under vigorous stirring for 1 hour. The resulting solution was filtered and washed several times with water followed by drying in air overnight. This sample is referred to as 50% loading, which corresponds to the percentage weight in the Pt-graphene composite if complete reaction between the added graphene and Pt occurred. Similarly 15% and 70%

loadings were also prepared. Electrode fabrication was then carried out by drop coating 5 μL of 1 mg mL^{-1} suspension of Pt-graphene onto the glassy carbon electrode.

The morphology of platinum graphene nanoplatelet composites (Pt-GNPs) produced from simultaneous reduction of hexachloro platinate (IV) complex and PDDA functionalized graphene oxide was elucidated by transmission electron microscopy in Fig. 1(b). The average Pt particle size was estimated to be around 8 nm with a narrow range of size distribution. Fig. 1(c) shows the typical AFM image of Pt-GNPs on Si substrate. The thickness of reduced graphene oxide sheet was found to be 2.4 nm on average corresponding to a few monolayers of graphene and is comparable to the thickness of PDDA functionalised graphene nanoplatelets reported elsewhere.[45]

To further characterize the surface chemical composition of the Pt-GNPs hybrid, XPS analysis was carried out. In the survey scan (data not shown), there are three main peaks observed at around 72 eV, 285 eV, and 531 eV in all Pt-graphene nanosheets samples, which correspond to Pt 4f, C 1s and O 1s respectively and one small peak at 402 eV due to N 1s from PDDA. The higher resolution spectra provide closer investigation of the chemical states on the surface of C 1s in Fig 1(d). The peak at binding energy of 284.5 eV can be assigned to the graphitic C–C bond, while those at higher binding energies correspond to C–OH (286.2 eV) and C=O (287.2 eV) bonds.[46] Additionally, a small peak observed at 285.7 eV is suggested which is consistent with the presence of C–N bonds from PDDA.

The relative surface composition of Pt to C in Pt-GNPs was estimated from XPS data with an atomic sensitivity factor of Pt 4f of 4.4[47] to be 5.5% for Pt deposited on GNPs, giving rise to 2.4 μg loading of Pt for 5 μg Pt-GNPs used for the sensor fabrication. This corresponds to 50% loading of Pt precursor to the reaction mixture in the synthesis step. The electrochemical active surface area of the Pt was estimated by the charge associated with the oxidation of adsorbed hydrogen[48] and found to be 2.6 cm^2. By contrast the Pt-BDD electrode showed a twofold higher Pt loading of 5.1 μg but a lower surface area of 0.98 cm^2. This trend is qualitatively consistent with the larger particle size of the Pt-BDD electrode.

3.2 Comparison of Pt-BDD and Pt-GNPs platform for detection of hydrogen peroxide and glutamate

As briefly explained in the introduction, the mechanism of the oxidation of L-glutamate catalysed by L-glutamate oxidase enzyme can be represented by the following two-step process.

$$L - glutamate + H_2O + GluOx/FAD \rightarrow \alpha - ketoglutarate + NH_3 + GluOx/FADH_2 \quad (2)$$

$$GluOx/FADH_2 + O_2 \rightarrow GluOx/FAD + H_2O_2 \quad (3)$$

The first step is the deamination process catalysed by glutamate oxidase enzyme with the cofactor flavin adenine dinucleotide (FAD) acting as a redox centre. Upon the uptake of the oxygen co-substrate in the second step, hydrogen peroxide is produced as a by-product. The production of H_2O_2 can then be directly detected at the electrode by the electrocatalytic oxidation as shown below.

$$H_2O_2 \rightarrow O_2 + 2H^+ + 2e^- \quad (4)$$

Measurement of H_2O_2 oxidation activity was therefore compared on the 2 types of electrode. The as-prepared platinum decorated diamond sensor (Pt-BDD) shows a reproducible oxidisation peak at about 0.5 V in the anodic scans of cyclic voltammograms measured in stirred H_2O_2 electrolyte. The chronoamperometric response of the electrode was thus measured at 0.5 V in PBS for varying hydrogen peroxide concentrations by addition of small amounts of 10 mM H_2O_2 solution using a RDE electrode rotated at 3000 rpm to accelerate the mixing.

A linear plot of current density against concentration of H_2O_2 was observed at least up to 20 μM as presented in Fig. 2(a). Similar experiments for the Pt-GNPs electrode were also carried out, for which data is also presented in Fig. 2(a). The Pt-BDD electrode exhibited the highest sensitivity of 2.5 μA μM^{-1} cm^{-2} while the 50wt% Pt-GNPs, despite being the highest amongst the Pt-GNPs samples, showed a sensitivity of 0.69 μA μM^{-1} cm^{-2}. Although lowering the amount of Pt added to the graphene to 15% reduced the response to H_2O_2 (curve (iv) in Fig. 2(a)) as expected it was found that increasing the amount of Pt had little effect, as can be seen in Fig. 2(a) where the 70% loaded sample produced even a smaller sensitivity than the 50% loading. It is not clear what differences between the differing electrodes are responsible for these differing responses. In comparing the performances of the differing sensors however, it should be borne in mind that the Pt-BDD electrode is significantly more responsive to H_2O_2 than is the Pt-GNPs modified electrode.

For control experiments, Pt-PDDA and GNPs-PDDA electrodes were also tested (traces (v) and (vi) in Fig. 2(a)). However the H_2O_2 sensitivity of both falls well below that of Pt-GNPs confirming that both graphene itself and also Pt nanoparticles coated with PDDA are relatively inactive towards H_2O_2 oxidation as expected.[32,49,50]

3.3 Detection of glutamate at low concentrations

The enzyme immobilisation step was carried out by drop evaporation of glutamate oxidase, using around 1 mg glutamate oxidase in 5 μL PBS solution,[22] on the

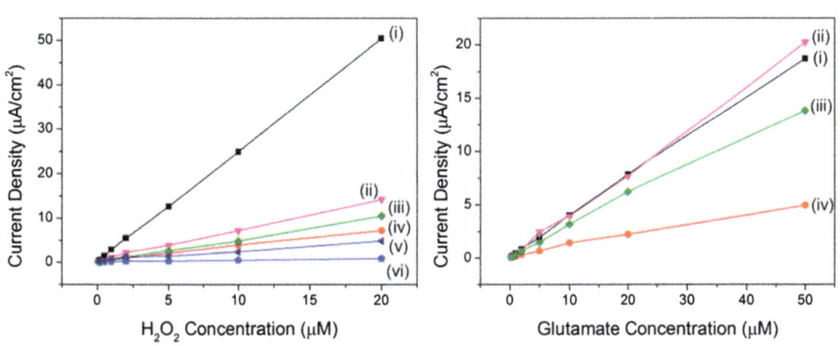

Fig. 2 Linear plot of current density response (a) against concentration of H_2O_2 and (b) glutamate measured at different electrodes; (i) Pt modified boron-doped diamond, (ii) 50wt%, (iii) 70wt%, (iv) 15wt% Pt loading on graphene nanosheets, (v) graphene nanosheets and (vi) Pt nanoparticles as controlled experiments.

Pt-BDD electrode and the 50 wt% Pt-GNPs. It was then stabilised on the surface by exposure to glutaraldehyde (GA) vapour for 10 minutes, which has been shown to efficiently immobilise glutamate oxidase on surfaces through cross-linking of amino groups.[51] The response of the electrodes to glutamate was then tested at a potential of 0.5 V for well-stirred PBS solutions containing differing concentrations of glutamate. The concentration of glutamate is around 10 μM in the brain extracellular fluid[52] and around 20 μM in serum and cerebrospinal fluids,[53] so concentrations in this range are of primary interest for practical applications.

The Pt-BDD/GluOx sensor shows a linear response at low glutamate concentrations (0.2 to 50 μM) as shown in Fig. 2(b), with a sensitivity of 370 nA μM^{-1} cm^{-2}, which leads to a detection limit (signal : noise (s–n) = 3) of 220 nM for the 3 mm diameter diamond electrode used, given the typical background noise level of 2 nA in our measurement system. Limits of detection for most electrochemical glutamate sensors are typically in the range 200 nM – 2μM so the performance of the diamond sensor is very competitive in this regard.[22,23,54–57] In comparison, the 50 wt% Pt-GNPs/GluOx sensor shows a slightly higher response of 400 nA μM^{-1} cm^{-2} for the same linear range measured with a detection limit of 320 nM (s–n = 3) due to higher background noise. It is interesting to note out that the glutamate sensitivity of Pt-BDD, despite possessing much higher hydrogen peroxide sensitivity, becomes similar to the glutamate response of Pt-GNPs. By contrast, for the graphene platform, the trend between different platinum loadings follows that observed for H_2O_2 response as seen in Fig. 2(b). This suggests a higher activity or more efficient enzyme immobilisation on the graphene surface compared to the diamond surface in the fabrication methodology employed.

3.4 Selectivity and effects of modifying the diamond sensor with poly-phenylenediamine

The operation of bio-sensors is susceptible to interference by endogenous electroactive species, especially ascorbic acid[22,52] which is normally present at significant concentrations and is easily oxidised. Interference of ascorbic acid, unsurprisingly, is therefore also observed in our diamond glutamate sensor, as shown in Fig. 3(a), which shows equal sensitivity to glutamate and ascorbic acid. A typical method to eliminate the interference of endogenous species is the use of a permselective membrane such as poly-phenylenediamine (PPD) to inhibit the diffusion of larger molecules such as ascorbic acid, while still allowing the diffusion of small molecules such as H_2O_2 to the electrode surface.[58] PPD was thus deposited as a barrier layer to enhance the selectivity. PPD polymerisation onto the electrode was conducted at 0.7 V for 1 minute in 100 mM phenylenediamine solution prepared in PBS using the Pt-decorated diamond electrode before functionalisation with the GluOx. The glutamate sensor was then characterized after at least 5 days storage in PBS at 4 °C in a fridge to achieve stable operating characteristics as discussed in the next section before being subject to systematic measurements.

The diamond sensor with a PPD layer exhibits a much improved selectivity to glutamate than to ascorbic acid, as shown in Fig. 3(b) and (c), although unsurprisingly the sensitivity is reduced by over a factor of 10. The diamond sensor incorporating the PPD layer shows a sensitivity to ascorbic acid of 4 nA μM^{-1} cm^{-2}, 6 times smaller than the sensitivity to glutamate (24 nA μM^{-1} cm^{-2}) as

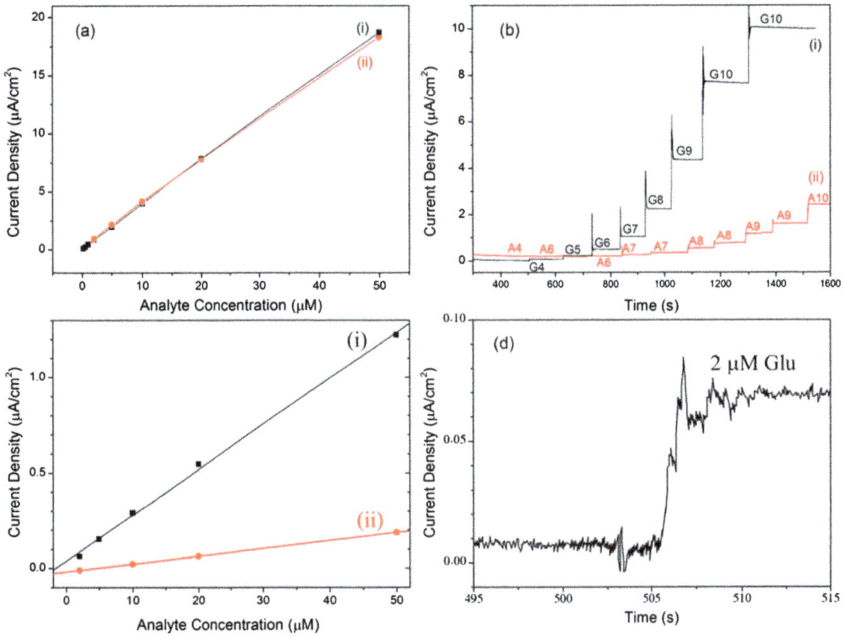

Fig. 3 (a) Calibration curve for the diamond glutamate sensor without a PPD layer for the response to glutamate (i) and ascorbic acid (ii) in a stirred PBS solution. (b) Amperometric response curve of the glutamate sensor with a PPD layer at 0.5 V upon successive injection of (i) glutamate or (ii) ascorbic acid solution in a stirred PBS buffer: 2, 5, 10, 20, 50, 100, 200 μM for G4 to G10 or A4 to A10 respectively. (c) The corresponding calibration plot and (d) background signal noise level of the amperometric response in (b).

shown in the linear plot of current density against analyte concentration in Fig. 3(c). The linear range is still up to 50 μM, but the glutamate detection limit is increased to 350 nM (s–n = 3) as a result of the change in sensitivity as illustrated in Fig. 3(d).

At high glutamate concentrations, the sensor current becomes non-linear as expected and shows characteristics of the Michaelis–Menten kinetics mechanism, as depicted in Fig. 4(a). The enzyme catalysed reaction can be described by the Michaelis–Menten equation,

$$I_{ss} = V_m[S]/(K_m + [S]) \tag{5}$$

where I_{ss} is the steady state current at a given reagent concentration $[S]$, V_m is the maximum current at enzyme saturation, and K_m is the Michaelis–Menten constant, corresponding to the reagent concentration when I_{ss} is equal to half of maximum current (V_m).

A higher V_m value and a lower K_m value are desired for the best sensitivity. The data suggests a K_m value for the diamond glutamate sensor of around 400 μM, greater than the value for the free enzyme which is 200 μM, but encouragingly lies to the lower end of the reported range of K_m values for surface bound glutamate oxidase (250–2800 μM).[17,52] The variation of K_m value is influenced by several aspects, such as steric and electrostatic hindrance by neighbouring GluOx

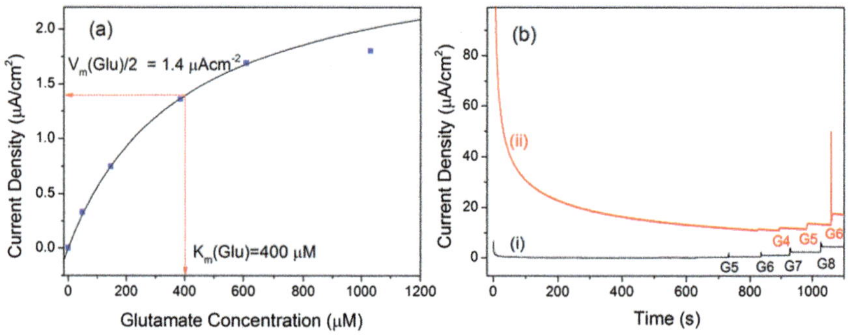

Fig. 4 (a) Calibration plot for diamond glutamate sensor with a PPD layer at high glutamate concentration. The solid line is the best curve fit to the Michaelis–Menten model (b) Amperometric response for (i) BDD based and (ii) glassy carbon based glutamate biosensor at 0.5 V upon successive addition of glutamate to stirred PBS buffer: 5, 10, 20, 50, 100 μM for G4 to G8 respectively.

enzymes.[17] The V_m value depends especially on the loading of GluOx enzyme, and the sensor fabricated here shows a V_m value of 2.8 μA cm^{-2}. As observed elsewhere, the PPD layer has little influence on K_m but reduces V_m.[17] It is expected that a thicker PPD layer could improve the selectivity further, but this is achieved at the cost of sensitivity. The work here therefore incorporates the PPD loading at this level, which represents a compromise between optimal sensitivity and selectivity.

Since differing fabrication procedures in different laboratories can produce varying performance levels making comparisons difficult, an identical sensor was fabricated, but using a glassy carbon electrode rather than diamond to compare the influence of the two substrates. Although qualitatively similar performance levels were noted, significant differences were also found. For both sensors some stabilisation when immersed in background PBS electrolyte is needed to achieve a sensitive analytical performance. Comparisons are made in Fig. 4(b) showing chronoamperometric response curves. The diamond sensor exhibits a small background current density when it is activated in blank solution and rapidly reaches a steady state, enabling fast monitoring of glutamate concentration changes. This must stem in part from the inert nature of the diamond electrode and the low interfacial capacitance, which is known to yield low electrochemical background current densities for this material.[59] In contrast the glassy carbon based sensor shows a large background signal when it is first activated in blank solution so a long incubation time is required before measurements can commence. Even after equilibration the background current is much larger than for the diamond sensor. Because of the varying and large background current, the detection limit was about 2.4 μM (s–n = 3), significantly higher than reported above for the diamond device, and comparable to what is observed for glutamate sensors based on carbon fiber[60] and glassy carbon.[20,22]

Since changes in the background signal are often the limiting issue in controlling the sensitivity of electrochemical measurements in this type of application, rather than signal : electrical noise ratios,[7,12] it is useful to consider the signal : background (S–B ratio) of an electrochemical sensor system. After the background current has stabilised, the glassy carbon based glutamate sensor exhibits an S–B ratio of 0.02 μM^{-1}, similar to the value observed previously for a similar sensor.[22] In comparison the same work shows a ratio of around 0.55 μM^{-1} for Pt, Au and Pd based glutamate biosensors. In contrast, an S–B ratio of 2.8 μM^{-1} is found for the diamond based glutamate sensor, around 140 times higher than that for the glassy carbon device. This stems solely from the low and stable background current seen for the diamond sensor.

3.5 Selectivity and effects of modifying the graphene sensor with poly-phenylenediamine

The glutamate biosensor based on Pt-GNPs was similarly modified with PPD to improve selectivity. The performance of the GC/Pt-GNPs/GluOx/PPD electrode in glutamate sensing was investigated also by amperometric sensing at 0.5 V with successive injections of increasing glutamate concentration into thoroughly stirred PBS solution as shown in Fig. 5(a). The electrode exhibited a good linear response range up to 100 μM with a high sensitivity of 174 nA μM^{-1} cm^{-2}, reflecting a smaller loss of sensitivity with addition of PPD than was observed for the diamond device. Good reproducibility was again seen with less than 10% response variation from sensor to sensor. The detection limit, LOD (s–n = 3) was found to be 0.75 μM, roughly two times higher than for diamond.

The fitted Michaelis–Menten curve of the GC/Pt-GNPs/GluOx/PPD electrode plotted in Fig. 5(b) revealed that K_m = 470 μM, which is comparable to that observed for diamond above and the maximum reaction rate, and V_m is 125.0 μA cm^{-2}, much higher than for the diamond electrode since the loss of activity due to addition of PPD is less.

As previously noted for the diamond sensor, several ubiquitous electrochemically active species in brain fluid may interfere with the sensor affecting the detection of glutamate. This nonetheless was minimised by addition of PPD as before. This permselective membrane provides cation-selective permeability[61] and hence successfully eliminated the interference from ascorbic acid as well as

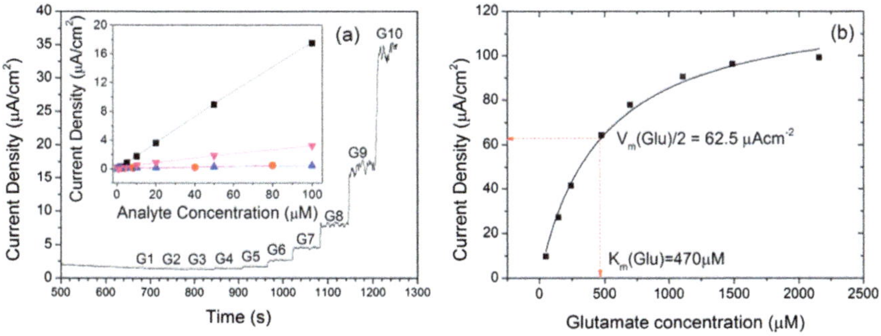

Fig. 5 (a) Amperometric response of 50% Pt-GNPs in stirred PBS buffer at 0.5V with successive additions of increasing concentration of glutamate: 0.1, 0.2, 0.5, 1, 2, 5, 10, 20, 50, 100 μM for G1 to G10 respectively with the inset showing the linear plot of current density against concentrations of glutamate (■) in comparison to commonly found interfering species in physiological conditions; uric acid (●), ascorbic acid (▲) and 4-acetamiophen (▼). (b) Michaelis–Menten plot of current density against concentration of glutamate at 0.5 V by 50% Pt-GNPs.

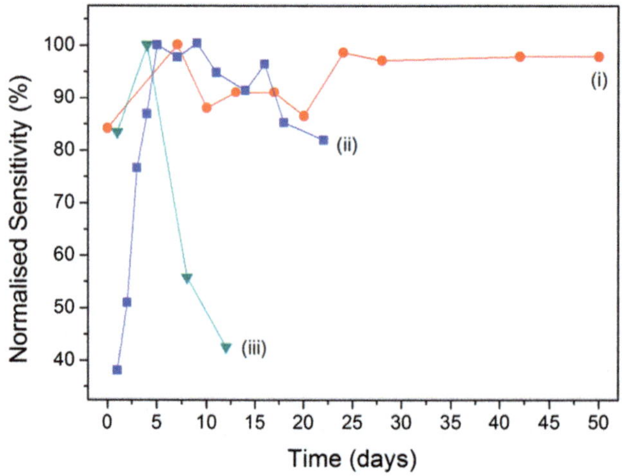

Fig. 6 Normalised sensitivity of (i) glutamate sensors based on Pt-GNPs (ii) Pt-BDD and (iii) Pt-glassy carbon as a function of storage time from fabrication.

uric acid and 4-acetamidophenol as shown as calibration plots in the inset of Fig. 5(a).

3.6 Long term stability

The reproducibility and stability of the fabricated diamond and graphene glutamate sensors were examined, and compared with that of glassy carbon based sensor, as shown in Fig. 6. All three sensors show an increasing sensitivity in the initial days after fabrication, which may be attributed to the opening of membrane channels in the PPD layer,[62] the swelling of PPD membrane[63,64] and conformational relaxation of the glutamate oxidase enzyme[65] when the electrode was stored in buffer solutions. After the relaxation period, the glassy carbon based sensor shows faster decay to less than 50% of its best performance in a week, as seen for similar devices such as SWCNT and Os-gel-HRP modified glassy carbon sensor for glutamate or NADH.[51,66] Frequent measurements and high accumulative concentration of glutamate also have shown adverse effect on the stability of glutamate biosensors, probably due to denaturation of enzyme by local decrease of pH value as a result of hydrogen peroxide oxidation.[63] Despite the high glutamate concentrations used in the stability test, the diamond based glutamate sensors still demonstrated rather stable performance with less than 20% decay in the next three weeks. The graphene based sensor on the other hand maintained around 90% of its highest sensitivity over a long period of 7 weeks. The good long term stability of diamond and graphene based glutamate sensor is attributed to good biocompatibility and the excellent electrochemical stability of the two substrates.[67]

4. Concluding remarks

Electrochemical glutamate biosensors for bioelectronic applications have been demonstrated using boron doped diamond and graphene for the first time. In

Table 1 Comparison of glutamate sensing performance by sensors developed in our laboratory, which emphasizes the largely enhanced sensitivity and linearity range by Pt-graphene hybrid electrode

	GC/Pt-GNPs/ GluOx/PPD	GC/Pt/ PPD/GluOx	BDD/Pt/ PPD/GluOx
Pt loading/μg	2.4	5.1	5.1
LoD/μM	0.75	2.4	0.35
Linearity/μM	0.2–100	0.5–50	0.5–50
Sensitivity/nA μM^{-1} cm^{-2}	174	91	24
Time response/s	4	5	4
V_m/μM cm^{-2}	125	-	2.8
K_m/μM	470	-	400

terms of limit of detection, stability and sensitivity, the devices exhibit comparable or in many cases significantly improved levels of performance compared to similar electrochemical devices produced on other platforms in the past.[23,55,57,68,69]

Since similar sensor fabrication approaches were used to produce devices on three different substrates, the suitability of the characteristics of these substrates with regard to the fabrication procedures used can be judged. Relevant parameters and overall sensor performance, for the devices fabricated, are summarised in Table 1. The mass of Pt was calculated for the diamond and glassy carbon electrodes from the electrolysis charge, whereas for the graphene electrodes from mass added during drop coating using the Pt content as deduced from XPS. The GC/Pt-GNPs/GluOx/PPD electrode structure exhibited superior overall performance particularly in terms of enhanced sensitivity and wide linear range, with the diamond platform showing poor sensitivity, largely because more activity seemed to be lost here when PPD was deposited in comparison to the other electrode materials. However the BDD/Pt/PPD/GluOx sensor on the other hand shows a more desirable detection limit due to very low background and electrical noise. All the sensors exhibited fast time response making them suitable for real-time measurement. Significant differences in stability were also observed. When freshly immersed in PBS buffer, the glassy carbon electrode required long stabilisation times (Fig. 4(b)) and in the stability testing rapid degradation were observed over a timescale of the order of 7 days. In contrast, the diamond sensor showed good stability for the three weeks it was tested for, whist the graphene device showed good stability over a period of seven weeks. Overall the present work illustrates how diamond and graphene materials can be used successfully for glutamate sensing.

Acknowledgements

We would like to acknowledge the financial support from Engineering and Physical Sciences Research Council (EPSRC) in the UK with grant no. EP/F025513/1. S. Wisetsuwannaphum also would like to thank the Royal Thai government for funding the PhD studentship. J. Hu acknowledges support from the Ramsay Memorial Fellowships Trust and Chinese National Thousand (Young) Talents Program.

References

1 G. C. Albert, C. M. Cook, F. S. Prato and A. W. Thomas, *Neurosci. Biobehav. Rev.*, 2009, **33**, 1042–1060.

2 A. L. Benabid, B. Wallace, J. Mitrofanis, C. Xia, B. Piallat, V. Fraix, A. Batir, P. Krack, P. Pollak and F. Berger, *C. R. Biol.*, 2005, **328**, 177–186.

3 M. Dankerl, S. Eick, B. Hofmann, M. Hauf, S. Ingebrandt, A. Offenhausser, M. Stutzmann and J. A. Garrido, *Adv. Funct. Mater.*, 2009, **19**, 2915–2923.

4 P. Fromherz, A. Offenhausser, T. Vetter and J. Weis, *Science*, 1991, **252**, 1290–1293.

5 S. Ingebrandt, C. K. Yeung, M. Krause and A. Offenhausser, *Biosens. Bioelectron.*, 2001, **16**, 565–570.

6 J. Park, V. Quaiserova-Mocko, B. A. Patel, M. Novotny, A. H. Liu, X. C. Bian, J. J. Galligan and G. M. Swain, *Analyst*, 2008, **133**, 17–24.

7 J. Park, Y. Show, V. Quaiserova, J. J. Galligan, G. D. Fink and G. M. Swain, *J. Electroanal. Chem.*, 2005, **583**, 56–68.

8 J. H. T. Luong, K. B. Male and J. D. Glennon, *Analyst*, 2009, **134**, 1965–1979.

9 L. Tang, C. Tsai, W. W. Gerberich, L. Kruckeberg and D. R. Kania, *Biomaterials*, 1995, **16**, 483–488.

10 R. G. Compton, J. S. Foord and F. Marken, *Electroanalysis*, 2003, **15**, 1349–1363.

11 J. Park, J. J. Galligan, G. D. Fink and G. M. Swain, *Anal. Chem.*, 2006, **78**, 6756–6764.

12 B. A. Patel, X. H. Bian, V. Quaiserova-Mocko, J. J. Galligan and G. M. Swain, *Analyst*, 2007, **132**, 41–47.

13 E. Popa, H. Notsu, T. Miwa, D. A. Tryk and A. Fujishima, *Electrochem. Solid-State Lett.*, 1999, **2**, 49–51.

14 S. T. Xie, G. Shafer, C. G. Wilson and H. B. Martin, *Diamond Relat. Mater.*, 2006, **15**, 225–228.

15 T. Bruhn, T. Christensen and N. H. Diemer, *J. Neurosci. Methods*, 1995, **59**, 169–174.

16 A. A. Karyakin, E. E. Karyakina and L. Gorton, *Anal. Chem.*, 2000, **72**, 1720–1723.

17 C. P. McMahon, G. Rocchitta, P. A. Serra, S. M. Kirwan, J. P. Lowry and R. D. O'Neill, *Analyst*, 2006, **131**, 68–72.

18 J. J. Burmeister, F. Pomerleau, M. Palmer, B. K. Day, P. Huettl and G. A. Gerhardt, *J. Neurosci. Methods*, 2002, **119**, 163–171.

19 J. Castillo, A. Blochl, S. Dennison, W. Schuhmann and E. Csoregi, *Biosens. Bioelectron.*, 2005, **20**, 2116–2119.

20 R. Maalouf, H. Chebib, Y. Saikali, O. Vittori, M. Sigaud and N. Jaffrezic-Renault, *Biosens. Bioelectron.*, 2007, **22**, 2682–2688.

21 J. H. Han, H. Boo, S. Park and T. D. Chung, *Electrochim. Acta*, 2006, **52**, 1788–1791.

22 R. D. O'Neill, S. C. Chang, J. P. Lowry and C. J. McNeil, *Biosens. Bioelectron.*, 2004, **19**, 1521–1528.

23 K. B. Male, S. Hrapovic and J. H. T. Luong, *Analyst*, 2007, **132**, 1254–1261.

24 C. E. Banks, T. J. Davies, G. G. Wildgoose and R. G. Compton, *Chem. Commun.*, 2005, 829–841.

25 I. Dumitrescu, P. R. Unwin and J. V. Macpherson, *Chem. Commun.*, 2009, 6886–6901.

26 K. S. Novoselov, A. K. Geim, S. V. Morozov, D. Jiang, Y. Zhang, S. V. Dubonos, I. V. Grigorieva and A. A. Firsov, *Science*, 2004, **306**, 666–669.
27 A. K. Geim and K. S. Novoselov, *Nat. Mater.*, 2007, **6**, 183–191.
28 M. Segal, *Nature Nanotechnology*, 2009, **4**, 611–613.
29 R. I. Jafri, T. Arockiados, N. Rajalakshmi and S. Ramaprabhu, *J. Electrochem. Soc.*, 2010, **157**, B874–B879.
30 S. Alwarappan, A. Erdem, C. Liu and C. Z. Li, *J. Phys. Chem. C*, 2009, **113**, 8853–8857.
31 C. S. Shan, H. F. Yang, J. F. Song, D. X. Han, A. Ivaska and L. Niu, *Anal. Chem.*, 2009, **81**, 2378–2382.
32 X. H. Kang, J. Wang, H. Wu, I. A. Aksay, J. Liu and Y. H. Lin, *Biosens. Bioelectron.*, 2009, **25**, 901–905.
33 N. G. Shang, P. Papakonstantinou, M. McMullan, M. Chu, A. Stamboulis, A. Potenza, S. S. Dhesi and H. Marchetto, *Adv. Funct. Mater.*, 2008, **18**, 3506–3514.
34 L. H. Tang, Y. Wang, Y. M. Li, H. B. Feng, J. Lu and J. H. Li, *Adv. Funct. Mater.*, 2009, **19**, 2782–2789.
35 H. Liu, J. Gao, M. Q. Xue, N. Zhu, M. N. Zhang and T. B. Cao, *Langmuir*, 2009, **25**, 12006–12010.
36 M. Zhou, Y. M. Zhai and S. J. Dong, *Anal. Chem.*, 2009, **81**, 5603–5613.
37 Y. X. Huang, X. C. Dong, Y. M. Shi, C. M. Li, L. J. Li and P. Chen, *Nanoscale*, 2010, **2**, 1485–1488.
38 C. Terashima, T. N. Rao, B. V. Sarada, N. Spataru and A. Fujishima, *J. Electroanal. Chem.*, 2003, **544**, 65–74.
39 F. Ricci and G. Palleschi, *Biosens. Bioelectron.*, 2005, **21**, 389–407.
40 Y. J. Teng, S. H. Zuo and M. B. Lan, *Biosens. Bioelectron.*, 2009, **24**, 1353–1357.
41 L. Zhang, *Biosens. Bioelectron.*, 2008, **23**, 1610–1615.
42 K. B. O'Brien, S. J. Killoran, R. D. O'Neill and J. P. Lowry, *Biosens. Bioelectron.*, 2007, **22**, 2994–3000.
43 N. Q. Jia, Q. Lian, Z. Y. Wang and H. B. Shen, *Sens. Actuators, B*, 2009, **137**, 230–234.
44 J. Hu, X. Lu and J. S. Foord, *Electrochem. Commun.*, 2010, **12**, 676–679.
45 K. Liu, J. Zhang, G. Yang, C. Wang and J.-J. Zhu, *Electrochem. Commun.*, 2010, **12**, 402–405.
46 S. Stankovich, R. D. Piner, X. Q. Chen, N. Q. Wu, S. T. Nguyen and R. S. Ruoff, *J. Mater. Chem.*, 2006, **16**, 155–158.
47 e. D. B. a. M. P. S. C. D. Wagner, *Practical Surface Analysis*, 2nd edn., J. Wiley and Sons, 1990.
48 A. Pozio, M. De Francesco, A. Cemmi, F. Cardellini and L. Giorgi, *J. Power Sources*, 2002, **105**, 13–19.
49 Y. Y. Shao, S. Zhang, C. M. Wang, Z. M. Nie, J. Liu, Y. Wang and Y. H. Lin, *J. Power Sources*, 2010, **195**, 4600–4605.
50 W. Qin and X. Li, *J. Phys. Chem. C*, 2010, **114**, 19009–19015.
51 Z. M. Liu, O. Niwa, T. Horiuchi, R. Kurita and K. Torimitsu, *Biosens. Bioelectron.*, 1999, **14**, 631–638.
52 M. R. Ryan, J. P. Lowry and R. D. ONeill, *Analyst*, 1997, **122**, 1419–1424.
53 R. L. Villarta, G. Palleschi, A. Suleiman and G. G. Guilbault, *Electroanalysis*, 1992, **4**, 27–31.
54 M. Ammam and J. Fransaer, *Biosens. Bioelectron.*, 2010, **25**, 1597–1602.

55 N. Hamdi, J. J. Wang, E. Walker, N. T. Maidment and H. G. Monbouquette, *J. Electroanal. Chem.*, 2006, **591**, 33–40.
56 V. M. Tolosa, K. M. Wassum, N. T. Maidment and H. G. Monbouquette, *Biosens. Bioelectron.*, 2013, **42**, 256–260.
57 T. T. C. Tseng, J. Yao and W.-C. Chan, *Biochem. Eng. J.*, 2013, **78**, 146–153.
58 L. T. Cai and H. Y. Chen, *Sens. Actuators, B*, 1999, **55**, 14–18.
59 M. Wei, Y. L. Zhou, J. F. Zhi, D. G. Fu, Y. Einaga, A. Fujishima, X. M. Wang and Z. Z. Gu, *Electroanalysis*, 2008, **20**, 137–143.
60 O. M. Schuvailo, S. Gaspar, A. P. Soldatkin and E. Csoregi, *Electroanalysis*, 2007, **19**, 71–78.
61 J. M. Elliott, L. M. Cabuche and P. N. Bartlett, *Anal. Chem.*, 2001, **73**, 2855–2861.
62 E. Ekinci, S. T. Ogunc and A. E. Karagozler, *J. Appl. Polym. Sci.*, 1998, **68**, 145–152.
63 K. S. Chang, W. L. Hsu, H. Y. Chen, C. K. Chang and C. Y. Chen, *Anal. Chim. Acta*, 2003, **481**, 199–208.
64 X. N. Zhang and G. A. Rechnitz, *Electroanalysis*, 1994, **6**, 361–367.
65 S. A. M. Marzouk, S. S. Ashraf and K. A. Al Tayyari, *Anal. Chem.*, 2007, **79**, 1668–1674.
66 L. Meng, P. Wu, G. X. Chen, C. X. Cai, Y. M. Sun and Z. H. Yuan, *Biosens. Bioelectron.*, 2009, **24**, 1751–1756.
67 H. L. Fan, L. L. Wang, K. K. Zhao, N. Li, Z. J. Shi, Z. G. Ge and Z. X. Jin, *Biomacromolecules*, 2010, **11**, 2345–2351.
68 M. Zhang, C. Mullens and W. Gorski, *Electrochim. Acta*, 2006, **51**, 4528–4532.
69 S. Chakraborty and C. R. Raj, *J. Electroanal. Chem.*, 2007, **609**, 155–162.

Faraday Discussions

PAPER

Comparison of carbon materials as electrodes for enzyme electrocatalysis: hydrogenase as a case study

Jonathan Quinson,*[ab] Ricardo Hidalgo,[b] Philip A. Ash,[b] Frank Dillon,[a] Nicole Grobert[a] and Kylie A. Vincent[b]

Received 4th April 2014, Accepted 11th April 2014
DOI: 10.1039/c4fd00058g

We present a study of electrocatalysis by an enzyme adsorbed on a range of carbon materials, with different size, surface area, morphology and graphitic structure, which are either commercially available or prepared *via* simple, established protocols. We choose as our model enzyme the hydrogenase I from *E. coli* (Hyd-1), which is an active catalyst for H_2 oxidation, is relatively robust and has been demonstrated in H_2 fuel cells and H_2-driven chemical synthesis. The carbon materials were characterised according to their surface area, surface morphology and graphitic character, and we use the electrocatalytic H_2 oxidation current for Hyd-1 adsorbed on these materials to evaluate their effectiveness as enzyme electrodes. Here, we show that a variety of carbon materials are suitable for adsorbing hydrogenases in an electroactive configuration. This unified study provides insight into selection and design of carbon materials for study of redox enzymes and different applications of enzyme electrocatalysis.

1. Introduction

Carbon materials with a variety of size, surface features, morphology and arrangement of graphitic structure can offer properties relevant for specific electrochemical applications, ranging from carbon fibre microelectrodes up to high surface area carbon blacks used as supports for metal electrocatalysts in fuel-cells.

In this paper, we focus on applications of carbon materials as supports for redox enzymes as electrocatalysts. A number of redox enzymes are amenable to immobilisation on carbon materials (either through covalent attachment or direct adsorption) allowing them to be electrochemically investigated over a wide potential range and in various solution conditions.[1] The electrode replaces

[a]*Department of Materials, University of Oxford, Parks Road, Oxford OX1 3PH, UK. E-mail: nicole.grobert@materials.ox.ac.uk*
[b]*Department of Chemistry, University of Oxford, Inorganic Chemistry Laboratory, South Parks Road, Oxford, OX1 3QR, UK. E-mail: kylie.vincent@chem.ox.ac.uk*

physiological donors and acceptors, and redox enzymes often show high rates of catalysis when immobilised on an electrode.[2] This has led to numerous demonstrations of technological applications of bioelectrocatalysis, including enzyme fuel cells[3] and electrochemical biosensors.[4]

Our model enzyme system in this study is a hydrogenase. This class of enzyme allows micro-organisms to oxidise dihydrogen, or reduce protons to generate H_2. Hydrogenases possess an electron relay chain of iron–sulphur redox centres which transport electrons between the surface of the protein framework and the buried catalytic centre, making them well equipped to function as electrocatalysts (Fig. 1). It is well established that immobilisation of hydrogenases onto an electrode makes it possible to probe a submonolayer film of the enzyme in a technique known as protein film electrochemistry. This approach has been used, for example, to understand potential-dependent features of hydrogenase electrocatalysis and their reactions with inhibitors and to compare features of hydrogenases from different organisms or with different genetic modifications.[5,6]

Pyrolytic graphite has been used most widely in fundamental studies of hydrogenase electrocatalysis, and it is often possible to adsorb enzymes directly without need for specialised attachment strategies, thus simplifying electrode preparation.[7] Pyrolytic graphite has ordered graphitic domains, is very soft, and a fresh surface for enzyme adsorption is readily prepared by mild abrasion.[8] Hydrogenases and other redox enzymes have been shown to interact preferentially with the rough 'edge' surface and at defects on the basal plane of graphitic materials.[7] It has been reported that protein adsorption is aided by fresh abrasion of the pyrolytic graphite prior to protein immobilisation[1] and this is likely to be due to changes in the chemical composition of the surface on contact with air or water. Hydrogenases have also been adsorbed onto plate-like particles of pyrolytic graphite, prepared freshly by abrading a piece of graphite with glass paper immediately prior to exposure to enzyme.[9–11] The preference for fresh preparation of the particles limits batch processing of larger quantities and the large proportion of basal surface on the plate-like particles probably means that the total surface area of pyrolytic graphite particles prepared in this way is not well utilised for enzyme loading.

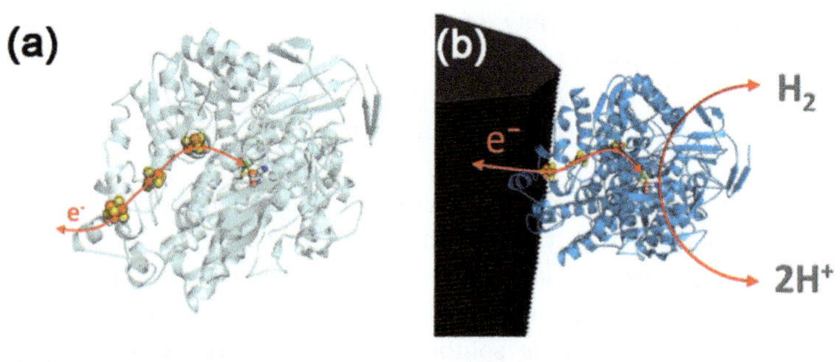

Fig. 1 (a) Representation of the crystal structure of a hydrogenase molecule showing how the enzyme is equipped for electron exchange with an electrode: the protein framework is shown in 'ribbon' format in grey, and metal clusters are shown in elemental colours; (b) schematic representation of direct adsorption of a hydrogenase molecule onto a carbon electrode material (exhibiting exposed graphitic carbon edges) giving rise to efficient electrocatalysis. Representations of the hydrogenase structure (PDB code: 3USC) were prepared using The PyMOL Molecular Graphics System, Version 1.3 Schrödinger, LLC.

The efficiency of hydrogenases as electrocatalysts led rapidly to possible applications for the enzymes in biological fuel-cells,[12,13] H_2-production devices[14] and enzyme-mediated chemical synthesis.[9,10] Identifying alternative carbon materials has now become an important challenge for developing and up-scaling applications of electrocatalysis by hydrogenases and other redox enzymes.[15–17] In order for enzymes to present an attractive alternative to metal catalysts for fuel cells, for example, high surface-area carbon supports are particularly important to facilitate high enzyme loading due to the relatively large footprint for enzymes (typically *ca* 5 nanometres in diameter).

A number of different carbon materials in addition to pyrolytic graphite have been explored for adsorption of hydrogenases, including carbon filaments,[18] carbon powders,[17,19] and various types of carbon nanotubes.[14,20–23] However these studies involve hydrogenases from many different organisms and there has not been a systematic study comparing electrocatalysis by the same enzyme on a range of carbon materials with different properties.[24] This work is an attempt to explore and compare various carbon materials for hydrogenase electrocatalysis. The particular enzyme we select is hydrogenase-1 (Hyd-1) from the common bacterium *Escherichia (E.) coli*, a robust O_2-tolerant enzyme already demonstrated for application in fuel cell devices.[17]

In this study we select and characterise eleven readily obtainable carbon-based materials displaying different shape, size, graphitic features, porosity and surface area, and we assess these as supports for H_2-oxidation electrocatalysis by Hyd-1. Due to the success demonstrated for pyrolytic graphite 'edge' surfaces in hydrogenase electrocatalysis, we selected several particle types with ordered graphitic character. Pyrolytic graphite was abraded with glass paper to produce plate-like particles (APG), already demonstrated in hydrogenase applications,[9] and micron-sized graphite particles (MG) were prepared by sonication of commercial graphite flakes to generate a material with similar graphitic character to the APG but with higher surface area and increased proportion of 'edge' sites. Multi-wall carbon nanotubes (MWCNT) were selected because they have been previously suggested as conductive networks for high surface area electrodes for hydrogenase immobilisation.[25,26] Nitrogen-doped multi-wall carbon nanotubes (N-MWCNT) were also tested because alteration of the carbon lattice by nitrogen doping confers defects and heteroatom sites that may improve hydrophilicity and provide anchor sites for protein attachment. Graphite platelet nanofibers (GNF) from Sigma-Aldrich, consisting of stacked graphitic layers perpendicular to the fibre axis (rather than parallel to the tube as observed in MWCNT) were selected due to the prevalence of graphitic 'edge' surface. In order to assess the suitability of amorphous carbon materials, we selected four commercially-available carbon blacks already used in fuel cell catalysis:[27] Mogul L (ML), Vulcan XC72R (VX) and Black Pearls 2000 (BP) from Cabot Corporation and carbon nanopowder (CNP) from Sigma-Aldrich. In addition to the commercially available amorphous carbon materials, carbon particles prepared by hydrothermal synthesis from a glucose precursor[28] (HNP) were tested because this versatile and well established synthetic approach can be adapted readily for various carbon-based materials and can therefore be tuned for specific applications. To test the applicability of composite materials that could lead to specialised applications, particles with a thin carbon shell surrounding a silicon core were prepared by polymerisation followed by carbonisation (Si@C).[29]

Here we focus on direct adsorption of the enzyme Hyd-1 onto these raw carbon supports. It may be possible to achieve further improvements to enzyme loading by additional treatments to the materials[16,22,26] but in the present study we limit our investigation to as-received or as-synthesised carbon materials in order to identify readily available or easily produced materials for hydrogenase applications and study.

2. Experimental

2.1 Commercially available carbon materials: ML, VX, BP, CNP and GNF

Carbon black powders: Mogul L (ML), Vulcan XC72R (VX) and Black Pearls 2000 (BP) were obtained from Cabot Corporation and used as received. Carbon nanopowder (less than 50 nm particle size, referred to as CNP) and graphite platelet nanofibers (GNF) were purchased from Sigma-Aldrich and used as received.

2.2 Abraded pyrolytic graphite: APG

Pyrolytic graphite blocks were obtained from Momentive Performance Materials. APG flakes were obtained by abrasion of a block of pyrolytic graphite with glass paper (Norton Abrasives, P400 grade). Contrary to previous methods reported, here APG was not freshly prepared prior to enzyme adsorption.[30]

2.3 Micron-sized graphite flakes: MG

Micron-sized (<1 μm) graphite flakes were prepared by modifying a method used to prepare graphene through exfoliation.[31] In a typical experiment 10 g of millimeter-sized graphite flakes (Aldrich, +100 mesh) were placed in 500 mL isopropanol (≥99.5%, Sigma-Aldrich) and exfoliated and broken down by sonication in an ultrasonic bath (FB11301, Fisherbrand, 600 W, 45kHz) for 200 hours. The mixture was then left for 12 hours to sediment, and a grey suspension containing smaller graphite flakes (<1 μm) was pipetted off from the remaining bulk graphite flakes (>1 μm) at the bottom of the flask and sedimented by centrifugation at 3100 × *g* (Hettich, Rotina 420R). After centrifugation the supernatant was decanted and the precipitated graphite flakes that were typically less than 1 μm in size were then collected and the solvent was left to evaporate overnight at 80 °C leading to a powder comprising micron-sized graphite flakes (MG) in yield of 0.5–1 g.

2.4 Multi-wall carbon nanotube synthesis: MWCNT and N-MWCNT

Multi-wall carbon nanotubes (MWCNT) and nitrogen doped multi-wall carbon nanotubes (N-MWCNT) were synthesised by aerosol-assisted chemical vapour deposition (AACVD) as reported in detail previously.[32] Briefly a mixture of 5 wt % ferrocene (98%, Aldrich, purified through sublimation at 90 °C prior to use) and toluene (99.9%, Fluka) for MWCNT, or benzylamine (≥99%, Fluka) for N-MWCNT were injected (for 10 and 20 minutes respectively) into a quartz tube located inside a tube furnace operated at 800 °C using argon (99.999%, BOC) at a flow of 2500 sccm (standard cubic centimetres per minute). The furnace was then left to cool down to <300 °C before the samples of *ca* 500 mg per experiment were collected from the centre of the quartz tube. Both MWCNT and N-MWCNT contained traces of iron impurities (typically 5 wt %)[33] but were used as-synthesised since these impurities were found not to show any electroactivity over the potential range relevant for studies of hydrogenase.

2.5 Hydrothermal carbon particle synthesis: HNP

Carbon particles with a size range 50–200 nm were obtained by hydrothermal synthesis, based upon the procedure of Sun *et al.*[28] In a typical experiment 1.27 g glucose (ACS reagent grade, Sigma-Aldrich) was added to distilled water (35 mL) in a PTFE-lined acid digestion vessel (45 mL, Parr Instrument Company) and the closed vessel was then placed into an oven at 180 °C for 8 hours. Parameters such as time, glucose concentration and temperature were selected to optimise the carbon particle size (100–120 nm) and overall yield. The oven was then left to cool down overnight before the sample consisting of the particles was collected from the vessel. The particles were washed and centrifuged at least 5 times using distilled water and 5 times with ethanol (99.8%, Fisher Scientific) for 10 to 15 minutes at 3100 × *g* per cycle. Subsequently, the particles were left to dry overnight at 80 °C. Typical carbon particle yields were *ca* 100 mg. In line with previous reports,[34,35] we observed that the electrochemical response of materials prepared by this route can be improved by annealing the particles at 1000 °C in argon (100 sccm) for 2 hours.

2.6 Silicon core–carbon shell nanoparticles: Si@C

Silicon core–carbon shell nanoparticles (Si@C) were generated through a polymerisation and carbonisation procedure reported by Gao *et al.*[29] Typically 44 mg 4,4′-sulfonyldiphenol (98%, Aldrich), 20 mg phosphonitrilic chloride trimer (99.95%, Aldrich) and 20 mg silicon nanopowder (≥98%, Aldrich) with a particle diameter of *ca* 100 nm were dispersed using an ultrasonic bath (U1250D from Ultrawave, 200 W, 40 kHz) for 20 hours in 30 mL ethanol (99.8%, Fisher Scientific) and tetrahydrofuran (anhydrous, ≥99.9%, Sigma-Aldrich) in a 1:1 mixture by volume to which 2 mL triethylamine (≥99%, Sigma-Aldrich) had been added. The resulting silicon nanoparticles with a poly(cyclotriphosphazene-4,4′-sulfonildiphenol) shell were washed 3 times with ethanol:tetrahydrofuran solution and left to dry at 80 °C overnight. The particles were then annealed under an argon atmosphere (100 sccm) at 900 °C for 2 hours to form the carbon shell.

2.7 Characterisation of materials

For the eleven carbon materials reported here, we investigate the following criteria: the size and shape of the materials, their graphitic structure, their surface area and their mesoporosity (porosity on a scale relevant for enzyme adsorption).

The overall morphology and structure of all materials presented here was assessed using scanning and transmission electron microscopy (SEM and TEM) in conjunction with Raman spectroscopy. Specific surface area and porosity were investigated by nitrogen adsorption measurements.

SEM images were recorded using a Jeol 840F microscope with a secondary electron detector operated at 5 kV and a Zeiss NVision FIB microscope equipped with an in-lens detector and operated at 5 kV. TEM images were taken with a Jeol 2000FX and a Jeol 2010 both operated at 200 kV.

The degree of graphitisation of the samples was investigated using a JY Horiba Labram Aramis imaging confocal Raman microscope (532 nm excitation) whereby the ratio of peak intensities of the D band (around 1340 cm^{-1}) and G band (around 1575 cm^{-1}) (I_D/I_G), and the presence or absence of 2D peak (also called G′, around 2675 cm^{-1}) were identified. In simple terms, for carbon materials the D peak intensity correlates with defects in the carbon lattice and is linked to the extent of sp^3 hybridisation while the G peak arises from the graphitic network and extent of sp^2 hybridisation. If I_D/I_G is smaller than 1 the structures are considered to show a higher concentration of sp^2 hybridisation (*i.e.* extended graphitic character). If the ratio is close to 1 then the material is considered to have a more amorphous structure with a mixture of sp^2 and sp^3 hybridisation. Even in cases where I_D/I_G is close to 1, the presence of a 2D peak, more related to interactions between graphene-like planes, suggests a strong graphitic component.

Specific surface areas of samples were evaluated using a Micromeritics Gemini V BET surface area analyser (nitrogen adsorption at 77 K). For data collection the samples were left at 200 °C in dinitrogen overnight prior to the adsorption measurements and the data were analysed using the BET model for relative pressure between 0.05 and 0.3 for all samples. For BP, HNP and Si@C the Langmuir model was more suitable as it gave a better fit to the data, therefore for these materials we calculate both the BET and Langmuir surface areas. The molecular cross sectional area of dinitrogen was taken as 0.1620 nm^2. The Barrett–Joyner–Halenda (BJH) model was used to estimate pore characteristics.[36]

2.8 Adsorption of hydrogenase

The enzyme Hyd-1 was purified from *E. coli* following a published procedure[37] and used at a concentration of 20 μM unless otherwise specified. Each of the 11 carbon materials was dispersed at 20 mg mL^{-1} by sonication in ultra-high purity water (Millipore MilliQ, 18 MΩ cm). Adsorption of Hyd-1 onto the carbon materials was carried out in a N_2-filled glove box (<1 ppm O_2, Glove Box Technology Ltd.). A 5 μL aliquot of Hyd-1 solution was added to 5 μL of each material dispersion and the mixture was left overnight at 4 °C to allow Hyd-1 to adsorb. Carbon materials were then washed 3 times with 20 μL potassium phosphate buffer (100 mM, pH 6.0) prepared from potassium phosphate dibasic (≥98%, Sigma-Aldrich) and monobasic (≥99%, Sigma). The enzyme-modified carbon materials were then separated by centrifugation (7000 × *g*, MiniSpin, Eppendorf). To each sample, 5 μL buffer was added.

2.9 Electrocatalytic activity of hydrogenase adsorbed on the different particles

The electrocatalytic activity of adsorbed hydrogenase was assessed in an anaerobic glove box using a water jacketed (25 °C) electrochemical cell sealed onto an electrode rotator (EG&G model 636) with a rotating disc electrode (RDE). A gas inlet and outlet in the cell make it possible to flow 1 bar H_2 (99.99%, BOC) at 1000 sccm through the headspace above the cell solution. Measurements were performed with the electrode rotating at 2000 rotations per minute (rpm) unless otherwise stated, with steps to higher rotation rates used to test for mass transport limitation. The working electrode was prepared by drop-casting 0.5 or 1 μL (as specified) of the various material dispersions with adsorbed Hyd-1 onto the surface of a pyrolytic graphite 'edge' electrode of geometrical area 0.02 or 0.04 cm^2 which had been first polished with 1 μm α-alumina powder and cleaned by sonication in MilliQ water for at least one minute. For each material, an electrode prepared without Hyd-1 showed no Faradaic current response in the presence of H_2 over the potential window used for experiments in this study.

The counter electrode was a platinum wire and the reference electrode a saturated calomel electrode (SCE, BAS). Potentials were converted to volts (V) against the standard hydrogen electrode (SHE) using the relation $E(\text{SHE}) = E(\text{SCE}) + 0.2412$ V at 25 °C. Potentials were applied with an EcoChemie Autolab PGSTAT 128N potentiostat controlled by GPES software. The hydrogenase was activated for at least 5 minutes at −0.559 V *vs* SHE prior to measurements and further activated by sequences of 5 minutes at this same potential if the current recorded was noticed to increase during cyclic voltammetry measurements indicative of incomplete hydrogenase activation.[30] Scans were acquired at 10 mV s^{-1} between −0.559 and +0.241 V *vs* SHE. The absence of mediators[26] in all experiments means that the catalytic current arises solely from enzyme molecules adsorbed in an electroactive configuration and in direct electronic contact with the carbon.[1]

3. Results

Structural characterisation using TEM and SEM was carried out to reveal the general morphology and size of the commercial materials (ML, VX, BP, CNP and GNF) and the carbon materials produced in-house following established recipes (APG, MG, MWCNT and N-MWCNT, HNP and Si@C). SEM (Fig. 2 and Fig. 3) and TEM images (Fig. 3 and Fig. 5), Raman spectra (Fig. 4) and surface area and porosity measurements using nitrogen adsorption (Fig. 6) for all 11 materials were found to be in agreement with data provided by suppliers or reported in the literature[9,28,29,32,38–42] and results are summarised in Table 1. All materials were tested using electrochemistry for Hyd-1 adsorption and electrocatalysis (Fig. 7 to Fig. 10).

3.1 SEM/TEM characterisation of material size and morphology

Fig. 2 shows SEM images of materials which can be classed as plate-like (APG and EG) or fibre-like (MWCNT, N-MWCNT and GNF). Images presented in the left hand panel were recorded at the same magnification and are presented to show differences in the overall morphology of the samples (Fig. 2(a), (c), (e), (g) and (i)), while images presented in the right hand panel show magnifications selected to highlight more specific aspects of the material shapes and sizes, (Fig. 2 (b), (d), (f), (h) and (j)). Further SEM and TEM images (not shown here) were also used in analysis of particle size and shape distributions.

We found that the diameter of the flat surface of the APG plates is in the range 3–5 μm whereas the MG plates are smaller, falling mostly in the range 0.5–1 μm. A rough estimate of the thickness (t) of the plates (Table 1) was also evaluated from examination of the SEM images (see Table 1). The MG plates are thinner (<300 nm) than APG (>500 nm). These observations confirm that exfoliation of graphite flakes is a relevant synthetic route to obtain small-size graphite plate-like materials (MG).

MWCNT and N-MWCNT have typical diameters below 100 nm, and a length up to 150 μm and 120 μm respectively. TEM images confirm that they are both hollow materials and qualify as tubes.[32] The hollow nature of the N-MWCNT can

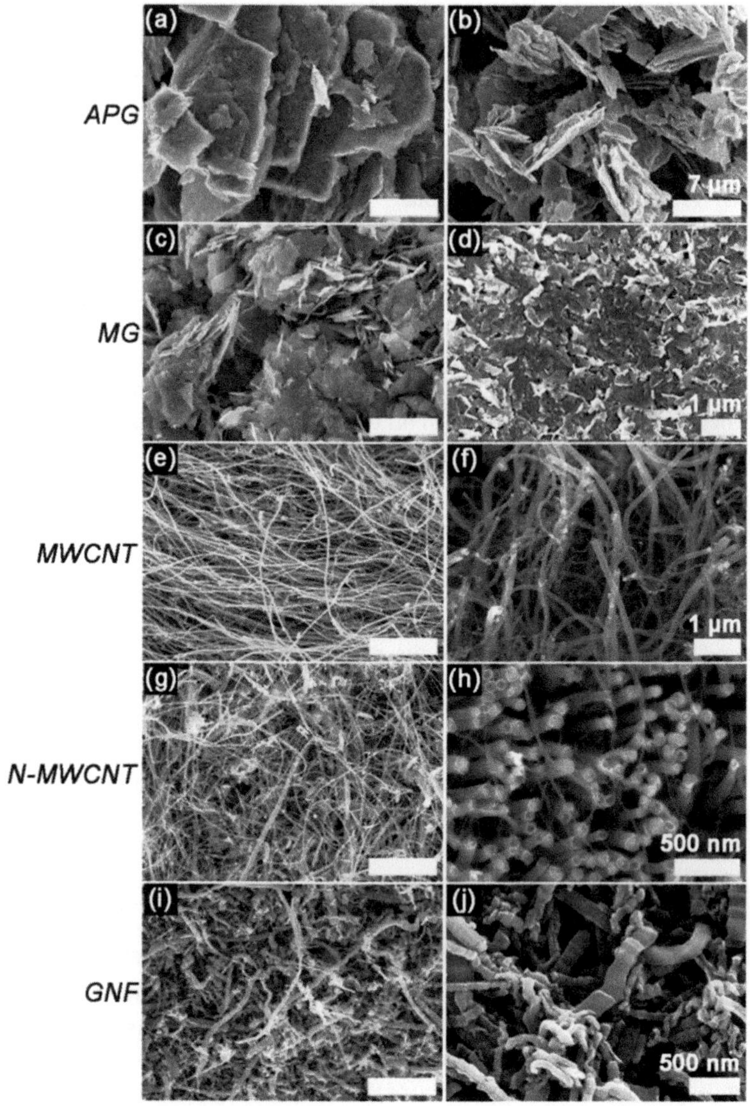

Fig. 2 Scanning electron micrographs of plate-like and fibre-like carbon materials. Images taken at the same magnification ((a), (c), (e), (g), (i); scale bars = 3 μm) reveal the overall morphology. In order to highlight specific morphological features of each material, images were also taken at different magnifications ((b), (d), (f), (h), (j); scale as indicated).

also be observed by looking at the catalyst-free tips of the N-MWCNT at higher magnification SEM images (see circular features in Fig. 2(h)). In contrast GNF are not hollow and consist of stacked layers (discussed further in section 3.2) with diameters in the range 20–250 nm, and are typically shorter than 3 μm.

The structural nature of the materials that have particle-like morphologies (CNP, ML, VX, BP, HNP and Si@C) is displayed in Fig. 3 which is split into SEM images in the left hand panel (Fig. 3(a), (c), (e), (g), (i) and (k)) and TEM images

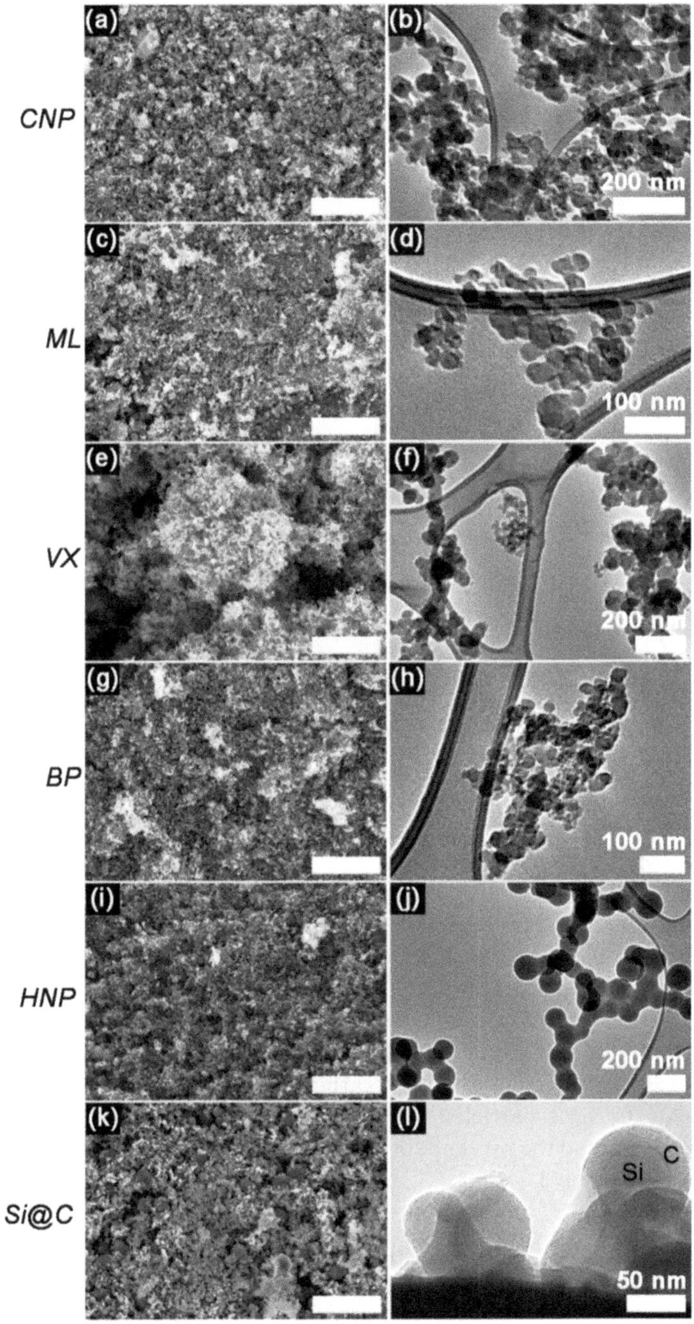

Fig. 3 Scanning and transmission electron micrographs of particle-like materials. Images taken at the same magnification ((a), (c), (e), (g), (i), (k); scale bars = 3 μm) reveal the overall morphology. In order to highlight specific morphological features of each material, images were also taken at higher magnifications ((b), (d), (f), (h), (j), (l); scale as indicated).

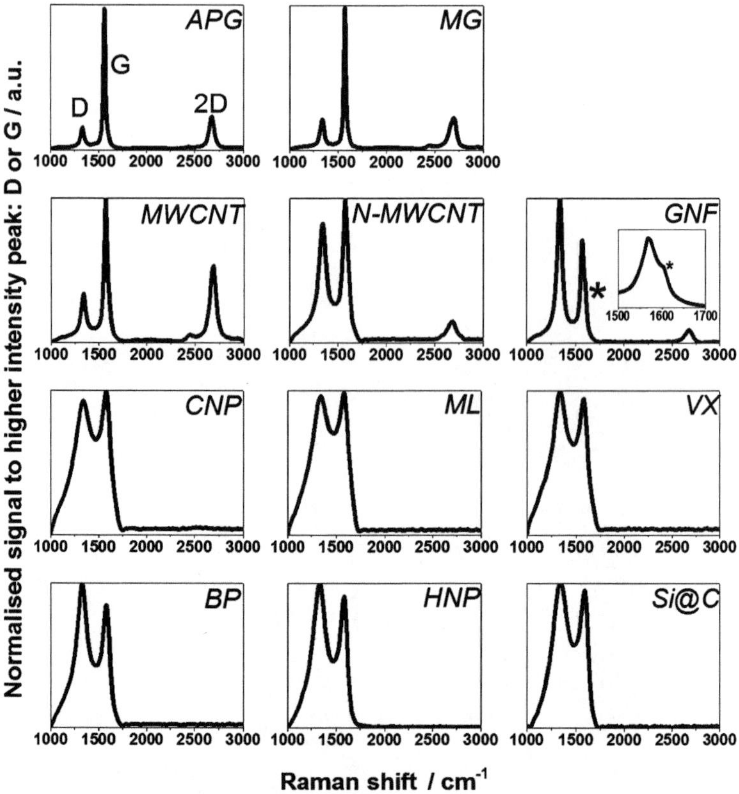

Fig. 4 Raman spectra of all 11 carbon samples investigated confirming the graphitic and amorphous nature of individual materials respectively. The * marks the D′ shoulder on GNF (see also the inset).

in the right hand panel (Fig. 3(b), (d), (f), (h), (j) and (l)). The electron microscopy studies showed that all the commercially available materials (CNP, ML, VX, BP) possess an average particle size less than 50 nm (Table 1) while the particle size for in-house synthesised HNP ranged between 100 nm and 120 nm. TEM images of the Si@C particles are consistent with a silicon core (50–150 nm) coated with a thin carbon shell (10–20 nm). The core-shell structure of the Si@C particles was indicated by the difference in contrast in the TEM images and the presence of the carbon shell was confirmed by Raman analysis (see section 3.2).

3.2 Characterisation of graphitic structure by Raman spectroscopy and high-resolution TEM

Raman spectra for each material are shown in Fig. 4, and I_D/I_G values are summarised in Table 1. Three main peaks in the Raman spectrum were considered in order to assess the degree of graphitisation in the materials: the D, G and 2D peaks at *ca* 1340, 1575 and 2675 cm^{-1}, respectively, as discussed in section 2.7. APG and MG display I_D/I_G less than 1 and show a strong 2D peak, confirming that the graphitic structure (sp^2 network) of the parent materials is retained during

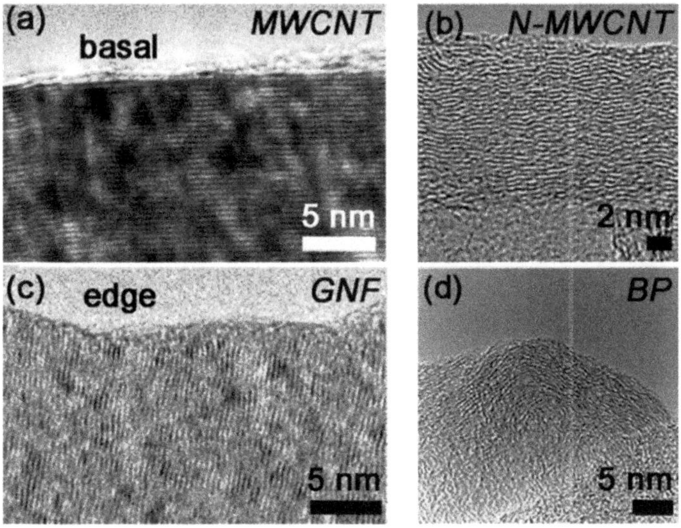

Fig. 5 High-resolution TEM images revealing the surface morphologies of (a) MWCNT, (b) N-MWCNT, (c) GNF and (d) BP. The image of BP is reprinted from ref. 40, with permission from Elsevier. The image of N-MWCNT is reprinted from ref. 32, with permission from Elsevier.

preparation of these plate-like particles. Similarly, the I_D/I_G of the MWCNT is less than 1 and the spectrum also shows a strong 2D peak. A high-resolution TEM (HRTEM) image of a MWCNT edge, Fig. 5(a), confirms an ordered graphitic structure and the majority of the exposed MWCNT surface is basal-plane type carbon. The presence of 'edge'-like structures can be expected at the open end of the tube only and are therefore not likely to be abundant in the material.[43] N-MWCNT shows I_D/I_G close to 1 and the presence of a 2D peak. Production of nitrogen-doped carbon nanotubes is known to introduce defects in the graphitic carbon network,[44] consistent with the observed increase in I_D/I_G and smaller 2D peak compared to MWCNT. A HRTEM image of N-MWCNT (Fig. 5(b)) confirms that this material has a less ordered graphic structured compared to the MWCNT. The presence of nitrogen sites has also been demonstrated.[32] The Raman spectrum of GNF shows I_D/I_G greater than 1 and a 2D peak confirming graphitic character. In contrast to MWCNT, HRTEM study of GNF, Fig. 5(c), shows the predominance of graphitic 'edge'-like features making up the majority of the GNF surface. The Raman spectrum of GNF also displays a D′ peak (shoulder on the D peak, see inset in Fig. 4) which is also related to an abundance of stacked platelets making up the fibre with their edges oriented perpendicular to the fibre axis and therefore responsible for the 'edge'-like features (rather than basal plane-like walls) along the fibre surface.[42]

As expected the remaining materials (CNP, ML, VX, BP, HNP, Si@C) have a Raman spectrum indicative of a rather amorphous structure without extended crystalline domains, with I_D/I_G close to 1 and no 2D peak. A HRTEM image of the BP particles reported by Zhu *et al.* (see Fig. 5(d)) confirms that these particles do not exhibit obvious 'edge'-like character.[40]

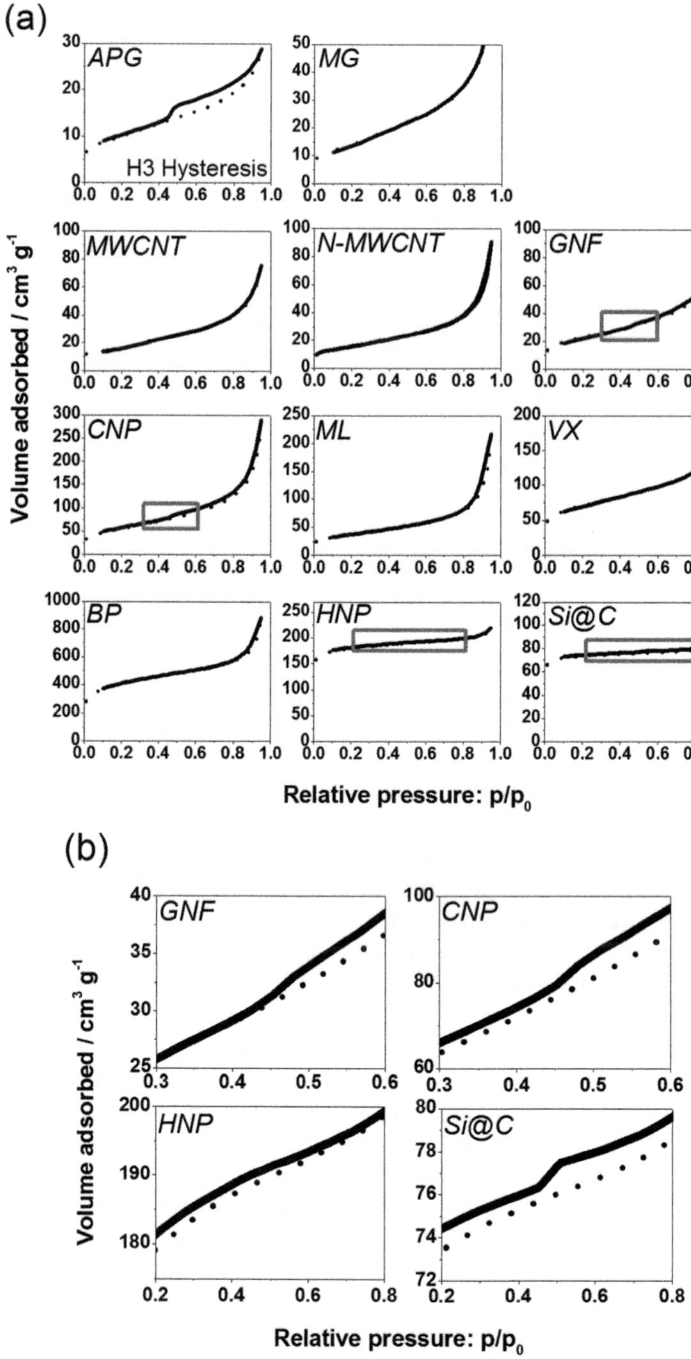

Fig. 6 (a) Nitrogen adsorption (dotted) and desorption (solid) isotherms at 77 K for each carbon material investigated. Grey boxes indicate the part of the isotherm which is enlarged in panel (b) for GNF, CNP, HNP and Si@C to highlight the H3 hysteresis for GNF, CNP, and the H4 hysteresis observed in the HNP and Si@C sample.

Table 1 Characteristics of carbon materials investigated

Material	Abbreviation	Source	Typical size	Raman I_D/I_G	Model used for specific surface area	Specific surface area ($m^2 g^{-1}$)	Pore diameter[a] (nm)	Pore volume[b] ($cm^3 g^{-1}$)
Abraded Pyrolytic Graphite	APG	Abrasion	*D*: 3–5 μm *t* > 500 nm	0.17 ± 0.04	BET	35.2 ± 0.1	4.6	0.05
Micron-sized Graphite	MG	Exfoliation	*D*: 500 nm–1μm *t* < 300 nm	0.19 ± 0.05	BET	52.5 ± 0.5	5.9	0.11
Graphite platelet NanoFibres	GNF	Sigma-Aldrich	*D*: 20–250 nm *L* < 3 μm	1.20 ± 0.11	BET	81.0 ± 0.4	5.7	0.14
Carbon Nanotubes	MWCNT	AACVD	*D*: 50–100 nm *L*: 50–150 μm	0.36 ± 0.04	BET	60.0 ± 0.6	6.1	0.12
Nitrogen doped Carbon Nanotubes	N-MWCNT	AACVD	*D*: 50–100 nm *L*: 40–120 μm	0.89 ± 0.11	BET	57.1 ± 0.2	7.2	0.14
Hydrothermal Nanoparticles	HNP	Hydrothermal synthesis	*D*: 100–120 nm	1.08 ± 0.01	Langmuir BET	816 ± 4 550 ± 14	3.6	0.34
Carbon NanoPowder	CNP	Aldrich	*D*: 30–50 nm	0.94 ± 0.08	BET	198.0 ± 0.5	6.9	0.45
Mogul L	ML	Cabot Corporation	*D*: 25–40 nm	0.96 ± 0.05	BET	127.6 ± 0.4	8.7	0.34
Vulcan XC72R	VX	Cabot Corporation	*D*: 12–63 nm	1.06 ± 0.05	BET	237 ± 3	5.1	0.28
Black Pearls 2000	BP	Cabot Corporation	*D*: 17–31 nm	1.18 ± 0.03	Langmuir BET	2053 ± 24 1332 ± 22	4.7	1.36
Silicon Core Carbon Shell Nanoparticles	Si@C	Polymerisation and Carbonisation	Core *D*: 50–150 nm Shell *D*: 10–20 nm	1.05 ± 0.03	Langmuir BET	328.9 ± 0.8 221 ± 5	3.8	0.13

[a] average pore diameter estimated by BJH desorption (4*V*/*A*). [b] single point measurement at relative pressure 0.95 for pores less than 40 nm; *t*: thickness; *L*: length; *D*: diameter, in the cases where the Langmuir model gives a better fit to the data than the BET model, values obtained with the Langmuir model are specified in addition to the results obtained using the BET model.

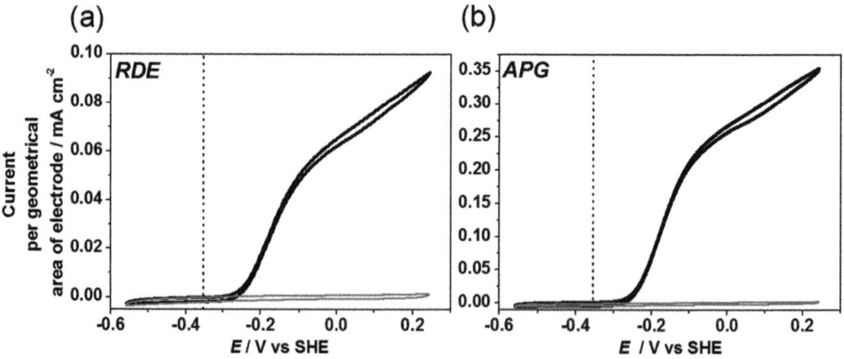

Fig. 7 Cyclic voltammograms of Hyd-1 (a) adsorbed directly on a pyrolytic graphite RDE with the 'edge' surface exposed or (b) on APG deposited on the RDE. The electrode was rotated at 2000 rpm in pH 6.0, 100 mM potassium phosphate buffer under 1 bar H_2 at 25 °C. The dotted vertical lines mark the thermodynamic potential of H^+/H_2 couple under the conditions of the experiment. The grey voltammograms are obtained under identical conditions with no hydrogenase deposited on the RDE.

3.3 BET characterisation: surface area

The surface area and porosity of the materials was assessed by nitrogen adsorption,[36] as shown in Fig 6(a), and respective values are summarised in Table 1. The observed porosity results from two factors: the surface properties of the material itself but also from aggregation of the carbon particles leading to porous networks in the material powder.

APG, MG, MWCNT, N-MWCNT, GNF, CNP, ML and VX all display a type II isotherm. This is an isotherm concave to the pressure axis, since at relative pressure 0 the volume of gas adsorbed must be 0, and displaying a B point marking the beginning of a linear relationship between adsorption and relative pressure. The presence of the type II isotherm is indicative of the predominance of rather large pores (macropores with diameter greater than 50 nm). Pores in this size range are likely to be due to aggregation of the materials, which can also be seen in the SEM and TEM images, Fig. 2 and Fig. 3.

Fitting the isotherms according to the BET model gives a specific surface area lower than 100 $m^2 g^{-1}$ for APG, MG, MWCNT, N-MWCNT and GNF. CNP, ML and VX have a specific surface area between 100 and 250 $m^2 g^{-1}$. Thus the plate-like and fibre-like particles fall into a lower surface-area range, consistent with the larger sizes of these materials. APG display a pronounced H3 hysteresis: the desorption isotherm does not follow exactly the adsorption isotherm at high relative pressure, Fig 6(a). GNF and CNP show a less pronounced H3 hysteresis, as evident from the enlargement of part of the isotherm in Fig 6(b). The combination of a type II isotherm and presence of H3 hysteresis is designated as a type II(b) isotherm. A type II(b) isotherm is usually observed in powders and aggregates, and thus is consistent with the nature of these materials.

The nitrogen adsorption isotherm for BP, HNP and Si@C differs slightly from the previously described type II isotherms. A more pronounced plateau at high relative pressure is observed indicative of a type I isotherm, suggesting the presence of micropores (pore diameter less than 2 nm). This is in agreement with the fact that for these three materials the Langmuir model gives a better fit than the BET model since the Langmuir model is known to give a better fit for materials with micropores.[36] The specific surface area of these materials is greater than 300 $m^2\ g^{-1}$ (as determined from the Langmuir model). Values obtained from the BET model are also included in Table 1 for comparison. HNP and Si@C show evidence of H4 hysteresis as shown in Fig 6(b). This behaviour is usually associated with activated carbons and materials displaying slit-shaped pores especially in the micropore range.[36] This is also consistent with the better fit of the Langmuir model for data from HNP and Si@C.

A convenient estimation of the average size of pores in a material can be inferred by dividing the pore volume (V) in the material by the surface area (A) of the material. Assuming the pores can be modelled by perfect cylinders, the average pore diameter is equal to $4V/A$.[36] For all materials considered in this study, the average mesopore size is within the range 3.5 to 9 nm (Table 1), although pores of smaller and larger diameter can also be found in the materials. Since all materials have a pronounced mesoporous structures (type II isotherm) and mesoporous volume (Table 1) most of the porosity of the whole material should be accessible to hydrogenase molecules (*ca* 5 nm diameter). The materials with smallest average pore size (<4 nm) are the synthesised materials, HNP and Si@C.

3.4 Electrocatalytic activity of adsorbed Hyd-1

A typical cyclic voltammogram for Hyd-1 adsorbed onto a rotating disc electrode (RDE) made of pyrolytic graphite with the edge surface exposed is reported in Fig. 7(a). The electrode is immersed in buffered aqueous solution (pH 6.0) equilibrated with 1 bar H_2 (*ca* 0.8 mM). Hyd-1 does not show significant activity for proton reduction under a H_2 atmosphere,[45] so no Faradaic current is observed at potentials more negative than −0.3 V. At higher potentials a clear oxidative catalytic wave is observed. No Faradaic current is observed under H_2 for a bare electrode without enzyme adsorbed. Each cycle of catalytic turnover by a molecule of hydrogenase leads to electron flow into the electrode, and thus the catalytic current is directly proportional to the number of electroactive enzyme molecules, provided that the current is not limited by availability of H_2 to the enzyme.[7] The electrode is rotated rapidly (2000 rpm) to maintain the H_2 concentration at the electrode surface close to that in the bulk solution. Hyd-1 exhibits an onset potential for H_2 oxidation just positive of −0.3 V, consistent with the slight overpotential requirement (50–60 mV) of this enzyme relative to the $E(H^+/H_2)$ potential at 1 bar H_2.[45] The thermodynamic potential for $E(H^+/H_2)$ under the experimental conditions is indicated by a vertical line in Fig. 7. The residual slope in the cyclic voltammogram at higher potentials is characteristic of immobilised enzyme electrocatalysts engaged in direct electron transfer, and has been explained by a dispersion of orientations of the enzyme on the electrode leading to a range of electron transfer rates.[46] A similar waveshape, although with higher electrocatalytic current, is obtained for Hyd-1 adsorbed on plate-like particles of pyrolytic graphite (APG) deposited onto the RDE (Fig. 7(b)). This is consistent with an earlier report in which Hyd-1 on APG was deposited on a RDE in a Nafion binder.[30]

In order to assess the potential benefit of alternative materials to APG we deposited an equal amount of each material, modified with Hyd-1, onto a RDE and recorded the associated catalytic current in a cyclic voltammetric sweep in solution equilibrated with H_2 (Fig. 8). For the results reported in this figure, electrodes were prepared by drop-casting 0.5 μL of a 20 mg mL^{-1} dispersion of the hydrogenase-modified particles onto a RDE of surface area 0.02 cm^2. Thus the current per geometric area of electrode is directly proportional to the current per gram of material. It is observed that the commercial materials with higher specific surface area (CNP, VX, ML and BP) give higher current densities. However the shape of the cyclic voltammogram for those materials yields a plateau at high potentials, characteristic of a film in which H_2 is not efficiently provided to the enzyme, and thus current is limited by mass transport. Rotation at a higher rate of 5000 rpm increases the observed current by improving the H_2 supply to the film (see the dashed line in Fig. 8 for BP). Further increase in rotation rate to 7000 rpm (not shown) provides negligible improvement in current and the plateau shape is still present, suggesting that fast rotation fails to provide efficient mass transport *through* the particle film to buried hydrogenase molecules.

For CNP, VX, ML and BP, to obtain conditions that are not limited by mass transport at the RDE we diluted the particle dispersions to 5 mg mL^{-1}, applying 0.5 μL to the RDE. This allowed us to assess the suitability of each carbon material for adsorption of Hyd-1 in an electroactive configuration. Results are shown in Fig. 9. Also shown on this figure are MWCNT and N-MWCNT modified with Hyd-1. Difficulty in dispersing the relatively hydrophobic nanotubes in water means that the mass of nanotubes deposited cannot be controlled reliably, so the absolute current is not comparable with the other materials. Importantly, the similarity in waveshape (general form and onset potential) for Hyd-1 on all the carbon materials suggests that the carbon support has little effect on enzyme

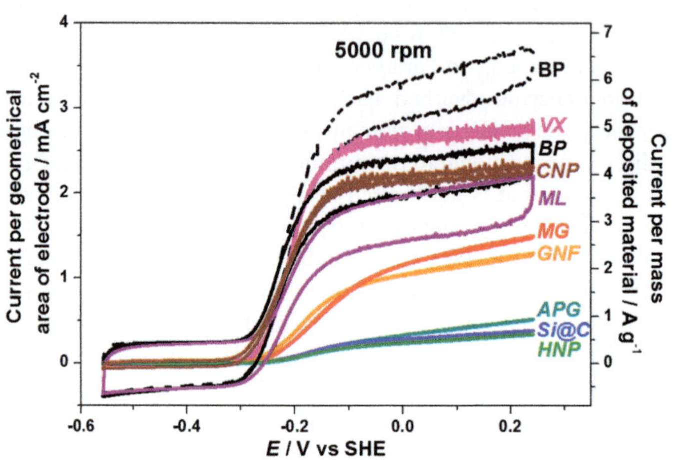

Fig. 8 Cyclic voltammograms of Hyd-1 adsorbed on various carbon materials in 100 mM potassium phosphate buffer, pH 6.0, 25 °C, equilibrated with 1 bar H_2. Materials were deposited on a 0.02 cm^2 pyrolytic graphite RDE with the 'edge' surface exposed, and the electrode was rotated at 2000 rpm. The left hand axis shows the current normalised to the geometric area of electrode and the right hand axis indicates the current per gram of deposited carbon material. In all cases 0.5 μL of 20 mg mL^{-1} dispersion of carbon material with adsorbed enzyme were deposited on the RDE. The current associated with BP material is reported at 2000 rpm (BP solid line) and 5000 rpm (BP dashed line).

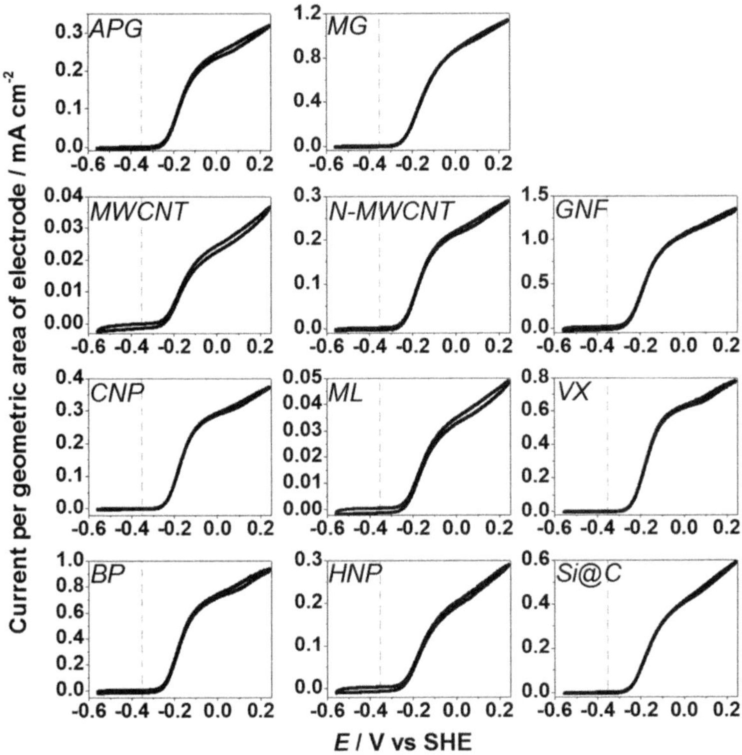

Fig. 9 Cyclic voltammograms of adsorbed Hyd-1 on the various carbon materials deposited onto a pyrolytic graphite RDE with the edge surface exposed, in 100 mM potassium phosphate buffer, pH 6.0, 25 °C, equilibrated with 1 bar H_2. Dilution of the dispersion of certain particles (CNP, ML, VX and BP, see text) means that currents are not significantly mass-transport limited. The dashed lines mark the thermodynamic potential of the H^+/H_2 couple under the experimental conditions.

electroactivity. Therefore, despite their intrinsic differences all materials investigated accommodate Hyd-1 in a suitable configuration to perform H_2 oxidation electrocatalysis.

The mass transport limitation observed for CNP, ML, VX and BP in Fig. 8 is likely to be a problem for realistic applications of hydrogenase electrocatalysis. We therefore investigated Hyd-1 activity on BP in a wall-jet electrochemical cell configuration where mass transport can be improved. In Fig. 10, scans (a) and (b) were recorded at an RDE electrode, whereas scan (c) was recorded with the same amount of material in the wall-jet cell. At high potentials a higher current is reached in the wall-jet cell and the plateau is absent suggesting that mass transport to enzyme molecules in the particle film is no longer significantly limiting the current. In order to test whether enzyme loading on BP could be improved by longer adsorption times from more concentrated enzyme solution, an electrode was prepared in the wall-jet cell using particles that had been incubated with 0.2 mM Hyd-1 for 48 hours at 4 °C, scan (d). In this case an improved current per mass of material of 12 A g^{-1} (at +0.24 V *vs* SHE) was recorded.

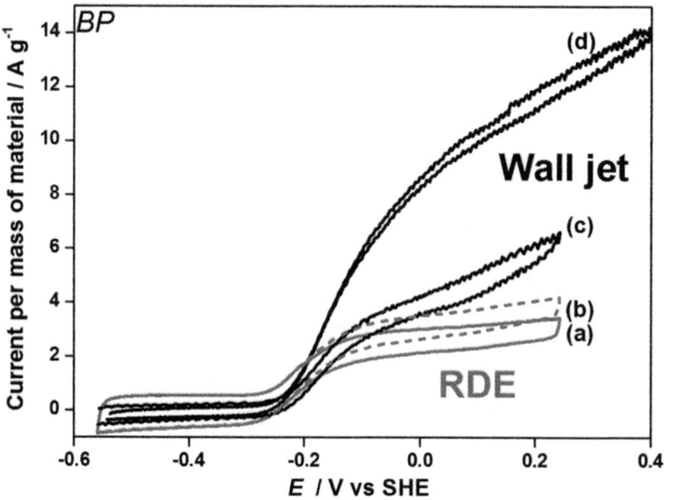

Fig. 10 Cyclic voltammograms of Hyd-1 adsorbed on BP assessed on a pyrolytic graphite RDE (grey) at (a) 2000 rpm or (b) 5000 rpm; and cyclic voltammograms for the same quantity of BP studied in a wall-jet cell configuration (with H_2-saturated solution flowing at 50 mL min^{-1}) for (c) the same set of particles as (a) and (b); and (d) for particles prepared with longer adsorption time (48 hours) from more concentrated Hyd-1 solution (0.2 mM). All were prepared by deposition of BP (1 μL) from a 20 mg mL^{-1} suspension.

4. Discussion and conclusions

The materials considered in this study fall into three broad categories with respect to their general shape and morphology. APG and MG are micron-sized platelets with low specific surface area compared to the other materials. They are highly graphitic, and it is expected that the thin edges of the plates expose graphite 'edge'-like regions that have been associated with effective protein adsorption. The fibre-like materials, MWCNT, N-MWCNT and GNF, have diameters in the nanometer range and up to 250 nm for GNF and have slightly higher surface area than the plate-like materials and are also graphitic. Whereas MWCNT and N-MWCNT show a predominance of basal-like surface, GNF has mainly 'edge'-like surface structure. CNP, ML, VX, BP, HNP and Si@C are particle-like materials that do not show an ordered graphitic structure but have the advantage of a high surface area. All materials show some extent of mesoporosity, on an appropriate scale for enzyme adsorption.

In this study we have investigated all of these materials as supports for electrocatalysis by the enzyme Hyd-1 from *E. coli* in order to present a unified comparison of different types of carbon with a single enzyme. The results demonstrate that a range of carbon materials can be used for adsorption of Hyd-1 in an electroactive configuration. The catalytic waveshapes in voltammograms recorded under conditions where mass transport is not significantly limiting (Fig. 9) are very similar to the well-characterised waveshape reported for Hyd-1 adsorbed on a pyrolytic graphite electrode (Fig. 7(a)). Importantly, the onset potential for H_2 oxidation is independent of material, showing that interfacial electron transfer to and from hydrogenase molecules is effective on each carbon material. The residual slope in the electrocatalytic voltammogram at high potential is similar for each material (Fig. 9), suggesting that there are not significant differences in the distribution of orientations of Hyd-1. The results of this study suggest that any conductive carbon-based material is likely to be a candidate for hydrogenase study or application.

All of the carbon materials give an increase in catalytic current per geometrical area of the electrode relative to Hyd-1 on a bare pyrolytic graphite electrode (Fig. 7(a)). In this set of experiments, the same mass of particles was applied to each electrode and so the magnitude of the current also gives a direct indication of the electrocatalytic activity per gram of carbon. A significant increase in current is observed for the commercially available carbons, BP, VX, ML and CNP but in each case the voltammogram is severely mass transport limited, showing a plateau at high potential. As demonstrated for BP (the highest surface area material with the most pronounced mesoporous structure), mass transport can be improved slightly by increasing the electrode rotation rate but the plateau remains. The mesopores, with an average diameter in the range 3.5–9 nm (Table 1) for all materials used in this study, are of similar dimension to Hyd-1 which has a diameter of roughly 5 nm. This is advantageous since the mesoporous network should be accessible to Hyd-1. Enzyme molecules inside the pores are likely to be almost completely surrounded by carbon, favouring contact of their electron relay centres with the electrode surface so that the majority of molecules are electroactive. Templated ordered mesoporous carbons and other inorganic materials have been developed specifically for the adsorption of enzymes in biotechnology, but usually involve use of a sacrificial template, multi-step synthesis, and may display relatively poor conductivity.[47,48] The success of Hyd-1 adsorption on BP and other commercial carbon blacks shows that it is possible to make use of readily available mesoporous carbons for enzyme electrocatalysis.

In general the materials which are most affected by mass transport limitations are those which also have the most developed mesoporous structures (Table 1), suggesting poor transport of H_2 within electrodes constructed from these materials. Mass transport limitation at mesoporous carbon electrodes modified with hydrogenase has been noted by other authors.[15] In Fig. 10 we compare the same preparation of Hyd-1 modified BP on a RDE (scans (a) and (b)) or in a cell with a wall-jet configuration (scan (c)). In the wall-jet cell, the current per gram of BP is substantially increased and the waveshape no longer shows a plateau, resembling that at the pyrolytic graphite electrode, Fig. 7(a). In the voltammograms of Fig. 8 electrodes formed from larger or less mesoporous particles, such as APG, MG and GNF show less evidence of mass transport limitation, although the relative currents are lower. This illustrates that in addition to selecting carbon materials to improve enzyme loading, another important consideration is optimisation of mass transport at the electrode.

Fig. 11 shows the voltammetric data from Fig. 8 recalculated to show the current density with respect to the surface area (BET) of each material. Fig. 11 gives an indication of how effectively the total surface area of each material is being used for electrocatalysis. In this case the highest current densities occur for materials with larger size or lower specific surface area. In particular APG, MG and GNF compare more favourably with the other materials when evaluated on the basis of current per surface area of material, showing that the surface of these materials is used efficiently for electrocatalysis. The fact that these materials with

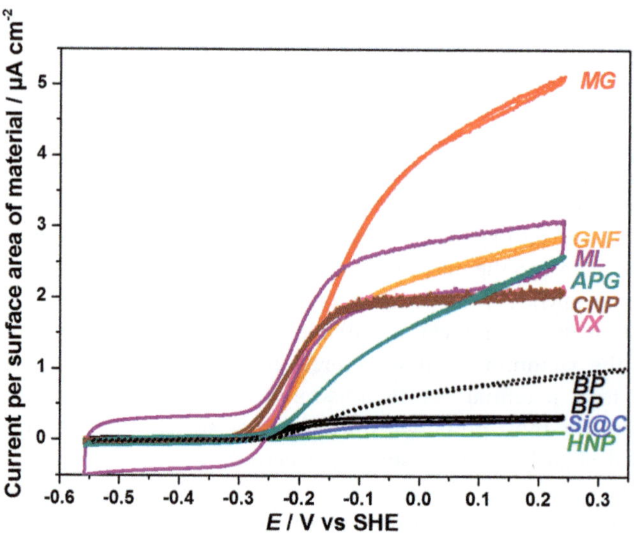

Fig. 11 Voltammetric data from Fig. 8 reported as current per surface area (BET) of material. For comparison, the scan obtained in the wall-jet configuration cell from Fig. 10(d) is also reported (dotted BP voltammogram).

exposed graphitic 'edge' surfaces show high currents per surface area of material is in line with the successful use of pyrolytic graphite 'edge' electrodes for protein film electrochemistry. Results associated with MG suggest that even more favourable current densities could be obtained from smaller particles of these graphitic materials. Conversely, materials with larger specific surface areas (BP, Si@C and HNP) show very low current densities in Fig. 11. This is particularly the case for BP which has a specific surface area (BET) of 1332 m^2 g^{-1}. Even after improving mass transport to Hyd-1 on BP by use of a wall-jet cell, the high surface area of the material is clearly being under-utilised (dashed line, Fig. 11). Slow diffusion of Hyd-1 into the mesoporous network of BP will lead to very long equilibration times for adsorption of the enzyme. Restricted or hindered diffusion of macromolecules through pores of a similar diameter is a well-documented phenomenon,[49] and is the basis of size exclusion chromatography. Enzyme loading on BP was improved by allowing Hyd-1 to adsorb over a longer time and from more concentrated solution (Fig. 10(d) and Fig. 11). In this case an improved current per surface area of material of 0.9 μA cm^{-2} was obtained. Hyd-1 is a particularly robust hydrogenase, and for some metalloproteins it would be unfeasible to allow such long adsorption times due to protein degradation. The fact that the current for BP in the wall-jet cell with longer adsorption time for Hyd-1 exceeds that obtained for the initial particle preparation (Fig. 10) at all potentials during H_2 oxidation implies that a significant amount of enzyme can adsorb within the mesopores of the material where efficient mass transport becomes particularly important.

Carbon nanotubes have been widely used in development of applications of hydrogenases.[25] Attractive features of MWCNT or N-MWCNT are the opportunity to develop relatively high surface area electrodes with a mesoporous network (Table 1, Fig. 2). We observed that direct adsorption of Hyd-1 is possible on

MWCNT and N-MWCMT, but their dispersion in aqueous solution is challenging due to their intrinsic hydrophobicity. A slight improvement in dispersion is observed for N-MWCNT compared to MWCNT, but we were still unable to make accurate comparisons of current per mass of material deposited for the nanotubes. It might be expected that the defects and heteroatom sites brought about by nitrogen incorporation within the carbon lattice in N-MWCNT should offer anchor-points for specific attachment of enzymes, but this is beyond the scope of the present study.

In other hydrogenase electrocatalysis studies, nanotubes have been handled in specific ways, for example, multi-wall carbon nanotubes grown directly on a gold substrate in a study by De Lacey and co-workers.[20] Carbon nanotubes have also been further modified for covalent attachment of hydrogenase, for example commercially available multi-wall carbon nanotubes modified by adsorption of a pyrene derivative which was covalently linked to Hyd-1 *via* an amide linkage were demonstrated in a high surface-area electrode for a hydrogenase H_2/O_2 fuel cell by Armstrong and co-workers.[16] Lojou and co-workers used chemically modified single and multi-wall carbon nanotubes for covalent attachment of hydrogenase from *Aquifex aeolicus*; stabilising the nanotubes network in the polymer electrolyte, Nafion.[26] These examples highlight the need for specialised handling of carbon nanotubes if they are to be used successfully for hydrogenase electrocatalysis.

The fact that high currents per gram of material can be achieved by Hyd-1 on commercial carbon blacks (BP, VX and ML) is promising for applications of hydrogenase electrocatalysis, for example in enzyme fuel cells or H_2-production devices, provided that mass transport issues can be properly addressed in the design of the device.

Successful adsorption of hydrogenase is not limited to commercially available materials. In-house generated particles HNP and Si@C investigated in this study are good examples of materials that can be easily tailored to accommodate hydrogenases for specific purposes. HNP material can be obtained by the versatile and simple hydrothermal synthesis:[50–52] a green synthetic route using glucose (a renewable feedstock) as starting material and requiring readily-available laboratory equipment. The high surface area material obtained has also been demonstrated in various other electrochemical devices[35,53] and so is a promising platform for tailoring of materials for enzyme electrocatalysis. Si@C may have interesting applications as an electrode material for coupling to *in situ* infrared spectroscopy as it has an IR-transparent silicon core but a sufficiently conductive outer carbon shell (10 to 20 nm) to support enzyme adsorption and electrochemistry. These examples illustrate opportunities to select and develop simple carbon materials for specific applications of hydrogenases.

Acknowledgements

We are grateful for financial support from Engineering and Physical Sciences Research Council [INSPIRE grant EP/J015202/1 (J.Q., N.G. and K.A.V.), EPSRC DTA award EP/J500495/1 (J.Q.), EPSRC Pathways to Impact Block Awards (N.G.) and EPSRC Small Equipment Award EP/K031503/1 (K.A.V.)], The Royal Society (N.G.), the European Research Council (ERC) [EnergyBioCatalysis-ERC-2010-StG-258600 (K.A.V. and P.A.A.) and DEDIGROWTH-ERC-2009-StG-240500 and DEVICE

ERC-2012-PoC-309786 (N.G.)]. J.Q. is also grateful for support from The Keble Association and Keble College Oxford. R.H. is supported by Ministerio de Ciencia y Tecnología and Univesidad de Costa Rica and Lincoln College Oxford. The authors thank Miss Judy Britton and Dr Antal A. Koós for their help in the acquisition of high resolution TEM images, and Prof. Fraser A. Armstrong and Elena Nomerotskaia for providing a sample of *E. coli* Hyd-1.

References

1 C. Léger and P. Bertrand, *Chem. Rev.*, 2008, **108**, 2379–2438.

2 F. A. Armstrong and J. Hirst, *Proc. Natl. Acad. Sci. U. S. A.*, 2011, **108**, 14049–14054.

3 J. A. Cracknell, K. A. Vincent and F. A. Armstrong, *Chem. Rev.*, 2008, **108**, 2439–2461.

4 W. Yang, K. R. Ratinac, S. P. Ringer, P. Thordarson, J. J. Gooding and F. Braet, *Angew. Chem., Int. Ed.*, 2010, **49**, 2114–2138.

5 S. V. Hexter, F. Grey, T. Happe, V. Climent and F. A. Armstrong, *Proc. Natl. Acad. Sci. U. S. A.*, 2012, **109**, 11516–11521.

6 A. A. Hamdan, P.-P. Liebgott, V. Fourmond, O. Gutiérrez-Sanz, A. L. D. Lacey, P. Infossi, M. Rousset, S. Dementin and C. Léger, *Proc. Natl. Acad. Sci. U. S. A.*, 2012, **109**, 19916–19921.

7 K. A. Vincent, A. Parkin and F. A. Armstrong, *Chem. Rev.*, 2007, **107**, 4366–4413.

8 C. F. Blanford and F. A. Armstrong, *J. Solid State Electrochem.*, 2006, **10**, 826–832.

9 K. A. Vincent, X. Li, C. F. Blanford, N. A. Belsey, J. H. Weiner and F. A. Armstrong, *Nat. Chem. Biol.*, 2007, **3**, 761–762.

10 H. A. Reeve, L. Lauterbach, P. A. Ash, O. Lenz and K. A. Vincent, *Chem. Commun.*, 2012, **48**, 1589–1591.

11 O. Lazarus, T. W. Woolerton, A. Parkin, M. J. Lukey, E. Reisner, J. Seravalli, E. Pierce, S. W. Ragsdale, F. Sargent and F. A. Armstrong, *J. Am. Chem. Soc.*, 2009, **131**, 14154–14155.

12 A. A. Karyakin, S. V. Morozov, E. E. Karyakina, S. D. Varfolomeyev, N. A. Zorin and S. Cosnier, *Electrochem. Commun.*, 2002, **4**, 417–420.

13 K. A. Vincent, J. A. Cracknell, J. R. Clark, M. Ludwig, O. Lenz, B. Friedrich and F. A. Armstrong, *Chem. Commun.*, 2006, 5033–5035.

14 D. Svedruzic, J. L. Blackburn, R. C. Tenent, J.-D. R. Rocha, T. B. Vinzant, M. J. Heben and P. W. King, *J. Am. Chem. Soc.*, 2011, **133**, 4299–4306.

15 A. de Poulpiquet, H. Marques-Knopf, V. Wernert, M. T. Giudici-Orticoni, R. Gadiou and E. Lojou, *Phys. Chem. Chem. Phys.*, 2014, **16**, 1366–1378.

16 S. Krishnan and F. A. Armstrong, *Chem. Sci.*, 2012, **3**, 1015–1023.

17 L. Xu and F. A. Armstrong, *Energy Environ. Sci.*, 2013, **6**, 2166–2171.

18 S. V. Morozov, E. E. Karyakina, N. A. Zorin, S. D. Varfolomeyev, S. Cosnier and A. A. Karyakin, *Bioelectrochemistry*, 2002, **55**, 169–171.

19 A. I. Yaropolov, A. A. Karyakin, S. D. Varfolomeev and I. V. Berezin, *Bioelectrochem. Bioenerg.*, 1984, **12**, 267–277.

20 M. A. Alonso-Lomillo, O. Rudiger, A. Maroto-Valiente, M. Velez, I. Rodriguez-Ramos, F. J. Munoz, V. M. Fernandez and A. L. De Lacey, *Nano Lett.*, 2007, **7**, 1603–1608.

21 J. Baur, A. Le Goff, S. Dementin, M. Holzinger, M. Rousset and S. Cosnier, *Int. J. Hydrogen Energy*, 2011, **36**, 12096–12101.
22 F. J. M. Hoeben, I. Heller, S. P. J. Albracht, C. Dekker, S. G. Lemay and H. A. Heering, *Langmuir*, 2008, **24**, 5925–5931.
23 E. Lojou, X. Luo, M. Brugna, N. Candoni, S. Dementin and M. T. Giudici-Orticoni, *JBIC, J. Biol. Inorg. Chem.*, 2008, **13**, 1157–1167.
24 S. V. Morozov, E. E. Karyakina, O. A. Zadvornyi, N. A. Zorin, S. D. Varfolomeev and A. A. Karyakin, *Russ. J. Electrochem.*, 2002, **38**, 97–102.
25 E. Lojou, *Electrochim. Acta*, 2011, **56**, 10385–10397.
26 A. De Poulpiquet, A. Ciaccafava, K. Szot, B. Pillain, P. Infossi, M. Guiral, M. Opallo, M.-T. Giudici-Orticoni and E. Lojou, *Electroanalysis*, 2013, **25**, 685–695.
27 M.-x. Wang, Q. Liu, H.-f. Sun, N. Ogbeifun, F. Xu, E. A. Stach and J. Xie, *Mater. Chem. Phys.*, 2010, **123**, 761–766.
28 X. M. Sun and Y. D. Li, *Angew. Chem., Int. Ed.*, 2004, **43**, 597–601.
29 P. Gao, J. Fu, J. Yang, R. Lv, J. Wang, Y. Nuli and X. Tang, *Phys. Chem. Chem. Phys.*, 2009, **11**, 11101–11105.
30 A. J. Healy, H. A. Reeve, A. Parkin and K. A. Vincent, *Electrochim. Acta*, 2011, **56**, 10786–10790.
31 Y. Hernandez, V. Nicolosi, M. Lotya, F. M. Blighe, Z. Sun, S. De, I. T. McGovern, B. Holland, M. Byrne, Y. K. Gun'ko, J. J. Boland, P. Niraj, G. Duesberg, S. Krishnamurthy, R. Goodhue, J. Hutchison, V. Scardaci, A. C. Ferrari and J. N. Coleman, *Nat. Nanotechnol.*, 2008, **3**, 563–568.
32 A. A. Koós, M. Dowling, K. Jurkschat, A. Crossley and N. Grobert, *Carbon*, 2009, **47**, 30–37.
33 S. S. Meysami, F. Dillon, A. A. Koós, Z. Aslam and N. Grobert, *Carbon*, 2013, **58**, 151–158.
34 Z. Wen, Q. Wang, Q. Zhang and J. Li, *Electrochem. Commun.*, 2007, **9**, 1867–1872.
35 X. XiaoHong, S. Lei, L. HongBo, Y. Li and H. YueDe, *J. Phys. Chem. Solids*, 2012, **73**.
36 F. Rouquerol, J. Rouquerol, K. S. W. Sing, P. Llewellyn and G. Maurin, *Adsorption by Powders and Porous Solids: Principles, Methodology and Applications*, Elsevier, 2nd edn, 2014.
37 M. J. Lukey, A. Parkin, M. M. Roessler, B. J. Murphy, J. Harmer, T. Palmer, F. Sargent and F. A. Armstrong, *J. Biol. Chem.*, 2010, **285**, 20421–20421.
38 D. Pantea, H. Darmstadt, S. Kaliaguine and C. Roy, *Appl. Surf. Sci.*, 2003, **217**, 181–193.
39 M. Li, W. Li and S. Liu, *Carbohydr. Res.*, 2011, **346**, 999–1004.
40 W. Z. Zhu, D. E. Miser, W. G. Chan and M. R. Hajaligol, *Carbon*, 2004, **42**, 1841–1845.
41 M. Maja, C. Orecchia, M. Strano, P. Tosco and M. Vanni, *Electrochim. Acta*, 2000, **46**, 423–432.
42 A. Ambrosi, T. Sasaki and M. Pumera, *Chem. – Asian J.*, 2010, **5**, 266–271.
43 C. E. Banks and R. G. Compton, *Analyst*, 2006, **131**, 15–21.
44 A. A. Koós, F. Dillon, E. A. Obraztsova, A. Crossley and N. Grobert, *Carbon*, 2010, **48**, 3033–3041.
45 M. J. Lukey, A. Parkin, M. M. Roessler, B. J. Murphy, J. Harmer, T. Palmer, F. Sargent and F. A. Armstrong, *J. Biol. Chem.*, 2010, **285**, 3928–3938.

46 C. Léger, A. K. Jones, S. P. J. Albracht and F. A. Armstrong, *J. Phys. Chem. B*, 2002, **106**, 13058–13063.
47 P. F. Fulvio, R. T. Mayes, X. Wang, S. M. Mahurin, J. C. Bauer, V. Presser, J. McDonough, Y. Gogotsi and S. Dai, *Adv. Funct. Mater.*, 2011, **21**, 2208–2215.
48 D. I. Fried, F. J. Brieler and M. Froeba, *ChemCatChem*, 2013, **5**, 862–884.
49 M. G. Davidson and W. M. Deen, *Macromolecules*, 1988, **21**, 3474–3481.
50 X. M. Sun and Y. D. Li, *J. Colloid Interface Sci.*, 2005, **291**, 7–12.
51 M. Liu, C. Wang and X. Wang, *J. Mater. Chem.*, 2011, **21**, 15197–15200.
52 Z. Liu, C. Zhang, L. Luo, Z. Chang and X. Sun, *J. Mater. Chem.*, 2012, **22**, 12149–12154.
53 M. Liu, Y. Yan, L. Zhang, X. Wang and C. Wang, *J. Mater. Chem.*, 2012, **22**, 11458–11461.

Faraday Discussions

DISCUSSIONS

Carbon electrode interfaces for synthesis, sensing and electrocatalysis: general discussion

Yury Gogotsi, Richard McCreery, Stephen M. Lyth, Robert Dryfe, John Foord, Matěj Velický, Julie Macpherson, Matteo Duca, Katherine Holt, Manuel Alvarez-Guerra, Heisi Kurig, Surbhi Sharma, Patrick R. Unwin, George Zheng Chen, Milo Shaffer, Taiwo Alaje, Robert Hamers, Mark Newton, Philip A. Ash, Keith Stevenson, Siegfried Waldvogel, Jingping Hu, Aleix Güell and Jonathan Quinson

DOI: 10.1039/C4FD90038C

Richard McCreery opened the discussion of the paper by Robert Hamers: You use "Conduction band of water" early in your presentation. Could you clarify the nature of this 'band'. Is it the LUMO of a water molecule?

Robert Hamers replied: One can think of water as being like an amorphous semiconductor, in that there are strong interactions between adjacent atoms but there is no long-range order. J.V. Coe published several important papers in which he describes the water conduction band. Terminology is important, as one can refer to an optical conduction band and an adiabatic conduction band. The optical conduction band is determined by finding the position of the valence band relative to vacuum, and then using the optical bandgap (onset of optical absorption) to define the position of the conduction band. For our purposes, the more important definition is that of the adiabatic conduction band, which is essentially determined by the electron affinity level of the water. That is, it refers to a state in which there is an excess electron, but the water molecules have not yet polarized around it to form a solvated electron. In this limit, the electron is essentially a plane wave and water behaves like a continuum dielectric, with the electron stabilized with respect to vacuum by ~0.1 V due to the long-range polarization forces that the electron exerts. A key aspect of the 'adiabatic conduction band' is that it has been used to explain how electrons created in water (usually created by high-energy ionizing radiation) are able to travel significant distances in water before the water finally polarizes around them to form solvated electrons, which then have much lower mobility. In our studies, the conduction band may be important in establishing how far the solvated electrons are located away from the electrode surface. If they enter the water with ~1 eV of kinetic energy (roughly the energy difference between the conduction band of diamond and the conduction band of water), then they may be able to travel some distance away from the electrode before becoming solvated. We have no direct evidence of this from our own studies, but electron mobility in such a band may be important in understanding the spatial distribution of electrons. A second aspect of the conduction band that may be relevant is in understanding the dynamics of how an electron is emitted from diamond into water. From a quantum mechanical viewpoint, we like to think about photoemission as arising from excitation from a bound state of the solid into an excited state of the solid that is degenerate with a plane wave state of the emitted electron, so that the 'final state' of the excitation process is a excited state that has some density of states in the solid and some density of states in the vacuum. Because photoemission is a nearly instantaneous process, one would imagine that in order to get into the water in the first place there must be an electronic state of the electron in water that can couple to the states in the diamond conduction band, meaning that they must have the same energy.

Richard McCreery continued: Has anyone done DFT or similar theory on the structure of a hydrated electron? Do they provide insights into the sequence of events following electron injection into water?

Robert Hamers responded: Yes, there are a number of calculations, it's a tough computational problem because the energetics involve long-range polarization forces and the short-range order of water. Yet, there are many attempts to look at these processes and there is continued progress in the level of treatments and their accuracy.

Katherine Holt said: You mention that you observe electron emission at longer wavelengths than the band gap of diamond. Is this attributed to band gap states in the diamond and if so are these bulk states or surface states?

Robert Hamers answered: There are studies in vacuum showing sub-bandgap emission of free electrons. In our experiments we also see emission at longer wavelengths, using several different types of diamond including very high purity 'electronic' grade diamond. There are computational studies showing the H-terminated diamond has unoccupied states in the bandgap that derive from the antibonding states of the C–H bonds formed during hydrogen termination. So, on high purity diamond we believe these C–H surface states may be playing a role. On lower purity samples, such as heavily boron-doped diamond, it is likely that mid-gap states associated with bulk nitrogen, with graphitic pi* states, or from the boron states can all possibly contribute. Certainly much more study is needed to understand the mechanism of sub-bandgap emission in water.

Katherine Holt continued: In your previous work on the generation of ammonia from nitrogen, the reaction is proposed to proceed *via* the reaction of an H atom with N_2, rather than direct reduction of the nitrogen molecule by addition of an electron. In the example of CO_2 reduction you propose that CO_2 is reduced to the CO_2^- anion by addition of an electron. Why in this case does CO_2 not react with an H atom?

Robert Hamers replied: Actually that could happen. The order of reduction potentials is such that the hydrogen atom (H) is about 0.4 eV higher in energy than the CO_2 anion. Consequently, if CO_2 encounters an H atom it will transfer an electron *via* $CO_2 + H \rightarrow CO_2^- + H^+$. So, solvated electrons could directly reduce CO_2 to the CO_2^- anion, or they could reduce H^+ to H atoms, when reacting with CO_2 to reform H^+ and CO_2^-. The net result is the same.

Katherine Holt added: So to ensure the CO_2 reduction reaction proceeds rather than the competing H_2 evolution, you have to use increased concentrations of CO_2?

Robert Hamers responded: Yes, that is correct. Using a higher concentration of CO_2 makes sure that the rate of formation of CO_2^- is higher than the rate of formation of H atoms.

Katherine Holt asked: Why does the CO_2^- radical not just couple or dimerise as is noted for this reaction at electrodes?

Robert Hamers replied: The CO_2^- anion is photochemically active when illuminated with UV light, and so dissociates to CO and O. The reactive species are present at relatively low concentration in solution, and consequently the simple dissociation is kinetically favored over bimolecular coupling.

John Foord enquired: One of the features of doing electrochemistry at hydrogen terminated surfaces, is that they are relatively unstable and so become oxidised. Why isn't there a similar problem in the present work?

Robert Hamers answered: H-terminated diamond is known to be very stable in water and air, and H-terminated samples can be kept in the atmosphere for long periods of time (months) without significant oxidation. With that said, when in direct contact with water and illuminated with ultraviolet light some oxidation might be expected, and in fact we do see this by taking samples and analyzing them *via* XPS. However, the process is slow, in part because diamond's valence band is very shallow, and not able to oxidize water (for example). In our experiments we usually separate the anodic (oxidation) and cathodic (reduction) reactions in two separate half-cells; in this configuration we ask diamond to do the reduction reaction, and we ask the Pt electrode in contact with sulfite ions to do the necessary oxidation reaction (necessary to maintain charge neutrality). After some time we do see some oxidation and reduction in catalytic activity, but at least over periods of 2 days or so (the longest we've done continuously) the electrode remains effective. Direct measurements of solvated electron yield show that the H-terminated surface gives high electron yield and the O-terminated surface gives lower electron yield, but reduced only perhaps by a factor of 2 or 3 compared with the H-terminated surface.

Julie Macpherson said: Can you comment on the effect of film thickness on your solvated electron emission *versus* photocatalytic activity? For example, in the paper you say that a growth time of 2 h gives a maximum electron yield, whilst a growth time of 1 h yields the best photocatalytic activity.

Robert Hamers explained: As diamond films grow there are several things that change in addition to the thickness. The microscopic morphology of the surface and the resulting microscopic surface area also change. In general, we see that both the electron yield and the catalytic activity toward ammonia formation (nitrogen reduction) exhibit a maximum. Given the error bars on both of these, I don't think the difference between 2 h for electron yield and 1 h for photocatalytic yield is particularly significant. The more important point is that both of them show a maximum at a similar thickness, and that thick diamond films show both lower electron emission and photocatalytic yield. Those facts are important to support our assertion that for thin diamond films grown on metals, the UV light is absorbed within the metallic substrate and is injected into the diamond film from the back. In this case the diamond film is not the absorbing medium, but is simply acting as a protective layer with a high-lying conduction band that shuttles electrons from the metal to the water, while keeping them at a high energy corresponding to the diamond conduction band edge.

Julie Macpherson continued: Is it best, in order to obtain the greatest solvated electron emission yields, to use diamond on a metal film, so that you get electron emission from both the diamond and the metal? If so (and also relating back to the previous question), can your diamond film ever be too thick?

Robert Hamers replied: So far we've seen that diamond films grown on metallic substrates have a higher yield of solvated electrons compared with either highly-doped B-doped substrates or high-quality electronic grade diamonds. We see a maximum in electron yield at a thickness where the diamond film just coalesces. So ideally, the best diamond film would be the thinnest film that is a complete film without pinholes. This is because the mobility of electrons and the lifetime make the diffusion length something on the order of a few hundred nanometers. If the film thickness comes into the micron range, then electrons injected from the metallic substrate would not make it to the front face. Also, there is always some absorbance within the film due to graphitic impurities, and the thicker film would have more photons lost before getting to the metallic backing.

Julie Macpherson added: Relating back to the previous question from John Foord on hydrogen termination stability, do you ever see this approach replacing current technology for catalytic processes such as the Haber process?

Robert Hamers responded: It would be very difficult to replace the Haber–Bosch process. However, I think this approach could be technologically relevant if we can find more effective ways of getting electrons up into the diamond conduction band. We know that electron emission from diamond is irreversible, so that every electron emitted induces a chemical reaction event. The question is, whether there are more efficient ways of getting electrons into the conduction band. That could potentially be done using n-type diamond and/or making more complicated kinds of devices that might be able to do this purely electrically.

Manuel Alvarez-Guerra asked: Regarding CO_2 reduction, I think the results you showed about avoiding the reduction of protons are really interesting. In this sense, do you think it could be possible to guide the reduction of CO_2 to other products apart from CO?

Robert Hamers answered: Quite possibly. If the CO_2 could be further reduced to methanol or to methane, it would be a very excellent result. Under the conditions of our current experiments, CO_2 becomes reduced to the radical anion CO_2^-, which is known to photodissociate into CO + O^- under UV illumination. Perhaps the CO could be further reduced. In our current experiments, the CO leaves the aqueous phase and goes into the gaseous headspace above the liquid. With agitation it might be possible to keep more CO in solution and further decompose it. Also, if solvated electrons can be created using longer-wavelength light, the CO_2^- anion might be induced to do further reactions.

Manuel Alvarez-Guerra continued: But do you think it would be possible to guide the reduction of CO_2 to other products apart from CO, keeping the very high efficiencies of more than 98% that you showed in your results for CO?

Robert Hamers replied: That is very difficult to predict. Perhaps with knowledge of the kinetic rates of reactions and possibly the use of physical separation devices such as gas-liquid separators, it might be possible.

Matteo Duca remarked: With respect to the photochemical cleavage of CO_2^- to yield CO, it is not particularly surprising that a non-electrochemical process is responsible for CO formation. In fact, it has been suggested that metal electrocatalysts for CO_2 reduction can be classified according to the dominant final product. Metals such as Sn produce $HCOO^-$ predominantly because, after forming CO_2^-, they are unable to adsorb it and to reduce it further to CO. On the contrary, metals such as Au or Ag mostly produce CO through adsorption of CO_2^- and further reduction to CO which, being weakly adsorbed, is released from the surface.[1]

1 Hori Y *et al.*, *Electrochim. Acta*, 1994, **39**, 1833–1839.

Robert Hamers said: Yes, that is correct. Thank you for that useful comment.

Katherine Holt opened the discussion of the paper by Siegfried Waldvogel: Do you always use constant current electrolysis to carry out these syntheses? Do you know what potential you are applying? Have you ever attempted using constant potential electrolysis, as this may give more control over the reaction conditions?

Siegfried Waldvogel responded: In most cases we carry out electrosyntheses in a constant current fashion. It is not only much more convenient, but rather gives better control of the organic reaction following the electrochemical step. In electroorganic transformations, the concentration of spin centers is very crucial since we have a whole cascade of steps. We did some studies with constant potential but we could not get a better result. The applied potential for the transformation is about 0.3 V higher than using platinum anodes. We attribute that to the over-potential.

Katherine Holt added: While I can see that constant current electrolysis is more useful for bulk synthesis, perhaps for mechanistic studies and for understanding product distribution some smaller scale constant potential electrolysis experiments might be useful.

Siegfried Waldvogel said: I absolutely agree!

Katherine Holt asked: Does the nature of the solvent affect the relative stability of the reactant and products?

Siegfried Waldvogel answered: Not really, it shifts the oxidation potential of the starting materials by solvate formation, but a significant stabilization effect for the product cannot be anticipated.

Katherine Holt continued: In this work I note that you are using a non-aqueous solvent, but how dry is your system and could electrogenerated hydroxyl radicals play a role in the reaction?

Siegfried Waldvogel replied: The electrogenerated hydroxyl radicals are formed at a +2.3 V or higher potential; the phenols are converted already at +1.0 V. Consequently, hydroxyl or methoxyl radicals can be excluded. The effect of methanol and water was studied in detail.[1]

1 A. Kirste, B. Elsler, G. Schnakenburg and S. R. Waldvogel, Efficient anodic and direct phenol-arene C,C cross-coupling – the benign role of water or methanol, *J. Am. Chem. Soc.*, 2012, **134**, 3571–3576.

Julie Macpherson enquired: In the paper you state you operate the electrode under constant current conditions (2.8 mA cm^{-2}). Do you know whether this current comes from the oxidation of the organic molecule only, or is it a combination of oxidation of the organic molecule and oxidation of the solvent?

Siegfried Waldvogel responded: Cyclic voltammetry revealed that the solvent 1,1,1,3,3,3-hexafluoroisopropanol cannot be oxidized under the given conditions as well as methanol. Indeed, the phenolic substrate is initially oxidized. We need that relatively low current density to allow the substrate to migrate in time to the anodic surface. The studies are in the supporting information of ref. 1.

1 A. Kirste, B. Elsler, G. Schnakenburg and S. R. Waldvogel, Efficient anodic and direct phenol-arene C,C cross-coupling – the benign role of water or methanol, *J. Am. Chem. Soc.*, 2012, **134**, 3571–3576.

Julie Macpherson continued: Can you comment on how the electrode(s) are experimentally arranged in your experiment?

Siegfried Waldvogel answered: Usually we have a co-planar arrangement when using BDD/BDD as a combination. The BDD is on silicon and the support is completely covered.

John Foord remarked: Could you give some further experimental details such as the electrode area needed to run the process at scale, and what control of mass transport you are employing? Do you need to clean the electrodes regularly due to fouling?

Siegfried Waldvogel replied: The big advantage of BDD is that there is no fouling because deposits on BDD anodes will be mineralized. Mass transport is usually achieved by vigorous stirring and operation at temperatures of approx. 50 °C. In the supporting information of our recent papers you will find detailed information.[1]

1 *e.g.* B. Elsler, D. Schollmeyer, K. M. Dyballa, R. Franke and S. R. Waldvogel, Metal- and Reagent-Free Highly Selective Anodic Cross-Coupling Reaction of Phenols, *Angew. Chem. Int. Ed.*, 2014, **53**, 5210–5213.

John Foord continued: Could you highlight the advantages of using diamond in this application by describing results you may have on alternative electrode materials?

Siegfried Waldvogel responded: BDD provides just the best results in terms of yield and selectivity. Consequently, we use this particular material. Glassy carbon is inferior in performance.

Julie Macpherson queried: Do you have an explanation for why you get different products forming when you use electrodes other than boron doped diamond?

Siegfried Waldvogel answered: Unfortunately, not yet a rational one! We are working on that.

Julie Macpherson continued: When you run the same experiment using alternative electrodes, do you keep the current density the same in all cases?

Siegfried Waldvogel replied: For all electrode materials, we screened the current densities in the range 1–15 mA cm^{-2} and used the best results. For the given electrolytic reaction, 2.8 mA cm^{-2} turned out to be the best for both BDD and glassy carbon. This underlines also that many steps in organic electrochemistry do not depend on the electrode material, since they happen in the electrolyte/bulk.

Katherine Holt said: As discussed previously,[1] diamond has a different catalytic activity to glassy carbon due to the differing ability to adsorb some molecules. It could be that synthesis and stability of certain species using a diamond anode comes about due to the increased overpotential for oxidation of some species at diamond electrodes in comparison to glassy carbon, and hence this prevents them from being overoxidised.

1 The many faces of carbon in electrochemistry: general discussion, *Faraday Discuss.*, 2014, DOI:10.1039/C4FD90025A.

Siegfried Waldvogel responded: I agree to that as well. I am just looking for somebody who is willing to address that in an electroanalytical fashion.

Richard McCreery addressed Siegfried Waldvogel and Julie Macpherson: What does the BDD surface look like to the solution? Has anybody demonstrated a real electrocatalytic system using diamond? I am not familiar with the relevant literature.

Siegfried Waldvogel answered: The key person for that chemistry is Christos Comninellis (EPF Lausanne). He studied basically the electrolysis of water and methanol at BDD anodes. He could demonstrate by kinetic studies that the conversion occurs with these substrates only by previous coordination to the diamond lattice. An initial voltage is necessary for this process to be induced. Consequently, BDD is a very good electrode material with low background currents. When using mediators based on metal ions, Ru or Fe based, these properties are not observed. I do not know a real example wherein boron-doped diamond plays an electrocatalytic role.

Julie Macpherson replied: The surface chemistry of the diamond electrode will really depend on what the user has done to the electrode before use. If it is fresh from the growth chamber it will be hydrogen terminated. If it has been polished, acid cleaned, or just left in air for a long period of time, the surface is likely to be oxygen terminated. The chemical nature of the oxygen termination depends on which crystal faces are present and the method of termination. XPS should be

Fig. 1 FE-SEM image of the surface of the diamond electrode. The dark areas indicate regions of the surface which contain a slightly higher boron dopant level.

ideally used to inform the user of the chemical state of their electrode surface. The sp^3 nature of the diamond electrode surface typically discourages electrocatalytic processes (hence the wide solvent window in aqueous solution) unless redox systems are used where some form of bond can be formed between them and the surface functional group presented to solution; however the number of systems which do this are limited. Typically researchers functionalise the surface of the diamond electrode with metal/inorganic nanoparticles to promote electrocatalysis of a species of interest.

Richard McCreery continued: You commented that the specific interactions between the BDD electrode and the organic molecules are generally unknown. Have you or anyone else considered chemical modification of the BDD to determine its effects on electrocatalysis? For example, does –OH termination of the surface affect catalysis?

Siegfried Waldvogel answered: As far as I know, there is a plethora of chemical modifications on BDD. The purpose is mostly for specific electroanalytical studies. The OH termination of BDD occurs automatically when this material is anodically immersed in an electrolyte containing water.

George Zheng Chen addressed Siegfried Waldvogel and Julie Macpherson: Was the surface of your electrode very heterogeneous in terms of distribution and composition of the metal and semi-metal elements?

Julie Macpherson replied: The diamond we use is 200 μm thick, polished to ~nm roughness and freestanding (*i.e.* removed from the growth substrate). It is polycrystalline in nature and boron doped such that all the crystallites are doped above the metallic threshold, *i.e.* $>10^{20}$ B atoms cm^{-3}. However, as different crystal faces take up boron differently, the surface will be heterogeneously doped. The surface doping heterogeneity can be simply revealed, on this material, using techniques such as field emission scanning electron microscopy (FE-SEM), conducting atomic force microscopy, or Raman microscopy. This has been previously described for similar growth material in ref. 1 and 2. Fig. 1 shows an FE-SEM image of the surface of our diamond electrode; the dark areas indicate regions of the surface which contain a slightly higher boron dopant level.

1 Hutton *et al.*, *Anal. Chem.*, 2013, **85**, 7230–7240.
2 Wilson *et al.*, *J. Phys. Chem. B*, 2006, **110**, 5639.

Siegfried Waldvogel added: As far as we know from our BDD provider, the diamond material contains approximately 1000 ppm boron and no traces of other metals. In terms of diamond, it was very homogeneous.

Mark Newton commented: Polycrystalline boron-doped diamond (pBDD) is a well known heterogeneous electrode material. The different facets that make up the crystallites in pBDD contain different amounts of boron, resulting in a heterogeneously doped electrode surface. The material is generally considered metal-like for $[B] > 10^{20}$ cm^{-3}, and semiconducting for $[B] < 10^{19}$ cm^{-3}, with a region between these limits characterized by hopping conduction.

Electrochemical mapping[1] revealed a direct correlation between heterogeneous electron-transfer kinetics and local density of states in diamond electrodes, and confirmed that the local density of states for BDD, even when doped with [B] > 10^{20} cm^{-3}, is two orders of magnitude less than a good metal.

1 H. V. Patten *et al.*, *Angew. Chem. Int. Ed.*, 2012, **51**, 7002–7006.

Yury Gogotsi opened the discussion of the paper by Mark Newton: Are there any effects, in addition to heating (*e.g.* injection of electrons into water) that are induced by IR laser irradiation?

Mark Newton replied: No, heating is the only effect we need to consider. The near infrared laser (915 nm) is focused onto the back side of the boron doped diamond, which has an absorptivity of ~1 at this wavelength, not the solution facing side. Furthermore, the laser energy (approximately 1.4 eV) is too small to excite electrons into the vacuum level and inject electrons from the diamond into solution.

Robert Hamers said: I think this work is very interesting. I think it's possible that the thermal pulse might be able to create an electrochemical gradient. The Dember effect, for example, due to the difference in mobilities of electrons and holes, can create a potential difference in response to a thermal gradient. Is it possible that could be occurring during your experiments?

Mark Newton responded: It is possible that a thermal pulse could create an electrochemical gradient, but we believe that it is highly unlikely in this situation.

George Zheng Chen asked: How high a temperature can you achieve? As high as that for molten salts (>400 °C)? Mo or W may be stable to resist the reaction with carbon at elevated temperatures.

Mark Newton answered: Boron doped diamond electrodes are not forever! Although at all temperatures diamond is thermodynamically the unstable form of carbon, at zero pressure the conversion to graphite, in an inert environment, is not detectable experimentally at temperatures below approximately 1500 °C. In oxidizing atmospheres, diamond will decompose at temperatures of 700–800 °C *via* oxidation to gaseous products which are either CO or CO_2, or a mixture of them. Diamond chemically reacts with ferrous metals at similar temperatures. Thus caution should be exercised when contemplating the use of diamond electrodes with molten salts, however there are a number of examples (*e.g.* Li2O–LiCl–KCl at 550 °C, ref. 1) in which the boron doped diamond has been shown to be chemically stable.

1 W. Park *et al.*, *J. Nucl. Mater.*, 2013, **432**(1–3), 175–181.

Robert Dryfe remarked: In the models, you have allowed for the temperature dependence (*via* a ΔT term) of the electrode potential, but temperature otherwise appears to be treated as constant, *e.g.* in the RT term of the denominator of the Butler-Volmer exponential factor. Is this a valid assumption?

Mark Newton replied: In the model we do account for the change in temperature; T in the denominator of the Butler-Volmer exponential factor (but not in the pre-factor) and in the calculation of the diffusion coefficient (molecule size is assumed temperature independent), as well as in the temperature dependence of the electrode potential. Hence, within the constraints of the model, we have taken full account of the variable temperature.

Robert Dryfe continued: Is it possible to use the laser-heated approach to induce phase transitions in electrochemical systems, rather than merely heating? For example, if you used an ionic liquid (or possibly polymer electrolyte) system, which melted slightly above room temperature, you could use the local, rapid heating to melt the system and induce an electrochemical process – or a chemical process coupled to the electron transfer – that would only occur in the molten state, and may therefore be able to 'freeze in' an unusual state to the system on re-solidification.

Mark Newton responded: This is a very interesting suggestion and I see no reason as to why this experiment is not possible. The shortest pulse length employed with the 30 W CW laser diode module (HLU30-F200-DL915, Laser2000 Ltd) used in the work is 10 mS, but a shorter pulse length laser could of course be employed for experiments needing to access shorter time scales. It is important to remember that the diamond is a very effective heat spreader, so after the thermal pulse the diamond (in contact with a heat sink) will rapidly cool any material in contact with it which may be useful for freezing in transient species.

John Foord said: You have shown the diamond substrates have ideal characteristics (damage threshold, absorption, thermal conductivity) for this application. Could the approach be easily extended to other materials by depositing them on the diamond, or are the thermal conductivity requirements too stringent?

Mark Newton answered: This is a very good suggestion. As long as the deposited layer was thin and in good thermal contact with the diamond (heated from the back side) then the diamond would rapidly transmit heat to the deposited layer, and rapidly cool, after the thermal pulse.

Katherine Holt addressed Julie Macpherson and Mark Newton: Do you see any change in the oxygen functional groups on the diamond surface as a result of the laser heating? If the functional groups present are stable and don't change during the experiment this would seem a good technique for probing the effect of different functional groups on the electrochemical response of different redox probes and the ordering of the double layer.

Mark Newton and **Julie Macpherson** replied: This is an interesting question; the couples we use are deemed 'outer-sphere' and are thus not sensitive to changes in the surface chemistry. Our electrochemical data obtained under heated conditions suggests that the change in the current response is driven primarily by thermodynamic (entropic, governed by the temperature coefficient term, β) and mass transport factors. In other work using diamond electrode materials of a very similar quality to those employed here, we performed high temperature XPS measurements on oxygen terminated electrodes subject to an alumina polish (electrodes in these experiments were alumina polished before use too)[1]. The XPS showed us that at room temperature the surface contained C–C, C–OH, C=O and C–O–C groups. At 300 °C the XPS showed that C–OH and C–O–C largely remained on the surface, however the C=O groups had decreased significantly. C=O is thought to be removed in the range 100–400 °C (ref. 2). In our experiments, the temperature of our current system achieves maximum heating temperatures of about 90 °C, so we don't think we are significantly perturbing the surface chemistry. However, we do have the capabilities to go to significantly higher temperatures by moving to higher power lasers. It would be of great interest to use a redox couple which is surface sensitive, especially to groups such as C=O, *e.g.* $Fe^{2+/3+}$, and explore the results obtained under higher temperature heating conditions. $Fe^{2+/3+}$, a couple which we find on our alumina polished conducting diamond, shows relatively slow electron transfer kinetics. By increasing the temperature of the electrode/electrolyte interface it would be interesting to consider what we see in terms of thermodynamic/kinetic effects and the effect of possible surface chemistry changes. Initial data has been collected with this redox couple, and we are currently looking at using simulations to understand the process.

1 L. A. Hutton *et al.*, *Anal. Chem.*, 2013, **85**, 7230–7240.
2 J. E. Field, *The Properties of Natural and Synthetic Diamond*, Academic Press Limited, London, 1992.

Katherine Holt continued: With your experimental setup, could you probe at the timescales required to measure reordering of the double layer on heating (ns)? This type of experiment would be a good way of probing the dynamics of double layer perturbations and the role of functional groups in charge ordering at the interface. It has relevance in the fundamental understanding of charge storage at the carbon interface, which is very important in understanding how carbon electrodes perform in supercapacitors.

Mark Newton and **Julie Macpherson** responded: We agree with you. However, as double charging/discharging happens on a much faster timescale (nanosecond) than we are probing with our diode based lasers (100s of microseconds), the current setup does not allow us to make this measurement. However, by switching to a more expensive, *e.g.* Q-switched Nd:YAG laser, this experiment now becomes possible. Double layer charging using a pulsed laser approach has been explored by others[1] at a gold disk electrode.

1 *e.g.* J. L. Brennan and R. J. Forster, *J. Phys. Chem. B*, 2003, **107**, 9344.

Yury Gogotsi opened the discussion of the paper by Aleix Güell: Is the solution in the double barrelled pipette flowing out of one channel and being sucked into the other channel, or is there just a flow of ions?

Aleix Güell replied: There is no flow between the double barrelled pipette, as for example in the push–pull probe of Momotenko *et al.*[1] In our scanning electrochemical cell microscopy (SECCM) setup, the use of a double barrelled pipette and the two electrodes inserted in each barrel (bias applied between them) is to obtain a conductance cell at the meniscus created at the end of the tapered pipette. The conductance current is used as a feedback parameter (in fact, the AC component of the conductance current), allowing us to scan over the surface at a constant tip–sample distance independently of any surface electrochemical current. A detailed description of SECCM can be found in ref. 2. In addition, the bias applied between electrodes in the barrels also provides an enhancement in the mass transport due to the extra migration component.[3]

1 D. Momotenko *et al.*, *Anal. Chem.*, 2011, **83**, 5275.
2 N. Ebejer *et al.*, *Annu. Rev. Anal. Chem.*, 2013, **6**, 329.
3 M. E. Snowden, A. G. Güell *et al.*, *Anal. Chem.*, 2012, **84**, 2483.

Yury Gogotsi continued: Do you understand well the dynamics of the electrolyte at the surface? Are there any efforts to model it?

Aleix Güell answered: In ref. 1 we provide a comprehensive model for mass transport in SECCM, including the ion conductance characteristics and the voltammetric response with a conducting electrode substrate, employing finite element method modelling.

1 M. E. Snowden, A. G. Güell *et al.*, *Anal. Chem.*, 2012, **84**, 2483.

Milo Shaffer asked: Your Raman shows an apparently low defect concentration. However, are you able to confirm that you do not have any dilute electrochemically relevant defects, say, at a concentration of around one per micron, which might be hard to resolve as discrete sites by either Raman or SECCM. Collins *et al.*[1] demonstrated that the formation of metal particles on individual nanotubes is reversible, but repeats at the same location, suggesting a means to locate this type of defect. Do you see any similar effects, for example during the silver electrodeposition step of your processing? The nature of these electrochemically active defects is unclear, but might be further probed, after 'tagging' in this way?

1 Y. Fan, B. R. Goldsmith and P. G. Collins, *Nat. Mater.*, 2005, **4**, 906–911.

Aleix Güell responded: SECCM is in fact able to resolve the presence of these point defects when surface sensitive reactions such as oxygen reduction (ORR) are being studied. As we describe in our communication,[1] we used SECCM to localize point defects (accumulated at the kink sites of pristine individual nanotubes) which exhibit an enhancement in electrocatalytic activity towards oxygen reduction. Furthermore, SECCM was also used to create defects (oxygenated points) and confirm that the these are much more active to ORR than pristine sidewalls. In addition, we have also shown in our other contributions[2,3] that the scan profile obtained when the SECCM probe scans over a nanotube can be analyzed (shape and current values) and simulated by FEM modelling. This was used to confirm the homogeneous activity of nanotube sidewalls, and discard that only defects are responsible for the carbon nanotube activity, since the shape of the scan profiles would differ significantly. In this paper, silver electrodeposition was used only to localize the nanotube, thus the shape/distribution of particles was not studied in depth.

1 J. C. Byers, A. G. Güell and P. R. Unwin, *J. Am. Chem. Soc.*, 2014, **136**, 11252.
2 A. G. Güell *et al.*, *Nano Lett.*, 2014, **14**, 220.
3 A. G. Güell *et al.*, *Proc. Natl. Acad. Sci. USA*, 2012, **109**, 11487.

Robert Dryfe remarked: You are measuring the electrochemical activity of individual single wall nanotubes, with a diameter of a few nm (see Fig. 2C of your paper, for example). However, diffusional broadening means the current maps are on the order of a micron or so (see Fig. 6A for example). Could you therefore comment on the spatial resolution/sensitivity of your technique?

Aleix Güell replied: With SECCM we are constantly acquiring data along with the tip displacement. In the case of this paper, EC current readings were acquired every 1 nm of tip displacement, obtaining scan profiles of 1 nm resolution. These profiles can be further analysed and compared with simulated scan profiles, as for example in ref. 1 and 2.

1 A. G. Güell *et al.*, *Nano Lett.*, 2014, **14**, 220.
2 A. G. Güell *et al.*, *Proc. Natl. Acad. Sci. USA*, 2012, **109**, 11487.

Robert Dryfe continued: You are able to measure electron transfer kinetics at high spatial resolution, *i.e.* at individual nanotubes. The work of Lemay and Dekker to which you refer (ref. 23 of your paper) suggests that it is possible to measure the difference in the heterogeneous rate constants for a given redox couple, due to the variations in electronic structure imposed by nanotubes of differing chirality. Is your method sensitive enough to be able to detect such differences experimentally?

Aleix Güell answered: This is definitely an interesting topic to investigate. In this paper and ref. 1 we distinguish between semiconductor and metallic nanotubes employing ruthenium hexaamine, since this mediator redox potential lies in the band gap of the semiconductor nanotube, hence giving a large difference in the EC activity; either ON or OFF for metallic or semiconductor, respectively. In order to distinguish between nanotubes of different chirality, the differences in EC activity will be due to the different level of overlap (in which the different position of the Van Hove singularities will definitely play a role, so selecting an adequate redox mediator to emphasize these changes will be also important). I don't think the changes in EC activity (electron transfer rate constant) between nanotubes will be enough (detectable) to establish a correlation with the chirality. Perhaps alternative methods can be used to determine the band structure or DOS, such as localized capacitance measurements (as Keith Stevenson suggested), instead of determining the EC activity with redox mediators.

1 A. G. Güell *et al.*, *Nano Lett.*, 2014, **14**, 220.

Matteo Duca enquired: With respect to the problem of determining the chirality of single-walled carbon nanotubes, I would like to suggest a possible experiment based on the observation that chiral single-crystal metal surfaces (in particular Au and Pt) are able to distinguish between enantiomers of an electroactive species (see ref. 1 for L- and D-glucose oxidation on chiral Pt surfaces, and ref. 2 for the same reaction on chiral gold surfaces). Glassy carbon is able to oxidise tyrosine,[3] and so should carbon nanotubes, which, after all, can oxidise serotonin, containing a phenol moiety. Therefore, one could attempt to discriminate the chirality of carbon nanotubes by comparing the voltammetric response towards the oxidation of L- and D-tyrosine at individual carbon nanotubes with SECCM.

1 A. Ahmadi *et al.*, *Langmuir*, 1999, **15**, 2420–2424.
2 A. Martins *et al.*, *Electrochem. Commun.*, 2003, **5**, 741–746.
3 A. Harriman, *J. Phys. Chem.*, 1987, **91**, 6102–6104.

Aleix Güell responded: That's a very interesting point that can be assessed with our platform of individual carbon nanotubes, described in our paper. However, I believe the curvature of the carbon nanotube and its dimensions (around 1.2 nm) are important factors to consider, compared to single crystal electrodes with large areas of atomically flat surface.

Matteo Duca continued: Would it be possible to exploit the fouling of the carbon surface by using products of serotonin oxidation to block the carbon surface in a controlled way with the SECCM setup? If a metal is then electrodeposited onto the non-fouled gaps, a very well-defined partially-blocked electrode could be obtained. Its electrochemical response could then be compared to theoretical predictions.

Aleix Güell answered: The experiment you described is certainly possible. As shown in ref. 1, Fig. 3a, the scanning parameters in SECCM can be changed at any time during imaging, it being therefore possible to apply different electrochemical potentials in different scan lines of the mapping. This can be used to selectively deposit byproducts of serotonin oxidation (or any insulating coating) on defined segments of the nanotubes and obtain the partially blocked electrode you are describing.

1 A. G. Güell *et al.*, *Nano Lett.*, 2014, **14**, 220.

Richard McCreery remarked: Fig. 5C of your paper shows that Fc is redox active on a semiconducting tube, but Ru(hexamine) is not. Does this correspond at all to the density of states predicted for the semiconducting tube? The low DOS in that case is about 0.3 V wide, yet you report slow kinetics over a wide potential range. Have there been any reports correlating CNT electrode behavior with the DOS?

Aleix Güell replied: For the reduction of ruthenium hexamine on a semiconducting nanotube (Fig. 5C), there are two facts to be considered: the low density of states at that energy range; and the Schottky barrier existing at the metal–semiconductor contact for electron injection. The slow kinetics exhibited in the experiment of Fig. 5C for such a wide potential range are a consequence of both, with a significant contribution from the ohmic drop. In contrast, in ref. 1, we used much smaller pipettes to confine the electrochemical measurements. This resulted in much lower EC currents (~pA range) and consequently lower ohmic drop effects. In that case, the slow kinetics are observed in a much more reduced range. I am not aware of any publication that correlates experimentally CNT behaviour with DOS, but there is the work from Heller *et al.*[2] where they assessed this from a theoretical perspective.

1 A. G. Güell *et al.*, *Nano Lett.*, 2014, **14**, 220.
2 I. Heller *et al.*, *J. Am. Chem. Soc.*, 2006, **128**, 7353

Richard McCreery continued: You report facile oxidation of serotonin on carbon nanotubes. Did you look for serotonin adsorption? Removal of the first electron from serotonin should be energetically difficult unless it is accompanied by ionization of the phenolic proton. That might be assisted by serotonin adsorption, which would generate a hydrogen-bonding site.

Aleix Güell responded: We didn't look at serotonin adsorption, but certainly SECCM would allow us to do this study. Zhang *et al.*[1] used SECCM to study the adsorption and electrochemistry of anthraquinone-2,6-disulfonate (AQDS) on HOPG.

1 G. Zhang *et al.*, *J. Am. Chem. Soc.*, 2014, **136**, 11444.

Richard McCreery added: You suggest that Raman might be a way to detect the adsorption of serotonin on carbon nanotubes. Possibly, but the signal would be very weak unless there is some sort of enhancement.

Aleix Güell answered: I agree. Perhaps atomic force microscopy tip-enhanced raman spectroscopy (AFM–TERS) could be one of the techniques to enhance the signal.[1]

1 T. Yano *et al.*, *Nat. Commun.*, 2013, **4**, 2592.

Heisi Kurig asked: A question about the measurement technique SECCM: how well is this technique adjustable for very rough surfaces, *e.g.* amorphous carbons? Does the electrolyte meniscus have an advantage as being a more flexible tip compared to tips used in STM?

Aleix Güell replied: SECCM can also be used on very rough surfaces. However, the imaging mode should be changed from scanning mode (lateral displacement of the probe) to a hopping mode (individual approaches per each pixel of the image). The hopping mode is widely used in scanning ion conductance microscopy for imaging protruding samples such as live cells.[1] An example of the SECCM platform being used on very rough samples (single point measurements) is ref. 2, where the electrochemical response of a high density of vertically aligned carbon nanotubes was studied, confining the electrochemical cell on the sidewalls and top ends by means of SECCM. Among of the advantages of SECCM compared to STM is that it does not require a conductive sample as a support, allowing the study of one single nanotube on an insulating substrate such as silicon/silicon oxide. This ensures the measured signal comes from the nanotube itself and there is no contribution from the substrate.

1 P. Novak *et al.*, *Nature Methods*, 2009, **6**, 279.
2 T. S. Miller *et al.*, *Chem. Commun.*, 2012, **48**, 7435.

Keith Stevenson said: Is it possible to measure the DOS electrochemically? You claim that you can map the defects of a bent tube, but if you measure the DOS electrochemically it may be that the DOS is higher at the bend and it may not be driven by defects, just a higher DOS at the bend. The same is claimed for twisted or folded single layer graphene sheets.

Aleix Güell responded: In our communication[1] we demonstrate that SECCM is able to resolve the presence of point defects, which are accumulated at kink sites, and exhibit an enhancement in surface sensitive reactions such as oxygen reduction. See also the above previous discussion on this topic. Measuring the LDOS with SECCM would be an alternative route to detect the presence of point defects, since these have a different density than pristine areas of nanotubes. We have not yet tried to carry out local capacitance measurements with the SECCM setup, which could be a possible experiment to measure the DOS. This could also be applied to step edges at HOPG, which are supposed to possess a different LDOS compared to the basal plane.[2]

1 J. C. Byers, A. G. Güell and P. R. Unwin, *J. Am. Chem. Soc.*, 2014, **136**, 11252.
2 Y. Niimi *et al.*, *Phys. Rev. B*, 2006, **73**, 085421.

Keith Stevenson addressed Aleix Güell and Patrick Unwin: You should be able to just measure the electrochemical impedance, and not probe the DOS using a redox couple, to measure the DOS. Redox mediators, for example (oxygen reduction *vs.* oxidation), really matter as some are more inner sphere than others. 'Defects' can have a positive effect on inner sphere redox reactions.

Aleix Güell answered: Agreed. This a very useful suggestion. However, I believe that outer sphere mediators, if chosen adequately,[1] can also be employed to emphasize relative differences in the LDOS within a sample/surface, and these, as outer sphere mediators, will be independent of surface functionalities.

1 R. McCreery *et al.*, *Anal. Chem.*, 1992, **64**, 2518.

Patrick Unwin commented: As highlighted earlier in this discussion, we have, in fact, previously used SECCM to estimate the density of states (DOS) in differently doped regions of a polycrystalline diamond electrode and correlated electron transfer kinetics with the DOS.[1] Related to this, scanning ion conductance microscopy (SICM) could be a powerful technique for probing surface charge and double layer properties at the nanoscale, and may be particularly valuable for studies of the electrode/electrolyte interface generally, and ultimately at defect sites on electrodes.

1 H. V. Patten *et al.*, *Angew. Chem. Int. Ed.*, 2012, **51**(28), 7002–7006.

Keith Stevenson added: Yes, we have included this reference in our manuscript. I do look forward to SICM measurements at defect sites such as were shown by your student.

Surbhi Sharma opened the discussion of the paper by Jingping Hu: In the experimental section, you say you start with GO, functionalise it with PDDA and then grow Pt nanoparticles using H_2PtCl_6 *via* the sodium borohydride reduction process. What is the need for the additional step of PDDA functionalisation? The large number of oxygen defects in well oxidised GO act as nucleation sites for metal nanoparticles in the standard borohydride reduction and also solvothermal processes, generating uniformly distributed metal nanoparticles on rGO.[1,2,3] Please explain the role of PDDA.

1 S. Sharma *et al.*, *J. Phys. Chem. C*, 2010, **114**, 19459–19466.
2 J. Wang *et al.*, *Energy Environ. Sci.*, 2012, **5**, 6885–6888.
3 S. Wang *et al.*, *Electrochim. Acta*, 2011, **56**, 3338–3344.

Jingping Hu replied: We found that if GO and hexPt were reduced in one pot, the reaction was slow, presumably due to electrostatic repulsion. By simply adding some PDDA, which adsorbs on the graphene and gives it a positive charge, the reaction was much faster, so the PDDA acts as a catalyst.

Surbhi Sharma continued: What is the size of the NPs? The Pt salt in the reduction system is already looking for defect sites on the GO sheets which allows the growth of well controlled small size nanoparticles; the need for the extra PDDA functionalisation step is not very clear.

Jingping Hu added: On graphene, a few nm. PDDA speeds up the reaction.

Surbhi Sharma asked: After Pt deposition on rGO, the XPS graph shows lots of residual oxygen functionalities. What effect do you think these have on the electrochemical and sensing behaviour of your system? Will you be doing any control studies with further reduction of GO to see the difference in performance of the two GOs reduced to different degrees?

Jingping Hu answered: If there was insufficient reduction, the device performance was poor. We are not intending further studies at present.

Surbhi Sharma added: Since you say that the residual oxygen groups seem to have some role to play in the biosensor performance, would you say you are investigating lightly oxidised graphene or partially reduced graphene oxide, as opposed to referring to it as reduced graphene oxide?

Jingping Hu responded: Some oxygen functionalities seemed to be generally present on the functioning device – it is not clear whether these originated from the GO or came from subsequent contamination.

Matěj Velický said: Experimental details in the paper suggest that there was perhaps only one layer of graphene nanosheets (GNS) prepared intentionally on the glassy carbon (GC) substrate (2.4 nm thickness of the nanosheets as determined by AFM)?

Jingping Hu replied: The quoted thickness relates to individual monolayer graphene platelets but actually the sensor was drop-coated to produce a significant graphene film, the thickness of which can be estimated from the mass loadings in section 3.1.

Matěj Velický continued: The question arises of how thick the layer of GNS is and how does it affect the electrochemical response and XPS measurement, *i.e.* is the underlying substrate (glassy carbon) interfering with the GNS signal? Only a limited comparison of the voltammetric response of the GNS and glassy carbon was provided (missing in Fig. 2). Furthermore, the XPS spectrum in Fig. 1d is much more likely to be interpreted as a strong sp^3 C–C bond signal, positioned around 285 eV with additional C–O and C=O signals. This would support the above comments, *i.e.* the penetration depth of the XPS beam is larger than the thickness of the GNS, and sp^3 signal comes from the underlying glassy carbon.

Jingping Hu answered: Due to the thickness of the graphene layer, all of the XPS signal must be derived from it, from which is derived the main C–C peak.

Matěj Velický added: Following the previous question, the comparison of the claimed catalytic performance with the bare underlying substate – glassy carbon – needs to be provided (which is missing in Fig. 2).

Jingping Hu said: There is a lot of voltammetric data from the three types of sensor, not all of which could be incorporated into the paper, but the bare glassy carbon electrode showed no useful response to peroxide or glutamate – the Pt and the enzyme were crucial components.

Matěj Velický asked: The oxygen content is very important for the conductivity of the electrode modified this way – stacking of GNS and high oxygen content will result in a decrease in conductivity and poor voltammetric performance. What role do these factors play in your case?

Jingping Hu replied: Reduction of the graphene oxide was crucial to get a useful response. As we had a significant graphene film, good conductivity was needed through the layer and graphene oxide was too insulating.

Matěj Velický added: Following the previous question, what is the maximum oxygen content still suitable for the use of these GNS as an electrode material? Do you know at what oxygen concentration the transition between 'good' and 'poor' electrodes occurs, *i.e.* do you lose the conductivity with 10% or perhaps 20% of oxygen?

Jingping Hu noted: We did not study this systematically, but for the graphene sensor, it seemed an oxygen content greater than 25% had a deleterious response.

Surbhi Sharma commented: The percentage of oxygen in rGO for an effective/ good electric and electrochemical response needs to be less than 18–20%.[1] The detailed studies on thermal de-oxygenation of GO in the reference provide a lot of information on the density of states and behaviour upon successive *in situ* reductions at temperatures between 200–1000 °C.

1 A. Ganguly *et al.*, *J. Phys. Chem. C*, 2011, **115**, 17009–17019.

Jingping Hu added: We would agree that >25% oxygen inhibited sensor performance.

Surbhi Sharma said: In the sodium borohydride method for the growth of Pt nanoparticles, the $NaBH_4$ enables reduction of the Pt salt and also acts as mild reducing agent for GO at the same time. Moreover, the process of growth of metal nanoparticles on the oxygen defect sites on GO seems to lead to simultaneous reduction of the GO. In order to reduce GO more and to study the effect of residual oxygen functional groups, you would need to determine a different reduction process which does not lead to simultaneous nucleation/growth of NPs and reduction of GO.

Jingping Hu replied: The graphene could be reduced in a prior step, but then it tends to aggregate in solution leading to poor functionalisation with Pt.

Julie Macpherson enquired: Was there a reason for using much larger Pt particle sizes (150 ± 30 nm) for the Pt electrodeposited boron doped diamond electrode compared to the 8 nm particles employed with the Pt-GNPs? Particle size can have an influence on electrocatalytic activity. We have shown previously it is possible to produce <10 nm sized Pt nanoparticles (mean size 2–3 nm) on boron doped diamond using an electrodeposition process.[1]

1 Hutton *et al.*, *Anal. Chem.*, 2009, **81**, 1023–1032.

Jingping Hu responded: The AFM data in the referenced paper seems to show lateral particle sizes of around 30 nm (albeit only a few nm in height), and if longer deposition times were used to get the loadings we needed, the particle size would be bigger. So we do believe it is difficult to produce small particles on diamond using constant potential deposition. The aim of the paper was to compare sensors which could be most obviously prepared on the differing platforms. Hence electrodeposition is the straightforward route to prepare adherent Pt on diamond, whereas wet chemical methods can be easily developed for Pt functionalisation on solution graphene. By comparing the response to peroxide or glutamate we could examine how differing components were functioning, and it appeared that the differences were more associated with the interaction of glutamate with the enzyme rather than the detection of the peroxide produced.

Julie Macpherson communicated: In response to your comment about lateral nanoparticle (NP) size, even though the NPs in our paper look bigger in the lateral direction compared to the measured height using AFM, we always use the height of the NP as an indicator of size. Imaging diamond with an AFM tip can be challenging, as the tip wears on diamond, so the AFM tip ends up getting bigger and hence introducing lateral tip size effects into the image. The height is, however, accurate, so we do believe we can make Pt nanoparticles on BDD using an electrodeposition approach.

Julie Macpherson continued: Do you think nanoparticle size effects played a role in the data seen?

Jingping Hu answered: This was discussed above, but we agree this would be an interesting angle for future investigation.

Julie Macpherson asked: In Fig. 2, for example, can you be sure that some of the GNPs are not removed from the surface of the electrode during rotation?

Jingping Hu replied: The final sensor configurations were stabilised to some extent by the addition of PPD as a barrier to promote selectivity, but regarding the data in Fig. 2, the response was stable over the measurement period. The graphene we were using was modified with PDDA – we are not sure how bare graphene would compare in stability.

Keith Stevenson commented: It would be interesting to measure the activity of the enzyme once it's mobilized, using a colorimetric assay, and see if the strength of enzyme adsorption affects the activity or effective turnover of the enzyme.

Philip A. Ash remarked: We have recorded infrared spectra of a variety of different proteins, both in solution and adsorbed on a range of carbon surfaces. In most cases, we find that adsorption does not affect the spectra in the amide regions, suggesting that any changes in the overall structure of the protein are minimal.

Jingping Hu responded: In the case of this paper, the Michaelis–Menten constant of the enzyme was raised, but comparable to that observed in solution, suggesting that the structure of the protein was not changed very much by binding to the surface.

Taiwo Alaje opened the discussion of the paper by Jonathan Quinson: You did mention in your paper that the materials APG, MG, MWCNT, N-MWCNT, GNF, CNP, ML and VX all display a type II isotherm. Now, type II isotherms are known to characterize materials that are non-porous but might be macroporous due to the way the particles aggregate. But on page 19 you went further to describe these materials have as having a well developed mesoporous structure. This description needs to be more specific, considering that from your isotherms the materials are clearly not mesoporous.

Jonathan Quinson replied: We acknowledge that other materials, such as ordered mesoporous carbons, will indeed have a far more developed mesoporous structure than those we have investigated. Whilst it is true that the materials we have presented are mainly macroporous, they do all show some degree of mesoporosity due to their pore size distribution. On page 19 we noted that BP2000 had the most mesoporous nature of all the materials we studied, and also showed the highest activity per unit mass following enzyme adsorption.

Taiwo Alaje continued: On page 19, line 33, you said "In general the materials which are most affected by mass transport limitations are those which also have the most developed mesoporous structures (Table 1), suggesting poor transport of H_2 within electrodes constructed from these materials."; this refers to Fig. 8 of the paper. My question is, does this presuppose that the lower curves, *i.e.* those whose end point does not hit a plateau, are not experiencing mass transfer limitations?

Jonathan Quinson responded: In the case of adsorbed enzymes, catalytic cyclic voltammograms often have complex shapes that do not exhibit a plateau at high overpotentials in the absence of mass transport limitation. Therefore, those which do not reach a plateau are not mass transport limited in our case. The waveshape we observe is characteristic of *E. coli* hydrogenase 1 under non-mass transport limited conditions. The hydrogenase active site is buried within the enzyme, and electrons are transported to and from the active site *via* a chain of iron-sulphur clusters. As a result, the electrons must enter and leave the enzyme at a well-defined point on its surface. Hydrogenase adsorbs to the electrode in a range of orientations, so the distance between this entrance point and the electrode will vary from molecule to molecule, resulting in a range of interfacial electron transfer rates. Léger *et al.*[1] reported a model that accounted for this distribution of electron transfer rates and accurately predicts experimental catalytic waveshapes, including the residual slope at high overpotential.

1 C. Léger, A. K. Jones, S. P. J. Albracht and F. A. Armstrong, *J. Phys. Chem. B*, 2002, **106**, 13058–13063.

Taiwo Alaje said: I wish to say that all the materials are experiencing mass transfer limitations. At the onset of the voltage sweep, the reaction rate is being governed by both mass transfer and kinetic limitations. Those that hit a plateau quickly have such a high activity or rapid kinetics which leads to the concentration of the substrate at the surface of the electrode becoming zero earlier on, hence the rate is being governed by how fast the substrate from the bulk gets to the surface of the electrode. Due to the small amount of substrate being consumed at the electrode, the bulk concentration remains constant. Hence, there exists a fixed concentration gradient between the bulk solution and the tip of the electrode, and hence a constant rate of reaction. This is what gives rise to the plateau.

Jonathan Quinson answered: In our case we do not have a planar electrode, rather we have a porous particle network deposited onto the surface of a rotating disc electrode. In terms of the voltammograms that show mass transport limitations and reach a plateau, this is indeed, as you suggest, due to the relative rates of consumption of H_2 within and diffusion of H_2 into the particle networks from bulk solution. Increasing the rotation rate for the mass transport limited films introduces a residual slope at high potential, as seen in Fig. 10 of the paper, which implies improved substrate supply. Similar effects have been reported in the very active membrane-bound hydrogenase from *A. vinosum* by Jones *et al.*,[1] where electron-transfer limitations are clearly observed at high rotation rates under 1 bar H_2. Fig. 10 of the paper also shows that mass transport can be improved further using a flow cell configuration, where H_2-saturated solution is forced through the particle network at a rate of approximately 2000 cell volumes per second. Under these conditions, the catalytic waveshape is independent of flow rate and the expected non-mass transport limited behaviour is reproduced.

1 A. K. Jones, E. Sillery, S. P. J. Albracht and F. A. Armstrong, *Chem. Commun.*, 2002, 866.

Taiwo Alaje added: I think they all have mass transport limitations, not just the ones that hit the plateau.

Jonathan Quinson replied: Armstrong and co-workers demonstrated that, at low coverages, enzyme films behave as microelectrode arrays, and used this fact to record non-mass transport limited voltammograms of the very active membrane-bound hydrogenase from *A. vinosum* on a rotating disc electrode.[1] The catalytic voltammograms that we present in Fig. 9 are all independent of rotation rate between 2000 rpm and 6000 rpm, and are consistent with reported non-mass transport limited voltammograms for *E. coli* hydrogenase 1 recorded at a planar rotating disc electrode.

1 K. A. Vincent, A. Parkin and F. A. Armstrong, *Chem. Rev.*, 2007, **107**, 4366–4413.

John Foord asked: Do you know what type of enzyme-support interaction – hydrophilic, hydrophobic, *etc.* – is operative in the enzyme–carbon materials you describe, and how controlled is the hydrogenase surface orientation as a consequence?

Jonathan Quinson responded: The hydrogenase surface orientation is not controlled directly in our experiments, although we do have spectroscopic evidence that all of the enzyme sample responds to an applied potential. In addition, the residual slope at high potential in our voltammograms is similar to that reported for other hydrogenases, implying that the sample must be oriented such that the 'entrance' to the chain of iron-sulphur clusters is close enough to a carbon surface for electron transfer to take place.[1] There are a number of reports that describe attachment strategies designed to favour specific orientation of hydrogenases at electrode surfaces, either through electrostatic interaction or direct covalent linkage. These strategies rely on knowledge of the charge, hydrophobicity or amino acid residues on the surface of the hydrogenase in the vicinity of the distal iron-sulphur cluster, and it is therefore likely that a range of interactions play a role when the enzyme adsorbs on an 'unmodified' carbon surface.

1 C. Léger, A. K. Jones, S. P. J. Albracht and F. A. Armstrong, *J. Phys. Chem. B*, 2002, **106**, 13058–13063.

John Foord continued: In your presentation, you have mainly described current–voltage curves, but other issues such as enzyme stability are important. How long do the supported hydrogenases stay active under what storage conditions, and does the stability vary with the carbon support?

Jonathan Quinson answered: The hydrogenase we selected for this study, hydrogenase 1 from *E. coli* (Hyd-1), is a robust oxygen-tolerant hydrogenase. We found that the stability of Hyd-1 did not vary with the carbon support under the conditions of our experiment. Previous studies have reported stability over several days of continuous operation on amorphous carbon electrodes,[1] and this is in line with our own observations that the hydrogenase remains stable over extended periods with little film loss or denaturation. Other hydrogenases are known to exhibit much lower stabilities, although some of these issues have been alleviated through the use of specific attachment strategies – for example Léger and co-workers covalently immobilised a [FeFe] hydrogenase to a carbon electrode.[2]

1 L. Xu and F. A. Armstrong, *Energy Environ. Sci.*, 2013, **6**(7), 2166–2171.
2 C. Baffert *et al.*, *Anal. Chem.*, 2012, **84**, 7999–8005.

Stephen Lyth said: In a paper published recently, Takahiro Matsumoto *et al.*[1] reported the hydrogen oxidation reaction on carbon-black supported, oxygen-tolerant hydrogenase with a mass activity ~600 times that of platinum. They also reported quite reasonable polymer electrolyte membrane fuel cell power densities using hydrogenase at the anode. What do you think of these results, and are hydrogenase-based electrodes ever likely to be utilised in hydrogen fuel cells?

1 T. Matsumoto *et al.*, [NiFe]Hydrogenase from *Citrobacter* sp. S-77 Surpasses Platinum as an Electrode for H_2 Oxidation Reaction, 2014, *Angew. Chem. Int. Ed.*, **53**, 8895.

Jonathan Quinson replied: The results of Matsumoto *et al.* are intriguing and serve to highlight possible applications of hydrogenases in fuel cells, particularly given the reported high activity of the [NiFe] hydrogenase from *Citrobacter* sp. S-77 relative to other [NiFe] hydrogenases.[1] Previous comparisons of Pt and the [NiFe] hydrogenase from *A. vinosum* found the H_2-oxidation rates to be indistinguishable, on a per active site basis, at a range of H_2 partial pressures.[2] Matsumoto *et al.* have used a highly purified sample of hydrogenase in their measurements, which is expensive. Although it is becoming increasingly common to prepare enzymes on an industrial scale that are extremely cost-effective, at present there are no examples of industrial-scale hydrogenase preparations to develop cheaper fuel cell devices. It is therefore difficult to assess the future technological impact of hydrogenases. Hydrogenases are, however, a source of inspiration since they use cheap, abundant metals like Fe and Ni. Understanding what makes the hydrogenase active site so efficient is therefore important in the search for alternative catalysts to platinum for fuel cell applications, and it is entirely conceivable that an optimised hydrogenase or hydrogenase mimic could be used in this way.

1 S. Eguchi, K.-S. Yoon and S. Ogo, *J. Biosci. Bioeng.*, 2012, **114**, 479–484.
2 A. K. Jones, E. Sillery, S. P. J. Albracht and F. A. Armstrong, *Chem. Commun.*, 2002, **8**, 866–867.

Faraday Discussions

PAPER

Concluding remarks: there's nowt so queer as carbon electrodes

Patrick R. Unwin*

Received 6th October 2014, Accepted 13th October 2014
DOI: 10.1039/c4fd00194j

This contribution provides a personal overview and summary of Faraday Discussion 172 on "Carbon in Electrochemistry", covering some of the key points made at the meeting within the broader context of other recent developments on carbon materials for electrochemical applications. Although carbon electrodes have a long history of use in electrochemistry, methods and techniques are only just becoming available that can test long-established models and identify key features for further exploration. This Discussion has highlighted the need for a better understanding of the impact of surface structure, defects, local density of electronic states, and surface functionality and contamination, in order to advance fundamental knowledge of various electrochemical processes and phenomena at carbon electrodes. These developments cut across important materials such as graphene, carbon nanotubes, conducting diamond and high surface area carbon materials. With more detailed pictures of structural and electronic controls of electrochemistry at carbon electrodes (and electrodes generally), will come rational advances in various technological applications, from sensors to energy technology (particularly batteries, supercapacitors and fuel cells), that have been well-illustrated at this Discussion.

I would like to thank Katherine [Holt] for inviting me to give the 'Concluding Remarks' at this Faraday Discussion. I was delighted to accept, not only because I have considerable interest in carbon electrodes, but also – as a proud Yorkshireman – it's always wonderful to come home to 'God's own county' (even, as a Leeds lad, to Sheffield)! I do have to warn you, however, that Yorkshire folk (or 'Tykes' as we are 'affectionately' known) are often stereotyped as *"friendly, but bloody-minded and argumentative"*; and that's according to Wikipedia,[1] so it must be true! On reflection, that may be the perfect attribute to participate in a Faraday Discussion on "Carbon in Electrochemistry" and give these particular 'Remarks'. As this has been a most interesting, but at times serious and intense, meeting, I aim to lighten the mood a little. Prepare for Yorkshire (and Sheffield) references to be sprinkled liberally! Indeed, as we have discussed spatial heterogeneity

Department of Chemistry, University of Warwick, Coventry CV4 7AL, UK. E-mail: p.r.unwin@warwick.ac.uk; Fax: +44 (0)2476 524112; Tel: +44 (0)2476 523264

Fig. 1 Late summer in upper Swaledale, North Yorkshire (photograph courtesy of Dr Tessa Mobbs).

extensively, let's set the scene with an appreciation of the characteristic topography of the world-famous Yorkshire Dales (Fig. 1).

We were naturally very pleased in the Warwick Electrochemistry & Interfaces Group that the Faraday Division of the RSC chose to use one of our images for the publicity and announcements for this Discussion (Fig. 2 and elsewhere[2]). This picture is not a map of some far off world or a distorted view of our own planet! Rather, it's data showing the electrochemical activity of a polycrystalline diamond electrode, which evidently varies spatially.[3] On a different scale, the surface of a diamond electrode, topographically, is not unlike the beautiful Limestone Pavement in Malham Cove, North Yorkshire, comprising facets and grain boundaries. This type of heterogeneity goes to the heart of many of the issues we have addressed in this meeting: carbon electrode surfaces often comprise such features and other defects, and this may impact massively on the electrochemical properties that have been considered extensively in this Discussion. The challenge is to be able to identify and characterise these surface features and demonstrate unambiguously how they influence electrochemical reactions.

Local character

Sheffield is a particularly appropriate location for this meeting on 'carbon', because we are situated here above the great South Yorkshire Coalfield, with coal mining in this area stretching back to Roman times, until Margaret Thatcher put a stop to it! And, of course, the city of Sheffield is world famous for cutlery manufacture, with a long history, as far back as the late 13th century, and a

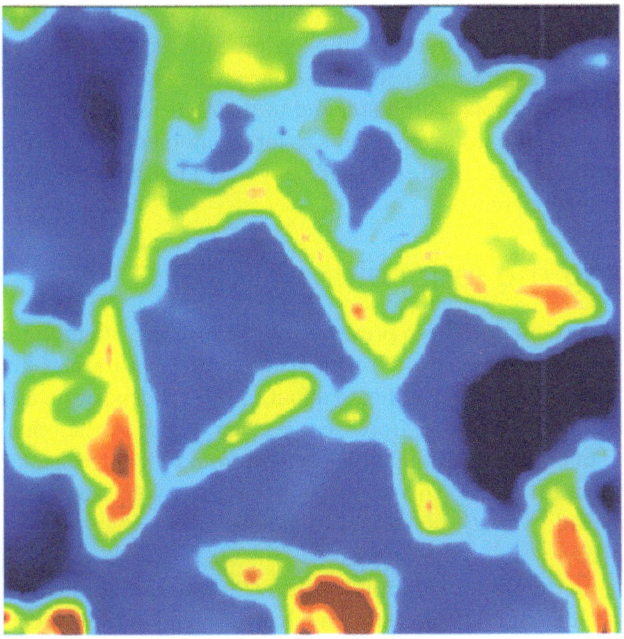

Fig. 2 Electrochemical map of a polycrystalline diamond electrode, adapted for the publicity graphic for this Discussion.[2]

mention in Chaucer's *Canterbury Tales*.[4] Crucible steel was manufactured extensively in this region during the Industrial Revolution, and stainless steel was discovered in Sheffield by Harry Brearley just over 100 years ago. In one of the papers on supercapacitors,[5] we heard about steel counter electrodes. So, there really is nowhere better than Yorkshire, and Sheffield, to have this meeting.

Yorkshire is famous for many unique and distinctive phrases, and I have drawn on one, 'There's nowt so queer as folk' or, to translate, 'There's nothing so peculiar as people' as the title for this talk. Why do I consider this appropriate? Well, as evident in some of the papers and debate in this meeting, and in work elsewhere,[6–13] carbon electrodes really do seem to have the most peculiar behaviour, with apparently the same material in different labs (and sometimes in different eras) giving rather different results. A major challenge in the field is to understand and rationalise these differences. For the present, it has proven difficult to identify the electrochemical behaviour of related new forms of carbon, such as graphene. Thus, is graphene more[14,15] or less[16] active than graphite, even for very simple outer sphere electron transfer processes? Similarly, for outer sphere redox couples, we may look at work in the field and ask, are single walled carbon nanotubes (SWNTs) only active at their ends,[17–19] and do such ends have to be open and oxygen functionalised?[18,19] Or do SWNTs have intrinsic activity on the sidewall?[20–24] We have had several papers on polycrystalline diamond electrodes, with applications spanning electroanalysis, electrocatalysis and electrodeposition. Yet, outside this meeting, there has been considerable debate on the active sites for electrochemistry at polycrystalline diamond, with high resolution electrochemical imaging providing insights into structural controls of electrochemical activity that are important for both fundamental understanding of charge transfer at this material,[3] and optimised applications. Although progress is being made in relating structure (and other surface properties) to activity, there is a long way to go, so with apologies for the anthropomorphism, and to the Arctic Monkeys, if carbon electrodes could speak they would probably say, for the moment, *"Whatever people say I am, that's what I'm not"*. And, for those not into English indie rock, that was another Sheffield reference!

Reviewing the rationale

A wide range of studies has been presented at this meeting, attempting to address the electroactivity of carbon electrodes at different length scales and under a diversity of conditions. Thus, we heard about the electrodeposition and stripping of carbon at high temperature, from George Chen,[25] and, on the other hand, the modification of graphene around point defects from Keith Stevenson[26] (a study I will mention again later). This Discussion has been very successful in highlighting the wide variety of studies that constitutes the use of carbon in electrochemistry.

In setting up the Discussion, the Scientific Committee outlined 4 points as a rationale, and it is useful to look at each in turn, in light of the papers presented. The first point was that *"Carbon is ubiquitous in electrochemical research..."* and we have certainly seen that in this meeting, which has been very successful in covering a wide range of topics in considerable depth. There is evidently much interest in the use of carbon electrodes in batteries,[27,28] and three papers in the session on supercapacitors[5,29,30] generated a huge amount of discussion and debate that could have gone on for much longer had the tea/coffee break not been called! There is clearly a lot of exciting work, and some controversy, in that area, and many new materials and concepts are coming through. The topic of electrochemical biosensors, likewise, was wide-ranging, from implant electrodes based on diamond microelectrode arrays[31] to the detection of biomolecules in solution,[32] and attempts to understand coupled electron transfer-proton transfer reactions, and the influence of electrode structure on such processes. Key properties of carbon electrodes that have come up in many of the discussions have been electrical conductivity, chemical functionality and biocompatibility. We even had a debate on the first afternoon about economics and carbon trading!

The second point was *"...the use of carbon in more novel forms (e.g. carbon nanotubes, graphene) has impacted on all areas of fundamental and applied electrochemistry..."*. In this respect, we have certainly covered many more types of carbon than I envisaged. One would have predicted that graphene,[26,32–34] highly oriented pyrolytic graphite (HOPG),[35] nanotubes,[36] glassy carbon[37] and diamond[31,38–40] would feature prominently, but we have also heard about nanoflakes,[34] nanodiamond,[39] a lot on carbon composites,[27,41–43] support effects[32,43] and porous materials.[30] This diversity and complexity in the types of carbon materials found in electrochemistry raises challenges as to how to look at these materials. This brings us to the third point which was *"Advanced electrochemical techniques such as scanning electrochemical microscopy, electrochemical AFM and spectroelectrochemistry have moreover increased our understanding of the interfacial properties of traditional carbon electrode materials"*. This is certainly true, as these techniques allow access to structure–function directly and simultaneously,[6,7,44–48] as highlighted above, and they can provide insight into chemical speciation during electrochemical reactions at carbon electrodes.[49,50] However, these types of methodologies did not feature as prominently in this Discussion as I expected, although they were evident.[26,36] Such approaches, and others, have an important role to play in the future, and I will return to this shortly.

The final aspect was that *"Carbon also plays an important role in technologically applied areas of electrochemistry – in energy generation and storage and catalyst support"*. This issue was covered admirably, for example, with sub-nanoporous carbide-derived carbons[30] and multi-functional structural composites for energy storage,[42] where the importance not only of electrochemical performance, but also structural integrity, was highlighted. We also heard about the use of redox active electrolytes for supercapacitors.[29] But, I think, the technological highlight was undoubtedly Rick's [McCreery] amazing 'electrochemical guitar pedal', based on an innovative carbon-based molecular junction and showcased with audio in the Introductory Lecture.[35] There is a future for this! Perhaps the hard-rocking Sheffield band, Def Leppard, might be interested? As Rick got this Discussion off to a musical start, I am happy to conclude in similar fashion!

Macroscopic to microscopic

An enduring theme of this meeting is that many papers tended to focus on average properties and *we had a lot of current–voltages curves*! As electrochemists, we know that current–voltage relationships have their place, but considerable care is needed when developing microscopic models from such measurements. We still rely heavily on correlations in electrochemistry, and particularly in studies of carbon electrodes, because (traditionally) they have been difficult to characterise microscopically, but it is important to bear in mind that just because one quantity appears to correlate with another does not mean there is necessarily a physical relationship between the two. Models derived in this way need to be tested continuously with new, emerging microscopic techniques. The electrochemical response of simple redox couples, notably $Fe(CN)_6^{4-/3-}$, has become a proxy for the surface quality (defect concentration) of graphitic materials and carbon nanotubes,[13,19,51–53] and is now widely used in that context,[54–58] but there are many other issues that can impact the response of (layered) carbon electrodes, such as the adsorption of adventitious material (impurities),[6,59,60] delamination,[6,61] surface oxidation, *etc.* So, without careful control of these parameters, macroscopic CV measurements, where one simply looks for 'sluggish kinetics' as an indicator of 'quality', do not provide an unambiguous view of (high) surface quality. Indeed, although $Fe(CN)_6^{4-/3-}$ was used as the main method to 'select' particular HOPG surfaces for further study in earlier work,[13,51,52] it is not a straightforward couple even on well-defined and easily prepared metal electrodes,[62,63] and recent work highlights complicated factors in its study at graphite (HOPG) electrodes.[6] Without a detailed understanding of structural controls of apparently simple electron transfer processes at well-defined graphitic electrodes, it will be difficult for the field to advance to understand more complex reactions and carbon materials. This situation contrasts markedly to the studies of precious metal electrodes, for example, platinum, where CV measurements can give considerable information on the structure and status of the electrode.[64–66] Challenges for the study of carbon electrodes would be: to identify similar unambiguous CV signatures indicative of surface quality; to make more widespread use of high resolution structural techniques to characterise materials used for electrochemistry, as highlighted in several papers in this Discussion; and to make more use of microscopic and nanoscopic electrochemical probes where one can target particular surface features of interest.

Capacitance measurements also featured strongly at the macroscopic level, along with related measurements such as the determination of active surface area from redox probe adsorption and gas adsorption.[30,37,43] There were interesting questions and debate about the limitations of these approaches. The importance of nanoscale properties featured a lot in the discussions, but there was a tendency to talk in very general terms about defects or dangling bonds, and much more needs to be known about these issues. We also need more dialogue between different groups about surface quality, functionality and reproducibility. We had some contributions to the debate along the lines that really nice super-clean, low defect graphene can be made very easily by chemical vapour deposition, whereas other groups recognise that there are many issues to making (and retaining) such materials with high surface integrity. Which is correct?

'The Full Monty'

So how do we move forward to improve our understanding? To begin with, I believe we have to focus on well-defined planar carbon surfaces (*e.g.* well-defined HOPG and conducting single crystal diamond), as these materials can now be characterised fairly easily in a variety of ways. But we have to ensure the surfaces are as close to pristine as possible. Individual point defects can be characterised with STM, as Keith Stevenson showed in this meeting for graphene.[26] Others have shown for HOPG that such defects are of low concentration and in a narrow range of 0.1–10 μm^{-2}.[67–69] In fact, a recent high resolution STM study of HOPG surface modification failed to find a single point defect in over 100 images (typically 25 nm^2 area per image).[70] Thus, we should be able to pin down the importance (or otherwise) of such features. Step edges, which are the main 'defect site' on such surfaces, are routinely quantified with AFM.[6,7,10,47] Thus, studying different grades of HOPG (see for example Fig. 3A), even by macroscopic electrochemistry, opens up the possibility of determining how step edge density impacts (or otherwise) various electrochemical processes.[10,71]

Well-characterised nanomaterials are now also amenable to study, such as single layer graphene, graphene bilayers, multiple layers, individual edges, and nanotubes. This meeting,[36] and related work,[23,72] has shown that it is now possible to understand multiple properties of an individual SWNT and relate these to the electrochemistry. Although challenging experimentally, defects in SWNTs can also be visualised with a range of techniques,[23,73–75] and it should thus be possible to assess the impact of defects on electrochemistry with high resolution electrochemical microscopy.

It is evidently important to understand the local electronic structure of carbon materials, and this was raised several times in the Discussion. Scanning tunnelling spectroscopy (STS) studies are revealing considerable information about the local density of states (LDOS) of various carbon materials.[61,76] For HOPG, a major outcome is that the LDOS at armchair step edges is similar to that at the basal surface away from the edge, and only enhanced at zigzag edges, which are the

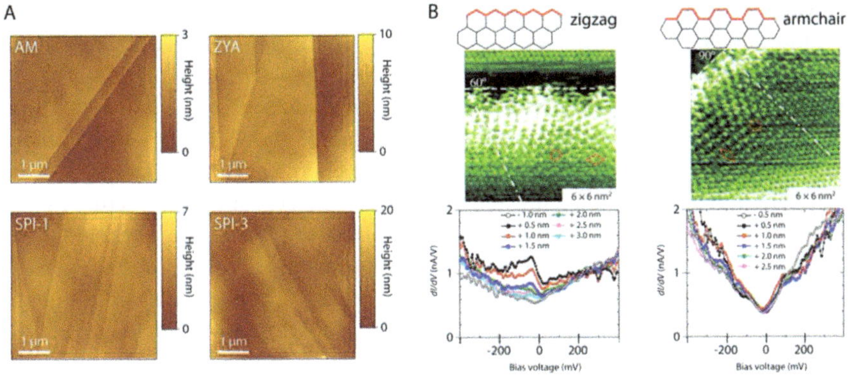

Fig. 3 (A) AFM images of different grades of HOPG (note the different height scales). (B) STM images of HOPG near a monoatomic step edge for zigzag (left) and armchair (right) edges. The d*I*/d*V* curves were measured at difference distances from the edge. Adapted with permission from ref. 10 and 77.

minority edge (see Fig. 3B). Thus, unless the density of such edges is very high (<10 nm spacing), the overall DOS at HOPG is essentially constant across the different grades of HOPG.[77] This is significant because it opens up the possibility – even on the macroscopic scale – of separating DOS effects from step edge density effects (and other defect effects, assuming that point defects, and other electronic disturbances that may occur in the basal surface[61,78–80] remain unimportant or can be quantified). Thus, for HOPG, and other carbon materials, it should be possible to build a very detailed picture of activity, by combining sophisticated surface structure probes and electrochemical measurements, while also taking precautions to avoid surface contamination.

So, in essence, I am arguing that to move the field forward we need the *"Full Monty"*! The title of this Great British film about 5 unemployed guys from Sheffield who form a striptease act most likely has its origin in the 'full Montague Burton (3-piece) suit' (from Leeds), and basically means 'everything' or 'the full works'. So who are the 'Full Monty electrochemists' at this Discussion? First, Keith [Stevenson] showed us how STM measurements revealed the importance of defects *vs.* basal activity during the functionalization of graphene.[26] Rob [Dryfe] tackled the issue of environment/impurity effects on graphene electroactivity, and used a microdroplet to probe a single layer, two layers, *etc.*, unambiguously, with knowledge of the environment of the graphene surface and how that impacted the electrochemistry. Laurence [Hardwick] told us about making *in situ* Raman measurements at micro-graphite electrodes during lithium intercalation and de-intercalation, with major implications for understanding the functioning of lithium ion batteries. This was a distinctive contribution, because while Raman spectroscopy is a natural complement to electrochemistry to enable understanding of the structure and properties of carbon,[81,82] and was mentioned a great deal at this meeting, *in situ* work in this type of system has been rare. Aleix [Güell] described a multi-technique platform, where many techniques, including high resolution electrochemical imaging, can be applied to individual nanomaterials.[36] A word of advice though guys – stick to the electrochemistry!

Outside of this meeting, there is evidence that other groups are embracing this philosophy. So Phil Collins' group (University of California, Irvine) has recently investigated the charge transfer resistance of individual SWNT composites.[83] This is a particularly nice study because it contributes to fundamental understanding of heterogeneous electron transfer, making use of individual pristine carbon nanotubes coated with a metal–oxide electrochemical-storage material. The experiments rule out any significant role of SWNT defects or band structure in the electrochemistry of this kind of composite. Moreover, the results contradict the expectation of enhanced out-of-plane electron transfer rates from defective (step edge) sp^2 carbon, based on older HOPG literature.[8,9,11,12,51,52,84] These studies found that multi-walled carbon nanotubes, platinum, HOPG and SWNTs had essentially the same charge transfer characteristics, even though they have very different electronic band structure. Investigations of this type are essential to challenge preconceptions and promote progress in this field.

Back to Yorkshire

As a beneficiary of an enlightened education environment in West Yorkshire in the early 1970's, under Sir Alec Clegg, I would like to close by quoting from an after dinner speech he gave at a workshop for school teachers in Bingley, West Yorkshire, just over 40 years ago. Entitled *"Making the whole world wonder"*,[85] his speech is illuminating and I encourage you to read it. In part of the speech he identified the characteristics he had seen in the greatest teachers (and by extension, I believe, these attributes make the best researchers): *"They were immersed in – and constantly excited by – what they were doing. They were innovators. They modified their views continuously as they learned from their experiences… and from the ideas that they picked up outside and wished to try out. Their professional life was a creative quest."*

While we (and the public) often link science with innovation and excitement, it can – at the same time – be very conservative, and hierarchical. Ideas can become entrenched (and personal), and the history of science tells us that change can be a painful, long and difficult process. Yet, science is not set in stone, and this is particularly true of electrochemistry. The challenge of studying complex electrochemical interfaces means that our views will change continuously, as new techniques and methods become available. This meeting has highlighted the tremendous potential of carbon materials for both fundamental and applied studies in electrochemistry, but we have much to learn, and our views will change. Carbon electrodes may have been around for a long time in electrochemistry, but this Discussion has highlighted that we are entering a new era of understanding.

I hope you have enjoyed your time in Yorkshire and this Discussion on "Carbon in Electrochemistry".

Acknowledgements

I would like to acknowledge many colleagues at Warwick with whom I have had interesting and illuminating discussions on carbon electrodes. I am particularly grateful to the electrochemical imaging team who have brought many new perspectives to bear on the local electrochemistry of carbon electrodes, aided by generous support from the European Research Council (ERC-2009-AdG247143-QUANTIF).

I would like to thank Aleix Güell and Guohui Zhang for their help and advice in the preparation of this manuscript.

References

1 http://en.wikipedia.org/wiki/Culture_of_Yorkshire, accessed on 5 October 2014.
2 http://www.rsc.org/conferencesandevents/rscconferences/fd/fd172/index.asp, accessed on 5 October 2014.
3 H. V. Patten, K. E. Meadows, L. A. Hutton, J. G. Iacobini, D. Battistel, K. McKelvey, A. W. Colburn, M. E. Newton, J. V. Macpherson and P. R. Unwin, *Angew. Chem., Int. Ed.*, 2012, **51**, 7002–7006.
4 G. Chaucer, The Reeve's Tale, in *The Canterbury Tales*, 'Ther was no man, for peril, dorste hym touche. A Sheffeld thwitel baar he in his hose. Round was his face, and camus was his nose'.
5 Q. Abbas, P. Ratajczak and F. Beguin, *Faraday Discuss.*, 2014, DOI: 10.1039/c4fd00056k.
6 A. N. Patel, M. G. Collignon, M. A. O'Connell, W. O. Hung, K. McKelvey, J. V. Macpherson and P. R. Unwin, *J. Am. Chem. Soc.*, 2012, **134**, 20117–20130.
7 S. C. S. Lai, A. N. Patel, K. McKelvey and P. R. Unwin, *Angew. Chem., Int. Ed.*, 2012, **51**, 5405–5408.
8 T. J. Davies, M. E. Hyde and R. G. Compton, *Angew. Chem., Int. Ed.*, 2005, **44**, 5121–5126.
9 C. E. Banks, T. J. Davies, G. G. Wildgoose and R. G. Compton, *Chem. Commun.*, 2005, 829–841.
10 G. Zhang, P. M. Kirkman, A. N. Patel, A. S. Cuharuc, K. McKelvey and P. R. Unwin, *J. Am. Chem. Soc.*, 2014, **136**, 11444–11451.
11 M. T. McDermott and R. L. McCreery, *Langmuir*, 1994, **10**, 4307–4314.
12 M. T. McDermott, K. Kneten and R. L. McCreery, *J. Phys. Chem.*, 1992, **96**, 3124–3130.
13 K. R. Kneten and R. L. McCreery, *Anal. Chem.*, 1992, **64**, 2518–2524.
14 W. Li, C. Tan, M. A. Lowe, H. D. Abruña and D. C. Ralph, *ACS Nano*, 2011, **5**, 2264–2270.
15 A. T. Valota, I. A. Kinloch, K. S. Novoselov, C. Casiraghi, A. Eckmann, E. W. Hill and R. A. W. Dryfe, *ACS Nano*, 2011, **5**, 8809–8815.
16 A. G. Güell, N. Ebejer, M. E. Snowden, J. V. Macpherson and P. R. Unwin, *J. Am. Chem. Soc.*, 2012, **134**, 7258–7261.
17 J. J. Gooding, R. Wibowo, J. Liu, W. Yang, D. Losic, S. Orbons, F. J. Mearns, J. G. Shapter and D. B. Hibbert, *J. Am. Chem. Soc.*, 2003, **125**, 9006–9007.
18 A. Chou, T. Bocking, N. K. Singh and J. J. Gooding, *Chem. Commun.*, 2005, 842–844.
19 A. F. Holloway, K. Toghill, G. G. Wildgoose, R. G. Compton, M. A. H. Ward, G. Tobias, S. A. Llewellyn, B. Ballesteros, M. L. H. Green and A. Crossley, *J. Phys. Chem. C*, 2008, **112**, 10389–10397.
20 I. Heller, J. Kong, H. A. Heering, K. A. Williams, S. G. Lemay and C. Dekker, *Nano Lett.*, 2004, **5**, 137–142.
21 A. G. Güell, N. Ebejer, M. E. Snowden, K. McKelvey, J. V. Macpherson and P. R. Unwin, *Proc. Natl. Acad. Sci. U. S. A.*, 2012, **109**, 11487–11492.

22 A. G. Güell, K. E. Meadows, P. V. Dudin, N. Ebejer, J. V. Macpherson and P. R. Unwin, *Nano Lett.*, 2014, **14**, 220–224.

23 J. C. Byers, A. G. Güell and P. R. Unwin, *J. Am. Chem. Soc.*, 2014, **136**, 11252–11255.

24 T. S. Miller, N. Ebejer, A. G. Guell, J. V. Macpherson and P. R. Unwin, *Chem. Commun.*, 2012, **48**, 7435–7437.

25 H. V. Ijije, R. C. Lawrence, N. J. Siambun, S. M. Jeong, D. A. Jewell, D. Hu and G. Z. Chen, *Faraday Discuss.*, 2014, DOI: 10.1039/c4fd00046c.

26 K. J. Stevenson, P. A. Veneman, R. I. Gearba, K. M. Mueller, B. J. Holliday, T. Ohta and C. K. Chan, *Faraday Discuss.*, 2014, DOI: 10.1039/c4fd00038b.

27 Y. Cao, S.-R. Cai, S.-C. Fan, W.-Q. Hu, M.-S. Zheng and Q.-F. Dong, *Faraday Discuss.*, 2014, DOI: 10.1039/c4fd00075g.

28 C. Sole, N. E. Drewett and L. J. Hardwick, *Faraday Discuss.*, 2014, DOI: 10.1039/c4fd00079j.

29 E. Frackowiak, M. Meller, J. Menzel, D. Gastol and K. Fic, *Faraday Discuss.*, 2014, DOI: 10.1039/c4fd00052h.

30 B. Dyatkin and Y. Gogotsi, *Faraday Discuss.*, 2014, DOI: 10.1039/c4fd00048j.

31 C. Hebert, E. Scorsone, A. Bendali, R. Kiran, M. Cottance, H. A. Girard, J. Degardin, E. Dubus, G. Lissorgues, L. Rousseau, P. Mailley, S. Picaud and P. Bergonzo, *Faraday Discuss.*, 2014, DOI: 10.1039/c4fd00040d.

32 J. Hu, S. Wisetsuwannaphum and J. S. Foord, *Faraday Discuss.*, 2014, DOI: 10.1039/c4fd00032c.

33 N. L. Ritzert, W. Li, C. Tan, G. G. Rodriguez-Calero, J. Rodriguez-Lopez, K. Hernandez-Burgos, S. Conte, J. J. Parks, D. C. Ralph and H. D. Abruna, *Faraday Discuss.*, 2014, DOI: 10.1039/c4fd00060a.

34 M. M. Lounasvuori, M. Rosillo-Lopez, C. G. Salzmann, D. J. Caruana and K. B. Holt, *Faraday Discuss.*, 2014, DOI: 10.1039/c4fd00034j.

35 R. McCreery, A. Bergren, A. Morteza-Najarian, S. Nagy and H. Yan, *Faraday Discuss.*, 2014, DOI: 10.1039/c4fd00172a.

36 A. G. Güell, K. E. Meadows, P. V. Dudin, N. Ebejer, J. C. Byers, J. V. Macpherson and P. R. Unwin, *Faraday Discuss.*, 2014, DOI: 10.1039/c4fd00054d.

37 A. Noofeli, P. J. Hall and A. J. R. Rennie, *Faraday Discuss.*, 2014, DOI: 10.1039/c4fd00057a.

38 L. Meng, J. G. Iacobini, M. B. Joseph, J. V. Macpherson and M. E. Newton, *Faraday Discuss.*, 2014, DOI: 10.1039/c4fd00044g.

39 T. S. Varley, M. Hirani, G. Harrison and K. B. Holt, *Faraday Discuss.*, 2014, DOI: 10.1039/c4fd00041b.

40 R. Hamers, J. N. Bandy, L. Zhang and D. Zhu, *Faraday Discuss.*, 2014, DOI: 10.1039/c4fd00039k.

41 S. A. Hodge, H. H. Tay, D. B. Anthony, R. Menzel, D. J. Buckley, P. L. Cullen, N. T. Skipper, C. A. Howard and M. S. P. Shaffer, *Faraday Discuss.*, 2014, DOI: 10.1039/c4fd00043a.

42 N. Shirshova, H. Qian, M. Houlle, J. H. G. Steinke, A. R. J. Kucernak, Q. P. V. Fontana, E. S. Greenhalgh, A. Bismarck and M. S. P. Shaffer, *Faraday Discuss.*, 2014, DOI: 10.1039/c4fd00055b.

43 J. Quinson, R. Hidalgo, P. A. Ash, F. Dillon, N. Grobert and K. A. Vincent, *Faraday Discuss.*, 2014, DOI: 10.1039/c4fd00058g.

44 N. Ebejer, A. G. Güell, S. C. S. Lai, K. McKelvey, M. E. Snowden and P. R. Unwin, *Annu. Rev. Anal. Chem.*, 2013, **6**, 329–351.

45 N. Ebejer, M. Schnippering, A. W. Colburn, M. A. Edwards and P. R. Unwin, *Anal. Chem.*, 2010, **82**, 9141–9145.
46 P. L. Frederix, P. D. Bosshart, T. Akiyama, M. Chami, M. R. Gullo, J. J. Blackstock, K. Dooleweerdt, N. F. de Rooij, U. Staufer and A. Engel, *Nanotechnology*, 2008, **19**, 384004.
47 A. N. Patel, S. Y. Tan, T. S. Miller, J. V. Macpherson and P. R. Unwin, *Anal. Chem.*, 2013, **85**, 11755–11764.
48 A. J. Wain, A. J. Pollard and C. Richter, *Anal. Chem.*, 2014, **86**, 5143–5149.
49 N. González-Diéguez, A. Colina, J. López-Palacios and A. Heras, *Anal. Chem.*, 2012, **84**, 9146–9153.
50 J. Garoz-Ruiz, D. Izquierdo, A. Colina, S. Palmero and A. Heras, *Anal. Bioanal. Chem.*, 2013, **405**, 3593–3602.
51 R. J. Rice and R. L. McCreery, *Anal. Chem.*, 1989, **61**, 1637–1641.
52 K. K. Cline, M. T. McDermott and R. L. McCreery, *J. Phys. Chem.*, 1994, **98**, 5314–5319.
53 C. E. Banks, R. R. Moore, T. J. Davies and R. G. Compton, *Chem. Commun.*, 2004, 1804–1805, DOI: 10.1039/b406174h.
54 M. Pumera, *Electrochem. Commun.*, 2013, **36**, 14–18.
55 A. Ambrosi, C. K. Chua, A. Bonanni and M. Pumera, *Chem. Rev.*, 2014, **114**, 7150–7188.
56 D. K. Kampouris and C. E. Banks, *Chem. Commun.*, 2010, **46**, 8986–8988.
57 D. A. C. Brownson, D. K. Kampouris and C. E. Banks, *Chem. Soc. Rev.*, 2012, **41**, 6944–6976.
58 S. Palanisamy, R. Madhu, S.-M. Chen and S. K. Ramaraj, *Anal. Methods*, 2014, **6**, 8368–8373.
59 Z. Li, Y. Wang, A. Kozbial, G. Shenoy, F. Zhou, R. McGinley, P. Ireland, B. Morganstein, A. Kunkel, S. P. Surwade, L. Li and H. Liu, *Nat. Mater.*, 2013, **12**, 925–931.
60 A. Kozbial, Z. Li, J. Sun, X. Gong, F. Zhou, Y. Wang, H. Xu, H. Liu and L. Li, *Carbon*, 2014, **74**, 218–225.
61 G. Li, A. Luican and E. Y. Andrei, *Phys. Rev. Lett.*, 2009, **102**, 176804.
62 C. M. Pharr and P. R. Griffiths, *Anal. Chem.*, 1997, **69**, 4673–4679.
63 J. López-Palacios, A. Heras, Á. Colina and V. Ruiz, *Electrochim. Acta*, 2004, **49**, 1027–1033.
64 J. Mostany, E. Herrero, J. M. Feliu and J. Lipkowski, *J. Phys. Chem. B*, 2002, **106**, 12787–12796.
65 N. Garcia-Araez, V. Climent and J. Feliu, *J. Phys. Chem. C*, 2009, **113**, 9290–9304.
66 V. Climent and J. Feliu, *J. Solid State Electrochem.*, 2011, **15**, 1297–1315.
67 H. Chang and A. J. Bard, *J. Am. Chem. Soc.*, 1991, **113**, 5588–5596.
68 H. Chang and A. J. Bard, *J. Am. Chem. Soc.*, 1990, **112**, 4598–4599.
69 J. V. Zoval, R. M. Stiger, P. R. Biernacki and R. M. Penner, *J. Phys. Chem.*, 1996, **100**, 837–844.
70 H. Ma, L. Lee, P. A. Brooksby, S. A. Brown, S. J. Fraser, K. C. Gordon, Y. R. Leroux, P. Hapiot and A. J. Downard, *J. Phys. Chem. C*, 2014, **118**, 5820–5826.
71 A. N. Patel, S. Y. Tan and P. R. Unwin, *Chem. Commun.*, 2013, **49**, 8776–8778.
72 P. V. Dudin, M. E. Snowden, J. V. Macpherson and P. R. Unwin, *ACS Nano*, 2011, **5**, 10017–10025.

73 K. Suenaga, H. Wakabayashi, M. Koshino, Y. Sato, K. Urita and S. Iijima, *Nat. Nanotechnol.*, 2007, **2**, 358–360.

74 Z. Osváth, G. Vértesy, L. Tapasztó, F. Wéber, Z. E. Horváth, J. Gyulai and L. P. Biró, *Phys. Rev. B: Condens. Matter Mater. Phys.*, 2005, **72**, 045429.

75 Y. Fan, B. R. Goldsmith and P. G. Collins, *Nat. Mater.*, 2005, **4**, 906–911.

76 W. O. Teri, H. Jin-Lin and M. L. Charles, *J. Phys.: Condens. Matter*, 2002, **14**, R145.

77 Y. Niimi, T. Matsui, H. Kambara, K. Tagami, M. Tsukada and H. Fukuyama, *Phys. Rev. B: Condens. Matter Mater. Phys.*, 2006, **73**, 085421.

78 A. Luican, G. Li and E. Y. Andrei, *Solid State Commun.*, 2009, **149**, 1151–1156.

79 Y. A. Eva, L. Guohong and D. Xu, *Rep. Prog. Phys.*, 2012, **75**, 056501.

80 G. Li, A. Luican, J. M. B. Lopes dos Santos, A. H. Castro Neto, A. Reina, J. Kong and E. Y. Andrei, *Nat. Phys.*, 2010, **6**, 109–113.

81 Y. Wang, D. C. Alsmeyer and R. L. McCreery, *Chem. Mater.*, 1990, **2**, 557–563.

82 M. S. Dresselhaus, A. Jorio and R. Saito, *Annu. Rev. Condens. Matter Phys.*, 2010, **1**, 89–108.

83 B. L. Corso, I. Perez, T. Sheps, P. C. Sims, O. T. Gül and P. G. Collins, *Nano Lett.*, 2014, **14**, 1329–1336.

84 R. J. Bowling, R. T. Packard and R. L. McCreery, *J. Am. Chem. Soc.*, 1989, **111**, 1217–1223.

85 http://www.educationengland.org.uk/documents/speeches/1972clegg.html, accessed on 5 October 2014.

Poster titles

Study of nitrogen doped mesoporous carbon as a catalyst for oxygen reduction in alkaline and buffer solutions, **T. Alaje, E. Yu and K. Scott**, *Newcastle University, UK*

Electroreduction of CO_2 to formate: effect of particle size and load of Sn powder deposited over carbon support, **A. Del Castillo, M. Alvarez-Guerra, J. Solla-Gullón, A. Sáez, V. Montiel and A. Irabien**, *Universidad de Cantabria, Spain*

Electrochemical reduction of CO_2 by Cu_2O/Zn-catalyzed carbon supports, **J. Albo, A. Sáez, J. Solla-Gullón, V. Montiel, P. Castaño, A. Irabien and M. Alvarez-Guerra**, *Universidad del País Vasco, Spain*

The application of exfoliated graphite-diamond composite electrode in the electrochemical degradation of trichloroethylene, **O. Arotiba, B. Ntsendwana, B. Mamba and S. Sampath**, *University of Johannesburg, South Africa*

Vertically aligned carbon nanotube/conducting polymer hybrid electrode, **D. Cakiroglu and F. C. Cebeci**, *Sabanci University, Turkey*

Electrocatalytic denitrification at neutral pH: a spectroelectrochemical study of activity and intermediates at carbon-supported metal nanoparticles, **M. Duca, I. McPherson and K. Vincent**, *University of Oxford, UK*

Surface chemistry and graphitic structure effects on electric double layer capacitance, **B. Dyatkin and Y. Gogotsi**, *Drexel University*

NMR studies of the carbon-electrolyte interface in ionic liquid supercapacitors, **A. Forse, J. Griffin, H. Wang, C. Merlet, D. Weingarth, V. Presser, Y. Gogotsi, P. Simon and C. Grey**, *University of Cambridge, UK*

Enhancement of capacitance by halide and hydroxybenzene additives to electrolyte, **D. Gastol, E. Frackowiak and J. Walkowiak**, *Poznan University of Technology, Poland*

Influence of nitrogen-containing groups on the capacitive properties of carbon nanofibers, **G. Gryglewicz, A. Śliwak and S. Suropek**, *Wroclaw University of Technology, Poland*

Carbon nanoparticle electrodes for coupling protein film electrochemistry with infrared spectroscopy: studies of hydrogenases, **R. Hidalgo, P. Ash and K. Vincent**, *University of Oxford, UK*

Using pair distribution function analysis to study the charge/discharge process of supercapacitors, **E. Howe, P. Allan, P. Chupas, K. Chapman, S. Clarke and C. Grey**, *University of Cambridge, UK*

Nitrogen doping onto reduced graphene oxide for controlling the mean electron transfer number of the oxygen reduction reaction, **C.-C. Hu, C.-L. Liu, S.-H. Wu and T.-H. Wu**, *National Tsing Hua University, Chinese Taipei*

Carbon nanoparticle electrodes for sensing and catalysis, **M. Jonsson-Niedziolka, M. Kundys, E. Rozniecka, K. Szot, A. Zloczewska, A. Celebanska, F. Marken and M. Opallo**, *Polish Academy of Sciences, Poland*

A long-term performance of μ-Si/graphite/pitch-coke composite in half- and full cell of Li-ion battery, **K. Kierzek**, *Wroclaw University of Technology, Poland*

Is the slit shape pore model appropriate for highly amorphous nanoporous carbons?, **H. Kurig, M. Russina, I. Tallo, M. Siebenbürger, S. Prevost and E. Lust**, *University of Tartu, Estonia*

Development of conjugated microporous polymers for supercapacitive energy storage, **J.-S. Lee, B. Alston, T. Hasell and A. Cooper**, *University of Liverpool, UK*

Synthesis and sorption property of magnetic Fe_3O_4/RGO nanocomposites on methylene blue, **K. Luo, P. Wang, Y. Xiang, X. Liu and L. Xing**, *Guilin University of Technology, China*

Platinum-free electrocatalysts for oxygen reduction reaction, **M. Longhi, S. Marzorati and L. Formaro**, *Università degli Studi di Milano, Italy*

Electrochemical characterisation of graphene nanoflakes with functionalised edges, **M. Lounasvuori, K. Holt, D. Caruana, M. Rosillo-Lopez and C. Salzmann**, *University College, London, UK*

Nitrogen-doped graphene foam electrocatalysts, **S. M. Lyth, J. Liu and K. Sasaki**, *Kyushu University, Japan*

Tuning the electronic valence structure of amorphous carbon surfaces: effects on the kinetics of electrochemical phenomena at the carbon/solution interface, **S. Marzorati, R. J. Cullen, R. L. Doyle, M. Longhi, M. E. G. Lyons and P. E. Colavita**, *Universita' degli Studi di Milano, Italy*

Single-walled carbon nanotubes as electrodes for the electrochemical characterisation of guest molecules, **R. McSweeney, T. Chamberlain, S. Davies and A. Khlobystov**, *University of Nottingham, UK*

Laser heated boron doped diamond electrode for lead deposition: stripping and microscopic analysis, **L. Meng, J. Iacobini, M. Joseph, M. Newton and J. Macpherson**, *The University of Warwick, UK*

Investigation of supercapacitors with an original lattice simulation method combined with *in situ* NMR experiments, **C. Merlet, A. C. Forse, J. M. Griffin, D. Frenkel and C. P. Grey**, *University of Cambridge, UK*

Plasma functionalised carbon nanomaterials for electrochemical sensing and energy applications, **H. Nolan, N. McEvoy, G. P. Keeley, N. Ashok Kumar and G. S. Duesberg**, *CRANN*, Ireland

Integrated carbon nanomaterials for enhanced electrochemical performance, **T. Palomäki, S. Sainio, K. Kordás, V. Singh, J. Koskinen and T. Laurila**, *Aalto University, Finland*

Continuous performance of double-layer technologies for water desalination and energy harvesting using flow electrodes, **S. Porada, D. Weingarth, B. Hamelers, M. Bryjak, V. Presser and M. Biesheuvel**, *INM - Leibniz Institute for New Materials, Germany*

Strong interaction between metal particles and graphitic carbon nitride/graphene sheet composites for facile oxygen reduction reaction, **K. Qiu and Z.-X. Guo**, *University College, London, UK*

Highly microporous carbons from olive pits for supercapacitors, **E. Redondo, J. Carretero-González, E. Goikolea, J. Ségalini and R. Mysyk**, *CIC Energigune, Spain*

Low Pt loading, CO tolerant manganese oxide - reduced graphene oxide hybrid (MnO_x-RGO) support, **S. Sharma, R. Kannan, G. Gupta and J. Oyekan**, *University of Birmingham, UK*

Aqueous supercapacitor on textiles with carbon electrode, **Y. Sheng, O. John, T. John and S. Beeby**, *University of Southampton, UK*

Preparation of carbon nanofibers grafted activated carbon for supercapacitor with enhanced capacitive performance, **A. Śliwak and G. Gryglewicz**, *Wroclaw University of Technology, Poland*

The electrochemical characterization of graphite felt, **R. Smith, T. Davies and R. Nichols**, *University of Liverpool, UK*

The electrochemical graphene surface force balance, **C. van Engers, N. Cousens, J. Britton, S. Perkin and N. Grobert**, *University of Oxford, UK*

Key parameters affecting cell voltage of aqueous EDLCs, **T.-H. Wu, C.-T. Hsu, C.-C. Hu and L. Hardwick**, *University of Liverpool, UK*

Pillared graphene nanocomposites for energy storage, **W. Yu**, *University of Cambridge, UK*

Quinone adsorption and electrochemistry at graphite electrodes: testing and revising the literature model, **G. Zhang, P. Kirkman, A. Cuharuc, K. McKelvey and P. R. Unwin**, *University of Warwick, UK*

Ion screening in carbon nanotubes - a density functional theory study, **F. Juarez, L. Mohammadzadeh, A. Goduljan, P. Quaino, E. Santos and W. Schmickler**, *University of Ulm, Germany*

High working voltage multilayer supercapacitor threads, **F. Qiu, D. Harrison and J. Fyson**, *Brunel University, UK*

CoS/CoS_2 nanomaterials hydrothermally decorated on CNT and reduced graphene oxide as electrocatalysts towards oxygen reduction reaction in alkaline medium, **S. M. S. Kumar, K. Selvakumar, R. Thangamuthu, X. Liu, T. Alaje and K. Scott**, *Newcastle University, UK*

List of participants

Mr Sadiki Lamari Abdelmajid, *Sultan Moulay Sliman University Faculty of Science and Technology, Morocco*
Professor Héctor Abruña, *Cornell University, USA*
Mr Taiwo Alaje, *Newcastle University, United Kingdom*
Dr Abdulaziz Alghamdi, *King Saud University, Saudi Arabia*
Dr Ali Alsalme, *King Saud University, Saudi Arabia*
Dr Manuel Alvarez-Guerra, *University of Cantabria, Department of Chemical and Biomolecular Engineering, Spain*
Professor Omotayo Arotiba, *University of Johannesburg, South Africa*
Dr Philip Ash, *University of Oxford, Department of Chemistry, United Kingdom*
Dr Klaus Bauer-Espindola, *Roche Diagnostics GmbH, Germany*
Dr Domnik Bayer, *BASF SE, Germany*
Dr Alisa Becker, *Royal Society of Chemistry, United Kingdom*
Professor François Béguin, *Poznan University of Technology, Poland*
Dr Philippe Bergonzo, *CEA-LIST, France*
Miss Dilek Cakiroglu, *Sabanci University, Turkey*
Dr Yong Cao, *Xiamen University, China*
Professor George Zheng Chen, *University of Nottingham, United Kingdom*
Miss Ming Chung, *Johnson Matthey, United Kingdom*
Dr Nico Cousens, *University of Oxford, United Kingdom*
Dr Keith Dawes, *EC Lab Ltd, United Kingdom*
Mr Son Dongrag, *Department of Chemistry, University of Ulsan, Korea*
Professor Robert Dryfe, *University of Manchester, United Kingdom*
Dr Matteo Duca, *University of Oxford, United Kingdom*
Mr Boris Dyatkin, *Drexel University, USA*
Professor John Foord, *University of Oxford, United Kingdom*
Mr Alexander Forse, *Department of Chemistry, University of Cambridge, United Kingdom*
Professor Elzbieta Frackowiak, *Poznan University of Technology, Poland*
Dr Dominika Gastol, *Poznan University of Technology, Poland*
Mr Yury Gogotsi, *Drexel University, USA*
Professor Dr Grażyna Gryglewicz, *Wroclaw University of Technology, Poland*
Dr Aleix Güell, *University of Warwick, United Kingdom*
Professor Dr Dirk Guldi, *University of Erlangen, Germany*
Professor Peter Hall, *Sheffield University, United Kingdom*
Professor Robert Hamers, *University of Wisconsin-Madison, USA*
Dr Laurence Hardwick, *University of Liverpool, United Kingdom*
Mr Ricardo Hidalgo, *University of Oxford, United Kingdom*
Miss Meetal Hirani, *University College, London, United Kingdom*
Dr Stephen Hodge, *Imperial College, London, United Kingdom*
Dr Katherine Holt, *University College, London, United Kingdom*
Miss Debbie Houghton, *Royal Society of Chemistry, United Kingdom*
Ms Elizabeth Howe, *University of Cambridge, United Kingdom*

Professor Dr Jingping Hu, *Huazhong University of Science and Technology*, *China*
Professor Dr Chi-Chang Hu, *National Tsing Hua University*, *Chinese Taipei*
Mrs Tanti Ruslini Hutapea, *PT DAFA JAYA*, *Indonesia*
Professor Graham Hutchings, *Cardiff University*, *United Kingdom*
Dr Abdelqader Imragaa, *University of Benghazi*, *Libya*
Dr Lyn Jennens, *Royal Society of Chemistry*, *United Kingdom*
Dr Martin Jonsson-Niedziolka, *Institute of Physical Chemistry, Polish Academy of Sciences*, *Poland*
Dr Fernanda Juarez, *Institute of Theoretical Chemistry*, *Germany*
Dr Krzysztof Kierzek, *Wroclaw University of Technology*, *Poland*
Mr Jeong Ho Kim, *University of Ulsan*, *South Korea*
Professor Ian Kinloch, *University of Manchester*, *United Kingdom*
Mr Cheol Wan Ko, *Department of Chemistry, University of Ulsan*, *South Korea*
Dr Heisi Kurig, *Univeristy of Tartu*, *Estonia*
Miss Sarah Latham, *Royal Society of Chemistry*, *United Kingdom*
Dr John Lee, *Camosun College*, *Canada*
Mr Jet-Sing Lee, *University of Liverpool*, *United Kingdom*
Dr Hong Ngee Lim, *Universiti Putra*, *Malaysia*
Dr Xiaoteng Liu, *Newcastle University*, *United Kingdom*
Miss Deborah Lomax, *The University of Manchester*, *United Kingdom*
Dr Mariangela Longhi, *Università degli Studi di Milano, Dipartimento di Chimica*, *Italy*
Ms Mailis Lounasvuori, *University College, London*, *United Kingdom*
Miss Sarah Lu, *Royal Society of Chemistry*, *United Kingdom*
Dr Stephen M. Lyth, *Kyushu University*, *Japan*
Professor Julie MacPherson, *University of Warwick*, *United Kingdom*
Ms Stefania Marzorati, *Universita' degli Studi di Milano*, *Italy*
Dr Richard McCreery, *University of Alberta/National Institute for Nanotechnology*, *Canada*
Mr Robert McSweeney, *University of Nottingham*, *United Kingdom*
Mr Lingcong Meng, *The University of Warwick*, *United Kingdom*
Dr Céline Merlet, *University of Cambridge*, *United Kingdom*
Dr Jonathan Midgley, *Royal Society of Chemistry*, *United Kingdom*
Dr Heather Montgomery, *Royal Society of Chemistry*, *United Kingdom*
Professor Mark Newton, *University of Warwick*, *United Kingdom*
Professor Richard Nichols, *The Department of Chemistry, University of Liverpool*, *United Kingdom*
Mr Hugo Nolan, *CRANN*, *Ireland*
Dr Andrew Osborne, *Metrohm UK*, *United Kingdom*
Mr Tommi Palomäki, *Aalto University*, *Finland*
Mr Rajesh Parishwad, *Royal Society of Chemistry*, *United Kingdom*
Dr Hollie Patten, *The University of Manchester*, *United Kingdom*
Dr Slawomir Porada, *INM - Leibniz Institute for New Materials, Energy Materials Group*, *Germany*
Mr Kaipei Qiu, *University College, London*, *United Kingdom*
Dr Fulian Qiu, *Department of Design, Brunel University*, *United Kingdom*
Mr Jonathan Quinson, *University of Oxford*, *United Kingdom*
Miss Edurne Redondo, *CIC Energigune*, *Spain*
Mr Andrew N. J. Rodgers, *University of Manchester*, *United Kingdom*
Professor Milo Shaffer, *Imperial College, London*, *United Kingdom*
Dr Surbhi Sharma, *Chemical Engineering, University of Birmingham*, *United Kingdom*
Mr Yong, Sheng, *University of Southampton*, *United Kingdom*

Dr Agata Śliwak, *Wroclaw University of Technology*, *Poland*
Mr Robert Smith, *University of Liverpool*, *United Kingdom*
Mr Christopher Sole, *University of Liverpool*, *United Kingdom*
Miss Claire Springett, *Royal Society of Chemistry*, *United Kingdom*
Professor Keith Stevenson, *The University of Texas at Austin*, *USA*
Mrs Titik Rahayu Sulardi, *PT DAFA JAYA*, *Indonesia*
Professor Patrick R. Unwin, *University of Warwick*, *United Kingdom*
Mr Christian van Engers, *University of Oxford*, *United Kingdom*
Dr Thomas Varley, *University College, London*, *United Kingdom*
Dr Matěj Velický, *University of Manchester*, *United Kingdom*
Professor Dr Siegfried Waldvogel, *Johannes Gutenberg University Mainz*, *Germany*
Mr Tom Warwick, *Blue Scientific*, *United Kingdom*
Mr Tzu-Ho Wu, *Stephenson Institute for Renewable Energy, Department of Chemistry, The University of Liverpool*, *United Kingdom*
Dr Wanjing Yu, *University of Cambridge*, *United Kingdom*
Mr Guohui Zhang, *University of Warwick*, *United Kingdom*